D1265731

Encyclopedia of Mobile Computing and Commerce

David Taniar
Monash University, Australia

Volume I
A–Mobile Hunters

INFORMATION SCIENCE REFERENCE
Hershey · London · Melbourne · Singapore

Acquisitions Editor: Kristin Klinger
Development Editor: Kristin Roth
Senior Managing Editor: Jennifer Neidig
Managing Editor: Sara Reed
Assistant Managing Editor: Diane Huskinson
Copy Editor: Maria Boyer and Alana Bubnis
Typesetter: Diane Huskinson
Support Staff: Sharon Berger, Mike Brehm, Elizabeth Duke, and Jamie Snavely
Cover Design: Lisa Tosheff
Printed at: Yurchak Printing Inc.

Published in the United States of America by
Information Science Reference (an imprint of Idea Group Inc.)
701 E. Chocolate Avenue, Suite 200
Hershey PA 17033
Tel: 717-533-8845
Fax: 717-533-8661
E-mail: cust@idea-group.com
Web site: http://www.idea-group-ref.com

and in the United Kingdom by
Information Science Reference (an imprint of Idea Group Inc.)
3 Henrietta Street
Covent Garden
London WC2E 8LU
Tel: 44 20 7240 0856
Fax: 44 20 7379 0609
Web site: http://www.eurospanonline.com

Library of Congress Cataloging-in-Publication Data

Encyclopedia of mobile computing and commerce / David Taniar, editor.
p. cm.
Summary: "Nowadays, mobile communication, mobile devices, and mobile computing are widely available. The availability of mobile communication networks has made a huge impact to various applications, including commerce. Consequently, there is a strong relationship between mobile computing and commerce. This book brings to readers articles covering a wide range of mobile technologies and their applications"--Provided by publisher.
Includes bibliographical references and index.
ISBN 978-1-59904-002-8 (hardcover) -- ISBN 978-1-59904-003-5 (ebook)
1. Mobile computing--Encyclopedias. 2. Mobile communication systems--Encyclopedias. 3. Mobile commerce--Encyclopedias. I. Taniar, David.
QA76.59.E47 2007
004.16503--dc22

2006039745

British Cataloguing in Publication Data
A Cataloguing in Publication record for this book is available from the British Library.

All work contributed to this encyclopedia set is new, previously-unpublished material. The views expressed in this encyclopedia set are those of the authors, but not necessarily of the publisher.

Editorial Advisory Board

List of Contributors

Koumaras, H. / *N.C.S.R., Demokritos, Greece* 758, 889
Kourtis, A. / *N.C.S.R., Demokritos, Greece* 758, 889
Kovács, Ferenc / *Budapest University of Technology and Economics, Hungary* 393
Kritzner, Jan / *Aachen University, Germany* 611
Kustov, Andrei L. / *First Hop Ltd., Finland* 15, 785
Kvasnica, Milan / *Tomas Bata University, Zlin, Czech Republic* 403, 651
Lalopoulos, George K. / *Hellenic Telecommunications Organization S.A., Greece* 581
Lang, Jia / *Nice Business Solutions Finland, Finland* 785
Lau, Chiew-Tong / *Nanyang Technological University, Singapore* 734
Le, Phu Dung / *Monash University, Australia* 227, 832, 1015
Lee, Cheon-Pyo / *Carson-Newman College, USA* 442
Lee, Dennis / *The University of Queensland, Australia & The Australian CRC for Interactive Design, Australia* 933
Lei, Pouwan / *University of Bradford, UK* 455
Leong, Hong Va / *The Hong Kong Polytechnic University, Hong Kong* 749
Leu, Huei / *Industrial Technology Research Institute, Taiwan* 178
Liberati, Diego / *Italian National Research Council, Italy* 68
Lim, Say Ying / *Monash University, Australia* 102, 154, 849
Lin, Chad / *Edith Cowan University, Australia* 178
Lin, Koong / *Taiwan National University of the Arts, Taiwan* 178
Liu, Chao / *Waseda University, Japan* 804, 820
Lívio Vasconcelos Guedes, Ádrian / *Federal University of Campina Grande, Brazil* 249
Lonthoff, Jörg / *Technische Universität Darmstadt, Germany* 510
Loureiro, Emerson / *Federal University of Campina Grande, Brazil* 71, 877
Luís do Nascimento, José / *Federal University of Campina Grande, Brazil* 341
Maamar, Zakaria / *Zayed University, UAE* 190
Mahatanankoon, Pruthikrai / *Illinois State University, USA* 461
Mahmoud, Qusay H. / *University of Guelph, Canada* 190
Malik, Haroon / *Acadia University, Canada* 328
Mamat, Ali Bin / *FSKTM – UPM, Malaysia* 693, 947
Maricar, Habeebur Rahman / *American University of Sharjah, UAE* 682
Marques, Stefânia / *Federal University of Campina Grande, Brazil* 978
Martakos, D. / *University of Athens, Greece* 758
Massimi, Michael / *University of Toronto, Canada* 291
McMeel, Dermott / *University of Edinburgh, Scotland* 516
Mecella, Massimo / *University of Rome, Italy* 1043
Menipaz, Ehud / *Ben-Gurion University, Israel* 413
Merten, Patrick S. / *University of Fribourg, Switzerland* 466
Misra, Manoj / *Indian Institute of Technology Roorkee, India* 172
Mitchell, Stella / *IBM T. J. Watson Research, USA* 644
Morais, Marcos / *Federal University of Campina Grande, Brazil* 260
Muhlberger, Ralf / *The University of Queensland, Australia & The Australian CRC for Interactive Design, Australia* 933
Muldoon, Conor / *University College Dublin, Ireland* 243
Nanopoulos, Alexandros / *Aristotle University, Greece* 660
Nishiyama, Mamoru / *Dentsu Communication Institute Inc., Japan* 639
O'Grady, Michael J. / *University College Dublin, Ireland* 243, 769, 1034
O'Hare, Gregory M. P. / *University College Dublin, Ireland* 243, 310, 769, 1034
O'Hare, Peter / *University College Dublin, Ireland* 310
Okazaki, Shintaro / *Autonomous University of Madrid, Spain* 296, 635, 639, 885
Oliveira, Loreno / *Federal University of Campina Grande, Brazil* 71, 621, 877
Olla, Phillip / *Madonna University, USA* 504
Olson, Andrew M. / *Indiana University Purdue University Indianapolis, USA* 436
Orosz, Mihály / *Budapest University of Technology and Economics, Hungary* 393

Contents
by Volume

VOLUME I

VOLUME II

Contents
by Topic

3G

Adhoc Network

Converging Technology

Human Factor

Location and Context Awareness

M-Business and M-Commerce

M-Entertainment

M-Health

M-Learning

Mobile Multimedia

Mobile Phone

Mobile Software Engineering

P2P

Security

Sensor Network

Service Computing

Wireless Networking

Foreword

Let us borrow this quote from the British humorist and cartoonist Ashleigh Brilliant to summarize the role of mobility in the development of the information society: "Unless you move, the place where you are is the place where you will always be." In more serious terms, it is fundamental to recognize that today's economic and societal progress is primarily dependent on the technological ability to sustain and facilitate the mobility of persons, physical goods (let us not forget, for instance, that the probably most critical component of global commerce today is deep sea shipping) and digital information (data and programs).

Recent years have witnessed a rapid growth of interest in mobile computing and communications. Indicators are the rapidly increasing penetration of the cellular phone market in Europe, and the mobile computing market is growing nearly twice as fast as the desktop market. In addition, technological advancements have significantly enhanced the usability of mobile communication and computer devices. From the first CT1 cordless telephones to today's Iridium mobile phones and laptops/PDAs with wireless Internet connection, mobile tools and utilities have made the life of many people at work and at home much easier and more comfortable. As a result, mobility and wireless connectivity are expected to play a dominant role in the future in all branches of economy. This is also motivated by the large number of potential users (a U.S. study reports of one in six workers spending at least 20 percent of their time away from their primary workplace, similar trends are observed in Europe). The addition of mobility to data communications systems has not only the potential to put the vision of "being always on" into practice;- but has also enabled new generation of services, for example, location-based services.

Mobile commerce leveraging the mobile Web and mobile multimedia is precisely the ability to deploy and utilize modern technologies for the design, development and deployment of a content rich, user and business friendly, integrated network of autonomous, mobile agents (here "agent" is to be taken in the sense of persons, goods and digital information).

I am delighted to write the foreword to this encyclopedia, as its scope, content and coverage provides a descriptive, analytical, and comprehensive assessment of factors, trends, and issues in the ever-changing field of mobile computing and commerce. This authoritative research-based publication also offers in-depth explanations of mobile solutions and their specific applications areas, as well as an overview of the future outlook for mobile computing.

I am pleased to be able to recommend this timely reference source to readers, be they researchers looking for future directions to pursue when examining issues in the field, or practitioners interested in applying pioneering concepts in practical situations and looking for the perfect tool.

Ismail Khalil Ibrahim,
Johannes Kepler University Linz, Austria
January 2007

Preface

Nowadays, mobile communication, mobile devices, and mobile computing are widely available. Everywhere people are carrying mobile devices, such as mobile phones. The availability of mobile communication networks has made a huge impact to various applications, including commerce. Consequently, there is a strong relationship between mobile computing and commerce. The *Encyclopedia of Mobile Computing and Commerce* brings to readers articles covering a wide range of mobile technologies and their applications.

Mobile commerce (m-commerce) is expanding, and consequently the impact to the overall economy is considerable. However, there are still many issues and challenges to be addressed, such as mobile marketing, mobile advertising, mobile payment, mobile authorization using voice, and so on. Providing users with more intelligent product catalogues for browsing on mobile devices and product brokering also plays an important role in m-commerce. Furthermore, the impact mobile devices give to the supply chain must be carefully considered. This includes the use of emerging mobile technology, such as RFID, sensor network, and so forth.

A wide range of mobile technology is available for m-commerce. Mobile phones are an obvious choice. Additionally, there are many different kinds of mobile phones sold in the market, some of which are labelled as smartphones. There is much research conducted in conjunction with the use of mobile phones. Mobile phone text messaging and SMS are common among mobile users. Subsequently, the use of text messaging and SMS enriches m-commerce, including the ability to support multilingual text messaging. Mobile phone supporting disability has also been a focus lately, which focuses on text messaging to disabled people. More advanced applications now require additional services, such as chatting using Bluetooth, mobile querying, and voice recognition. Mobile privacy issues are also still an important topic.

Apart from mobile phones, there is a wide variety of mobile technology, some of which are mobile robots, RFID, pen-based mobile computing, and so forth. Many advanced applications have been developed utilizing these technologies. Current research has been focusing on man-machine interfaces and sensory systems, particularly for mobile robots, biometric and voice based authentication, traffic infractions, and so forth. The context of smart spaces also gives a new dimension to mobile technology.

The use of mobile technology in entertainment is growing rapidly. Some examples include mobile phone gambling, mobile collaborative games, mobile television, mobile sport videos, and mobile hunting incorporating location-based information. The list is expanding as the technology is advancing. Understanding the success factors for mobile gaming and other entertainment is equally important as the technical aspects of the technology itself.

Videos and multimedia undoubted play an important role in mobile entertainment. Video technologies, such as mobile video sequencing, mobile video transcoding, and mobile video communications, have been studied extensively. One of the main limitations of mobile devices is the limited memory capacity, which has to be carefully addressed, especially in the context of mobile multimedia, because these kinds of applications generally require large amount of spaces. Beside videos, radio technology should not be neglected either.

There are many other applications of mobile technology. For example, the use of mobile technology in health, called m-health, is expanding. Mobile medical imaging is made possible thru the use of 3G wireless network. Another example is the use of mobile technology in learning, called m-learning, such as the use of SMS and text messaging, although some still argue whether m-learning is the way to go in learning, while others are still looking at how to combine the infrastructures and tools with pedagogy.

Developing mobile applications requires a novel software engineering approach. The design for mobile information systems is still maturing. Some researchers are still formulating design patterns for mobile applications, while others are focusing on the user interface aspects. Programming for handheld devices is quite common to use various programming languages and tools, including Java micro edition, J2ME, Corba, and Extreme programming. Since the device generally has a small screen, content transformation and content personalization need to be examined. Other forms of interfaces, including brain computing interfacing, are also interesting. Mobile databases and XML-based mobile technology have received some degree of attention as well.

Other issues that have been incorporated into mobile technologies include mobile agents, service-oriented computing, and various forms of caching, such as peer-to-peer, cooperative, and semantic caching. Service delivery and resource discovery are gaining their popularities too. Security—especially in a mobile environment—should not be neglected. Some work on mobile PKI and limited key generations has been carried out by a number of researchers in order to contribute to advancing m-commerce.

The impact of mobile technology in commerce needs to be evaluated, including its socio-psychological influence and technological adoption and diffusion, as well as readiness and transformation. We need to understand the adoption, barrier, and influencing factors of m-commerce. Some gender issues have been pointed out by some researchers.

All of the abovementioned applications will not be made possible without addressing the advancement of mobile networks. Most of the articles in this encyclopedia may be categorized into the mobile network and communication category. 3G architectures have made their entries lately. Mobile ad-hoc network, IPv6 and P2P are also maturing. Some new work in wireless sensor network is presented.

Last but not least, mobile technology and its applications will not be complete without mentioning location-aware and context-aware. New technologies in positioning; either indoor or outdoor, as well as tracking of moving objects, are presented. Some applications of location-aware include ad-hoc mobile querying, use of iPod as a tourist guide, location-based multimedia for monitoring purposes, and location-based multimedia for tourists. Some notable context-aware applications are notification services, context-aware mobile GIS, and semantic mobile agents for context-aware applications.

As a final note, the *Encyclopedia of Mobile Computing and Commerce* covers a broad range of aspects pertaining to mobile computing, mobile communication, mobile devices, and various mobile applications. These technologies and applications will shape mobile computing and commerce into a new era of the 21st century whereby mobile devices are not only pervasive and ubiquitous, but also widely accepted as the main tool in commerce.

David Taniar
Melbourne, Australia
January 2007

Acknowledgments

I would like to acknowledge the help of all involved in the collation and review process of the encyclopedia, without whose support the project could not have been satisfactorily completed.

I would like to thank all the staff at IGI, whose contributions throughout the whole process, from inception of the initial idea to final publication, have been invaluable. In particular, our thanks go to Kristin Roth, who kept the project on schedule by continuously monitoring the progress on every stage of the project, and to Mehdi Khosrow-Pour and Jan Travers, whose enthusiasm initially motivated me to accept their invitations to take on this project. I am also grateful to my employer Monash University for supporting this project.

A special thank goes to Mr. John Goh of Monash University, who assisted me in almost the entire process of the encyclopedia: from collecting and indexing the proposals, distributing chapters for reviews and re-reviews, constantly reminding reviewers and authors, liaising with the publisher, to many other housekeeping duties, which are endless.

I would also like to acknowledge the assistance and advice from the editorial board members. In closing, I wish to thank all of the authors for their insights and excellent contributions to this encyclopedia, in addition to all those who assisted us in the review process.

David Taniar
Melbourne, Australia
January 2007

About the Editor

David Taniar received a PhD degree in computer science from Victoria University, Australia, in 1997. He is now a senior lecturer at Monash University, Australia. He has published more than 100 research articles and co-authored a number of books in the mobile technology series. He is on the editorial board of a number of international journals in the fields of data warehousing and mining, business intelligence and data mining, mobile information systems, mobile multimedia, Web information systems, and Web and grid services.

Academic Activities Based on Personal Networks Deployment

Vasileios S. Kaldanis
NTUA, Greece

Charalampos Z. Patrikakis
NTUA, Greece

Vasileios E. Protonotarios
NTUA, Greece

INTRODUCTION

Personal networking has already become an increasingly important aspect of the unbounded connectivity in heterogeneous networking environments. Particularly, personal networks (PNs) based on mobile ad-hoc networking have seen recently a rapid expansion, due to the evolution of wireless devices supporting different radio technologies. Bluetooth can be considered as the launcher of the self-organizing networking in the absence of fixed infrastructure, forming pico nets or even scatternets. Similar other wireless technologies (e.g., WiFi) attract a lot of attention in the context of mobile ad hoc networks, due to the high bandwidth flexibility and QoS selection ranges they feature, leveraging the path to develop advanced services and applications destined to the end user and beyond. Furthermore, personal networks are expected to provide a prosperous business filed for exploitation to third-party telecom players such as service and content providers, application developers, integrators, and so forth.

In this article, a personal-to-nomadic networking case is presented. Academic PN (AcPN) is a generic case that aims to describe several situations of daily communication activities within a university campus or an extended academic environment through the support of the necessary technological background in terms of communication technologies. The concept is straightforward: a number of mobile users with different characteristics and communication requirements ranging from typical students to instructors and lecturers, researchers and professors, as well as third parties (e.g., visitors, campus staff), are met, work, interact, communicate, educate, and are being educated within such an environment. This implies the presence of a ubiquitous wireless personal networking environment having nomadic characteristics. Several interesting scenarios and use cases are analyzed, along with a number of proposed candidate mobile technology solutions per usage case.

The article is organized as follows: first, a general description of the academic case is presented identifying examples of typical communication activities within an academic environment; the technical requirements necessary for a successful deployment of personal area network (PAN)/PN technologies within the academic environment are also listed. Next, specific deployment scenarios are presented, followed by a business analysis. The article closes with a concluding section.

ACADEMIC CASE DESCRIPTION

The AcPN case describes several situations of daily communication activities, taking place within a typical university campus environment. Members of the academic community, such as students, make use of personal networking concepts and related technologies to acquire and maintain constant connectivity among them or with local or remote networks, and utilize offered services—applications discovered at their point of presence. In this fashion, they may exchange files on the move, interact with each other in different ways (e.g., messaging, audio/videoconference), connect to a home desktop PC to download a missing file, or configure remotely a project installation located in a lab.

The AcPN case aims to support a number of communication activities known in an academic environment. Typical examples of such activities include:

- entering the campus, and making inquiries for local information (maps, buildings, etc.);
- monitoring information updates (announcements, urgent notices, deadlines, events);
- meeting with a colleague/friend/other student mates, exchanging data with others (docs, mp3, video clips, etc.), work management, and so on;
- seeking a friend/colleagues somewhere on campus;
- communicating with a professor/tutor/technical supervisor;
- reporting project results to colleagues and real-time discussion;

- borrowing/returning a book from/to the local library;
- performing remote home/office network setup (upon returning home);
- monitoring and controlling a lab experiment/project installation; and
- responding to emergency situations within the campus area (fire drill, medical assistance, etc).

The objective of developing the AcPN case is to provide the academic users with an easy way to perform their everyday work as efficiently as possible—in the least time and with the least cost. The academic entity concept-model used here is very general and includes all different types of academics existing in a typical university environment. These are undergraduates/postgraduates/PhD students, tutors/lecturers/professors, research associates, and third-party entities such as visitors and permanent/temporary staff. The campus infrastructure is supposed to support as many communication technologies as possible to the academic entities roaming on campus, in order to provide a variety of services, featuring flexibility in constructing different networking configurations. These technologies could range from short-distance wireless protocols (Bluetooth, infrared) to large-scale networking solutions such as WLAN or GSM/GPRS and 3G/UMTS.

In any case, academic users can benefit from PN concepts such as P-PAN, PAN/PN, W-PAN, and so forth in order to acquire access to other networks or services. Each user is equipped with a number of wireless communicating devices such as mobile phones, PDAs, laptops, headsets, and mobile storage devices, featuring GSM/GPRS/UMTS Bluetooth and WiFi technologies. These devices can detect and interact with each other in various ways, providing new communication capabilities and fields for different networking configurations.

For example, a student is able to form his own personally attached network or private PAN (P-PAN) by interconnecting his wearable short-range devices (e.g., headset, mp3 player, mobile hard disc, PDA) via Bluetooth or infrared protocol. On a larger scale, the user can also connect to a local network of short-range devices (other users' devices or local wireless printer) becoming part of the existing personal area network, and interact with users in his or her close vicinity who belong to the same network. The student may use his or her mobile device as a GSM/GPRS or UMTS terminal to extend his or her current P-PAN and PAN configuration in order to connect to his or her home DSL network to download an important file from the remote desktop PC. In this case, the student establishes a personal network that can be further used for numerous other remote actions. In the same way, the ubiquitous campus network provider can interconnect all PANs within the campus area and form a "personal"-like network: the campus PN.

Similarly, any other academic user can form one or more PNs dependent on the following parameters:

- the number of interconnecting devices,
- the inherent characteristics of used wireless technologies,
- the connection capabilities per technology in terms of bandwidth and QoS, and
- the requirements imposed by each service used on a particular PN.

Finally, administration of the campus PN is a very important issue for the successful management of attached users in terms of resources and security and successful service provision. Different security levels can be used, according to the trust policy followed when a foreign user (e.g., visitor) is accepted locally in a PAN or globally in the campus PN.

PN CONCEPT IN ACADEMIC CASE

PNs in our case comprise potentially all of a person's devices capable to detect and connect each other in the real or virtual vicinity. Connection is performed via any known and applicable wireless access technology (Bluetooth, infrared, WiFi, MAGNET low/high data rate, WLAN/GSM/GPRS/UMTS, and so on). PN establishment requires an extension of the present and locally detected PAN by the person's attached network (set of person's devices) called private PAN. The physical architecture of the networks and devices (for the AcPN case) has already been mentioned, while all interactions among them is illustrated in the Figure 1.

PNs are configured in an ad hoc fashion, establishing any possible peer-to-peer (P2P) connection among users belonging to the same local PAN and other remote PANs or PNs as well, in order to support a person's private and professional applications. Such applications may be installed and executed on a user's personal device, but also on foreign devices in the same way. PNs consist of communicating clusters of personal digital devices, possibly shared with others and connected through different communication technologies remaining reachable and accessible via at least a PAN/PN. Obviously, PANs have a limited geographical coverage, while PNs have unrestricted geographical span, incorporating devices into the personal environment, regardless of their physical or geographical location. In order to extend their access range, they need the support of typical infrastructure-based and ad-hoc mobile networks.

Strict security policies determine PNs' performance. Any visiting (foreign to the local PAN) mobile user bearing his or her own P-PAN may acquire trust and become a member of the locally detected PAN, as long as another member of the same PAN can guarantee his or her proper behavior in

Figure 1. Academic PN concept topology and interactions

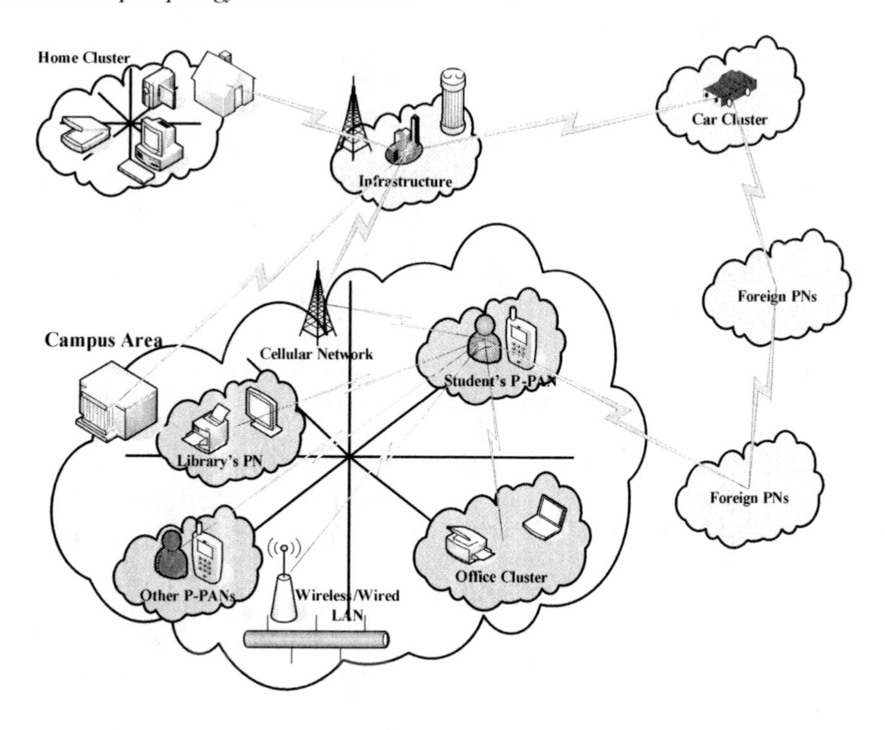

Table 1. Devices used in AcPN case

	P-PAN	Home PAN	Office PAN	Campus PN	Laboratory PAN / PN
Desktop PC	-	√	√	√	√
Laptop	Optional	Optional	Optional	Optional	√
PDA	√	√ (Optional)	√ (Optional)	√ (Optional)	√ (Optional)
Mobile Phone	√	Optional	Optional	√	√
MP3 Player	√	√	Optional	Optional	-
Wireless Headset	√	-	-	-	√
Printer	-	√	√	-	√
Scanner	-	√	√	-	√
Mobile Hard Drive	-	√ (Optional)	√ (Optional)	-	√
Camera	-	√	√	-	√
DVD R/Player	-	√	-	-	√
Wall Screen	-	√	√	-	√
Sensor Tx/Rx	-	-	-	-	√

this network. In this way, the new user can become trusted and behave as any other existing user in the PAN. Similar mechanisms exist for AAA functionalities in other clusters and PN domains as well.

A list of important devices for the use cases listed formerly is summarized in Table 1.

- **Desktop PC and Laptop:** High processing power, unlimited power supply, high storage capacity, graphical UI, support of 802.11/Ethernet/Bluetooth, HDR, Internet connectivity (TCP/IP, UDP, etc.), database software, and so forth.
- **PDA:** Low processing power, unlimited power supply, high storage capacity, graphical UI, HDR/LDR,

support 802.11/Ethernet/Bluetooth/56K, support of TCP/IP, security software integrated, low weight, and so forth.

- **High-Featured Mobile Phone Device:** Low processing power, low power consumption, moderate storage space, support of wireless protocols (Bluetooth/GSM/CDMA), IrDA support, cellular connectivity (GSM/GPRS/UMTS), WiFi/WLAN connectivity, portability, synchronization with other devices.
- **Printer:** Support of various wireless technologies (Bluetooth, IrDA, etc), wired networking, and so forth.
- **MP3 Player:** Weak power supply, moderate storage capacity, support of basic wireless access technologies (Bluetooth/WiFi), HDR/LDR, large battery power consumption, low recharging time, handy UI and control, high sound quality, and so forth.
- **Wireless Headset:** Support of basic wireless access technologies (Bluetooth, IrDA), LDR, and so forth.
- **Wireless Sensors:** Low power consumption, wireless interconnectivity (Bluetooth, IrDA, etc.), LDR, large operation life flexible functionality, light weight device, low volume, remotely controllable, and so forth.

It should be noted that currently, there is ongoing work on specifying devices that support new protocols (especially in the wireless physical layer), and the expansion of the use cases to current networking technologies is also still under development.

SCENARIOS AND USE CASES

The scenario generation procedure has been based on the obtained results from an end user workshop held at the NTUA campus. The workshop participants were academic people coming from different knowledge backgrounds and professions (undergraduates/MSc students, PhD candidates/research associates, tutors, lecturers, professors, and visitors). During the workshop all participants had the chance to exchange thoughts and express their own needs regarding communication solutions and services they wish or expect to have within a typical campus area environment.

Login to the Ubiquitous Campus PN Network

This is a fundamental case for the AcPN, since it presents the most important thing an AcPN user must do if he wants to utilize services and applications available in the university domain (single campus or a set of campuses belonging to the same organization).

According to this case, the AcPN user must login to the campus network via his mobile device mainly in two cases: whenever he reaches the real campus area physically (e.g., by car, by bus, or by foot) via a locally detected campus PAN or remotely via a PN which he has previously established dynamically with the campus PN. The AcPN could be a registered user to the campus network (e.g., student, lecturer, researcher, or permanent staff) or a foreign (third-party) user (e.g., visitor) who should follow a registration procedure before attaching to the local network. The login procedure is required for the AcPN case in order to maintain a certain level of security, which is higher for locally connected users in contrast to remotely connected ones. After successful login, the AcPN users can immediately be informed by the campus PN administrator for urgent messages from their colleagues, reminders, scheduled power outages, and other important messages of general importance.

Information Update and Real-Time P2P Interaction

In this use case, an AcPN member, after logging into the campus PN network, wishes to have access to any available local services and applications according to his or her educational activity (e.g., student, researcher). At the same time, he or she can be informed about course announcements, important notices (e.g., deadline extensions, change of lecture classrooms, etc.) from the student office or from any other local online source related to his or her studies. Furthermore, using a mobile device he or she may directly connect to a course database to download important files such as handouts, past papers, presentations, or any other material in electronic form. In P2P fashion, the student may have the chance to see on his or her device who is currently roaming into the campus area from among his contact people (friends, colleagues, tutors, etc.) and to interact with them in various ways. He may also publish hello messages everywhere he wants to, arrange a meeting (physical or not) on the fly, be informed by other people who also "see" him on their devices, exchange files with friends (mp3s, pictures, video clips), send an important file to a colleague or to his or her technical supervisor, setup an audio/video conference, and so forth.

Using a Trusted PAN to Connect to Other Networks

In this scenario, a mobile user who is not a member in the campus PN currently lies within the campus and wishes to get an Internet connection or to acquire access to the local network for several reasons (e.g., utilize local services, get library access, view local events, etc.). This user is considered a foreign user, since he does not belong to the campus PN or to any other local PAN, as privileged campus PN members do. Obviously, the foreign user is considered by the campus network as a third party-user or a visitor and in some way has to be accepted by the campus PN administrator into the ubiquitous local network. This can be done directly or indirectly. In the direct way, the user can be connected to a locally detected PAN at its point of presence if another registered user of the same PAN can guarantee his or her proper and safe behavior. In other words, the foreign user may be attached to any PAN and consequently to the campus PN if another user of the same PAN can verify him or her as a trusted entity and provide him or her with access rights characterized by the basic required security level. In case the foreign user violates the invitation policy agreement, he or she may be warned or even banned by any other PAN user reporting the event to the campus PN administrator. Then the user who signed his or her trustworthiness may lose credits on his or her membership to the campus PN, or his or her authorization provision to other users in the future may be suspended for some period. Following the indirect way, the foreign user may use the local wide area network (e.g., WLAN) to ask for a temporary registration from the campus PN administrator. For example he or she may use a credit card to register to the ubiquitous campus PN network; buy connection time duration; service access rights, bandwidth, and QoS; and so on. In this way, the registered foreign user may be accepted by any other PAN anywhere in the campus, gaining access to the allowed local services in general. This type of user cannot access individual department resources and services (e.g., engineering department database, ftp software, etc.) but only allowed services for third-party users (e.g., library access, local knowledge base intranets, projects, etc.).

Remote Laboratory Monitoring and Control

This is the case where a remote monitoring and controlling of a procedure taking place in a location is required using PAN/PN technologies. Particularly, a group of scientists (students, researchers, professors, etc.) is performing a lab experiment that is long lasting, and the overall progress and results need to be monitored continuously on a 24-hour basis. Furthermore, it is required that according to the collected ongoing results, some experiment parameters may be changed dynamically (locally or remotely). The scientific group must have continuous communication using their mobile devices independent of their point of presence, in order to discuss the change of parameters whenever needed to do so. In this case we consider that there is no physical presence by any member to the lab location and the procedure runs remotely using PAN/PN.

The experiment consists of a number of wireless sensors attached on the examined sample under test, forming a P-PAN which sends reports to a report collector. The report collector enriches the raw report signals and forwards them to the central processing device (high processing power desktop PC) where the experiment software is running. The central processing device sends formatted reports to a local database for data warehousing purposes, while reporting results to the scientific group using the lab PAN as well. Each member of the scientific group has been attached to the lab PAN forming individual PNs and also maintains a direct online connection with the other members for results discussion. Depending on the results, if a parameter change is decided, the user responsible for the experiment sends the required commands to the command executor device, which runs an external application controlling the interaction functionality with the sample under test. The change is verified and archived wirelessly into the database, again using the lab PAN, while a report is sent back to the group about its successful command execution.

Future Library Loaning and Reservation

This scenario presents a proposed loaning and reservation system for academic libraries in the future. In this case, the reservation and loaning of a book title may be performed based on the well-known Web service (via the library Web site) and the campus PN infrastructure. The campus PN consists of all PAN/PN clusters in different departments (or offices/labs, etc.) or smaller departmental libraries and the on-campus users equipped with mobile devices.

According to this scenario, a requestor for a book is an on-campus entity (normal/MSc/PhD student, research fellow), who is using his or her mobile device and the campus networking infrastructure to get access to the local online library database. The requestor should also be a registered member of the campus PN with a stored profile in the university database already logged in. This profile entry automatically enables a number of useful privileges according to the AcPN user type (user profession) that allows him or her to access specific applications and services.

An example use scenario is the following: a requestor gets informed by the library service on his mobile that a requested book is currently loaned and has been delayed to return (i.e., for a day). He is also notified about the priority in the request queue (if any exists) for that title. After that, the system generates an urgent message and forwards it to the loaner of the book using the campus PN. The system, using a tracing mechanism regarding the user status-location, is aware that the loaner is currently active and able to receive notifications via the campus PN, so it prefers to notify the user in this way. The loaner must provide as soon as possible a new book returning date to the library system if he does not want his membership to be blacklisted or in the worst case banned from the campus PN database. Hence, the loaner provides as the new returning date a specific time during the same day. The system forwards the new returning date to the requestor and provides a validity period for his request. After that period, his request is no longer valid and a next requestor (on the queue) gets the right to reserve that book. When finally the book returns to the library desk, the system via the campus PN notifies the active requestor about the book availability and his validity period to come and collect it. The requestor may provide himself as the collecting person or another registered PN user.

Remote Course Exam Participation and Distant Learning

Finally, using this case, a student currently away from the campus area for several reasons (urgent reasons, recuperation in hospital, etc.) has the option to participate remotely in her course exams using the PN technology. At her current location, she as to scan for a local PAN to attach or to search for another local wireless Internet connection means (e.g., WLAN, UMTS, WiFi). Then she must setup a PN with the campus network, logon to the campus PN using her student account, and connect to the local examinations server who has privately published an exams-related session link for such cases. Then after authorizing and authenticate herself, she must download the support software for this online session or any other auxiliary utility supplied by the exam center administrator, install it properly, and directly connect to the exam server before the actual start time of the exams. It is supposed that she has already applied for a remote exam participation by sending an e-mail to the exam administrator, and on reply she has received all the relevant details—information of that session according to the course requirements (e.g., multiple-choice form), connection bandwidth, QoS, and personal mobile device capabilities (e.g., large viewable display, keyboard, memory, etc). The student using this exam PN session participates remotely in the same way she would if she was present in the real exam center location for the required time period of the exam. It is required that she has an interruptible connection with the campus PN network and particularly with the exam center local server. The student provides her answers to the exam paper questions by ticking the appropriate box in each online XML Web interface, presses the "SEND" button to proceed to the next question, and so on. Each provided answer cannot be changed or undone since it has already been sent to the server and saved to the database system. If any problem occurs (e.g., connection is lost or service application fails), the session state is continuously monitored by the exam administrator and resumed when the problem is solved. At the end, the session is closed and a message informs the student that the application has already completed successfully. The service will later inform the student of her achieved results.

In the same way any possible distant learning activity can be supported using similar PN setups and configurations as long as the remote users can create any possible type of PN with the distant network of interest where a relevant service can run reliably.

BUSINESS PROSPECTS

Many players in mobile business may find PN technology to be a prosperous field to extend the market in many dimensions, ranging from high data rate connectivity solutions to advanced services and Web-based applications. The value chain of the mobile market can be dynamically expanded including more than one network and service providers, integrators, service and application developers, or even small-to-medium network operators.

The AcPN case exploits PN concepts in a very efficient way, allowing the use of well-known wireless technologies and common networking configurations of the present and the future to be used and easily applied. Target users are people actively involved in educational activities who present high expectations from communication technologies such as increased bandwidth, connection flexibility (among different technologies), use of a wide range of services and applications, more personalized devices, large mobile storage capability, interoperability, friendly user-device interface, and so forth.

Based on the collected results from the AcPN end user workshop held in Athens, Greece, a number of important requirements have been identified. These requirements have led to several conclusions regarding the new players in the value chain and the business aspects of PAN/PN concepts within the academic environment. The most important conclusions are:

- **Regarding Network Infrastructure:** The network infrastructure should include the normal mobile networks (GSM/GPRS/UMTS), as well as additional networking infrastructure such as WLAN/WiFi on a single or multi-operator environment and the ubiquitous campus PN operator. The campus PN infrastructure must include networking configurations among all campus PANs (different departments, labs, offices) and possibly other PNs (other campuses of the same organization).
- **Regarding Security:** The campus PN operator is responsible for network security in the supported connections of the wired/wireless domain, user login/logout functionality, mobility support within the campus (or campuses of the same university), and other required PAN/PN operations. If the particular university operates more than one campus, then a university PN is required to interconnect the different campus PNs and support the previous on a higher administrative level, securing of course the communication between the PNs.

- **Regarding Service Aggregators:** In this case, the role of service aggregation and provision to the AcPN users is performed primarily by the campus PN operator and partially by third-party service operators who may have agreements with the campus operator. Any service provided to the campus PN is expected to be controlled and maintained by the unique campus PN operator, which plays the twofold role of the service aggregator and the provider. Other services can be provided on the campus by typical mobile operators through the use of voice, e-mail, SMS, or MMS, but PN services and relevant interconnections must be realized via the campus PN network or service operator.
- **Regarding Terminal Equipment:** This requirement takes into account all the different vendors and manufacturers who provide the terminal devices to the end users. The fact that any AcPN user is supposed to be equipped with his or her own P-PAN requires a number of different featured portable devices coming from different vendors to be used. This is feasible as long as the PAN-proposed standards are supported. (It should be noted that for the air interface, the MAGNET LDR/HDR standard has been proposed.)
- **Regarding End Users:** These can be divided into two types. The first one includes all the normal students (undergraduates, postgraduates, etc.) who wish to use typical (low QoS) applications and services (Web browsing, chat, e-mail, voice, SMS, MMS, etc.) within the campus PN at a low cost. The second user type includes any other academic person or third party (visitors, temporary staff) who wish to have (and are willing to pay for) a higher bandwidth wireless connection or access to QoS demanding services such as (real-time) audio/videoconference, streaming applications, and so on. Such users could be professors, tutors, researchers, associates, or general university employees who use telecom technology to communicate with their work contacts for several reasons.

CONCLUSION

The academic case is very promising for the future deployment of PN technologies for many important reasons. First of all, it attempts to combine and reuse efficiently almost any wireless access technologies of the present with proposed ones for the future in many scalable configurations according to the case. Secondly, it provides the option to choose which type of PN could better serve its purposes in terms of connection bandwidth and cost. The user may choose the most efficient way (in terms of cost) to construct his or her own PN; for example, he or she may prefer a relatively cheap WLAN to connect to his or her office rather than a UMTS. Finally, since the use of PN technology might not be possible in some cases without the existence of PAN or P-PAN, the definition of clusters eases the PAN or P-PAN formation as a set of preferable devices, but not all.

ACKNOWLEDGMENTS

The AcPN case was presented, developed, and analyzed in detail within the IST-MAGNET framework (http://www.ist-magnet.org). Specific documents referenced include: MAGNET WP1 Task 1.1, D.1.1.1a, March 2004; MAGNET, WP1 Task 1.4, D1.4.1a, September 2004; MAGNET WP1 Task 1.1, D.1.1.1b, December 2004; MAGNET WP1 Task 1.1, D.1.1.1b, December 2005; and Academic Case Workshop Results, Internal Report D-1.3.1b.

The work acceptance by the academic community is very encouraging and promising for the future. Currently the project group is implementing, based on the previous use cases and scenarios, a number of services.

KEY TERMS

Academic PN (AcPN): Use case descriptive name for a PN exploitation into a typical academic environment.

Cluster: A network of personal devices and nodes located within a limited geographical area (such as a house or a car) which are connected to each other by one or more network technologies and characterized by a common trust relationship between each other.

Context: The information that characterizes a person, place, or object. In that regard, there exist user, environment, and network context. The context information is used to enable context-aware service discovery.

Foreign Device: A device that is not personal and cannot be part of the PN. The device can be either trusted, having an ephemeral trust relationship with another device in the PN, or not trusted at all.

Private Personal Area Network (P-PAN): A dynamic collection of personal nodes and devices around a person.

Personal Area Network (PAN): A network that consists of a set of mobile and wirelessly communicating devices that are geographically close to a person but which may not belong to him.

Personal Device: A device related to a given user or person with a pre-established trust attribute. These devices are typically owned by the user. However, any device exhibiting the trust attribute can be considered as a personal device. The same remarks as those for the personal nodes definition hold for devices.

Personal Network (PN): Network including the P-PAN and a dynamic collection of remote personal nodes and devices in clusters that are connected to each other via interconnecting structures.

Accessibility of Mobile Applications

Pankaj Kamthan
Concordia University, Canada

INTRODUCTION

The increasing affordability of devices, advantages associated with a device always being handy while not being dependent on its location, and being able to tap into a wealth of information/services has brought a new paradigm to mobile users. Indeed, the *mobile Web* promises the vision of universality: access (virtually) anywhere, at any time, on any device, and to *anybody*.

However, with these vistas comes the realization that the users of the mobile applications and their context vary in many different ways: personal preferences, cognitive/neurological and physiological ability, age, cultural background, and variations in computing environment (device, platform, user agent) deployed. These pose a challenge to the ubiquity of mobile applications and could present obstacles to their proliferation.

This article is organized as follows. We first provide the motivation and background necessary for later discussion. This is followed by introduction of a framework within which accessibility of mobile applications can be systematically addressed and thereby improved. This framework is based on the notions from semiotics and quality engineering, and aims to be practical. Next, challenges and directions for future research are outlined. Finally, concluding remarks are given.

BACKGROUND

The issue of accessibility is not new. However, the mobile Web with its potential flexibility on both the client-side and the server-side presents new challenges towards it.

Figure 1 illustrates the dynamics within which the issue of accessibility of a mobile application arises.

We define a *mobile application* as a domain-specific application that provides services and means for interactivity in the mobile Web. For example, education, entertainment, or news syndication are some of the possible domains. The issue of accessibility is intimately related to the user and user context that includes client-side computing environment. To that regard, we define *accessibility* in context of a mobile application as access to the mobile Web by everyone, regardless of their human or environment properties. A *consumer* (user) is a person that uses a mobile application. A *producer* (provider) is a person or an organization that creates a mobile application.

The Consumer Perspective of Mobile Accessibility

The accessibility concerns of a consumer are of two types, namely human and environment properties, which we now discuss briefly.

Human Properties

Human properties are issues relating to the differences in properties among people. One major class of these properties is related to a person's ability, and often the degree of absence of such properties is termed as a disability. We will use the term "disability" and "impairment" synonymously.

The statistics vary, but according to estimates of the United Nations, about 10% of the world's population is considered disabled. The number of people with some form of disability that do have access to the Internet is in the millions.

Figure 1. The interrelationships between a consumer, a producer, accessibility, and a mobile application

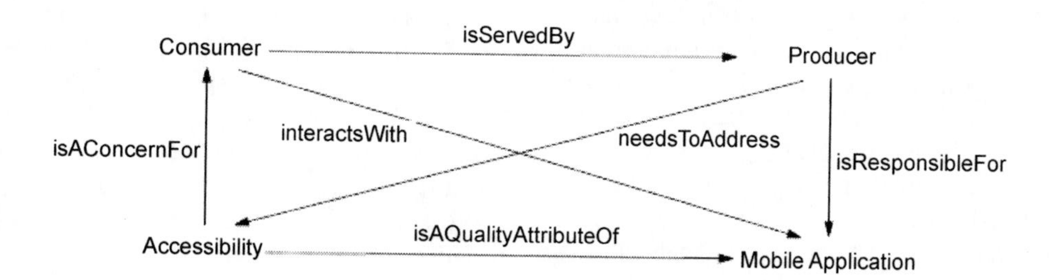

Table 1. A semiotic framework for accessibility of mobile applications

Semiotic Level	Quality Attributes	Means for Accessibility Assurance and Evaluation	Decision Support
Pragmatic	**Accessibility** [T4;E]	• Training in Primary and Secondary Notation • "Expert" Knowledge (Principles, Guidelines, Patterns) • Inspections • Testing • Metrics • Tools	Feasibility
	Comprehensibility, Interoperability, Performance, Readability, Reliability, Robustness [T3;E]		
Semantic	Completeness and Validity [T2;I]		
Syntactic	Correctness (Primary Notation) and Style (Secondary Notation) [T1;I]		

There are several types of disabilities that a producer of a mobile application needs to be concerned with. These can include visual (e.g., low visual acuity, blindness, color blindness), neurological (e.g., epilepsy), auditory (e.g., low hearing functionality, deafness), speech (e.g., difficulties in speaking), physical (e.g., problems using an input device), cognitive (e.g., difficulties of comprehending complex texts and complex structures), cultural/regional (e.g., differences in the use of idioms, metaphors leading to linguistic problems).

Environment Properties

Environment properties are issues relating to different situations in which people find themselves, either temporarily or permanently. These situations could be related to their connectivity, the location they are in, or the device/platform/user agent they are using. For example, a user using a computer in a vehicle shares many of the issues that some people have permanently due to a disability in hand motorics. Or, for example, a user may be accessing the *same* information using a personal digital assistant (PDA) or a cellular phone.

The Producer Perspective of Mobile Accessibility

The motivation for accessibility for a business is to reach as many users as possible and in doing so reduce concerns over customer alienation.

It is the producer of the mobile application that needs to adjust to the user context (and address the issue of accessibility), not the other way around. It is not reasonable for a producer to expect that the consumer environment will be conducive to *anything* that is delivered to him/her. In certain cases, when a consumer has a certain disability, such adaptation is not even possible.

If the success of a mobile application is measured by the access to its services, then improving accessibility is critical for the producers. Still, any steps that are taken by a producer related to a mobile application have associated costs and trade-offs, and the same applies to improvements towards accessibility.

Initiatives for Improving Accessibility in Mobile Contexts

There are currently only a few efforts in systematically addressing accessibility issues pertaining to mobile applications.

There are guidelines available for addressing accessibility (Chisholm, Vanderheiden, & Jacobs, 1999; Ahonen, 2003) in general and language-specific techniques (Chisholm et al., 2000) in particular.

ADDRESSING THE ACCESSIBILITY OF MOBILE APPLICATIONS

To systematically address the accessibility of mobile applications, we take the following steps:

1. View accessibility as a qualitative aspect and address it indirectly via quantitative means.
2. Select a theoretical basis for communication of information (semiotics), and place accessibility in its setting.
3. Address semiotic quality in a practical manner.

Based on this, we propose a framework for accessibility of mobile applications (see Table 1). The external attributes (denoted by E) are extrinsic to the mobile application and are directly the consumer's concern, while internal attributes (denoted by I) are intrinsic to the mobile application and are directly the producer's concern. Since not all attributes corresponding to a semiotic level are on the same echelon, the different tiers are denoted by "Tn."

We now describe each component of the framework in detail.

Semiotic Levels

The first column of Table 1 addresses semiotic levels. Semiotics (Stamper, 1992) is concerned with the use of symbols to convey knowledge.

From a semiotics perspective, a representation such as a mobile resource can be viewed on six interrelated levels: physical, empirical, syntactic, semantic, pragmatic, and social, each depending on the previous one in that order. The physical and empirical levels are concerned with the physical representation of signs in hardware and communication properties of signs, and are not of direct concern here. The syntactic level is responsible for the formal or structural relations between signs. The semantic level is responsible for the relationship of signs to what they stand for. The pragmatic level is responsible for the relation of signs to interpreters. The social level is responsible for the manifestation of social interaction with respect to signs, and is not of direct concern here.

We note that none of the layers in Table 1 is sufficient in itself for addressing accessibility and intimately depends on other layers. For example, it is readily possible to create a document in XHTML Basic, a markup language for small information appliances such as mobile devices, that is syntactically correct but is semantically non-valid. This, for instance, would be the case when the elements are (mis)used to create certain user-agent-specific presentation effects. Now, even if a mobile resource is syntactically and semantically acceptable, it could be rendered in such a way that it is unreadable (and therefore violates an attribute at the pragmatic level). For example, this could be the case by the use of very small fonts for some text, or the colors chosen for background and text foreground being so close that the characters are hard to discern.

Quality Attributes

The second column of Table 1 draws the relationship between semiotic levels and corresponding quality attributes. We contend that the quality attributes we mention are necessary but make no claim of their sufficiency.

The internal quality attributes for syntactic and semantic levels are inspired by Lindland, Sindre, and Sølvberg (1994). At the semantic level, we are only concerned with the conformance of the mobile application to the domain it represents (that is, semantic correctness or completeness) and at the syntactic level the interest is in conformance with, with respect to the languages used to produce the mobile application (that is, syntactic correctness).

Accessibility belongs to the pragmatic level and depends on the layers beneath it. It in turn depends upon the other external quality attributes, namely comprehensibility, interoperability, performance, readability, reliability, and robustness, which are also at the pragmatic level. Since these are perceived as necessary conditions, violations of any of these lead to a deterioration of accessibility.

Means for Accessibility Assurance and Evaluation

The third column of Table 1 lists the direct and indirect (and not necessarily mutually exclusive) means for assuring and evaluating accessibility:

- **Training in Primary and Secondary Notation:** The knowledge of the *primary notation* of all technologies (languages) is necessary for guaranteeing conformance to tier T1. The Cognitive Dimensions of Notations (CDs) (Green, 1989) are a generic framework for describing the utility of information artifacts by taking the system environment and the user characteristics into consideration. Our main interest here is in the CD of *secondary notation.* This CD is about appropriate use (that is, *style*) of primary notation in order to assist in interpreting semantics. It uses the notions of *redundant recoding* and *escape from formalism* along with spatial layout and perceptual cues to clarify information or to give hints to the stakeholder, both of which aid the tiers T2 and T3. Redundant Recoding is the ability to express information in a representation in more than one way, each of which simplifies different cognitive tasks. It can be introduced in a textual mobile resource by making effective use of orthography, typography, and white space. Escape from Formalism is the ability to intersperse natural language text with formalism. Mobile resources could be complemented via natural language annotations (metadata) to make the intent or decision rationale of the author explicit, or to aid understanding of stakeholders that do not have the necessary technical knowledge. Incidentally, many of the language-specific techniques for accessibility (Chisholm et al., 2000) are in agreement with this CD.
- **"Expert" Body of Knowledge:** The three types of knowledge that we are interested are principles, guidelines, and patterns. Following the basic principles (Ghezzi, Jazayeri, & Mandrioli, 2003; Bertini, Catarci, Kimani, & Dix, 2005) underlying a mobile application enables a provider to improve quality attributes related to tiers T1-T3 of the framework. However, principles tend to be abstract in nature which can lead to multiple interpretations in their use and not mandate conformance. The guidelines encourage the use of conventions and good practice, and could serve as a checklist with respect to which an application could be heuristically or otherwise evaluated. The guidelines

available for addressing accessibility (Chisholm et al., 1999; Ahonen, 2003) when tailored to mobile contexts can be used as means for both assurance and evaluation of accessibility of mobile applications. However, guidelines tend to be more useful for those with an expert knowledge than for a novice to whom they may seem rather general to be of much practical use. The problems in using tools to automatically check for accessibility have been outlined in Abascal, Arrue, Fajardo, Garay, and Tomás (2004). Patterns (Alexander, 1979) are reusable entities of knowledge and experience aggregated by experts over years of "best practices" in solving recurring problems in a domain including mobile applications (Roth, 2002; Van Duyne, Landay, & Hong, 2003). They are relatively more structured compared to guidelines and, if represented adequately, provide better opportunities for sharing and reuse. There is, however, an associated cost of learning and adapting patterns to new contexts.

- **Inspections:** Inspections (Wiegers, 2002) are a rigorous form of auditing based upon peer review that can address quality concerns for tiers T1, T2, and most of T3, and help improve the accessibility of mobile applications. Inspections could, for example, use the guidelines and decide what information is and is not considered "comprehensible" by consumers at-large, or whether the choice of labels in a navigation system enhances or reduces readability. Still, inspections, being a means of static verification, cannot completely assess interoperability, performance, reliability, or robustness. Furthermore, inspections do involve an initial cost overhead from training each participant in the structured review process, and the logistics of checklists, forms, and reports.
- **Testing:** Some form of testing is usually an integral part of most development models of mobile applications (Nguyen, Johnson, & Hackett, 2003). However, due to its very nature, testing addresses quality concerns only at of some of the tiers (T1, subset of T2, subset of T3). Interoperability, performance, reliability, and robustness would intimately depend on testing. Unlike inspections, tool support is imperative for testing. Therefore, testing *complements* but does not replace inspections.
- **Metrics:** In a resource-constrained environment of mobile devices, efficient use of time and space is critical. Metrics (Fenton & Pfleeger, 1997) provide a quantitative means for making qualitative judgments about quality concerns at technical levels. There is currently limited support for metrics for mobile applications in general and for their accessibility (Arrue, Vigo, & Abascal, 2005) in particular. Any dedicated effort of deploying metrics for accessibility measurement would inevitably require tool support, which at present is lacking.
- **Tools:** Tools that have help improve quality concerns at all tiers. For example, tools can help report violations of accessibility guidelines, or find non-conformance to markup or scripting language syntax. However, at times tools cannot address some of the stylistic issues (such as an "optimal" distance between two text fragments that will improve readability) or semantic issues (like semantic correctness of a resource included in a mobile application). Therefore, the use of tools as means for automatic accessibility evaluation should be kept in perspective.

Decision Support

A systematic approach to a mobile application must take a variety of constraints into account: organizational constraints (personnel, infrastructure, schedule, budget, and so on) and forces (market value, competitors, and so on).

A producer would need to, for example, take into consideration the cost of an authoring tool vs. the accessibility support it provides; since complete accessibility testing is virtually impossible, determine a stopping criteria that can be attained within the time constraints before the application is delivered; and so on.

Indeed, the last column of Table 1 acknowledges that with respect to any assurance and/or evaluation, and includes feasibility as an all-encompassing consideration on the layers to make the framework practical. There are well-known techniques such as analytical hierarchy process (AHP) and quality function deployment (QFD) for carrying out feasibility analysis, and further discussion of this aspect is beyond the scope of this article.

FUTURE TRENDS

Much of the development of mobile applications is carried out on the desktop. The tools in the form of software development toolkits (SDK) and simulators such as Nokia Mobile Internet Toolkit, Openwave Phone Simulator, and NetFront Mobile Content Viewer assist in that regard. However, explicit support for accessibility in these tools is currently lacking.

The techniques for accessibility for mobile technologies such as XHTML Basic/XHTML Mobile Profile (markup of information) and CSS Mobile Profile (presentation of information) would be of interest. This is especially an imperative considering that the widely used traditional representation languages such as Compact HTML (cHTML), an initiative of the NTT DoCoMo, and the Wireless Markup Language (WML), an initiative of the Open Mobile Alliance (OMA), have evolved towards XHTML Basic or its extensions such as XHTML Mobile Profile.

Identification of appropriate CDs, and an evaluation of the aforementioned languages for presentation or representation of information in a mobile context with respect to them, would also be of interest.

As mobile applications increase in size and complexity, a systematic approach to developing them arises. Indeed, accessibility needs to be a part of the *entire* lifecycle of a mobile application—that is, in the typical workflows of planning, modeling, requirements, design, implementation, and verification and validation. To that regard, integrating accessibility into "lightweight" process methodologies such as Extreme Programming (XP) (Beck & Andres, 2005) that is adapted for a systematic development of small-to-medium scale mobile applications would be useful. A similar argument can be made for the "heavyweight" case, for example, by instantiating the Unified Process (UP) (Jacobson, Booch, & Rumbaugh, 1999) for medium-to-large scale mobile applications.

Finally, a natural extension of the issue of accessibility is to the next generation of mobile applications, namely mobile applications on the semantic Web (Hendler, Lassila, & Berners-Lee, 2001). The mobile applications for the Semantic Web present unique accessibility issues such as inadequacy of current searching techniques (Church, Smyth, & Keane, 2006) and a promising avenue for potential research.

CONCLUSION

This article takes the view that accessibility is not only a technical concern, it is also a social right. In that context, the issues of credibility and legality are particularly relevant as both are at a higher echelon (social level) than accessibility within the semiotic framework.

Credibility is considered to be synonymous to (and therefore interchangeable with) believability (Hovland, Janis, & Kelley, 1953). Indeed, improvement of accessibility is necessary for a demonstration of *expertise,* which is one of the dimensions (Fogg, 2003) of establishment of credibility of the producer with the consumer.

Accessibility is now a legal requirement for public information systems of governments in Canada, the U.S., Australia, and the European Union. The producers need to be aware of the possibility that, as mobile access becomes pervasive in society, the legal extent could expand to mobile applications.

As is well known in engineering contexts, preventative measures such as addressing the problem *early* are often better than curative measures at late stages when they may just be prohibitively expensive or simply infeasible. If accessibility is to be considered as a first-class concern by the producer, it needs to be more than just an afterthought; it needs to be integral to mobile Web engineering.

REFERENCES

Abascal, J., Arrue, M., Fajardo, I., Garay, N., & Tomás, J. (2004). The use of guidelines to automatically verify Web accessibility. *Universal Access in the Information Society, 3*(1), 71-79.

Ahonen, M. (2003, September 19). Accessibility challenges with mobile lifelong learning tools and related collaboration. *Proceedings of the Workshop on Ubiquitous and Mobile Computing for Educational Communities: Enriching and Enlarging Community Spaces (UMOCEC 2003)*, Amsterdam, The Netherlands.

Alexander, C. (1979). *The timeless way of building.* Oxford, UK: Oxford University Press.

Arrue, M., Vigo, M., & Abascal, J. (2005, July 26). Quantitative metrics for Web accessibility evaluation. *Proceedings of the 1st Workshop on Web Measurement and Metrics (WMM05)*, Sydney, Australia.

Beck, K., & Andres, C. (2005). *Extreme programming explained: Embrace change* (2nd ed.). Boston: Addison-Wesley.

Bertini, E., Catarci, T., Kimani, S., & Dix, A. (2005). A review of standard usability principles in the context of mobile computing. *Studies in Communication Sciences, 1*(5), 111-126.

Chisholm, W., Vanderheiden, G., & Jacobs, I. (1999). *Web content accessibility guidelines 1.0.* W3C Recommendation, World Wide Web Consortium (W3C).

Chisholm, W., Vanderheiden, G., & Jacobs, I. (2000). *Techniques for Web content accessibility guidelines 1.0.* W3C Note, World Wide Web Consortium (W3C).

Church, K., Smyth, B., & Keane, M.T. (2006, May 22). Evaluating interfaces for intelligent mobile search. *Proceedings of the International Cross-Disciplinary Workshop on Web Accessibility 2006 (W4A2006)*, Edinburgh, Scotland.

Fogg, B.J. (2003). *Persuasive technology: Using computers to change what we think and do.* San Francisco: Morgan Kaufmann.

Jacobson, I., Booch, G., & Rumbaugh, J. (1999). *The unified software development process.* Boston: Addison-Wesley.

Ghezzi, C., Jazayeri, M., & Mandrioli, D. (2003). *Fundamentals of software engineering* (2nd ed.). Englewood Cliffs, NJ: Prentice-Hall.

Fenton, N. E., & Pfleeger, S. L. (1997). *Software metrics: A rigorous & practical approach.* International Thomson Computer Press.

Green, T. R. G. (1989). Cognitive dimensions of notations. In V. A. Sutcliffe & L. Macaulay (Ed.), *People and computers* (pp. 443-360). Cambridge: Cambridge University Press.

Hendler, J., Lassila, O., & Berners-Lee, T. (2001). The semantic Web. *Scientific American, 284*(5), 34-43.

Hovland, C. I., Janis, I. L., & Kelley, J. J. (1953). *Communication and persuasion.* New Haven, CT: Yale University Press.

Lindland, O. I., Sindre, G., & Sølvberg, A. (1994). Understanding quality in conceptual modeling. *IEEE Software, 11*(2), 42-49.

Nguyen, H. Q., Johnson, R., & Hackett, M. (2003). *Testing applications on the Web: Test planning for mobile and Internet-based systems* (2nd ed.). New York: John Wiley & Sons.

Paavilainen, J. (2002). *Mobile business strategies: Understanding the technologies and opportunities.* Boston: Addison-Wesley.

Van Duyne, D. K., Landay, J., & Hong, J. I. (2003). *The design of sites: Patterns, principles, and processes for crafting a customer-centered Web experience.* Boston: Addison-Wesley.

KEY TERMS

Cognitive Dimensions of Notations: A generic framework for describing the utility of information artifacts by taking the system environment and the user characteristics into consideration.

Delivery Context: A set of attributes that characterizes the capabilities of the access mechanism, the preferences of the user, and other aspects of the context into which a resource is to be delivered.

Mobile Accessibility: Access to the Web by everyone, regardless of their human or environment properties.

Mobile Resource: A mobile network data object that can be identified by a URI. Such a resource may be available in multiple representations.

Mobile Web Engineering: A discipline concerned with the establishment and use of sound scientific, engineering, and management principles, and disciplined and systematic approaches to the successful development, deployment, and maintenance of high-quality mobile Web applications.

Quality: The totality of features and characteristics of a product or a service that bear on its ability to satisfy stated or implied needs.

Semantic Web: An extension of the current Web that adds technological infrastructure for better knowledge representation, interpretation, and reasoning.

Semiotics: The field of study of signs and their representations.

A

Acoustic Data Communication with Mobile Devices

Victor I. Khashchanskiy
First Hop Ltd., Finland

Andrei L. Kustov
First Hop Ltd., Finland

INTRODUCTION

One of the applications of m-commerce is mobile authorization, that is, rights distribution to mobile users by sending authorization data (a token) to the mobile devices. For example, a supermarket can distribute personalized discount coupon tokens to its customers via SMS. The token can be a symbol string that the customers will present while paying for the goods at the cash desk. The example can be elaborated further—using location information from the mobile operator, the coupons can only be sent to, for example, those customers who are in close vicinity of the mall on Saturday (this will of course require customers to allow disclosing their location).

In the example above, the token is used through its manual presentation. However, most interesting is the case when the service is released automatically, without a need for a human operator validating the token and releasing a service to the customer; for example, a vending machine at the automatic gas station must work automatically to be commercially viable.

To succeed, this approach requires a convenient and uniform way of delivering authorization information to the point of service—it is obvious that an average user will only have enough patience for very simple operations. And this presents a problem.

There are basically only three available local (i.e., short-range) wireless interfaces (LWI): WLAN, IR, and Bluetooth, which do not cover the whole range of mobile devices. WLAN has not gained popularity yet, while IR is gradually disappearing. Bluetooth is the most frequently used of them, but still it is not available in all phones.

For every particular device it is possible to send a token out using some combination of LWI and presentation technology, but there is no common and easy-to-use combination. This is a threshold for the development of services.

Taking a deeper look at the mobile devices, we can find one more non-standard simplex LWI, which is present in all devices—acoustical, where the transmitter is a phone ringer. Token presentation through acoustic interface along with general solution of token delivery via SIM Toolkit technology (see 3GPP TS, 1999) was presented by Khashchanskiy and Kustov (2001). However, mobile operators have not taken SIM Toolkit into any serious use, and the only alternative way of delivering sound tokens into the phone-ringing tone customization technology was not available for a broad range of devices at the time the aforementioned paper was published.

Quite unexpectedly, recent development of mobile phone technologies gives a chance for sound tokens to become a better solution for the aforementioned problem, compared with other LWI. Namely, it can be stated that every contemporary mobile device supports either remote customization of ringing tones, or MMS, and in the majority of cases, even both, thus facilitating sound token receiving over the air.

Most phone models can playback a received token with only a few button-clicks. Thus, a sound token-based solution meets the set criteria better than any other LWI. Token delivery works the same way for virtually all phones, and token presentation is simple.

In this article we study the sound token solution practical implementation in detail. First, we select optimal modulation, encoding, and recognition algorithm, and we estimate data rate. Then we present results of experimental verification.

ACOUSTIC DATA CHANNEL

We consider the channel being as follows. The transmitter is a handset ringer; information is encoded as a sequence of sine wave pulses, each with specified frequency and amplitude.

Multimedia message sounds and most ringing tones are delivered as sequences of events in MIDI (musical instrument digital interface) format. A basic pair of MIDI events (note on and note off) defines amplitude, frequency, duration of a note, and the instrument that plays this note. MIDI events can be used to produce information-bearing sound pulses with specified frequency and amplitude.

Widely used support of polyphonic MIDI sequences allows playback of several notes simultaneously. Nonetheless, this has been proved worthless because in order to get

Figure 1. Frequency response measured with test MIDI sequence in hold-max mode

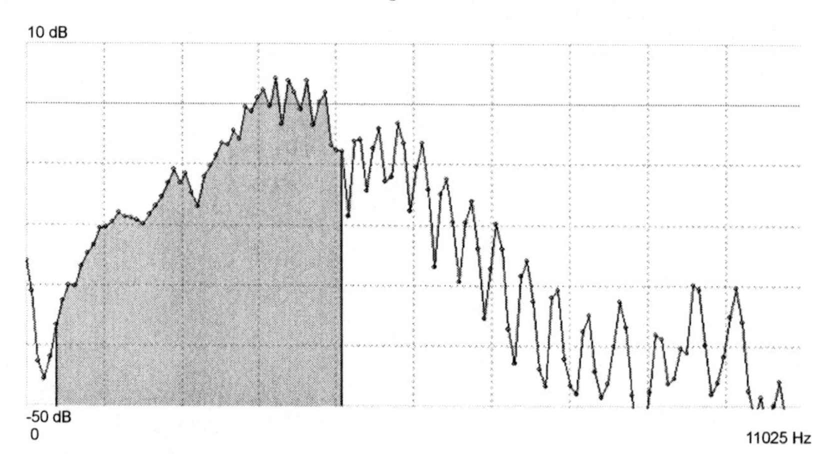

distinguished, these notes have to belong to different non-overlapping frequency ranges. Then the bit rate that can be achieved would be the same as if wider frequency range was allocated for a single note.

The receiver is a microphone; its analog sound signal is digitized and information is decoded from the digital signal by recognition algorithm, based on fast fourier transform (FFT) technique. FFT is, in our opinion, a reasonable trade-off between efficiency and simplicity.

We investigated acoustics properties of mobile devices. After preliminary comparison of a few mobile phone models, we found that ringer quality is of approximately the same level. All handsets have a high level of harmonic distortions and poor frequency response. The results shown in Figures 1 and 2 are obtained for a mid-class mobile phone SonyEricsson T630 and are close to average.

MIDI-based sound synthesis technology applies limitations on pulse magnitude, frequency, and duration. At the same time, ringer frequency response is not linear and the level of harmonic distortions is very high. Figure 1 shows frequency response measured with a sweeping tone or, to be precise, a tone leaping from one musical note to another. To obtain this, the phone played a MIDI sequence of non-overlapping in-time notes that covered a frequency range from 263 to 4200 Hz (gray area).

The frequency response varies over a 40 dB range, reaching its maximum for frequencies from approximately 2.5 to 4 kHz. Moreover, spectral components stretch up to 11 KHz, which is caused by harmonic distortions. This is illustrated also by Figure 2.

Horizontal axis is time; overall duration of the test sequence is 15 seconds. Vertical axis is sound frequency, which is in range from 0 to 11025 Hz. Brightness is proportional to sound relative spectral density; its dynamic range is 60 dB, from black to white.

We also found that frequency of the same note may differ in different handsets. Nevertheless, the ratio of note frequencies (musical intervals) remains correct, otherwise melodies would sound wrong.

For a simplex channel with such poor parameters, as reliable a data encoding method as possible is to be used. Frequency shift keying (FSK) is known as the most reliable method which finds its application in channels with poor signal-to-noise ratio (SNR) and non-linear frequency response.

It is not possible to negotiate transfer rate or clock frequency, as it is usually done in modem protocols because acoustic channel is simplex. To make the channel as adaptive as possible, we have chosen to use differential FSK (DFSK), as it requires no predefined clock frequency. Instead, frequency leaps from one pulse to another provide the channel clocking. The difference between frequencies of consecutive pulses determines the encoded value.

Once encoding scheme is selected, let us estimate possible transfer rate before we can find the balance between data

Figure 2. A spectrogram of the test MIDI sequence

transfer rate and channel reliability. Suppose the transmitter generates a sequence of pulses of duration τ, which follow without gaps with repetition frequency f. If each frequency leap between two consequent pulses carries N bits of information, the overall bit rate p is obviously:

$$p = N \bullet f. \tag{1}$$

In DFSK, for each frequency leap to carry N bits, we must be able to choose pulse frequencies from a set of 2^N+1 values. If a pulse frequency can have n values, we will have

$$p = [log_2(n-1)] \bullet f, \tag{2}$$

where by $[]$ we denote integer part. It follows from (1, 2), that to increase p, we must increase pulse repetition frequency f and the amount of possible values for pulse carrier frequencies n. However, if the recognition is based on spectral analysis, we cannot increase n and f independently. Let us show it. Assume for simplicity that pulse frequency can have any value within frequency range F. Then the number of available values of coding frequencies will be

$$n = [log_2(F/\Delta f - 1)], \tag{3}$$

where Δf is the minimal shift of pulse frequency between two consecutive pulses. Maximum n is achieved with maximum F and minimum Δf. Both parameters have their own boundaries. Bandwidth is limited by the ringer capabilities, and frequency shift is dependant on pulse repetition frequency f, due to the fundamental rule of spectral analysis (Marple, 1987), which defines frequency resolution δf to be in reverse proportionality to observation time T:

$$\delta f = 1 / T. \tag{4}$$

How can (4) be understood in our case of a sequence of pulses? Having converted the signal into frequency domain, we will get the sequence of spectra. As information is encoded in the frequency pulses, we must determine the pulse frequency for every spectrum. This can only be done with certain accuracy δf called frequency resolution. The longer time T we observe the signal, the better frequency resolution is. So for given pulse duration τ, equation (4) sets the lower limit for frequency difference Δf between two consecutive pulses:

$$\Delta f \geq 1 / \tau = f. \tag{5}$$

This means, that if we increase pulse repetition rate f, then we have to correspondingly increase frequency separation Δf for the consecutive pulses; otherwise the spectral analysis-based recognizing device will not principally be able to detect signal.

Let us now try to estimate the data rate for the system we studied earlier. Figures 1 and 2 show that harmonic distortions are very high, and second and third harmonics often have higher magnitudes than the main tone. Consequently, the coding frequencies must belong to the same octave. Their frequency separation should be no less than defined by (5).

An octave contains 12 semitones, so possible frequency values f_i are defined by the following formula:

$$f_i = f_0 \bullet 2^{i/12}, i=0...11. \tag{6}$$

The minimum spacing between consecutive notes is for i=1; maximum for i=11.

In our case, we decided to use the fourth octave—as the closest to the peak area of phone ringer frequency response—in order to maximize SNR and thus make recognition easier. For it, f_0= 2093 Hz, and minimum spacing between notes is 125 Hz. Taking the maximum amount of N = 3 (9 coding frequencies), we can estimate transfer rate as:

$$p_{max} = 3 \bullet 125 = 375\ bps. \tag{7}$$

Recognition Algorithm (Demodulation)

The following algorithm was developed to decode information transferred through audio channel. Analog audio signal from the microphone is digitized with sampling frequency Fs satisfying Nyquist theorem (Marple, 1987). A signal of duration Ts is then represented as a sequence of Ts/Fs samples. FFT is performed on a sliding vector of M signal samples, where M is a power of 2.

- First, sequence of instant power spectra is obtained from the signal using discrete Fourier transform with sliding window (vector) of M samples. To get consecutive spectra overlapped by 50%, the time shift between them was taken $M/2Fs$. Overlapping is needed to eliminate the probability of missing the proper position of a sliding window corresponding to the pulse existence interval, when the pulse duration is not much longer than analysis time significantly (at least twice).
- Second, the synchronization sound is found as sine wave with a constant, but not known in advance frequency, and a certain minimum duration.
- Third, the spectrum composed of maximum values over the spectra sequence (so-called hold-max spectrum) is used to find the pulse carrier frequencies. This step relies on the assumption that used frequency range does not exceed one octave. In other words, the highest frequency is less than twice the value of the lowest one.
- Forth, time cross-sections of spectra sequence at found carrier frequencies are used to recognize moments of sound pulse appearances.

- The last step is reconstruction of encoded bit sequence having the time-ordered set of frequency leaps.

Such an algorithm does not need feedback and can work with unknown carrier frequencies in unknown but limited frequency range. Recognizing the beginning of the transmission is critical for the correct work, so we added "synchronization header" in the beginning of the signal. The length of this header is constant, so the throughput of the system will rise with the message length.

Recognizer Parameters

Here we explain how the parameters of analyzer (***Fs, M***) are defined from that of signal (***f***, Δ***f***). After FFT, we have ***M/2*** of complex samples in frequency domain, corresponding to frequency range from 0 to ***Fs/2***. So for this particular case, frequency resolution obviously equals the difference between the consecutive samples in the frequency domain; namely,

$$df = Fs / M. \tag{7}$$

According to (4), minimum required time of analysis is

$$T = M / Fs. \tag{8}$$

It is obvious, that T must not exceed burst duration τ. Combining (8) and (5), we get:

$$M / Fs \leq 1 / f \tag{9}$$

On the other hand, frequency resolution df must not exceed spacing Δf between carrier frequencies:

$$Fs / M \leq \Delta f \tag{10}$$

Combining (9) and (10), we will finally get:

$$f \leq Fs / M \leq \Delta f \tag{11}$$

which shows that values of analyzer parameters may be restricted when (5) is close to the equation. This imposes requirements on the sound recognition algorithm to work reliably nearby the "critical points," where the recognition becomes principally impossible.

EXPERIMENTAL RESULTS

We implemented a prototype of acoustic data channel with the mobile phone SonyEricsson T630, whose characteristics are seen in Figures 1 and 2.

For encoding, we developed software that encoded symbol strings in ASCII to melody played by an electric organ. The instrument was chosen from 127 instruments available in MIDI format, because its sound is the closest to the sine wave pulses model we used in calculations. It is maintained at approximately the same level over the whole note duration.

Figure 3. Encoded "hello world"; note the leading synchronization header. Overall duration is approximately 2 seconds.

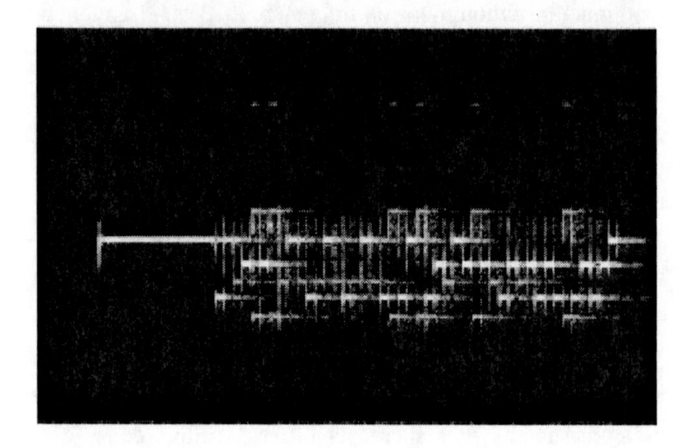

The recognizer consisted of a Sony ECM-MS907 studio microphone for signal recording, and a conventional PC with a sound card was used for signal analysis. FFT processing was done by our own software.

In the beginning of our experiments, we used the parameters described in the theoretical section. Later we found that at the highest possible transfer rate, data recognition is not reliable. So we gradually increased pulse duration until recognition became reliable. Eventually we selected the following modulation parameters: ***n=5*** (each frequency leap carries two data bits), notes were evenly distributed over the octave (C, D#, F, G, A in musical notation, and they correspond to frequencies 2093, 2489, 2794, 3136, and 3520 Hz), and pulse duration was 46 ms.

Figure 3 shows a spectrogram of recognizable signal from the microphone.

Horizontal axis is time, and overall signal duration is 2 seconds. Vertical axis is frequency, and one can see the leaps between consecutive pulses. Brightness is proportional to the signal intensity.

This example signal carries 88 bits of information (a string "hello world," coded as 11 ASCII characters), which makes the data transfer rate approximately 40 bps. Overhead from the synchronization header is ca. 25%; for longer messages the average transfer rate would be higher.

DISCUSSION

We have managed to implement a reliable data channel from the phone; the advantage of the proposed recognition algorithm is that it can work in the same way for every mobile device, independent on acoustic properties of different brands and models, although encoding frequencies are different.

The channel is principally one way: the handset cannot receive any feedback that can be used, for example, for error correction. Nevertheless, developed recognition algorithm provided good reliability. For a handset placed 30 cm from the microphone, in a room environment, recognition was 100% reliable. This condition corresponds to the output of the average phone in a "normal" room environment.

Ensuring reliability does not seem to be a very difficult task. First of all, SNR can be improved by increasing the number of receiving microphones. On the other hand, in practical systems simple shielding is very easy to implement. And finally, even one error in recognition is not fatal: the user can always have another try. A recognizing device can easily identify cases of unsuccessful recognition and indicate the former case for the user to retry.

The recognition system can be implemented on any PC equipped with a sound card. The algorithm is so simple that the system can also be implemented as an embedded solution based on digital signal processors. Microphone requirements are not critical either: both the frequency response and SNR of entry level microphones are much better than those of mobile device ringers. This means that cheap stand-alone recognizers can be implemented and deployed at the points of service.

It is interesting to note that other devices capable of playing MIDI sequences (e.g., PDAs) can be used as well as mobile phones.

Measured transfer rate (40 bps) was considerably less than the estimation, obtained in our simple model—375 bps. We think that the reason for this was slow pulse decay rate in combination with non-linear frequency response. Amplitude of the note with frequency close to a local frequency response maximum might remain higher than amplitude of the consecutive note through the whole duration of the latter. Thus, the weaker sound of the second note might be not recognized.

However, we consider even such relatively slow transmission still suitable for the purposes of mobile authorization applications, because authorization data is usually small and its transmission time is not critical.

Our example (Figure 3) seems to be a quite practical situation—transmitting 11-symbol password during 2s is definitely not too long for a user. Typing the same token on the vending machine keyboard would easily take twice as long.

The acoustic presentation method might be an attractive feature for teenagers (e.g., mobile cinema tickets being one conceivable application).

ACKNOWLEDGMENTS

The authors would like to thank Petteri Koponen for the original idea.

REFERENCES

Khashchanskiy, V., & Kustov, A. (2001). Universal SIM Toolkit-based client for mobile authorization system. *Proceedings of the 3rd International Conference on Information Integration and Web-Based Applications & Services (IIWAS 2001)* (pp. 337-344).

Marple, S. Lawrence Jr. (1987). *Digital spectral analysis with applications.* Englewood Cliffs, NJ: Prentice-Hall.

3GPP TS 11.14. (1999). *Specification of the SIM application toolkit for the Subscriber Identity Module-Mobile Equipment (SIM-ME) interface.* Retrieved from http://www.3gpp.org/ftp/Specs/html-info/1114.htm

KEY TERMS

Fast Fourier Transform (FFT): An optimized form of the algorithm that calculates a complex spectrum of digitized signals. It is most widely used to obtain a so-called power spectrum as a square of a complex spectrum module. Power spectrum represents energy distribution along frequency axis.

Frequency Resolution: The minimum difference in frequencies which can be distinguished in a signal spectrum.

Frequency Response: For a device, circuit, or system, the ratio between output and input signal spectra.

Frequency Shift Keying (FSK): The digital modulation scheme that assigns fixed frequencies to certain bit sequences. Differential FSK (DFSK) uses frequency differences to encode bit sequences.

Harmonic Distortions: Alteration of the original signal shape caused by the appearance of higher harmonics of input signal at the output.

IR: Short-range infrared communication channel.

Musical Instrument Digital Interface (MIDI): A standard communications protocol that transfers musical notes between electronic musical instruments as sequences of events, like 'Note On', 'Note Off', and many others.

Sampling Frequency: The rate at which analogue signal is digitized by an analogue-to-digital converter (ADC) in order to convert the signal into numeric format that can be stored and processed by a computer.

Adaptive Transmission of Multimedia Data over UMTS

Antonios Alexiou
Patras University, Greece

Dimitrios Antonellis
Patras University, Greece

Christos Bouras
Patras University, Greece

INTRODUCTION

As communications technology is being developed, users' demand for multimedia services raises. Meanwhile, the Internet has enjoyed tremendous growth in recent years. Consequently, there is a great interest in using the IP-based networks to provide multimedia services. One of the most important areas in which the issues are being debated is the development of standards for the universal mobile telecommunications system (UMTS). UMTS constitutes the third generation of cellular wireless networks which aims to provide high-speed data access along with real-time voice calls. Wireless data is one of the major boosters of wireless communications and one of the main motivations of the next-generation standards.

Bandwidth is a valuable and limited resource for UMTS and every wireless network in general. Therefore, it is of extreme importance to exploit this resource in the most efficient way. Consequently, when a user experiences a streaming video, there should be enough bandwidth available at any time for any other application that the mobile user might need. In addition, when two different applications run together, the network should guarantee that there is no possibility for any of the above-mentioned applications to prevail against the other by taking all the available channel bandwidth. Since Internet applications adopt mainly TCP as the transport protocol, while streaming applications mainly use RTP, the network should guarantee that RTP does not prevail against the TCP traffic. This means that there should be enough bandwidth available in the wireless channel for the Internet applications to run properly.

BACKGROUND

Chen and Zachor (2004) propose a widely accepted rate control method in wired networks which is the equation-based rate control, also known as TFRC (TCP-friendly rate control). In this approach the authors use multiple TFRC connections as an end-to-end rate control solution for wireless streaming video. Another approach is presented by Fu and Liew (2003). As they mention, TCP Reno treats the occurrence of packet losses as a manifestation of network congestion. This assumption may not apply to networks with wireless channels, in which packet losses are often induced by noise, link error, or reasons other than network congestion. Equivalently, TCP Vegas uses queuing delay as a measure of congestion (Choe & Low, 2003). Thus, Fu and Liew (2003) propose an enhancement of TCP Reno and TCP Vegas for the wireless networks, namely TCP Veno.

Chen, Low, and Doyle (2005) present two algorithms that formulate resource allocation in wireless networks. These procedures constitute a preliminary step towards a systematic approach to jointly design TCP congestion control algorithms, not only to improve performance, but more importantly, to make interaction more transparent. Additionally, Xu, Tian, and Ansari (2005) study the performance characteristics of TCP New Reno, TCP SACK, TCP Veno, and TCP Westwood under the wireless network conditions and they propose a new TCP scheme, called TCP New Jersey, which is capable of distinguishing wireless packet losses from congestion.

Recent work provides an overview of MPEG-4 video transmission over wireless networks (Zhao, Kok, & Ahmad, 2004). A critical issue is how we can ensure the QoS of video-based applications to be maintained at an acceptable level. Another point to consider is the unreliability of the network, especially of the wireless channels, because we observe packet losses resulting in a reduction of the video quality. The results demonstrate that the video quality can be substantially improved by preserving the high-priority video data during the transmission.

THE TCP-FRIENDLY RATE CONTROL PROTOCOL

TFRC is not actually a fully specified end to-end transmission protocol, but a congestion control mechanism that is designed

Figure 1. Typical scenario for streaming video over UMTS

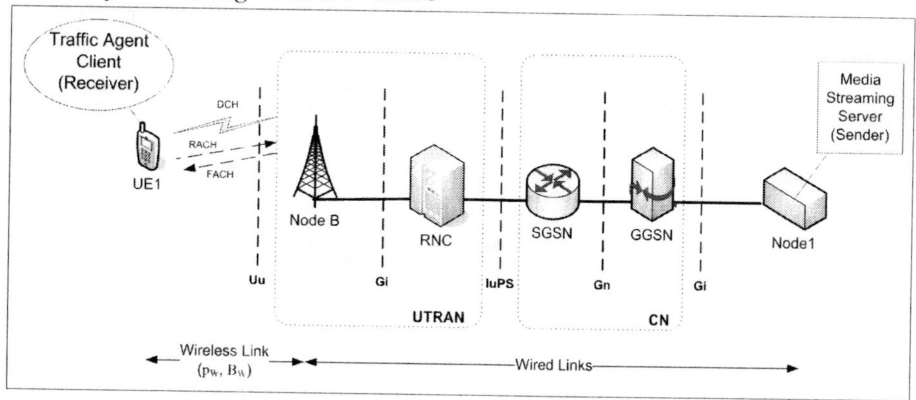

to operate fairly along with TCP traffic. Generally TFRC should be deployed with some existing transport protocol such as UDP or RTP in order to present its useful properties (Floyd, Handley, Padhye, & Widmer, 2000). The main idea behind TFRC is to provide a smooth transmission rate for streaming applications. The other properties of TFRC include slow response to congestion and the opportunity of not aggressively trying to make up with all available bandwidth. Consequently, in case of a single packet loss, TFRC does not halve its transmission rate like TCP, while on the other hand it does not respond rapidly to the changes in available network bandwidth. TFRC has also been designed to behave fairly when competing for the available bandwidth with concurrent TCP flows that comprise the majority of flows in today's networks. A widely popular model for TFRC is described by the following equation (Floyd & Fall, 1999):

$$T = \frac{kS}{RTT\sqrt{p}} \tag{1}$$

T represents the sending rate, *S* is the packet size, *RTT* is the end-to-end round trip time, *p* is the end-to-end packet loss rate, and *k* is a constant factor between 0.7 and 1.3 (Mahdavi & Floyd, 1997) depending on the particular derivation of equation (1).

The equation describes TFRC's sending rate as a function of the measured packet loss rate, round-trip time, and used packet size. More specifically, a potential congestion in the nodes of the path will cause an increment in the packet loss rate and in the round trip time according to the current packet size. Given this fluctuation, it is easy to determine the new transmission rate so as to avoid congestion and packet losses. Generally, TFRC's congestion control consists of the following mechanisms:

1. The receiver measures the packet loss event rate and feeds this information back to the sender.
2. The sender uses these feedback messages to calculate the round-trip-time (*RTT*) of the packets.
3. The loss event rate and the *RTT* are then fed into the TRFC rate calculation equation (described later in more detail) in order to find out the correct data sending rate.

ANALYSIS OF THE TFRC MECHANISM FOR UMTS

The typical scenario for streaming video over UMTS is shown in Figure 1, where the server is denoted by Node1 and the receiver by UE1. The addressed scenario comprises a UMTS radio cell covered by a Node B connected to an RNC. The model consists of a UE connected to DCH, as shown in Figure 1. In this case, the DCH is used for the transmission of the data over the air. DCH is a bi-directional channel and is reserved only for a single user. The common channels are the forward access channel (FACH) in the downlink and the random access channel (RACH) in the uplink.

The wireless link is assumed to have available bandwidth B_W, and packet loss rate p_w, caused by wireless channel error. This implies that the maximum throughput that could be achieved in the wireless link is $B_W(1 - p_W)$. There could also be packet loss caused by congestion at wired nodes denoted by $p_{node\,name}$ (node name: GGSN, SGSN, RNC, Node B). The end-to-end packet loss rate observed by the receiver is denoted as *p*. The streaming rate is denoted by *T*. This means that the streaming throughput is $T(1 - p)$. Under the above assumptions we characterize the wireless channel as underutilized if $T(1 - p) < B_W(1 - p_W)$. Given the above described scenario, the following are assumed:

1. The wireless link is the long-term bottleneck. This means that there is no congestion due to streaming traffic to the nodes GGSN, SGSN, and RNC.
2. There is no congestion at Node B due to the streaming application, if and only if the wireless bandwidth is underutilized—that is, $T(1 - p) < B_W(1 - p_W)$. This also implies that no queuing delay is caused at Node

B and hence, the round trip time for a given route has the minimum value (i.e., RTT_{min}). Thus, this assumption can be restated as follows: for a given route, $RTT = RTT_{min}$ if and only if $T(1 - p) \leq B_W(1 - p_W)$. This in turn implies that if $T(1 - p) > B_W(1 - p_W)$ then $RTT \geq RTT_{min}$.

3. The packet loss rate caused by wireless channel error (p_W) is random and varies from 0 to 0.16.
4. The backward route is error-free and congestion-free.

The communication between the sender and the receiver is based on RTP/RTCP sessions; and the sender, denoted by Node 1 (Figure 1), uses the RTP protocol to transmit the video stream. The client, denoted by UE1 (Figure 1), uses the RTCP protocol in order to exchange control messages. The mobile user in recurrent time space sends RTCP reports to the media server. These reports contain information about the current conditions of the wireless link during the transmission of the multimedia data between the server and the mobile user. The feedback information contains the following parameters:

- **Packet Loss Rate:** The receiver calculates the packet loss rate during the reception of sender data, based on RTP packets sequence numbers.
- **Timestamp of Every Packet Arrived at the Mobile User:** This parameter is used by the server for the RTT calculation of every packet.

The media server extracts the feedback information from the RTCP report and passes it through an appropriate filter. The use of filter is essential for the operation of the mechanism in order to avoid wrong estimations of the network conditions. On the sender side, the media server using the feedback information estimates the appropriate rate of the streaming video so as to avoid network congestion. The appropriate transmission rate of the video sequence is calculated from equation (1), and the media server is responsible for adjusting the sending rate with the calculated value. Obviously, the media server does not have the opportunity to transmit the video in all the calculated sending rates. However, it provides a small variety of them and has to approximate the calculated value choosing the sending rate from the provided transmission rates.

This extends the functionality of the whole congestion control mechanism. More specifically, the sender does not have to change the transmission rate every time it calculates a new one with a slight difference from the previous value. Consequently, it changes the transmission rate of the multimedia data to one of the available sending rates of the media server as has already been mentioned. In this approach, the number of the changes in the sending rate is small and the mobile user does not deal with a continually different transmission rate.

As mentioned above, it is essential to keep a history of the previous calculated values for the transmission rate. Having this information, the media server can estimate the smoothed transmission rate, using the m most recent values of the calculated sending rate from the following equation:

$$T^{Smoothed} = \frac{\sum_{i=1}^{m} w_i \cdot T_{m+1-i}^{Smoothed}}{\sum_{i=1}^{m} w_i} \quad (2)$$

The value m, used in calculating the transmission rate, determines TFRC's speed in responding to changes in the level of congestion (Handley, Floyd, Padhye, & Widmer, 2003). The weights w_i are appropriately chosen so that the most recent calculated sending rates receive the same high weights, while the weights gradually decrease to 0 for older calculated values.

Equivalently to the calculation of the transmission rate, the mobile user (client) measures the packet loss rate p_l based on the RTP packets sequence numbers. This information is sent to the media server via the RTCP reports. In order to prevent a single spurious packet loss having an excessive effect on the packet loss estimation, the server smoothes the values of packet loss rate using the filter of the following equation, which computes the weighted average of the m most recent loss rate values (Vicisiano, Rizzo, & Crowcroft, 1998).

$$p_1^{Smoothed} = \frac{\sum_{i=1}^{m} w_i \cdot p_{1,m+1-i}^{Smoothed}}{\sum_{i=1}^{m} w_i} \quad (3)$$

The value of $p_1^{Smoothed}$ is then used by equation (1) for the estimation of the transmission rate of the multimedia data. The weights w_i are chosen as in the transmission rate estimation.

FUTURE TRENDS

The most prominent enhancement of the adaptive real-time applications is the use of multicast transmission of the multimedia data. The multicast transmission of multimedia data has to accommodate clients with heterogeneous data reception capabilities. To accommodate heterogeneity, the server may transmit one multicast stream and determine the transmission rate that satisfies most of the clients (Byers et al., 2000). Additionally, Vickers, Albuquerque, and Suda (1998) present different approaches where the server transmits multiple multicast streams with different transmission rates allocating the client at these streams, as well as using layered encoding and transmitting each layer to a different multicast stream. An interesting survey of techniques for multicast multimedia data over the Internet is presented in Li, Ammar, and Paul (1999).

Single multicast stream approaches have the disadvantage that clients with a low-bandwidth link will always get a high-bandwidth stream if most of the other members are connected via a high-bandwidth link, and the same is true the other way around. This problem can be overcome with the use of a multi-stream multicast approach. Single multicast stream approaches have the advantages of easy encoder and decoder implementation and simple protocol operation, due to the fact that during the single multicast stream approach, there is no need for synchronization of clients' actions (as the multiple multicast streams and layered encoding approaches require).

The subject of adaptive multicast of multimedia data over networks with the use of one multicast stream has engaged researchers all over the world. During the adaptive multicast transmission of multimedia data in a single multicast stream, the server must select the transmission rate that satisfies most the clients with the current network conditions. Totally, three approaches can be found in the literature for the implementation of the adaptation protocol in a single stream multicast mechanism: equation based (Rizzo, 2000; Widmer & Handley, 2001), network feedback based (Byers et al., 2000), or a combination of the above two approaches (Sisalem & Wolisz, 2000).

CONCLUSION

An analysis of the TCP friendly rate control mechanism for UMTS has been presented. The TFRC mechanism gives the opportunity to estimate the appropriate transmission rate of the video data for avoiding congestion in the network. The three goals of this rate control could be stated as follows. First, the streaming rate does not cause any network instability (i.e., congestion collapse). Second, TFRC is assumed to be TCP Friendly, which means that any application that transmits data over a network presents friendly behavior towards the other flows that coexist in the network and especially towards the TCP flows that comprise the majority of flows in today's networks. Third, it leads to the optimal performance—that is, it results in the highest possible throughput and lowest possible packet loss rate. Furthermore, an overview of video transmission over UMTS using real-time protocols such as RTP/RTCP has been presented.

REFERENCES

Byers, J., Frumin, M., Horn, G., Luby, M., Mitzenmacher, M., Roetter, A., & Shaver, W. (2000). FLID-DL: Congestion control for layered multicast. *Proceedings of the International Workshop on Networked Group Communication* (pp. 71-81), Palo Alto, CA.

Chen, M., & Zachor, A. (2004). Rate control for streaming video over wireless. *IEEE INFOCOM,* Hong Kong, China, (pp. 1181-1190).

Chen, L., Low, S., & Doyle, J. (2005). Joint congestion control and media access control design for ad hoc wireless networks. *IEEE INFOCOM,* Miami, FL.

Choe, H., & Low, S. (2003). Stabilized Vegas. *IEEE INFOCOM, 22*(1), 2290-2300.

Floyd, S., Handley, M., Padhye, J., & Widmer, J. (2000). Equation-based congestion control for unicast applications. *Proceedings of ACM SIGCOMM,* Stockholm, Sweden, (pp. 43-56).

Floyd, S., & Fall, K. (1999). Promoting the use of end-to-end congestion control in the Internet. *IEEE/ACM Transactions on Networking, 7*(4), 458-472.

Fu, C. P., & Liew, S. C. (2003). TCP Veno: TCP enhancement for transmission over wireless access networks. *IEEE Journal on Selected Areas in Communications, 21*(2), 216-228.

Handley, M., Floyd, S., Padhye, J., & Widmer, J. (2003). TCP Friendly Rate Control (TFRC). *RFC, 3448.*

Li, X., Ammar, M., & Paul, S. (1999). Video multicast over the Internet. *IEEE Network Magazine, 12*(2), 46-60.

Mahdavi, J., & Floyd, S. (1997). *TCP-Friendly unicast rate-based flow control.* Retrieved from http://www.psc.edu/networking/papers/tcp_friendly.html

Rizzo, L. (2000). pgmcc: A TCP-friendly single-rate multicast congestion control scheme. *Proceedings of ACM SIGCOMM,* Stockholm, Sweden, (pp. 17-28).

Sisalem, D., & Wolisz, A. (2000). LDA+ TCP-Friendly adaptation: A measurement and comparison study. *Proceedings of the International Workshop on Network and Operating Systems Support for Digital Audio and Video (NOSSDAV),* Chapel Hill, NC.

Vicisiano, L., Rizzo, L., & Crowcroft, J. (1998). TCP-like congestion control for layered multicast data transfer. *IEEE INFOCOM,* San Francisco, CA, (pp. 996-1003).

Vickers, J., Albuquerque, N., & Suda, T. (1998). Adaptive multicast of multi-layered video: Rate-based and credit-based approaches. *IEEE INFOCOM,* San Francisco, CA, (pp. 1073-1083).

Widmer, J., & Handley, M. (2001). Extending equation-based congestion control to multicast applications. *Proceedings of ACM SIGCOMM,* San Diego, CA, (pp. 275-286).

Xu, K., Tian, Y., & Ansari, N. (2005). Improving TCP performance in integrated wireless communications networks. *Computer Networks, Science Direct, 47*(2), 219-237.

Zhao, J., Kok, C., & Ahmad, I. (2004). MPEG-4 video transmission over wireless networks: A link level performance study. *Wireless Networks, 10*(2), 133-146.

KEY TERMS

Adaptive Real-Time Application: An application that has the capability to transmit multimedia data over heterogeneous networks and adapt media transmission to network changes.

Delay Jitter: The mean deviation (smoothed absolute value) of the difference in packet spacing at the receiver compared to the sender for a pair of packets.

Frame Rate: The rate of the frames, which are encoded by video encoder.

Multimedia Data: Data that consist of various media types like text, audio, video, and animation.

Packet Loss Rate: The fraction of the total transmitted packets that did not arrive at the receiver.

RTP/RTCP: Protocol used for the transmission of multimedia data. The RTP performs the actual transmission, and the RTCP is the control and monitoring transmission.

Addressing the Credibility of Mobile Applications

Pankaj Kamthan
Concordia University, Canada

INTRODUCTION

Mobile access has opened new vistas for various sectors of society including businesses. The ability that anyone using (virtually) any device could be reached anytime and anywhere presents a tremendous commercial potential. Indeed, the number of mobile applications has seen a tremendous growth in the last few years.

In retrospect, the fact that almost *anyone* can set up a mobile application claiming to offer products and services raises the question of credibility from a consumer's viewpoint. The obligation of establishing credibility is essential for an organization's reputation (Gibson, 2002) and for building consumers' trust (Kamthan, 1999). If not addressed, there is a potential for lost consumer confidence, thus significantly reducing the advantages and opportunities the mobile Web as a medium offers. If a mobile application is not seen as credible, we face the inevitable consequence of a product, however functionally superior it might be, rendered socially isolated.

The rest of the article is organized as follows. We first provide the motivational background necessary for later discussion. This is followed by introduction of a framework within which different types of credibility in the context of mobile applications can be systematically addressed and thereby improved. Next, challenges and directions for future research are outlined. Finally, concluding remarks are given.

BACKGROUND

In this section, we present the fundamental concepts underlying credibility, and present the motivation and related work for addressing credibility within the context of mobile applications.

Basic Credibility Concepts

For the purposes of this article, we will consider credibility to be synonymous to (and therefore interchangeable with) believability (Hovland, Janis, & Kelley, 1953). We follow the terminology of Fogg and Tseng (1999), and view credibility and trust as being slightly different. Since trust indicates a *positive* belief about a person, object, or process, we do not consider credibility and trust to be synonymous.

It has been pointed out in various studies (Fogg, 2003; Metzger, 2005) that credibility consists of two primary dimensions, namely *trustworthiness* and *expertise* of the source of some information. Trustworthiness is defined by the terms such as well-intentioned, truthful, unbiased, and so on. The trustworthiness dimension of credibility captures the *perceived* goodness or morality of the source. Expertise is defined by terms such as knowledgeable, experienced, competent, and so on. The expertise dimension of credibility captures the *perceived* knowledge and skill of the source. Together, they suggest that "highly credible" mobile applications will be perceived to have high levels of *both* trustworthiness and expertise.

We note that trustworthiness and expertise are at such a high level of abstraction that direct treatment of any of them is difficult. Therefore, in order to improve credibility, we need to find quantifiable attributes that can improve each of these dimensions.

A Classification of Credibility

The following taxonomy helps associating the concept of credibility with a specific user class in context of a mobile application. A user could consider a mobile application to be credible based upon direct interaction with the application (*active credibility*), or consider it to be credible in absence of any direct interaction but based on certain pre-determined notions (*passive credibility*). Based on the classification of credibility in computer use (Fogg & Tseng, 1999) and adapting them to the domain of mobile applications, we can decompose these further.

There can be two types of *active credibility:* (1) *surface credibility,* which describes how much the user believes the mobile application is based on simple inspection; and (2) *experienced credibility,* which describes how much the user believes the mobile application is based on first-hand experience in the past.

There can be two types of *passive credibility:* (1) *presumed credibility,* which describes how much the user believes the mobile application because of general assumptions that the user holds; and (2) *reputed credibility,* which describes how much the user believes the mobile application because of a reference from a third party.

Finally, credibility is not absolute with respect to users and with respect to the application itself (Metzger, Flanagin, Eyal, Lemus, & McCann, 2003). Also, credibility can be associated with a whole mobile application or a part of a mobile application. For example, a user may question the credibility of information on a specific product displayed in a mobile application. We contend that for a mobile application to be labeled non-credible, there must exist at least a part of it that is labeled non-credible based on the above classification by at least one user.

The Origins and Significance of the Problem of Mobile Credibility

The credibility of mobile applications deserves special attention for the following reasons:

- **Delivery Context:** Mobile applications are different from the desktop or Web environments (Paavilainen, 2002) where context-awareness (Sadeh, Chan, Van, Kwon, & Takizawa, 2003) is a unique challenge. The delivery context in a changing environment of mobile markup languages, variations in user agents, and constrained capabilities of mobile devices presents unique challenges towards active credibility.
- **Legal Context:** Since the stakeholders of a mobile application need not be co-located (different jurisdictions in the same country or in different countries), the laws that govern the provider and the user may be different. Also, the possibilities of fraud such as computer domain name impersonation (commonly known as "pharming") or user identity theft (commonly known as "phishing") with little legal repercussions for the perpetrators is relatively high in a networked environment. These possibilities can impact negatively on presumed credibility.
- **User Context:** Users may deploy mobile devices with varying configurations, and in the event of problems with a mobile service, may first question the provider rather than the device that they own. In order for providers of mobile portals to deliver user-specific information and services, they need to know details about the user (such as profile information, location, and so on). This creates the classical dichotomy between personalization and privacy, and striking a balance between the two is a constant struggle for businesses (Kasanoff, 2002). The benefits of respecting one can adversely affect the other, thereby impacting their credibility in the view of their customers. Furthermore, the absence of a human component from non-proximity or "facelessness" of the provider can shake customer confidence and create negative perceptions in a time of crisis such as denial of service or user agent crash. These instances can lead to a negative passive credibility.

Initiatives for Improving Mobile Credibility

There have been initiatives to address the credibility of Web applications such as a user survey to identify the characteristics that users consider necessary for a Web application to be credible (Fogg et al., 2001) and a set of guidelines (Fogg, 2003) for addressing *surface, experienced, presumed,* and *reputed credibility* of Web applications.

However, these efforts are limited by one or more of the following issues. The approach towards ensuring and/or evaluating credibility is not systematic, the proposed means for ensuring credibility is singular (only guidelines), and the issue of feasibility of the means is not addressed. Moreover, these guidelines are not specific to mobility, are not prioritized and the possibility that they can contradict each other is not considered, can be open to broad interpretation, and are stated at such a high level that they may be difficult to realize by a novice user.

ADDRESSING THE CREDIBILITY OF MOBILE APPLICATIONS

In this section, we consider approaches for understanding and improving active credibility of mobile applications.

A Framework for Addressing Active Credibility of Mobile Applications

To systematically address the active credibility of mobile applications, we take the following steps:

1. View credibility as a qualitative aspect and address it indirectly via quantitative means.
2. Select a theoretical basis for communication of information (semiotics), and place credibility in its setting.
3. Address semiotic quality in a practical manner.

Based on this and using the primary dimensions that affect credibility, we propose a framework for active credibility of mobile applications (see Table 1). The external attributes (denoted by E) are extrinsic to the software product and are directly a user's concern, while internal attributes (denoted by I) are intrinsic to the software product and are directly an engineer's concern. Since not all attributes corresponding to a semiotic level are at the same echelon, the different tiers are denoted by "Tn."

Table 1. A semiotic framework for active credibility of mobile applications

Semiotic Level	Quality Attributes	Means for Credibility Assurance and Evaluation	Decision Support
Social	**Credibility**	• "Expert" Knowledge (Principles, Guidelines, Patterns) • Inspections • Testing • Metrics • Tools	Feasibility
	Aesthetics, Legality, Privacy, Security, (Provider) Transparency [T5;E]		
Pragmatic	Accessibility, Usability [T4;E]		
	Interoperability, Portability, Reliability, Robustness [T3;E]		
Semantic	Completeness and Validity [T2;I]		
Syntactic	Correctness [T1;I]		

We now describe each of the components of the framework in detail.

Semiotic Levels

The first column of Table 1 addresses semiotic levels. Semiotics (Stamper, 1992) is concerned with the use of symbols to convey knowledge.

From a semiotics perspective, a representation can be viewed on six interrelated levels: physical, empirical, syntactic, semantic, pragmatic, and social, each depending on the previous one in that order. The physical and empirical levels are concerned with the physical representation of signs in hardware and communication properties of signs, and are not of direct concern here. The syntactic level is responsible for the formal or structural relations between signs. The semantic level is responsible for the relationship of signs to what they stand for. The pragmatic level is responsible for the relation of signs to interpreters. The social level is responsible for the manifestation of social interaction with respect to signs.

Quality Attributes

The second column of Table 1 draws the relationship between semiotic levels and corresponding quality attributes.

Credibility belongs to the social level and depends on the layers beneath it. The external quality attributes *legality, privacy, security,* and *(provider) transparency* also at the social level depend upon the external quality attributes *accessibility* and *usability* at the pragmatic level, which in turn depend upon the external quality attributes *interoperability, performance, portability, reliability,* and *robustness* also at the pragmatic level. (We note here that although *accessibility* and *usability* do overlap in their design and implementation, they are not identical in their goals for their user groups.)

We discuss in some detail only the entries in the social level. Aesthetics is close to human senses and perception, and plays a crucial role in making a mobile application "salient" to its customers beyond simply the functionality it offers. It is critical that the mobile application be legal (e.g., is legal in the jurisdiction it operates and all components it makes use of are legal); takes steps to respect a user's privacy (e.g., does not use or share user-supplied information outside the permitted realm); and be secure (e.g., in situations where financial transactions are made). The provider must take all steps to be transparent with respect to the user (e.g., not include misleading information such as the features of products or services offered, clearly label promotional content, make policies regarding returning/exchanging products open, and so on).

The internal quality attributes for syntactic and semantic levels are inspired by Lindland, Sindre, and Sølvberg (1994) and Fenton and Pfleeger (1997). At the semantic level, we are only concerned with the conformance of the mobile application to the domain(s) it represents (that is, semantic correctness or completeness) and vice versa (that is, semantic validity). At the syntactic level the interest is in conformance with respect to the languages used to produce the mobile application (that is, syntactic correctness).

The definitions of each of these attributes can vary in the literature, and therefore it is important that they be adopted and followed consistently. For example, the definition of usability varies significantly across ISO/IEC Standard 9126 and ISO Standard 9241 with respect to the perspective taken in their formulation.

Means for Credibility Assurance and Evaluation

The third column of Table 1 lists (in no particular order, by no means complete, and not necessarily mutually exclusive) the means for assuring and evaluating active credibility.

- **"Expert" Body of Knowledge:** The three types of knowledge that we are interested in are principles, guidelines, and patterns. Following the basic principles (Ghezzi, Jazayeri, & Mandrioli, 2003) underlying a mobile application enables a provider to improve quality attributes related to (T1-T3) of the framework. The guidelines encourage the use of conventions and good practice, and could also serve as a checklist with respect to which an application could be heuristically or otherwise evaluated. There are guidelines available for addressing accessibility (Chisholm, Vanderheiden, & Jacobs, 1999; Ahonen, 2003), security (McGraw & Felten, 1998), and usability (Bertini, Catarci, Kimani, & Dix, 2005) of mobile applications. However, guidelines tend to be more useful for those with an expert knowledge than for a novice to whom they may seem rather general to be of much practical use. Patterns are reusable entities of knowledge and experience aggregated by experts over years of "best practices" in solving recurring problems in a domain including that in mobile applications (Roth, 2001, 2002). They are relatively more structured compared to guidelines and provide better opportunities for sharing and reuse. There is, however, a lack of patterns that clearly address quality concerns in mobile applications. Also, there is a cost of adaptation of patterns to new contexts.
- **Inspections:** Inspections (Wiegers, 2002) are a rigorous form of auditing based upon peer review that can address quality concerns at both technical and social levels (T1-T5), and help improve the credibility of mobile applications. Inspections could, for example, decide what information is and is not considered "promotional," help improve the labels used to provide cues to a user (say, in a navigation system), and assess the readability of documents. Still, inspections do involve an initial cost overhead from training each participant in the structured review process, and the logistics of checklists, forms, and reports.
- **Testing:** Some form of testing is usually an integral part of most development models of mobile applications (Nguyen, Johnson, & Hackett, 2003). There are test suites and test harnesses for many of the languages commonly used for representation of information in mobile applications. However, due to its very nature, testing addresses quality concerns of only some of the technical and social levels (T1, subset of T2, T3, T4, subset of T5). Therefore, testing *complements* but does not replace inspections. Accessibility or usability testing that requires hiring real users, infrastructure with video monitoring, and subsequent analysis of data can prove to be prohibitive for small-to-medium-size enterprises.
- **Metrics:** In a resource-constrained environment of mobile devices, efficient use of time and space is critical. Metrics (Fenton & Pfleeger, 1997) provide a quantitative means for making qualitative judgments about quality concerns at technical levels. For example, metrics for a document or image size can help compare and make a choice between two designs, or metrics for structural complexity could help determine the number of steps required in navigation, which in turn could be used to estimate user effort. However, well-tested metrics for mobile applications are currently lacking. We also note that a dedicated use of metrics on a large scale usually requires tool support.
- **Tools:** Tools that have help improve quality concerns at technical and social levels. For example, tools can help engineers detect security breaches, report violations of accessibility or usability guidelines, find non-conformance to markup or scripting language syntax, suggest image sizes favorable to the small devices, or detect broken links. However, at times, tools cannot address some of the technical quality concerns (like complete semantic correctness of the application with respect to the application domain), as well as certain social quality concerns (like provider intent or user bias). Therefore, the use of tools as means for automatic quality assurance or evaluation should be kept in perspective.

Decision Support

A mobile application project must take a variety of constraints into account: organizational constraints of time and resources (personnel, infrastructure, budget, and so on) and external forces (market value, competitors, and so on). These compel providers to make quality-related decisions that, apart from being sensitive to credibility, must also be feasible.

For example, the provider of a mobile application should carry out intensive accessibility and usability evaluations, but ultimately that application must be delivered on a timely basis. Also, the impossibility of complete testing is well known.

Indeed, the last column of Table 1 acknowledges that with respect to any assurance and/or evaluation, and includes feasibility as an all-encompassing consideration on the layers to make the framework practical. There are well-known techniques such as analytical hierarchy process (AHP) and quality function deployment (QFD) for carrying out feasibility analysis, and further discussion of this aspect is beyond the scope of this article.

Limitations of Addressing Credibility

We note here that credibility, as is reflected by its primary dimensions, is a socio-cognitive concern that is not always amenable to a purely technological treatment. However, by decomposing it into quantifiable elements and approaching them in a systematic and feasible manner, we can make improvements towards its establishment.

We assert that the quality attributes we mention in pragmatic and social levels are necessary but make no claim of their sufficiency. Indeed, as we move from bottom to top, the framework gets less technically oriented and more human oriented. Therefore, finding sufficient conditions for establishing credibility is likely to be an open question, and it may be virtually impossible to provide complete guarantees for credibility.

FUTURE TRENDS

In the previous section, we discussed active credibility; the issue of passive credibility poses special challenges and is a potential area of future research. We now briefly look at the case of reputed credibility.

In case of Web applications, there have been two notable initiatives in the direction of addressing reputed credibility, namely WebTrust and TRUSTe. In response to the concerns related to for business-to-consumer electronic commerce and to increase consumer confidence, the American Institute of Certified Public Accountants (AICPA) and Canadian Institute of Chartered Accountants (CICA) have developed WebTrust Principles and Criteria and the related WebTrust seal of assurance. Independent and objective certified public accountant or chartered accountants, who are licensed by the AICPA or CICA, can provide assurance services to evaluate and test whether a particular Web application meets these principles and criteria. The TRUSTe program enables companies to develop privacy statements that reflect the information gathering and dissemination practices of their Web application. The program is equipped with the TRUSTe "trustmark" seal that takes users directly to a provider's privacy statement. The trustmark is awarded only to those that adhere to TRUSTe's established privacy principles and agree to comply with ongoing TRUSTe oversight and resolution process. Admittedly, not in the realm of pure academia, having similar quality assurance and evaluation programs for mobile applications, and perhaps even the use of ISO 9001:2000 as a basis for a certification, would be of interest.

A natural extension of the preceding discussion on credibility could be in the context of the next generation of mobile applications such as semantic mobile applications (Alesso & Smith, 2002) and mobile Web services (Salmre, 2005). For example, ontological representation of information can present certain human-centric challenges (Kamthan & Pai, 2006) that need to be overcome for it to be a credible knowledge base.

Finally, viewing a mobile application as an information system, it would of interest to draw connections between credibility and ethics (Johnson, 1997; Tavani, 2004).

CONCLUSION

Although there have been significant advances towards enabling the technological infrastructure (Coyle, 2001) for mobile access in the past decade, there is much to be done in addressing the social challenges. Addressing credibility of mobile applications in a systematic manner is one step in that direction.

The organizations that value credibility of their mobile applications need to take two aspects into consideration: (1) take a *systematic* approach to the development of the mobile applications, and (2) consider credibility as a first-class concern *throughout* the process. The former need to particularly include support for modeling a user's environment (context, task, and device) (Gandon & Sadeh, 2004) and mobile user interface engineering. The latter implies that credibility is viewed as a *mandatory* non-functional requirement during the analysis phase and treated as a central design concern in the synthesis phase.

In a user-centric approach to engineering, mobile applications belong to an *ecosystem* that includes both the people and the product. If the success of a mobile application is measured by use of its services, then establishing credibility with the users is critical for the providers. By making efforts towards improving the criteria that directly or indirectly affect credibility, the providers can meet user expectations and change the user perceptions in their favor.

REFERENCES

Ahonen, M. (2003, September 19). Accessibility challenges with mobile lifelong learning tools and related collaboration. *Proceedings of the Workshop on Ubiquitous and Mobile Computing for Educational Communities (UMOCEC 2003)*, Amsterdam, The Netherlands.

Alesso, H. P., & Smith, C. F. (2002). *The intelligent wireless Web.* Boston: Addison-Wesley.

Bertini, E., Catarci, T., Kimani, S., & Dix, A. (2005). A review of standard usability principles in the context of mobile computing. *Studies in Communication Sciences, 1*(5), 111-126.

Chisholm, W., Vanderheiden, G., & Jacobs, I. (1999). *Web content accessibility guidelines 1.0.* W3C Recommendation, World Wide Web Consortium (W3C).

Coyle, F. (2001). *Wireless Web: A manager's guide.* Boston: Addison-Wesley.

Fenton, N. E., & Pfleeger, S. L. (1997). *Software metrics: A rigorous & practical approach.* International Thomson Computer Press.

Fogg, B. J. (2003). *Persuasive technology: Using computers to change what we think and do.* San Francisco: Morgan Kaufmann.

Fogg, B. J., Marshall, J., Laraki, O., Osipovich, A., Varma, C., Fang, N., et al. (2001, March 31-April 5). What makes Web sites credible?: A report on a large quantitative study. *Proceedings of the ACM CHI 2001 Human Factors in Computing Systems Conference,* Seattle, WA.

Fogg, B. J., & Tseng, S. (1999, May 15-20). The elements of computer credibility. *Proceedings of the ACM CHI 99 Conference on Human Factors in Computing Systems,* Pittsburgh, PA.

Gandon, F. L., & Sadeh, N. M. (2004, June 1-3). Context-awareness, privacy and mobile access: A Web semantic and multiagent approach. *Proceedings of the 1st French-Speaking Conference on Mobility and Ubiquity Computing,* Nice, France (pp. 123-130).

Gibson, D. A. (2002). *Communities and reputation on the Web.* PhD Thesis, University of California, USA.

Ghezzi, C., Jazayeri, M., & Mandrioli, D. (2003). *Fundamentals of software engineering* (2nd ed.). Englewood Cliffs, NJ: Prentice-Hall.

Hovland, C. I., Janis, I. L., & Kelley, J. J. (1953). *Communication and persuasion.* New Haven, CT: Yale University Press.

Johnson, D. G. (1997). Ethics online. *Communications of the ACM, 40*(1), 60-65.

Lindland, O. I., Sindre, G., & Sølvberg, A. (1994). Understanding quality in conceptual modeling. *IEEE Software, 11*(2), 42-49.

Kamthan, P. (1999). *E-commerce on the WWW: A matter of trust.* Internet Related Technologies (IRT.ORG).

Kamthan, P., & Pai, H.-I. (2006, May 21-24). Human-centric challenges in ontology engineering for the semantic Web: A perspective from patterns ontology. *Proceedings of the 17th Annual Information Resources Management Association International Conference (IRMA 2006)*, Washington, DC.

Kasanoff, B. (2002). *Making it personal: How to profit from personalization without invading privacy.* New York: John Wiley & Sons.

McGraw, G., & Felten, E. W. (1998). Mobile code and security. *IEEE Internet Computing, 2*(6).

Metzger, M. J. (2005, April 11-13). Understanding how Internet users make sense of credibility: A review of the state of our knowledge and recommendations for theory, policy, and practice, *Proceedings of the Internet Credibility and the User Symposium,* Seattle, WA.

Metzger, M. J., Flanagin, A. J., Eyal, K., Lemus, D., & McCann, R. (2003). Bringing the concept of credibility into the 21st century: Integrating perspectives on source, message, and media credibility in the contemporary media environment. *Communication Yearbook, 27,* 293-335.

Nguyen, H. Q., Johnson, R., & Hackett, M. (2003). *Testing applications on the Web: Test planning for mobile and Internet-based systems* (2nd ed.). New York: John Wiley & Sons.

Paavilainen, J. (2002). *Mobile business strategies: Understanding the technologies and opportunities.* Boston: Addison-Wesley.

Roth, J. (2001, September 10). Patterns of mobile interaction. *Proceedings of the 3rd International Workshop on Human Computer Interaction with Mobile Devices (Mobile HCI 2001)*, Lille, France.

Roth, J. (2002). Patterns of mobile interaction. *Personal and Ubiquitous Computing, 6*(4), 282-289.

Sadeh, N. M., Chan, T.-C., Van, L., Kwon, O., & Takizawa, K. (2003, June 9-12). A semantic Web environment for context-aware m-commerce. *Proceedings of the 4th ACM Conference on Electronic Commerce,* San Diego, CA (pp. 268-269).

Salmre, I. (2005). *Writing mobile code: Essential software engineering for building mobile applications.* Boston: Addison-Wesley.

Stamper, R. (1992, October 5-8). Signs, organizations, norms and information systems. *Proceedings of the 3rd Australian Conference on Information Systems*, Wollongong, Australia.

Tavani, H. T. (2004). *Ethics and technology: Ethical issues in an age of information and communication technology.* New York: John Wiley & Sons.

Wiegers, K. (2002). *Peer reviews in software: A practical guide.* Boston: Addison-Wesley.

KEY TERMS

Delivery Context: A set of attributes that characterizes the capabilities of the access mechanism, the preferences of the user, and other aspects of the context into which a resource is to be delivered.

Mobile Web Engineering: A discipline concerned with the establishment and use of sound scientific, engineering, and management principles, and disciplined and systematic approaches to the successful development, deployment, and maintenance of high-quality mobile Web applications.

Personalization: A strategy that enables delivery that is customized to the user and user's environment.

Quality: The totality of features and characteristics of a product or a service that bear on its ability to satisfy stated or implied needs.

Semantic Web: An extension of the current Web that adds technological infrastructure for better knowledge representation, interpretation, and reasoning.

Semiotics: The field of study of signs and their representations.

User Profile: A information container describing user needs, goals, and preferences.

Adoption and Diffusion of M-Commerce

Ranjan B. Kini
Indiana University Northwest, USA

Subir K. Bandyopadhyay
Indiana University Northwest, USA

INTRODUCTION

Mobile commerce (or in short, m-commerce) is currently at the stage where e-commerce was a decade ago. Many of the concerns consumers had regarding e-commerce (such as security, confidentiality, and reliability) are now directed towards m-commerce. To complicate the matter further, the lack of a standardized technology has made m-commerce grow in multiple directions in different parts of the world. Thus, the popularity of m-commerce-based services varies by country, by culture, and by individual user. For example, in Europe the most popular application is SMS (short message service) or text messaging, in Japan interactive games and picture exchange via NTT DoCoMo i-mode, and in North America e-mail via interactive pagers (such as RIM BlackBerry) and wireless application protocol-based (WAP-based) wireless data portals providing news, stock quotes, and weather information. It is safe to predict that these applications will take on different forms as the technologies mature, devices become more capable in form and functionality, and service providers become more innovative in their business models.

It is true that m-commerce has witnessed spectacular growth across the globe. It is also encouraging that several factors are expected to accelerate the pace of adoption of m-commerce. Notable among these drivers is convergence in the voice/data industry, leaping improvements in related technology and standards, adoptive technology culture in many parts of the world, and governmental and regulatory initiatives.

Despite the undisputed promise of m-commerce, there are several barriers that are slowing the pace of adoption of m-commerce. The major barriers include: (a) lack of good business models to generate revenues, (b) perception of lack of security, (c) short product lifecycle due to rapidly changing technology, (d) non-convergence of standards, (e) usability of devices, (f) limitation of bandwidth, and (g) cost.

Many of the aforesaid were common to e-commerce also at its introduction and growth stage. We strongly believe it is worthwhile to investigate how e-commerce has been able to overcome these barriers so that we can incorporate some of the successful strategies to m-commerce. In our study, we will first compare and contrast e-commerce and m-commerce with respect to a set of common criteria such as: (1) hardware requirement, (2) software requirement, (3) connection or access, and (4) content. In the process, we will identify the principal barriers to the development of m-commerce as outlined in the above list.

The Growth in E-Commerce

Electronic commerce or e-commerce is the mode of commerce wherein the communication and transactions related to marketing, distributing, billing, communicating, and payment related to exchange of goods or services is conducted through the Internet, communication networks, and computers. Since the Department of Defense opened up the Internet for the public to access in 1991, there has been exponential growth in the number of Web sites, users on the Web, commerce through the Web, and now change of lifestyle through the Web (Pew, 2006).

The chronology of events shows that as the Internet became easier and cheaper to use, and as the applications (such as e-mail and Web interaction) became necessary or useful to have, the rate of adoption of the Internet accelerated. In fact, the rate of adoption of the Internet surpassed all projections that were made based on the traditional technology adoption rates that were documented for electricity, automobile, radio, telephone, and television (Pew, 2006). Unfortunately, the over-enthusiastic media hyped up the growth rate to an unsustainable level, leading to unprecedented growth of investment in the Internet technologies and followed by a melt-down in the stock market. This shattered the confidence in Internet technologies in the investment market. Although there was a significant deceleration in IT investment, e-commerce has rebounded to a large extent since the dot.com bust. It has been growing at about 30% compound rate per year (Pew, 2006).

In the last 10 years, the adoption of e-commerce has been extensively studied both by academicians as well as practitioners. During this period e-commerce and the scope of its definition also went through various iterations. For example, people may not buy a car on the Internet, but it is documented that 65% of car buyers have done extensive research on the Web about the car they eventually buy. Is this e-commerce? Should we restrict the e-commerce definition to financial exchange for goods or services? We have various such examples in the marketplace where extensive research about the product or service is conducted on the Internet, but the final purchase is made in the physical environment. Hence, although the number of consumer financial transactions has not grown to the level industry projected initially, there has been a significantly high rate of adoption of the activities supporting e-commerce.

In addition, there has been a very high rate of adoption of business-to-business (B2B) commerce both in terms of financial and supporting transactions. In this article, we are interested in business-to-consumer (B2C) commerce. Hence, the comparison and contrast is made between e-commerce and m-commerce. All our discussion henceforth will be on B2C commerce using desktop and/or mobile technologies.

The Growth Potential of M-Commerce

Mobile commerce is the model of commerce that performs transactions using a wireless device and data connection that result in the transfer of value in exchange for information, services, or goods. Mobile commerce is facilitated generally by mobile phones and newly developed handheld devices. It includes services such as banking, payment, ticketing, and other related services (DEVX, 2006; Kini & Thanarithiporn, 2005).

Currently, most m-commerce activity is performed using mobile phones or handsets. This type of commerce is common in Asian countries led by Japan and South Korea. Industry observers are expecting that the United States will catch up soon, with mobile phones replacing existing devices such as ExxonMobil's Speedpass (eMarketer, 2005; Kini & Thanarithiporn, 2005).

Although the U.S. is lagging behind many countries in Asia and Europe in m-commerce, a UK-based research firm projects North American m-commerce users to total 12 million by 2009, with two-thirds of them using the devices to buy external items such as tickets and goods, and a third of them using it to make smaller transactions through vending machines (eMarketer, 2005). The firm also notes that there is a large potential number of the 95 million current American teens who are already making purchases on the Web that will adopt m-commerce. However, the study also remarks that generating widespread user interest in m-commerce and addressing security fears of mobile payment technologies and m-commerce services are critical in achieving a high level of adoption (eMarketer, 2005).

While the Asia Pacific Research Group (APRG, 2006) projected in 2002 that global m-commerce would reach US$10 billion 2005, Juniper Research currently projects that the global mobile commerce market, comprising mobile entertainment downloads, ticket purchases, and point-of-sale (POS) transactions, will grow to $88 billion by 2009, largely on the strength of micro-payments (e.g., vending machine type purchases). See eMarketer (2005) for more details.

Today, a large percentage of mobile phone users use mobile phones to download ring tones and play games; hence content-based m-commerce is expected to make up a small percentage of m-commerce. One recent study, however, projects that in the future mobile phone users will move up the value chain from purchases that are used and enjoyed on the mobile phone to external items such as tickets, snacks, public transportation, newspapers, and magazines (eMarketer, 2005).

Diffusion Models of Technology Adoption

There are many models that have been formulated and studied with regard to technology adoption, acceptance, diffusion, and continued adoption. These theories identify factors that are necessary to support different levels of adoption of information and communication technologies (ICTs). Notable among these models are the innovation-diffusion theory (Roger, 1995), technology acceptance model (or TAM) based on the theory of reasoned action (Davis, 1989; Fishbein & Ajzen, 1975), extended TAM2 model that incorporates social factors (Venkatesh & Davis, 2000), technology adoption model based on the theory of planned behavior (Ajzen & Fishbein, 1980), post acceptance model based on marketing and advertising concepts (Bhattacherjee, 2001), and SERVQUAL (Parasuraman, Berry, & Zeithaml, 1988) for service quality. These models have been extensively used to predict and evaluate online retail shopping and continued acceptance of ICTs. In addition, varieties of integrated models have been developed to measure the success of information systems, ICT, and Internet adoption and diffusion. Currently, many of these models are being tested in the context of mobile technology (primarily mobile phone services).

The integration models mentioned above have been empirically tested in the e-commerce area. The models have been authenticated and proven to be extremely useful in predicting behavior of users of ICT and e-commerce. In the case of m-commerce, the results have been slightly inconsistent. Primarily these inconsistencies have been found because of the differing market maturity levels or the usage pattern of mobile devices. For example, in a South Korean study where mobile phones have been in use for quite some time, the results of testing an integrative m-commerce adoption model yielded different results for actual use than in a similar study conducted in Thailand where mobiles devices were introduced much later in the market. South Koreans were not influenced much by advertising, unlike Thai people in the initial adoption phase of m-commerce. Conversely, Thai people were not influenced by word-of-mouth to the extent South Koreans were influenced in the initial adoption (Thanarithiporn, 2005). According to Thanarithiporn

(2005), this is due to the fact South Koreans are at a more advanced level of adoption for ICTs. Furthermore, Thanarithiporn (2005) found that, unlike in South Korea where content availability had no influence in the continued use of mobile phones, it had a strong influence in Thailand on mobile usage rate. Also, in both countries self-efficacy had no influence one way or the other in the initial adoption of the mobile phone.

Key Factors that Affect the Adoption and Diffusion of E-Commerce and M-Commerce

As expected, many factors influence the rate of adoption and diffusion of technological innovations. We reviewed the extant literature, as outlined above, to identify those factors. In particular, we were interested in a set of factors that have significant influence in the adoption and diffusion of both e-commerce and m-commerce. These include: (a) hardware requirement, (b) software requirement, (c) connection or accessibility, and (d) content. In the following paragraphs, we will outline how these factors have influenced the development of e-commerce, and are currently influencing the adoption and diffusion of m-commerce.

Hardware Requirement

E-Commerce

Computer users were used to the QWERTY keyboard (of typewriters), thus they easily adapted to the standardized desktop of the first personal computers (PCs) in the 1980s. The development of graphical user interface (GUI), mice, and various other multimedia-related accessories has made PCs and variations thereof easy to use. With the introduction of open architecture, the adoption and diffusion of PCs proliferated. The introduction of the Internet to the common public, and the introduction of the GUI browser immediately thereafter, allowed PC users to quickly adopt the Web browsers and demand applications in a hurry. The limitation of hardware at the user level was only restricted by the inherent rendering capability of a model based on the processors, configuration, and accessories that supported them. Since the Web and e-commerce server technologies that serve Internet documents or Web pages are also based on open architecture, limitations were similar to that of desktops.

M-Commerce

The hardware used for mobile devices are complex. The evolution of the hardware technology used in mobile devices is diverse because of the diversity in fundamental architecture. These architectures are based on diverse technology standards such as TDMA, CDMA, GPRS, GSM, CDMA/2000, WCDMA, and i-mode. In addition, these architectures have gone through multiple generations of technology such as 1G (first generation – analog technology); 2G (second generation – digital technology, including 2.5G and 2.75G); and 3G, to meet the demands of customers in terms of bandwidth speed, network capabilities, application base, and corresponding price structures. The lack of uniform global standards and varied sizes and user interfaces to operate the devices has further disrupted the smoother adoption process. While the U.S. still suffers from a lack of uniform standard, Europe is moving towards uniformity through some variation of TDMA technology, and China is modifying CDMA technology to develop its own standard. Other countries are currently working towards a uniform standard based on a variation of base TDMA or CDMA technology (Keen & Mackintosh, 2001).

The innovation in the changing standards, devices, applications, and cultural temperament have constantly maintained a turbulent environment in the adoption and diffusion of commerce through mobile devices. For example, if the device is WAP-enabled, then Web services can be delivered using standardized WML, CHTML, or J2ME development tools. But the WAP enabling has not given scale advantages because hardware standards have not converged, at least not in the U.S. where consumers use a multitude of devices such as Palm, different Web-enabled phones, and different pocket phones.

Software Requirement

E-Commerce

The standardization and open architecture of PCs, along with the high degree of penetration of PCs in the office and home environment, allowed for standardization of client devices. This allowed for the development of text browsers, and subsequently the development of the graphical interface through Web browsers. Apples, PCs, and other UNIX-based workstations were able to use the device-independent Web browsers, thus leading to rapid adoption and expansion in the usage of Web browsers. The low price of earlier browsers such as Mosaic and Netscape, and the distribution of Internet Explorer with the Windows Operating System by Microsoft allowed the diffusion of the browsing capability in almost every client in the market.

Standardized browser software and interface, along with market dominant operating systems such as the Windows family of desktop operating systems and server platforms, facilitated the exponential growth of Internet users and applications. The availability, integration, and interoperability of application development tools, and the reliance on open systems concept and architecture, fueled further changes in the interactivity of the Web and indirectly boosted the commerce on the Web. The development of hardware-independent Java (by Sun Microsystems) and similarly featured tools allowed growth in the interactivity of the Web and application integration both at the front end and backend of the Web. The interoperability of Web applications to communicate with a wide variety of organizational systems initiated a concern for security of the data while in transit and storage. In the early stages of e-commerce, major credit card companies did not trust the methodologies that were used, although they allowed the transactions. Beginning in 1999, they started protecting the online customers just as they protected off-line customers (namely, a customer is only responsible for $50 if she reports the card stolen within 24 hours). The technology companies and financial service organizations collaboratively created and standardized methodologies for online secure transactions, and originated the concept of third-party certification of authority. This certification practice further strengthened the security of online commerce and established a strong basis for consumers to trust and online commerce to grow.

M-Commerce

Software for mobile technologies is dependent on the technology standard used and type of applications suitable for the mobile device. In most nations, like in the U.S., the use of mobile devices started with the use of analog cellular phones. These required proprietary software and proprietary networks. The digitization of handheld devices started with personal digital assistants (PDAs) for personal information management. The transformation of the PDA as a digital communication tool was made possible by private networks, operating systems, and applications developed by companies such as Palm. However, as Microsoft's Windows CE (Compact Edition) and BlackBerry started offering e-mail, information management tools, and Web surfing using micro-browsers, the growth in the use of handheld devices for Web applications started growing. The handheld industry responded with a variety of applications and made WAP a standard for applications development.

Concurrently, the telecom industry brought out digital phones and devices that could offer voice, personal information management (PIM), and data applications. However, until now, operating systems, servers, and Web applications are not standardized in the handheld market. The diversity of server software and client operating systems, and the availability of applications have not made these devices interoperable. In addition, with each player offering its own network and original content or converted content (i.e., content originally developed for the desktop computers), the interest in commerce using mobile devices has not been too enthusiastic. Furthermore, the lack of common security standards has made mobile commerce adoption very slow.

Connection or Access

E-Commerce

In the United States, where telephone wire lines have been in existence for over 100 years, it was natural for the telecom companies to focus on offering Internet connectivity through the existing telephone network. In the early stages of pubic offering of the Internet, it was easy for people to adopt the Internet using their modem from a private network. As the Internet evolved into the World Wide Web, and innovation brought faster modems to the market, more Internet service providers (ISPs) started providing ramps to the Internet. When the Windows98 Operating System with its integrated Internet Explorer was introduced to the marketplace, the Internet adoption was growing in triple digits per year. The major infrastructural components were already in place. The telecom sector invested heavily into building the bandwidth and router network to meet the insatiable demand for Web surfing. Worldwide Internet adoption and use was growing exponentially. The ICT industry responded with innovative technologies, software and services using standardized PCs, modems, support for (Internet protocol suite) TCP/IP protocol of Internet, and highly competitive pricing. The e-tail industry subsequently started growing rapidly, and the financial service industry introduced innovative products and services while collaboratively designing secure electronic payment mechanisms with ICT industry players.

The drop in pricing, availability of bandwidth, security, and quality of products and services bolstered the commerce activity on the Internet until the 'dot.com bust' of May 2000. Although the bust slowed the growth rate of e-commerce, in reality e-commerce continuously grew despite the bust. Support for e-commerce from the U.S. government to fuel the e-commerce growth through moratorium on taxes by two administrations considerably helped the diffusion of e-commerce. The concern about the security in e-commerce shown by laggards was eased by a variety of security and encryption tools, and the creation of the certification of authority concept by strong security services offered by companies such as Verisign, TRUSTe, and others.

Lately, the demand for highly competitive broadband service availability, and the availability and delivery of media-rich content, has brought media and entertainment industry to the Web with greater force. These technological advances in the e-commerce sector have received increased attention, thus ensuring a strong global growth rate in e-commerce.

M-Commerce

In the mobile arena, customers may have been using analog cellular phones (1G) for a long of time. During the era of analog cellular phones, the common mobile commerce activity was the downloading of ring tones. This type of commerce activity is still quite prevalent in developing nations. In addition to this type of commerce, other types of commerce conducted using these devices are the same as the ones that can be performed using a standard desk phone, such as ordering tickets for an event, ordering catalog items, and similar tasks.

With the introduction of digital devices (2G), mobile phones quite suddenly have become the lifeline for many transactions, such as e-mail, voicemail, and text messaging. With 2.5G, 2.75G, and now with 3G devices, more varied and complex applications such as photo transfers, interactive games, and videos have become the norm. The capabilities of these devices are determined by technical ability of the devices and the support of terrestrial tower structures by the vendors offering these services. In addition, the content availability and their desirability by the customers also determine the adoption of such services. The technology, standards, and competition have left U.S. vendors in the distance in rolling out new technology and services. While Asia's (South Korea, Japan, and China) mobile penetration growth is three times that of the United States, Europe is closely behind Asia, with England (87%) and Finland (75%) achieving very high penetration rates (Shim, 2005). In the U.S., the major players in the telecom industry are collaborating to achieve the 3G-standard Universal Mobile Telecommunication System (UMTS) to provide penetration and support rollout of new technology and services. Several countries including South Korea were planning to offer a more advanced technology called the Digital multimedia broadband (DMB) or wire broadband (WiBro) by the end of 2006 (Shim, 2005). According to Shim (2005), it will take a while to obtain DMB cellular phone services in the U.S., since technical standards and logistical barriers will have to be overcome first.

The private networks built by the wireless service providers through the customized devices will determine the access and speed available in the future in the United States. The investment in the network, along with the rollout of new technology and methods used to price the services, will be strong factors in building the capacity. Government policies are also vital in this respect. According to Shim (2005), the government commitment and push for IT strategy and long-term goals are among the most important factors to advance a country's cellular mobile business, particularly for less-developed countries.

Content

E-Commerce

Identifying the most preferred method for delivery of any content has always been a thorny issue. In electronic commerce, the complete digital conversion of all media into technology mandated by the FCC by 2008 would be much easier (FCC, 2006). Voice, as well as radio and television signals, will be broadcast digitally. The Internet has built capacity to deliver rich media content at high speed using the fiber network in the U.S. The convergence of devices such as TV monitors and PC monitors has already brought down the prices for such devices due to scale effects. The stumbling blocks to achieve a greater level of broadband adoption (from the current 53% in the U.S.) are pricing and quality of content (Pew, 2006). In e-commerce, content can be provided by anyone using standardized development tools and can be served on the standardized server software since most desktops can handle all the content delivered through the Web. The diffusion of such innovations is constrained by the pricing and the investment made by consumers at the client level. The industry has converged in standardizing hardware, software, and protocols. Globally as well as in the U.S., there is a clear trend to make the technology affordable throughout the world through the open systems concept. This has helped tremendously, especially in developing countries, in the adoption and diffusion of the Internet and generalized applications.

M-Commerce

In mobile commerce, the content such as data, text, audio, video, and video streaming can be delivered through the devices provided by service providers through their network infrastructure. As the service providers rollout new network technologies with greater capabilities to adapt to the new generation of hardware and software technologies, consumers can expect more media-rich content. Any content that is available in the e-commerce world will be specially modified for mobile delivery using specific development tools for WAP-enabled devices such as WML, CHTML, and J2ME.

Depending on the type of device, the content will have to be delivered in device-specific configuration—for example, the content has to be delivered differently to a PocketPC, WAP-enabled mobile phone, and WAP-enabled PDAs. This type of dynamic configuration in the content delivery requires investment from service providers and/or value-added intermediaries. The special intermediaries provide enormous value-added services in converting the e-commerce content for different mobile devices and become consolidators of content and applications and essentially become data portals for mobile devices. The diversity of devices available in the market will require a significant amount of investments in the U.S. to offer it nationwide, unless it focuses only on high-population density regions to maximize returns.

CONCLUSION

Based on the foregoing discussion, we can say that the introduction of e-commerce has been comparatively smoother than m-commerce. The development of the hardware capability (from PC to GUI to other multimedia-related accessories such as printers, camera, etc.), the software capability (such as browsers, open operating systems, payment schemes, secure systems, etc.), better accessibility (such as phone lines, cables, etc.), and more varied content (such as voice, radio, and television signals) ensured a fast adoption and diffusion of e-commerce throughout the world.

It is true that m-commerce also enjoys many advantages similar to e-commerce. For example, the mobile phone—the principal mode of m-commerce—is witnessing a spectacular growth throughout the world. Unfortunately, unlike e-commerce, m-commerce does not enjoy an open architecture that can accommodate varied standards in hardware, software, connection technology, and the content. Several countries (such as Japan and South Korea) are further ahead of the U.S. in solving this issue of incompatible technologies. It is heartening to see a sincere effort in many countries, including the U.S., to achieve convergence in technologies so that m-commerce is able to grow true to its full potential.

REFERENCES

Ajzen, I., & Fishbein, M. (1980). *Understanding attitudes and predicting social behavior.* Englewood Cliffs, NJ: Prentice Hall.

APRG. (2006). Retrieved from http://www.aprg.com

Bhattacherjee, A. (2001). Understanding information systems continuance: An expectation-confirmation model. *MIS Quarterly, 25*(3), 351-370.

Cassidy, J. (2002). *dot.con: The greatest story every sold.* New York: Harper Collins.

Davis, F. D. (1989). Perceived usefulness, perceived ease of use and user acceptance of information technology. *MIS Quarterly, 13*(2), 319-339.

DEVX. (2006). Retrieved from http://www.devx.com/wireless/Door/11297

eMarketer. (2005). Mobile marketing and m-commerce: Global spending and trends. *eMarketer,* (February 1).

FCC. (2006). Retrieved from http://www.fcc.gov/cgb/consumerfacts/digitaltv.html

Fishbein, M., & Ajzen, I. (1975). *Belief, attitude, intention and behavior: An introduction to theory and research.* Reading, MA: Addison-Wesley.

Keen, P., & Mackintosh, R. (2001). *The freedom economy: Gaining the m-commerce edge in the era of the wireless Internet.* Berkeley, CA: Osborne/McGraw-Hill.

Kini, R. B., & Thanarithiporn, S. (2004). M-commerce and e-commerce in Thailand—A value space analysis. *International Journal of Mobile Communications, 2*(1), 22-37.

Parasuraman, A., Berry, L. L., & Zeithaml, V. A. (1988). SERVQUAL: A multiple-item scale for measuring customer perceptions of service quality. *Journal of Retailing, 64*(1), 12-40.

Pew. (2006). Retrieved from http://www.pewinternet.org

Rogers, E. M. (1995). *Diffusion of innovations.* New York: The Free Press.

Schifter, D. E., & Ajzen, I. (1985). Intention, perceived control, and weight loss: An application of the theory of planned behavior. *Journal of Personality and Social Psychology, 49*(3), 843-851.

Shim, J. P. (2005). Korea's lead in mobile cellular and DMB phone services. *Communications of the Association for Information Systems, 15,* 555-566.

Thanarithiporn, S. (2004). *A modified technology acceptance model for analyzing the determinants affecting initial and post intention to adopt mobile technology in Thailand.* Unpublished dissertation, Bangkok University, Thailand.

Venkatesh, V., & Davis, F. D. (2000). A theoretical extension of the technology acceptance model: Four longitudinal field studies. *Management Science, 46*(2), 186-204.

Adoption of M-Commerce Devices by Consumers

Humphry Hung
Hong Kong Polytechnic University, Hong Kong

Vincent Cho
Hong Kong Polytechnic University, Hong Kong

INTRODUCTION

The Internet has undoubtedly introduced a significant wave of changes. The increased electronic transmission capacity and technology further paves a superhighway towards unrestricted communication networks (Chircu & Kauffman, 2000; Cowles, Kiecker, & Little, 2002). It is estimated that by 2007, the total number of Internet users in the world will be over 1.4 billion and the percentage of wireless users is projected to take up about 57% of the vast number (Magura, 2003). Most people anticipate that the next-generation commerce will emerge from traditional commerce to PC-based e-commerce, and eventually to mobile commerce (Ellis-Chadwick, McHardy, & Wiesnhofer, 2000, Miller, 2002, Watson, Pitt, Berthon, & Zinkhan, 2002).

Mobile commerce (m-commerce) is an extension, rather than a complete replacement, of PC-based electronic commerce. It allows users to interact with other users or businesses in a wireless mode, anytime and anywhere (Balasubramanian, Peterson, & Jarvenpaa, 2002; Samuelsson & Dholakia, 2003). It is very likely that PC-based e-commerce will still prevail for a relatively long period of time in spite of the trend that more and more people will choose to adopt m-commerce for their purchases (Miller, 2002).

The focus of our article is on the consumers' adoption of m-commerce devices (MCDs), which are equipment and technologies that facilitate users to make use of m-commerce. MCDs include mobile phones, personal digital assistants (PDA), portable computer notebooks, Bluetooth, WAP, and other facilities that can have access to the wireless networks. We expect that the heading towards a world of mobile networks and wireless devices, which will present a new perspective of time and space, is definitely on its way.

Several basic questions about m-commerce devices will be addressed in this article. First, why should consumers adopt MCDs? What will be the influencing factors for consideration? Are these MCDs easy to use and proven to be useful? Second, how do the MCDs compare with the devices for other types of commerce, such as e-commerce or traditional mail order? Consumers will only adopt MCDs when there are some potential significant advantages when comparing to old devices for other types of commerce. There is still a lack of comprehensive framework within which the adoption of MCDs can be evaluated. Traditional viewpoints regarding this issue, especially those that are based on technology acceptance models, will need to be revisited and revised when consumers are considering such an adoption.

In this article, we propose a framework for identifying the various influencing factors of the adoption of MCD, as well as the antecedents of these influencing factors. Because of the need of the standardization of the application, interface, and inter-connectivity of all hardware and software relevant to the adoption and usage of MCDs, our proposed framework will have some global implications (Zwass, 1996). Our conceptual framework can, therefore, make significant contributions to a more in-depth understanding in the spread and acceptability of m-commerce through knowing why and how relevant MCDs are adopted.

While using technology acceptance models (TAMs) as our primary reference, we also incorporate the important implications of an options model into our basic framework of analyzing consumers' adoption of MCDs. Based on our theoretical framework, we identify four influencing factors—merits, maturity, maneuverability, and mentality—which we consider to be relevant to the decision of consumers in adopting MCDs. We also identify two generic antecedents of these influencing factors—mobility and matching. We plan to investigate the extent of influence of these influencing factors and their antecedents, which will affect consumers' adoption decisions of MCDs. Figure 1 is a graphical representation of our conceptual model of the adoption of MCDs by consumers.

INFLUENCING FACTORS BASED ON TECHNOLOGY ACCEPTANCE MODEL

The technology acceptance model is an information systems theory that models how users come to accept and use a new technology, with reference to two major considerations, perceived usefulness and perceived ease of use (Venkatesh & Davis, 2000). The former is about the degree to which a person believes that using a particular system will make

Figure 1. A conceptual model of the adoption of m-commerce devices

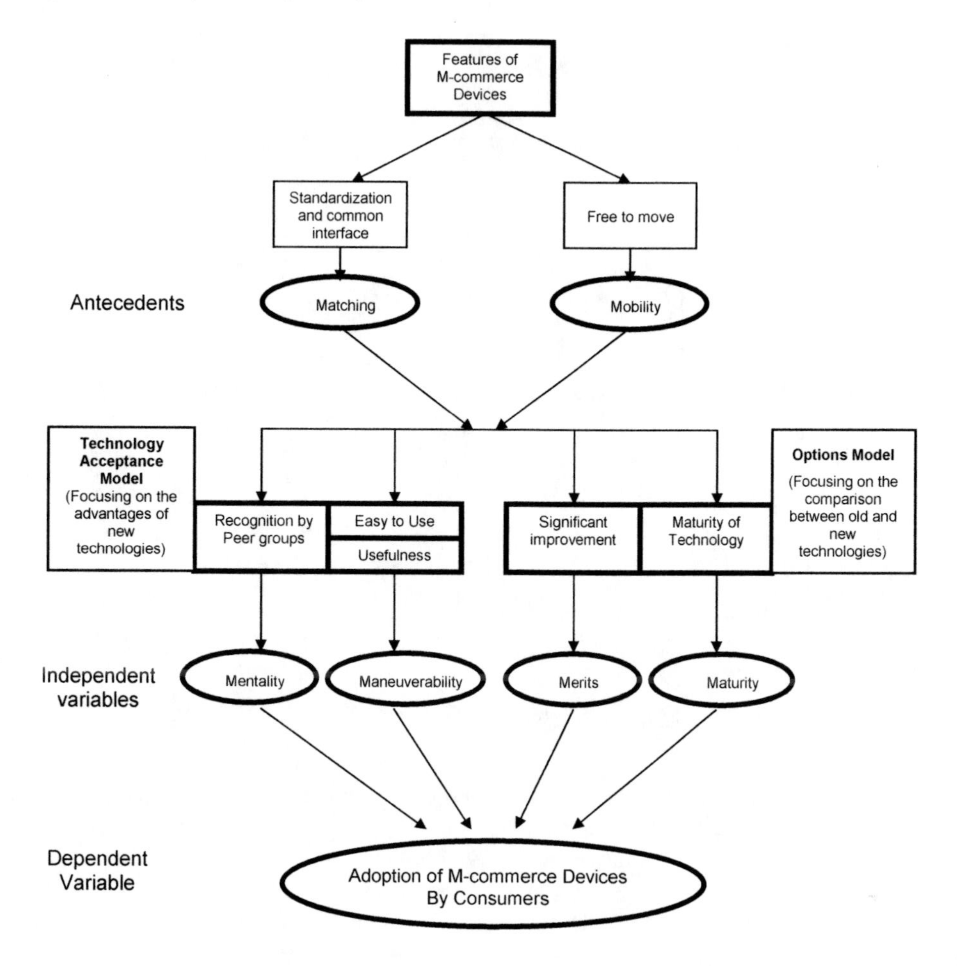

his or her life easier, for instance, by enhancing his or her job performance or reducing the workload, while the latter is the degree to which a person believes that it is not difficult to actually use a particular system (Venkatesh & Davis, 2000).

With reference to TAM, we consider whether the adoption of MCDs will bring advantages to consumers. We identify two Ms, maneuverability and mentality, for relating the acceptability of MCDs to users.

The first influencing factor, maneuverability, is related to the perceived usefulness in the adoption of MCDs and the degree to which a person can make the best use of such MCDs. Consumers will tend to adopt devices that are user friendly and do not require some intensive training of adoption (Prasanna et al., 1994).

The second influencing factor, mentality, is concerned with the match between the new technology and consumers' own mindsets, as well as the appropriate recognition of their peer groups (Bessen, 1999; Venkatesh & Davis, 2000). General acceptance by the consumers, especially by their peer groups, will be very important to consumers when they consider using MCDs for matching the devices of other people.

INFLUENCING FACTORS BASED ON OPTIONS MODEL

While mainstream literature on the adoption of new technologies is primarily based on the technology acceptance model, we consider that, in the context of m-commerce, we also need to think about some other aspects.

The options model demonstrates that a new technology with a moderate expected improvement in performance can experience substantial delays in adoption and price distortions even in a competitive market (Bessen, 1999; Sheasley, 2000). Rather than adopting a new technology that demonstrates only marginal improvement, consumers have the option of not adopting until the new technology, in terms of performance, is substantially better than the old technology. Consumers contemplating the adoption of a new technology are, of course, aware of the possibility of

Figure 2. A diagrammatic representation of the role of the four influencing factors of MCDs

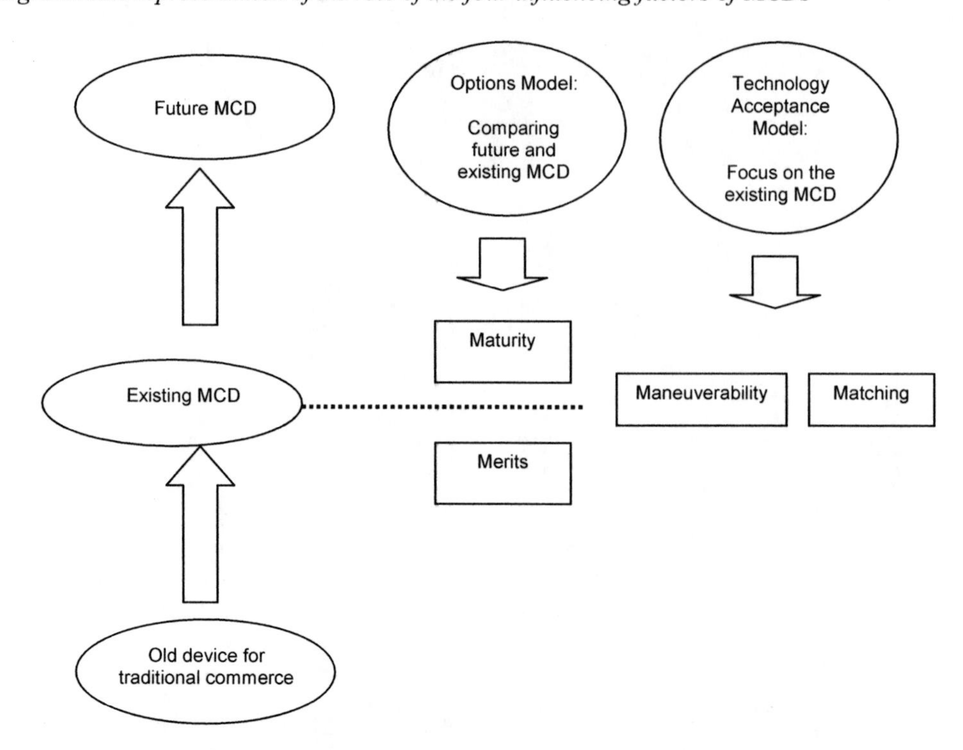

sequential improvement. They consider not only the current technical level of the new technology, but also their expectations of possible upgrades and changes in the future of the new technology (Sheasley, 2000).

With regards to the options model, we consider the comparison between MCDs and devices for other types of commerce, and in particular, the comparative advantages of MCDs to consumers. Based on the options model, we identify two Ms, merits and maturity, in relation to the comparison.

We identify the third influencing factor, merits, which is about the degree to which a buyer believes that the MCD can provide significant improvement in the purchase process. Handheld mobile devices, such as PDAs and other enhanced alphanumeric communicators, have supplemented mobile telephones, thus expanding the range of MCDs available for m-commerce transactions. With the abilities to be connected to digital communication networks, MCDs are considered to be in possession of important comparative advantage of mobility.

The fourth influencing factor, maturity, is the possibility that the technology of the MCD is mature enough so that there will not be any possible significant improvements at a later stage. While academic researchers and business practitioners recognize that the electronic market will penetrate and replace a traditional type of commerce, there are still some reservations that will likely cause the early adopters of new technologies some problems in terms of the obsolescence of devices (Samuelsson & Dholakia, 2003). Most consumers will prefer adopting MCDs with more mature technologies so that there is no need for a high level of subsequent upgrading of devices.

In essence, the option model focuses on the comparison between existing and old MCDs, while TAM places emphasis on the generic attributes and utility of MCDs. Figure 2 shows the inter-relationship among the four influencing factors. Based on the four factors that we have identified, we propose the followings:

Proposition 1: Maneuverability, mentality, merits, and maturity are the influencing factors when consumers consider adopting MCDs for purchases.

GENERIC ATTRIBUTES OF MCD

In addition to the identification of the influencing factors of the adoption of MCDs, we also consider their antecedents, which are related to the very basic and essential characteristics of MCDs.

We start our analysis by considering two generic attributes of MCDs, mobility and matching. Mobility is the most fundamental aspect of m-commerce because the name m-commerce arises from the mobile nature of the wireless environment that supports mobile electronic transactions (Coursaris, Hassanein, & Head, 2003). Mobile wireless devices, such as mobile phones, PDAs, and portable computer notebooks, can have the ability to help users gain access to the Internet. Based on these wireless devices, m-commerce is a natural extension of e-commerce but can provide some additional advantages of mobility for consumers. Mobility is a major prerequisite for the adoption of MCDs. It is an antecedent of the influencing factors of the adoption of MCDs because people will consider adopting a wireless connection because it can allow significant improvement compared with traditional device (i.e., merits), and is perceived to be useful and convenient (i.e., maneuverability).

Matching describes the need for the standardized and common interface of MCDs (Coursaris et al., 2003). The unique characteristic of m-commerce very often requires both ends of this new type of commerce to have a common interface. M-commerce applications have the challenging task of discovering services in a dynamically changing environment. Effective mechanisms need to be in place for the interface between various types of MCDs. Matching is an important antecedent of the influencing factors of consumers' adoption of MCDs because the need for standardization (i.e., matching) is important for m-commerce technology, which allows for the connection of MCDs with the wireless networks and the connections among different MCDs. This standardized interface (i.e., matching) also reflects that the MCD is mature (i.e., maturity). Moreover, the standardized interface (matching) will also help to promote the universal acceptance of MCDs by people (i.e., mentality). Based on these arguments, we develop the second proposition:

Proposition 2: The generic attributes of m-commerce, mobility and matching, are the antecedents of the influencing factors when consumers adopt MCDs for purchases.

RESEARCH IMPLICATIONS

Based on our conceptual framework, we identify the various influencing factors (i.e., 4 Ms) which can affect consumers' decisions about the adoption of MCDs in their purchases. It is possible to collect data on whether consumers will consider the adoption of MCDs, and at the same time, researchers can also investigate the reasons why they adopt or do not adopt MCDs, in terms of timing, opportunities, changing trends, and applications.

In our conceptual framework, the dependent variable is the intention of consumers to adopt MCDs. We identify four Ms as the primary influencing factors of the adoption of new technologies in m-commerce (maneuverability, mentality, merits, and maturity). These are independent variables in our framework. We also identify the antecedents of these influencing factors, mobility and matching.

First, maneuverability will be measured by the usability of the MCD. Mentality can be evaluated by the perceived peer groups' acceptance of MCDs. Merits can be measured by the comparative advantages of the MCD in relation to the old devices for other types of commerce. Maturity can be assessed by the perception that the relevant MCD can or cannot be upgraded. The first antecedent, mobility, can be measured by the extent of access to wireless networks. Matching can be measured by the degree that MCDs can be compatible with each other.

In addition to the primary independent variables, we suggest measuring some important control or moderating variables, such as price, and completing a demographic profile such as sex, age, and education levels, as well as occupations and incomes of consumers.

CONCLUSION AND IMPLICATIONS

In our proposed model, we are exploring new insights and new adoption behavior in the ubiquitous world of m-commerce, which we believe is not yet fully understood by most marketers and scholars (Stevens & McElhill, 2000; Struss, El-Ansary, & Frost, 2003). Our proposed model will be of interest to academics in the IT field, who may be keen to know how they can perform further relevant research in m-commerce.

Our proposed framework represents a theory-driven examination of the adoption of MCDs by consumers in their purchase processes. The powerful tool of m-commerce can allow for faster and easier response to market demand, and at the same time consumers can obtain relevant information as well as purchasing goods and services at any time and anywhere as they prefer.

It is expected that our proposed framework can provide important guidelines for pointing the way towards some relevant research on the significance of the adoption of MCDs. Our conceptual framework contributes to literature by ascertaining the most significant independent variables from among all those key variables that we have identified based on our literature review, which can determine which and how new technologies are likely to be adopted in m-commerce.

REFERENCES

Balasubramanian, S., Peterson, R.A., & Jarvenpaa, S.L. (2002). Exploring the implications of m-commerce for markets and marketing. *Academy of Marketing Science Journal, 30*(4), 348-361.

Bessen, J. (1999). *Real options and the adoption of new technologies.* Retrieved from http://www.researchoninnovation.org/online.htm#realopt

Bisbal, J., Lawless, D., Wu, B., & Grimson, J. (1999). Legacy information system migration: A brief review of problems, solutions and research issues. *IEEE Software, 16,* 103-111.

Chircu, A., & Kauffman, R. (2000). Reintermediation strategies in business-to-business electronic commerce. *International Journal of Electronic Commerce, 4*(4), 7-42.

Coursaris, C., Hassanein, K., & Head, M. (2003). M-commerce in Canada: An interaction framework for wireless privacy. *Canadian Journal of Administrative Sciences, 20*(1), 54-73.

Cowles, D.L., Kiecker, P., & Little, M.W. (2002). Using key informant insights as a foundation for e-retailing theory development. *Journal of Business Research, 55,* 629-636.

Ellis-Chadwick, F., McHardy, P., & Wiesnhofer, H. (2000). Online customer relationships in the European financial services sector: A cross-country investigation. *Journal of Financial Services Marketing, 6*(4), 333-345.

Magura, B. (2003). *What hooks m-commerce customers? MIT Sloan Management Review, 44*(3), 9-10.

Miller, A.I. (2002). *Einstein, Picasso: Space, time, and the beauty that causes havoc.* New York: Basic Books.

Samuelsson, M., & Dholakia, N. (2003). Assessing the market potential of network-enabled 3G m-business services. In S. Nansi (Ed.), *Wireless communications and mobile commerce.* Hershey, PA: Idea Group Publishing.

Sheasley, W.D. (2000). Taking an options approach to new technology development. *Research Technology Management, 43*(6), 37-43.

Stevens, G.R., & McElhill, F. (2000). A qualitative study and model of the use of e-mail in organizations. *Electronic Networking Applications and Policy, 10*(4), 271-283.

Struss, J., El-Ansary, A., & Frost, R. (2003). *E-marketing* (3rd ed.). Englewood Cliffs, NJ: Prentice Hall.

Venkatesh, V., & Davis, F.D. (2000). A theoretical extension of the technology acceptance model: Four longitudinal field studies. *Management Science, 46*(2), 186-204.

Watson, R.T., Pitt, L.F., Berthon, P., & Zinkhan, G.M. (2002). U-commerce: Extending the universe of marketing. *Journal of the Academy of Marketing Science, 30*(4), 329-343.

Zwass, V. (1996). Electronic commerce: Structures and issues. *International Journal of Electronic Commerce, 1*(1), 3-23.

KEY TERMS

Maneuverability: The perceived usefulness in the adoption of MCDs and the degree to which a person can make the best use of such MCDs; one of the influencing factors when consumers consider adopting MCDs for purchases.

Matching: The need for the standardized and common interface of MCDs.

Maturity: The possibility that the technology of the MCD is mature enough so that there will not be any possible significant improvements at a later stage; one of the influencing factors when consumers consider adopting MCDs for purchases.

MCD: M-commerce device.

Mentality: The match between the new technology and consumers' own mindsets, as well as the appropriate recognition of their peer groups; one of the influencing factors when consumers consider adopting MCDs for purchases.

Merits: The degree to which a buyer believes that MCDs can provide significant improvement in the purchase process; one of the influencing factors when consumers consider adopting MCDs for purchases.

Options Model: A model that proposes that consumers have the option of not adopting until the new technology, in terms of performance, is substantially better than the old technology, and as a result of such options, a new technology with a moderate expected improvement in performance can experience substantial delays in adoption and price distortions even in a competitive market.

Technology Acceptance Model (TAM): An information systems theory that models how users come to accept and use a new technology, with reference to two major considerations, perceived usefulness and perceived ease of use.

A

Advanced Resource Discovery Protocol for Semantic-Enabled M-Commerce

Michele Ruta
Politecnico di Bari, Italy

Tommaso Di Noia
Politecnico di Bari, Italy

Eugenio Di Sciascio
Politecnico di Bari, Italy

Francesco Maria Donini
Università della Tuscia, Italy

Giacomo Piscitelli
Politecnico di Bari, Italy

INTRODUCTION

New mobile architectures allow for stable networked links from almost everywhere, and more and more people make use of information resources for work and business purposes on mobile systems. Although technological improvements in the standardization processes proceed rapidly, many challenges, mostly aimed at the deployment of value-added services on mobile platforms, are still unsolved. In particular the evolution of wireless-enabled handheld devices and their capillary diffusion have increased the need for more sophisticated service discovery protocols (SDPs).

Here we present an approach, which improves Bluetooth SDP, to provide m-commerce resources to the users within a piconet, extending the basic service discovery with semantic capabilities. In particular we exploit and enhance the SDP in order to identify generic resources rather than only services.

We have integrated a "semantic layer" within the application level of the standard Bluetooth stack in order to enable a simple interchange of semantically annotated information between a mobile client performing a query and a server exposing available resources.

We adopt a simple piconet configuration where a stable networked zone server, equipped with a Bluetooth interface, collects requests from mobile clients and hosts a semantic facilitator to match requests with available resources. Both requests and resources are expressed as semantically annotated descriptions, so that a *semantic distance* can be computed as part of the ranking function, to choose the most promising resources for a given request.

STATE OF THE ART

Usually, resource discovery protocols involve a requester, a lookup or directory server and finally a resource provider. Most common SDPs, as service location protocol (SLP), Jini, UPnP (Universal Plug aNd Play), Salutation or UDDI (universal description discovery and integration), include registration and lookup of resources as well as matching mechanisms (Barbeau, 2000).

All these systems generally work in a similar manner. Basically a client issues a query to a directory server or to a specific resource provider. The request may explicitly contain a resource name with one or more attributes. The lookup server—or directly the resource provider—attempts to match the query pattern with resource descriptions stored in its database, then it replies to the client with discovered resources identification and location (Liu, Zhang, Li, Zhu, & Zhang, 2002).

These discovery architectures are based on some common assumptions about network infrastructure under the application layer in the protocol stack. In particular, current SDPs usually require a continuous and robust network connectivity, which may not be the case in wireless contexts, and especially in the ad-hoc ones. In fact in such environments, network consistence varies continuously and temporary disconnections occur frequently, so bringing to a substantial decrease traditional SDP performances (Chakraborty, Perich, Avancha, & Joshi, 2001).

Actually, there are several issues that restrain the expansion of advanced wireless applications, among them, the variability of scenarios. An ad-hoc environment is based on short-range, low power technologies like Bluetooth

(Bluetooth, 1999), which grant the peer-to-peer interaction among hosts. In such a mobile infrastructure there could be one or more devices providing and using resources but, as a MANET is a very unpredictable environment, a flexible resource search system is needed to overcome difficulties due to the host mobility. Furthermore, existing mobile resource discovery methods use simple string-matching, which is largely inefficient in advanced scenarios as the ones related to electronic commerce. In fact, in these cases there is the need to submit articulate requests to the system to obtain adequate responses (Chakraborty & Chen, 2000).

With specific reference to the SDP in the Bluetooth stack, it is based on a 128-bit universally unique identifier (UUID); each numeric ID is associated to a single service class. In other words, Bluetooth SDP is code-based and consequently it can handle only exact matches. Yet, if we want to search and retrieve resources whose description cannot be classified within a rigid schema (e.g., the description of goods in a shopping mall), a more powerful discovery architecture is needed (Avancha, Joshi, & Finin, 2002). SDP should be able to cope with non-exact matches (Chakraborty & Chen, 2000), and to provide a ranked list of discovered resources, computing a distance between each retrieved resource and the request after a matchmaking process.

To achieve these goals, we exploit both theoretical approach and technologies of semantic Web vision and adapt them to small ad-hoc networks based on the Bluetooth technology (Ruta, Di Noia, Di Sciascio, Donini, & Piscitelli, 2005).

In a semantic-enabled Web—what is known as the semantic Web vision—each available resource should be annotated using RDF (RDF Primer, 2004), with respect to an OWL ontology (Antoniou & van Harmelen, 2003). There is a close relation between the OWL-DL subset of OWL and description logics (DLs) (Baader, Calvanese, McGuinness, Nardi, & Patel-Schneider, 2002) semantics, which allows the use of DLs-based reasoners in order to infer new information from the one available in the annotation itself.

In the rest of the article we will refer to DIG (Bechhofer, 2003) instead of OWL-DL because it is less verbose and more compact: a good characteristic in an ad-hoc scenario. DIG can be seen as a syntactic variant of OWL-DL.

THE PROPOSED APPROACH

In what follows we outline our framework and we sketch the rationale behind it. We adopt a mobile commerce context as reference scenario.

In our mobile environment, a user contacts via Bluetooth a zone resource provider (from now on *hotspot*) and submits her semantically annotated request in DIG formalism. We assume the zone server—which classifies resource contents by means of an OWL ontology—has previously identified shopping malls willing to promote their goods and it has already collected semantically annotated descriptions of goods. Each resource in the m-marketplace owns an URI and exposes its OWL description.

The *hotspot* is endowed with a *MatchMaker* [in our system we adapt the MAMAS-tng reasoner (Di Noia, Di Sciascio, Donini, & Mongiello, 2004)], which carries out the matchmaking process between each compatible offered resource and the requested one measuring a "semantic distance." The provided result is a list of discovered resources matching the user demand, ranked according to their degree of correspondence to the demand itself.

By integrating a semantic layer within the OSI Bluetooth stack at service discovery level, the management of both syntactic and semantic discovery of resources becomes possible. Hence, the Bluetooth standard is enriched by new functionalities, which allow to maintain a backward compatibility (handheld device connectivity), but also to add the support to matchmaking of semantically annotated resources. To implement matchmaking and ontology support features, we have introduced a *semantic service discovery* functionality into the stack, slightly modifying the existing Bluetooth discovery protocol.

Recall that SDP uses a simple request/response method for data exchange between SDP client and SDP server (Gryazin, 2002). We associated unused classes of 128-bit UUIDs in the original Bluetooth standard to mark each specific ontology and we call this identifier *OUUID* (*ontology universally unique identifier*). In this way, we can perform a preliminary exclusion of supply descriptions that do not refer to the same ontology of the request (Chakraborty, Perich, Avancha, & Joshi, 2001). With OUUID matching we do not identify a single service, but directly the context of resources we are looking for, which can be seen as a class of similar services. Each resource semantically annotated is stored within the *hotspot* as resource record. A 32-bit identifier is uniquely associated to a semantic resource record within the *hotspot*, which we call *SemanticResourceRecordHandle*. Each resource record contains general information about a single semantic enabled resource and it entirely consists of a list of resource attributes. In addition to the *OUUID* attribute, there are *ResourceName*, *ResourceDescription*, and a variable number of *ResourceUtilityAttr_i* attributes (in our current implementation 2 of them). *ResourceName* is a text string containing a human-readable name for the resource, the second one is a text string including the resource description expressed in DIG formalism and the last ones are numeric values used according to specific applications. In general, they can be associated to context-aware attributes of a resource (Lee & Helal, 2003), as for example its price or the physical distance it has from the *hotspot* (expressed in metres or in terms of needed time to get to the resource). We use them as parameters of the overall *utility function* that computes matchmaking results.

Table 1. List of PDU IDs with corresponding descriptions

PDU ID	Description
0x00	Reserved
0x01	SDP_ErrorResponse
0x02	SDP_ServiceSearchRequest
0x03	SDP_ServiceSearchResponse
0x04	SDP_ServiceAttributeRequest
0x05	SDP_ServiceAttributeResponse
0x06	SDP_ServiceSearchAttributeRequest
0x07	SDP_ServiceSearchAttributeResponse
0x08	SDP_OntologySearchRequest
0x09	SDP_OntologySearchResponse
0x0A	SDP_SemanticServiceSearchRequest
0x0B	SDP_SemanticServiceSearchResponse
0x0C-0xFF	Reserved

In particular, to allow the representation and the identification of a semantic resource description we introduced in the data representation of the original Bluetooth standard two new *data element type descriptor*: OUUID and DIG text string. The first one is associated to the type descriptor value 9 whereas to the second one corresponds the type descriptor value 10 (both reserved in the original standard). We will associate 1, 2, 4 byte as valid size for the first one and 5, 6, 7 for the DIG text string.

Since the communication is referred to the peer layers of the protocol stack, each transaction is represented by one request Protocol Data Unit (PDU) and another PDU as response. If the SDP request needs more than a single PDU (this case is frequent enough if we use semantic service discovery) the SDP server generates a partial response and the SDP client waits for the next part of the complete answer.

By adding two SDP features *SDP_OntologySearch* (request and response) and *SDP_SemanticServiceSearch* (request and response) to the original standard (exploiting not used PDU ID) we inserted together with the original SDP capabilities further semantic-enabled resource search functions (see Table 1).

The transaction between service requester and *hotspot* starts after ad-hoc network creation. When a user becomes a member of a MANET, she is able to ask for a specific service/resource (by submitting a semantic-based description). The generic steps, up to response providing, for a service request are detailed in the following:

1. The user searches for a specific ontology identifier by submitting one or more $OUUID_R$ she manages by means of her client application
2. The *hotspot* selects OUUIDs matching each $OUUID_R$ and replies to the client
3. The user sends a service request (*R*) to the *hotspot*
4. The *hotspot* extracts descriptions of each resource cached within the *hotspot* itself, which is classified with the previously selected $OUUID_R$
5. The *hotspot* performs the matchmaking process between *R* and selected resources it shares. Taking into account the matchmaking results, all the resources are ranked with respect to *R*
6. The *hotspot* replies to the user.

Table 2. SDP_OntologySearchRequest PDU parameters

PDU ID	parameters
0x08	- ***OntologySearchPattern*** - *ContinuationState*

Table 3. SDP_OntologySearchResponse PDU parameters

PDU ID	parameters
0x09	- ***TotalOntologyCount*** - ***OntologyRetrievedPattern*** - *ContinuationState*

It is important to remark that basically all the previous steps are based on the original SDP in Bluetooth. No modifications are made to the original structure of transactions, but simply we differently use the SDP framework. In what follows we outline the structure of the SDP PDUs we added within the original framework to allow semantic resource discovery.

The first one is the *SDP_OntologySearchRequest* PDU. Their parameters are shown in Table 2.

The *OntologySearchPattern* is a data element sequence where each element in the sequence is a OUUID. The sequence must contain at least 1 and at most 12 OUUIDs, as in the original standard. The list of OUUIDs is an ontology search pattern. The *ContinuationState* parameter maintains the same purpose of the original Bluetooth (Bluetooth, 1999).

The *SDP_OntologySearchResponse* PDU is generated by the previous PDU. Their parameters are reported in Table 3.

The *TotalOntologyCount* is an integer containing the number of ontology identifiers matching the requested ontology pattern. Whereas the *OntologyRetrievedPattern* is a data element sequence where each element in the sequence is a OUUID matching at least one sent with the *OntologySearchPattern*. If no OUUID matches the pattern,

Table 4. SDP_SemanticServiceSearchRequest PDU parameters

PDU ID	parameters
0x0A	- ***SemanticResourceDescription*** - ***ContextAwareParam1*** - ***ContextAwareParam2*** - *MaximumResourceRecordCount* - *ContinuationState*

Table 5. SDP_SemanticServiceSearchResponse PDU parameters

PDU ID	parameters
0x0B	- *TotalResourceRecordCount* - *CurrentResourceRecordCount* - ***SemanticResourceRecordHandleList*** - *ContinuationState*

the *TotalOntologyCount* is set to 0 and the *OntologyRetrievedPattern* contains only a specific OUUID able to allow the browsing by the client of all the OUUIDs managed by the *hotspot* (see the following *ontology browsing* mechanism for further details). Hence the pattern sequence contains at least 1 and at most 12 OUUIDs.

The *SDP_SemanticServiceSearchRequest* PDU follows previous PDU. Their parameters are shown in Table 4.

The *SemanticResourceDescription* is a data element text string in DIG formalism representing the resource we are searching for; *ContextAwareParam1* and *ContextAwareParam2* are data element unsigned integers. In our case study, which models an m-marketplace in an airport terminal, we use them respectively to indicate a reference price for the resource and the hour of the scheduled departure of the flight. Since a generic client interacting with a *hotspot* is in its range, using the above PDU parameter she can impose—among others—a proximity criterion in the resource discovery policy.

The *SDP_SemanticServiceSearchResponse* PDU is generated by the previous PDU. Their parameters are reported in Table 5.

The *SemanticResourceRecordHandleList* includes a list of resource record handles. Each of the handles in the list refers to a resource record potentially matching the request. Note that this list of service record handles does not contain header fields, but only the 32-bit record handles. Hence, it does not have the data element format. The list of handles is arranged according to the relevance order of resources, excluding resources not compatible with the request. The other parameters maintain the same purpose of the original Bluetooth (Bluetooth, 1999).

In all the previous cases, the error handling is managed with the same mechanisms and techniques of Bluetooth standard (Bluetooth, 1999).

Notice that each resource retrieval session starts after settling between client and server the same ontology identifier (OUUID).

Nevertheless if a client does not support any ontology or if the supported ontology is not managed by the *hotspot*, it is desirable to discover what kind of merchandise class (and then what OUUIDs) are handled by the zone server without any a priori information about resources. For this purpose we use the *service browsing* feature (Bluetooth, 1999) in a slightly different fashion with respect to the original Bluetooth standard, so calling this mechanism *ontology browsing*. It is based on an attribute shared by all semantic enabled resource classes, the *BrowseSemanticGroupList* attribute which contains a list of OUUIDs. Each of them represents the browse group a resource may be associated with for browsing.

Browse groups are organized in a hierarchical fashion, hence when a client desires to browse a *hotspot* merchandise class, she can create an *ontology search pattern* containing the OUUID that represents the *root browse semantic group*. All resources that may be browsed at the top level are made members of the *root browse semantic group* by having the root browse group OUUID as a value within the *BrowseSemanticGroupList* attribute.

Generally a *hotspot* supports relatively few merchandise classes, hence all of their resources will be placed in the root browse group. However, the resources exposed by a provider may be organised in a browse group hierarchy, by defining additional browse groups below the root browse group.

Having determined the goods category and the corresponding reference ontology, the client can also download a DIG version of it from the *hotspot* as *.jar* file [such a file extension—among other things—also allows a total compatibility with the Connected Limited Device Configuration (CLDC) technology].

Also notice that since the proposed approach is fully compliant with semantic Web technologies, the user exploits the same semantic enabled descriptions she may use in other Semantic Web compliant systems (e.g., in the Web site of a shopping mall). That is, there is no need for different customized resource descriptions and modelling, if the user employs different applications either on the Web or in mobile systems. The syntax and formal semantics of the descriptions is unique with respect to the reference ontology and can be shared among different environments.

In e-commerce scenarios, the match between demand and supply involves not only the description of the good but also data-oriented properties. It would be quite strange to have a commercial transaction without taking into account price, quantity, and availability, among others. The demander usually specifies how much she is willing to pay, how many items she wants to buy, and the delivery date. Hence, the overall match value depends not only on the distance between the (semantic-enabled) description of the demand and of the supply. It has to take into account the description distance with the difference of (the one asked by the demander and the other proposed by the seller), quantity, and delivery date. The overall utility function combines all these values to give a global value representing the match degree.

Also notice that, in m-commerce applications, in addition to "commercial" parameters also context-aware variables should influence matching results. For example, in our airport case study, we consider the price difference but also the physical distance between requester and seller to weigh the match degree. The distance becomes an interesting value since a user has a temporal deadline for shopping: the scheduled hour of her flight. Hence, a resource might be chosen also according to its proximity to the user.

We will express this distance in terms of time to elapse for reaching the shop where a resource is, leaving from the *hotspot* area. In such a manner the *hotspot* will exclude resources not reachable by the user while she is waiting for boarding and it will assign to resources unlikely reachable (farther) a weight smaller than one assigned to easily reachable ones.

The above approach can be further extended to other data-type properties.

The utility function we used depends on:

- p_D : price specified by the demander
- p_O : price specified by the supplier
- t_D : time interval available to the client
- t_O : time to reach the supplier and come back, leaving from the *hotspot* area
- *s_match*: score computed during the semantic matchmaking process, computed through *rankPotential* (Di Noia, Di Sciascio, Donini, & Mongiello, 2004) algorithm.

$$u(s\_match, p_D, p_O, t_D, t_O) = \frac{s\_match}{2} + \frac{\tanh\frac{t_D - t_O}{\beta}}{3} + \frac{(1+\alpha)p_D - p_O}{6(1+\alpha)p_D} \qquad (1)$$

Notice that p_D is weighted by a $(1+\alpha)$ factor. The idea behind this weight is that, usually, the demander is willing to pay up to some more than what she originally specified on condition that she finds the requested item, or something very similar. In the tests we carried out, we find α=0.1 and β=10 are values in accordance with user preferences. These values seem to be in some accordance with experience, but they could be changed according to different specific considerations.

RUNNING EXAMPLE

A simple example can clarify the rationale of our setting. Here we will present a case study analogous to the one presented in Avancha, Joshi, and Finin (2002), and we face it by means of our approach.

Let us suppose a user is in a duty free area of an airport, she is waiting for her flight to come back home and she is equipped with a wireless-enabled PDA. She forgot to buy a present for her beloved little nephew and now she wants to purchase it from one of the airport gift stores.

In particular she is searching for a learning toy strictly suitable for a kid (she dislikes a child toy or a baby toy) and possibly the toy should not have any electric power supply.

Clearly this request is too complex to be expressed by means of standard UUID Bluetooth SDP mechanism. In addition, non-exact matches between resource request and offered ones is highly probable and the on/off matching system provided by the original standard in this case could be largely inefficient.

Hence both the semantic resource request and offered ones can be expressed in a DIG statement exploiting DL semantics and encapsulated in an SDP PDU.

The *hotspot* equipped with MAMAS reasoner collects the request and initially selects supplies expressed by means of the same ontology shared with the requester. Hence a primary selection of suitable resources is performed. In addition, the matchmaker carries out the matchmaking process between each offered resource in the m-marketplace and the requested one measuring a "semantic distance" (Colucci, Di Noia, Di Sciascio, Donini, & Mongiello, 2005). Finally the matchmaking results are ranked and returned to the user.

A subset of the ontology used as a reference in the examples is reported in Figure 1. For the sake of simplicity, only the class hierarchy and disjoint relations are represented.

Let us suppose that after the *hotspot* selects supplies, its knowledge base is populated with the following individuals whose description is represented using DL formalism:

- *Alice_in_wonderland.* Price 20$. 5 min from the *hotspot*:
  book ⊓ ∀has_genre.fantasy
- *Barbie_car.* Price 80$. 10 min from the *hotspot*:
  car ⊓ ∀suggested_for.girl ⊓ ∀has_power_supply.battery
- *classic_guitar.* Price 90$. 17 min from the *hotspot*:
  musical_instrument ⊓ ∀suitable_for.kid ⊓ (£ 0 has_power_supply)
- *shape_order.* Price 40$. 15 min from the *hotspot*:
  educational_tool⊓∀suitable_for.child⊓∀stimulates_to_learn.shape_and_color
- *Playstation.* Price 160$. 28 min from the *hotspot*:
  video_game ⊓ ∀has_power_supply.DC

Figure 1. The simple toy store ontology used as reference in the example

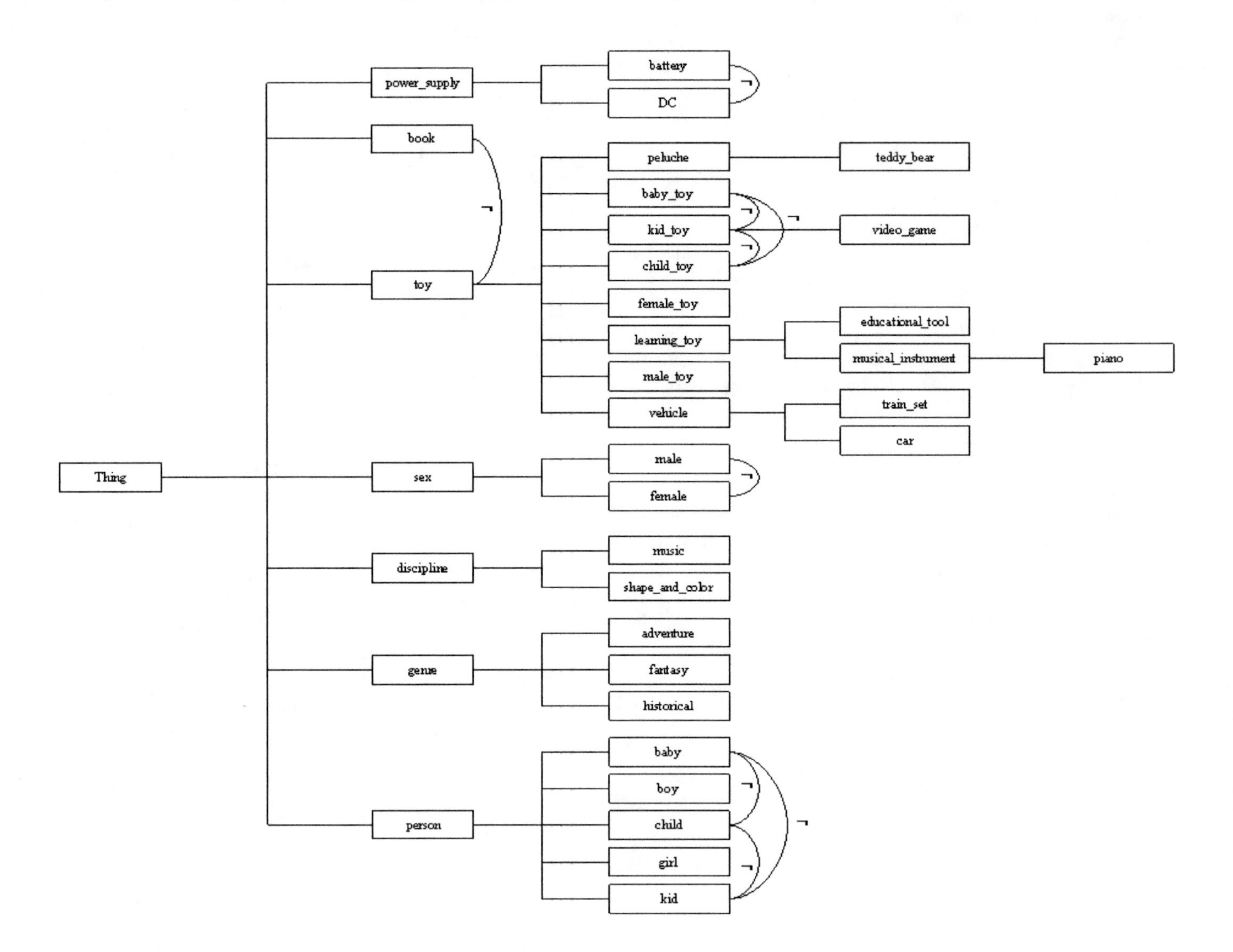

- *Winnie_the_pooh.* Price 30$. 15 min from the *hotspot*:
  teddy_bear ⊓ ∀suitable_for.baby

On the other hand, the request ***D*** submitted to the system by the user can be formalized in DL syntax as follows:

learning_toy ⊓ ∀suggested_for.boy ⊓ ∀suitable_for.kid ⊓ (£ 0 has_power_supply)

In addition she imposes a reference price of 200$ (p_D=200) as well as the scheduled departure time as within 30 minutes (t_D=30).

In Table 6 matchmaking results are presented. The second column shows whether each retrieved resource is compatible or not with request *D* and, in case, the *rankPotential* computed result. In the fourth column, matchmaking results are also expressed in a relative form between 0 and 1 to allow a more immediate semantic comparison among requests and different resources and to put in a direct correspondence various rank values.

Finally in the last column results of the overall utility function application are shown.

Notice that the semantic distance of the individual *classic_guitar* from *D* is the smaller one; then the system will recommend to the user this resource first. Hence the ranked list returned by the *hotspot* is a strict indication for the user about best available resources in the airport duty free piconet in order of relevance with respect to the request. Nevertheless

Table 6. Matchmaking results

demand – supply	compatibility (y/n)	score	s_match	u(·)
D - Alice_in_wonderland	n	-	-	-
D - Barbie_car	y	7	0.364	0.609
D - classic_guitar	y	3	0.727	0.748
D - shape_order	n	-	-	-
D - Playstation	y	5	0.546	0.378
D - Winnie_the_pooh	n	-	-	-

a user can choose or not a resource according to her personal preferences and her initial purposes.

After having selected the best resource, the server of the chosen virtual shop will receive a connection request from the user PDA with its connection parameters and in this manner the transaction may start. The user can provide her credit card credentials, so that when she reaches the store, her gift will be already packed. This final part of the application is not yet implemented, but it is trivially achievable exploiting the above SDP infrastructure.

CONCLUSION AND FUTURE WORK

In this article we have presented an advanced semantic-enabled resource discovery protocol for m-commerce applications. The proposed approach aims to completely recycle the basic functionalities of the original Bluetooth service discovery protocol by simply adding semantic capabilities to the classic SDP ones and without introducing any change in the regular communication work of the standard. A matchmaking algorithm is used to measure the semantic similarity among demand and resource descriptions.

Future trends of the proposed framework aim to create a more advanced DSS to help a user in a generic m-marketplace. Under investigation is the support to creation of P2P small communities of mobile hosts where goods and resources are advertised and opinions about shopping are exchanged (Avancha, D'Souza, Perich, Joshi, & Yesha, 2003). If a user decides to "open" her shopping trolley sharing information she owns (purchased goods, discounts, opinion about specific vendors or products) the system will insert her in a buyer mobile community where she can exchange information with other users.

Another future activity focuses on strict control of the good advertising. In an m-marketplace, the system will send to various potential buyers best proposals about their interests.

We intend to implement a mechanism to advertise goods or services in a more direct and personalized fashion. From this point of view, an additional feature of the system is oriented to the user profiling extraction and management (Prestes, Carvalho, Paes, Lucena, & Endler, 2004; Ruta, Di Noia, Di Sciascio, Donini, & Piscitelli, 2005; von Hessling, Kleemann, & Sinner, 2004). Without imposing any explicit profile submission to the user, the system could collect her preferences by means of previously submitted requests (Ruta, Di Noia, Di Sciascio, Donini, & Piscitelli, 2005); that is, by means of the "history" of the user in the m-marketplace.

REFERENCES

Antoniou, G., & van Harmelen, F. (2003). Web ontology language: OWL. In *Handbook on Ontologies in Information Systems.*

Avancha, S., D'Souza, P., Perich, F., Joshi, A., & Yesha, Y. (2003). P2P m-commerce in pervasive environments. *ACM SIGecom Exchanges*, *3*(4), 1-9.

Avancha, S., Joshi, A., & Finin, T. (2002). Enhanced service discovery in Bluetooth. *IEEE Computer*, *35*(6), 96-99.

Baader, F., Calvanese, D., McGuinness, D., Nardi, D., & Patel-Schneider, P. (2002). *The description logic handbook*. Cambridge: Cambridge University Press.

Barbeau, M. (2000). Service discovery protocols for ad hoc networking. *Workshop on Ad-hoc Communications (CASCON '00).*

Bechhofer, S. (2003). *The DIG description logic interface: DIG/1.1.* Retrieved from http://dlweb.man.ac.uk/dig/2003/02/interface.pdf.

Bluetooth specification document. (1999). Retrieved from http://www.bluetooth.com.

Chakraborty, D., & Chen, H. (2000). Service discovery in the future for mobile commerce. *ACM Crossroads*, *7*(2), 18-24.

Chakraborty, D., Perich, F., Avancha, S., & Joshi, A. (2001). Dreggie: Semantic service discovery for m-commerce applications. In *Workshop on Reliable and Secure Applications in Mobile Environment*.

Colucci, S., Di Noia, T., Di Sciascio, E., Donini, F. M., & Mongiello, M. (2005). Concept abduction and contraction for semantic-based discovery of matches and negotiation spaces in an e-marketplace. *Electronic Commerce Research and Applications*, *4*(4), 345-361.

Di Noia, T., Di Sciascio, E., Donini, F. M., & Mongiello, M. (2004). A system for principled matchmaking in an electronic marketplace. *International Journal of Electronic Commerce*, *8*(4), 9-37.

Gryazin, E. (2002). *Service discovery in Bluetooth*. Retrieved from http://www.hpl.hp.com/techreports/2002/HPL-2002-233.pdf.

Lee, C., & Helal, S. (2003). Context attributes: An approach to enable context awareness for service discovery. In *Symposium on Applications and the Internet (SAINT '03)* (pp. 22-30).

Liu, J., Zhang, Q., Li, B., Zhu, W., & Zhang, J. (2002). A unified framework for resource discovery and QoS-aware provider selection in ad hoc networks. *ACM Mobile Computing and Communications Review*, *6*(1), 13-21.

Prestes, R., Carvalho, G., Paes, R., Lucena, C., & Endler, M. (2004). Applying ontologies in open mobile systems. In *Workshop on Building Software for Pervasive Computing OOPSLA'04*.

RDF Primer-W3C Recommendation. (2004, February 10). Retrieved from http://www.w3.org/TR/rdf-primer/

Ruta, M., Di Noia, T., Di Sciascio, E., Donini, F.M., & Piscitelli, G. (2005). Semantic based collaborative P2P in ubiquitous computing. In *IEEE/WIC/ACM International Conference Web Intelligence 2005 (WI '05)* (pp. 143-149).

von Hessling, A., Kleemann, T., & Sinner, A. (2004). Semantic user profiles and their applications in a mobile environment. In *Artificial Intelligence in Mobile Systems 2004*.

KEY TERMS

Description Logics (DLs): A family of logic formalisms for knowledge representation. Basic syntax elements are concept names, role names, and individuals. Intuitively, concepts stand for sets of objects, and roles link objects in different concepts. Individuals are used for special named elements belonging to concepts. Basic elements can be combined using constructors to form concept and role expressions, and each DL has its own distinct set of constructors. DL-based systems are equipped with reasoning services: logical problems whose solution can make explicit knowledge that was implicit in the assertions.

M-Marketplace: Virtual environment where demands and supplies (submitted or offered by users equipped with mobile devices) encounter each other.

Ontology: An explicit and formal description referred to concepts of a specific domain (classes) and to relationships among them (roles or properties).

Piconet: Bluetooth-based short-range wireless personal area network. A Bluetooth piconet can host up to eight mobile devices. More piconets form a *scatternet*.

Service Discovery Protocol (SDP): It identifies the application layer of an OSI protocol stack and manages the automatic detection of devices with joined services.

Semantically Annotated Resource: any kind of good, tangible or intangible (e.g., a document, an image, a product or a service) endowed of a description that refers to a shared ontology.

Semantic Matchmaking: The process of searching the space of possible matches between a request and several resources to find those best matching the request, according to given semantic criteria. It assumes that both the request and the resources are annotated according to a shared ontology.

Anycast-Based Mobility

István Dudás
Budapest University of Technology and Economics, Hungary

László Bokor
Budapest University of Technology and Economics, Hungary

Sándor Imre
Budapest University of Technology and Economics, Hungary

INTRODUCTION

We have entered the new millennium with two great inventions, the Internet and mobile telecommunication, and a remarkable trend of network evolution toward convergence of these two achievements. It is an evident step to combine the advantages of the Internet and the mobile communication methods together in addition to converge the voice and data into a common packet-based and heterogeneous network infrastructure. To provide interworking, the future systems have to be based on a universal and widespread network protocol, such as Internet protocol (IP) which is capable of connecting the various wired and wireless networks (Macker, Park, & Corson, 2001).

However, the current version of IP has problems in mobile wireless networks; the address range is limited, IPv4 is not suitable to efficiently manage mobility, support real-time services, security, and other enhanced features. The next version, IPv6 fixes the problems and also adds many improvements to IPv4, such as extended address space, routing, quality of service, security (IPSec), network autoconfiguration and integrated mobility support (Mobile IPv6).

Today's IP communication is mainly based on unicast (one-to-one) delivery mode. However it is not the only method in use: other delivery possibilities, such as broadcast (one-to-all), multicast (one-to-many) and anycast (one-to-one-of-many) are available. Partridge, Mendez, and Milliken (1993) proposed the host anycasting service for the first time in RFC 1546. The basic idea behind the anycast networking paradigm is to separate the service identifier from the physical host, and enable the service to act as a logical entity of the network. This idea of anycasting can be achieved in different layers (e.g., network and application layers) and they have both strengths and weaknesses as well. We focus on network-layer anycasting in this article, where a node sends a packet to an anycast address and the network will deliver the packet to at least one, and preferably only one of the competent hosts. This approach makes anycasting a kind of group communication in that a group of hosts are specified for a service represented by an anycast address and underlying routing algorithms are supposed to find out the appropriate destination for an anycast destined packet.

OVERVIEW OF IPV6 ANYCASTING

RFC 1546 introduced an experimental anycast address for IPv4, but in this case the anycast addresses were distinguishable from unicast addresses. IPv6 adopted the paradigm of anycasting as one of the basic and explicitly included services of IP and introduced the new anycast address besides the unicast and multicast addresses (Deering & Hinden, 1998). IPv6 anycast addresses were designed to allow reaching a single interface out of a group of interfaces. The destination node receiving the sent packets is the "nearest" node. The distance is dependent on the metric of the underlying routing protocol. In case of IPv6, an anycast address is defined as a unicast address assigned to more than one interface, so anycast addresses can not be distinguished from unicast addresses: they both share the same address space. Therefore the beginning part of any IPv6 anycast address is the network prefix. The longest P prefix identifies the topological region in which all interfaces are belonging to that anycast address reside. In the region identified by P, each member of the anycast membership must be handled as a separate entry of the routing system. Based on the length of P, IPv6 anycast can be categorized into two types: subnet anycast and global anycast. Hashimoto, Ata, Kitamura, and Murata (2005) summarized all that issues and defined the main terminology of IPv6 anycasting (Figure 1).

Hinden and Deering (2003) declared some restrictions concerning the further usage of the anycast addressing paradigm. The main purpose for setting these limitations was to keep the usage of anycast addresses under control until enough experience has been gathered in order to fit this new scheme to the existing structure of the Internet. These restrictions are now being eased that research could find appropriate solution for them (Abley, 2005). The biggest concern that had to be dealt with was routing since anycast packets (packets with an anycast address in the destination

Figure 1. IPv6 anycast terminology basics

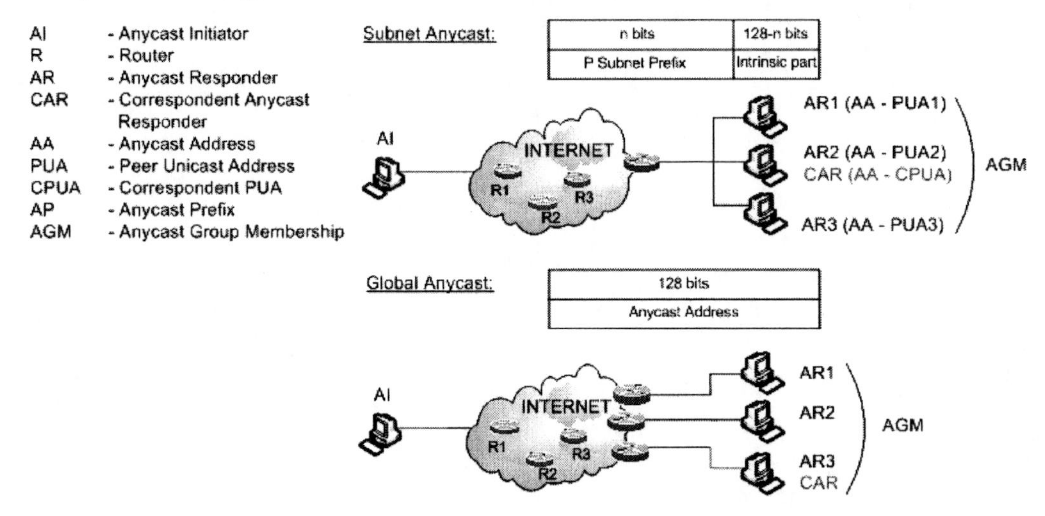

field) might be forwarded to domains with different prefixes, as anycast receivers might be distributed all over the Internet. As a result a scalable and stable routing solution for anycasting is necessary.

Routing Protocols for IPv6 Anycasting

The current IPv6 standards do not define the anycast routing protocol, although the routing is one of the most important elements of network-layer anycasting. There is a quite small amount of literature about practical IPv6 anycasting. Park and Macker (1999) proposed and evaluated anycast extensions of link-state routing algorithm and distance-vector routing algorithm. Xuan, Jia,, Zhao, and Zhu (2000) proposed and compared several routing algorithms for anycast. Eunsoo Shim (2004) proposed an application load sensitive anycast routing method (ALSAR) and analyzed the existing routing algorithms in his PhD thesis. Doi, Ata, Kitamura, and Murata (2004) summarized the problems and possible solutions regarding the current specifications for IPv6 anycasting and proposed an anycast routing architecture based on seed nodes, gradual deployment and the similarities to multicasting. Based on their work, Matsunaga, Ata, Kitamura,and Murata (2005) designed and implemented three IPv6 anycast routing protocols (AOSPF—anycast open shortest path first, ARIP—anycast routing information protocol and PIA-SM—protocol independent anycast - sparse mode) based on existing multicast protocols.

The recent studies are focusing on subnet anycast routing protocols since they offer various possibilities for research while global anycast routing still faces scalability problems to be solved. The recently introduced anycast routing protocols all share a common ground as they are all based on multicast routing protocols because of the similarities of the two addressing schemes.

Unfortunately it does not fit the scope of this document to introduce each anycast routing protocol one-by-one although it is important to present the main idea that lies beneath all these protocols. The principal task to be performed is to discover all the anycast capable routers and nodes in the network: this can happen by flooding (as in case of AOSPF) or discovery methods (e.g., PIA-SM). The next, and maybe the most important step, is to maintain an up-to-date anycast routing table so all possible receivers could be reached in case of need. The easiest way to keep the routing entries up-to-date is to maintain a so-called Anycast Group Membership (Figure 1) where the anycast hosts can sign in or out when joining or leaving a certain anycast group designated by its anycast address.

APPLICATIONS OF ANYCASTING

Since the introduction of IPv6 anycast only a few applications have emerged using these addresses. It is mainly because the flexibility of the anycasting paradigm has not yet been widespread in the public. An excellent survey of the IPv6 anycast characteristics and applications was made by Weber and Cheng, 2004; Doi, Ata, Kitamura, and Murata, 2004; Matsunaga, Ata, Kitamura, and Murata (2005), where the authors describe many advantages and possible applications of anycasting. These applications can be classified into the following main types.

Main Application Schemas

The most popularly known application of anycast technology is helping the communicating nodes in selection of service providing servers. In the *server selection* approach the client host can choose one of many functionally identical servers.

The anycast server location and selection method could be a simple and transparent technique since the same address can be used from anywhere in the network, and the anycast routing would automatically choose the best destination for the client.

Anycast addresses can also be useful in discovering and locating services. In case of *service discovery,* the clients just need to know only one address: they can communicate with an optimal (e.g., minimum delay) host selected from the anycast group and easily discover the closest provider. This is especially beneficial in case of dynamically and frequently changing environments such as mobile ad-hoc systems. Services based on this characteristic can be acquired easily and optimally by the mobile clients through network-layer anycasting.

Application Scenarios

The most important advantage of network-layer anycasting is its ability to provide a simple mechanism where the anycast initiator (Figure 1) can receive a specific service without exact information about the server nodes and networks. Moreover the whole procedure is totally transparent: the clients do not need to know whether the server's address is unicast or anycast, because anycast addresses are syntactically indistinguishable from unicast addresses. Only servers have additional knowledge about their explicitly configured anycast addresses. The main application schemas and the application scenarios below are demonstrating the possibilities of the anycasting communication paradigm.

With the help of IPv6 anycasting *local information services* (e.g., emergency calls) can be given by getting each node to communicate with the appropriate server to the node's actual location. This kind of application is very useful in a mobile environment where nodes move from one network to another while resorting a given service.

By assigning a well-known anycast address to widespread applications, we can achieve *host auto-configuration.* The clients can use these services without knowing the appropriate unicast address of the server. The clients can utilize these applications everywhere only by specifying the service's well-known anycast address. For example, DNS resolvers no longer have to be configured with the unicast IP addresses for every host in every network if a standardized anycast address is built in the hardware or software, end users can get the service without configuration.

Improving the system reliability is another good example of IPv6 anycasting. Anycast communication grants multiple numbers of hosts with the same address and by increasing the number of hosts *load balancing, service redundancy* and *DoS attack avoidance* can be achieved based on the routing mechanism where anycast requests are fairly forwarded.

In a widely distributed environment (like a peer-to-peer architecture) services can construct a logical topology above the physical network. This logical topology can be based on anycast addresses. When a client wants to participate, it specifies the anycast address of the logical level in order to join in the logical network. In such a way, one of the participating nodes will become the *gate of the logical network* determined by the underlying anycast routing protocol.

As we can see, there are some real promising application scenarios of IPv6 anycasting. However there is only one standardized anycast application these days, called *dynamic home agent address discovery*. In Mobile IPv6 the home agents (HA) have an anycast address, since the HA may change while the mobile terminal is not attached to its home network. Therefore a mobile node should use the anycast address of the home agents to reach one HA out of the set of home agents on its home link.

EMERGING APPLICATION: ANYCAST-BASED MICROMOBILITY

In Mobile IPv6 every mobile node (MN) is identified by its home address (HA), totally independently of where it is located in the network. When a MN is away from its home network (HN), it gets a new care-of-address (CoA). The IPv6 packets sent to the mobile node's HA will be routed to the mobile node's new CoA (Johnson, Perkins, & Arkko, 2004). Although Mobile IPv6 is capable of handling global mobility of users, it has shortcomings in supporting low latency and packet loss—required by real time multimedia services—during handover. To improve handover performance, the movement of a mobile node inside a subnet has to be dealt locally, by hiding intra-domain movements. As a result, the number of signaling messages reduced and the handover performance improved. Inside such a local subnet—called micromobility domain—the terminal receives a temporal IP address, which is valid throughout the subnet, and can be used a temporal CoA for the HA while the mobile terminal is located in the micromobility domain. Inside the micromobility domain, micromobility protocols are responsible for the proper routing of packets intended to the mobile hosts (Saha, Mukherjee, Misra, & Chakraborty, 2004). Leaving the micromobility domain, Mobile IPv6 provides global mobility management.

We have developed a new type of anycast application based on the main characteristics and the new research achievements of IPv6 anycasting in order to provide micromobility support in a standard IPv6 environment. In our proposed scheme, anycast addresses are used to identify mobile IPv6 hosts entering a micromobility domain while the underlying anycast routing protocol is used to maintain the anycast address routing information exchange. As a result the care-of-address obtained if the mobile terminal moves into a micromobility area is an anycast address. According to our proposal an anycast address identifies a single mobile

Figure 2. Entering a foreign micromobility domain

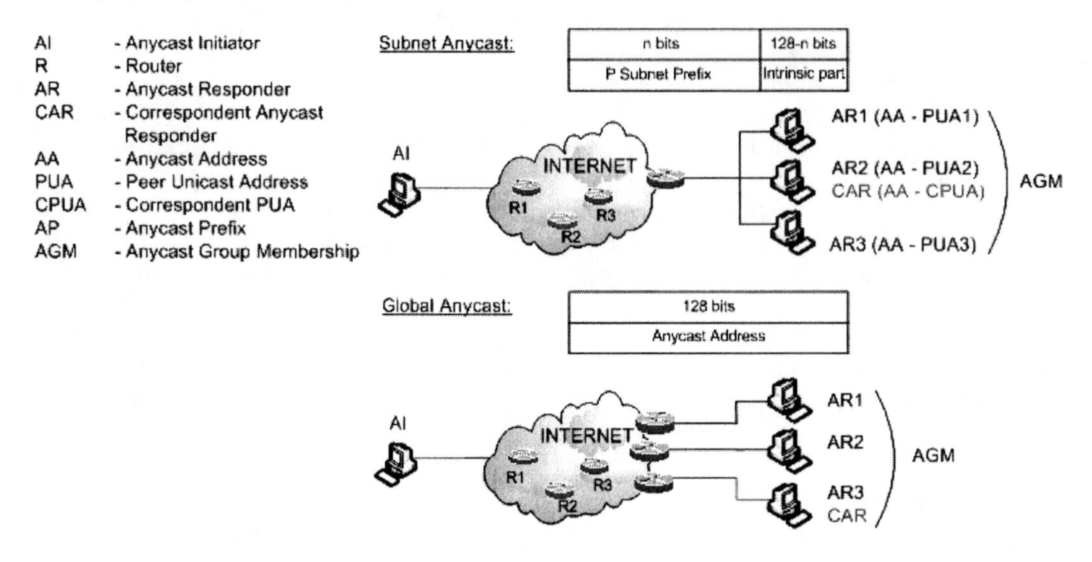

node. Therefore IP packets sent to the CoA of the mobile terminal have no chance to reach another "nearest" mobile node, since in this sense anycast addresses identifying mobile nodes are unique. The mobile node with a unique anycast care-of-address matches the correspondent anycast responder (CAR) in anycasting terminology. Also it has to be noted that in case of anycast address-based mobility there is no need for a peer unicast address since the CoA obtained is unique. The reason why unique anycast address is used instead of unicast address is the fact that anycast addresses are valid in the whole micromobility domain. Therefore the same anycast address can not be assigned to a second mobile node in a given micromobility domain.

The mobile node with a unique anycast address forms a virtual group. The members of this virtual group are the possible positions of the mobile node in the micromobility domain (that equals the validity area of the anycast address defined by the anycast *P* prefix) and the "nearest" mobile equipment is at the actual position of the mobile node. Therefore the mobile node remains reachable at any time (Figure 2). The purpose of using anycast address as an identifier for mobile nodes is that routing and handover management can be simplified with the help of changing the routing metrics. With the proper selection of the *P* prefix, the size of the virtual anycast group (VAG) can be adjusted easily. The virtual anycast group equals anycast group membership (AGM), while the virtual copies of the mobile node match the anycast responders. The operation of the anycast addressing-based mobility has to be investigated in case of different scenarios.

In the first scenario the mobile terminal leaves its current domain (e.g., its home network) and enters (1) another local administrative mobility domain (a new micromobility domain), as seen in Figure 2. In such case the mobile node first of all obtains (2)—with the help of IPv6 address autoconfiguration method—a unique anycast address that is valid in the whole area due to the properly set *P* prefix of the anycast address. As a result, the source address can be a unique anycast address since the source of a packet can be identified unequivocally. After getting the unique anycast care-of-address, the mobile node has to build the binding towards its home agent; therefore a binding procedure (3) is started by sending a binding update message. Next the mobile terminal has to initiate its membership in the virtual anycast group (VAG) of the new micromobility domain by having its anycast CoA (4). On receiving an anycast group membership report message the anycast access-router starts to propagate the new routing information by creating special routing information messages and sending it towards its adjacent routers. Based upon the underlying anycast routing protocol, each router in the new micromobility domain will get an entry in their routing table on how to reach the mobile terminal. Since each routing entry has a timeout period, thus the mobile node should send the membership report mes-

Figure 3. Moving in a given micromobility domain

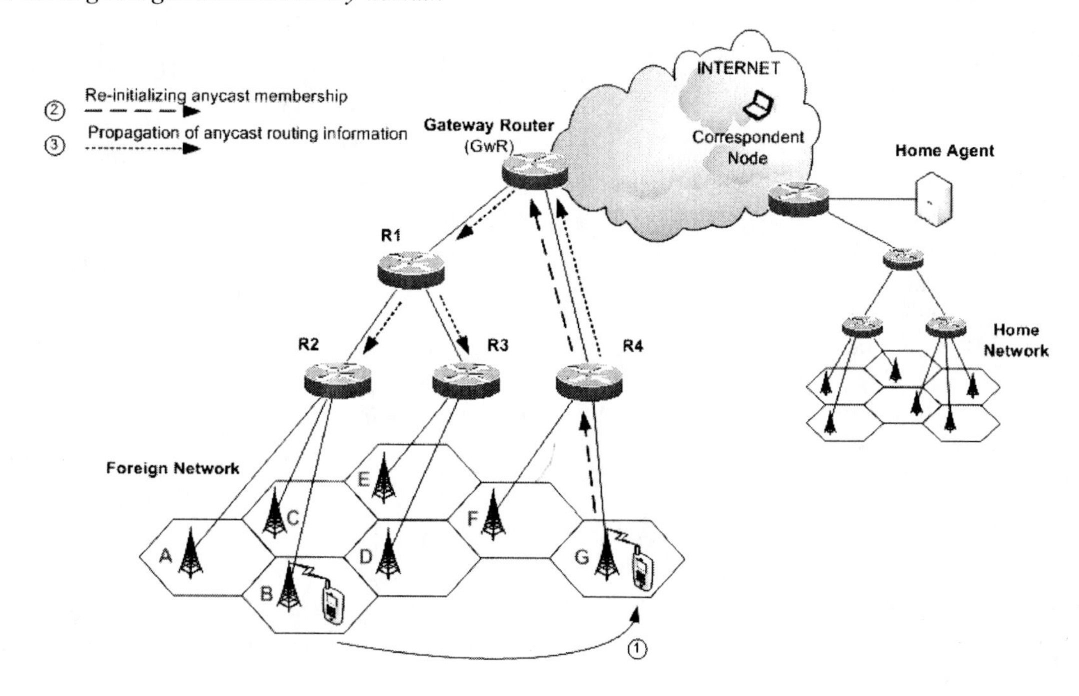

sage periodically to maintain its routing entry. The updating time of the routing entry should be defined according to the refresh interval of the routing entries.

In the second scenario (Figure 3), the mobile node moves in a given micromobility area (1). At the new wireless point of attachment the mobile terminal has to notify the new access router about its new location. This updating process can be done, for example, with the help of data packets of an active communication. In this case the new access router notices that packets with the anycast address in the source address field are being sent over one of its interfaces (2) (the access router checks the direction where it receives the anycast-sourced packets). According to the anycast routing protocol the access router has an entry in its routing table regarding this source anycast address. Therefore the router modifies the entry regarding the anycast address of the mobile node so that the new entry forwards the packets towards their new destination (the interface from which it has received the packet with the anycast address in the source address field), the actual location of the mobile terminal. The access router also has to initiate anycast routing information exchange (3).

Our approach gives a unique viewpoint on applications of IPv6 anycasting: introduces a new solution for micromobility management based on the IPv6 anycast addresses. The proposed method fits to the Mobile IPv6 standard and works efficiently in micromobility environment, while reducing the volume of control messages during the mobile operation and resulting in more seamless handover. The procedure can be realized without new protocol stacks, because the method is based only on the built-in features of IPv6 standard. The anycast-based micromobility can work on any mobility-supporting IPv6 system.

FUTURE TRENDS

In this article we have tried to give you an overview of the main issues that are being tackled by the ongoing research. The focus of the recent research is to construct an anycast routing protocol that is capable of handling large amounts of anycast hosts with reasonably low overhead generated by the routing system. At the moment there are multiple candidate protocols that could fulfill all the requirements set for the anycast routing protocol, while the standardization of these protocols is on the way.

It is also important to take a look on the trends that can be found among the applications. One can easily see that various applications could benefit from the properties of anycast addressing, therefore more and more applications emerge for exploit these possibilities.

First of all, it should be highlighted that application of the anycast addressing scheme is closely related to introduction of IPv6 into today's network, therefore until the usage of IPv6 gets more widespread, the scope of anycasting is also limited. In accordance with the present trends the vision for the anycasting looks bright, since as soon as there will be standardized routing protocols more and more application will be able to use the advanced services of the anycast addressing. In our view micromobility management could be one of the driving applications using anycast addresses.

CONCLUSION

Our aim in this article was to present an overview of the usage of anycast addressing paradigm and also show a possible new usage of the anycast address introduced in IPv6. The proposed anycast-based micromobility scheme is fairly simple: the mobile node after joining a foreign network obtains a unique anycast care-of-address that is valid until the mobile terminal stays inside the micromobility area, no matter if the mobile node moves around. Our method uses the services of anycast routing protocols that are capable of routing the traffic towards the "nearest" node from the set of nodes having the same anycast address. Currently none of the existing anycast routing protocols have been widely adopted, due to the lack of standardization.

REFERENCES

Abley, J. (2005). *Anycast addressing in IPv6*. draft-jabley-v6-anycast-clarify-00.txt

Deering, S., & Hinden, R. (1998), *Internet Protocol Version 6 (IPv6)*. IETF RFC 2460.

Doi, S., Ata, S., Kitamura, H., & Murata M. (2004). IPv6 anycast for simple and effective service-oriented communications. *IEEE Communications Magazine*, 163-171.

Doi, S., Ata, S., Kitamura, H., & Murata, M. (2005). *Design, implementation and evaluation of routing protocols for IPv6 anycast communication*. In *19th International Conference on Advanced Information Networking and Applications AINA '05* (Vol. 2, pp. 833-838). Taiwan.

Eunsoo, S. (2004). *Mobility management in the wireless Internet*. PhD Thesis, Columbia University.

Hashimoto, M., Ata, S., Kitamura, H., & Murata, M. (2005). *IPv6 anycast terminolgy definition*. draft-doi-ipv6-anycast-func-term-03.txt

Hinden, R., & Deering, S. (2003). *.IP Version 6 Addressing Architecture*. IETF RFC 3513.

Johnson, D., Perkins, C., & Arkko, J. (2004). *Mobility support in IPv6*. IETF RFC 3775.

Macker, J. P., Park, V. D., & Corson, S. M. (2001). Mobile and wireless Internet services: Putting the pieces together. *IEEE Communications Magazine*, *39*(6), 148-155

Matsunaga, S., Ata, S., Kitamura, H., & Murata, M. (2005). *Applications of IPv6 Anycasting*. draft-ata-ipv6-anycast-app-01.txt.

Matsunaga, S., Ata, S., Kitamura, H., & Murata, M. (2005). Design and implementation of IPv6 anycast routing protocol: PIA-SM. In *19th International Conference on Advanced Information Networking and Applications (AINA '05)*, (Vol. 2, pp. 839-844). Taiwan.

Partridge, C., Mendez, T., & Milliken, W. (1993). *Host anycasting service*. IETF RFC 1546.

Saha, D., Mukherjee, A., Misra, I.S., & Chakraborty, M. (2004). Mobility support in IP: A survey of related protocols. *IEEE Network*, *18*(6), 34-40.

Park, V. D., & Macker, J.P. (1999). Anycast routing for mobile networking. In *MILCOM '99 Conference Proceedings*.

Weber, S., & Cheng, L. (2004). A survey of anycast in IPv6 Networks. *IEEE Communications Magazine*, 127-133

Xuan, D., Jia, W., Zhao, W., & Zhu, H. (2000). A routing protocol for anycast messages. *IEEE Transactions on Parallel and Distributed Systems*, *11*(6), 571-588

Applications Suitability on PvC Environments

Andres Flores
University of Comahue, Argentina

Macario Polo Usaola
Universidad de Castilla-La Mancha, Spain

INTRODUCTION

Pervasive computing (PvC) environments should support the *continuity* of users' daily tasks across dynamic changes of operative contexts. Pervasive or ubiquitous computing implies computation becoming part of the environment. Many different protocols and operating systems, as well as a variety of heterogeneous computing devices, are interrelated to allow accessing information anywhere, anytime in a secure manner (Weiser, 1991; Singh, Puradkar, & Lee, 2005; Ranganathan & Campbell, 2003).

According to the initial considerations by Weiser (1991), a PvC environment should provide the feeling of an enhanced natural human environment, which makes the computers themselves vanish into the background. Such a disappearance should be fundamentally a consequence not of technology but of human psychology, since whenever people learn something sufficiently well, they cease to be aware of it.

This means that the user's relationship to computation changes to an implicit human-computer interaction. Instead of thinking in terms of doing explicit tasks "*on the computer*"—creating documents, sending e-mail, and so on—on PvC environments individuals may behave as they normally do: moving around, using objects, seeing and talking to each other. The environment is in charge of facilitating these actions, and individuals may come to expect certain services which allow the feeling of "*continuity*" on their daily tasks (Wang & Garlan, 2000).

Users should be allowed to change their computational tasks between different operative contexts, and this could imply the use of many mobile devices that help moving around into the environment. As a result, the underlying resources to run the required applications may change from wide memory space, disk capacity, and computational power, to lower magnitudes. Such situations could make a required service or application inappropriate in the new context, with a likely necessity of supplying a proper adjustment. However, users should not perceive the surrounding environment as something that constraints their working/living activities. There should be a continuous provision of proper services or applications. Hence the environment must be provided with a mechanism for *dynamic applications suitability* (Flores & Polo, 2006).

PERVASIVE COMPUTING ENVIRONMENTS

In the field of PvC there is still a misuse of some related concepts, since often PvC is used interchangeably with ubiquitous computing and mobile computing. However, nowadays consistent definitions are identified in the literature as follows (Singh et al., 2005).

Mobile computing is about elevating computing services and making them available on mobile devices using the wireless infrastructure. It focuses on reducing the size of the devices so that they can be carried anywhere or by providing access to computing capacity through high-speed networks. However, there are some limitations. The computing model does not change considerably as we move, since the devices cannot seamlessly and flexibly obtain information about the context in which the computing takes place and adjust it accordingly. The only way to accommodate the needs and possibilities of changing environments is to have users manually control and configure the applications while they move—a task most users do not want to perform.

PvC deals with acquiring context knowledge from the environment and dynamically building computing models dependent on context. That is, providing dynamic, proactive, and context-aware services to the user. It is invisible to human users and yet provides useful computing services (Singh et al., 2005). Three main aspects must be properly understood (Banavar & Bernstein, 2002). First is the way people view mobile computing devices and use them within their environments to perform tasks. A device is a portal into an application/data space, not a repository of custom software managed by the user. Second is the way applications are created and deployed to enable such tasks to be performed. An application is a means by which a user performs a task, not a piece of software that is written to exploit a device's capabilities. And third is the environment and how it is enhanced by the emergence and ubiquity of new information and functionality. The computing environment is the user's information-enhanced physical surroundings, not a virtual space that exists to store and run software.

Ubiquitous computing uses the advances in mobile computing and PvC to present a *global computing environment* where seamless and invisible access to computing resources

Figure 1. Vision of an enhanced physical environment by ubiquitous computing

On the other side, there are approaches particularly concerned with providing higher level services more related to users tasks, in order to help them reduce the working effort (Roman, Ziebart, & Campbell, 2003; Becker & Schiele, 2003; Chakraborty, Joshi, Yesha, & Finin, 2006; Gaia Project, 2006; Aura Project, 2006). Most of them have been conceptualized with some sort of self-adjusted applications or by applications relying on basic services provided by the underlying platform (e.g., CORBA).

No matter how users need a transparent delivery of functionality, so they could have a sense of continued presence of the environment. Therefore, any unavailability of a required service implies that a user understand that the underlying environment cannot provide all that is needed, thus destroying the aspiration of transparency.

is provided to the user. It aims to provide PvC environments to a human user as s/he moves from one location to another. Thus, it is created by sharing knowledge and information between PvC environments (Singh et al., 2005). Figure 1 shows the vision a user may have of a physical environment that is enhanced by ubiquitous computing.

Some approaches for PvC are concerned with interconnecting protocols from different hardware artifacts and devices, or solving problems of intermittent network connections and fluctuation on bandwidth. Therefore, their applications are quite general or low level, yet mainly related to communication tools which still requires a big effort for a user to accomplish a working task. Other approaches are focused on solving problems of prohibited access to information or even to a closed or restricted environment. If we consider that the environment is populated with an enormous amount of users, each intending accesses to different hardware and software resources, the security concerns increase proportionally (Kallio, Niemelä, & Latvakoski, 2004).

SUITABILITY FOR PERVASIVE APPLICATIONS

Functionality on a PvC environment is usually shaped as a set of aggregated components that are distributed among different computing devices. On changes of availability of a given device, the involved component behavior still needs to be accessible in the appropriate form according to the updated technical situation. This generally makes users be involved on a dependency with the underlying environment and increases the complexity of its internal mechanisms (Iribarne, Troya, & Vallecillo, 2003; Warboys et al., 2005).

Applications composed of dynamically replaceable components imply the need of an appropriate integration process according to component-based software development (CBSD) (Cechich, Piattini, & Vallecillo, 2003; Flores, Augusto, Polo, & Varea, 2004). For this, an application model may provide the specification of a required functionality in the form of the aggregation of component models, as can be seen in Figure 2. A component model provides a definition to

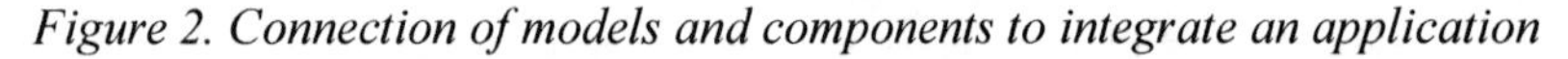

Figure 2. Connection of models and components to integrate an application

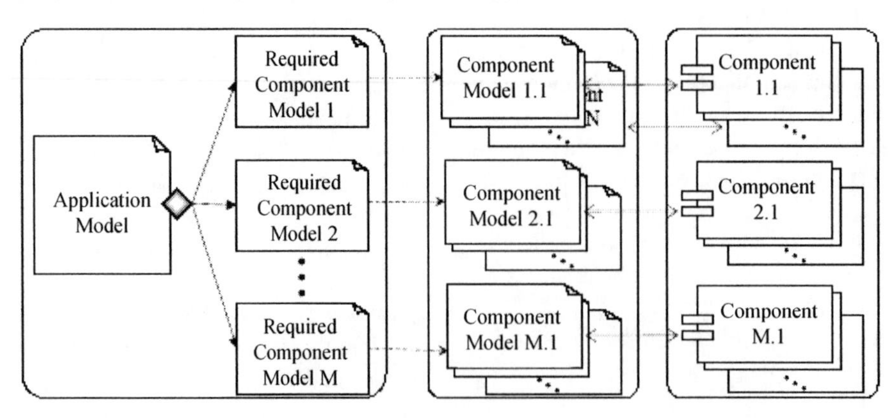

instantiate a component and its composition aspects through standard interactions and unambiguous interfaces (Cechich et al., 2003; Iribarne et al., 2003; Warboys et al., 2005). In order to assure the adequacy of a given component with respect to an application model, there is a need to evaluate its component model. Hence we present an assessment procedure which can be applied both on a development stage and also at runtime. The latter becomes necessary when the current technical situation makes unsuitable a given component demanding that a surrogate be provided.

The assessment procedure compares functional aspects from components against the specification provided by the application model, which is component oriented. Besides analyzing component services at a syntactic level, its behavior is also inspected, thus embracing semantic aspects. The latter is done by abstracting out the black box functionality hidden on components in the form of *assertions,* and also exposing its likely interactions by means of the *protocol of use,* which describes the expected order of use for its services (Flores & Polo, 2005)—also called choreography (Iribarne et al., 2003).

So far we have been experimenting with the addition of metadata for comparing behavioral aspects from components. Metadata has been used in several approaches as a technique to easy verification procedures (Cechich et al., 2003; Cechich & Polo, 2005; Orso et al., 2001). By adding meta-methods we may then retrieve detailed information concerning *assertions* and the *protocol of use,* which somehow implies a component adaptation, particularly referred to as the instrumentation mechanism (Flores & Polo, 2005).

The assessment procedure is described by means of a set of conditions which must be satisfied according to certain thresholds. Different techniques are applied to achieve the required evaluations. Compatibility on both assertions and the usage protocol is carried out by generating Abstract Syntax Trees and applying some updated algorithms, which were originally developed to detect similar pieces of code (*clones*) on existing programs. Such compatibility analysis is based on the following consideration. Post-conditions (for example) on services from two similar components necessarily should relate to a similar structure and semantic. Hence, they could be thought of as one being a clone of the other (Flores & Polo, 2005).

All such techniques applied on our assessment procedure allow the accomplishment of a consistent mechanism to assure a fair component integration. As PvC environments imply many challenges, the whole integration process is based on considering all aspects concerning reliability.

We may make use of a simple example to illustrate the way a functionality is composed from distributed disparate components and understand how the assessment procedure may help to assure the suitability of a certain involved component.

Suppose we represent a PvC environment for a museum, which includes a tour guide application for proposing different paths according to the user's dynamic choices. When the user enters the museum, s/he may carry a computing device (a PDA or a smart phone), and an automatic detection is done in order to identify and connect the device to the environment. As the user walks by each art piece (e.g., painting or sculpture), descriptions and information of particular interest to the user are displayed on the PDA or spoken through the phone. Figure 3 shows a likely scenario of the presented case study.

A related application could allow creating an album with images of some art pieces the user has visited. To obtain the images the user will probably have to pay a fee, for example, when s/he intends to leave the museum. The album organizer application—maybe downloaded into a user's notebook recognized by the environment—may allow creating a sort of document with images and some notes written by the user. Notes could be stored on separated text files and bind to the document by means of hypertext links. Thus every time the user needs to write or edit a note, a proper editor is provided.

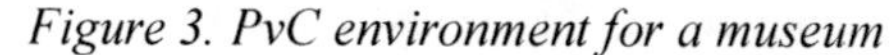

Figure 3. PvC environment for a museum

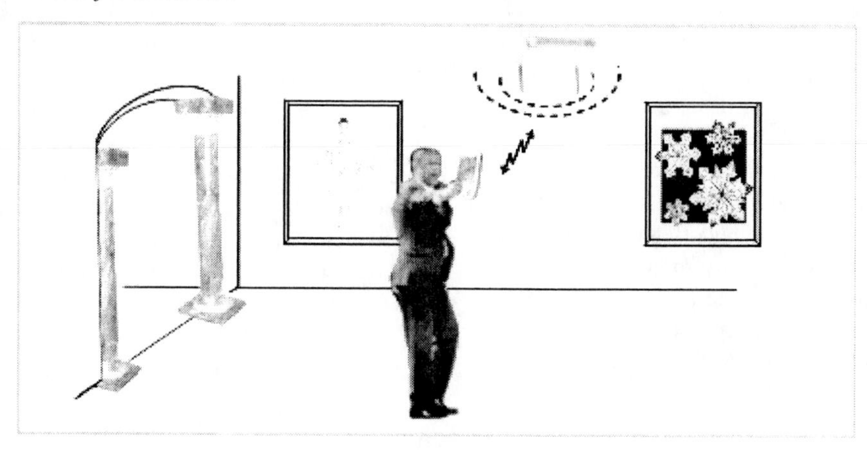

Figure 4. Distributed components for the album organizer application

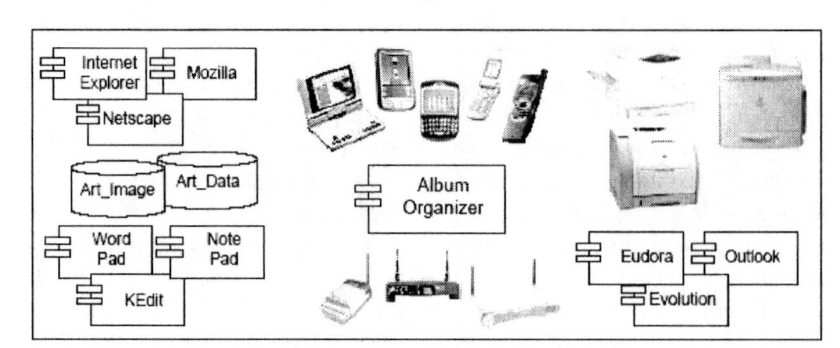

The user may also be allowed to print a selection of pages of the document, or even send the created album by e-mail in case s/he does not have a device to carry those files.

If we focus on the album organizer application, we might analyze the potential required components. There could be an album organizer component to represent the main logic of the application. This component could have an ad-hoc sophisticated album visual editor or it could be a Web-style editor in which is additionally required a generic Web browser. The visual editor also depends on the actual used device. In order to make notes, different components could be used as a simple sort of Notepad, or other replacements like WordPad or similar applications according to the underlying software platform. For sending e-mail, applications like Outlook or Eudora could be used, and to provide a printer service, different kinds of printers and ad hoc wireless sensors should be available. Another component is concerned with the database for images and descriptions of art pieces. Figure 4 shows a diagram with the likely comprised components and devices for the album organizer application.

Suppose a user needs to write a note by using a notebook which runs a Linux platform. One available text editor is KEdit. The environment then evaluates this component so to ensure it is appropriate to fulfill the task. Following can be seen the interfaces of both the KEdit component and the required component model named TextEditor. In order to assure the compatibility with the surrogate, the assessment procedure is applied to analyze if the degree of similarity raises a proper level.

Since implementation alternatives are fairly important, we have properly explored some of them. For example both Microsoft .NET and Enterprise Java Beans allow the addition of metadata to components. Information about components—their structure and the state of their corresponding instances—can be retrieved at runtime by using *reflection* on both environments as well. Particularly .NET includes the possibility of adding *attributes,* which is a special class intended to provide additional information about some design element as a class, a module, a method, a field, and so on (Flores & Polo, 2005).

Hence, we have implemented on the .Net technology the current state of our approach (Flores & Polo, 2006). Though simple, this prototype gives us rewarding data on possibilities to make our proposals concrete. All the applied techniques are selected according to our goal of achieving consistent mechanisms to assure a fair component integration. As we proceed with our work, reliability is mainly considered, since we focus the whole integration process for those challenging systems as PvC environments.

FUTURE TRENDS

Our work continues by completing the coverage of functional aspects for components, describing the components replacement mechanism and then focusing on the non-functional aspects, particularly on quality of service. For this we are analyzing the use and extending the schemas from Iribarne et al. (2003) which provide a consistent format for specifications of components by means of XML. The approach covers all of the aspects from components: functional, non-functional, and commercial.

As some authors have pointed out (Chen, Finin, & Joshi, 2005; Ranganathan & Campbell, 2003), temporal aspects could also be helpful to achieve a more accurate component integration process. This may give the chance to analyze whether a component can fit the requirements when time conditions are also included in the set of updated requirements upon a change on the user's context of operation.

Selection of appropriate methods, techniques, and languages must be accurately accomplished upon the concern of a reliable mechanism. This is the emphasis of our next development in this area.

REFERENCES

Aura Project. (2006). *Aura Project Web site.* Retrieved from http://www-2.cs.cmu.edu/ aura/

Banavar, G., & Bernstein, A. (2002). Issues and challenges in ubiquitous computing: Software infrastructure and design challenges for ubiquitous computing applications. *Communications of the ACM, 45*(12).

Becker, C., & Schiele, G. (2003). Middleware and application adaptation requirements and their support in pervasive computing. *Proceedings of IEEE ICDCSW* (pp. 98-103), Providence, RI.

Cechich, A., & Polo, M. (2005). COTS component testing through aspect-based metadata. In S. Beydeda & V. Gruhn (Eds.), *Building quality into components—testing and debugging.* Berlin: Springer-Verlag.

Cechich, A., Piattini, M., & Vallecillo, A. (Eds.). (2003). *Component-based software quality: Methods and techniques* (LNCS 2693). Berlin: Springer-Verlag.

Chakraborty, D., Joshi, A., Yesha, Y., & Finin, T. (2006). Toward distributed service discovery pervasive computing environments. *IEEE Transactions on Mobile Computing, 5*(2).

Chen, H., Finin, T., & Joshi, A. (2004). Semantic Web in the context broker architecture. *Proceedings of IEEE PerCom,* Orlando, FL, (pp. 277-286).

Flores, A., & Polo, M. (2005, June 27-30). Dynamic component assessment on PvC environments. *Proceedings of IEEE ISCC,* Cartagena, Spain, (pp. 955-960).

Flores, A., & Polo, M. (2006, May 23). An approach for applications suitability on pervasive environments. *Proceedings of IWUC* (held at ICEIS). Paphos, Cyprus: INSTICC Press.

Flores, A., Augusto, J. C., Polo, M., & Varea, M. (2004, October 10-13). Towards context-aware testing for semantic interoperability on PvC environments. *Proceedings of IEEE SMC,* The Hague, The Netherlands, (pp. 1136-1141).

Gaia Project. (2006). *Gaia Project Web site.* Retrieved from http://www.w3.org/2001/sw

Iribarne, L., Troya, J., & Vallecillo, A. (2003). A trading service for COTS components. *The Computer Journal, 47*(3).

Kallio, P., Niemelä, E., & Latvakoski, J. (2004). UbiSoft—pervasive software. *Research Notes, 2238.* Finland: VTT Electronics. Retrieved from www.vtt.fi/inf/pdf/tiedotteet/2004/T2238.pdf

Orso, A., Harrold, M.J., Rosenblum, D., Rothermel, G., Do, H., & Sofía, M.L. (2001). Using component metacontent to support the regression testing of component-based software. *Proceedings of IEEE ICSM,* Florence, Italy, (pp. 716-725).

Ranganathan, A., & Campbell, R. (2003). An infrastructure for context-awareness based on first order logic. *Personal and Ubiquitous Computing, 7,* 353-364.

Roman, M., Ziebart, B., & Campbell, R. (2003). Dynamic application composition: Customizing the behavior of an active space. *Proceedings of IEEE PerCom.*

Singh, S., Puradkar, S., & Lee, Y. (2005). Ubiquitous computing: Connecting pervasive computing through semantic Web. *Journal of ISeB.* Berlin: Springer-Verlag.

Wang, Z., & Garlan, D. (2000). *Task-driven computing.* Technical Report No. CMU-CS-00-154, School of Computer Science, Carnegie Mellon University, USA. Retrieved from http://reports-archive.adm.cs.cmu.edu/anon/2000/abstracts/00-154.html

Warboys, B., Snowdon, B., Greenwood, R. M., Seet, W., Robertson, I., Morrison, R., Balasubramaniam, D., Kirby, G., & Mickan, K. (2005). An active-architecture approach to COTS integration. *IEEE Software, 22*(4), 20-27.

Weiser, M. (1991). The computer for the 21st century. *Scientific American, 265*(3), 94-104. Retrieved from http://www.ubiq.com/hypertext/weiser/SciAmDraft3.html

KEY TERMS

Component-Based Software Development (CBSD): A development paradigm where software systems are developed from the assembly or integration of software components.

Component Model: A specification that describes how to instantiate or build a software component, and gives guidelines for its binding to other software components by means of standard interactions or communication patterns and unambiguous interfaces.

Mobile Computing: Small wireless devices that can be carried anywhere, allowing a computing capacity through wireless networks. Also known as nomadic computing.

Pervasive Computing (PvC): Enhancement of the physical surroundings by providing and adapting mobile computing according to the user's needs. Also known as ambient intelligent.

Software Component: A unit of independent deployment that is ready "off-the-shelf" (OTS), from a commercial source (COTS) or reused from another system (in-house or legacy systems). It is usually self-contained, enclosing a collection of cooperating and tightly cohesive objects, thus providing a significant aggregate of functionality. It is used "as it is found" rather than being modified, may possibly execute independently, and can be integrated with other components to achieve a required bigger system functionality.

Ubiquitous Computing: Provides PvC environments to a human user as s/he moves from one location to another. It allows sharing knowledge and information between PvC environments. Also known as Global Computing.

Wireless Device: A small computer that is reduced in size and in computing power, can be carried everywhere, and provides voice, data, games, and video applications. The most familiar is the mobile or cell phone, then Palm Pilot and its handheld descendent, the PDA (personal digital assistant), a great evolution because of its large amount of new applications. Laptop or notebook and tablet PCs are also well-known wireless devices.

Wireless Network: Offers mobility and elimination of unsightly cables, by the use of radio waves and/or microwaves to maintain communication channels between computers. It is an alternative to wired networking, which relies on copper and/or fiber optic cabling between network devices. Popular wireless local area networking (WLAN) products conform to the 802.11 "Wi-Fi" standards.

A Bio-Inspired Approach for the Next Generation of Cellular Systems

Mostafa El-Said
Grand Valley State University, USA

INTRODUCTION

In the current 3G systems and the upcoming 4G wireless systems, *missing neighbor pilot* refers to the condition of receiving a high-level pilot signal from a Base Station (BS) that is not listed in the mobile receiver's neighbor list (LCC International, 2004; Agilent Technologies, 2005). This pilot signal interferes with the existing ongoing call, causing the call to be possibly dropped and increasing the handoff call dropping probability. Figure 1 describes the missing pilot scenario where BS1 provides the highest pilot signal compared to BS1 and BS2's signals. Unfortunately, this pilot is not listed in the mobile user's active list.

The horizontal and vertical handoff algorithms are based on continuous measurements made by the user equipment (UE) on the Primary Scrambling Code of the Common Pilot Channel (CPICH). In *3G systems,* UE attempts to measure the quality of all received CPICH pilots using the Ec/Io and picks a dominant one from a cellular system (Chiung & Wu, 2001; El-Said, Kumar, & Elmaghraby, 2003). The UE interacts with any of the available radio access networks based on its memorization to the neighboring BSs. As the UE moves throughout the network, the serving BS must constantly update it with neighbor lists, which tell the UE which CPICH pilots it should be measuring for handoff purposes. In *4G systems*, CPICH pilots would be generated from any wireless system including the 3G systems (Bhashyam, Sayeed, & Aazhang, 2000). Due to the complex heterogeneity of the 4G radio access network environment, the UE is expected to suffer from various carrier interoperability problems. Among these problems, the missing neighbor pilot is considered to be the most dangerous one that faces the 4G industry.

The wireless industry responded to this problem by using an inefficient traditional solution relying on using antenna downtilt such as given in Figure 2. This solution requires shifting the antenna's radiation pattern using a mechanical adjustment, which is very expensive for the cellular carrier. In addition, this solution is permanent and is not adaptive to the cellular network status (Agilent Technologies, 2005; Metawave, 2005).

Therefore, a self-managing solution approach is necessary to solve this critical problem. Whisnant, Kalbarczyk, and Iyer (2003) introduced a system model for dynamically reconfiguring application software. Their model relies on considering the application's static structure and run-time behaviors to construct a workable version of reconfiguration software application. Self-managing applications are hard to test and validate because they increase systems complexity (Clancy, 2002). The ability to reconfigure a software application requires the ability to deploy a dynamically hardware infrastructure in systems in general and in cellular systems in particular (Jann, Browning, & Burugula, 2003).

Figure 1. Missing pilot scenario

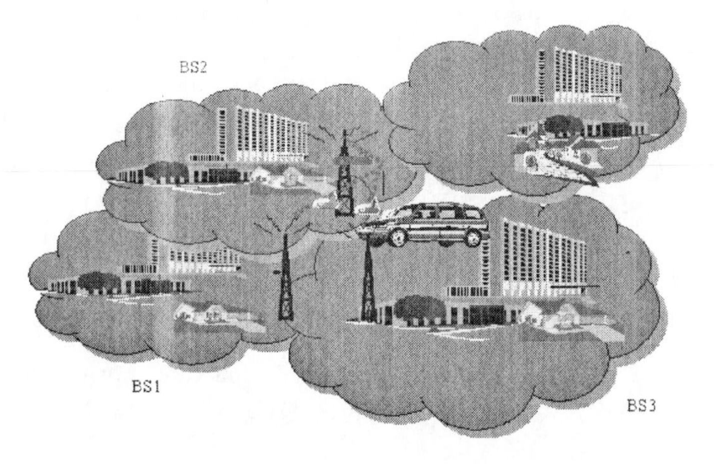

Figure 2. Missing pilot solution: Antenna downtilt

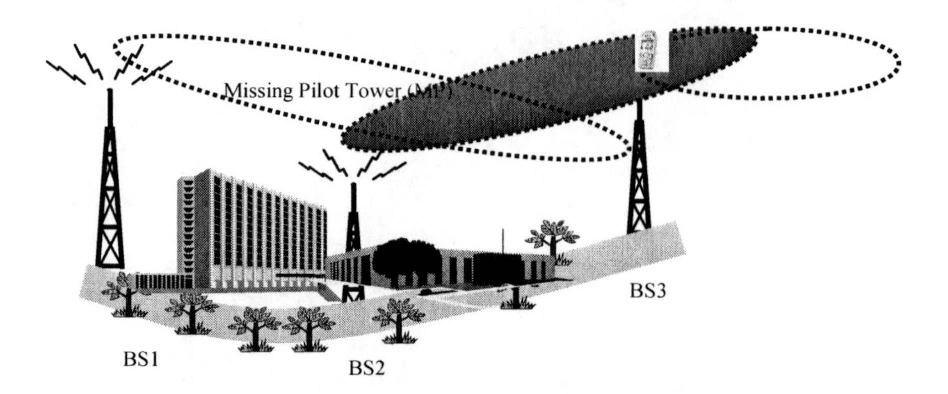

Konstantinou, Florissi, and Yemini (2002) presented an architecture called NESTOR to replace the current network management systems with another automated and software-controlled approach. The proposed system is inherently a rule-based management system that controls change propagation across model objects. Vincent and May (2005) presented a decentralized service discovery approach in mobile ad hoc networks. The proposed mechanism relies on distributing information about available services to the network neighborhood nodes using the analogy of an electrostatic field. Service requests are issued by any neighbor node and routed to the neighbor with the highest potential.

The autonomic computing system is a concept focused on adaptation to different situations caused by multiple systems or devices. The IBM Corporation recently initiated a public trail of its Autonomic Toolkit, which consists of multiple tools that can be used to create the framework of an autonomic management system. In this article, an autonomic engine system setting at the cellular base station nodes is developed to detect the missing neighbor (Ganek & Corbi, 2003; Haas, Droz, & Stiller, 2003; Melcher & Mitchell, 2004). The autonomic engine receives continuous feedback and performs adjustments to the cell system's neighboring set by requiring the UE to provide signal measurements to the serving BS tower (Long, 2001).

In this article, I decided to use this toolkit to build an autonomic rule-based solution to detect the existence of any missing pilot. The major advantage of using the IBM autonomic toolkit is providing a common system infrastructure for processing and classifying the RF data from multiple sources regardless of its original sources. This is a significant step towards creating a transparent autonomic high-speed physical layer in 4G systems.

PROPOSED SOLUTION

The proposed AMS relies on designing an autonomic high-speed physical layer in the smart UE and the BS node. *At the UE side,* continuous CPICH pilot measurements will be recorded and forwarded to the serving BS node via its radio interface. *At the BS node,* a scalable self-managing autonomic engine is developed using IBM's autonomic computing toolkit to facilitate the mobile handset's vertical/horizontal handover such as shown in Figure 3. The proposed engine is cable of interfacing the UE handset with different wireless technologies and detects the missing pilot if it is existed.

The autonomic engine relies on a generic log adapter (GLA), which is used to handle any raw measurements log file data and covert it into a standard format that can be understood by the autonomic manager. Without GLA, separate log adapters would have be coded for any system that the autonomic manager interfaced with. The BS node will then lump all of the raw data logs together and forward them to the Generic Log Adapter for data classification and restructuring to the common base event format. Once the GLA has parsed a record in real time to common base event format, the autonomic manager will see the record and process it and take any action necessary by notifying the BS node to make adjustments to avoid the missing pilot and enhance the UE devices' quality of service.

PERFORMANCE MEASUREMENTS AND KEY FINDINGS

To test the applicability of the proposed solution, we decided to use the system's response time, AS's service rate for callers experiencing missing pilot problem, and the performance

Figure 3. Autonomic base station architecture

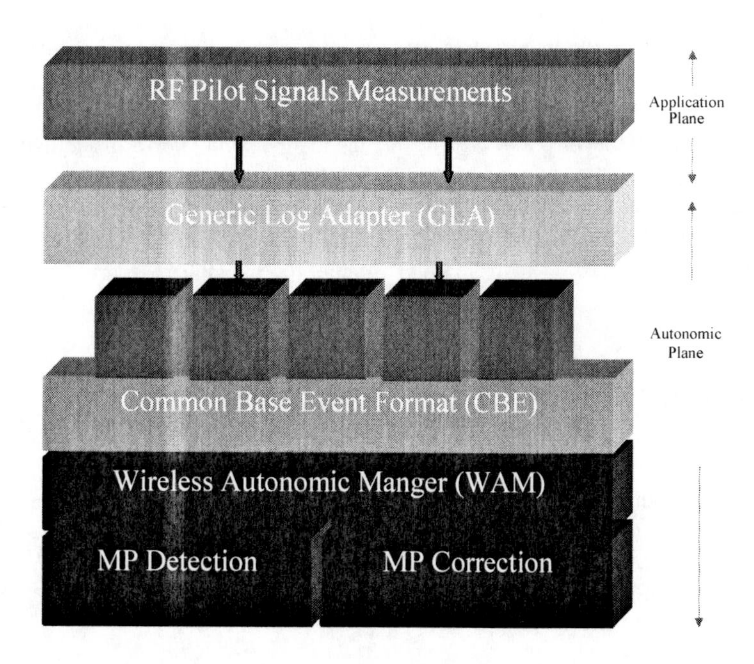

Table 1. Summary of the system performance analysis

	Log File Size in (# Records)	System Response Time in (Sec)	Processing Rate by the Base Station in (Records/Sec)
Trial Experiment 1	985	145	6.793103448
Trial Experiment 2	338	95	3.557894737
Trial Experiment 3	281	67	4.194029851
Trial Experiment 4	149	33	4.515151515
Average Processing Rate by the Base Station in (Records/Sec)	**4.765044888**		
Base Station Service Rate For callers experiencing missing pilot problem (Records/Sec)	**5.3**		
Performance Gain	**1.112266542**		

gain as performance metrics. Also, we developed a Java class to simulate the output of a UE in a heterogeneous RF access network. Table 1 summarizes the simulation results for four simulation experiments with different log files size.

The results shown in Table 1 comply with the design requirements for the current 3G system. This is illustrated in the following simple example.

DESIGN REQUIREMENTS FOR 3G SYSTEMS

- The 3G cell tower's coverage area is divided into three sectors, with each sector having (8 traffic channel * 40 call/channel = 320 voice traffic per sector) and (2 control channels * 40 callers/channels = 80 control traffic per sector).
- The overlapped area between towers (handoff zone) occupies 1/3 of the sector size and serves (1/3 of 320

= 106 callers (new callers and/or exciting ones)). If we consider having the UE report its status to the tower every 5 seconds, we could potentially generate 21.2 records in 1 second.

- It is practical to assume that 25% of the 21.2 reports/second accounts for those callers that may suffer from the missing pilot problem—that is, the tower's service rate for missing neighbor pilot callers is 21.2/4 = 5.3 records/second. This is the threshold level used by the tower to accommodate those callers suffering from the missing pilot problem.

ANALYSIS OF THE RESULTS

- Response time is the time taken by the BS to process, parse the incoming log file and detect the missing neighbor pilot. It is equal to (145, 95, 67, and 33) for the four experiment trials. All values are in seconds.
- Processing rate by the base station is defined as the total number of incoming records divided by the response time in (records/second). It is equal to (6.7, 3.5, 4.1, and 4.5) for the four experiment trials.
- The UE reports a missing pilot problem with an average rate of 4.7 records/second.
- The base station's service rate for callers experiencing the missing pilot problem = 5.3 records/second.
- The performance gain is defined as:

$$\frac{\text{Base Station Service Rate For callers experiencing missing pilot problem in (Records/Sec)}}{\text{Average Processing Rate by the Base Station in (Records/Sec)}}$$

$= 5.3/4.7 = 1.1$

- Here it is obvious that the service rate (5.3 records/second) is greater than the UE's reporting rate to the base station node (4.7 records/second). Therefore, the above results prove that the proposed solution does not overload the processing capabilities of the BS nodes and can be scaled up to handle a large volume of data.

FUTURE TRENDS

An effective solution for the interoperability issues in 4G wireless systems must rely on an adaptive and self-managing network infrastructure. Therefore, the proposed approach in this article can be scaled to maintain continuous user connectivity, better quality of service, improved robustness, and higher cost-effectiveness for network deployment.

CONCLUSION

In this article, we have developed an autonomic engine system setting at the cellular base station (BS) nodes to detect the missing neighbor. The autonomic engine receives continuous feedback and performs adjustments to the cell system's neighboring set by requiring the user equipment (UE) to provide signal measurements to the serving BS tower. The obtained results show that the proposed solution is able to detect the missing pilot problem in any heterogeneous RF environment.

REFERENCES

Agilent Technologies. (2005). Retrieved October 2, 2005, from http://we.home.agilent.com

Bhashyam, S., Sayeed, A., & Aazhang, B. (2000). Time-selective signaling and reception for communication over multipath fading channels. *IEEE Transaction on Communications, 48*(1), 83-94.

Chiung, J., & Wu, S. (2001). Intelligent handoff for mobile wireless Internet. *Journal of Mobile Networks and Applications, 6,* 67-79.

Clancy, D. (2002). *NASA challenges in autonomic computing. Almaden Institute 2002, IBM Almaden Research Center,* San Jose, CA.

El-Said, M., Kumar, A., & Elmaghraby, A. (2003). Pilot pollution interference cancellation in CDMA systems. *Special Issue of Wiley Journal: Wireless Communication and Mobile Computing on Ultra Broadband Wireless Communications for the Future, 3*(6), 743-757.

Ganek, A., & Corbi, T. (2003). The dawning of the autonomic computing era. *IBM Systems Journal, l42*(1), 5-19.

Haas, R., Droz, P., & Stiller, B. (2003). Autonomic service deployment in networks. *IBM Systems Journal, 42*(1), 150-164.

Jann, L., Browning, A., & Burugula, R. (2003). Dynamic reconfiguration: Basic building blocks for autonomic computing on IBM pSeries servers. *IBM Systems Journal, 42*(1), 29-37.

Konstantinou, A., Florissi, D., & Yemini, Y. (2002). Towards self-configuring networks. *Proceedings of the DARPA Active Networks Conference and Exposition* (pp. 143-156).

LCC International. (2004). Retrieved December 10, 2004, from http://www.hitech-news.com/30112001-MoeLLC.htm

Lenders, V., May, M., & Plattner, B. (2005). Service discovery in mobile ad hoc networks: A field theoretic approach. *Special Issue of Pervasive and Mobile Computing, 1,* 343-370.

Long, C. (2001). *IP network design.* New York: McGraw-Hill Osborne Media.

Melcher, B., & Mitchell, B. (2004). Towards an autonomic framework: Self-configuring network services and developing autonomic applications. *Intel Technology Journal, 8*(4), 279-290.

Metawave. (2005). Retrieved November 10, 2005, from http://www.metawave.com

Whisnant, Z., Kalbarczyk, T., & Iyer, R. (2003). A system model for dynamically reconfigurable software. *IBM Systems Journal, 42*(1), 45-59.

KEY TERMS

Adaptive Algorithm: Can "learn" and change its behavior by comparing the results of its actions with the goals that it is designed to achieve.

Autonomic Computing: An approach to self-managed computing systems with a minimum of human interference. The term derives from the body's autonomic nervous system, which controls key functions without conscious awareness or involvement.

Candidate Set: Depicts those base stations that are in transition into or out of the active set, depending on their power level compared to the threshold level.

Missing Neighbor Pilot: The condition of receiving a high-level pilot signal from a base station (BS) that is not listed in the mobile receiver's neighbor list.

Neighbor Set: Represents the nearby serving base stations to a mobile receiver. The mobile receiver downloads an updated neighbor list from the current serving base station. Each base station or base station sector has a unique neighbor list.

Policy-Based Management: A method of managing system behavior or resources by setting "policies" (often in the form of "if-then" rules) that the system interprets.

Virtual Active Set: Includes those base stations (BSs) that are engaged in a live communication link with the mobile user; they generally do not exceed three base stations at a time.

Brain Computer Interfacing

Diego Liberati
Italian National Research Council, Italy

INTRODUCTION

In the near future, mobile computing will benefit from more direct interfacing between a computer and its human operator, aiming at easing the control while keeping the human more free for other tasks related to displacement.

Among the technologies enabling such improvement, a special place will be held by brain computer interfacing (BCI), recently listed among the 10 emerging technologies that will change the world by the *MIT Technology Review* on January 19, 2004.

The intention to perform a task may be in fact directly detected from analyzing brain waves: an example of such capability has been for instance already shown trough artificial neural networks in Babiloni et al. (2000), thus allowing the switch of a bit of information in order to start building the control of a direct interaction with the computer.

BACKGROUND

Our interaction with the world is mediated through sensory-motor systems, allowing us both to acquire information from our surroundings and manipulate what is useful at our reach. Human-computer interaction ergonomically takes into account the psycho-physiological properties of such interaction to make our interactions with computers increasingly easy. Computers are in fact nowadays smaller and smaller without significant loss of power needed for everyday use, like writing an article like this one on a train going to a meeting, while checking e-mail and talking (via voice) to colleagues and friends.

Now, the center of processing outside information and producing intention to act consequently is well known to be our brain. The capability to directly wire neurons on electronic circuits is not (yet?) within our reach, while interesting experiments of compatibility and communication capabilities are indeed promising at least in vitro. At the other extreme, it is not hard to measure non-invasively the electromagnetic field produced by brain function by positioning small electrodes over the skull, as in the standard clinical procedure of electroencephalography.

Obviously, taking from outside a far-field outside measure is quite different than directly measuring the firing of every single motor neuron of interest: a sort of summing of all the brain activity will be captured at different percentages. Nonetheless, it is well known that among such a messy amount of signal, when repeating a task it is not hard to enhance the very signal related to task, while reducing—via synchronized averaging—the overwhelming contribution of all the other neurons not related to the task of interest. On this premise, Deecke, Grozinger, and Kornhuber (1976) have been able to study the so-called event-related brain potential, naming the onset of a neural activation preceding the task, in addition to the neural responses to the task itself.

Statistical pattern recognition and classification has been shown to improve such event-related detection by Gevins, Morgan, Bressler, Doyle, and Cutillo (1986).

A method to detect such preparatory potential on a single event basis (and then not needing to average hundreds of repetitions, as said before) was developed and applied some 20 years ago by Cerutti, Chiarenza, Liberati, Mascellani, and Pavesi (1988). One extension of the same parametric identification approach is that developed by del Millan et al. (2001) at the European Union Joint Research Center of Ispra, Italy, while a Bayesian inference approach has been complementary proposed by Roberts and Penny (2000).

BRAIN COMPUTER INTERFACING

Autoregressive with exogenous input parametric identification (Cerutti et al., 1988) is able to increase by some 20 dB the worst signal-to-noise ratio of the event-related potential with respect to the overwhelming background brain activity. Moreover, it provides a reduced set of parameters that can be used as features to perform post-processing, should it be needed.

A more sophisticated, though more computing-demanding, time variant approach based on an optimal so-called Kalman filer has been developed by Liberati, Bertolini, and Colombo (1991a). The joint performance of more than one task has also been shown to evoke more specific brain potential (Liberati, Bedarida, Brandazza, & Cerutti, 1991b).

Multivariable joint analysis of covariance (Gevins et al., 1989), as well as of total and partial coherence among brain field recordings at different locations (Liberati, Cursi, Locatelli, Comi, & Cerutti, 1997), has also improved the capability of discriminating the single potential related to a particular task (Liberati, 1991a).

Artificial neural networks (Liberati, 1991b; Pfurtscheller, Flotzinger, Mohl, & Peltoranta, 1992; Babiloni et al., 2000) offered the first approach to the problem of so-called artificial intelligence, whose other methods of either soft computing, like the fuzzy sets made popular by Lofthi Zadeh, and the rule extraction like the one proposed by Muselli and Liberati (2002), are keen to be important post-processing tools for extracting real commands from the identified parameters.

In particular, the rule extraction approach proposed by Muselli and Liberati (2000) has the nice properties of processing the huge amount of data in a very fast quadratic time (and even in terms of binary operations), yielding both pruning of the redundant variables for discrimination (like not necessary recording points or time windows) and an understandable set of rules relating the residual variable of interest: this is thus quite useful in learning the BCI approach.

When the space of the feature is then confined to the few really salient, the Piece-Wise Affine identification recently developed (Ferrari-Trecate, Muselli, Liberati, & Morari, 2003) and also applied in a similar context for instance to hormone pulse or sleep apnea detection (Ferrari-Trecate & Liberati, 2002), is keen to be a good tool to help refine the detection of such mental decision on the basis of the multivariate parametric identification of a multiple set of dynamic biometric signals. Here the idea is to cluster recorded data or features obtained by the described pre-processing in such a way to identify automatically both approximate linear relations among them in each region of interest and the boundaries of such regions themselves, thus allowing quite precise identification of the time of the searched switching.

FUTURE TRENDS

Integration of more easily recorded signals is even more promising for automatic processing of the intention of interacting with the computer, both in a context of more assisted performance in everyday life, as well as in helping to vicariate lost functions because of handicaps.

CONCLUSION

The task is challenging, though at a first glance it would even appear not so complex: it wants to discriminate at least a bit of information (like opening and closing a switch), and then sequentially, it would be possible to compose a word of any length.

The point is that every single bit of intention should be identified with the highest accuracy, in order to avoid too many redundancies, demanding time while offering safety.

REFERENCES

Babiloni, F., Carducci, F., Cerutti, S., Liberati, D., Rossini, P., Urbano, A., & Babiloni, C. (2000). Comparison between human and ANN detection of Laplacian-derived electroencephalographic activity related to unilateral voluntary movements. *Comput Biomed Res, 33,* 59-74.

Cerutti, S., Chiarenza, G., Liberati, D., Mascellani, P., & Pavesi, G. (1988). A parametric method of identification of the single trial event-related potentials in the brain. *IEEE Transactions of Biomedical Engineering, 35*(9), 701.

Deecke, L., Grözinger, B., & Kornhuber, H. (1976). Voluntary finger movements in man: Cerebral potentials and theory. *Biological Cybernetics, 23,* 99-119.

del Millan et al. (2001). Brain computer interfacing. In D. Liberati (Ed.), *Biosys: Information and control technology in health and medical systems.* Milan: ANIPLA.

Ferrari-Trecate, G., & Liberati, D. (2002). Representing logic and dynamics: The role of piecewise affine models in the biomedical field. *Proceedings of the EMSTB Math Modeling and Computing in Biology and Medicine Conference,* Milan.

Ferrari–Trecate, G., Muselli, M., Liberati, D., & Morari, M. (2003). A clustering technique for the identification of piecewise affine systems. *Automatica, 39,* 205-217.

Gevins, A., Morgan, N., Bressler, S., Doyle, J., & Cutillo, B. (1986). Improved event-related potential estimation using statistical pattern classification. *Electroenceph. Clin. Neurophysiol, 64,* 177.

Gevins, A., Bressler, S. L., Morgan, N. H., Cutillo, B., White, R. M., Greer, D. S., & Illes, J. (1989). Event-related covariances during a bimanual visuomotor task: Methods and analysis of stimulus and response-locked data. *Electroenceph. Clin. Neurophysiol, 74,* 58.

Liberati, D. (1991a), Total and partial coherence analysis of evoked brain potentials. *Proceedings of the 4th International Symposium on Biomedical Engineering,* Peniscola, Spain, (pp. 101-102).

Liberati, D. (1991b). A neural network for single sweep brain evoked potential detection and recognition. *Proceedings of the 4th International Symposium on Biomedical Engineering,* Peniscola, Spain, (pp. 143-144).

Liberati, D., Bertolini, L., & Colombo, D. C. (1991a). Parametric method for the detection of inter and intra-sweep variability in VEP's processing. *Med Biol Eng Comput, 29,* 159-166.

Liberati, D., Bedarida, L., Brandazza, P., & Cerutti, S. (1991b). A model for the cortico-cortical neural interaction in multisensory evoked potentials. *IEEE T Bio-Med Eng, 38*(9), 879-890.

Liberati, D., Cursi, M., Locatelli, T., Comi, G., & Cerutti, S. (1997). Total and partial coherence of spontaneous and evoked EEG by means of multi-variable autoregressive processing. *Med Biol Eng Comput, 35*(2), 124-130.

Muselli, M., & Liberati, D. (2000). Training digital circuits with hamming clustering. *IEEE T Circuits I, 47,* 513-527.

Muselli, M., & Liberati, D. (2002). Binary rule generation via hamming clustering. *IEEE T Knowl Data En, 14*(6), 1258-1268.

Pfurtscheller, G., Flotzinger, D., Mohl, W., & Peltoranta, M. (1992). Prediction of the side of hand movements from single-trial multi-channel EEG data using neural networks. *Electroenceph. Clin Neurophysiol, 82,* 313.

Roberts, S. J., & Penny, W. D. (2000). Real-time brain-computing interfacing: A preliminary study using Bayesian learning. *Medical & Biological Engineering & Computing, 38*(1), 56-61.

KEY TERMS

Artificial Neural Networks: Non-linear black box input-output relationships built on a regular structure of simple elements, loosely inspired to the natural neural system, even in learning by example.

Bayesian Statistics: Named after its developer, Bayes, it takes into account conditional probabilities in order to describe variable relationships.

Brain Computer Interfacing: Ability, even if only partial, of outside controlling by detecting intention through brain wave analysis.

Event-Related Potential: Electro-magnetic brain activity related to a specific event: it may be evoked from the outside, or self-ongoing.

Fuzzy Logic: Set theory mainly developed and made popular by Lofthi Zadeh at the University of California at Berkeley where belonging of elements is not crisply attributed to only one disjoint subset.

Parametric Identification: Black box mathematical modeling of an input-output relationship via simple, even linear equations depending on a few parameters, whose values do identify the system dynamics.

Piecewise Affine Identification: Linearization of a non-linear function, automatically partitioning data in subsets whose switching identifies state commuting in a hybrid dynamic-logical process.

Rule Induction: Inference from data of "if…then…else" rules describing the logical relationships among data.

Bridging Together Mobile and Service-Oriented Computing

Loreno Oliveira
Federal University of Campina Grande, Brazil

Emerson Loureiro
Federal University of Campina Grande, Brazil

Hyggo Almeida
Federal University of Campina Grande, Brazil

Angelo Perkusich
Federal University of Campina Grande, Brazil

INTRODUCTION

The growing popularity of powerful *mobile devices,* such as modern cellular phones, smart phones, and PDAs, is enabling *pervasive computing* (Weiser, 1991) as the new paradigm for creating and interacting with computational systems. Pervasive computing is characterized by the interaction of mobile devices with embedded devices dispersed across *smart spaces,* and with other mobile devices on behalf of users. The interaction between user devices and smart spaces occurs primarily through services advertised on those environments. For instance, airports may offer a notification service, where the system registers the user flight at the check-in and keeps the user informed, for example, by means of messages, about flight schedule or any other relevant information.

In the context of smart spaces, *service-oriented computing* (Papazoglou & Georgakopoulos, 2003), in short SOC, stands out as the effective choice for advertising services to mobile devices (Zhu, Mutka, & Ni, 2005; Bellur & Narendra, 2005). SOC is a computing paradigm that has in services the essential elements for building applications. SOC is designed and deployed through *service-oriented architectures* (SOAs) and their applications. SOAs address the flexibility for dynamic binding of services, which applications need to locate and execute a given operation in a pervasive computing environment. This feature is especially important due to the dynamics of smart spaces, where resources may exist anywhere and applications running on mobile clients must be able to find out and use them at runtime.

In this article, we discuss several issues on bridging mobile devices and service-oriented computing in the context of smart spaces. Since smart spaces make extensive use of services for interacting with personal mobile devices, they become the ideal scenario for discussing the issues for this integration. A brief introduction on SOC and SOA is also presented, as well as the main architectural approaches for creating SOC environments aimed at the use of resource-constrained mobile devices.

BACKGROUND

SOC is a distributed computing paradigm whose building blocks are distributed services. Services are self-contained software modules performing only pre-defined sets of tasks. SOC is implemented through the deployment of any software infrastructure that obeys its key features. Such features include loose coupling, implementation neutrality, and granularity, among others (Huhns & Singh, 2005). In this context, SOAs are software architectures complying with SOC features.

According to the basic model of SOA, service providers advertise service interfaces. Through such interfaces, providers hide from service clients the complexity behind using different and complex kinds of resources, such as databanks, specialized hardware (e.g., sensor networks), or even combinations of other services. Service providers announce their services in service registries. Clients can then query these registries about needed services. If the registry knows some provider of the required service, a reference for that provider is returned to the client, which uses this reference for contacting the service provider. Therefore, services must be described and published using some machine-understandable notation.

Different technologies may be used for conceiving SOAs such as grid services, Web services, and Jini, which follow the SOC concepts. Each SOA technology defines its own standard machineries for (1) service description, (2) message format, (3) message exchange protocol, and (4) service location.

In the context of pervasive computing, services are the essential elements of smart spaces. Services are used for interacting with mobile devices and therefore delivering personalized services for people. Owning to the great benefits that arise with the SOC paradigm, such as interoperability, dynamic service discovery, and reusability, there is a strong and increasing interest in making mobile devices capable of providing and consuming services over wireless networks (Chen, Zhang, & Zhou, 2005; Kalasapur, Kumar, & Shirazi, 2006; Kilanioti, Sotiropoulou, & Hadjiefthymiades, 2005). The dynamic discovery and invocation of services are essential to mobile applications, where the user context may change dynamically, making different kinds of services, or service implementations, adequate at different moments and places.

However, bridging mobile devices and SOAs requires analysis of some design issues, along with the fixing of diverse problems related to using resources and protocols primarily aimed at wired use, as discussed in the next sections.

INTEGRATING MOBILE DEVICES AND SOAS

Devices may assume three different roles in a SOA: service provider, service consumer, or service registry. In what follows, we examine the most representative high-level scenarios of how mobile devices work in each situation.

Consuming Services

The idea is to make available, in a wired infrastructure, a set of services that can be discovered and used by mobile devices. In this context, different designs can be adopted for bridging mobile devices and service providers. Two major architectural configurations can be derived and adapted to different contexts (Duda, Aleksy, & Butter, 2005): direct communication and proxy aided communication. In Figure 1 we illustrate the use of direct communication.

Figure 1. Direct communication between mobile client and SOA infrastructure

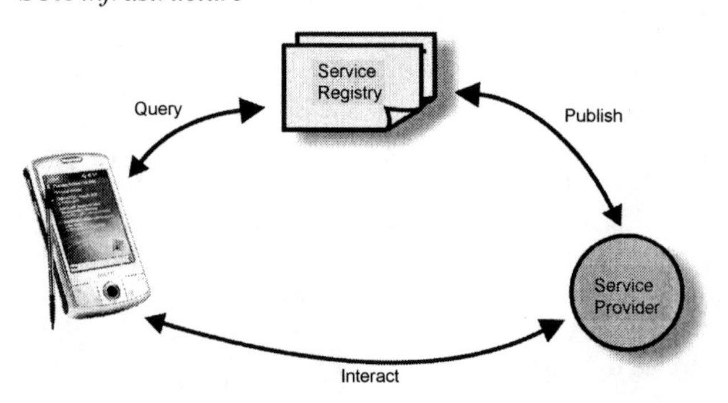

In this approach, applications running at the devices directly contact service registries and service providers. This approach assumes the usage of fat clients with considerable processing, storage, and networking capabilities. This is necessary because mobile clients need to run applications coupled with SOA-defined protocols, which may not be suited for usage by resource-constrained devices.

However, most portable devices are rather resource-constrained devices. Thus, considering running on mobile devices applications with significant requirements of processing and memory footprint reduces the range of possible client devices. This issue leads us to the next approach, proxy-aided communication, illustrated in Figure 2.

In this architectural variation, a proxy is introduced between the mobile device and the SOA infrastructure, playing the role of mobile device proxy in the wired network. This proxy interacts via SOA-defined protocols with registries and service providers, and may perform a series of content adaptations, returning to mobile devices results using lightweight protocols and data formats.

This approach has several advantages over the previous one. The proxy may act as a cache, storing data of previous service invocations as well as any client relevant information, such as bookmarks and profiles. Proxies may also help client devices by transforming complex data into lightweight formats that could be rapidly delivered through wireless channels and processed by resource-constrained devices.

Advertising Services

In a general way, mobile devices have two choices for advertising services (Loureiro et al., 2006): the push-based approach and the pull-based approach. In the first one, illustrated in Figure 3, service providers periodically send the descriptions of the services to be advertised directly to potential clients, even if they are not interested in such services (1). Clients update local registries with information about available services (2), and if some service is needed, clients query their own registries about available providers (3).

In the pull-based approach, clients only receive the description of services when they require a service discovery. This process can be performed in two ways, either through centralized or distributed registries. In the former, illustrated in Figure 4, service descriptions are published in central servers (1), which maintain entries about available services (2). Clients then query this centralized registry in order to discover the services they need (3).

In the distributed registry approach, illustrated in Figure 5, the advertisement is performed in a registry contained in each provider (1). Therefore, once a client needs to discover a service, it will have to query all the available hosts in the environment (2) until discovering some service provider for the needed service (3).

Figure 2. Proxy intermediating communication between mobile client and SOA

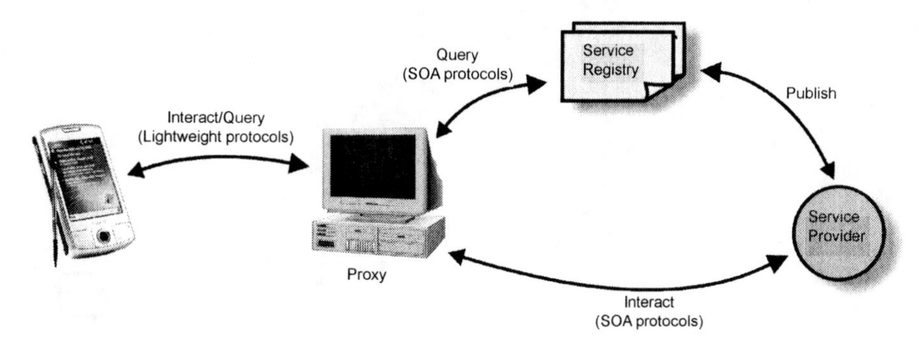

Figure 3. Push-based approach

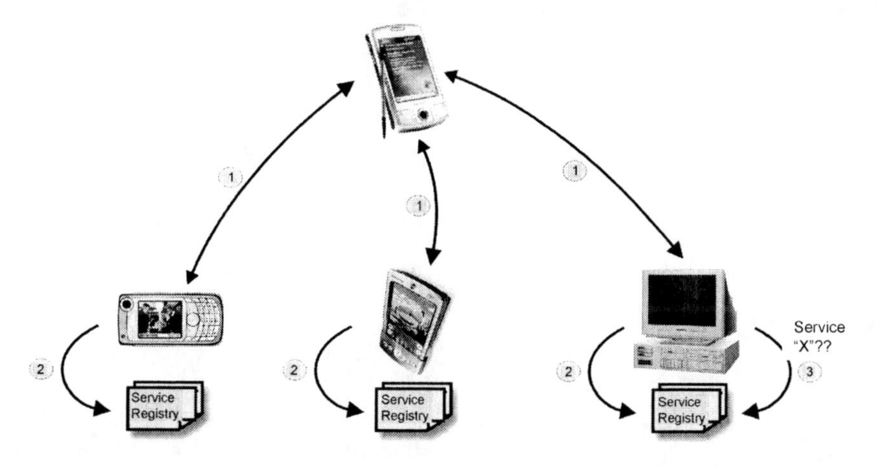

Figure 4. Pull-based approach with centralized registry

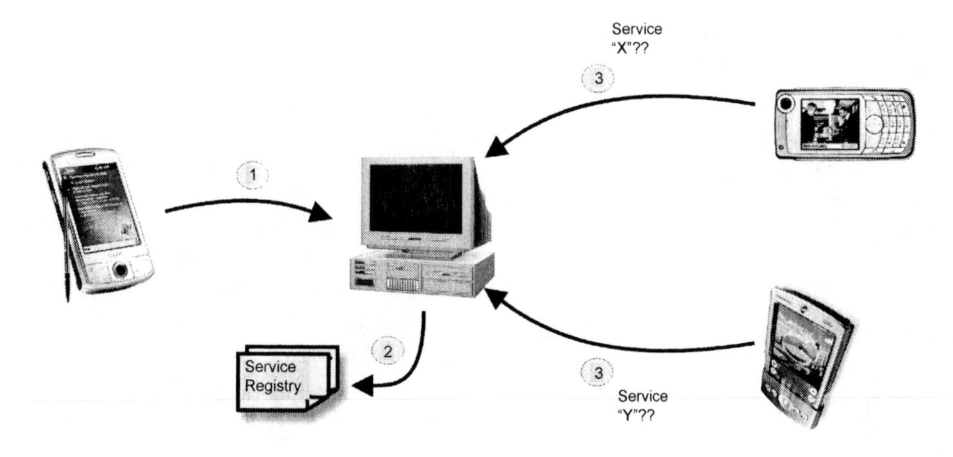

ISSUES ON INTEGRATING MOBILE DEVICES AND SOAs

Regardless of using mobile devices for either consuming or advertising services in SOAs, both *mobility* and the *limitations* of these devices are raised as the major issues for this integration. Designing and deploying effective services aimed at mobile devices requires careful analysis of diverse issues related to this kind of service provisioning.

Next, we depict several issues that arise when dealing with mobile devices in SOAs. This list is not exhaustive, but rather representative of the dimension of parameters that should be balanced when designing services for mobile use.

Figure 5. Pull-based approach with distributed registries

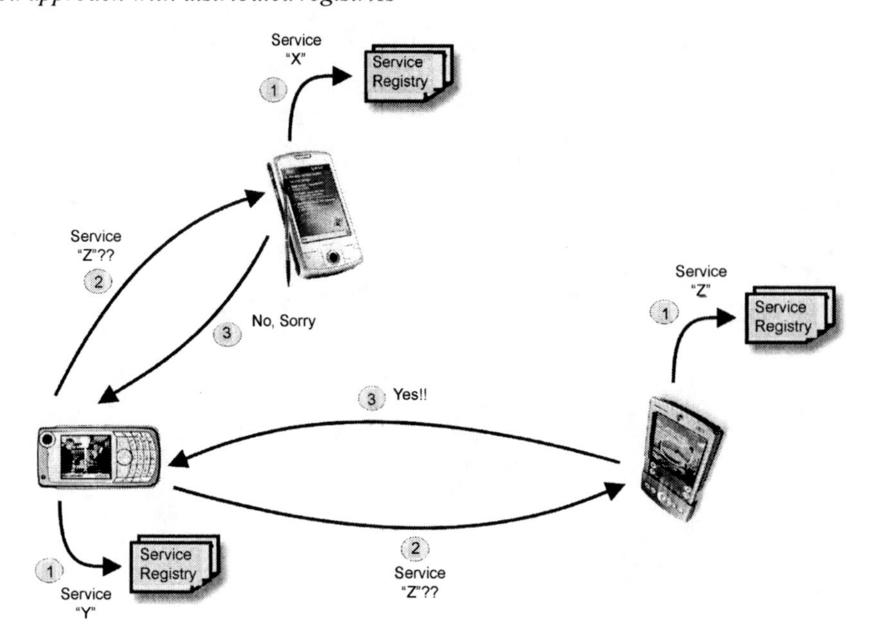

Suitability of Protocols and Data Formats

SOAs are primarily targeted at wired infrastructures. Conversely, small mobile devices are known by their well-documented limitations. Thus, protocols and formats used in conventional SOAs may be inadequate for use with resource-constrained wireless devices (Pilioura, Tsalgatidou, & Hadjiefthymiades, 2002; Kilanioti et al., 2005).

For instance, UDDI and SOAP are, respectively, standard protocols for service discovery and messaging in Web services-based SOAs. When using UDDI for service discovery, multiple costly network round trips are needed. In the same manner, SOAP messages are too large and require considerable memory footprint and CPU power for being parsed. Hence, these two protocols impact directly in the autonomy of battery-enabled devices.

Disconnected and Connected Services

In the scope of smart spaces, where disconnections are the norm rather than the exception, we can identify two kinds of services (Chen et at., 2005): disconnected and connected services. The first ones execute by caching the inputs of users in the local device. Once network connectivity is detected, the service performs some sort of synchronization. Services for messaging (e.g., e-mail and instant messages) and field research (e.g., gathering of data related to the selling of a specific product in different supermarkets) are some examples of services that can be implemented as disconnected ones.

Connected services, on the other hand, are those that can only execute when network connectivity is available in the device. Some examples of connected services include price checking, ordering, and shipment tracking. Note, however, that these services could certainly be implemented as disconnected services, although their users will generally need the information when demanded, neither before nor later. Therefore, there is no precise categorization of what kind of services would be connected or disconnected, as this decision is made by the system designer.

User Interface

User interfaces of small portable devices are rather limited in terms of screen size/resolution and input devices, normally touch screens or small built-in keyboards. This characteristic favors services that require low interaction to complete transactions (Pilioura et al., 2002). Services requiring many steps of data input, such as long forms, tend to: stress users, due to the use of non-comfortable input devices; reduce device autonomy, due to the extra time for typing data; and increase the cost of data transfer, due to larger amounts of data being transferred.

A possible alternative for reducing data typing by clients is the use of context-aware services (Patterson, Muntz, & Pancake, 2003). Context-aware services may reduce data input operations of mobile devices by inferring, or gathering through sensors, information about a user's current state and needs.

Frequent Temporary Disconnections

Temporary disconnections between mobile device and service provider are common due to user mobility. Thus, both client applications and service implementations must consider the design of mechanisms for dealing with frequent disconnections.

Different kinds of services require distinct solutions for dealing with disconnections. For instance, e-business applications need machineries for controlling state of transactions and data synchronization between mobile devices and service providers (Sairamesh, Goh, Stanoi, Padmanabhan, & Li, 2004). Conversely, streaming service requires seamless reestablishment and transference of sessions between access points as the user moves (Cui, Nahrstedt, & Xu, 2004).

Security and Privacy

Normally, mobile devices are not shared among different users. Enterprises may benefit from this characteristic for authenticating employees, for instance. That is, the system knows the user and his/her access and execution rights based on profiles stored in his/her mobile devices. However, in commercial applications targeted at a large number of unknown users, this generates a need of anonymity and privacy of consumers. This authentication process could cause problems, for example in case of device thefts, because the device is authenticated and not the user (Tatli, Stegemann, & Lucks, 2005).

Security also has special relevance when coping with wireless networks (Grosche & Knospe, 2002). When using wireless interfaces for information exchange, mobile devices allow any device in range, equipped with the same wireless technology, to receive the transferred data. At application layer, service providers must protect themselves from opening the system to untrusted clients, while clients must protect themselves from exchanging personal information with service providers that can use user data for purposes different than the ones implicit in the service definition.

Device Heterogeneity and Content Adaptation

Modern mobile devices are quite different in terms of display sizes, resolutions, and color capabilities. This requires services to offer data suitable for the display of different sorts of devices. Mobile devices also differ in terms of processing capabilities and wireless technologies, which makes harder the task of releasing adequate data and helper applications to quite different devices.

Therefore, platform-neutral data formats stand out as the ideal choice for serving heterogeneous sets of client devices. Another possible approach consists of using on-demand data adaptation. Service providers may store only one kind of best-suited data format and transform the data, for example, using a computational grid (Hingne, Joshi, Finin, Kargupta, & Houstis, 2003), when necessary to transfer the data to client devices. Moreover, dynamic changes of conditions may also require dynamic content adaptation in order to maintain pre-defined QoS threshold values. For instance, users watching streamed video may prefer to dynamically reduce video quality due to temporary network congestion, therefore adapting video data, and to maintain a continuous playback instead of maintaining quality and experiencing constant playback freezing (Cui et al., 2004).

Consuming Services

As discussed before, system architects can choose between two major approaches for accessing services of SOAs from mobile devices: direct communication and proxy-aided communication. The two approaches have some features and limitations that should be addressed in order to deploy functional services. Direct communication suffers from the limitations of mobile devices and relates to other discussions presented in this article, such as adequacy of protocols and data formats for mobile devices and user interface.

If, on the one hand, proxy-aided communication seems to be the solution for problems of the previous approach, on the other hand it also brings its own issues. Probably the most noted is that proxies are single points of failures.

Furthermore, some challenges related to wired SOAs are also applicable to both approaches discussed. Service discovery and execution need to be automated to bring transparency and pervasiveness to the service usage. Moreover, especially in the context of smart spaces, services need to be personalized according to the current user profile, needs, and context. Achieving this goal may require describing the semantics of services, as well as modeling and capturing the context of the user (Chen, Finin, & Joshi, 2003).

Advertising Services

A number of issues and technical challenges are associated with this scenario. The push-based approach tends to consume a lot of bandwidth of wireless links according to the number of devices in range, which implies a bigger burden over mobile devices.

Using centralized registries creates a single point of failure. If the registry becomes unreachable, it will not be possible to advertise and discover services. In the same manner, the discovery process is the main problem with the approach of distributed registries, as it needs to be well designed in order to allow clients to query all the hosts in the environment.

Regardless of using centralized or distributed registries, another issue rises with mobility of service providers. When service providers move between access points, a new address is obtained. This changing of address makes service providers inaccessible by clients that query the registry where it published its services. Mechanisms for updating the registry references must be provided in order for services to continue to be offered to their requestors.

FUTURE TRENDS

The broad list of issues presented in this article gives suggestions about future directions for integrating SOC and mobile devices. Each item depicted in the previous section is already an area of intensive research. Despite this, both SOC and mobile computing still lack really functional and mature solutions for the problems presented.

In particular, the fields of context-aware services and security stand out as present and future hot research fields. Besides, the evolution itself of mobile devices towards instruments with improved processing and networking power, as well as better user interfaces, will reduce the complexity of diverse challenges presented in this article.

CONCLUSION

In this article we have discussed several issues related to the integration of mobile devices and SOC. We have presented the most representative architectural designs for integrating mobile devices to SOAs, both as service providers and service consumers.

While providing means for effective integration of mobile devices and service providers, SOC has been leveraging fields such as mobile commerce and pervasive computing. Nonetheless, several issues remain open, requiring extra efforts for designing and deploying truly functional services.

REFERENCES

Bellur, U., & Narendra, N. C. (2005). Towards service orientation in pervasive computing systems. *Proceedings of the International Conference on Information Technology: Coding and Computing (ITCC'05)* (Vol. II, pp. 289-295).

Chen, H., Finin, T., & Joshi, A. (2003). An ontology for context-aware pervasive computing environments. *The Knowledge Engineering Review, 18*(3), 197-207.

Chen, M., Zhang, D., & Zhou, L. (2005). Providing Web services to mobile users: The architecture design of an m-service portal. *International Journal of Mobile Communications, 3*(1), 1-18.

Cui, Y., Nahrstedt, K., & Xu, D. (2004). Seamless user-level handoff in ubiquitous multimedia service delivery. *Multimedia Tools Applications, 22*(2), 137-170.

Duda, I., Aleksy, M., & Butter, T. (2005). Architectures for mobile device integration into service-oriented architectures. *Proceedings of the 4th International Conference on Mobile Business (ICBM'05)* (pp. 193-198).

Grosche, S.S., & Knospe, H. (2002). Secure mobile commerce. *Electronics & Communication Engineering Journal, 14*(5), 228-238.

Hingne, V., Joshi, A., Finin, T., Kargupta, H., & Houstis, E. (2003). Towards a pervasive grid. *Proceedings of the 17th International Parallel and Distributed Processing Symposium (IPDPS'03)* (p. 207.2).

Huhns, M.N., & Singh, M.P. (2005). Service-oriented computing: Key concepts and principles. *IEEE Internet Computing, 9*(1), 75-81.

Kalasapur, S., Kumar, M., & Shirazi, B. (2006). Evaluating service oriented architectures (SOA) in pervasive computing. *Proceedings of the 4th IEEE International Conference on Pervasive Computing and Communications (PERCOMP'06)* (pp. 276-285).

Kilanioti, I., Sotiropoulou, G., & Hadjiefthymiades, S. (2005). A client/intercept based system for optimized wireless access to Web services. *Proceedings of the 16th International Workshop on Database and Expert Systems Applications (DEXA'05)* (pp. 101-105).

Loureiro, E., Bublitz, F., Oliveira, L., Barbosa, N., Perkusich, A., Almeida, H., & Ferreira, G. (2007). Service provision for pervasive computing environments. In D. Taniar (Ed.), *Encyclopedia of mobile computing and commerce.* Hershey, PA: Idea Group Reference.

Papazoglou, M. P., & Georgakopoulos, D. (2003). Service-oriented computing: Introduction. *Communications of the ACM, 46*(10), 24-28.

Patterson, C. A., Muntz, R. R., & Pancake, C. M. (2003). Challenges in location-aware computing. *IEEE Pervasive Computing, 2*(2), 80-89.

Pilioura, T., Tsalgatidou, A., & Hadjiefthymiades, S. (2002). Scenarios of using Web services in m-commerce. *ACM SIGecom Exchanges, 3*(4), 28-36.

Sairamesh, J., Goh, S., Stanoi, I., Padmanabhan, S., & Li, C. S. (2004). Disconnected processes, mechanisms and architecture for mobile e-business. *Mobile Networks and Applications, 9*(6), 651-662.

Tatli, E. I., Stegemann, D., & Lucks, S. (2005). Security challenges of location-aware mobile business. *Proceedings of the 2nd IEEE International Workshop on Mobile Commerce and Services (WMCS'05)* (pp. 84-95).

Weiser, M. (1991). The computer for the 21st century. *Scientific American, 265*(3), 66-75.

Zhu, F., Mutka, M. W., & Ni, L. M. (2005). Service discovery in pervasive computing environments. *IEEE Pervasive Computing, 4*(4), 81-90.

KEY TERMS

Grid Service: A kind of Web service. Grid services extend the notion of Web services through the adding of concepts such as statefull services.

Jini: Java-based technology for implementing SOAs. Jini provides an infrastructure for delivering services in a network

Mobile Device: Any low-sized portable device used to interact with other mobile devices and resources from smart spaces. Examples of mobile devices are cellular phones, smart phones, PDAs, notebooks, and tablet PCs.

Proxy: A network entity that acts on behalf of another entity. A proxy's role varies since data relays to the provision of value-added services, such as on-demand data adaptation.

Streaming Service: One of a number of services that transmit some sort of real-time data flow. Examples of streaming services include audio streaming or digital video broadcast (DVB).

Web Service: Popular technology for implementing SOAs built over Web technologies, such as XML, SOAP, and HTTP.

Browser-Less Surfing and Mobile Internet Access

Gregory John Fleet
University of New Brunswick at Saint John, Canada

Jeffery G. Reid
xwave Saint John, Canada

INTRODUCTION

Lately, we have seen the use of a number of new technologies (such as Javascript, XML, and RSS) used to show how Web content can be delivered to users without a traditional browser application (e.g., Microsoft Explorer). In parallel, a growing number of PC applications, whose main job previously was to manage local resources, now are adding Internet connectivity to enhance their role and use (e.g., while iTunes started as a media player for playing and managing compressed audio files, it now includes Web access to download and purchase music, video, podcasts, television shows, and movies).

While most attempts at providing Internet access on mobile devices (whether wireless phones or personal digital assistants) have sought to bring the traditional browser, or a mobile version of the browser, to these smaller devices, they have been far from successful (and a far cry from the richer experience provided by browsers on the PC using standard input and control devices of keyboards and a mouse). Next, we will highlight a number of recent trends to show how these physical and use-case constraints can be significantly diminished.

BACKGROUND

Mobile telephony and mobile computing continue to display unprecedented growth worldwide. Zee News (2005) reports that in some parts of the world, such as India, mobile phones are now more popular than traditional landline phones. Since 2000, many developed countries have spent large amounts of money on the installation and deployment of wireless communication infrastructure (Kunz & Gaddah, 2005). And this growth trend is not confined to the mobile phone handset market. It is also being experienced across other mobile devices. In fact, in 2004 more mobile phones shipped than both automobiles and personal computers (PCs) combined, making them the fastest adopted consumer product of all time (Clarke & Flaherty, 2005). Further, Wiberg (2005) points out that this increase in mobile device usage spans across business and non-business usage. Therefore, this growth is not simply due to increased consumer demand; businesses are continually seeing new value in equipping employees with mobile computing and communication devices.

There has also been steady growth in the use of the Internet, as well as in the nature of Internet usage. The size of the Internet, measured in terms of the number of users, is more than 800 million users (Global Reach, 2004). While the majority of the users are English, other languages are experiencing significant growth in the number of users, and this growth is expected to continue, given the large numbers of non-English-speaking populations.

Some of the drivers for the increase in Internet usage include the growth in Web-enabled applications and the availability of high-speed, always-on Internet (Bink, 2004).

Kunz and Gaddah (2005) identify two broad technological developments that are converging to enable mobile computing (the use of the Internet through mobile devices). The first of these technological developments is the accessibility to the Internet regardless of location, as evidenced by the growth in wireless *hotspots*. Now users can connect to the Internet from various locations and access Internet content without being connected to a physical local area network (LAN) connection or other type of landline connection.

The second technological development is the drive to reduce the size of computer hardware (Kunz & Gaddah, 2005). This size reduction increases the portability of these devices, leading to the mobile nature of the devices as well as the desire to connect these devices to the Internet.

Unfortunately, being *able* to provide Internet access to mobile devices has not *ensured* a quality Web experience. The next section will profile the current mobile Web experience.

USER EXPERIENCE OF WEB ON MOBILE DEVICES

The Web Browser on a PC

Let us start with the typical experience of the Web. The most common way to navigate the Internet is through the use of a browser, a software application that allows the user to locate and display Web pages (Webopedia, 2006). On the personal computer (PC), there are a variety of browsers available, including Microsoft Internet Explorer, Mozilla Firefox, Opera, Netscape, Apple Safari, and Konqueror (Wikipedia, 2006a).

A cross-section of definitions from the Web outlines the basic functionality of these browsers (http://www.google.com/search?hl=en&lr=&q=define%3A+web+browser&btnG=Search); the Web browser is a graphical interface (i.e., icons, buttons, menu options) that:

- interprets HTML files (resources, services) from Web servers, and formats them into Web pages; and
- provides the ability to both view and interact with Web content (including download and upload of media content).

Yet most modern browsers also include additional functionality, assisting with the management of the tool's functionality and the content to which they provide access. This functionality includes:

- **Bookmarking:** The ability to save and manage Web addresses.
- **Cookies and Form-Filling:** The ability of the browser to pre-fill form fields (e.g., address or contact information), or provide the Web server with identifying information in order to customize the content received from the server.
- **Searching:** The ability to conduct a Web or local file search.
- **History:** The automatic cataloging of previously visited Web sites.
- **Display Modification:** The ability to customize the way Web content is displayed (e.g., size of text, types of media files that can be viewed, etc.)

It is also important to note that in the typical use of a Web browser, the user searches for information on the Web, often starting with a broad search and successively narrowing that search to meet his or her information goal (i.e., to go from the general to the specific).

The Web Browser on a Mobile Device

For mobile devices, such as cell phones or personal digital assistants (PDAs), the Web browser application is often referred to as a microbrowser (also minibrowser and mobile browser; see Wikipedia, 2006b). The difference between a *full* browser and the microbrowser is that the code in the microbrowser application has been optimized to accommodate the smaller screens, memory, and bandwidth limitations of mobile devices. In addition, the Web servers often communicate with these microbrowsers using variations on the standard HTML (hypertext markup language), again to accommodate the screen, memory, and bandwidth restrictions.

Internet usage on mobile devices poses a number of challenges that are different than those found on a traditional computing device such as a PC (Becker, 2005). As mentioned previously, the computing power (processor and memory configuration), the transmission bandwidth, and screen size on the mobile device are really just a fraction of what users have available to them on a PC. More importantly, the limitations in screen size and physical interface often require users to restrict the activities they might otherwise seek to accomplish on the Web.

The physical restrictions (that being the telephone keypad and four-way scroll and navigation keys) can be quite significant. On a PC, we have a full-sized QWERTY keyboard and mouse interface for entering searches and addresses, or navigating Web pages. On the mobile device, in particular the cell phone, these physical input and control devices are replaced with a keypad designed for dialing phone numbers (not entering text strings), and horizontal/vertical navigation keys that significantly slow simple scrolling and selection of content.[1] In user studies, Chen, Xie, Ma, and Zhang (2005) report that users, when browsing the Web on a phone, handheld computer, or personal digital assistant, spend the majority of their time scrolling the screen to locate and select the content of interest.

Despite these real challenges, Nugent (2005) expects that the need for mobile Web browsing will increase, and people will want these devices to stay small, weigh less, cost less, run cooler and longer on one charge, but continue to do more than today's devices. Lawton (2001) believes that meeting these needs will require faster wireless connections, larger displays, as well as new usage paradigms and/or content that fits these smaller devices.

This is the environment mobile users are operating in today. A user can either struggle with a small screen and content that does not fit within that screen, or lug around a larger device that has an adequately sized screen but more limited connectivity options.

Technology and Service Barriers

There are a number of technological hurdles that need to be overcome for widespread adoption of mobile Internet usage. Chan and Fang (2005) identify a number of technological barriers, which range from connectivity and bandwidth issues to the lack of standards and broad use of proprietary tools and languages. Kuniavsky (2006) also notes the numerous and often complex relationships that exist between the multiple service, application, and technology providers currently needed to deliver mobile computing to the user, and how none of these players is wholly responsible for the resulting user experience.

Identification of the challenges associated with mobile Internet usage is only part of the problem. After the issues have been identified, developing solutions to deal with the problems is the next step in the process. Many manufacturers are presently making moves to deal with some of these identified issues. One common move is to increase the screen size of the mobile device. One example of this trend is the Sony Ericsson P800 SmartPhone, which has increased the size of its display area while attempting to maintain the overall size of the device itself. Another more recent example is the Sony Ericsson P910i, with its larger screen, miniature QWERTY keyboard, and pen-based interface.

Another design approach to deal with limited screen space is to focus on the content rather than the size of the screen, as is attempted with standards such as WAP, WML, HDML, as well as services such as i-mode (Chen et al., 2005).

What is needed, though, is the development of Web content and mobile applications that can be viewed, navigated, and controlled from small devices (Nayak, 2005), because, at this time, consumers find the small screen display and small buttons on these devices difficult to use. Chan and Fang (2005) believe that these technologies need to mature, and until that time, the mobile Internet will be geared toward applications requiring limited bandwidth, short exchange of data and text, and simple functionality. Therefore, using smaller mobile devices to perform tasks similar to those carried out on a traditional computing platform poses challenges for users and manufacturers alike.

FUTURE TRENDS

Interestingly, there are a number of new technologies and trends that might suggest an evolution of the mobile computing Web experience. This evolution comes from a number of different places. In this section of the article, a few specific trends are highlighted in order to demonstrate this potential.

Web-Enabled Desktop Clients

In recent years, we are seeing more and more desktop clients (or applications) reaching beyond the processes of the PC and the content on the local drive to networked and Internet resources and content. Apple's iTunes media player was first released as a desktop application for playing music files from one's hard drive. Since that first release, it has grown to not only allow streaming of music libraries over networks, but now has a built-in Web browser tied to one of the most successful online music stores today.

There are additional examples of traditional desktop clients that have added Web connectivity to their functional specifications; these include desktop applications with built-in version checking; address book applications that communicate with LDAP servers; Google desktop™, providing the ability to search and find information not only on your local drives, but also on e-mail and Web servers; and cataloguing programs that match your own library of CDs, books, or DVDs with online databases such as Internet Movie Database (imd.com) or Amazon.com. I suspect this trend is only just beginning, and we will continue to see additional examples as software companies add both Internet connectivity and imbedded browsers into desktop applications in order to add new and unique value for users.

Webtop Clients

At the same time, there are also some exciting examples of Web applications (or services) that only require a standard Web browser. Web services have been around since the beginning of the Web, but what differentiates these newer Web applications is their attention to usability and responsiveness, resulting in a *Webtop client* that responds and behaves in ways similar to a desktop client. For example, Flickr.com allows individuals to upload photos from their cameras and hard drives to the Flickr Web servers. Then, in desktop-client fashion, they allow us to arrange the order with simple drag-and-drop, or name, edit, and tag photo labels by directly clicking on the titles within the Web browser.

Other examples include the Web services of Google, MyYahoo, and MyMSN, as well as the excellent services from 37signals.com (Basecamp, Backpack, Writeboard, Ta-Da List, and the growing number of services built using the Ruby on Rails development environment). In all these examples, the responsiveness of these Web clients is quite impressive, mimicking the behavior of their desktop counterparts.

The Changing Mental Model of Web Access

These examples of desktop and Webtop clients demonstrate a blurring of the lines between Web browsing or surfing and running local applications. I believe this is a good thing, since it suggests that Web connectivity is not limited to what is accomplished and viewed through a browser. And for mobile devices, this lack of distinction should also be a good thing—allowing users to think about Internet content separate from traditional browsers. This could also suggest that users might adapt their user model of expecting Internet content (especially on small devices) to be only through an Internet browser.

Seeing the Internet on mobile devices as separate from a browser is only one (significant) step in producing a better user experience. It is also important to recognize the other constraints that limit a quality Internet experience on these small devices: the constrained visual and physical interface.

Reproducing Web sites onto small screens, at best, requires the ability to visualize content beyond the screen, and, at worst, produces a frustrating, unacceptable experience.

SOLVING THE VISUAL LIMITS OF MOBILE DEVICES

A variety of technologies (XML, ATOM, Javascript, WebKit) have been used of late to create a number of useful Web services. One of the most common is RSS feeds, where the user can *subscribe* to the content found on a Web server. The current implementation of RSS satisfies two user goals: to filter Web content to only those topics of interest, and to provide real-time notification of updates to the Web site. Therefore, RSS provides a technology to allow users to *browse* the Web in a more focused manner, providing personalized views of self-selected content.

Another example of viewing self-selected Web content is found with Yahoo's Konfabulator (also known as the Yahoo! Widget Engine—see http://widgets.yahoo.com). This desktop client is a real-time Javascript compiler that can execute small Javascript files (called *widgets*) to accomplish whatever task they have been programmed to accomplish. The result is a small, windowless, (and in the case of scripts that communicate with Web servers) browser-less view of live Web content. Figure 1 shows an example using the Weather widget, which displays live weather conditions and the five-day forecast for a particular location configured by the user. Additional Web information is available with mouse-over or clicks on the widget.

The latest operating system from Apple (known as Tiger or 10.4) has also added similar functionality for displaying self-selected Web content. More appropriately named, these Dashboard™ widgets organize and/or present Web content in a way that is easy to read. Presently, thousands of widgets are available for download, whether from Yahoo's Web site or Apple.com (as of January 2006, there were more than 4,000 widgets available for download). What is interesting about widgets is the fact that most are designed to present their Web content in a fairly constrained visual space, separate from large resource requirements or visual real estate needs. In other words, these widgets provide what could be a perfect example of self-selected, rich Internet content for small screens.

Figure 1. The Weather widget, using Yahoo's Widget Engine, showing current and forecasted weather for Palo Alto, California

Another interesting example comes from some software developers in Japan. They have demonstrated the ability to create a Dashboard™ widget of a full Web browser, only miniaturized to dimensions that could easily work on a standard PDA-size screen (see http://hmdt-web.net/shiira/mini/en). Therefore, with a high-quality display such as that found on today's mobile devices and iPods, it is quite easy to imagine using this miniaturized view of Web pages to surf the Web, especially on Web sites where the format and layout is familiar.

Now we turn to the problem of the physical (input and control) interface on mobile devices. If you have this view of a miniaturized Web page, how do you move around and select the buttons and links on the page? Using a four-way scroll key might work for the limited content in most Widgets, but is a very poor substitute for a keyboard and mouse when browsing a full (though miniaturized) Web page.

Solving the Physical Interface of Mobile Devices

A keyboard and mouse is not just the standard input and control device for Web surfing, but provides a rich interaction for control and text input. A telephone keypad and four-way scroll key does not even come close to that user experience, therefore making the Web experience on mobile devices very constrained indeed.

Yet we have an excellent example of one specific mobile device that has, over the past four years, shown that navigation of large hierarchies of data can be very quickly accomplished with only a finger or thumb. The iPod music player provides both a simple and extremely intuitive interface for moving through and selecting from vast playlists, photos, folders, and files—using the scroll wheel or click wheel[2] design. The click wheel interface is currently only available on Apple's iPod music and video players, but there has been much discussion of the possibility of this interface being used on other small devices, such as cell phones or PDAs (see Shortflip, 2006; Baig, 2005). After experiencing how easy it is to use an iPod to navigate and select music, photos, or videos, it is not difficult to imagine the same physical interface being used to select phone numbers from an address book, and even select and navigate content on a Web page.

More recently, there has been discussion of some patents filed at the U.S. Patent Office site (see http://tinyurl.com/8zxuv). The patent application document demonstrates a device where the whole front is a display screen, and the control interface comes from a touch-activated, touch-sensitive click wheel that is available from the visual display wherever the user touches the screen. Therefore, the combination of a large screen, with a touch-activated telephone keypad and/or click wheel, provides a compelling possibility for a high-fidelity Web experience on a mobile device.

CONCLUSION

In this article, we have argued that today's mobile devices provide a poor user experience when presenting Internet and Web content. These devices have two major physical constraints: a visual display that is too small to present the typical browser-based view of the Web, and a physical interface (telephone keypad and four-way scroll key with center select) that is not easily adapted to text entry or interface control. Users, therefore, are unable to reproduce the familiar encyclopedic browsing of Web content on these miniature visual and physical interfaces.

At the same time, a number of new trends has been highlighted, demonstrating the possibility of producing a much stronger and more compelling user experience of the Web on small mobile devices.

REFERENCES

Baig, E. C. (2005). New iTunes phone a snazzy device. *USA Today.com.* Retrieved February 1, 2006, from http://www.usatoday.com/tech/columnist/edwardbaig/2005-09-07-itunes-phone_x.htm

Becker, S. A. (2005). Web usability. In M. Khosrow-Pour (Ed.), *Encyclopedia of information science and technology* (Vol. 5, pp. 3074-3078). Hershey, PA: Idea Group Reference.

Bink, S. (2004). Browserless Net use on the rise. *Bink.nu.* Retrieved February 1, 2006, from http://bink.nu/Article798.bink

Chan, S. S., & Fang, X. (2005). Interface design issues for mobile commerce. In M. Khosrow-Pour (Ed.), *Encyclopedia of information science and technology* (Vol. 3, pp. 1612-1616). Hershey, PA: Idea Group Reference.

Chen, Y., Xie, X., Ma, W., & Zhang, H. (2005). Adapting Web pages for small-screen devices. *IEEE Internet Computing.* Retrieved February 1, 2006, from http://research.microsoft.com/~xingx/tic1.pdf

Clarke, I., & Flaherty, T. (2005). Portable portals for m-commerce. In M. Khosrow-Pour (Ed.), *Encyclopedia of information science and technology* (Vol. 4, pp. 2293-2296). Hershey, PA: Idea Group Reference.

Fleet, G. J. (2003, October 16-18). The devolution of the Web browser: The fracturing of Internet Explorer. *Proceedings of the Atlantic Schools of Business Conference,* Halifax, Nova Scotia, Canada.

Global Reach. (2004). *Global Internet statistics by language.* Retrieved February 1, 2006, from http://global-reach.biz/globstats/index.php3

Kuniavsky, M. (2006). User experience and HCI. In J. Jacko & A. Sears (Eds.), *The human-computer interaction handbook* (2nd ed.). Retrieved February 1, 2006, from http://www.orangecone.com/hci_UX_chapter_0.7a.pdf

Kunz, T., & Gaddah, A. (2005). Adaptive mobile applications. In M. Khosrow-Pour (Ed.), *Encyclopedia of information science and technology* (Vol. 1, pp. 47-52). Hershey, PA: Idea Group Reference.

Lawton, G. (2001). Browsing the mobile Internet. *IEEE Computer, 35*(12), 18-21.

Nugent, J. H. (2005). Critical trends in telecommunications. In M. Khosrow-Pour (Ed.), *Encyclopedia of information science and technology* (Vol. 1, pp. 634-639). Hershey, PA: Idea Group Reference.

Nayak, R. (2005). Wireless technologies to enable electronic business. In M. Khosrow-Pour (Ed.), *Encyclopedia of information science and technology* (Vol. 5, pp. 3101-3105). Hershey, PA: Idea Group Reference.

Shortflip.com. (2006). *The future of the iPod.* Retrieved February 1, 2006, from http://www.shortflip.com/article/The-Future-of-the-iPod-149.html

Webopedia. (2004). *Browser.* Accessed February 1, 2006, from http://www.webopedia.com/TERM/B/browser.html

Wiberg, M. (2005). "Anytime, anywhere" in the context of mobile work. In M. Khosrow-Pour (Ed.), *Encyclopedia of information science and technology* (Vol. 1, pp. 131-134). Hershey, PA: Idea Group Reference.

Wikipedia. (2006a). *Web browser.* Accessed February 1, 2006, from http://en.wikipedia.org/wiki/Web_browser

Wikipedia. (2006b). *Microbrowser.* Accessed February 1, 2006, from http://en.wikipedia.org/wiki/Microbrowser

Zee News. (2005). *Mobile phones outpace landline but with grey calls.* Retrieved February 1, 2006, from http://www.zeenews.com/znnew/articles.asp?aid=194056&sid=ZNS

KEY TERMS

Atom: One of a number of Web formats that supports user subscription to online content.

Click Wheel: The physical interface on Apple's iPod for moving through the directories and selecting items.

HDML: Handheld Device Markup Language.

i-mode: A popular wireless Internet service initially available only in Japan.

Javascript: Scripting programming language.

LDAP: Lightweight directory access protocol.

Podcast: The distribution of audio or video content over the Web using Atom or RSS.

RSS: Rich site summary or really simple syndication.

Ruby on Rails: An new open-source Web application framework.

WAP: Wireless application protocol.

WebKit: Application framework for Apple's Safari Web browser.

WML: Wireless Markup Language.

XML: eXtensible Markup Language.

ENDNOTES

[1] It is true that some PDAs and cell phones are using miniature QWERTY keyboards for an input device, though the tiny size is only marginally better than the keypad.

[2] The click wheel interface allows the user to navigate a vertical array of items or folders by rotating the wheel either clockwise or counterclockwise. Selecting an item in the list or moving deeper into the folder structures can be accomplished with the center button or the four buttons placed 90 degrees apart.

Building Web Services in P2P Networks

Jihong Guan
Tongji University, China

Shuigeng Zhou
Fudan University, China

Jiaogen Zhou
Wuhan University, China

INTRODUCTION

Nowadays peer-to-peer (P2P) and Web services are two of the hottest research topics in computing. Roughly, they appear as two extremes of distributed computing paradigm. Conceptually, P2P refers to a class of systems and applications that employ distributed resources to perform a critical function in a decentralized way. A P2P distributed system typically consists of a large number of nodes (e.g., PCs connected to the Internet) that can potentially be pooled together to share their resources, information, and services. These nodes, taking the roles of both consumer and provider of data and/or services, may join and depart the P2P network at any time, resulting in a truly dynamic and ad-hoc environment. Apart from improving scalability by avoiding dependency on centralized servers, the distributed nature of such a design can eliminate the need for costly infrastructure by enabling direct communication among clients, along with enabling resource aggregation, thus providing promising opportunities for novel applications to be developed (Ooi, Tan, Lu, & Zhou, 2002).

On the other hand, Web services technologies provide a language-neutral and platform-independent programming model that can accelerate application integration inside and outside the enterprise (Gottschalk, Graham, Kreger, & Snell, 2002). It is convenient to construct flexible and loosely coupled business systems by application integration under a Web services framework. Considering Web services are easily applied as wrapping technology around existing applications and information technology assets, new solutions can be deployed quickly and recomposed to address new opportunities. With the acceleration of Web services adoption, the pool of services will grow, fostering development of more dynamic models of just-in-time application and business integration over the Internet.

However, current proposals for Web services infrastructures are mainly based on centralized approaches such as UDDI: a central repository is used to store services descriptions, which will be queried to discover or, in a later stage, compose services. Such centralized architecture is prone to introducing single points of failure and hotspots in the network, and exposing vulnerability to malicious attacks. Furthermore, making full use of Web services capabilities using a centralized system does not scale gracefully to a large number of services and users. This difficulty is severe by the evolving trend to ubiquitous computing in which more and more devices and entities become services, and service networks become extremely dynamic due to constantly arriving and leaving service providers.

We explore the techniques of building Web services systems in a P2P environment. By fitting Web services into a P2P environment, we aim to add more flexibility and autonomy to Web services systems, and alleviate to some degree the inherent limitations of these centralized systems. As a case study, we present our project *BP-Services*. BP-Services is an experimental Web services platform built on BestPeer (http://xena1.ddns.comp.nus.edu.sg/p2p/)—a generic P2P infrastructure designed and implemented collaboratively by the National University of Singapore and Fudan University of China (Ng, Ooi, & Tan, 2002).

FITTING WEB SERVICES INTO A P2P FRAMEWORK

A *Web service* can be seen as an interface that describes a collection of operations that are network accessible through standardized XML messaging (Gottschalk et al., 2002). Web services consist of three roles and three operations: the roles are *providers, requesters,* and *registrars* of services, and the operations are *publish, find,* and *bind.* The service providers are responsible for creating Web services and corresponding service definitions, and then publishing the services with a service registry based on UDDI specification. The service requesters first find the services requested via the UDDI interface, and the UDDI registry provides the requesters with WSDL service descriptions and URLs pointing to these services themselves. With the information obtained, the requesters can then bind directly with the services and *invoke* them.

Figure 1. Network topology of BestPeer

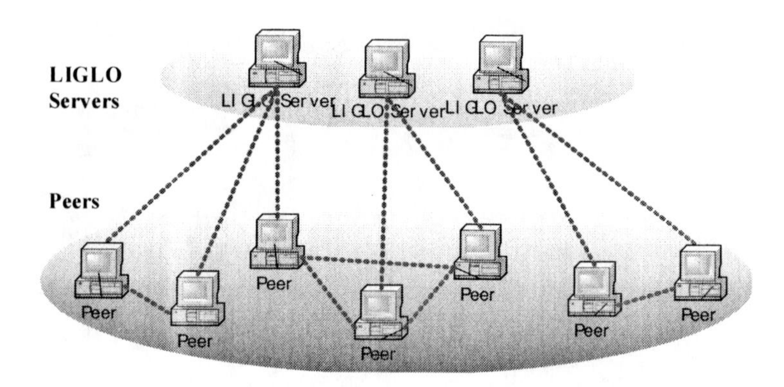

Over the last few years, many P2P systems have been developed and deployed for different purposes and with different technologies, such as Napster (http://www.napster.com/), Gnutella (http://gnutella.wego.com/), and Freenet (http://freenet.sourceforge.com/), to name a few. The architecture of these systems can be categorized into three groups mainly based on their network topologies: centralized P2P, pure P2P, and hybrid P2P systems (Yang & Garcia-Molina, 2001). In a centralized P2P network, there is a central server responsible for maintaining indexes on the metadata for all the peers in the network. Pure P2P is simply P2P systems with fully autonomous peers—that is, all nodes are equal, no functionality is centralized, and the communication between peers is also symmetric. Hybrid P2P is a kind of tradeoff between centralized P2P and pure P2P, which is structured hierarchically with a supernode layer and a normal peers layer.

Fitting Web services into P2P framework is to adapt Web services to P2P environment, which results in the so-called *P2P Web services,* or simply *P2P services*. Here P2P service is different from the ordinary Web services at least in three aspects. First, typically a peer in P2P services takes all three roles of services provider, consumer, and registrar, whereas in ordinary Web services, a node can typically be a producer and/or a consumer, but not a registrar at the same time. Second, generally speaking, servers in ordinary Web services systems are well-known hosts, with static IP addresses and on the outside of a firewall. However, this is not usually the case in the P2P world. A services node may join or depart the P2P services network at any time. Third, the preferred method of finding Web services in the ordinary architecture is currently through a central repository known as a UDDI operator. Nevertheless, P2P services systems have no central server to hold UDDI registry; each peer node manages its own UDDI registry locally. So, new and efficient mechanisms for services discovery in P2P services environment are required.

Corresponding to the architecture of P2P systems, there may also be three schemes for building P2P services applications: centralized P2P services, pure P2P services, and hybrid P2P services. For centralized P2P services, there is a central server in P2P services systems. However, the central server is not used as a central UDDI registry server; instead it is used for storing metadata of services to facilitate services discovery, which includes business names, services types, URLs, and so forth. In pure P2P services systems, services UDDI registry is distributed on every services node, so there is no need for services publication of the ordinary sense, and UDDI registry maintenance is also simplified because all services information is published and maintained locally. And in hybrid P2P services systems, the supernodes will be used for storing services metadata. It is useful for services discovery to cluster services nodes based on metadata, and then register the nodes in the same cluster under the same supernode.

BP-SERVICES: BESTPEER-BASED WEB SERVICES

As mentioned, the BP-Services project aims to develop an experimental P2P-based Web services platform as a test-bed for further P2P and Web services research.

BestPeer (Ng et al., 2002) is a generic P2P system with an architecture more pure P2P than hybrid P2P. The BestPeer system consists of two types of nodes: a large number of normal computers (i.e., peers), and a relatively fewer number of *Location-Independent Global names Lookup* (LIGLO) servers. Every peer in the system runs the BestPeer software, and will be able to communicate and share resources with any other peers. There are two types of data in each peer: private data and public (or sharable) data. For a certain peer, only its public data can be accessed by and shared with other

Figure 2. The internals of a BP-Services peer node

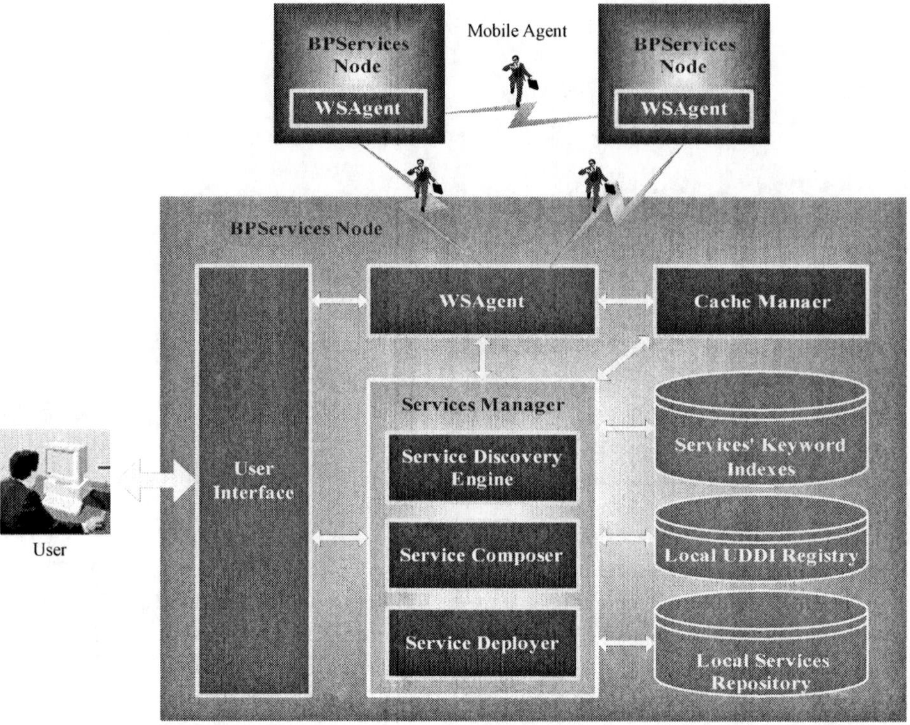

peers. Figure 1 shows the network topology of BestPeer. In the top layer are LIGLO servers, and in the bottom layer are normal peers.

The Architecture of BP-Services

In BP-Services, except for the LIGLO servers adhering to BestPeer, each peer node takes both the roles of a services provider and a services consumer, as well as a services registrar. That is to say, there is no central UDDI registry in BP-Services; all services and their definitions are distributed over the peer nodes. Figure 2 illustrates the internals of a BP-Services peer node. There are essentially seven components that are loosely integrated.

The first component, also the most important component, is the *Services Manager* that facilitates services discovery, services composition, and services deploying. Corresponding to its functionalities, the services manager consists of three sub-components: the *services discovery engine,* the *services composer,* and the *services deployer.* The services discovery engine is responsible for the publication and location of services. The services composer provides facilities for defining new composite services from existing services, and editing existing services (local). The services deployer facilitates the binding and invocation of requested services, as well as coordination of composite services.

The second component is the *Web Services Agent System,* or simply WSAgent. The WSAgent provides the environment for mobile agents to operate on. Each BP-Services node has a master agent that manages the services discovery and services description retrieval. In particular, it will clone and dispatch worker agents to neighboring nodes, receive results, and present to the user. It also monitors the statistics and manages the network reconfiguration policies.

The third component is a *Cache Manager,* which is used for caching the results of services discovery and retrieval. Furthermore, by collaboration among the cache managers, a P2P cache subsystem can be formed under the BP-Services framework so that all peers can share the caching results among themselves.

The fourth component is the *User Interface,* which consists of several interface modules, corresponding to services discovery and retrieval, services composition, and deploying.

The other three components are *Services Indexes*, *Local UDDI Registry,* and *Local Services Repository* respectively. The services repository keeps the services provided locally. Local UDDI registry holds the description (or publication) information of local services. And services indexes are simply inverted lists of services keywords extracted from the description information of local services, mainly business names and service types. Extracting and keeping services keywords speeds up services discovery.

Neighbor Nodes Finding in BP-Services

In BP-Services, given a participant node, its neighbor nodes are defined as those nodes that can provide as many services as possible similar to that in the given node. Here we use the information retrieval method to find neighbor nodes.

We can treat each node in BP-Services as a document, whose content is the services description information (UDDI registry) contained in that node. Thus we can cluster the nodes in BP-Services by using documents clustering methods (Baesa-Yates & Ribeiro-Neto, 1999). Roughly speaking, nodes in the same cluster may provide more similar services than those from different clusters. However, traditional documents clustering methods are based on a global data view, which is not realistic because it is not easy, if not impossible, to gather data of all nodes in a dynamic P2P network. In BP-Services, we adopt a simple local clustering strategy. We use a Boolean model to represent a peer services node, for the Boolean model is easier to evaluate than the vector space model (VSM), and it is difficult to set the document vector space without deterministic global data view in a P2P environment.

Given a peer services node *p*, there exists a set of keywords extracted from the services description document of *p*. We denote the set of keywords K_p, and treat it equal to the node *p* itself. For two services nodes *p* and *q*, suppose their keyword sets are K_p and K_q; the similarity of the two nodes are defined as follows. Here, |•| indicates the cardinality of a set.

$$sim(K_p, K_q) = \frac{|K_p \cap K_q|}{|K_p \cup K_q|}. \quad (1)$$

As in BestPeer, when a services node would like to become a participant of BP-Services, then it first registers with a LIGLO server, and the LIGLO server will issue the node with a global and unique identifier (i.e., BPID, BestPeerID), and meanwhile the LIGLO server will also send the node a list of peer nodes that have already registered in the network (i.e., the initial direct peers of the node). We term the links between these initial direct peers and the new participant node the *initial links* of the node.

After joining the network, the node (say *p*) can begin to find its neighbors by the following steps: (1) Through the ping-pong messages, it contacts the set of peers within a certain number (say *k*) of hops away from it. Let denote the set of peers as Peer(*p*, *k*)=$\{q_1,q_2,...q_n\}$, and get these peers' keywords sets $\{K_i|i=1\sim n\}$. (2) Calculate the similarity of *p* and each peer in *Peer*(*p*, *k*)—that is, $\{sim(p, q_i)| i=1\sim n\}$. (3) Suppose *q* is the peer in *Peer*(*p*, *k*) that has largest similarity with *p*, then take *q* as *p*'s neighbor node, and connect *p* and *q* by a direct link, which is termed *neighbor link* of *p* and *q*.

Through the process of neighbor finding, the peers that share services tend to be connected together by neighbor links, and consequently form clusters of services peers. Considering the dynamism of the P2P system, the peers should update their neighbors regularly.

Services Discovery in BP-Services

Services discovery is the key process of P2P services. Because P2P services' UDDI registries are distributed on the peer nodes, it is inefficient to search the targeted services by traversing all peers one by one. Note that service discovery in P2P is different from P2P information retrieval. In service discovery, once a service that satisfies the requester's requirements is found, the discovery process can be stopped. It is not necessary to find a lot of similar services for a certain requester's specific service requirements.

In BP-Services, once a requester submits his or her service requirement, say a service query *Q*, the following process will be launched:

1. Extracting keywords from *Q*, the service search process is equal to carrying out keywords matching in information retrieval.
2. First, search at the local peer. The searching task is done by using the local services indexes as in traditional IR. If there are services matching the query, then go to (3); otherwise, go to (4).
3. Return the matched services' descriptions to the user, and the user browses the services descriptions to see whether there are services (s)he wants. If there is at least one service (s)he wants, then the process of service discovery is over; otherwise, go to (5).
4. Select randomly an *initial link* of the local peer, then clone a working agent and dispatch it with the service query to the peer at the other end of the selected initial link. At that remote peer, do the searching as at the local peer.
5. Clone a working agent and dispatch it with the service query to the local peer's neighbor. At the neighbor peer, do the searching as at the local peer.
6. At the remote peer, once there are services matching the query, then return the matching services' descriptions to the user, who decides whether the returned results contain the target service. If the target service is found, then the search task is over and the working agent would return the source peer or be destroyed at the remote peer. If no target service is found, the working agent has to continue the search target until the target service is found or the working agent's TTL is 0.

Note in the above process, when the working agent gets to a peer along a *neighbor link,* its TTL will not decrease; only walking along *initial link,* its TTL will decrease.

RELATED WORK

Recently, combining P2P and Web services is gaining importance both in industry and academia. From the industry, two ambitious projects were launched, Sun Microsystems' JXTA (Li, 2001) and Microsoft's .net, more recently Hailstorm. Both JXTA and Hailstorm are trying to provide a general, language/environment-independent P2P services environment by putting forward a set of protocols for communication among peers.

From research institutions, Hoschek (2002) proposed a unified peer-to-peer database framework for scalable service discovery; Schlosser Sinteck, Decker, and Nejdl (2002) put forward a scalable and ontology-based infrastructure for semantic Web services; Sheng, Benatallah, Dumas, and Mak (2002) developed a platform for rapid composition of Web services in peer-to-peer environment; and Abiteboul, Benjelloun, Manolescu, Milo, and Weber (2002) designed a kind of active XML document to integrate peer-to-peer data and Web services.

Unlike the projects above, BP-Services is based on the BestPeer platform. We use an information retrieval method for services discovery, which is quite different from other P2P services projects. BP-Services is easy to implement because, except for the ordinary Web services protocols, it does not need any additional and complex protocols.

CONCLUSION

To overcome the limitations of Web services systems caused by their centralized architecture, we explore the techniques of building Web services applications under a P2P environment. The ongoing project, BP-Services, is presented as a case study to demonstrate our approach. BP-Services is an experimental Web services platform developed on the propriety BestPeer infrastructure. Future work will focus on developing some concrete applications on BP-Services put on a campus network as a test-bed for future research on P2P and Web services. And semantic Web service will also be considered in BP-Services in the future.

ACKNOWLEDGMENTS

This work was supported by grants numbered 60573183 and 60373019 from the NSFC, grant No. 20045006071-16 from the Chenguang Program of Wuhan Municipality, grant No. WKL(04)0303 from the Open Researches Fund Program of LIESMARS, and the Shuguang Scholar Program of Shanghai Education Development Foundation.

REFERENCES

Abiteboul, S., Benjelloun, O., Manolescu, I., Milo, T., & Weber, R. (2002). Active XML: Peer-to-peer data and Web services integration. *Proceedings of the 28th International Conference on Very Large Databases* (pp. 1087-1090), Hong Kong, China.

Baesa-Yates, R., & Ribeiro-Neto, B. (1999). *Modern information retrieval* (pp. 124-127). Boston: Addison-Wesley/ACM Press.

Christensen, E., Curbera, F., & Meredith, G. (2001). *Web Services Description Language (WSDL) 1.1.* W3C Note 15. Retrieved from http://www.w3.org/TR/wsdl

Gottschalk, K., Graham, S., Kreger, H., & Snell, J. (2002). Introduction to Web services architecture. *IBM Systems Journal, 41*(2), 170-177.

Hoschek, W. (2002). A unified peer-to-peer database framework and its application for scalable service discovery. *Proceedings of the 3rd International IEEE/ACM Workshop on Gird Computing,* Baltimore, MD (pp. 126-144).

Li, G. (2001). JXTA: A network programming environment. *IEEE Internet Computing,* (May-June), 88-95.

Ng, W. S., Ooi, B. C., & Tan, K.L. (2002). BestPeer: A self-configurable peer-to-peer system. *Proceedings of the 18th International Conference on Data Engineering* (pp. 272-272).

Ooi, B. C., Tan, K-L., Lu, H., & Zhou, A. (2002). P2P: Harnessing and riding on peers. *Proceedings of National Database Conference*, Zhengzhou, China, (pp. 1-5).

Schlosser, M., Sinteck, M., Decker, S., & Nejdl, W. (2002). A scalable and ontology-based infrastructure for semantic Web services. *Proceedings of the 2nd International Workshop on Agents and Peer-to-Peer Computing*, Linköping, Sweden, (pp. 104-111).

Sheng, Q., Benatallah, B., Dumas, M., & Mak, E. (2002). SELF-SERV: A platform for rapid composition of Web services in a peer-to-peer environment. *Proceedings of the 28th International Conference on Very Large Databases*, Hong Kong, China, (pp. 1051-1054).

SOAP. (2000). *Simple Object Access Protocol (SOAP) 1.1.* W3C Note 8. Retrieved from http://www.w3.org/TR/soap

Yang, B., & Garcia-Molina, H. (2001). Comparing hybrid peer-to-peer systems. *Proceedings of the 27th International Conference on Very Large Databases*, Roma, Italy, (pp. 561-570).

Web Services Conceptual Architecture. (n.d.). Retrieved from *http://*www.ibm.com/software/solutions/webservices/documentation.html

KEY TERMS

Centralized P2P: In a centralized P2P network, there is a central server responsible for maintaining indexes on the metadata for all the peers in the network.

Hybrid P2P: A kind of tradeoff between centralized P2P and pure P2P, which is structured hierarchically with a supernode layer and a normal peers layer.

Peer-to-Peer (P2P): A class of systems and applications that employ distributed resources to perform a critical function in a decentralized way. A P2P distributed system typically consists of a large number of nodes that can share resources, information, and services, taking the roles of both consumer and provider, and may join or depart the network at any time, resulting in a truly dynamic and ad-hoc environment.

Pure P2P: A P2P system with fully autonomous peers—that is, all nodes are equal, no functionality is centralized, and the communication between peers is also symmetric.

Service Discovery: An operation of finding Web services. After Web services are created and published in Web services registries such as UDDI, the service users or consumers need to search Web services manually or automatically. The implementation of UDDI servers should provide simple search APIs or Web-based GUI to help find Web services.

Universal Description, Discovery and Integration (UDDI): The protocol for Web service publishing. It should enable applications to look up Web services information in order to determine whether to use them.

Web Service: Can be seen as an interface that describes a collection of operations that are network accessible through standardized XML messaging. Software applications written in various programming languages and running on various platforms can use Web services to exchange data over computer networks due to the interoperability of using open standards.

Web Services Description Language (WSDL): An XML language for describing Web services.

eXtensible Markup Language (XML): A meta-language written in SGML that allows one to design a markup language, used to allow for the easy interchange of documents on l is misd Wide Web.

Business and Technology Issues in Wireless Networking

David Wright
University of Ottawa, Canada

INTRODUCTION

A major development in the enabling technologies for mobile computing and commerce is the evolution of wireless communications standards from the IEEE 802 series on local and metropolitan area networks. The rapid market growth and successful applications of 802.11, WiFi, is likely to be followed by similar commercial profitability of the emerging standards, 802.16e, WiMAX, and 802.20, WiMobile, both for network operators and users. This article describes the capabilities of these three standards and provides a comparative evaluation of features that impact their applicability to mobile computing and commerce. In particular, comparisons include the range, data rate in Mbps and ground speed in Km/h plus the availability of quality of service for voice and multimedia applications.

802.11 WiFi

WiFi (IEEE, 1999a, 1999b, 1999c, 2003) was originally designed as a wireless equivalent of the wired local area network standard IEEE802.3, Ethernet. In fact there are many differences between the two technologies, but the packet formats are sufficiently similar that WiFi packets can easily be converted to and from Ethernet packets. Access points can therefore be connected using Ethernet and can communicate with end stations using WiFi.

WiFi can transport both real-time communications such as voice and video plus non-real time communications such as Web browsing, by providing quality of service, QoS, using 802.11e (IEEE, 2005). There are 2 QoS options. One provides four priority levels allowing real-time traffic to be transmitted ahead of non-real-time traffic, but with no guarantee as to the exact delay experienced by the real-time traffic. The other allows the user to request a specific amount of delay, for example, 10 msecs., which may then be guaranteed by the access point. This is suited to delay sensitive applications such as telephony and audio/video streaming.

WiFi has a limited range of up to 100 metres, depending on the number of walls and other obstacles that could absorb or reflect the signal. It therefore requires only low powered transmitters, and hence meets the requirements of operating in unlicensed radio spectrum at 2.4 and 5 GHz in North America and other unlicensed bands as available in other countries.

WiFi is deployed in residences, enterprises and public areas such as airports and restaurants, which contain many obstacles such as furniture and walls, so that a direct line of sight between end-station and access point is not always possible, and certainly cannot be guaranteed when end stations are mobile. For this reason the technology is designed so that the receiver can accept multipath signals that have been reflected and/or diffracted between transmitter and receiver as shown in Figure 1(a). WiFi uses two technologies that operate well in this multipath environment: DSSS, Direct Sequence Spread Spectrum, which is used in 802.11b, and OFDM, Orthogonal Frequency Division Multiplexing, which is used in 802.11a and g (Gast, 2002). A key distinguishing factor between these alternatives, which is important to users, is spectral efficiency, that is, the data rate that can be achieved given the limited amount of wireless spectrum available in the unlicensed bands. DSSS as implemented in 802.11b uses 22 MHz wireless channels and achieves 11 Mbps, that is, a spectral efficiency of 11/22 = 0.5. OFDM achieves a higher spectral efficiency and is therefore making more effective use of the available wireless spectrum. 802.11g has 22 MHz channels and delivers 54 Mbps, for a spectral efficiency of 54/22 = 2.5 and 802.11a delivers 54 Mbps in 20 MHz channels, with a spectral efficiency of 54/20 = 2.7. A recent development in WiFi is 802.11n (IEEE, 2006a), which uses OFDM in combination with MultiInput, MultiOutput, MIMO, antennas as shown in Figure 1(b). MIMO allows the spectral efficiency to be increased further by exploiting the multipath environment to send several streams of data between the multiple antennas at the transmitter and receiver. At the time of writing the details of 802.11n are not finalized, but a 4x4 MIMO system (with 4 transmit and 4 receive antennas) will probably generate about 500 Mbps in a 40 MHz channel, that is, a spectral efficiency of 500/40 = 12.5. 802.11n is suited to streaming high definition video and can also support a large number of users per access point.

The data rates in WiFi are shared among all users of a channel, however some users can obtain higher data rates than others. Network operators may choose to police the data rate of individual users and possibly charge more for higher rates, or they may let users compete so that their data rates vary dynamically according to their needs and the priority levels

Figure 1. (a) Receiver recovers a single signal from multiple incoming signals; (b) MIMO receiver recovers multiple signals using multiple antennas

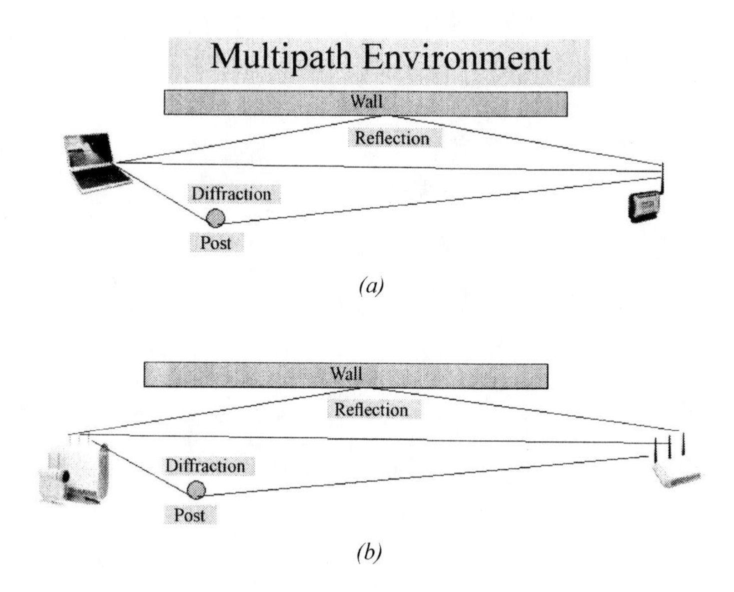

of their traffic. This provides considerable flexibility allowing many users to spend much of their time with low data rate applications such as VoIP, e-mail and Web browsing, with occasional high data rate bursts for audio/video downloads and data-intensive mesh computing applications.

Many deployments of WiFi use multiple access points to achieve greater coverage than the range of a single access point. When the coverage of multiple access points overlaps they should use different radio channels so as not to interfere with each other, as shown in Figure 2. For instance, in the North American 2.4 GHz band there is 79 MHz of spectrum available and the channels of 802.11b and g are 22 MHz wide. It is therefore possible to fit 3 non-overlapping channels into the available 79 MHz, which are known as channels 1, 6 and 11. Other intermediate channels are possible, but overlap with channels 1, 6 and 11. In Figure 2, the top three access points are shown connected by Ethernet implying that they are under the control of a single network operator, such as an airport. As an end-station moves among these access points the connection is handed off from one access point to another using 802.11r (IEEE, 2006b), while maintaining an existing TCP/IP session. Movement can be up to automobile speeds using 802.11p (IEEE, 2006c). Standard technology, 802.21 (IEEE, 2006d), is also available to handoff a TCP/IP session when a mobile end-station moves from an access point of one network operator to that of another, and this requires a business agreement between the two operators.

802.11 networks can therefore span extensive areas by interconnecting multiple access points, and city-wide WiFi networks are available in, for example, Philadelphia in the U.S., Adelaide in Australia, Fredericton in Canada and Pune in India. The broad coverage possible in this way greatly expands the usefulness of WiFi for mobile computing and electronic commerce. Enterprise users can set up secure virtual private networks from laptops to databases and maintain those connections while moving from desk to conference room to taxi to airport. A VoIP call over WiFi can start in a restaurant, continue in a taxi and after arriving at a residence.

The features of WiFi, IEEE 802.11, that are of particular importance for mobile computing and commerce are:

- Broad coverage achieved by handing off calls between access points, using 802.11r and 802.21, in cities where there are sufficient access points.
- Multimedia capability achieved by QoS, 802.11e.
- Flexibility in data rates achieved by allowing the total data rate of an access point to be shared in dynamically changing proportions among all users.
- Low cost achieved by using unlicensed spectrum, low power transmitters and mass produced equipment.

The downside to WiFi, IEEE 802.11, is limited coverage in cities that do not have extensive access point deployment.

802.16E WIMAX

802.16E (IEEE, 2006e) has a greater range than 802.11, typically 2-4 km and operates between base stations and subscriber stations. The initial IEEE standard 802.16 is for fixed applications, which compete with DSL and cable

Figure 2. WiFi handoff among access points

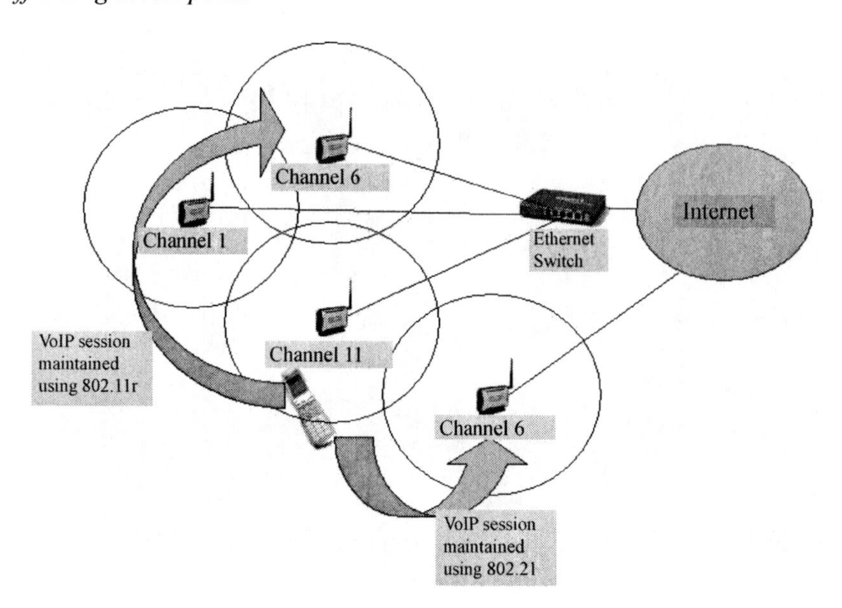

modems. Mobile applications including handoff capability among base stations, which we deal with here, are provided by 802.16E, and are based on similar but incompatible technology.

In 802.16E, WiMAX, mobility is limited to automobile speeds, up to about 100 Km/h so that it has limited use in high speed trains and aircraft. WiMAX uses the terminology "subscriber" stations, implying that customers are paying for a public service. Since the geographic range extends well into public areas, this is certainly one application. Another mobile application is a private campus network in which a central base station serves a business park or university campus. Initial deployment of WiMAX uses licensed spectrum, although low power applications in unlicensed spectrum are also specified in the standard.

WiMAX has sophisticated QoS capabilities, which allow customers to reserve capacity on the network including a reserved data rate plus quality of service. The data rate is specified by a minimum reserved traffic rate, MRTR, on which quality of service is guaranteed (Figure 3). The customer is allowed to send at a higher rate, up to a maximum sustainable traffic rate, MSTR, without necessarily receiving QoS, and above that rate, traffic will be policed by the network operator, that is, it may be discarded. The QoS parameters that can be specified by the customer are latency and jitter, plus a priority level, which is used by the base station to distinguish among service flows that have the same latency and jitter requirements. The combination of latency and jitter can be used to distinguish among service flows, and further detail on the performance of WiMAX is given by Ghosh et al. (2005).

Combinations of QoS parameters and data rates make WiMAX highly suited to mobile computing and commerce. Each subscriber can set up multiple service flows, for example, for Web browsing during a multimedia conference, and use data rates that are quite different from those of other customers. The service provider can charge based on a combination of data rate and QoS.

WiMAX is based on OFDM, thus achieving a high spectral efficiency. There are a number of options within 802.16E for the channel widths and modulation techniques, resulting in a corresponding range of data rates and spectral efficiencies. It is important to recognize that the spectral efficiency depends on the distance between the base station and the subscriber station (Figure 4). As the signal degrades with distance it is not possible to encode so many bps within each Hz and 802.16E assigns encodings that take this into account. Closer to the base station the data rate is therefore higher. The exact distance depends on the operating environment

Figure 3. WiMAX traffic rate guarantees

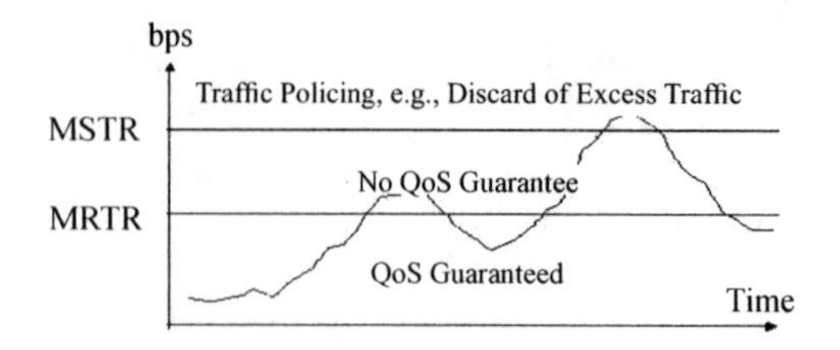

Figure 4. Spectral efficiency and maximum data rates for WiMAX

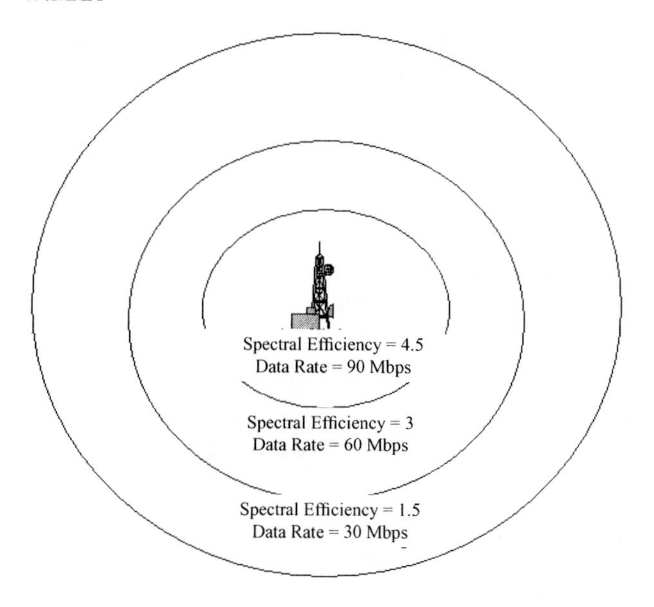

since 802.16E uses multipath signals involving reflections and diffractions. The data rates shown in Figure 4 are the maximum achievable with the highest channel bandwidth allowed according to the standard—20 MHz—and can vary not only with distance but also according to how much forward error correction is used.

The features of 802.16E that are of particular importance for mobile computing and commerce are:

- Good range, enabling city-wide coverage with a reasonable number of base stations.
- Multimedia capability achieved by QoS, and guaranteed data rates.
- Flexibility in data rates achieved by allowing the total data rate of a base station to be shared in dynamically changing proportions among all users.

The downside to 802.16E is the cost of licensed spectrum.

802.20 WIMOBILE

At the time of writing, (1Q06), the specification of 802.20, (IEEE, 2006, f), is under development, so that less detail is available than for 802.11 and 802.16e. The key features of 802.20 are:

- It operates in licensed spectrum below 3.5 GHz.
- It is designed from the start for an all-IP environment and interfaces to IP DiffServ QoS service classes, (Grossman, 2002) which provide for prioritization of users' traffic.
- It interfaces to "Mobile IP" (Montenegro, 2001) as part of its mobility capability. Mobility includes not just automobile speed, but also high speed trains at up to 250 Km/h.
- It uses OFDM with MIMO antennas to achieve a very high spectral efficiency, so that large numbers of users can share access to a single base station.

COMPARATIVE EVALUATION

Mobile computing and commerce involves communicating from mobile devices for a variety of purposes including: data transfer for processing intensive applications and for Web browsing; voice and multimedia calls between human users;

Table 1. Comparative evaluation of technologies for mobile computing and commerce

	802.11, WiFi	802.16e, WiMAX	802.20, WiMobile
Range	100 metres	2-4 Km	2-4 Km
Coverage	Hot spots. Some city-wide deployments.	Designed for city-wide deployment	Designed for national deployment
Data Rate	11, 54, 500 Mbps flexibly shared among all users	Up to 90 Mbps flexibly shared among all users	> 1 Mbps per user
QoS	(a) Prioritization mechanism (b) data rate and QoS guarantees	Data rate and QoS guarantees	Data rate guarantees and QoS prioritization
Mobility Speed	100 Km/h	100 Km/h	250 Km/h
Cost	Very low unit cost access points. End-station interfaces built into phones, laptops, PDAs. Large number of access points required. Unlicensed spectrum.	Medium unit cost access points. End-station interfaces built into phones, laptops, PDAs. Licensed or unlicensed spectrum.	Medium unit cost access points. End-station interfaces built into phones, laptops, PDAs. Licensed spectrum.

downloading audio, video and multimedia from a server, (a) streaming for real-time playout to human users and (b) file transfer for subsequent access on the mobile device. Each of these requires appropriate data rate and quality of service. Cost is also an important factor, since subscription may be required to a public network operator or an enterprise may need to build its own wireless network. Employees using mobile computing devices within a building require mobility only at pedestrian speeds. In public areas such as city streets, automobile speeds are required and between cities high speed trains may be used. The type of mobile computing application determines which speed is appropriate. Table 1 provides a comparison among the three technologies described in this paper.

CONCLUSION

Mobile computing and commerce users have a wide range of emerging wireless communication technologies available: WiFi, WiMAX and WiMobile. Each of them offers high data rates and spectral efficiencies, and will therefore likely be available at low cost. They are the major enabling telecommunication technologies for mobile computing and are likely to be deployed in public areas and private campuses for in-building and outdoor use. WiFi is already extensively deployed and WiMAX is being deployed in Korea in 2006 and can be expected in many other countries in 2007. The WiMobile standard has not yet been specified (as of the time of writing 1Q06) and commercial equipment can be expected after WiMAX.

REFERENCES

Gast, M. (2002). *802.11 wireless networks: The definitive guide*. O'Reilly.

Ghosh, A., Wolter, D. R., Andrews, J. G., & Chen, R. (2005). Broadband wireless access with WiMax/8O2.16: Current performance benchmarks and future potential. *IEEE Communications Magazine*, *43*(2), 129-136.

Grossman, D. (2002). *New terminology and clarifications for Diffserv*. RFC3260. Internet Engineering Task Force.

IEEE. (1999a). *802.11 wireless LAN: Medium access control (MAC) and physical layer (PHY) specifications*. New York: IEEE Publications.

IEEE. (1999b). *802.11a high-speed physical layer in the 5 GHz band*. New York: IEEE Publications.

IEEE. (1999c). *802.11b higher-speed physical layer (PHY) extension in the 2.4 GHz band*. New York: IEEE Publications.

IEEE. (2003). *802.11g further higher-speed physical layer extension in the 2.4 GHz band*. New York: IEEE Publications.

IEEE. (2005). *802.11e wireless LAN: Quality of service enhancements*. New York: IEEE Publications.

IEEE. (2006a). *802.11n wireless LAN: Enhancements for higher throughput* (In progress). Retrieved March 2006, from http://standards.ieee.org/board/nes/projects/802-11n.pdf.

IEEE. (2006 b). *802.11r wireless LAN: Fast BSS transition* (In progress). Retrieved March 2006, http://standards.ieee.org/board/nes/projects/802-11n.pdf.

IEEE. (2006c). *802.11p wireless LAN: Wireless access in vehicular environments*. (In progress). Retrieved March 2006, from http://standards.ieee.org/board/nes/projects/802-11p.pdf.

IEEE. (2006d). *802.21 media independent handover services*. (In progress). Retrieved March 2006, from http://grouper.ieee.org/groups/802/21/.

IEEE. (2006e). *802.16E-2005 air interface for fixed and mobile broadband wireless access systems: Amendment for physical and medium access control layers for combined fixed and mobile operation in licensed bands*. New York: IEEE Publications.

IEEE. (2006f). 802.20 mobile broadband wireless access systems. (In progress). Retrieved March 2006, from http://grouper.ieee.org/groups/802/20/.

Montenegro, G. (2001) *Reverse tunneling for mobile IP*. RFC3024. Internet Engineering Task Force.

KEY TERMS

Direct Sequence Spread Spectrum (DSS): A transmission technique in which data bits are multiplied by a higher frequency code sequence, so that the data are spread over a wide range of frequencies. If some of these frequencies fade, the data can be recovered from the data on the other frequencies together with a forward error correction code.

Mobile IP: An Internet standard that allows a mobile user to move from one point of attachment to the network to another while maintaining an existing TCP/IP session. Incoming packet to the user are forwarded to the new point of attachment.

Multipath: A radio environment in which signals between transmitter and receiver take several different spatial paths due to reflections and diffractions.

Orthogonal Frequency Division Multiplexing (OFDM): A transmission technique in which data bits are transmitted on different frequencies. The data transmitted on one frequency can be distinguished from those on other frequencies since each frequency is orthogonal to the others.

Quality of Service (QoS): Features related to a communication, such as delay, variability of delay, bit error rate and packet loss rate. Additional parameters may also be included, for example, peak data rate, average data rate, percentage of time that the service is available, mean time to repair faults and how the customer is compensated if QoS guarantees are not met by a service provider.

WiFi: A commercial implementation of the IEEE 802.11 standard in which the equipment has been certified by the WiFi Alliance, an industry consortium.

WiMAX: A commercial implementation of the IEEE 802.16 standard in which the equipment has been certified by the WiMAX Forum, an industry consortium.

WiMobile: Another name for the IEEE 802.20 standard which is in course of development at the time of writing (1Q06).

Business Strategies for Mobile Marketing

Indranil Bose
University of Hong Kong, Hong Kong

Chen Xi
University of Hong Kong, Hong Kong

INTRODUCTION

With the appearance of advanced and mature wireless and mobile technologies, more and more people are embracing mobile "things" as part of their everyday lives. New business opportunities are emerging with the birth of a new type of commerce known as mobile commerce or m-commerce. M-commerce is an extension to electronic commerce (e-commerce) with new capabilities. As a result, marketing activities in m-commerce are different from traditional commerce and e-commerce. This chapter will discuss marketing strategies for m-commerce. First we will give some background knowledge about m-commerce. Then we will discuss the pull, push, and viral models in m-marketing. The third part will be the discussion about the future developments in mobile marketing. The last part will provide a summary of this article.

BACKGROUND

Popularity of Mobile Services

From the research done by Gartner Dataquest (Business Week, 2005), there will be more than 1.4 billion mobile service subscribers in the Asia-Pacific region by 2009. Research analysts of Gartner Dataquest also estimated that China will have over 500,000 subscribers, and more than 39% of the people will use mobile phones at that time. In India, the penetration rate of mobile phones is expected to increase from 7% in 2005 to 28% in 2008. The Yankee Group has also reported a growing trend of mobile service revenues from 2003 to 2009. Although the revenue generated by traditional text-based messaging service will not change much, revenue from multimedia messaging services will rise to a great extent. Other applications of mobile services, such as m-commerce-based services and mobile enterprise services, will continue to flourish. One thing that is very important in driving Asia-Pacific mobile service revenue is mobile entertainment services. Revenue from mobile entertainment services will make up almost half of the total revenues from all kinds of mobile data services from now on. Not only in the region of Asia-Pacific, but mobile services will increase in popularity in other parts of the world as well. In the United States, it is expected that the market for m-commerce will reach US$25 billion in 2006.

The Development of Mobile Technologies

Two terms are frequently used when people talk about mobile information transmission techniques: the second-generation (2G) and the third-generation (3G) wireless systems. These two terms actually refer to two generations of mobile telecommunication systems. Three basic 2G technologies are time division multiple access (TDMA), global system for mobile (GSM), and code division multiple access (CDMA). Among these three, GSM is the most widely accepted technology. There is also the two-and-a-half generation (2.5G) technology of mobile telecommunication, such as general packet radio service (GPRS). 2.5G is considered to be a transitional generation of technology between 2G and 3G. They have not replaced 2G systems. They are mostly used to provide additional value-added services to 2G systems. The future of mobile telecommunication network is believed to be 3G. Some standards in 3G include W-CDMA, TD-SCDMA, CDMA 2000 EV-DO, and CDMA EV-DV. The advancement in mobile telecommunication technology will bring in higher speed of data transmission.. The speed of GSM was only 9.6 kilobits per second (kbps), while the speed of GPRS can reach from 56 to114 kbps. It is believed that the speed of 3G will be as fast as 2 Megabits per second (mbps). The acceptance of 3G in this world began in Japan. NTT DoCoMo introduced its 3G services in 2001. Korea soon followed the example of Japan. In 2003, the Hutchison Group launched 3G commercially in Italy and the UK, and branded its services as '3'. '3' was later introduced in Hong Kong, China in 2004. Mainland China is also planning to implement 3G systems. Some prototypes or experimental networks have been set up in the Guangdong province. It is expected that 3G networks will be put into commercial use in 2007 using the TD-SCDMA standard that has been indigenously developed in China. Mobile information transmission can also be done using other technical solutions such as wireless local area network (WLAN) and Bluetooth. The interested reader may refer to Holma and Toskala (2002) for a fuller description of 3G systems, and to Halonen, Romero, and Melero (2003) for details of 2G and 2.5G systems.

The most popular mobile devices currently in use include mobile phones, wireless-enabled personal digital assistants (PDAs), and wireless-enabled laptops (Tarasewich, Nickerson, & Warkentin, 2002). Smartphones are also gaining favor from customers. Mobile phones are the most pervasive mobile devices. Basically, mobile phones can make phone calls, and can send and receive short text messages. More advanced mobile phones have color screens so that they can send or receive multimedia messages, or have integrated GPRS modules so that they can connect to the Internet for data transmission. PDAs are pocket-size or palm-size devices which do limited personal data processing such as recording of telephone numbers, appointments, and notes on the go. Wireless-enabled PDAs have integrated Wi-Fi (wireless fidelity)—which is the connection standard for W-LAN or Bluetooth—which helps them access the Internet. Some PDAs can be extended with GPRS or GSM modules so that they can work as a mobile phone. PDAs nowadays usually have larger screens than that of mobile phones and with higher resolution. They are often equipped with powerful CPUs and large storage components so that they can handle multimedia tasks easily. Smartphones are the combination of mobile phones and PDAs. Smartphones have more complete phoning function than PDAs, while PDAs have more powerful data processing abilities. However, the boundary between smartphones and PDAs are actually becoming more and more fuzzy.

The Need for Mobile Marketing

The rapid penetration rate of mobile devices, the huge amounts of investment from industries, and the advancement of mobile technologies, all make it feasible to do marketing via mobile devices. Mobile commerce refers to a category of business applications that derive their profit from business opportunities created by mobile technologies. Mobile marketing, as a branch of m-commerce (Choon, Hyung, & Kim, 2004; Varshney & Vetter, 2002), refers to any marketing activities conducted via mobile technologies. Usually m-commerce is regarded as a subset of e-commerce (Coursaris & Hassanein, 2002; Kwon & Sadeh, 2004). That is true, but due to the characteristics of mobile technologies, mobile marketing is different from other e-commerce activities. The first difference is caused by mobile technologies' ability to reach people anywhere and anytime; therefore mobile marketing can take the advantage of contextual information (Zhang, 2003). Dey and Abowd (2001) defined context as "any information that characterizes a situation related to the interaction between users, applications, and the surrounding environment." Time, location, and network conditions are three of the key elements of context. The second difference is caused by the characteristics of mobile devices. Mobile devices have limited display abilities. The screens are usually small, and some of the devices cannot display color pictures or animations. On the other hand, mobile devices have various kinds of screen shapes, sizes, and resolutions. Thus, delivering appropriate content to specific devices is very important. Mobile devices also have limited input abilities, and this makes it difficult for customers to respond. Mobile marketing shares something in common with e-commerce activities. An important aspect of e-commerce is to deliver personalized products/services to customers. Mobile marketing inherits this feature. Mobile marketing also inherits some of the problems from e-commerce, especially the problem of spamming. Personalization in mobile marketing is to conduct marketing campaigns which can meet the customer's needs by providing authorized, timely, location-sensitive, and device-adaptive advertising and promotion information (Scharl, Dickinger, & Murphy, 2005).

MOBILE MARKETING

Benefits of Mobile Marketing

There are two main approaches to advertise and promote products in industry—mass marketing and direct marketing. The former uses mass media to broadcast product-related information to customers without discrimination, whereas the latter is quite different in this regard. Mobile marketing takes a direct marketing approach. Using mobile marketing, marketers can reach customers directly and immediately. Similarly, customers can also respond to marketers rapidly. This benefit makes the interaction between marketers and customers easy and frequent. Compared to direct marketing using mail or catalogs, mobile marketing is comparatively cost effective and quick. Compared to telephone direct marketing, mobile marketing can be less interruptive. Compared to e-mail direct marketing, mobile marketing can reach people anytime and anywhere, and does not require customers to sit in front of a computer. Therefore, to some extent, mobile marketing can be a replacement for other types of marketing channels such as mail, telephone, or e-mail. Advertisement or promotion information sent via the Internet can be sent via a mobile device. Mobile marketing can enhance marketing by adding new abilities like time-sensitive and location-sensitive information. On the other hand, mobile commerce can generate new customers' data, like mobile telecommunication usage data and mobile Internet surfing data. Mobile marketing is the first choice for conducting marketing activities for m-commerce applications. However, due to limited size of screens of mobile devices, only brief information can be provided in mobile marketing solicitations, while e-mail or mail marketing can provide very detailed information. On the other hand telephone marketing requires the good communication skill of telesales. Once telemarketers have acquired this skill, the interaction between marketers and customers is quicker and more effective. It is not clear if mobile marketing is as effective or as popular as mass marketing agents like television and newspaper, but it can be said that it is indeed a powerful medium that is likely to gain in popularity in the future.

Models for Mobile Marketing

Mobile marketing usually follows one of the three kinds of models—push, pull (Haig, 2002; Zhang, 2003), and viral (Ahonen, 2002; Ahonen, Kasper, & Melkko, 2004; Haig, 2002), as illustrated in Figure 1.

Push Model

The push model sends marketing information to customers without the request of the customer. If the push model is used, besides knowing the targeted customers' interests, understanding the context of customers at the time the marketing activities are to be carried out is very critical. The timing in sending mobile information should also be appropriate. The content delivered to customers should also be displayable on their mobile devices. Permissions from customers are necessary before any solicitations can be sent. In the push model, marketers like to make it easier for customers to respond because of the poor input ability of mobile devices. G2000, a Hong Kong clothing chain, launched a mobile marketing campaign in November 2004. Mobile coupons in the format of SMS were sent to the mobile phones of selected customers. Customers could then use these coupons stored in their mobile phones when purchasing items in designated G2000 stores in order to get discounts. The campaign was considered to be a success because a number of customers responded to this program and used mobile coupons at G2000 stores.

Figure 1. Push, pull and viral models for mobile marketing

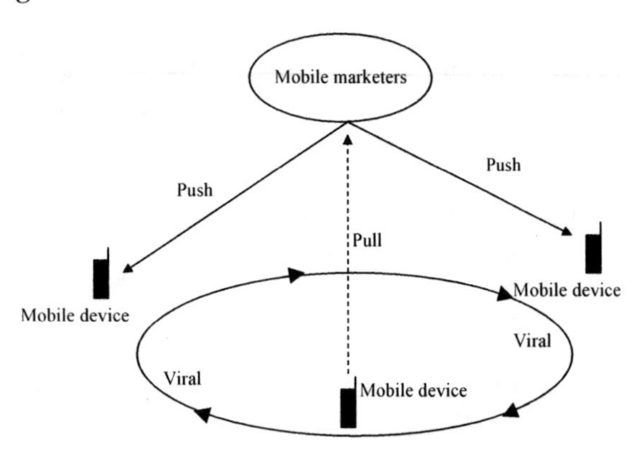

Pull Model

In the pull model, the marketer waits for the customer to send a request for a solicitation. The marketer prepares the marketing information in a format that is displayable in all possible mobile devices and scalable for various connection speeds. The significance of the pull model is that the information from customers is very useful for understanding customers' preferences, such as the preferred marketing time and interests. Mobile marketing following a pull model can be conducted in many ways. One possible approach is to let the customers select and download coupons to their mobile devices. Mobile service providers can build a Web site using a mobile Internet protocol such as wireless application protocol (WAP), and place various text-based coupons on this Web site. Customers can use their GPRS-enabled phone to browse the Web site and download coupons they like in the format of SMS. Each coupon will have a unique identity number. When the coupon is redeemed, related information, such as the phone number of the customer who downloaded the coupon and the time it was redeemed, is recorded. This kind of information is later used to analyze the behaviors of customers and build a profile for the customer. China Mobile, a mobile telecommunication operator in China, had established such a Web site for customers in Xiamen (a city located in southeast China) in 2005. The customers of China Mobile could download coupons displayed on this Web site onto their mobile phones using text messaging.

Viral Model

The phrase viral marketing was created by Steve Jurvetson in 1997 to describe the burgeoning use of Hotmail (Jurvetson & Draper, 1997). The principle of the viral model is based on the fact that customers forward information about products/services to other customers. The viral model enlarges the effect of other marketing activities while it costs the marketers very little in monetary terms. The viral model enables customer-to-customer communication. Like the pull model, the format of information that is delivered by viral model should be displayable in different devices and scalable for different connection speeds. Actually mobile marketing has the ability to be viral inherently because it is quite easy for people to forward mobile advertising or promotion information to their friends. However, viral marketing information has to be interesting and attractive enough to make the customers willing to forward it to other people. For example, "reply to this message in order to win $5000" may be a very attractive viral marketing message. Usually, viral marketing begins with push marketing activities to customers. According to Linner (2003), when the movie "2 Fast 2 Furious" was running in movie theatres, marketers tried to create a viral promotion using a mobile marketing strategy. Fans were asked to send SMS to enter a certain film-related competition. Besides inviting fans on every major phone network through advertisements on television, newspapers, and also through posters, a special code was designed and a low fee was offered to customers in order to encourage them to forward promotion information to their friends. Exciting gifts were offered as prizes in this competition (such as a replica of the vehicle, the EvO VII, that was used in the movie) to spur the enthusiasm of customers.

These three models of direct marketing can be complimentary to each other. Push-based mobile marketing can be used to stimulate pull-based marketing activities. For example, book marketers can send a short introduction to customers via SMS with a remark at the end saying "for more details, please reply to XXXXX." Once a customer responds to this by replying using SMS, more promotion or advertising information on this book can be sent to him. All three models—push, pull, and viral—can even be integrated together in a mobile marketing campaign. An example of an integrated mobile marketing approach was adopted by Fox Txt Club for the movie "phone booth" (Linner, 2003). At the beginning of the marketing campaign, members of the Fox Txt Club were sent invitations via SMS to a preview. The aim of this was to pull customers to the campaign. A competition that invited people to send SMS about questions pertaining to various details in the film was set up. The forwarding of SMS about the movie and the competition among club members and their friends, together with other media such as entertainment and event listings magazines and city-center posters, made the marketing campaign viral. The details of those who responded were recorded by Fox Txt Club, and this helped in building the database of customers for future release promotions. This could be used for push marketing for another movie in the future.

Strategies for Mobile Marketing

The most fundamental task for marketing activities following the push model is to send advertising or promotion information about products or services that the targeted customers once bought. This is the most direct and easy way to decide what is to be offered to customers in a solicitation. However, just marketing products already existing in customers' transaction records is not enough for marketers. It is necessary for marketers to explore the needs of customers. Two of the most commonly used marketing strategies are cross-selling and up-selling. Cross-selling is the practice of suggesting similar products or services to a customer who is considering buying something, such as showing a list of ring tones on a mobile Internet Web page that are similar to the one a customer has downloaded. Up-selling is the practice of suggesting higher priced, better versions of products or services to a customer who is considering a purchase, such as a mobile phone plan with higher fees and additional features. Two approaches can be used to find opportunities for up-selling or cross-selling. One is to find products or services that are similar to the ones a customer has bought. The other is to find people who have characteristics that are similar to a targeted customer. Products or services those people have bought and the targeted customer has not can be recommended to the target customers.

Pull-based marketing is relatively passive compared to push-based marketing. Usually in pull marketing, customers are responsible for searching for useful advertising or promotion information. The marketers' responsibility is to help customers find what they want more efficiently. Therefore, knowing what customers may request is very important in pull marketing. Instead of sending related information to customers like push marketing, marketers doing pull marketing can make information about products or services available on their mobile Internet Web site or ordinary Internet Web site. In viral marketing, marketers stand in a more passive position than even in pull marketing. However, for both pull and push marketing, some push activities should be carried out to start the marketing.

Whatever model one may use when carrying out mobile marketing activities, one issue must always be kept in mind and that is the necessity of obtaining explicit permission from customers (Bayne, 2002). Mobile technology makes connections so direct that it can interfere with customers' privacy very easily. Therefore, sending advertising or promotion information to people will cause trouble if permissions are not sought before solicitations or customers' wishes about not receiving a solicitation are not respected.

Understanding Customers in Mobile Marketing

All of the three models require good understanding of customers' needs. Marketing information that is not well designed will be regarded as spam by customers. Once a customer identifies some information from a company as spam, he or she will pay very little attention to or simply discard any information from that company. If a customer cannot find useful information on the Web site a company provides, it may be ok for the first time, a pity for the second time, but for the third time it will mean business lost forever. If information sent to customers is not interesting, customers may not want to forward them to their friends. All these situations may lead to failure of a marketing campaign. To avoid these situations, marketers need to understand customers well enough in order to send personalized marketing information. Customer profiling is a necessary approach to understand customers better. Customer profiling aims to find factors that can characterize customers. These factors are found by comparing customers to each other in order to discover similarities and differences among customers. Customer profiling encompasses two tasks—customer clustering and customer behavior pattern recognition. Customer clustering aims to classify customers into different groups. Customers within the same group are said to be more similar to each other than to customers in different groups. Marketers cluster customers using various data. Traditionally, customers are clustered according to their geographic locations, demographic characteristics, and the industries they are working for. They can also be clustered based on information about their purchasing history, such as what they bought, when they bought, and how much they spent. With the appearance of mobile services and m-commerce, usage data of new customer data services can also be used for clustering. For example, messaging services that customers subscribed to, GPRS surfing and download records, the type of mobile devices the customers use, and monthly mobile phone usage including use of IDD and roaming can yield many interesting information about the customers.

Aside from these hard facts, marketers may also want to infer some soft knowledge about customers' behaviors as well. To recognize customer behavior the marketers must discover relationships between hard facts. For example, customers that download ring tones of game music may download games-related screensavers later on. Since mobile technologies can enable context-sensitive marketing activities, marketers should gather knowledge about customers' location preferences and time preferences. For example, when does a customer usually go shopping and which place does he/she visit on the shopping trips? Marketers can find this kind of soft knowledge from various mobile network usage data. Again, collecting information on location and time requires permission from customers. Based on customer profiling, more sophisticated personalized advertising or promotion information can be sent to customers.

FUTURE TRENDS

Mobile technologies will advance further in the future. New technologies will enable new kinds of marketing activities. For example, the implementation of fourth-generation (4G) wireless systems will make the bandwidth much larger than that in current networks. On the other hand, the mobile device will have larger screens with higher resolutions. These two factors together will make interactive audio and even interactive video marketing possible. Generally speaking, the limitation of current mobile technologies will be weakened or removed in the future. As a result, more emphasis may be put on time- and location-related marketing, as well as on better understanding customers' interests. The principle is not only to know what customers want, but also to know when and where they may have a certain kind of need. Data mining techniques can be used in the future to find customer behavior patterns with time and location factors. Data mining techniques have been used widely in direct marketing for targeting customers (Ling & Li, 1998). There are also data mining techniques for clustering customers such as self-organizing-map (SOM—Kohonen, 1995) and techniques for discovering customer behavior such as association rules mining (Agrawal & Srikant, 1994). In the future, the availability of huge amounts of data about customers will compel marketers to adopt strong data mining tools to delve deep into customers' nature.

CONCLUSION

Equipped with advanced mobile technologies, more sophisticated marketing activities can be conducted now and in the future. In this article, we have discussed the benefits of mobile marketing, the role of mobile marketing in m-commerce, and the models used in mobile marketing. Although mobile marketing is powerful, it cannot replace other methods of marketing and should only be used as a powerful complement to traditional marketing. Mobile marketing should be integrated into the whole marketing strategy of a firm so that it can work seamlessly with other marketing approaches.

REFERENCES

Agrawal, R., & Srikant, R. (1994). Fast algorithms for mining association rules. In *Proceedings of the 20th International Conference on Very large Databases* (pp. 487-499), Santiago, Chile.

Ahonen, T. T. (2002). *M-profits: Making money from 3G services.* West Sussex, UK: John Wiley & Sons.

Ahonen, T. T., Kasper, T., & Melkko, S. (2004). *3G marketing: Communities and strategic partnerships.* West Sussex, UK: John Wiley & Sons.

Bayne, K. M. (2002). *Marketing without wires: Targeting promotions and advertising to mobile device users.* New York: John Wiley & Sons.

BusinessWeek. (2005). Special advertising section: 3G the mobile opportunity. *BusinessWeek* (Asian ed.), (November 21), 92-96.

Choon, S. L., Hyung, S. S., & Kim, D. S. (2004). A classification of mobile business models and its applications. *Industrial Management & Data Systems, 104*(1), 78-87.

Coursaris, C., & Hassanein, K. (2002). Understanding m-commerce. *Quarterly Journal of Electronic Commerce, 3*(3), 247-271.

Dey, A. K., & Abowd, G. D. (2001). A conceptual framework and a toolkit for supporting the rapid prototyping of context-aware applications. *Human-Computer Interaction, 16*(2-4), 97-166.

Haig, M. (2002). *Mobile marketing: The message revolution.* London: Kogan Page.

Halonen, T., Romero, J., & Melero, J. (2003). *GSM, GPRS and EDGE performance: Evolution towards 3G/UMTS.* West Sussex, UK: John Wiley & Sons.

Holma, H., & Toskala, A. (2002). *WCDMA for UMTS* (2nd ed.). West Sussex, UK: John Wiley & Sons.

Jurvetson, S., & Draper, T. (1997). *Viral marketing.* Retrieved from http://www.dfj.com/cgi-bin/artman/publish/steve_tim_may97.shtml

Kohonen, T. (1995). *Self-organizing maps.* Berlin: Springer-Verlag.

Kwon, O.B., & Sadeh, N. (2004). Applying case-based reasoning and multi-agent intelligent system to context-aware comparative shopping. *Decision Support Systems, 37*(2), 199-213.

Ling, C. X., & Li, C.-H. (1998). Data mining for direct marketing: Problems and solutions. *Proceedings of the 4th International Conference on Knowledge Discovery and Data Mining* (pp. 73-79), New York.

Linner, J. (2003). *Hitting the mark with text messaging.* Retrieved from http://wireless.sys-con.com/read/41316.htm

Scharl, A., Dickinger, A., & Murphy, J. (2005). Diffusion and success factors of mobile marketing. *Electronic Commerce Research and Applications, 4,* 159-173.

Tarasewich, P., Nickerson, R.C., & Warkentin, M. (2002). Issues in mobile e-commerce. *Communications of the Association for Information Systems, 8,* 41-84.

Varshney, U., & Vetter, R. (2002). Mobile commerce: Framework, applications and networking support. *Mobile Networks and Applications, 7,* 185-198.

Zhang, D. (2003). Delivery of personalized and adaptive content to mobile devices: A framework and enabling technology. *Communications of the Association for Information Systems, 12,* 183-202.

KEY TERMS

Bluetooth: Used mostly to connect personal devices wirelessly like PDAs, mobile phones, laptops, PCs, printers, and digital cameras.

Code Division Multiple Access (CDMA): A kind of 2G technology that allows users to share a channel by encoding data with channel-specified code and by making use of the constructive interference properties of the transmission medium.

Enhanced Data rates for GSM Evolution (EDGE): A kind of 2.5G technology. A new modulation scheme is implemented in EDGE to enable transmission speed of up to 384 kbps within the existing GSM network.

General Packet Radio Service (GPRS): Belongs to the family of 2.5G. GPRS is the first implementation of packet switching technology within GSM. The speed of GPRS can reach up to 115 Kbps.

Global System for Mobile (GSM) Communications: One of the 2G wireless mobile network technologies and the most widely used today. It can now operate in the 900 MHz, 1,800 MHz, and 1,900 MHz bands.

3G: The third generation of mobile telecommunication technologies. 3G refers to the next generation of mobile networks which operate at frequencies as high as 2.1 GHz, or even higher. The transmission speeds of 3G mobile wireless networks are believed to be able to reach up to 2 Mbps.

Time Division Multiple Access (TDMA): Divides each network channel into different time slots in order to allow several users to share the channel.

Time Division Synchronous Code Division Multiple Access (TD-SCDMA): A 3G mobile telecommunications standard developed in China.

2G: The second generation of mobile telecommunication technologies. It refers to mobile wireless networks and services that use digital technology. 2G wireless networks support data services.

2.5G: The second-and-a-half generation of mobile telecommunication technologies. 2.5G wireless system is built on top of a 2G network. 2.5G networks have the ability to conduct packet switching in addition to circuit switching. 2.5G supports higher transmission speeds compared to 2G systems.

W-CDMA: Developed by NTT DoCoMo as the air interface for its 3G network called FOMA. It is now accepted as a part of the IMT-2000 family of 3G standards.

Wireless Local Area Network (WLAN): Connects users wirelessly instead of using cables. WLAN is not a kind of mobile telecommunication technology. The coverage of WLAN may vary from a single meeting room to an entire building of a company.

Cache Invalidation in a Mobile Environment

Say Ying Lim
Monash University, Australia

INTRODUCTION

The rapid development, as well as recent advances in wireless network technologies, has led to the development of the concept of mobile computing. A mobile computing environment enables mobile users to query databases from their mobile devices over the wireless communication channels (Cai & Tan, 1999). The potential market for mobile computing applications is projected to increase over time by the currently increasingly mobile world, which enables a user to satisfy their needs by having the ability to access information anywhere, anytime. However, the typical nature of a mobile environment includes low bandwidth and low reliability of wireless channels, which causes frequent disconnection to the mobile users. Often, mobile devices are associated with low memory storage and low power computation and with a limited power supply (Myers & Beigl, 2003). Thus, for mobile computing to be widely deployed, it is important to cope with the current limitation of power conservation and low bandwidth of the wireless channel. These two issues create a great challenge for fellow researchers in the area of mobile computing.

By introducing data caching into the mobile environment, it is believed to be a very useful and effective method in conserving bandwidth and power consumptions. This is because, when the data item is cached, the mobile user can avoid requests for the same data if the data are valid. And this would lead to reduced transmissions, which implies better utilization of the nature of the wireless channel of limited bandwidth. The cached data are able to support disconnected or intermitted connected operations as well. In addition, this also leads to cost reduction if the billing is per KB data transfer (Lai, Tari, & Bertok, 2003). Caching has emerged as a fundamental technique especially in distributed systems, as it not only helps reduce communication costs but also offloads shared database servers. Generally, caching in a mobile environment is complicated by the fact that the caches need to be kept consistent at all time.

In this article, we describe the use of caching that allows coping with the characteristics of the mobile environment. We concentrate particularly on cache invalidation strategy, which is basically a type of caching strategy that is used to ensure that the data items that are cached in the mobile client are consistent in comparison to the ones that are stored on the server.

BACKGROUND

Caching at the mobile client helps in relieving the low bandwidth constraints imposed in the mobile environment (Kara & Edwards, 2003). Without the ability to cache data, there will be increased communication in the remote servers for data and this eventually leads to increased cost and, with the nature of an environment that is vulnerable to frequent disconnection, may also lead to higher costs (Leong & Si, 1997). However, the frequent disconnection and the mobility of clients complicate the issue of keeping the cache consistent with those that are stored in the servers (Chand, Joshi, & Misra, 2004).

Thus, when caching is used, ensuring data consistency is an important issue that needs considerable attention at all times (Lao, Tari, & Bertok, 2003). This is because the data that has been cached may have been outdated and no longer valid in comparison to the data from the corresponding servers or broadcast channel.

Figure 1 shows an illustration of a typical mobile environment that consists of mobile clients and servers, which are also know as mobile host (MH) and mobile support system (MSS) respectively. The mobile clients and servers communicate via a wireless channel within a certain coverage, known as cell (Chand, Joshi, & Misra, 2003; Cai & Tan, 1999). There are two approaches for sending a query in a mobile environment, which are: (a) The mobile clients are free to request data directly from the server via the wireless channel and the server will process and pass the desired data items back and (b) the mobile clients can tune into the broadcast channel to obtain the desired data items and download it to his/her mobile device. This can be illustrated in Figure 1a and Figure 1b respectively. The assumption is that updates are only able to occur at the server side and mobile clients can only have a read only feature.

CACHE INVALIDATION

Due to the important issue in the mobile environment, which is the ability to maintain data consistency, cache invalidation strategy is of utmost significance to ensure that the data items cached in the mobile client are consistent with those that are stored on the server. In order to ensure that data that are about to be used is consistent, a client must validate its cache prior to using any data from it.

Figure 1. Mobile environment architecture

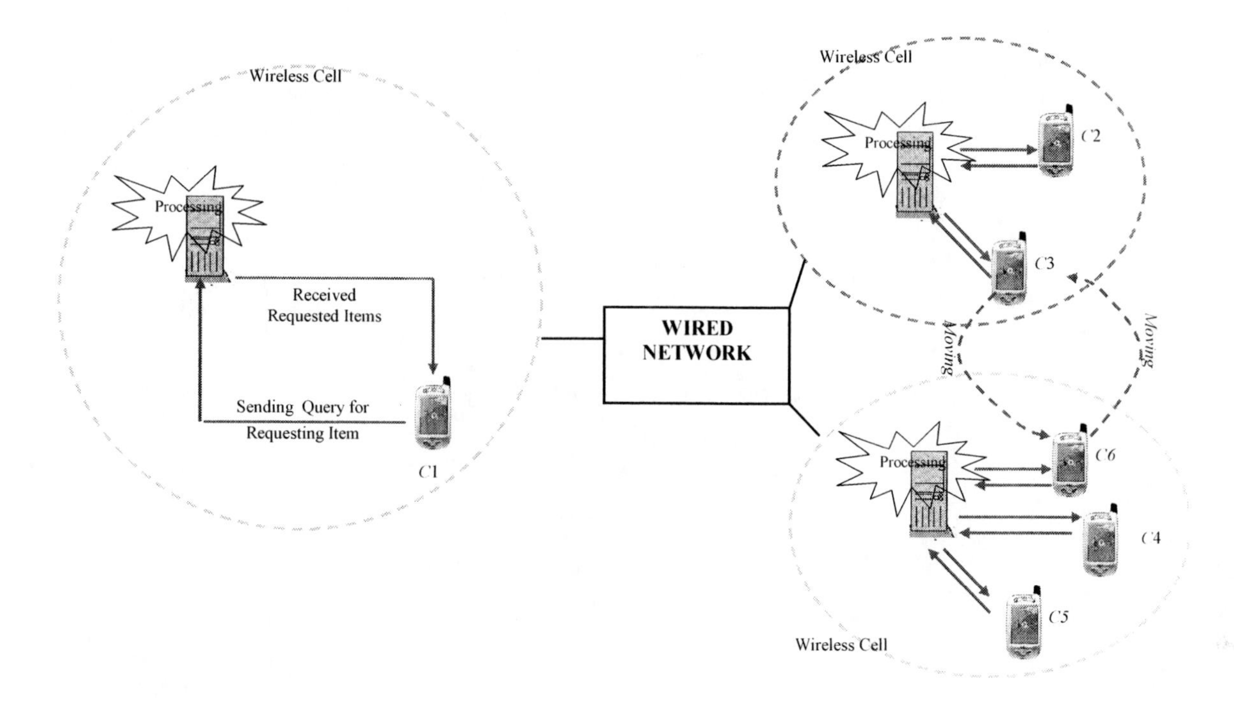

There are several distinctive and significant benefits that cache invalidation brings to a mobile computing environment. If cache data are not validated to check for consistency, it will become useless and out-of-date. However, if one can utilize the cache data then the benefits it may bring include energy savings—that is, by reducing the amount of data transfer—and in return result in cost savings.

Using Cache Invalidation in a Mobile Environment

This can be done by using the broadcasting concept in communicating cache validation information to mobile clients. The server broadcasts the cache information, which is known as cache invalidation report (IR), periodically on the air to help clients validate their cache to ensure they are still consistent and can be used. It appears that the broadcast mechanism is more appropriate for the mobile environment due to its characteristic of salability, which allows it to broadcast data to an arbitrary number of clients who can listen to the broadcast channel anytime (Lai, Tari, & Bertok, 2003). By using the broadcasting approach, whereby the server periodically broadcasts the IR to indicate the change data items, it eliminates the need to query directly to the server for a validation cache copies. The mobile clients would be able to listen to the broadcast channel on the IR and use them to validate their local cache respectively (Cao, 2002).

Although cache invalidation strategy is important in a mobile environment, it will be vulnerable to disconnection and the mobility of the clients. One of the main reasons that cause mobile clients frequent disconnection is the limited battery power, and that is why mobile clients often disconnect to conserve battery power. It may appear to be very expensive at times to validate the cache for clients that experience frequent disconnection, especially with narrow wireless links. Other drawbacks would include long query latency, which is associated with the need of the mobile client to listen to the channel for the next IR first before he is able to conclude whether the cache is valid or not before answering a query. Another major drawback is the unnecessary data items in the IR that the server keeps. This refers to data items that are not cached by any mobile clients. This is thereby wasting a significant amount of wireless bandwidth.

Example 1: A mobile client in a shopping complex denoted as *C*1 in Figure 2 wanted to know which store to visit by obtaining a store directory. The client has previously visited this store and already has a copy of the result in his cache. In order to answer a query, the client will listen to the IR that are broadcasted and use it for validation against its local cache to see if it is valid or not. If there is a valid cached copy that can be used in answering the query, which is getting the store directories, then the result will be returned immediately. Otherwise, if the store directories have changed and now contain new shops, then the invalid caches have to be refreshed via sending a query to the server (Elmagarmid et al., 2003). The server would keep track of the recently updated data and broadcast the up-to-date IR every now and

Figure 2. Using cache invalidation in a mobile environment

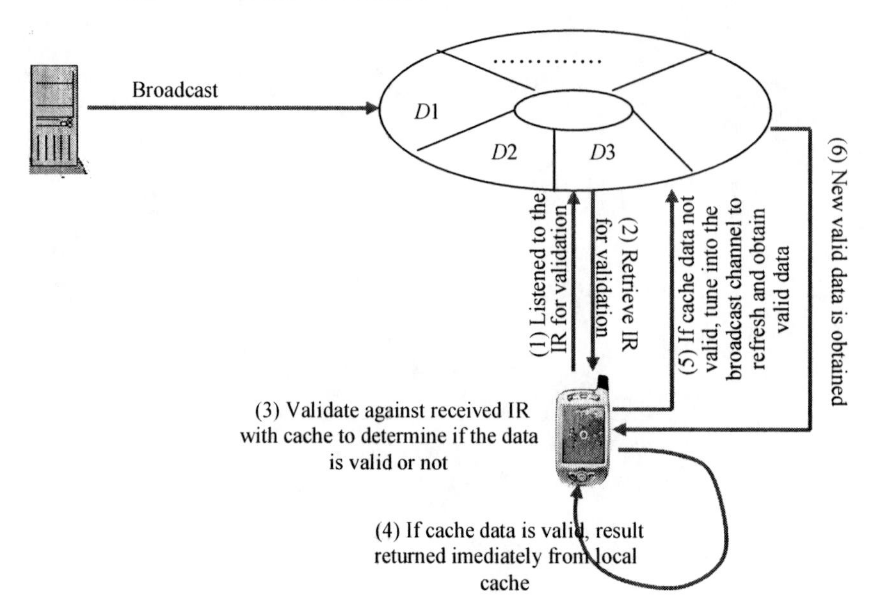

then for the clients to tune in. This can be done either by sending a request directly to the server (pull-based system) or tune into the broadcast channel (push-based system). Figure 2 illustrates an example of a push-based system.

In summary, effective cache invalidation strategies must be developed to ensure consistency between cached data in the mobile environment and the original data that are stored on the server (Hu & Lee, 1998).

Designing Cache Invalidation Strategies

It is important to produce an effective cache invalidation strategy to maintain a high level of consistency between cached data in the mobile devices with those that are stored on the server. In general, there are three possible basic ways in designing the invalidation strategies that described as follows (Hu & Lee, 1998).

Assuming the server is stateful, whereby it knows which data are cached and by which particular mobile clients. Whenever there are changes in the data item in the server, the server would send a message to those clients, which has cached that particular item that has been updated or changed. In this way, the server would be required to locate the mobile clients. However, there is a major limitation in this method, that is, particularly in cases of disconnection. This is because mobile clients that are disconnected cannot be contacted by the server and thus its cache would have turned into invalid upon reconnection. Another aspect is if the mobile client moves to a new location, it will have to notify the server of the relocation. And all these issues, such as disconnection and mobility, have to be taken into account because it incurs costs from sending the messages to and from the server via the uplink and downlink messages.

The second possible way is to have the mobile client query the server directly in order to verify the validity of the cache data prior to using it. This appears to be straightforward and easy, but one has to bear in mind that this method would generate a lot of uplink traffic in the network.

In contrast to stateful method, another way that can be taken into account in designing the invalidation is using a stateless method. This method is in direct opposite from the first possible way, which is the stateful method. In this method, the server is not aware of the state of the client's cache and the client location and disconnected status. The server would not care about all these but just periodically broadcasts an IR containing the data items that have been updated or changed in comparison to its previous state. Thus the server just keeps track of which item is recently updated and broadcasts them in an IR. Then only the client determines whether its cache is valid or not by validating it against the IRs that are broadcasted on the wireless channel.

Another challenging issue that involves determining an efficient invalidation strategy is to optimize the organization of the IR. Commonly, a large-sized report provides more information and appears to be more effective. But publishing a large report also brings drawbacks, such as implying a long latency for mobile clients to listen to the report due to the low bandwidth wireless channel. There have been several methods proposed in addressing the report optimization issue in other works, such as using the dual report scheme and bit sequence scheme (Tan, Cai, & Ooi, 2001; Elmagarmid et al., 2003).

Figure 3. Architecture of location dependent query processing

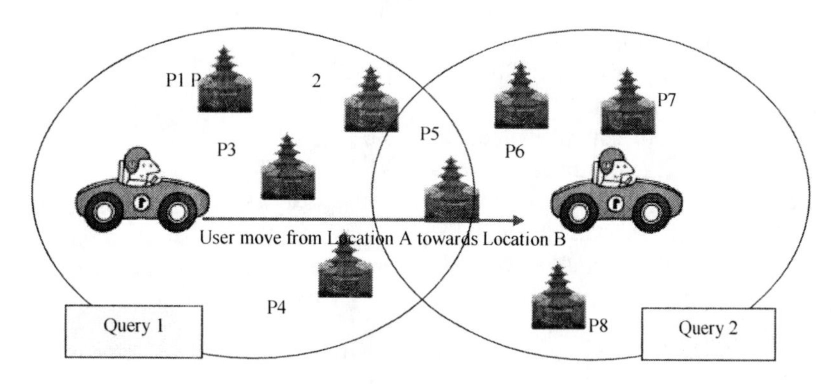

Location-Dependent Cache Invalidation

Due to the fact that mobile users in a typical mobile environment move around frequently by changing location has opened up a new challenge of answering queries that is dependent on the current geographical coordinates of the users (Barbara, 1999; Waluyo, Srinivasan, & Taniar, 2005). This is known as location dependent queries (Kottkamp & Zukunft, 1998). In this location dependent query, the server would produce answers to a query based on the location of the mobile client issuing the query. Thus, a different location may sometimes yield a different result even though the query is taken from a similar source.

Figure 3 depicts an illustration of a location dependent query processing. This shows that when the mobile client is in Location *A*, the query would return a set of results and when the mobile client moves towards a new Location *B*, another set of results will be returned. However there are cases of results overlapping between nearby locations. An example of a location dependent query can be: "Find the nearest restaurants from where I am standing now." This is an example of static object whereby restaurants are not moving. An example of a dynamic object would be: "What is the nearest taxi that will pass by me" (Lee et al., 2002).

With the frequent movement of mobile users, very often the mobile clients would query the same server to obtain results, or with the frequent movement of mobile users from location to location, very often the mobile clients would suffer from scarce bandwidth and frequent disconnection, especially when suddenly moved towards a secluded area (Jayaputera & Taniar, 2005). Hence, is essential to have data caching that can cope with cases of frequent disconnections. And often data may have become invalid after a certain point of time, especially in the area of location dependent.

Example 2: A mobile user who is in Location *A*, cached the results of the nearby vegetarian restaurants in Location *A*. As he moves to Location *B*, he would like another list of nearby vegetarian restaurants. The user is sending a query to the same source, but the results returned are different due to location dependent data. And because there are data previously cached, this data—which is the result obtained when he is in Location *A*—would become invalid since he has now moves to Location *B*.

Hence, location dependent cache invalidation serves the purpose of maintaining the validity of the cached data when the mobile client moves from one location to another location (Zheng, Xu, & Lee, 2002). The emergence of this location dependent cache invalidation is due to mobile client's movement and thus the data value for a data item is actually dependent on the geographical location. Hence, traditional caching that does not consider geographical location is inefficient for location dependent data.

There are both advantages and disadvantages of location dependent cache invalidation. The major benefit that the attached invalidation information provided is that it provides a way for the client to be able to check for validity of cached data in respect to a certain location. These are necessary, especially in cases of when the mobile client wishes to issue the same query later when he/she moves to a new location. Another situation for the importance in checking the validity of the cached data is that because mobile clients keep on moving even right after they submit a query, they would have arrived to a new location when the results are returned. This may occur if there is a long delay in accessing data. Thus, if this two situations occur, then it is significant to validate the cached data because it may have become invalid (Zheng, Xu, & Lee, 2002).

FUTURE TRENDS

There have been several researches done in the area of exploring cache invalidation in a mobile environment. The usage of cache invalidation has obviously provoked extensive complicated issues. There are still many limitations of the nature of the mobile environment as well as mobility of the users that generate a lot of attention from research in finding a good cache strategy that can cope well with frequent disconnection and low power consumption.

In the future, it is critical to build an analytical model to get a better understanding of how cache invalidation works and how well it can cope in the mobile environment. Developing caching strategies that support cache invalidation for a multiple channel environment is also desirable, whereby a mixture of broadcast and point-to-point channels are being used. Including a dynamic clustering is also beneficial in order to allow the server to group data items together as their update changes. Besides these, further investigation on other cache replacement policies, as well as granularities issues, is also beneficial.

Due to the non-stop moving clients, further research on adapting cache invalidation into location dependent data is favorable. Another possible issue that could open up for future work may involve minimizing the waiting time for the mobile client in acquiring the IR, since the mobile client has to obtain an IR prior to their cache being validated. Thus, it is essential to be able to reduce waiting time. Another aspect is due to wireless channels that are often error prone due to their instability, bandwidth, and so on. Thereby, having techniques to handle errors in a mobile environment is definitely helpful. Last but not least, having further study on integrating several different strategies to obtain a more optimal solution in coping with mobile environment is advantageous.

CONCLUSION

Although there is a significant increase in the popularity of mobile computing, there are still several limitations that are inherent, be it the mobile device itself or the environment itself. These include limited battery power, storage, communication cost, and bandwidth problems. All these have become present challenges for researchers to address.

In this article, we have described the pros and cons of adopting cache invalidation in a mobile environment. We include adapting cache invalidation strategy in both location and non-location dependent queries. Discussion regarding the issue in designing cache invalidation is also provided in a preliminary stage. This article serves as a valuable starting point for those who wish to gain some introductory knowledge about the usefulness of cache invalidation.

REFERENCES

Barbara, D., & Imielinski, T. (1994). Sleepers and workaholics: Caching strategies in mobile environments. *MOBIDATA: An Interactive Journal of Mobile Computing*, *1*(1).

Cai, J., & Tan, K. L. (1999). Energy efficient selective cache invalidation. *Wireless Networks*, *5*(6), 489-502.

Chan, B. Y., Si, A., & Leong, H. V. (1998). Cache management for mobile databases: Design and evaluation. In *Proceedings of the International Conference on Data Engineering (ICDE)* (pp. 54-63).

Chand, N., Joshi, R., & Misra, M. (1996). Energy efficient cache invalidation in a disconnected mobile environment. In *Proceedings of the Twelfth International Conference on Data Engineering* (pp.336-343).

Cao, G. (2003). A scalable low-latency cache invalidation strategy for mobile environment. *IEEE Transaction on Knowledge and Data Engineering (TKDE)*, *15*(2), 1251-1265.

Cao, G. (2002). On improving the performance of cache invalidation in mobile environment. *Mobile Networks and Applications*, *7*(4), 291-303.

Deshpande, P. M., & Ramasamy, K. (1998). Caching multidimensional queries using chunks. In *Proceedings of the ACM SIGMOD Conference on Management of Data* (pp. 259-270).

Dong Jung, Y. H., You, J., Lee, W., & Kim, K. (2002). Broadcasting and caching policies for location dependent queries in urban areas. In *Proceedings of the 2nd International Workshop on Mobile Commerce* (pp. 54-60).

Elmagarmid, A., Jing, J., Helal, A., & Lee, C. (2003). Scalable cache invalidation algorithms for mobile data access. *IEEE Transaction on Knowledge and Data Engineering (TKDE)*, *15*(6), 1498-1511.

Hu, Q., & Lee, D. (1998). Cache algorithms based on adaptive invalidation reports for mobile environment. *Cluster Computing*, pp. 39-48.

Hurson A.R., & Jiao, Y. (2005). Data broadcasting in mobile environment. In D. Katsaros, A. Nanopoulos, & Y. Manolopaulos (Eds.), *Wireless information highways* (Chapter 4). Hershey, PA: IRM Press.

Imielinski, T., & Badrinath, B. (1994). Mobile wireless computing: Challenges in data management. *Communications of the ACM, 37*(10), 18-28.

Jayaputera, J., & Taniar, D. (2005). Data retrieval for location-dependent queries in a multi-cell wireless environment. *Mobile Information Systems*, *1*(2), 91-108.

Kara, H., & Edwards, C. (2003). A caching architecture for content delivery to mobile devices. In *Proceedings of the 29th EUROMICRO Conference: New Waves in System Architecture (EUROMICRO'03)*.

Kottkamp, H.-E., & Zukunft, O. (1998). Location-aware query processing in mobile database systems. In *Proceedings of ACM Symposium on Applied Computing* (pp. 416-423).

Lai, K.Y., Tari, Z., & Bertok, P. (2003). Cost efficient broadcast based cache invalidation for mobile environment. In *Proceedings of the 2003 ACM symposium on Applied Computing* (pp. 871-877).

Lee, G., Lo, S-C., & Chen, A. L. P. (2002). Data allocation on wireless broadcast channels for efficient query processing. *IEEE Transactions on Computers, 51*(10), 1237-1252.

Leong, H. V., & Si, A. (1997). Database caching over the air-storage. *The Computer Journal, 40*(7), 401-415.

Lee, D-L., Zhu, M., & Hu, H. (2005). When location-based services meet databases. *Mobile Information Systems, 1*(2), 81-90.

Lee, D. K., Xu, J., Zheng, B., & Lee, W-C. (2002). Data management in location-dependent information services. *IEEE Pervasive Computing, 2*(3), 65-72.

Prabhajara, K., Hua, K. A., & Oh, J.H. (2000). Multi-level, multi-channel air cache designs for broadcasting in a mobile environment. In *Proceedings of the 16th International Conference on Data Engineering* (pp. 167-186).

Park, K., Song, M., & Hwang, C. S. (2004). An efficient data dissemination schemes for location dependent information services. In *Proceedings of the First International Conference on Distributed Computing and Internet Technology (ICDCIT 2004)* (Vol. 3347, pp.96-105). Springer-Verlag.

Tan, K. L., Cai, J., & Ooi, B. C. (2001). An evaluation of cache invalidation strategies in wireless environment. *IEEE Transactions on Parallel and Distributed Systems, 12*(8), 789-807.

Waluyo, A. B., Srinivasan, B., & Taniar, D. (2005). Research on location-dependent queries in mobile databases. *International Journal on Computer Systems: Science and Engineering, 20*(3), 77-93.

Waluyo, A. B., Srinivasan, B., & Taniar, D. (2005). Research in mobile database query optimization and processing. *Mobile Information Systems, 1*(4).

Xu, J., Hu, Q., Tang, X., & Lee, D. L. (2004). Performance analysis of location dependent cache invalidation scheme for mobile environments. *IEEE Transaction on Knowledge and Data Engineering (TKDE), 15*(2), 125-139.

Xu, J., Hu, Q., Lee, D. L., & Lee, W.-C. (2000). SAIU: An efficient cache replacement policy for wireless on-demand broadcasts. In *Proceedings of the 9th International Conference on Information and Knowledge Management* (pp. 46-53).

Xu, J., Hu, Q., Lee, W.-C., & Lee, D. L. (2004). Performance evaluation of an optimal cache replacement policy for wireless data dissemination. *IEEE Transaction on Knowledge and Data Engineering (TKDE), 16*(1), 125-139.

Yajima, E., Hara, T., Tsukamoto, M., & Nishio, S. (2001). Scheduling and caching strategies for correlated data in push-based information systems. *ACM SIGAPP Applied Computing Review, 9*(1), 22-28.

Zheng, B., Xu, J., & Lee, D.L. (2002, October). Cache invalidation and replacement strategies for location-dependent data in mobile environments. *IEEE Transactions on Computers, 51*(10), 1141-1153.

KEY TERMS

Caching: Techniques of temporarily storing frequently accessed data designed to reduce network transfers and therefore increase speed of download

Cache Invalidation Strategy: A type of caching strategy that is used to ensure that the data items that are cached in the mobile client are consistent in comparison to the ones that are stored on the server.

Caching Management Strategy: A strategy that relates to how client manipulates the data that has been cached in an efficient and effective way by maintaining the data items in a client's local storage.

Invalidation Report (IR): An informative report in which the changed data items are indicated; it is used for mobile clients to validate against their cache data to check if it is still valid or not.

Location-Dependent Cache Invalidation: maintaining the validity of the cached data when the mobile client changes locations.

Mobile Environment: Refers to a set of database servers, which may or may not be collaborative with one another, that disseminate data via wireless channels to multiple mobile users.

Pull-Based Environment: Also known as an on demand system, which relates to techniques that enable the server to process request that are sent from mobile users.

Push-Based Environment: Also known as a broadcast system where the server would broadcast a set of data to the air for a population of mobile users to tune in for their required data.

Communicating Recommendations in a Service-Oriented Environment

Omar Khadeer Hussain
Curtin University of Technology, Australia

Elizabeth Chang
Curtin University of Technology, Australia

Farookh Khadeer Hussain
Curtin University of Technology, Australia

Tharam S. Dillon
University of Technology, Sydney, Australia

INTRODUCTION

The Australian and New Zealand Standard on Risk Management, AS/NZS 4360:2004 (Cooper, 2004), states that risk identification is the heart of risk management. Hence risk should be identified according to the context of the transaction in order to analyze and manage it better. Risk analysis is the science of evaluating risks resulting from past, current, anticipated, or future activities. The use of these evaluations includes providing information for determining regulatory actions to limit risk, and for educating the public concerning particular risk issues. Risk analysis is an interdisciplinary science that relies on laboratory studies, collection, and exposure of data and computer modeling.

Chan, Lee, Dillon, and Chang (2002) state that the advent of the Internet and its development has simplified the way transactions are carried out. It currently provides the user with numerous facilities which facilitate transaction process. This process evolved into what became known as e-commerce transactions. There are two types of architectures through which e-commerce transactions can be conducted. They are: (a) client-server business architecture, and (b) peer-to-peer business architecture.

In almost all cases, the amount of risk involved in a transaction is important to be understood or analyzed before a transaction is begun. This also applies to the transactions in the field of e-commerce and peer-to-peer business. In this article we will emphasize transactions carried out in the peer-to-peer business architecture style, as our aim is to analyze risk in such transactions carried out in a service-oriented environment.

Peer-to-peer (P2P) architecture is so called because each node has equivalent responsibilities (Leuf, 2002). This is a type of network in which each workstation or peer has equivalent capabilities and responsibilities. This differs from client/server architecture, in which some computers or central servers are dedicated to serving others. As mentioned by Oram (2001), the main difference between these two architectures is that in peer-to-peer architecture, the control is transferred back to the clients from the servers, and it is the responsibility of the clients to complete the transaction. Some of the characteristics of peer-to-peer or decentralized transactions are:

1. There is no server in this type of transaction between peers.
2. Peers interact with each other directly, rather than through a server, as compared to a centralized transaction where the authenticity can be checked.
3. Peers can forge or create multiple identities in a decentralized transaction, and there is no way of checking the identity claimed by the peer to be genuine or not.

The above properties clearly show that a decentralized transaction carries more risks and hence merits more detailed investigation. Similarly, in a service-oriented peer-to-peer financial transaction, there is the possibility of the trusted agent engaging in an untrustworthy manner and in other negative behavior at the buyer's expense, which would result in the loss of the buyer's resources. This possibility of failure and the degree of possible loss in the buyer's resource is termed as risk. Hence, risk analysis is an important factor in deciding whether to proceed in an interaction or not, as it helps to determine the likelihood of loss in the resources involved in the transaction.

Risk analysis by the trusting agent before initiating an interaction with a trusted agent can be done by:

- determining the possibility of failure of the interaction, and
- determining the possible consequences of failure of the interaction.

Figure 1. The riskiness scale and its associated levels

Riskiness Levels	Magnitude of Risk	Riskiness Value	Star Rating
Unknown Risk	-	-1	Not Displayed
Totally Risky	91-100% of Risk	0	Not Displayed
Extremely Risky	71-90% of Risk	1	From to
Largely Risky	70% of Risk	2	From to
Risky	26-50% of Risk	3	From to
Largely Unrisky	11-25% of Risk	4	From to
Unrisky	0-10% of Risk	5	From to

The trusting agent can determine the possibility of failure in interacting with a probable trusted agent either by:

a. considering its previous interaction history with the trusted agent, if any, in the context of its future interaction, or
b. soliciting recommendations for the trusted agent in the particular context of its future interaction, if it does not have any previous interaction history with it.

When the trusting agent solicits for recommendations about a trusted agent for a particular context, then it should consider replies from agents who have previous interaction history with the trusted agent in that particular context. The agents replying back with the recommendations are called the *recommending agents*. But it is possible that each recommending agent might give its recommendation in its own way, and as a result of that, it will be difficult for the trusting agent to interpret and understand what each element of the recommendations mean. Hence, a standard format for communicating recommendations is needed so that it is easier for the trusting agent to understand and assimilate them. Further, the trusting agent has to determine whether the recommendation communicated by the recommending agent is trustworthy or not before considering it.

In this article we propose a methodology by which the trusting agent classifies the recommendation according to its trustworthiness. We also define a standard format for communicating recommendations, so that it is easier for the trusting agent to interpret and understand them.

BACKGROUND

Security is the process of providing sheltered communication between two communicating agents (Singh & Liu, 2003; Chan et al., 2002). We define *risk* in a peer-to-peer service-oriented environment transaction as the likelihood that the transaction might not proceed as expected by the trusting agent in a given context and at a particular time once it begins resulting in the loss of money and the resources involved in it. The study of risk cannot be compared with the study of security, because securing a transaction does not mean that there will be no risk in personal damages and financial losses. Risk is a combination of:

a. the uncertainty of the outcome; and
b. the cost of the outcome when it occurs, usually the loss incurred.

Analyzing risk is important in e-commerce transactions, because there is a whole body of literature based on rational economics that argues that the decision to buy is based on the risk-adjusted cost-benefit analysis (Greenland, 2004). Thus it commands a central role in any discussion of e-commerce that is related to a transaction. Risk plays a central role in deciding whether to proceed with a transaction or not. It can broadly be classified as an attribute of decision making that reflects the variance of its possible outcomes.

Peer-to-peer architecture-type transactions are being described as the next generation of the Internet (Orlowska, 2004). Architectures have been proposed by researchers (Qu & Nejdl, 2004; Schmidt & Parashar, 2004; Schuler, Weber, Schuldt, & Schek, 2004) for integrating Web services with peer-to-peer communicating agents like Gnutella. However, as discussed earlier, peer-to-peer-type transactions suffer from some disadvantages, and risk associated in the transactions is one of them. Hence, this disadvantage has to be overcome so that they can be used effectively with whatever service they are being integrated with.

Through the above discussion, it is evident that risk analysis is necessary when a transaction is being conducted in a peer-to-peer architecture environment. As mentioned before, risk analysis by the trusting agent can be done by determining the possibility of failure and the possible consequences of failure in interacting with a probable trusted agent. In order for the trusting agent to determine and quantify the possibility of failure of an interaction, we define the term riskiness. Riskiness is defined as the numerical value that is assigned by the trusting agent to the trusted agent after the interaction, which shows the level of possibility of failure of an interaction on the riskiness scale. The numerical value corresponds to a level on the riskiness scale, which gives an indication to other agents about the level of possibility of failure in interacting with a particular trusted agent. The riskiness scale as shown in Figure 1 depicts different levels of possibility of failure that could be present in an interaction.

The riskiness value to the trusted agent is assigned by the trusting agent after assessing the level of un-commitment in its actual behavior with respect to the promised commitment. The promised commitment is the expected behavior by which the trusted agent was supposed to behave in the interaction. The expected behavior is defined by the trusting agent according to its criteria, before starting its interaction with the trusted agent. The actual behavior is the actual commitment that the trusted agent showed or behaved in the interaction. Criteria are defined as the set of factors or bases that the trusting agent wants in the interaction and later against which it determines the un-committed behavior of the trusted agent in the interaction.

If the trusting agent has interacted previously with the trusted agent in the same context as its future interaction, then it can determine the possibility of failure in interacting with it by analyzing the riskiness value that it assigned to the trusted agent in their previous interaction. If a trusting agent has not interacted previously with a trusted agent in a particular context, then it can determine the possibility of failure in their future interaction, by soliciting for its recommendation from other agents who have dealt with the same trusted agent previously in the same context as that of the trusting agent's future interaction. As mentioned earlier the agents giving recommendations are called *recommending agents*.

But it would be difficult for the trusting agent to assimilate the data that it gets from the recommending agents and draw a conclusion if each agent gives its recommendation in its own format. It would rather be easier for the trusting agent if the recommendations came in a standard set or format that enables the trusting agent to ascertain the meaning of each element in the recommendations.

But even in the same context, each recommending agent might have different criteria in its interaction with the trusted agent. Consequently the riskiness value that it recommends for the trusted agent depends on its assessment of un-commitment in the trusted agent's actual behavior with respect to its expected behavior in those criteria. It would be baseless for the trusting agent to consider recommendations for a trusted agent in criteria of assessment which are not similar to those in its future interaction with that particular trusted agent. Additionally it is highly unlikely that the recommendations provided by the recommending agents would be completely reliable or trustworthy. Some agents might be communicating un-trustworthy recommendations. The trusting agent has to consider all these scenarios before it assimilates the recommendations from the recommending agents to assess the risk in dealing with a trusted agent.

In order to propose a solution to these issues, in the next sections we will define a methodology by which the trusting agent can classify the recommendations according to its trustworthiness. We also define a standard format for communicating recommendations so that the trusting agent can ascertain the meaning of each element of the recommendations before assimilating them, and consider only those whose criteria are of interest to it in its future interaction.

CLASSIFYING THE RECOMMENDATIONS AS TRUSTWORTHY OR UN-TRUSTWORTHY

As stated earlier, it is possible that the recommendation communicated by a recommending agent might not be trustworthy. The recommending agent might be communicating recommendations that the trusting agent finds to be incorrect or misleading after its interaction with the trusted agent. So the trusting agent has to determine whether the recommendation is trustworthy or not before assimilating it. To achieve that, we propose that each recommending agent is assigned a *riskiness value* while giving recommendations called *riskiness of the recommending agent (RRP).*

The riskiness value of the recommending agent is determined by the difference between:

- the riskiness value that the trusting agent found out for the trusted agent after interacting with it, and
- the riskiness value that the recommending agent recommend for the trusted agent to the trusting agent when solicited for.

When the trusting agent broadcasts a query soliciting for recommendations about a trusted agent in a particular context, it will consider replies from those agents who have interacted with that particular trusted agent previously in that same context. Hence, whatever riskiness value the recommending agents recommend to the trusting agent will be greater than -1, as -1 on the riskiness scale represents the riskiness value as *Unknown Risk,* which cannot be assigned to any agent after an interaction. After an interaction a value only within the range of (0, 5) on the riskiness scale can be assigned. So, the maximum range for the riskiness value of the recommending agent (RRP) is between (-5, 5), since this is the maximum possible range of difference between the riskiness value that the trusting agent might determine for the trusted agent after its interaction with it and the riskiness value recommended by the recommending agent for the trusted agent to the trusting agent.

We adopt the approach mentioned by Chang, Dillon, and Hussain (2006) which states that a recommending agent is said to be communicating trustworthy recommendations if its riskiness value while giving recommendations (RRP) is in the range of (-1, 1). A value within this range will state that there is a difference of one level in the riskiness value that the trusting agent found out after the interaction and what the recommending agent suggested for the trusted agent. If the riskiness value of the recommending agent is beyond those levels, then it hints that the recommending agent is giving recommendations that the trusting agent finds to vary a lot after the interaction, and there is at least a difference of two levels on the riskiness scale between what the trusting agent found and what the recommending agent recommended. An agent whose Riskiness value while giving recommendation (RRP) is beyond the level of (-1, 1) is said to be an *Un-trustworthy* recommending agent. Chang et al. (2006) mention that the trusting agent should only consider recommendations from agents who are either *Trustworthy* or *Unknown* in giving recommendations and leave the recommendation from agents who are *Un-trustworthy* in giving them. Hence the recommendation from agents with riskiness values beyond the levels of (-1, 1) will not be considered.

If the recommending agent gives more than one recommendation in an interaction, then its riskiness value while giving recommendation can be determined by taking the average of the difference of each recommendation.

Hence riskiness of the recommending agent (RRP) =

$$\frac{1}{N}\sum_{i=1}^{N}(Ti - Ri)$$

where Ti is the riskiness value found out by the trusting agent after the interaction, Ri is the riskiness value recommended by the recommending agent for the trusted agent, and

N is the number of recommendations given by a particular agent.

DEFINING A STANDARD FORMAT FOR COMMUNICATING RECOMMENDATIONS

Whenever a trusting agent interacts with a trusted agent, a risk relationship is formed between them. The risk relationship consists of a number of factors. These include the trusting agent:

1. considering its previous experience with the trusted agent in the context of its future interaction with it, or soliciting recommendations for the trusted agent in the context of its future interaction if they have not interacted before;
2. determining the riskiness value of the trusted agent according to its previous interactions or recommendations;
3. predicting the future riskiness value of the trusted agent, within the time period of its interaction with the trusted agent; or
4. taking into consideration the cost of the interaction and assigning a riskiness value to the trusted agent after completing the interaction.

A risk relationship exists between a trusting agent and a trusted agent only if they interact with each other. Between a trusting agent and a trusted agent, there might exist one or more risk relationships depending on the number of times they interact with each other. For each interaction a new risk relationship is formed. Hence, the trusting agent and the trusted agent are in a ternary association (Eriksson & Penker, 2000) with the risk relationship as shown in Figure 2. But the risk relationship exists only if the trusting agent interacts with the trusted agent, and hence it is realized by a transaction between them. The risk relationship in turn is dependent on a number of factors. Figure 2 shows the risk relationship and the factors on which it is dependent.

As mentioned earlier, the trusting agent will consider recommendations from agents who have previous interaction history with the particular trusted agent in question in context similar to that of its future interaction with the trusted agent. A recommending agent when solicited for recommendation by a trusting agent for a particular trusted agent in a particular context will give its recommendation depending on its previous interaction with the particular trusted agent in the particular context. In other terms, it gives the risk relationship that it had formed with the trusted agent in that particular interaction as its recommendation. We propose that when any trusting agent solicits for recommendations for a trusted agent, then the recommending agents should give their replies in a standard format so that it is easier for the trusting agent to interpret their recommendation. The standard format is represented by a *risk set.*

The risk set is formed from the risk relationship that the recommending agent had from the last time it interacted with the particular trusted agent. Alternately, a risk set exists between any two agents only if there is a risk relationship between them, and hence it is dependent on the risk relationship as shown in Figure 2.

Once the risk relationship between any two agents has been established, then a risk set can be defined. The risk set contains the same elements as that of the risk relationship but in an ordered way. The order of appearance of the

Figure 2. Risk relationship that exists between any two agents

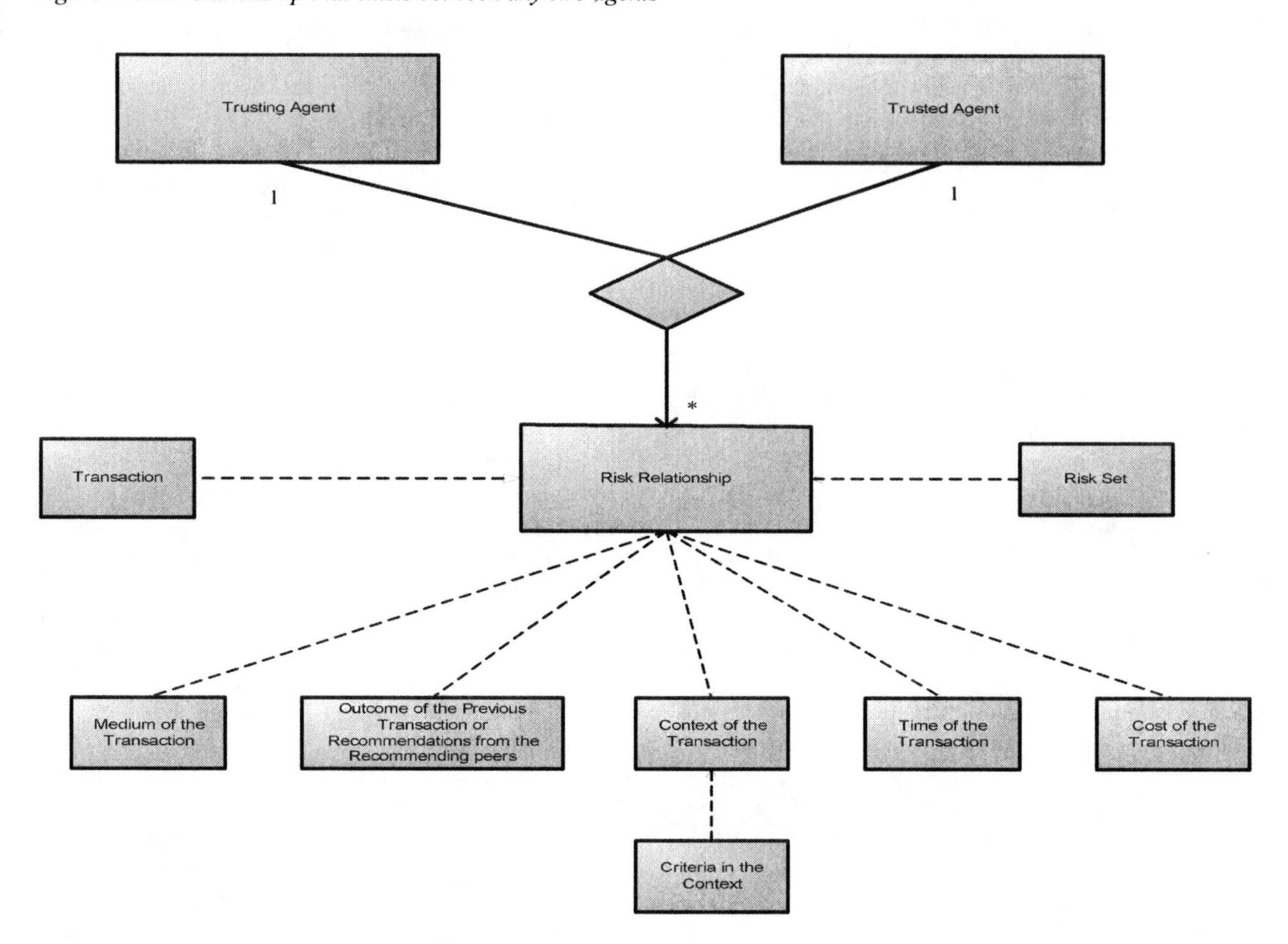

elements of the risk set is: {TP1, TP2, Context, CR, R', (Criteria, Commitment level), R, Cost, Start time, End time, RRP}, where:

- **TP1** denotes the trusting agent in the interaction. This is also the recommending agent while communicating recommendations.
- **TP2** denotes the trusted agent in the interaction.
- **Context** represents the context of the interaction.
- **CR** represents the 'current riskiness' value of the trusted agent before its interaction with the recommending agent. This is achieved either by the previous interaction history between the recommending agent and the trusted agent in the same context or by soliciting recommendations for the trusted agent by the recommending agent before its interaction,
- **R'** shows the predicted riskiness value of the trusted agent as determined by the recommending agent within the time slot of its interaction.
- **(Criteria, Commitment Level)** shows the factors or bases that the recommending agent used in its interaction with the trusted agent to assign it a riskiness value. These criteria are necessary to mention while giving recommendations, so that a trusting agent who asks for recommendation knows the factors on which this particular trusted agent has assigned the recommended riskiness value and only considers those recommendations which are of interest to it according to its criteria. Commitment level specifies whether the particular criterion was fulfilled by the trusted agent or not. A value of either 0 or 1 assigned here is based on its commitment. A value of 0 signifies that the criterion was not fulfilled by the trusted agent according to the expected behavior, whereas a value of 1 signifies that the criterion was fulfilled according to the expected behavior.
- **R** is the riskiness value assigned by the recommending agent to the trusted agent after its interaction. As discussed earlier, the riskiness value is determined after the interaction by assessing the level of un-commitment in the trusted agent's actual behavior with respect to the expected behavior.

- **Cost** represents the cost of the interaction.
- **Start Time** is the time at which the recommending agent started the interaction with the trusted agent.
- **End Time** is the time at which the interaction of the recommending agent ended with the trusted agent.
- **RRP** is the riskiness value of the recommending agent while giving recommendations. This value determines whether the recommendation is trustworthy or not.

To highlight the advantages of communicating the recommendations in a standard format by risk set and the usefulness of its elements, let us consider a scenario in which Bob wants to interact with a logistic company 'LC'. The context of its interaction with the logistic company is to 'transport its goods' on April 15, 2006. Let us represent the context as 'Transport'. The goods are worth $1,500. The criteria put up by Bob in its interaction with the logistic company 'LC' are:

1. Packing the goods properly.
2. Pickup of the goods on time by the logistic company.
3. Delivering the goods to the correct address on time as promised.

For explanation sake the criteria in the interaction are represented by C1, C2, and C3 respectively. The trusting agent Bob does not have any previous dealings with the logistic company 'LC', and in order to analyze the risk before proceeding in a business transaction with it, Bob solicits for recommendations from other agents who have previously dealt with logistic company 'LC' in a context similar to that in this interaction. The agents who had interacted previously in the same context give their recommendations to Bob in the form of a risk set, which relates to their previous interactions with the trusted agent 'LC'. Let us suppose that Bob receives recommendations from agents 'A', 'B', 'C', and 'D' in the form of risk set.

The recommendation from agent 'A' is:

{Agent 'A', Logistic Company 'LC', Transport, 5, 5, ((C3, 1) (C1, 0), (C2, 1)), 5, $5000, 15/07/2005, 22/07/2005, 0.8}

Similarly, recommendation from agent 'B' is:

{Agent 'B', Logistic Company 'LC', Transport, 3, 3, ((C5, 1) (C6, 0)), 3, $1000, 1/02/2006, 22/02/2006, -1}

Recommendation from agent 'C' is:

{Agent 'C', Logistic Company 'LC', Transport, 2, 3, ((C1, 0) (C2, 0), (C3, 1)), 2, UNKNOWN, 01/04/2006, 03/04/2006, -2.5}

Recommendation from agent 'D' is:

{Agent 'D', Logistic Company 'LC', Transport, 4, 4, ((C1, 1) (C2, 1), (C3, 1)), 5, UNKNOWN, 07/04/2006, 10/04/2006, 0}

The properties to be followed while forming or representing the risk set are:

1. The elements should be represented in the same order as defined above.
2. Each element of the risk set is mandatorally to be defined except the element 'cost'.
3. Each criteria and its commitment level should be represented inside a single '(', ')' bracket, separated by a comma, so as to differentiate it from the other criteria and the elements of the risk set.
4. If the cost is not represented, then it should be written as UNKNOWN.
5. If the riskiness value of the recommending agent (RRP) is not known, then it should be represented as UNKNOWN.
6. The elements of the risk set should be separated by a comma ','.

From the above recommendations it can be seen that:

- The recommendation from the recommending agent 'A' is trustworthy and exactly according to the criteria of the trusting agent's future interaction with the trusted agent. But there is a huge gap in time between the recommending agent's interaction with the trusted agent and the future interaction of the trusting agent with the trusted agent.
- The recommendation from recommending agent 'B' is trustworthy, but the criteria of its recommendation does not match with those of the trusting agent's future interaction with the trusted agent and so it is baseless for it to consider this recommendation.
- The criteria of recommending agent 'C' is similar to those of the trusting agent, but the riskiness value of the recommending agent 'C' (RRP) is not within the range of (-1,1). So it can be concluded that this is an un-trustworthy recommendation and it will not be considered by the trusting agent.
- The recommendation from recommending agent 'D' is trustworthy and in the criterions that the trusting agent wants in its interaction.

Hence, as can be seen, the trusting agent, by making use of the risk set, can interpret the meaning of each element of the recommendation that would help it to understand the recommendation better and assimilate it easily.

The advantages of communicating the recommendations in the form of a risk set are:

1. The recommendations come in a standard format and it is easier for the trusting agent to understand them.
2. Even if the context of two interactions is the same, the criteria might differ considerably, and the riskiness value assigned to each interaction is in accordance with its corresponding criteria. Therefore, while giving recommendations, the recommending agent must specify the criteria apart from the context of the interaction. By doing so, the trusting agent who is soliciting for recommendations might know the exact criteria in which the trusted agent was assigned the riskiness value recommended by the recommending agent and consider only those recommendations that are of interest to it. The risk set communicates the criteria along with the recommendations and also specifies the commitment level of the trusted agent in those criteria.
3. The risk set specifies the riskiness value of the trusted agent as determined by recommending agent before starting an interaction with it (CR), the predicted future riskiness value of the trusted agent within the time space of its interaction (R'), and the actual riskiness value of the trusted agent determined by the recommending agent (R) after its interaction with it depending on the level of un-commitment in the actual behavior of the trusted agent with respect to the expected behavior.
4. The risk set specifies the time of the interaction between the recommending agent and the trusted agent. As defined in the literature, risk is dynamic and it keeps on changing. It is not possible for an agent to have the same impression of another agent that it had at a given point of time. Hence, the trusting agent should give more weight to those recommendations which are near to the time slot of its interaction as compared to the far recent ones while assimilating them. This is achieved by using the proposed risk set, which specifies the context along with the accessing criteria and the riskiness values of the trusted agent according to the time assigned, in an ordered way.

CONCLUSION

In this article we discussed the need to analyze risk that could be associated in a peer-to-peer financial transaction. Further we discussed how a trusting agent can assess the possible risk beforehand that could be present in interacting with a particular trusted agent. We proposed a methodology of classifying the recommendations according to its trustworthiness. Further we discussed the risk relationship that exists between a trusting agent and a trusted agent in the post-interaction phase, and ascertain the factors on which the risk relationship is dependent. From that relationship we defined the risk set, which is an ordered way of representing the details of the transaction between the agents. This risk set is utilized by the recommending agents while communicating recommendations to the trusting agents, so that it can be interpreted and understood easily.

REFERENCES

Chan, H., Lee, R., Dillon, T. S., & Chang, E. (2002). *E-commerce: Fundamentals and applications* (1st ed.). New York: John Wiley & Sons.

Chang, E., Dillon, T., & Hussain F. K. (2006). *Trust and reputation in service oriented environments* (1st ed.). New York: John Wiley & Sons.

Cooper, D. F. (2004). *The Australian and New Zealand standard on risk management, AS/NZS 4360:2004, tutorial notes: Broadleaf Capital International Pty Ltd.* Retrieved from http://www.broadleaf.com.au/tutorials/Tut_Standard.pdf

Eriksson, H., & Penker, M. (2000). *Business modeling with UML: Business patterns at work.* New York: John Wiley & Sons.

Greenland, S. (2004). Bounding analysis as an inadequately specified methodology. *Risk Analysis, 24*(5), 1085-1092.

Leuf, B. (2002). *Peer to peer collaboration & sharing on the Internet.* Pearson Education.

Oram, A. (2004). *Peer-to-peer: Harnessing the power of disruptive technologies.* Retrieved February 16, 2004, from http://www.oreilly.com/catalog/peertopeer/chapter/ch01.html

Orlowska, M. E. (2004). The next generation messaging technology—makes Web services effective. *Proceedings of the 6th Asia Pacific Web Conference* (pp. 13-19). Berlin/Heidelberg: Springer-Verlag.

Qu, C., & Nejdl, W. (2004). Interacting the Edutella/JXTA peer-to-peer network with Web services. *Proceedings of the International Symposium on Applications and the Internet (SAINT'04),* Tokyo, (pp. 67-73).

Schmidt, C., & Parashar, M. (2004). A peer-to-peer approach to Web service discovery. *World Wide Web Journal, 7*(2), 211-229.

Schuler, C., Weber, R., Schuldt, H., & Schek, H. (2004). Scalable peer-to-peer process management—the OSIRIS approach. *Proceedings of the 2nd International Conference on Web Services*, San Diego, (pp. 26-34).

Singh, A., & Liu, L. (2003). TrustMe: Anonymous management of trust relationships in decentralized P2P systems. *Proceedings of the 3rd IEEE International Conference on P2P Computing* (pp. 142-149), Linköping, Sweden.

KEY TERMS

Recommending Agent: An agent who gives its recommendation about a trusted agent to a trusting agent, when solicited for.

Risk Set: A standard format for giving recommendations by the recommending agents.

Riskiness Scale: A scale that represents different levels of risk that could be possible in an interaction.

Riskiness Value: A value that is assigned to the trusted agent by the trusting agent after its interaction with it. This value specifies a level of risk on the riskiness scale that the trusted agent deserves according to the level of un-committed behavior in its interaction with the trusting agent.

RRP: Stands for riskiness value of the recommending agent. This value is used to determine if the recommending agent is communicating trustworthy recommendations or not.

Trusted Agent: An agent with whom the trusting agent deals with and reposes its faith in.

Trusting Agent: An agent who controls the resources and interacts with another agent after reposing its faith in it.

Content Personalization for Mobile Interfaces

Spiridoula Koukia
University of Patras, Greece

Maria Rigou
University of Patras, Greece and
Research Academic Computer Technology Institute, Greece

Spiros Sirmakessis
Technological Institution of Messolongi and
Research Academic Computer Technology Institute, Greece

INTRODUCTION

The contribution of context information to content management is of great importance. The increase of storage capacity in mobile devices gives users the possibility to maintain large amounts of content to their phones. As a result, this amount of content is increasing at a high rate. Users are able to store a huge variety of content such as contacts, text messages, ring tones, logos, calendar events, and textual notes. Furthermore, the development of novel applications has created new types of content, which include images, videos, MMS (multi-media messaging), e-mail, music, play lists, audio clips, bookmarks, news and weather, chat, niche information services, travel and entertainment information, driving instructions, banking, and shopping (Schilit & Theimer, 1994; Schilit, Adams, & Want, 1994; Brown, 1996; Brown, Bovey, & Chen, 1997).

The fact that users should be able to store the content on their mobile phone and find the content they need without much effort results in the requirement of managing the content by organizing and annotating it. The purpose of information management is to aid users by offering a safe and easy way of retrieving the relevant content automatically, to minimize their effort and maximize their benefit (Sorvari et al., 2004).

The increasing amount of stored content in mobile devices and the limitations of physical mobile phone user interfaces introduce a usability challenge in content management. The physical mobile phone user interface will not change considerably. The physical display sizes will not increase since in the mobile devices the display already covers a large part of the surface area. Text input speed will not change much, as keyboard-based text input methods have been the most efficient way to reduce slowness. While information is necessary for many applications, the human brain is limited in terms of how much information it can process at one time. The problem of information management is more complex in mobile environments (Campbell & Tarasewich, 2004).

One way to reduce information overload and enhance content management is through the use of *context metadata*. Context metadata is information that describes the context in which a content item was created or received and can be used to aid users in searching, retrieving, and organizing the relevant content automatically. Context is any information that can be used to characterize the situation of an entity. An entity is a person, place, or object that is considered relevant to the interaction between a user and an application, including the user and the applications themselves (Dey, 2001). Some types of context are the *physical context,* such as time, location, and date; the *social context,* such as social group, friends, work, and home; and the *mental context,* which includes users' activities and feelings (Ryan, Pascoe, & Morse, 1997; Dey, Abowd, & Wood, 1998; Lucas, 2001).

By organizing and annotating the content, we develop a new way of managing it, while content management features are created to face efficiently the usability challenge. Context metadata helps the user find the content he needs by enabling single and multi-criteria searches (e.g., find photos taken in Paris last year), example-based searches (e.g., find all the video clips recorded in the same location as the selected video clip), and automatic content organization for efficient browsing (e.g., location-based content view, where the content is arranged hierarchically based on the content capture location and information about the hierarchical relationships of different locations).

DATE, TIME, LOCATION, AND PROXIMITY

While context can be characterized by a large number of different types of attributes, the contribution of context attributes to content management is of great importance. We focus on a small number of attributes, which are considered the most important in supporting content management and also have the most practical implementations in real products, such as date, time, location, and proximity (nearby Bluetooth devices). Bluetooth is a short-range wireless technology used to create personal area networks among user mobile devices and with other nearby devices.

The first two attributes, date and time, are the most common in use in a wide range of applications. They are used to organize both digital and analog content, and offer an easy way of searching and retrieving the relevant content automatically. For example, many cameras automatically add the date and time to photographs. Furthermore, the location where content is created is another useful attribute for searching the content (e.g., home, workplace, summer cottage). Mobile devices give users the possibility to create content in many different locations. Users can associate the location with the equivalent content in order to add an attribute to it that will enable them to find it easier. Finally, proximity also plays an important role in content management, as nearby Bluetooth devices can provide information both in social and physical context. While each Bluetooth device can be uniquely identified, information can be provided on nearby people by identifying their mobile phones. An example for physical context is the case of a Bluetooth-based hands-free car kit that can be used to identify that the user is in a car.

USABILITY ISSUES AND PROBLEMS

The expansion of the dimension of context information in order to include location, as well as proximity context, can be of benefit to users while they are able to store, access, and share with others their own location-based information such as videos and photos, and feel the sense of community growing among them (Kasinen, 2003; Cheverist, Smith, Mitchell, Friday, & Davies, 2001). But when it comes to proximity to be included in context information, the problem of *privacy* emerges. It appears that users are willing to accept a loss of privacy when they take into account the benefits of receiving useful information, but they would like to control the release of private information (Ljungstrand, 2001; Ackerman, Darrel, & Weitzner, 2001).

While context metadata is attached to content, when users share content, they have to decide if they share all the metadata with the content or they filter out all or some part of them. The cost for memory and transmission of metadata, as it is textual information, is not an important factor to influence this decision. When the user receives location and proximity information attached to content, he or she may also find out where and with whom the creator of the content was when the content was created. As a result, both the location of the content creator and the location of nearby people are shared along with the content information. If this information is private, the sharing of it could be considered as a privacy violation. This violation may be 'multiplied' if the first recipient forwards the content and the metadata to other users.

However, users seem to be willing to share context metadata attached to content, as it would be convenient if context metadata were automatically available with the content (so that users do not have to add this information manually). Furthermore, it would be very helpful for the recipient if the received content was annotated with context metadata so that the recipient does not have to annotate it manually and be able to manage the content more easily. For example, in the case of image and video content, the filtering of context metadata such as location and people could be useless, since these same items appearing in the image or video can be identified visually from the image content itself.

But what is meaningful information to the end user? It seems that users want meaningful information, but they are not willing to put too much effort in creating it, unless this information is expected to be very useful. In the case of location, it would be difficult for users to type the name of the place and other attributes manually, since it would require their time and effort. Thus it would be important if meaningful context metadata, which include the required information, are automatically generated.

Proximity information also needs to be meaningful. In this way, meaningfulness is important when attaching information on nearby devices in the form of metadata. If the globally unique Bluetooth device address and the real name of the owner of the device could be connected, this functionality would give meaningful information to the user.

It is hard to determine which information is useful, while what is useful information in one situation might be totally useless in another. For example, when looking at photo albums, what is thought to be useful information varies a lot. When one is looking at family pictures taken recently, it is needless to write down the names of the people, since they were well known and discernable. But it is different looking at family pictures taken many years ago: the same people may not be that easily recognizable.

It appears that useful information depends on a user's location, what the information is used for, and in which time span. In order to create meaningful information, users need to put much effort into getting the data, organizing it, and annotating it with context metadata. Ways to minimize their effort and maximize their benefit should be developed.

CONCLUSION

The increasing amount of stored content in mobile devices and the limitations of physical mobile phone user interfaces introduce a usability challenge in content management. The efficient management of large amounts of data requires developing new ways of managing content. Stored data are used by applications which should express information in a sensible way, and offer users a simple and intuitive way of organizing, searching, and grouping this information. Inadequate design of user interface results in poor usability and makes an otherwise good application useless. Therefore, it is necessary to design and built context-aware applications.

Issues of usefulness and meaningfulness in utilizing context metadata need to be further investigated. Usefulness depends on the type of metadata. As far as location and proximity are concerned, it appears that the more time has passed since the recording of the data, the more accurate the information needs to be. Furthermore, in the case of location information, the closer to one's home or familiar places the data refers to, the more detailed the information needs to be. A main usability challenge is the creation of meaningful context metadata automatically, without users having to add this information manually. There exist many ways for automatic recording of information about a user's context, but the generated information is not always meaningful.

Another field that requires further research is privacy. It seems that users are willing to accept a loss of privacy, provided that the information they receive is useful and they have control over the release of private information. Content management provides users with a safe, easy-to-use, and automated way of organizing and managing their mobile content, as well as retrieving useful information efficiently.

REFERENCES

Ackerman, M., Darrel, T., & Weitzner, D. J. (2001). Privacy in context. *Human Computer Interaction, 16,* 167-176.

Brown, P. J. (1996). The stick-e document: A framework for creating context-aware applications. *IFIP Proceedings of Electronic Publishing '96,* Laxenburg, Austria, (pp. 259-272).

Brown, P. J., Bovey, J. D., & Chen, X. (1997). Context-aware applications: From the laboratory to the marketplace. *IEEE Personal Communications, 4*(5), 58-64.

Campbell, C., & Tarasewich, P. (2004). What can you say with only three pixels? *Proceedings of the 6th International Symposium on Mobile Human-Computer Interaction*, Glasgow, Scotland, (pp. 1-12).

Cheverist, K., Smith, G., Mitchell, K., Friday, A., & Davies, N. (2001). The role of shared context in supporting cooperation between city visitors. *Computers & Graphics, 25,* 555-562.

Dey, A. K., Abowd, G. D., & Wood, A. (1998). CyberDesk: A framework for providing self–integrating context-aware services. *Knowledge Based Systems, 11*(1), 3-13.

Dey, A. K. (2001). Understanding and using context. *Personal & Ubiquitous Computing, 5*(1), 4-7.

Kaasinen, E. (2003). User needs for location-aware mobile services. *Personal Ubiquitous Computing, 7,* 70-79.

Kim, H., Kim, J., Lee, Y., Chae, M., & Choi, Y. (2002). An empirical study of the use contexts and usability problems in mobile Internet. *Proceedings of the 35th Annual International Conference on System Sciences* (pp. 1767-1776).

Ljungstrand, P. (2001). Context-awareness and mobile phones. *Personal and Ubiquitous Computing, 5,* 58-61.

Lucas, P. (2001). Mobile devices and mobile data—issues of identity and reference. *Human-Computer Interaction, 16*(2), 323-336.

Ryan, N., Pascoe, J., & Morse, D. (1997). Enhanced reality fieldwork: The context-aware archaeological assistant. In V. Gaffney, M. v. Leusen, & S. Exxon (Eds.), *Computer applications in archaeology.*

Schilit, B., & Theimer, M. (1994). Disseminating active map information to mobile hosts. *IEEE Network, 8*(5), 22-32.

Schilit, B., Adams, N., & Want, R. (1994). Context-aware computing applications. *IEEE Proceedings of the 1st International Workshop on Mobile Computing Systems and Applications*, Santa Cruz, CA, (pp. 85-90).

Sorvari, A., Jalkanen, J., Jokela, R., Black, A., Kolil, K., Moberg, M., & Keinonen, T. (2004). Usability issues in utilizing context metadata in content management of mobile devices. *Proceedings of the 3rd Nordic Conference on Human-Computer Interaction*, Tampere, Finland, (pp. 357-363).

KEY TERMS

Bluetooth: A short-range wireless technology used to create personal area networks among user devices and with other nearby devices.

Content Management: Ways of organizing and annotating content in order to retrieve and search it more efficiently.

Context: Any information that can be used to characterize the situation of an entity.

Context Metadata: Information that describes the context in which a content item was created or received.

Entity: A person, place, or object that is considered relevant to the interaction between a user and an application, including the user and the applications themselves.

Location: The place where content is created by the user.

Usability: The effectiveness, efficiency, and satisfaction with which users can achieve tasks in the environment of mobile devices.

Content Transformation Techniques

Ioannis Antonellis
Research Academic Computer Technology Institute, Greece
University of Patras, Greece

Christos Bouras
Research Academic Computer Technology Institute, Greece
University of Patras, Greece

Vassilis Poulopoulos
Research Academic Computer Technology Institute, Greece
University of Patras, Greece

INTRODUCTION

The expansion of the Web is enormous and, more and more, people everyday access its content trying to make their life easier and their informational level complete. One can realize that lately the advances in computers are such that many appliances exist in order to offer to its users the chance to access any type of information. The use of microcomputers, such as PDAs, laptops, palmtops, mobile phones and generally mobile devices, has lead to a situation where a way had to be found in order to offer to the users the same information as if they had a normal screen device. Almost all the mobile devices offer "Web-ready" functionality, but it seems that few of the Web sites are considering offering to the mobile users the opportunity to access their pages from the mobile devices.

On the one hand, the widespread use of mobile devices introduces a new big market and many chances for research and development. On the other hand, the use of small screen devices introduces a basic constraint both to the constructors of the devices and to the users: the small screen limitation. This is making difficult for the users to establish a mental model of the data, often leading to user disorientation and frustration (Albers & Kim, 2000). Many other restrictions have to be taken under consideration when using small devices, especially the low resolution, the amount of the memory, and the speed of the processor. Additionally, when using such devices the users are often in places with distractions of noise, interruptions and movement of the handheld device (Jameson et al., 1998).

Many companies exist in order to offer to the users of small screen devices the opportunity to access Web pages by doing syntactic translation (AvantoGo, DPWeb, Palmscape, and Eudora). Syntactic translation recodes the Web content in a rote manner, usually tag-for-tag or following some predefined templates or rules. This method seems to be successful especially for the devices that have graphical display. But, in order to achieve this, the Web pages are scaled down and small devices like mobile phone (very small screen and low resolution) are problematic. This happens because either the graphics are too small or the letters and links cannot be explored.

Another major problem of the use of small screen devices is that users often migrate from device to device during a day and they demand to be able to work in the same way whether they work on their personal computer or their mobile phone. This is the main issue that is going to be analyzed in this article: the way of migrating data from device to device without damaging the integrity of the data and without distracting the user.

Migration is the process of taking data originally designed for display on a large screen and transforming it to be viewed on the small screen (Jameson et al., 1998). The main techniques that exist and are used for data migration are direct migration, data modification, data suppression and data overview. The first one, direct migration, is a very simple. The data are sent directly to the small screen device and the user navigates to the data by scrolling horizontally and vertically on the page. The second method is more complicated and data is shortened and minimized in order to be viewable in a small screen device. Data suppression technique removes parts of the data and presents parts of them and the latest technique is based on the focus and context model (Spence, 2001).

All the aforementioned techniques are useful and any of them can be used efficiently for different types of data. This is a difficult part for the construction of the small screen devices. The constructors of the devices cannot include all the implementations of the techniques or, even if they do, the user has to be asked which one to choose or try the different implementations while viewing a source of data. The differences between the aforementioned techniques are focused on the quality of the information shown to the user and the range of information that is shown. This means that in some techniques the quality of the information shown is high but the amount of information shown is quite poor. One can think that the quality of information is more important while another can think that the amount of information is more important. This is a question that cannot be answered simply. What we can safely note is that the answer depends on the type of data that we want to present to the users.

The rest of this article is structured as follows. In the next section we present the efforts of some companies that offer to the users of small screen devices the opportunity to access Web pages by doing syntactic translation. The first method of transforming information, its use, its advantages and disadvantages are presented to the third section. The fourth section presents the data modification technique and how it is implemented, and the fifth section the data suppression technique. The next section covers the issues concerning data overview technique and the last section presents a summarization and general overview of the techniques.

DIRECT MIGRATION TECHNIQUE

The most simple and most often used technique is the direct migration technique. It is used mostly for Web pages and its scope is to send to the users exactly the same data regardless of the device in use. The users are free to interact with the data and they are actually responsible for making themselves comfortable with the amount of data that they are presented. We cannot say that it is a user-centric technique but it is very easy to be implemented, very fast and does not require much effort either for machine or human. The main problem, which is actually a failure of the technique, is that it produces data that needs horizontal scrolling in order to be accessed and that way the user is much distracted.

Some additional techniques are used together with the data migration technique in order to reduce or remove the horizontal scrolling problem. The additional technique is mainly the wrapping technique, which removes the horizontal scrolling by putting the extra data under the main page that is shown to the small screen. The problem is not solved but it becomes minor, because it does not lessen the amount of data but transforms the horizontal scrolling to vertical.

Another additional technique requires duplicate creation of the data. It is used very often for Web sites and the method is creating two kinds of pages for the same data: one for large screen devices and one for small screen. Surely, this technique has major problems. One is that someone has to create two totally different pages for the same content. The other and more crucial problem is the size of the World Wide Web and the fact that almost nobody has made any effort to create two types of Web pages makes the technique difficult to be applied.

Research has shown that the users react better when they are confronting vertical scrolling rather than horizontal (Nielsen, 1999). However even vertical scrolling—generally any kind of scrolling—affects negatively the completion of any task (Albers & Kim, 2000; Dyson & Haselgrove, 2001; Jones et al., 1999). The above implies that this technique can be suitable only for situations where the user just wants to access and read some kind of information and the interaction level between the user and the data remains low.

Summarizing, we can say that this technique is very suitable for short text, sequential text, lists and menus that can be displayed within the width constraints of small screens (the impact of migration). It is not recommended to be used when the data include big tables and images (big, high resolution) because these types of data add horizontal scrolling that cannot be transformed.

DATA MODIFICATION

In this section we will analyze the second method for data migration, which is the data modification technique. Its main idea approaches the direct migration technique, but the data modification technique has countered the problem of big images and tables. When the data are to be presented to a small device, the size of the images, tables and lists is reduced and some parts of the text are summarized. In this way the users can save in download time and device memory (Mani, 2001).

The text summarization is the difficult part of the technique and it introduces a whole new theme for discussion. Many approaches have been proposed (Buyukkokten et al., 2000; Fukushima, 2001, Mani, 2001; Amitay & Paris, 2000). Some of them require a human expert to create the summaries while some others are based on machines.

The data that is presented to the users is a reduced form of the actual data. The user has the option to scroll vertically through the data that he comes up with. He can also select a part of the reduced data in order to "open" in another page of his small screen device the real text, which is hidden behind. This procedure can be algorithmic. When data are presented in this way to the user, then the procedure is to read the summarized, reduced data, select a specific topic that suits the user's needs, read the whole data that is hidden behind the summarized and then go back. The procedure then starts from the beginning.

We can say that this technique is very similar to the aforementioned direct migration technique but it goes one step further. It is used mainly for Web browsing where the data are already reduced and offer the user a style of navigation. The summarization that is included, whether it is for images (lower size, resolution) or text (summary), is very helpful for the end-user as it lessens the scrolling either vertical or horizontal. Actually this method does not have horizontal scrolling at all except for some specific, very rare conditions (very large images or tables).

Summarizing, we can say that this technique is very useful when users are determined of the information and can easily understand what they are looking for, from a summary of text or simple keywords. It cannot be useful for very specialized texts with difficult and mannered terminology. In general, the summaries have to be very specific and represent accurately the meaning of the text. The main problem of all the summarization techniques is that they do not succeed very often and cannot replace numerical data like financial information, weather information and dates. If one can think that some users want their small screen devices for accessing their bank accounts, watching the weather in a place that they visit or finding the financial exchange then this technique cannot be recommended.

DATA SUPPRESSION

As we are able to figure out from the name of this technique, what it actually does is to remove parts of the data that "seem" to be unimportant. What is presented to the user is the basic frame of the data. Displaying only skeleton information can simplify navigation and may reduce disorientation (Spence, 2001).

The data is not removed randomly, but there exist several techniques that help in this direction. Some methods for suppressing data is to select only some of the keywords (that are produce from text summarization), present only a specific number of words from each sentence or Z-thru mapping that imposes selective display (Spence, 2001).

This approach is very similar to the previous but it seems to be more compact. Very few data is presented to the user and most of the time there is no scrolling at all. The absence of scrolling has advantages and disadvantages. When there is no scrolling the user is not distracted from completing his task, but no scrolling means that the data is extremely reduced in order to fit the screen and it may be difficult for a simple user to locate the information he/she wants.

Navigating through the data in this technique is like a file system. The user has a list of words (like a folder) for each amount of data and by selecting an element of the list the information is expanded (files, subfolders) and shown to the user. Every time the user is able to return to the starting frame of data and start exploring from the beginning.

Like the previous, this technique has applications where the users know the exact information that they are looking for and they can figure it out from just a heading or a set of keywords. Searching through this type of data is almost impossible because the little amount of data that is presented is often not representative of the data that it comes from. However it is very useful for browsing through news portal when just a title or part of the title is enough for the user to understand the meaning of the whole article. It is used for structured data, which include information hierarchically structured. Sequential data with little or no structure could be less compatible to manipulate into categories for suppression (impact of migration).

DATA OVERVIEW

The last technique that is used for data migration is data overview technique. In reverse to the aforementioned techniques, which reduce parts of the data, this technique creates an overview of the whole data and presents it to the users. The whole data is minimized and the whole information is presented to the user minimized in order to fit the small screen of the device. It is based on the "focus and expand" method. When the user points a specific set of data that is contiguous then it is expanded and shown bigger to the screen in order to fit the screen and be readable.

The approach makes it easier for the users to access at once a very large amount of data without losing or not seeing any part of it and, in this way, the disorientation is lessened (Spence, 2001; Storey et al., 1999). Some methods that are used concerning the data overview technique are:

- Focus and context (Spence, 2001; Buyukkokten et al., 2000; Bjork, 2000)
- Fisheye Techniques (Spence, 2001; Storey et al., 1999)
- Zoom and pan (Good et al., 2002; Spence, 2001)
- Content lens (Dieberger et al., 2002)

In general, the technique seems to be problematic as the user is presented with a large amount of data in a small screen. The data is shrunk in order to fit the screen and may be difficult for the user even to see it and figure out what he is looking for. Movements while using a small screen device could create further distortion, or could make it difficult to discern what has been distorted (MacKay & Watters, 2003).

The nature of this approach produces both positive and negative points for the end-users. The point that the user is presented the whole information can be both positive and negative depending on the amount of data. However, it is very useful for the users to have full observation of the information they are looking for. The navigation is easy and is based at presenting in large the parts that are focused from the user, but the user can focus only on a part of information and he is not able to combine parts of data.

In general this method seems to be the best when the information that is accessed by the user includes large images, big tables, maps, graphs and in general everything that a "focus" method cannot distort but help.

OVERVIEW OF THE TECHNIQUES

In the previous sections we have discussed and analyzed the most common methods for data migration from large screen displays to small screen. As we can obviously see, each method has its advantages and disadvantages making difficult the selection of only one of them in order to cope with every type of data.

Direct migration cannot preserve scrolling and it is the fastest and easiest way to present data that are for reading. Its simplicity is its power but we can admit that is not user friendly.

Data modification technique solves many problems of the previous technique but still scrolling is an issue. At least paging of the data is preserved and the user can see a large part of the information in only one screen. The matter that rises from this technique is the summarization of the information, which may be distracting or not useful depending on the type of information. It could be seen as a good method for Web browsing.

Data suppression goes one step further than the previous technique by removing parts of data and summarizing the rest. It is named as the best method for browsing news portals where just the keywords of a news title can represent successfully the whole article. It is very weak for textual data and for searching, as it provides in a hierarchic manner only some keywords and often distracts a user that does not know exactly what he is looking for.

Data overview has a different angle of view than the three previous methods. It is based on the idea "focus and expand" and the philosophy is to present to the user all the information. When the data include large images, big tables and graphs, data overview is the best method for migrating data because it does not lessen or break into many pages all this information, which is by nature connected. On the contrary, when the user wants to read a text or browse in a big portal then this technique seems to be weak, as it provides to the user all the information in one screen and the data are often unreadable.

Summarizing, all the techniques offer to the users the opportunity to access any kind of information through their small screen devices like they would do to big screen ones. It is not fair to select one of them as the best one because each one is created for coping with different types of information and data. A device that could combine the implementation of all the aforementioned techniques could be a solution, but the complexity of modern life would prevent us to permute to the users the effort of data migration and thus make modern life more complex.

REFERENCES

Albers, M. J., & Kim, L. (2000). User Web browsing characteristics using Palm handhelds for information retrieval. In *Proceedings Of IPCC/SIGDOC Technology & Teamwork* (pp. 125-135). September, 2000, Cambridge, MA: IEEE.

Amitay, E., & Paris, C. (2000, November). Automatically summarizing Web sites: Is there a way around it? In *Proceedings of the 9th Internet Conference on Information and Knowledge Management* (pp. 173-179). McLean, VA.

AvantGo, Inc. (n.d.). *AvantGo*. Retrieved from http://www.avantgo.com

Bjork, S. (2000, May). Hierarchical flip zooming: Enabling parallel exploration of hierarchical visualization. In *Proceedings of the Working Conference on Advanced Visual Interfaces* (pp. 232-237). Palermo, Italy.

Buyukkokten, O., Garcia-Molina, H., & Paepcke, A. (2001). *Seeing the whole in parts: Text summarization for Web browsing on handheld devices.* Retrieved from http://wwwconf.ecs.soton.ac.uk/archive/00000067/01/index.html

Dieberger, A., & Russell, D. M. (2002, January). Exploratory navigation in large multimedia documents using context lenses. In *Proceedings of 35th Hawaii International Conference on System Sciences* (pp. 1462-1468). Big Island, Hawaii.

Digital Paths LLC. DPWeb. Http://www.digitalpaths.com/prodserv/dpwebdx.htm

Dyson, M., & Haselgrove, M. (2001). The influence of reading, speed and line length and effectiveness of reading from screen. *International Journal Human Computer Studies, 54*(4), 585-612.

Fukusima, T., & Okumura, M. (2001, June). Text summarization challenge: Text summarization evaluation in Japan. In *Proceedings North American Association for Computational Linguistics* (pp. 51-59). ittsburgh, Philadelphia, Association of Computational Linguistics.

Good, L., Bederson, B., Stefik, M., & Baudisch, P. (2002). Automatic text reduction for changing size constraints. In *Proceedings of Conference on Human Factors in Computer Systems, Extended Abstracts* (pp. 798-799). April 2001, Minneapolis, MN.

ILINX, Inc. (n.d.). *Palmscape*. Retrieved from http://www.ilinx.co.jp/en/products/ps.html

Jameson A., Schafer, R., Weis, T., Berthold A., & Weyrath, T. (1998). Making systems sensitive to the user's time and working memory constraints. In *Proceedings of 4th international Conference on Intelligent User Interfaces* (pp. 79-86). December 1998, Los Angeles, CA: ACM Press.

Jones, M., Marsden, G., Mohd-Nasir, N., Boone, K., & Buchanan, G. (1999). Improving Web interaction on small displays. In *Proceedings of the 8th International WWW*

Conference. May 1999, Toronto, Canada. Retrieved from http://www8.org/w8-papers/1b-multimedia/improving/improving.html

MacKay, B., & Watters, C. (2003, Winter). The impact of migration of data to small screens on navigation. *IT&Society, 1*(3), 90-101.

Mani, I. (2001, October). Text summarization and question answering: Recent developments in text summarization. In *Proceedings of the 10th International Conference on Information and Knowledge Management* (pp. 529-531).

Nielsen, J. (1999, December). *Changes in usability since 1994.*

QUALCOMM, Inc. (n.d.). *Eudora Internet Suite.* Retrieved from www.eudora.com/internetsuite/eudoraweb.html

Spence, Robert. (2001). *Information visualization.* New York: *ACM Press.*

Storey, M. D., Fraachia, F., Davic, M., & Hausi, A. (1999, June). Customizing a fisheye view algorithm to preserve the mental map. *Journal of Visual Languages and Computing,* 254-267.

KEY TERMS

Content Transformation: The procedure that leads to changes to content in order to make it interoperable.

Migration: Migration is the process of taking data originally designed for display on a large screen and transforming it to be viewed on the small screen.

Small Screen Devices: Devices with small screen size where it is difficult to access large-sized blocks of information.

Syntactic Translation (of WWW Data): The recoding of the Web content in a rote manner, usually tag-for-tag or following some predefined templates or rules.

Context-Adaptive Mobile Systems

Christian Kaspar
University of Goettingen, Germany

Thomas Diekmann
University of Goettingen, Germany

Svenja Hagenhoff
University of Goettingen, Germany

INTRODUCTION

Even though a major part of the industrialized world works with computers on a daily basis and operating computers became much easier since the introduction of graphic interfaces, many users do not experience their computers as work relief, but rather as an increased burden in their everyday lives. One of the most important reasons for this attitude is the unnatural mode of communication between user and computer: the natural interpersonal communication takes information from the communication situation (e.g., the location of the interacting communicators, their personal preferences, or their relationship with each other) implicitly into account. On the other side, despite the development of new interfaces—such as voice and character recognition, which are much closer to interpersonal communication than keyboard terminals—communication between user and computer is still complex and characterized by little intuition. This is where the objectives of the context-adaptive systems come into play: it is the aim of context-adaptive systems to implicitly collect information about the situation of a system request (context) in order to enable more efficient communication between user and computer.

Currently, the concept of context-awareness and context-adaptation has attracted particular attention in the area of mobile communication. This is largely due to the fact that the obligatory requirement of the devices' portability leads to certain constraints of mobile devices. Small-sized screens, low data processing capacity, and inconvenient ways of navigation and data entry are some examples for these constraints. To overcome these limitations is particularly relevant in the area of multimedia Internet content and therefore requires the communication to be as efficient as possible. One possible option to reduce the resulting problem of presentation and selection of content on mobile devices is to automatically offer the user only those contents relevant for the concrete situation of the service request. Such services require that the computer can sense the particular situation of the service request and autonomously respond with appropriate actions.

AUTOMATED CONTEXT-AWARENESS

In order for real situations to be sensed automatically by computing devices, the situations of these system requests have to be computed as abstract, automatically understood events, so-called contexts. A context is any information that is used to characterize relevant situations of people, locations, or objects that are important for the interaction between application and user (Dey, 2001). A common classification of context information traces back to Schilit, Adams, and Want (1994). They distinguish between the technical context of participating and available computing resources, the social context of users that are involved in the system interaction, and the physical context of the location of the system interaction.

Computing context describes available network connections and network bandwidth. Additionally, computing context includes accessible peripherals such as printers, screens, or additional terminals. For example, if a multimedia application knows the user's available network bandwidth, it is able to adapt a video stream to its capacity and ensures streaming without jerks and with the highest possible resolution. Furthermore, if the multimedia application is aware of a high-resolution display close to the user, it can suggest this device as an alternative screen for displaying the video stream. To be able to identify each other, mobile computing devices must have radio or infrared sensors. A computing device equipped with radio or infrared sensors spans a distinct logical space (a so-called "smart space") within its sensor coverage. If a device enters another device's sensor space, the device will identify itself and send its network address or appropriate commands for application requests.

The social context contains information about the users involved in the interaction. The user's information, such as identity, age, gender, and preferences, can be gathered either explicitly from surveys or implicitly by observing the user's behavior. Surveying each user's personal characteristics and preferences is the most common form of gathering user information. Most often, surveying user information is directly linked with service registration. Because the provider

Figure 1. Components of an agent system

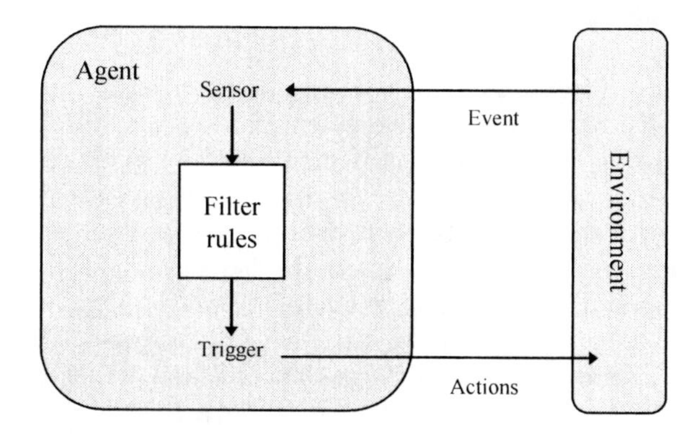

has little means to control (usually voluntary) submitted information, information from user surveys is often of poor quality. Additionally, profiles that were gathered from a one-time survey remain static over time. Therefore, apart from voluntary authentication information on a specific Web site, the user can be additionally identified on the basis of his or her behavior. Every Web server has a protocol component that logs every server activity and stores these logs chronologically into different application-oriented protocol files. Analyzing these server protocols, it can be determined what requests for which resources have been completed during a specific unique Web site visit. To link recorded requests with an individual user, the IP address of the user's device or identification data stored as cookie on the user's device can be used.

Information about the physical context can be collected from a multitude of data sources such as contact-, thermo-, humidity-, acceleration-, torsion-, or photo-sensors, cameras, and microphones. Sensors that are equipped with processors not only collect data but also pre-process this data. Additionally, they can identify specific patterns such as fingerprints. The user's interaction location is of particular importance for the perception of the physical interaction context. "Location-based-services" are services that take the location of the user into consideration. These services promise to have great chances on the market (Lehner, 2003). The geographical location of a user that is required for services of this kind can either be determined by terminal-locating or by external network-locating.

Terminal-locating is carried out by an especially designed device that autonomously executes location measurements. The global positioning system (GPS) operated by the U.S. military is the best-known technology for self-locating. Receiving positioning signals that are beamed down by GPS satellites, a GPS device can accurately triangulate its position for up to 10 meters. Techniques that can position a location or object from a photo are more sophisticated than the GPS system. Yet they strongly resemble human orientation. Using photo cameras, these methods can calculate the angle and distance to a specific object (such as a building) with the help of a stored three-dimensional model.

Network-locating fixes a device's position using network information. The best-known method for network-locating is the cell identity technique (or cell of origin technique). This technique locates a mobile device within a cellular radio network using the network's cell-ID. Other network-locating techniques fix a particular position based on time differences of signals arriving at different base stations, the angle of arrival, or attenuation of signals from different base stations.

While Schilit et al. (1994) differentiated between three forms of context, Dey (2001) adds the primary and secondary context to these categories (Conlan, Power, & Barrett, 2003). The location, the type of device, the behavior of the user, and the time of inquiry represent primary request contexts. On the other hand, secondary request contexts are composed of a combination of primary context data. By taking the location and other people in the vicinity into consideration to form the secondary social context, the social situation of the user can be determined; for example, the user might be in his office with his colleagues, or out with friends. Furthermore, Chen and Kotz (2000) differentiate between the active and the passive context. The active context determines a change in behavior for the present application (e.g., by defining sections of interest for an adaptive online newspaper). A passive context shows the change in context conditions for the system inquiry only as extra information for the user. For example, a recommendation system can suggest a specific purchase in an online shop, or the user can see his location when using a navigation system.

CONTEXT-RELATED SYSTEM ADAPTION

A system is considered context-adaptive if it uses context information to offer its users relevant information or, rather, services (Dey & Abowd, 2000). Generally speaking, a context-adaptive system is characterized by a certain degree of autonomy when fulfilling its tasks. Therefore, adaptive systems are also referred to as agent systems (Russel & Norvig, 2003). Agents are software systems that, with the help of sensors, identify their environment as application-related events and by using pre-defined rules that activate respective events (see Figure 1). The actions that have been triggered by the agent can refer to information or they can contain control commands for other systems. The agent can either perform the action directly and autonomously (active context-awareness), or these actions are only a suggestion to the user (passive context-awareness).

A context-adaptive application consists of software objects that are automatically requested when the system senses a certain event in the systems environment. If the application identifies such a key event, it executes the respective object, or rather, the relevant methods that are part of this object. Usually, the adaptive application needs to combine raw data from its sensory perception of its environment to compute such a key event (e.g., GPS-coordinates, entries in a Web-server's log file, or identified, closely related network resources) since an individual piece of raw data is only able to hint at the relevant situation. An adaptive traffic information service, for instance, needs to compute information about the current date and time (system clock), the user's current position (e.g., GPS), the user's destination, and his or her preferred routes and means of transportation (user profile). In this context, we can distinguish different adaptation methods, depending on the type and extent of combined raw data collected while computing such a key event.

Raw sensor data that closely related to the system (such as time, temperature, or the device's data processing capacity)—whose values are, in accordance with the closed-world-assumption, known ex ante—are usually directly linked to the respective application-specific event. Therefore, dynamic key-value-pairs are composed, whereby the application-event contains the key and the sensor system contributes the key value (Chen & Kotz, 2000).

With regard to raw sensor data that cannot be directly linked to application events, such as user preferences or information about the location, however, the process is different. In this case, values from various sensor systems need to be aggregated. For instance, a standardized locating technique that is able to locate a user in closed rooms and outside does not yet exist. In order to locate users accurately inside and outside of buildings, data from different locating systems need to be combined. Since different locating systems utilize different measures (e.g., distance, angular separation, or geometrical position), these values need to be translated into standardized measures first and then have to be transferred into a joint data model (Hightower, Brumitt, & Borriello, 2002). This process is commonly known as "sensor fusion" (Chen, Li, & Kotz, 2004).

Adaptive systems that process contexts whose quality rating is subject to a high level of changeability (e.g., a user's movement in a room or outside of a building) need a high number of key values to be computed. Generally speaking, it does not make much sense to compose one key-value-pair for each of these values. Instead, artificial intelligence techniques, such as artificial neural networks that have been developed to process knowledge, can be utilized to compute these values. An artificial neural network (ANN) is a system that consists of a multitude of identical, networked computing elements, so-called neurons. ANNs are thus able to simultaneously process extensive and changeable raw data in an efficient manner (Van Laerhoven, Aidoo, & Lowette, 2001).

EXAMPLES FOR CONTEXT-ADAPTIVE MOBILE SERVICES

Scholars focusing on mobile systems were among the first to publish research on adaptive systems. Olivetti and XEROX, two of the leading producers of copying machines at that time, were pioneers in this new research area of context-adaptive computing. In the early 1990s, their research facilities introduced the first prototypical adaptive systems. These early applications include automatic call-forwarding systems that are based on where in the office building the person who receives the call is located (Want, Hopper, Falcao, & Gibbons, 1992; Wood, Richardson, Bennett, Harter, & Hopper, 1997), browser software that can be adapted to specific locations (Voelker & Bershad, 1994), as well as location-aware shopping assistants (Asthana, Cravatts, & Krzyzanowski, 1994). These groundbreaking applications are not so much derived from data communication that is supported by mobile radio technology, but from communication that is connected to ubiquitous or, rather, pervasive computing. They primarily use infrared or radio transponders (so-called active badges). Special room sensors are able to detect users, who carry these transponders with them at all times. In later years, scholars developed various kinds of adaptive systems: these new developments include tour guides (Bederson, 1995; Long, Kooper, Abowd, & Atkeson, 1996; Davies, Cheverst, Mitchell, & Friday, 1999), software assistants for conference participants (Dey, Futakawa, Salber, & Abowd, 1999), and field researchers (Pascoe, 1998).

In the late 1990s, researchers proposed adaptive service concepts for commercial, content-related mobile radio services. One of the first studies offered a solution to the problem of content recipients' partially bound attention: this study developed an adaptive screen for GSM terminals which is able to change the screen's font and brightness in accordance with the room conditions and the user's activity. As a reaction to the deregulation of the cell-based user locating services in GSM-networks for commercial purposes in 2001, scholars proposed a number of other options for location-specific mobile services (Lehner, 2004). These suggestions include gas station search services that take locating and vehicle data into consideration, and adaptive multimedia applications for cars that are able to switch between different reception options (e.g., GSM or DVB), depending on the specific reception quality (Herden, Rautenstrauch, Zwanziger, & Planck, 2004). In addition to these cell-based location services, researchers have also discussed adaptive services based on GPS (Diekmann & Gehrke, 2003). Furthermore, some authors suggest individualizing concepts that are attuned to the special features of mobile terminals. These concepts would be able to adapt contents to individual preferences. In this context, it has to be distinguished between those individualization concepts that carry out the adaptation on the aggregation level (or rather, on a mobile portal—Smyth & Cotter, 2003; Kaspar & Hagenhoff, 2004) and those that aim at the individual service (Anderson, Domingos, & Weld, 2001). For the most part, scholars have discussed individualization techniques that are based on explicit information from users and on limited potentialities. Currently, however, research is also trying to find ways to distribute contents to a multitude of different types of mobile devices, a development that has become necessary because of the growing diversity of mobile devices that are equipped with varying hardware and software. One way to achieve this is to use markup transformations that are based on schematic libraries for different devices and standards such as WML, XHTML, und HTML. In order to identify a mobile device, various standards that allow users and providers to exchange the respective configuration during each data communication process have been developed. The best-known standards include the "Composite Capabilities/Preference Profiles" (CC/PP), which was developed by W3C in 2004 or its earlier implementation—the "User Agent Profile" (U-AProf)—by the WAP Forum in 2003. After identifying the respective device configuration, it is possible to adapt the syntax, for example on the basis of the style sheet transformation language, XSLT.

CONCLUSION

Most of the current examples of context-adaptive systems represent isolated solutions that are based on a closed-world assumption. At this point, these systems have little commercial value. This lack of commercial relevance is basically rooted in two main problems that have to be overcome in the future. On the one hand, the development of adaptive systems is very cost intensive. In addition, currently existing solutions are usually based on proprietary data models, which keep them from interacting with different systems and do not allow them to add additional contexts. On the other hand, an adaptive system is dependent on a comparatively large set of personal information that has to be gathered and processed automatically. This causes user concerns about possible abuses of personal data and intrusions of privacy. So far, neither legal measures nor technical control instruments have been able to eliminate users' apprehensions of permanent surveillance by a "big brother."

REFERENCES

Anderson, C., Domingos, P., & Weld, D. (2001). *Personalizing Web sites for mobile users.* Retrieved May 31, 2005, from http://www.cs.washington.edu/ai/proteus/www10.pdf

Asthana, A., Cravatts, M., & Krzyzanowski, P. (1994). An indoor wireless system for personalized shopping assistance. *Proceedings of the IEEE Workshop on Mobile Computing Systems and Applications*, Santa Cruz, CA, (pp. 69-74).

Bederson, B. (1995). Audio augmented reality: A prototype automated tour guide. *Proceedings of the Conference on Human Factors and Computing Systems*, Denver, (pp. 210-211).

Chen, G., & Kotz, D. (2000). *A survey of context-aware mobile computing research.* Dartmouth Computer Science Technical Report TR2000-381. Retrieved October 31, 2005, from http://www.cs.dartmouth.edu/~dfk/papers/chen:survey-tr.pdf

Chen, G., Li, M., & Kotz, D. (2004, August). Design and implementation of a large-scale context fusion network. *Proceedings of the 1st Annual International Conference on Mobile and Ubiquitous Systems: Networking and Services,* Boston.

Davies, N., Cheverst, K., Mitchell, K., & Friday, A. (1999). Caches in the air: Disseminating tourist information in the GUIDE system. *Proceedings of the 2nd IEEE Workshop on Mobile Computing Systems and Applications,* New Orleans.

Dey, A., & Abowd, G. (2000, June). The context toolkit: Aiding the development of context-aware applications. *Proceedings of the Workshop on Software Engineering for Wearable and Pervasive Computing,* Limerick, Ireland, (pp. 434-441).

Dey, A. (2001). Understanding and using context. *Personal and Ubiquitous Computing Journal, 5*(1), 4-7.

Dey, A., Futakawa, M., Salber, D., & Abowd, G. (1999). The conference assistant: Combining context-awareness with wearable computing. *Proceedings of the 3rd International Symposium on Wearable Computers (ISWC '99)*, San Francisco, (pp. 21-28).

Diekmann, T., & Gehrke, N. (2003). Ein framework zur nutzung situationsabhängiger dienste. In K. Dittrich, W. König, A. Oberweis, K. Rannenberg, & W. Wahlster (Eds.), *Lecture notes in informatics, informatik 2003. Innovative anwendungen* (Vol. 1, pp. 217-221). Bonn: Gesellschaft fuer Informatik.

Herden, S., Rautenstrauch, C., Zwanziger, A., & Planck, M. (2004). Personal information guide. In K. Pousttchi & K. Turowski (Eds.), *Mobile Economy: Proceedings of the 4th Workshop on Mobile Commerce*, Augsburg, (pp. 86-102).

Hightower, J., Brumitt, B., & Borriello, G. (2002, June). The location stack: A layered model for ubiquitous computing. *Proceedings of the 4th IEEE Workshop on Mobile Computing Systems & Applications,* Callicoon, New York, (pp. 22-28).

Kaspar, C., & Hagenhoff, S. (2004). Individualization of a mobile news service—a simple approach. In S. Jönsson (Ed.), *Proceedings of the 7th SAM/IFSAM World Congress,* Gothenburg.

Lehner, F. (2003). *Mobile und drahtlose informationssysteme.* Berlin: Springer-Verlag.

Lehner, F. (2004). Lokalisierungstechniken und location based services. *WISU, 2,* 211-219.

Long, S., Kooper, R., Abowd, G., & Atkeson, C. (1996). Rapid prototyping of mobile context-aware applications: The Cyberguide case study. *Proceedings of the 2nd Annual International Conference on Mobile Computing and Networking* (pp. 97-107), White Plains, NY.

Pascoe, J. (1998). Adding generic contextual capabilities to wearable computers. *Proceedings of the 2nd International Symposium on Wearable Computers,* Pittsburgh, PA.

Russel, S., & Norvig, P. (2003). *Artificial intelligence, a modern approach.* NJ: Pearson Education.

Schilit, B., Adams, N., & Want, R. (1994, December). *Proceedings of the IEEE Workshop on Mobile Computing Systems and Applications,* pp. 85-90.

Smyth, B., & Cotter, P. (2003). Intelligent navigation for mobile Internet portals. *Proceedings of the 18th International Joint Conference on Artificial Intelligence (IJCAI-03),* Acapulco.

Van Laerhoven, K., Aidoo, K., & Lowette, S. (2001). Realtime analysis of data from many sensors with neural networks. *Proceedings of the 5th IEEE International Symposium on Wearable Computers 2001,* (p. 115).

Want, R., Hopper, A., Falcao, V., & Gibbons, J. (1992). The active badge location system. *ACM Transactions on Information Systems, 10*(1), 91-102.

Wood, K., Richardson, T., Bennett, F., Harter, A., & Hopper, A. (1997). Global tele-porting with Java: Toward ubiquitous personalized computing. *Computer, 30*(2), 53-59.

KEY TERMS

Agent: A software system that is able to perceive its surroundings as events that are relevant for the application and that, in accordance with previously defined filter rules, causes actions in accordance with its perceptions.

Context: Any piece of information that can be used to characterize the situation of a person, a place, or an object in a way that is significant for the interaction between user and application.

Context-Adaptive System: An application that changes its behavior in accordance with information about the respective situation.

Location-Based Service: A service that takes the location of the user into consideration. A user's location can therefore be detected by using network-locating or terminal-locating.

Network-Locating: Detects an end device's position on the basis of network information.

Sensor Fusion: Refers to the translation of values from different locating systems using different measures (e.g., distance, angular separation, or geometrical position) into standardized measures and the transfer of these standardized measures into a joint data model.

Smart Space: The logical space that is covered by a computing device equipped with radio or infrared sensors within its sensor coverage.

Terminal-Locating: A procedure that enables a specially designed end device to locate its own position.

Context-Aware Mobile Geographic Information Systems

Slobodanka Djordjevic-Kajan
University of Nis, Serbia

Dragan Stojanović
University of Nis, Serbia

Bratislav Predić
University of Nis, Serbia

INTRODUCTION

A new breed of computing devices is taking more and more ground in the highly dynamic market of computer hardware. We refer to smart phones and PocketPCs, which redefine typical usage procedures we are all familiar with in traditional, desktop information systems. Dimensions of this class of computing devices allow users to keep them at hand virtually at all times. This omnipresence allows development of applications that will truly bring to life the motto: "availability always and everywhere."

Hardware and software characteristics of the aforementioned devices require a somewhat modified approach when developing software for them. Not only technical characteristics should be considered in this process, but also a general set of functionalities such an application should provide. Equally important is the fact that the typical user will be on the move, and his attention will be divided between the application and events occurring in his environment. Fundamentally new and important input to mobile applications is constantly changing the user environment. The term that is used most frequently and describes the user environment is a context, and applications that are able to independently interpret a user's context and autonomously adapt to it are named context-aware applications.

Recent developments in wireless telecommunications, ubiquitous computing, and mobile computing devices allowed extension of geographic information system (GIS) concepts into the field. Contemporary mobile devices have traveled a long way from simple mobile phones or digital calendars and phonebooks to powerful handheld computers capable of performing a majority of tasks, until recently reserved only for desktop computers. Advancements in wireless telecommunications, packet data transfer in cellular networks, and wireless LAN standards are only some of technological advancements GIS is profiting from. This mobile and ubiquitous computing environment is perfect incubation grounds for a new breed of GIS applications, mobile GIS. Advances in mobile positioning have given a rise to a new class of mobile GIS applications called location-based services (LBS). Such services deliver geographic information and geo-processing services to the mobile/stationary users, taking into account their current location and references, or locations of the stationary/mobile objects of their interests.

But the location of the user and the time of day of the application's usage are not the only information that shapes the features and functionalities of a mobile GIS application (Hinze & Voisard, 2003). Like other mobile and ubiquitous applications, mobile GIS completely relies on context in which the application is running and used. The full potential of mobile GIS applications is demonstrated when used in the geographic environment they represent (Raento, Oulasvirta, Petit, & Toivonen, 2005). Thus, development of mobile GIS applications requires thorough analysis of requirements and limitations specific to the mobile environment and devices. Practices applied to traditional GISs are usually not directly applicable to mobile GIS applications. Limitations shaping future mobile applications, including mobile GISs, are ranging from hardware limitations of client devices to physical and logical environment of the running application. Considering the fact that mobile applications are used in open space and in various situations, the ability of the application to autonomously adapt itself to a user's location and generally a user's context significantly increases the application's usability. Regardless of the type of LBS and mobile GIS application, the part of the system that is handling context is fairly independent and can be separately developed and reused. The proper management of contextual data and reasoning about it to shape the characteristics and functionalities of mobile GIS applications leads to a full context-aware mobile GIS.

The second section presents concepts of mobile GISs and context awareness, and the principles of how context can be incorporated into traditional GIS features adapted to mobile devices. Data structures and algorithms supporting context awareness are also given. The third section presents GinisMobile, a mobile GIS and LBS application framework developed at Computer Graphics and GIS Lab, University of Nis, which demonstrates the concepts proposed in this article. The last section concludes the article, and outlines future research and development directions.

CONTEXT AWARENESS IN MOBILE GIS

Even though the concept of mobile GIS is in its infancy, technologies that were prerequisite for development of this niche of GIS applications are today widely available and well known to GIS developers. It is reasonable to expect that there are prototypes available demonstrating all the advantages mobile GIS offers to field fork personnel. ESRI, as one of the leading companies in the GIS field in its palette of products, offers a mobile GIS solution targeting the PocketPC platform. It is called ArcPad (http://www.esri.com/software/arcgis/ bout/arcpad.html). It is a general type of mobile GIS solution with open architecture allowing easy customization and tailoring according to a specific customer's needs. It therefore offers a set of basic GIS functionalities and tools that are used to extend application with functionalities needed for specific usage scenarios. ESRI bases its ArcPad on four basic technologies: mobile computing device (PocketPC), basic set of spatial analysis and manipulation tools, global positioning system (GPS), and wireless network communication interface.

Basic GIS functionality understandably supported by ArcPad is geographic maps visualization in the form of raster images. In order to avoid the need for maps conversion into some highly specialized proprietary raster map format, ArcPad supports usage of all of today's widely used raster image formats, like JPEG, JPEG 2000, and BMP, as well as MrSID, which is common in GIS applications. Thematically different maps in the form of raster images can be grouped into layers. Apart from raster type, layers can also contain vector data. Also, standard vector type data formats are supported, most importantly the shapefile format. That is the most common vector data format in use in GIS today and is also well supported by other ESRI GIS software like ArcInfo, ArcEditor, ArcView, ArcIMS, and others. Other optimizations which enable sufficient speed in handling spatial data include spatial indexing schemes. Spatial indexing significantly increases speed of spatial objects visualization and search, especially on portable devices with limited processing power. Indexes are prepared on other desktop-type ESRI applications, and afterwards are transferred to a mobile device and used by ArcPad. In order to support usage of ArcPad throughout the world, a majority of map projections are included.

ArcPad is conceived as an integral part of the ESRI GIS platform consisting of other products, so there is the possibility of ArcPad functioning as a client for ArcIMS or Geography Network (http://www.geographynetwork.com/). Data is transferred to ArcPad using TCP/IP protocol and any sort of packet-based wireless networking technology (wireless LAN, GSM, GPRS, EDGE, 3G, etc.). Possibly the strongest advantage of ArcPad is its extensibility and adaptability. Forms used for thematic data input and manipulation are created and customizes independently using ArcPad Studio and Application builder development tools. Application toolbars can be adapted to specific user needs. More importantly, specific interfaces can be developed and added to ArcPad, enabling it to acquire data from different database types and sensors (GPS location devices, laser rangefinders, magnetic orientation sensors, etc.).

One academic project that encompasses the development of mobile GIS is "Integrated Mobile GIS and Wireless Image Servers for Environmental Modeling and Management," developed at San Diego State University (2002). The project includes an integrated GIS platform where, in the field, data collection must be performed using a mobile GIS client platform. Effectiveness of the developed system is tested in three different services: campus security, national park preservation service, and sports events. The development group's decision was not to develop a mobile GIS solution from scratch, but to upgrade and customize ArcPad. Similarly to other mobile GIS solutions, this project is based on modified client/server architecture. Fieldwork personnel are using a PocketPC device with a customized ArcPad version installed. Customization includes components developed specifically for testing on campus. PocketPC is connected with an external GPS device, and therefore it has constant access to user location information. Considering wireless communications, campus grounds are covered with a wireless LAN, and all client PocketPCs are equipped with WLAN adapters. The server side of this system includes a typical set of servers and tools from ESRI including ArcIMS and ArcGIS.

When this system is employed by the campus security service, field units use mobile GIS components to locate a reported incident location more easily and swiftly. Mobile GIS is also used to report new incidents to central. Following report-in, information about a new event taking place is momentarily available to all units. Therefore, reaction time is shortened and all patrolling units within campus are synchronized more easily.

Demonstration use case shows the field unit receiving a warning about a fire reported at the specified site. The closest field unit is being notified. Using the campus WLAN, the central ArcIMS server is contacted and a map of that part of the campus is acquired, as well as blueprints of buildings endangered by fire. The central server also contains thematic data about the estimated number of people in these buildings, evacuation plans, and similar information. Simultaneously, units on site can update fire reports with more detailed information and therefore shorten response time of other units enroute. The ArcPad application customized for this use and being used in this scenario is shown in Figure 1.

Figure 1. Mobile GIS implemented in San Diego State University campus security

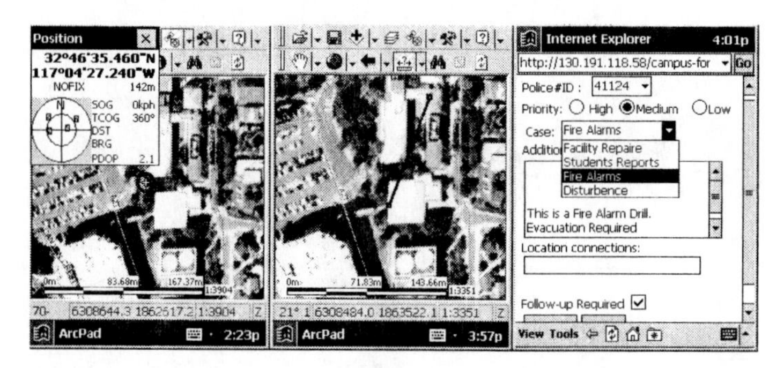

Besides the location of the user, contemporary mobile GIS applications, such as the one previously described, lack the support for context awareness. Such support must be developed and integrated into the basic framework or platform on top of which the mobile GIS application is developed. But first we must define the context and basic principles of context awareness.

In interpersonal communication, a significant amount of information is transmitted without explicit communication of such information. If we take verbal communication as an example, nonverbal signs will significantly influence the completeness of verbally communicated data. We are referring to facial expressions, body posture, voice tone, and nearby objects and persons included in the past history of communications. All this is helping the process of interpretation of verbally transmitted data. In a typical human-machine communication, there is very little context information available in a form that can be interpreted by machine. Therefore our first step should be to define the context. No matter how obvious this may seem, the definition of the context influences significantly all the decisions in the further process of context-aware application development. Dey and Abowd (2000) give a relatively abstract definition of context influenced by their work on "context toolkit" architecture:

We define context as any information that can be used to characterize the situation of an entity, where an entity can be a person, place, or physical or computational object.

Schilit and Theimer (1994) give a very concrete context definition which is therefore rather local in its application:

Context refers to location, identity of spatially nearby individuals and objects and changes that are relevant to aforementioned individuals and objects.

Summarizing numerous definitions of context, we can notice three aspects of context that are standing out:

- **Technical Characteristics of the Environment:** Hereby we are mainly referring to technical characteristics of the client device, processing power, available memory capacity, display characteristics, as well as characteristics of network connections available to the device (bandwidth, latency, price, etc.).
- **Logical Characteristics of the User's Environment:** This group contains geographic location, identity of individuals and objects nearby, and general social situation.
- **Physical Characteristics of the User's Environment:** This group contains levels of noise, light, and movement parameters (speed, direction, etc.).

In the process of context modeling and management, the system can use information that is both automatically collected or manually entered by the user. Although the first approach is attractive and seems to be the only true manner of handling contextual data, we believe that manual input should not be excluded. Also, some characteristics of the context (e.g., user preferences, history, and predictions of actions) are much more easily acquired by manual input at the current level of advancements in context management algorithms.

The important step in development of a context-aware LBS and mobile GIS is to define the set of functionalities the application should provide to the user, implicitly or explicitly. Numerous types of contextual information produce adequately numerous potential functionalities. We can group them as follows:

- **Display of Information and Services:** In order to reduce user workload, the system adjusts the set of offered information and functions according to detected and deduced environment of the user. For a typical mobile GIS, a section of the map surrounding the current user's location is displayed. According to the user's speed and heading, the central point of the map view is chosen and the speed vector displayed. Also, font and color scheme are adjusted to the situ-

Figure 2. XML scheme describing profile

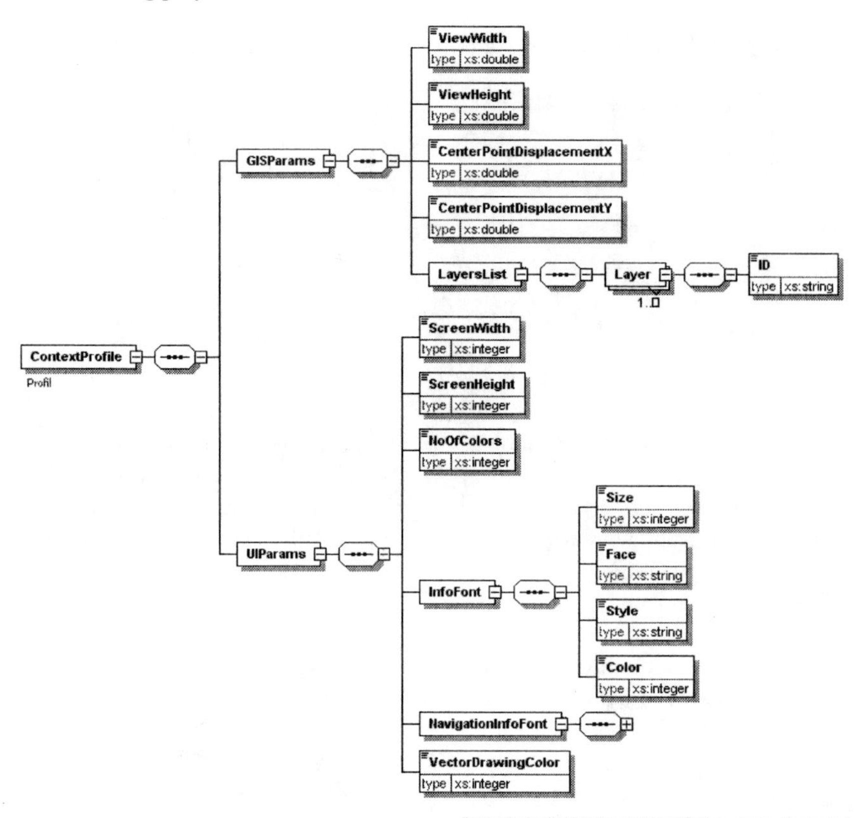

ation the user is in (e.g., the user is steering a vehicle at night).

- **Automated Execution of Commands:** An example would be a navigation GIS application that detects the user has missed the intersection and automatically initiates rerouting to find the new shortest path to destination.
- **Storage of Contextual Information:** Potential use of stored contextual information would be to enable application to autonomously extract user preferences from previous actions using data mining techniques.

The user context that is of interest in LBS and mobile GIS applications is classified into specific classes. Each class of contextual information is assigned a context variable. Often, in other papers published by researchers in this field, authors have noticed a hierarchical structure of context information, so some sort of graph structure is used for context representation (Meissen, Pfennigschmidt, Voisard, & Wahnfried, 2004). Since one class contains contextual information of various levels of generality, the most appropriate data structure for representing contextual information is directed acyclic graph. This data structure is the closest match to human cognition of structure and connections existing within a context data class. Another advantage of the hierarchical context model is the possibility to narrow down the choice of possible actions induced by a detected context. In this manner, a set of rules used by the rule-based expert system is kept to a minimum of candidate rules (Biegel & Cahill, 2004). The rule-based expert system is used to perform generalization of raw contextual data acquired from sensors and contained in leaf nodes. In this manner a "vertical" structure within each context class is built. As an example of a rule-based expert system that is widely used in literature, we have opted for the C Language Integrated Production System (CLIPS, 2006). The main advantage of CLIPS in our case is the existence of jCLIPS, a library that enables Java programs to use the CLIPS engine, embedding it in a Java code (jCLIPS, 2005).

The typical context data flow path in a context-aware application is as follows: raw data is collected by connected sensors, and the software interface associated with each of the sensors converts the data into facts and stores the facts into the expert system. After each modification of a rule set, CLIPS executes a generalization process and generates the new facts at higher levels of generality. Also, the possibility of performing action is tested. The action is represented by forming an XML file containing configuration parameters for the client device. This XML file generally describes a profile that a context-aware application will use as a response to a context change. The XML scheme of such a profile is shown in Figure 2.

Figure 3. The XML file representing the user profile according to the user's context


```xml
<?xml version="1.0" encoding="UTF-8"?>
<ContextProfile xmlns="http://gislab.elfak.ni.ac.yu/bpredic" xmlns:xsi="http://www.w3.org/2001/XMLSchema-instance" xsi:schemaLocation="http://gislab.elfak.ni.ac.yu/bpredic:\Dragan\CONTEXT\sema.xsd">
  <GISParams>
    <ViewWidth>1000</ViewWidth>
    <ViewHeight>700</ViewHeight>
    <CenterPointDisplacementX>850</CenterPointDisplacementX>
    <CenterPointDisplacementY>0</CenterPointDisplacementY>
    <LayersList>
        <Layer>
            <ID>Gas_Stations</ID>
        </Layer>
        <Layer>
            <ID>Fast_Food_Restaurants</ID>
        </Layer>
    </LayersList>
  </GISParams>
  <UIParams>
    <ScreenWidth>200</ScreenWidth>
    <ScreenHeight>320</ScreenHeight>
    <NoOfColors>4092</NoOfColors>
    <InfoFont>
        <Size>12</Size>
        <Face>Courier</Face>
        <Style>Normal</Style>
        <Color>Yellow</Color>
    </InfoFont>
    <NavigationInfoFont>
        <Size>24</Size>
        <Face>Arial</Face>
        <Style>Bold</Style>
        <Color>Red</Color>
    </NavigationInfoFont>
    <VectorDrawingColor>Blue</VectorDrawingColor>
  </UIParams>
</ContextProfile>
```


The particular profile transferred to the client is represented as an XML file described in Figure 3.

GinisMobile: CONTEXT-AWARE MOBILE GIS FRAMEWORK

To support efficient development of mobile GIS solutions, GinisMobile as a component mobile GIS framework is developed in the Laboratory for Computer Graphics and Geographic Information Systems (CG&GIS) Lab at the University of Nis. This framework represents the result of continuous development and advancement of GIS frameworks intended for rapid development of desktop and Web GIS applications (Ginis and GinisWeb). Well-tested GIS concepts of Ginis and GinisWeb have been supplemented with application components from the mobility domain (Predic & Stojanovic, 2004).

Mobile GIS architecture is somewhat similar to architectures used for Web or WAP GIS solutions (Fangxiong & Zhiyong, 2004). An extended client/server approach is used as the basis. The most frequent modification of this architecture, which is used in commercial solutions, is the three-tier model. It consists of the presentation layer, and the GIS logic layer including mobile database components and external GIS services layer. The third layer encompasses GIS services like Web map service (WMS) and Web feature service (WFS) (OGC, 2003). The main advantage of this approach is clear separation of functionalities into independent modules, easily upgradeable and substitutable. The grouping of layers in the case of mobile GIS is shown in Figure 4.

The presentation layer is tasked with information visualization (spatial data in the form of maps and attributes in alphanumerical tabular form). This layer also receives users' requests and queries, interprets them, and activates corresponding functions in the application logic layer. Using adapted HTTP protocol, the client is supplied with static

Figure 4. Layered architecture of mobile GIS

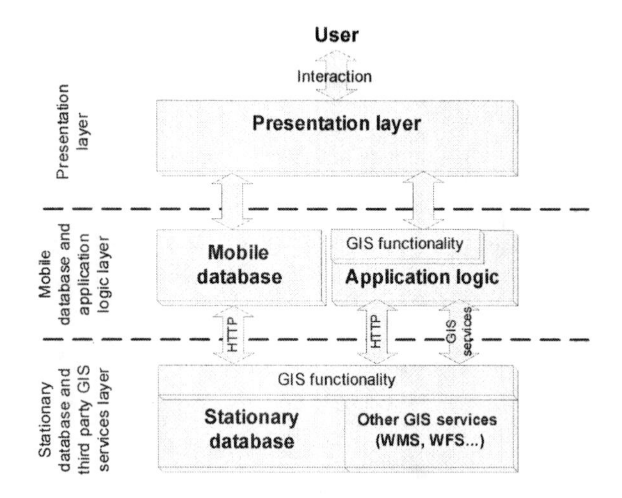

sections of maps in the form of raster images and vector type spatial data which is XML encoded. XML encoding also contains attribute data. All data the user interacts with and can change are classified and transferred in the vector form. All other data, regardless of their form when stored on the server side, are rasterized and transferred to the client in the form of raster images. Since the data's role is only auxiliary, this is the most appropriate form of visualization.

Database component must contain the functionality of partial replication of the data subset that is of interest to the individual user and data synchronization with local data storage located on the mobile device (Huang & Garcia-Molina, 2004). In our usage scenario (mobile GIS), the client possesses significant processing power and memory capacity which can therefore be used for performance improvement of this typical layered architecture. That is the reason the client-side component in the case of mobile GIS is usually named 'rich client' (Predic, Milosavljevic, & Rancic, 2005).

The mobile database and application logic layer is physically located at the client device according to its significant processing capacities. Application logic, which is also physically located at the client device, contains a portion of GIS functionalities, basic functionalities which can be performed on the client side solely without data transfer to/from the server side. The stationary database and third-party GIS services layer (e.g., Web map server) is physically located on the server side. The stationary database contains the complete set of data available. The mobile GIS client is supplied only with a subset of available data, a subset that is of direct interest to the individual client (fieldwork team). Determining the scope and volume of this subset is the task of the GIS functionalities component located in this layer. Other GIS services belonging to this layer (WMS, WFS) are also controlled by GIS functionalities of this layer. WMS is tasked with supplying raster map segments which are used at the client to form a continuous geo-referenced map. WFS provides data about geographic features in the vector format encoded in Geography Markup Language (GML, 2004).

To support development of context-aware LBS and mobile GIS applications, we have developed and integrated context-aware support and components into GinisMobile, an LBS and mobile GIS application framework (Predic et al., 2005). GinisMobile is a mobile extension of GinisWeb, a Web GIS application framework (Predic & Stojanovic, 2004). As such it includes support for management and presentation of raster and vector spatial data, as well as dynamic data about mobile objects (Stojanovic, Djordjevic-Kajan, & Predic, 2005). The first obstacle encountered when developing LBS and mobile GIS applications is a highly constrained mobile client platform. Therefore, these applications are already aware of the hardware characteristics of the device it is running on and able to automatically adapt to it, enabling full utilization of the device's capabilities. The type of sensors relevant to context-aware LBS and mobile GIS applications are widely available, and either are already integrated in modern devices (GPS sensors, Bluetooth radios, level of light sensors, etc.) or are available as add-on devices connected via PAN (Bluetooth). Each of these sensors implements its internal data format. Therefore, each sensor has a software interface attached to it. Its task is to convert data from the format used internally by the sensor into a format appropriate to the application. Contextual data concerning technical characteristics of the device are accessible directly by the application and therefore do not require a separate software interface. A compiled set of contextual data is encoded according to a defined XML scheme and transferred to the server for analysis and storage. The proposed architecture of GinisMobile, a context-aware LBS and mobile GIS framework, is illustrated in Figure 5.

This model requires minimal changes to starting a mobile GIS architecture and minimizes processing requirements on the client side. On the server side, context information is handled separately from the user commands. It is inserted into the rules and facts database, and is analyzed by a rule-based expert system. Every change in the rules and facts database can lead to either insertion of new context information at the higher logical level into the database or entering a new state. Reaching a new state results in picking out a profile that most adequately fits new change in the user's context. Handling of other spatial data is the same as in Predic and Stojanovic (2005) and will not be further discussed in this article. Profile, packed with static spatial layers (rasterized to a single map layer) and a dynamic map component (e.g., moving objects), is transferred to the client.

On the client side, the XML profile is parsed and used to customize the user interface. Rasterized layers are stored on internal cache and displayed. Static objects are presented on the background map according to display settings, profile,

Figure 5. Mobile GIS architecture with context-aware support

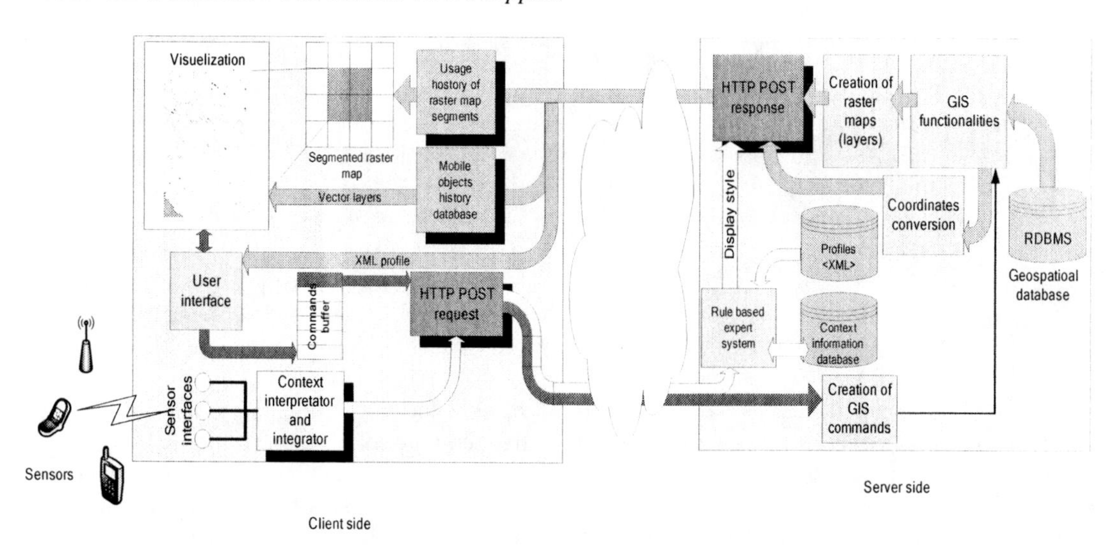

and with appropriate symbols. Finally, moving objects are superimposed on the map display. Since raster segments are static in nature and change rarely, we keep a local cache of frequently used segments. This approach speeds up the visualization process significantly. The adopted least recently used (LRU) algorithm is used to keep memory requirements minimal.

To test the context awareness support and context-aware components built into the GinisMobile framework, we have developed a mobile GIS application for vehicle navigation and fleet tracking on top of the GinisMobile. The application setting assumes that the sensors connected to the user's device are able to determine speed and direction, time of a day, and levels of noise and light. The higher level of contextual information is deduced based on basic contextual data. Based on deduced facts and rules within the knowledge-based engine, the server recognizes that the user drives a vehicle during the evening/night hours. According to this information, an appropriate profile is constructed which describes the user interface with night colors. The navigation data are displayed on the screen with appropriate font size according to the speed of the user's vehicle. Also, appropriate zoom level is chosen with the user's location displaced from the view center. In this manner the user is enabled to see more of the map in front of him. Finally, the view contains vector speed as a reference.

We assume the existence of a GPS receiver attached to the mobile device since this is a very common type of sensor today. It provides data on a user's geographic location as well as motion data (speed and direction). Another "sensor" relies on time of day to detect light conditions (day/night). More specifically, a level-of-light sensor that is present on mobile devices available on the market could be used for this purpose for additional accuracy.

The role of context detection and interpretation subsystem is to decrease the workload needed to operate a typical vehicle navigation system and therefore increase safety. In the demonstration application the following scenario is employed: a mobile user drives a vehicle in the urban environment during the day and at night. Screenshots are taken at different levels of adaptation that an LBS application has performed autonomously in response to changing user context. Figure 6(a) shows a user traveling at 20km/h

Figure 6. Screenshots taken from the sample mobile GIS application

(a)

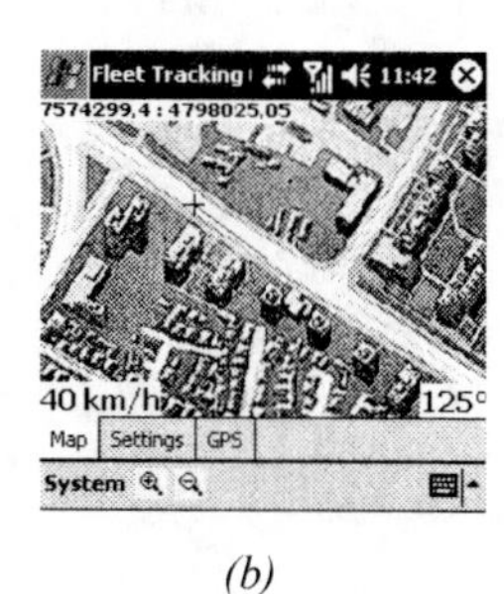

(b)

(c)

Figure 6. Continued

(d)

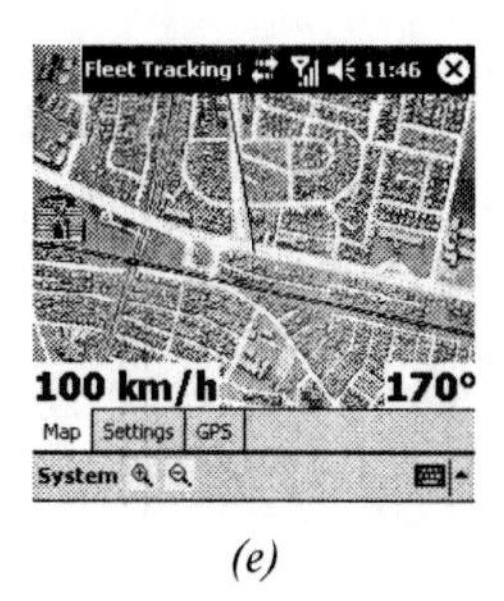

(e)

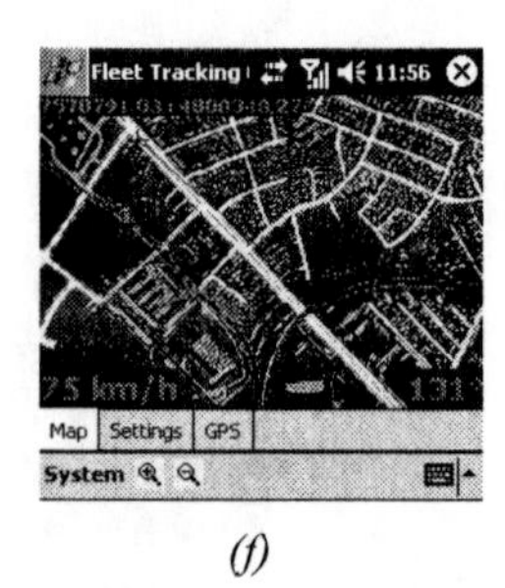

(f)

along a city street. It is worth noticing that the user's location (indicated by the cross symbol) is decentralized and a velocity vector is drawn on the map view. The map view also includes speed and heading.

As the user increases his speed, the font size for displayed motion data (speed and heading) is increased, the amount of map view decentralization is increased, and the velocity vector is updated accordingly. This is illustrated in Figure 6(b). As the speed further increases above a certain threshold (Figure 6(c)), the map view zoom scale is changed (decreased). This, along with additional decentralization of a map view, allows the user to see more of the map laying in front of him, in the direction of the velocity vector. The effects of further increase of speed and change in direction of velocity vector are shown in Figures 6(d) and 6(e). When the application detects night conditions, it switches to using a set of colors customized to night conditions. The map view customized to night conditions is shown in Figure 6(f). This choice of colors minimizes the distraction effect for the user (driver).

CONCLUSION

Considering the general trend in development of information technologies and the computer industry in general, we can notice a constant migration into mobile and ubiquitous computing (Hinze & Voisard, 2003). With further development of wireless communication technologies, data stored in heterogeneous distributed databases will be available at any specific instance in time and at any location. The most important beneficiaries of this newly introduced concept of mobile GIS will be professional users whose job descriptions include a lot of field work with spatial data. Considering the current state of the art, we believe that mobile GIS applications are perfect testing grounds for the context-aware concept. Being used in unconstrained free space, context awareness considerably enhances usability of the mobile GIS applications. Hereby, context-aware applications are a super set of location-based services. As this article has stressed, location information is only one class, although very frequently used, of context information. This information will be used to automate many procedures and decisions, and relieve the user of the repeatable and tedious tasks of frequent reconfiguration.

REFERENCES

Biegel, G., & Cahill, V. (2004, March 14-17). A framework for developing mobile, context-aware applications. *Proceedings of the 2nd IEEE Conference on Pervasive Computing and Communications (Percom 2004)*, Orlando, FL.

CLIPS. (2006). *Version 6.24: A tool for building expert systems.* Retrieved from http://www.ghg.net/clips/CLIPS.html

Dey, A. K., & Abowd, G. D. (2000, April 1-6). Towards a better understanding of context and context-awareness. *Proceedings of the Workshop on the What, Who, Where, When and How of Context-Awareness* (affiliated with CHI 2000), The Hague, The Netherlands.

Fangxiong, W., & Zhiyong, J. (2004, July 12-23). Research on a distributed architecture of mobile GIS based on WAP. *Proceedings of the ISPRS Congress,* Istanbul, Turkey.

GML. (2004). *Geography Markup Language (version 3.1.1)—encoding specification.* Retrieved from http://portal.opengeospatial.org/files/?artifact_id=4700

Hinze, A., & Voisard, A. (2003, July 24-27). Location- and time-based information delivery. *Proceedings of the 8th International Symposium on Spatial and Temporal Databases,* Santorini Island, Greece.

Huang, Y., & Garcia-Molina, H. (2004). Publish/subscribe in a mobile environment. *Wireless Networks, 10*(6), 643-652.

JCLIPS. (2005). Retrieved from http://www.cs.vu.nl/~mrmenken/jclips/#developer

Meissen, U., Pfennigschmidt, S., Voisard, A., & Wahnfried, T. (2004). Context and situation awareness in information logistics. *Proceedings of the Workshop on Pervasive Information Management* (held in conjunction with EDBT 2004).

Open GIS® Reference Model. (2003). *Version 0.1.2, Open GIS Consortium, Reference Number: OGC 03-040.* Retrieved from http://www.opengis.org/specs/?page=orm

Predic, B., & Stojanovic, D. (2004, March 8-12). XML integrating location based services with Web based GIS. *Proceedings of YU INFO 2004,* Kopaonik, Serbia, and Montnegro.

Predic, B., & Stojanovic, D. (2005, May 26-28). Framework for handling mobile objects in location based services. *Proceedings of the 8th Conference on Geographic Information Science (AGILE 2005)* (pp. 419-427), Estoril, Lisbon, Portugal.

Predic, B., Milosavljevic, A., & Rancic, D. (2005, June 5-10). RICH J2ME GIS client for mobile objects tracking. *Proceedings of the XLIX ETRAN Conference,* Budva.

Raento, M., Oulasvirta, A., Petit, R., & Toivonen, H. (2005). Context phone: A prototyping platform for context-aware mobile applications. *IEEE Pervasive Computing—Mobile and Ubiquitous Systems,* (April-June), 51-59.

San Diego State University. (2002). *Integrated mobile GIS and wireless Internet image servers for environmental monitoring and management.* Retrieved from http://map.sdsu.edu/mobilegis/photo_mtrp.htm

Schilit, B., & Theimer, M. (1994). Disseminating active map information to mobile hosts. *IEEE Network, 8*(5), 22-32.

Stojanovic, D., Djordjevic-Kajan, S., & Predic, B. (2005, December 15-16). Incremental evaluation of continuous range queries over objects moving on known network paths. *Proceedings of the 5th International Workshop on Web and Wireless Geographical Information Systems* (LNCS 3833, pp. 168-182), Lausanne, Switzerland. Berlin: Springer-Verlag.

KEY TERMS

Context-Aware Application: Application that posses the ability to autonomously and independently detect and interpret a user's environment parameters, and adapts its performance and functionalities according to detected context.

Geographic Information System (GIS): Information system that stores, analyzes, and presents data about geographic entities.

Geography Markup Language (GML): The XML grammar defined by the Open Geospatial Consortium (OGC) to express geographical features.

Global Positioning System (GPS): Satellite-based system that is using radio triangulation techniques to determine geographic position time and speed of user.

Location-Based Service (LBS): Service that delivers geographic information and geo-processing services to a mobile/stationary user, taking into account the user's current location and references, or locations of the stationary/mobile objects of the user's interests.

Web Feature Service (WFS): Web-accessible service that provides data about geographic entities encoded in GML using HTTP protocol.

Web Map Service (WMS): Web-accessible service that provides geo-referenced raster map data using HTTP protocol.

Context-Aware Systems

Chin Chin Wong
British Telecommunications (Asian Research Center), Malaysia

Simon Hoh
British Telecommunications (Asian Research Center), Malaysia

INTRODUCTION

Fixed mobile convergence is presently one of the crucial strategic issues in the telecommunications industry. It is about connecting the mobile phone network with the fixed-line infrastructure. With the convergence between the mobile and fixed-line networks, telecommunications operators can offer services to users irrespective of their location, access technology, or terminal.

The development of hybrid mobile devices is bringing significant impact on the next generation of mobile services that can be rolled out by mobile operators. One of the visions for the future of telecommunication is for conventional services such as voice call to be integrated with data services like e-mail, Web, and instant messaging. As all these new technologies evolve, more and more efforts will be made to integrate new devices and services. New markets for services and devices will be created in this converged environment. Services become personalized when they are tailored to the context and adapted to changing situation.

A context-aware network system is designed to allow for customization and application creation, while at the same time ensuring that application operation is compatible not just with the preferences of the individual user, but with the expressed preferences of the enterprise or those which own the networks. In a converged world, an extended personalization concept is required. The aspects covered include user preferences, location, time, network, and terminal; these must be integrated and the relationships between these aspects must be taken into consideration to design business models. Next-generation handsets are capable of a combination of services available on a personal digital assistant (PDA), mobile phone, radio, television, and even remote control. This kind of information and communications technology and mobile services together form one of the most promising business fields in the near future.

The voice average revenue per user (ARPU) is declining, the competition is getting fiercer, and voice over Internet protocol (VoIP) is entering the market with aggressive pricing strategies. Fixed mobile convergence should help in this context by providing converged services to both consumer and small-business users. For telecommunication companies it is now crucial to attempt to identify concrete applications and services for commercial offerings based on fixed mobile convergence which go beyond the current hype. Market scenarios and business models for such fixed mobile convergence solutions will be required and are therefore valuable for future strategy decisions.

This article examines market aspects, user requirements, and usage scenarios to come up with a roadmap and suggestions on how to deal with this matter.

CURRENT AND FUTURE TRENDS

In the past, user movement has often implied interruption of service. With the advent of pocket-size computers and wireless communication, services can be accessed without interruption while the entity using the services is moving (Floch, Hallsteinsen, Lie, & Myrhaug, 2001). There is a strong need for seamless access. Convergence has been taking place for years now. A study performed by the European Commission (1997) defines convergence as allowing both traditional and new communication services, whether voice data, sound, or pictures to be provided over many different networks. An excellent example of convergence in the telecommunications industry is the IP multimedia subsystem (IMS).

Similar to other emerging industries, fixed mobile convergence is characterized by a continuously changing and complex environment, which creates uncertainties at technology, demand, and strategy levels (Porter, 1980). Porter (1980) asserts that it is possible to generalize about processes that drive industry evolution, even though their speed and direction vary. According to Ollila, Kronzell, Bakos, and Weisner (2003), these processes are of different types and are related to:

- market behavior;
- industry innovation;
- cost changes;
- uncertainty reduction; and
- external forces, such as government policy and structural change in adjacent industries.

Each evolutionary process recognizes strategic key issues for the companies within the industry, and their effects are usually illustrated as either positive or negative from an industry development viewpoint. For example, uncertainty reduction is an evolutionary process that leads to an increased diffusion of successful strategies among companies and the entry of new types of companies into the industry. Both of these effects are believed to contribute to industry development with regards to the fixed mobile convergence value Web.

The technological uncertainties are usually caused by fast technological development and the battles for establishing standards, which are common in the beginning stages of the lifecycle of a specific industry as a result of a technological innovation (Camponovo, 2002). Concerning demand, regardless of the generalized consensus about the huge potential of fixed mobile convergence, there are many uncertainties about what services will be developed, whether the users are ready to pay for them, and the level and timeframe of their adoption (Camponovo, 2002).

While the wireless industry is often cited as an example of a rapidly changing sector, the period from 2001-2005 could (in some respects) be regarded as relatively stable (Brydon, Heath, & Pow, 2006). Mobile operators have made the vast majority of their service revenue from simple voice telephony and text messaging, while their value chain has remained largely undisturbed (Brydon et al., 2006). However, new services, alternative technologies, and an evolving competitive landscape mean that the possibility of substantial industry change over the course of the next five to 10 years cannot be discounted (Brydon et al., 2006).

The telecommunication industry has experienced several waves of changes from the introduction of wired telephony to wireless telephony, and it is currently heading towards fixed-mobile convergence. Users become more demanding: a "user-centric" and not "network-centric" approach is needed.

According to Hellwig (2006), many fixed operators lose their market dominance and merge units (fixed and mobile). New technologies and new actors (e.g., VoIP, Wi-Fi operators) coming into the picture are driving the adoption of fixed mobile convergence. The formation of new roles in the communication industry—including brokers, aggregators, alliances, and cooperation—have further pushed the stakeholders to take aggressive strategies to gain competitive advantage.

However, since new roles have been introduced, it is unclear how the market acceptance in the near future will be. Existing business models might not be applicable in the new business environment. The lack of terminal devices at the moment also hinders the diffusion.

In markets where there are high levels of fixed-mobile substitution and where broadband penetration and wireless local area network (WLAN) diffusion in the home are accelerating, it is most likely that consumers will be drawn to fixed mobile convergence, provided the cost savings and added convenience of carrying one device are apparent to the consumer (McQuire, 2005). According to the Yankee Group (2005), almost one-third of users make more calls within the home using their mobile phone than their landline. The trend is stronger among younger respondents. Figure 1 shows the number of fixed mobile convergence households in Spain, the UK, and France from 2006 to 2010.

The increasing need for a personal communication device that can connect to any type of network—a mobile network, IP network, or even public switched telephone network (PSTN)—and that supports all voice and text-based communication services drives the development of context-aware systems. The primary objective of the system is to facilitate acquisition, translation, and representation of context information in a structured and extensible form, in order to enable the development and enhancement of functionality of network

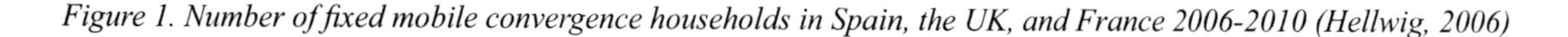

Figure 1. Number of fixed mobile convergence households in Spain, the UK, and France 2006-2010 (Hellwig, 2006)

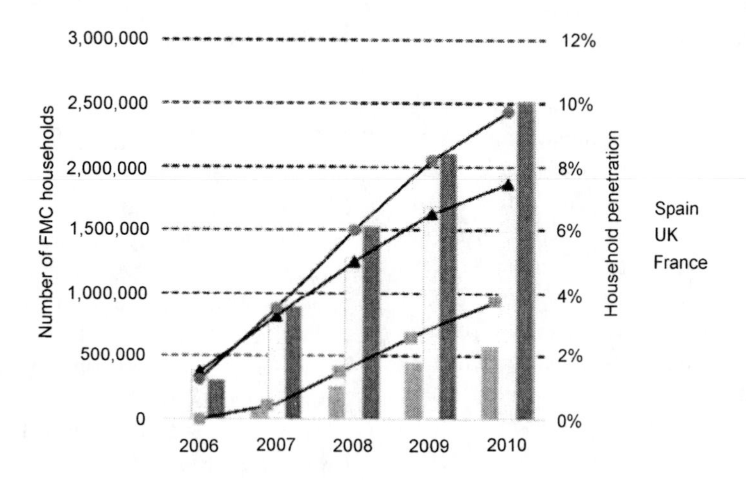

resources, personalized according to each individual's needs. The secondary goal is to facilitate rapid development and deployment of services and applications through a defined framework, which can maintain interoperability between different services and domains.

An example of a context-aware system would be BT's proposed Context Aware Service Platform (CASP). In June 2004, NTT DoCoMo, BT, and a number of other incumbent operators from around the world formed the Fixed Mobile Convergence Alliance (2006), with the objective of developing common technology standards and low-cost devices for integrated fixed-mobile services. The CASP middleware mentioned is the interpretation and one of BT's visions for the development of a converged platform. The salient features of the proposed product include:

- **User-Centered Operability:** One important requirement for a heterogeneous network environment is the ability to instantaneously optimize services for individual users without the need for them to perform any annoying operations. CASP aims to provide transparent connectivity between users with devices and surrounding communication resources. It is able to recognize users' situations and environmental information automatically.
- **Ease of Service Provisioning:** The proposed platform and generic framework guidelines in respect to security, data integrity, non-repudiation, registration, subscription, and quality of service (QoS) for all services will be made available. It offers standard interfaces for all services which enable easier access to a less complex network, with common operation and management, maintenance and training, as well as a common environment for services development and delivery.
- **Interoperability of Shared Services:** The proposed platform provides a common specification for services to guarantee the interoperability between shared services in the communication networks. Specific context information with respect to specific aspects characterizing a service or entity can be expressed in an eXtensible Markup Language (XML)-based instance document.
- **Unified Identity:** In a true seamless access communication world, every user or communication object is represented by a unified identity. A session initiation protocol (SIP) address (e.g., simon.hoh@bt.com) can be used to uniquely identify a user or communication object even when it moves across different networks or between different devices. By having identity management, it simplifies mobility management, security management, and unified user profile management.
- **Dynamic User Interface (UI) on Shared Device:** Through the proposed platform, the user can have a shared device that can connect and interact with the ubiquitous communication objects nearby. Each networked object or entity such as cameras, scanners, printers, video players, and so forth can be represented by a different UI based on its own dynamic profile and thus can react intelligently to events in the communication space.
- **Context-Enabled Adaptive Service:** The heterogeneity of the converged networks, in terms of network capacity and terminal capabilities, is expected to cause unpredictable changes of network condition. The traditional QoS mechanisms, which do not take the presence of mobility and seamless connectivity into consideration, are not sufficient to guarantee a stable service. Thus, the use of adaptive services being able to change their settings to adapt to the available network resources is a must. CASP enables dynamic selection of the settings used by multimedia services and applications during a multimedia session based on the context of the surrounding environments.

In the near future, stakeholders in the industry will move towards IP-based transport, call control, and service creation and delivery platform functionality. They will follow and adopt developments in the 3rd Generation Partnership Project (3GPP) (http://www.3gpp.org/), European Telecommunications Standards Institute (ETSI) (TISPAN) (http://www.etsi.org/), and Internet Engineering Task Force (http://www.ietf.org/) Next-Generation Network (http://www.ngni.org/) to support open interfaces and avoid interconnection and co-operation incompatibilities. In addition, the will: (1) support IP-based signaling and addressing, media negotiation, QoS, and security mechanisms; (2) support a very large variety of multimedia, banking, and mobile office applications seamlessly across different networks; and (3) adopt seamlessly to the network characteristics and device used. The players in the industry must also consider entering new positions in the value network by taking on new roles and working in cooperation with other telecommunication companies (Hellwig, 2006).

Figure 2 shows how and when future enterprise telephony services will embrace convergence.

USAGE SCENARIOS

Cheryl subscribes to an Internet VoIP service to save money on calls to her family and university friends who are now spread around the globe, but since her mobile operator utilizes unlicensed mobile access (http://www.umatechnology.org/) technology, she is now able to enjoy cheaper calls by using her mobile phone and connecting to her home WLAN or public hotspots. Her mobile device also enables a number of rich services that enable her to communicate with her friends via voice, video, as well as text.

Figure 2. Future enterprise telephony services will embrace convergence

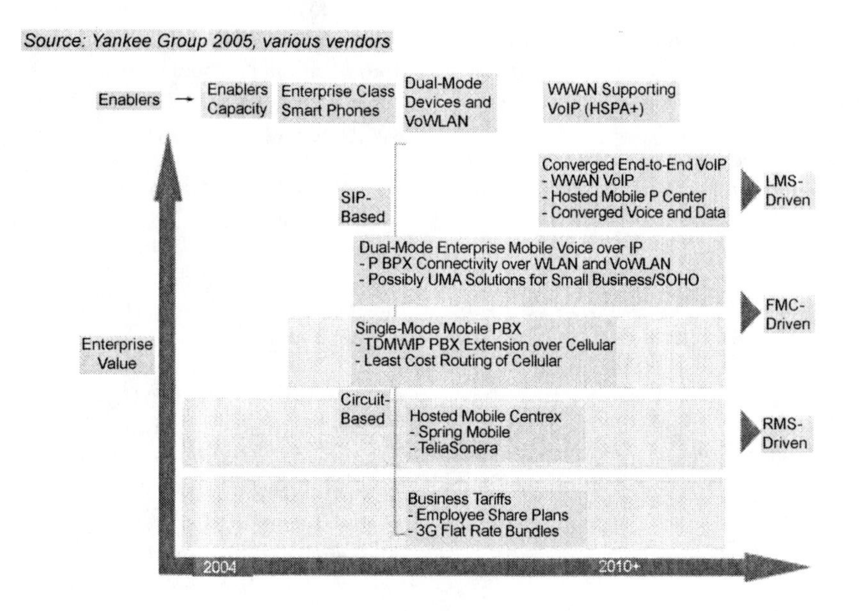

When she was on vacation in Hawaii recently, she was able to show the pictures she had taken with her mobile device to a colleague while he was talking to her over a VoIP call, and later sent a "wish you were here" video message to her parents.

Since all her communications are unified in a single device, Cheryl's friends and family can always reach her, either by voice, text, instant message, video call, or any other means, while Cheryl can use presence to broadcast her availability to her contacts (such as "in a meeting" or "traveling"), as well as manage the incoming communication depending on the context of what she is doing.

Now, she can keep her long-distance bills lower by using VoIP, but since it is in her mobile device, she does not have to be sitting in front of her personal computer (PC) to use it. And whereas some of the PC-based services Cheryl previously used were cumbersome to set up, and she had separate providers for her telecom services, she now gets all these functionalities in a bundled offering from a single operator, and all technology takes care of itself, working invisibly to her through a user-friendly interface.

Figure 3.

Cheryl is a recent university graduate whose work as a researcher means she has a busy travel schedule. She often travels to several places to attend international conferences and workshops. She uses a multi-radio mobile device and a mobile VoIP service subscription to keep in touch with her friends and family whether she is at home or on the move.

CONCLUSION

Context-aware systems, when made available to the end users, will be greatly valued by them. Consumers will be in a more interactive environment that could help them to take care of small yet related matters automatically. Any possible devices around them could be used to bridge any services offered, giving them the familiarity they preferred. Users would always have the option to alter the service execution or switch it off anytime they like.

Meanwhile from the network perspective, knowing the situation of the network and each network node's role could enable an adaptive and intelligent network. The capabilities such as self-healing, autonomous utilization optimization, and self-reconfiguration to adapt for changes could also be enabled with context-sensitive service logic.

The CASP provides a stable and robust environment for the context-aware service developers and operators. This stable environment is extremely important for them to have the accurate anticipated outcome and have the flexibility on changes. On top of the stable environment, the platform will be assisting the service execution to reduce the complexity for the service creation.

REFERENCES

Brydon, A., Heath, M., & Pow, R. (2006). *Scenarios for the evolution of the wireless industry in Europe to 2010 and beyond.* Analysys Research.

Camponovo, G. (2002). *Mobile commerce business models.* Paper presented at the International Workshop on Business Models, Lausanne, Switzerland.

European Commission. (1997). *Towards an information society approach.* Green paper on the convergence of the telecommunications, media and information technology sectors, and the implications for regulation. European Union.

Floch, J., Hallsteinsen, S., Lie, A., & Myrhaug, H. I. (2001, November 26-28). *A reference model for context-aware mobile services.* Tromsø, Norway: Norsk Informatikkonferanse.

FMCA. (2006). Retrieved from http://www.thefmca.com/

Hellwig, C. (2006). *New business and services by converging fixed and mobile technologies and applications.* T-Systems.

McQuire, N. (2005). *Residential fixed-mobile convergence heats up but SME Is next frontier.* Yankee Group.

Ollila, M., Kronzell, M., Bakos, M., & Weisner, F. (2003). *Mobile entertainment industry and culture: Barriers and drivers.* UK: MGAIN.

Porter, M. (1980). *Competitive strategy.* New York: The Free Press.

Yankee Group. (2005). *European mobile user survey.* Yankee Group.

KEY TERMS

eXtensible Markup Language (XML): A specification developed by the W3C, XML is a pared-down version of the Standard Generalized Markup Language (SGML), designed especially for Web documents. It allows designers to create their own customized tags, enabling the definition, transmission validation, and interpretation of data between applications and between organizations.

Internet Protocol Multimedia Subsystem (IMS): A standardized next-generation networking (NGN) architecture for telecommunication companies that want to provide mobile and fixed multimedia services. It uses a VoIP implementation based on a 3GPP standardized implementation of session initialization protocol (SIP), and runs over the standard Internet protocol (IP). Existing phone systems (both packet-switched and circuit-switched) are supported.

Public-Switched Telephone Network (PSTN): The international telephone system based on copper wires carrying analog voice data. This is in contrast to newer telephone networks based on digital technologies, such as the integrated services digital network (ISDN) and fiber distributed data interface (FDDI).

Quality of Service (QoS): A networking term that specifies a guaranteed throughput level. One of the biggest advantages of asynchronous transfer mode (ATM) over competing technologies such as frame relay and fast ethernet is that it supports QoS levels. This allows ATM providers to guarantee to their customers that end-to-end latency will not exceed a specified level.

Session Initiation Protocol (SIP): An application-layer control protocol; a signaling protocol for Internet telephony. SIP can establish sessions for features such as audio/videoconferencing, interactive gaming, and call forwarding to be deployed over IP networks, thus enabling service providers to integrate basic IP telephony services with Web, e-mail, and chat services. In addition to user authentication, redirect, and registration services, the SIP server supports traditional telephony features such as personal mobility, time-of-day routing, and call forwarding based on the geographical location of the person being called.

Unlicensed Mobile Access (UMA): The technology that provides access to global system for mobile communications (GSM) and general packet radio service (GPRS) mobile services over unlicensed spectrum technologies, including Bluetooth and 802.11. By deploying UMA technology, service providers can enable subscribers to roam and handover between cellular networks and public and private unlicensed wireless networks using dual-mode mobile handsets.

Voice Over Internet Protocol (VoIP): The routing of voice conversations over the Internet or any other IP-based network. The voice data flows over a general-purpose packet-switched network, instead of traditional dedicated, circuit-switched telephony transmission lines. Voice over IP traffic might be deployed on any IP network, including those lacking a connection to the rest of the Internet, for instance on a private building-wide LAN.

Contractual Obligations between Mobile Service Providers and Users

Robert Willis
Lakehead University, Canada

Alexander Serenko
Lakehead University, Canada

Ofir Turel
McMaster University, Canada

INTRODUCTION

The purpose of this chapter is to discuss the effect of contractual obligations between users and providers of mobile services on customer loyalty. One of the unique characteristics of mobile commerce that distinguishes it from most other goods and services is the employment of long-term contractual obligations that users have to accept to utilize the service. In terms of over-the-counter products, sold in one-time individual transactions in well-established markets, a strong body of knowledge exists that suggests that businesses may enhance loyalty through the improvement of quality and customer satisfaction levels. With respect to mobile commerce, however, this viewpoint may not necessarily hold true given the contractual nature of business-customer relationships.

In the case of mobile computing, it is suggested that loyalty consists of two independent yet correlated constructs that are influenced by different factors: repurchase likelihood and price tolerance. Repurchase likelihood is defined as a customer's positive attitude towards a particular service provider that increases the likelihood of purchasing additional services or repurchasing the same services in the future (e.g., after the contract expires). For example, when people decide to purchase a new mobile phone, they are free to choose any provider they want. In other words, repurchase likelihood is not affected by contractual obligations. In contrast, price tolerance corresponds to a probability of staying with a current provider when it increases or a competitor decreases service charges. In this situation, individuals have to break the existing contractual obligations. Currently, there is empirical evidence to suggest that the discussion above holds true in terms of mobile computing. However, there are few well-documented works that explore this argument in depth. This article attempts to fill that void.

This article will present implications for both scholarship and practice. In terms of academia, it is believed that researchers conducting empirical investigations on customer loyalty with mobile services should be aware of the two independent dimensions of the business-customer relationship and utilize appropriate research instruments to ensure the unidimensionality of each construct. With regards to practice, it is suggested that managers and marketers be aware of the differences between repurchase likelihood and price tolerance, understand their antecedents, and predict the consequences of manipulating each one. It is noted that overall loyalty is not the only multidimensional constuct in mobile commerce. Recently, it was emperically demonstrated that perceived value of short messaging services is a second-order construct that consists of several independent yet correlated dimensions (Turel et al., 2007).

Theoretical separation of the overall loyalty construct into two dimensions has been already empirically demonstrated in three independent mobile commerce investigations. First, Turel and Serenko (2006) applied the American customer satisfaction model (ACSM) to study mobile services in North America. By utilizing the original instrument developed by Fornell, Johnson, Anderson, Cha, and Bryant (1996), they discovered a low reliability of the overall satisfaction construct, and found that the correlation between two items representing price tolerance and one item reflecting repurchase likelihood was only 0.21 ($p<0.01$, $N=204$). Second, Turel et al. (2006) adapted the ACSM to study the consequences of customer satisfaction with mobile services in four countries (Canada, Finland, Israel, and Singapore), and reported that the correlation between price tolerance and repurchase likelihood was 0.20 ($p<0.01$, $N=736$). Third, Yol, Serenko, and Turel (2006) analyzed the ACSM with respect to mobile services in the U.S. and again found the same correlation to be 0.45 ($p<0.01$, $N=1,253$). All these correlations fall into the small-to-medium range, and two of them are beyond the lowest cut-off value of 0.35 for item-to-total correlation (Nunnally & Bernstein, 1994). The statistical significance of these correlations is explained by large sample sizes. Therefore, it is impossible to design a single unidimensional construct in mobile commerce research consisting of both price tolerance

Figure 1. The American Customer Satisfaction Model (adapted from Fornell et al., 1996)

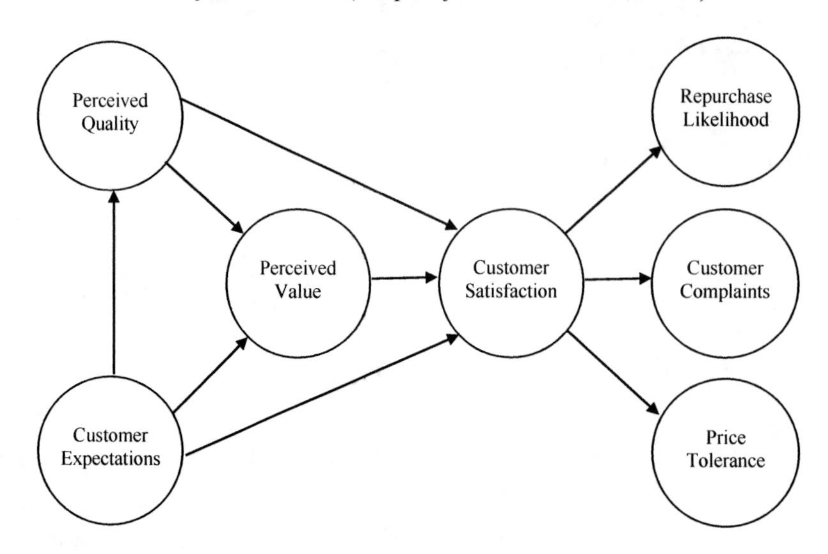

and repurchase likelihood. In all of these studies, most users had long-term contractual obligations with their respective mobile service provider that confirms the validity of the aforementioned conceptual discussion.

To better understand the customer loyalty concept in light of contractual obligations, this article briefly describes the American customer satisfaction model (ACSM), and then discusses the concepts of price tolerance and repurchase likelihood. Finally, it presents a summary which outlines implications for research and practice.

THE AMERICAN CUSTOMER SATISFACTION MODEL

The mobile telephony market continues to be one of the fastest growing service sector markets, creating a fiercely competitive industry environment (Kim & Yoon, 2005). As has happened in other, subscription-based mobile service industries, the nature of this competition has changed from the acquisition of new customers to the retention of existing customers and the luring away of competitors' customers. This last strategy is known in the industry as *outbound churn* or, more simply, as *churn*. Given the increasing penetration of mobile computing devices and the maturation of the market, avoiding churn and maximizing customer loyalty has become a primary concern for wireless providers. The first step in minimizing churn in a company's customer base is to understand its root causes.

The determinants of churn may be estimated by the adapted American Customer Satisfaction Model (see Figure 1). The original model suggests that satisfaction affects overall customer loyalty, where loyalty is a unidimensional construct that consists of price tolerance (i.e., the probability of staying with the current provider if it increases prices or if a competitor decreases prices) and repurchase likelihood (i.e., the probability of purchasing the same service again). At the same time, several recent works suggest that these loyalty dimensions are distinct yet correlated because of the contractual nature of the customer-service provider relationship.

Customer loyalty is one of the major constructs in marketing, and a large part of a marketing manager's effort is aimed at creating and maintaining loyalty among an organization's customer base. The significance of loyalty comes from the positive impact it has on the operations of the company in terms of customer retention, repurchase, long-term customer relationships, and company profits (Caruana, 2004). In other words, loyalty is a primary factor in reducing churn.

The notion of switching costs affecting loyalty has been recognized and researched by several professional and academic disciplines, including marketing, economics, and strategy. "Switching costs are generally defined as costs that deter customers from switching to a competitor's product or service" (Caruana, 2004, p. 256). For managers and researchers, it is important to understand the concepts of switching costs and customer loyalty, and to clearly identify both their dimensions and their interaction.

PRICE TOLERANCE

Switching costs are generally defined as one-time costs facing the consumer/buyer of switching from one supplier to another (Porter, 1980; Burnham, Frels, & Mahajan, 2003). Several researchers have identified various attributes or types of switching costs (e.g., Thibault & Kelly, 1959; Klemperer, 1987; Guiltnan, 1989; Burnham et al., 2003; Hu & Hwang,

2006); however, for the purposes of this article, switching costs are broadly categorized as three types: transaction, learning, and contractual. Transaction costs are costs incurred when a consumer begins a relationship with a provider and includes the costs associated with ending that relationship or terminating an existing relationship. Learning costs are associated with the effort required by the consumer to achieve the same level of knowledge and comfort acquired using a particular supplier's product, but which may not be transferable to similar/same products of other suppliers. Additionally, the notion of learning costs incorporates the implicit switching costs associated with decision biases, risk aversion, and market knowledge/familiarity. Learning what one's options are, what the relative competitive position of all suppliers are, and other such knowledge involves learning costs that will be differentially valued by individuals. In the case of mobile services, the switching costs are created by a service provider that requires customers to sign a long-term contract. If a customer wants to switch to another provider, he or she will have to pay a penalty to the current provider. As such, contractual costs are those costs that are directly provider induced in order to penalize churn and which are intended to prevent poaching of customers by other suppliers. With respect to the American customer satisfaction model, switching costs directly affect price tolerance. The ACSM survey instrument presents two questions: (1) by how much their current provider should increase its current prices in order for them to switch to a competitor, and (2) by how much a competitor should reduce its prices in order for them to switch. Peoples' answers to these questionnaire items are greatly affected by the direct switching costs they incur, such as a penalty.

Consistent with this proposition, Weiss and Anderson (1992) found that switching costs are a major consideration when consumers are making a churn decision, and that these costs (barriers) tend to reduce customers' churn behavior. These findings were further supported by research done by Jones, Mothersbaugh, and Beatty (2000). Burnham et al. (2003) suggested that switching costs are negatively correlated with a customer's intention to churn: the higher the costs, the lower the intention to switch. As Hu and Hwang (2006) point out, "the industry remains in a state of dynamic competition" (p. 75) and providers continue implementing flexible offerings that are aimed at reducing consumers' churn behavior. Shapiro and Varian (1999) found that *perceived* switching costs—which incorporate all of the explicit costs as well as the implicit costs discussed above—act as barriers to churn behavior. They suggest consumers will weigh the benefits of switching against the actual and psychological costs when considering churning.

Overall, the discussion above demonstrates that the concept of switching barriers has its own unique dimensions. In terms of the American customer satisfaction model applied in the context of mobile services, it is believed that two items pertaining to the customer switching behavior (conceptualized as price tolerance in the model) reflect a unique latent variable entitled *price tolerance.*

REPURCHASE LIKELIHOOD

The notion of overall customer loyalty has changed in both breadth and depth over the years in which it has been studied by academics and practitioners alike. The breadth of its definition is demonstrated by the multiplicity of areas that are examined, such as brand, product, vendor, or service loyalty. Initial research was primarily focused on brand loyalty, and mostly examined the behavioral aspects of the construct. In this view, Newman and Werbel (1973) defined customer loyalty as the repurchase of a brand that only considered that brand and which involved no brand-related information seeking.

Day (1969) was one of the first researchers to highlight the role of a positive attitude in the construct of loyalty. Following this line of reasoning, which incorporated both the behavioral and attitudinal conceptions of loyalty, operationalization of the construct of customer loyalty involved combining the aspects of purchasing a particular brand together with an affective attitudinal measure, whether that measure used a single scale or multi-scale items. With regards to the American customer satisfaction model, the discussion above relates to the unique dimension of loyalty as *repurchase likelihood,* or the probability of buying new services from the current provider when these purchases are not affected by prior contractual obligations, for example, when a contract has expired.

PRICE TOLERANCE AND REPURCHASE LIKELIHOOD

The literature—and intuition—suggests that higher switching costs are positively related to price tolerance—that is, that higher switching costs compel customers to remain loyal. Fornell et al. (1996) were among the first to include switching costs by adding them to the construct of customer satisfaction in the reflection of customer loyalty. In the ACSM, all items (i.e., two pertaining to price tolerance and one relating to repurchase likelihood) were believed to reflect overall loyalty. A number of subsequent studies demonstrated the unidimensionality of this construct. However, in the context of mobile services when high switching costs exist, unidimensionality does not apply. As such, it is suggested that, based on the theoretical rationale as well as empirical studies cited earlier, loyalty should be analyzed along two distinct dimensions: price tolerance and repurchase likelihood.

In terms of prior empirical research, Jones and Sasser (1995) included switching costs as one factor or competitiveness: since high switching costs discourage churning, they reduce the incentive for firms to compete. Bateson and Hoffman (1999) similarly suggest that customer satisfaction and switching costs are the primary influencers of loyalty. More recent studies have shown that switching costs have a direct and strong influence on the re-purchase decision (customer loyalty) in all markets, for example France (Lee, Lee, & Feick, 2001), Korea (Kim & Yoon, 2005), Australia (Caruana, 2004), Taiwan (Hu & Hwang, 2006), and Turkey (Aydin, Özer, & Arasil, 2005).

Jones et al.'s (2000) study examined the role of switching costs (barriers) in customer retention for services. They found that although core-service satisfaction was a primary issue in retention, switching factors in the form of interpersonal relationships, direct and indirect costs, and the perceived benefits of potential alternatives were also important. As such, these factors represented different unique dimensions of the overall loyalty concept. This supports the notion, outlined above, that loyalty of mobile service users must be considered as multidimensional and not simply as direct, contractual costs.

IMPACTS FOR MANAGERS AND RESEARCHERS

The findings of the many studies in the area show support for the intuitive link between higher switching costs and greater levels of customer loyalty (or at least, retention). More importantly, they also provide a greater understanding of the interaction between switching costs and loyalty, and refine the model that has, to date, served as a guide to management of mobile phone companies.

Management of mobile phone companies must understand the complexity and multidimensionality of the concepts of switching costs that directly influence price tolerance and repurchase likelihood that is not affected by contractual obligations. They must also understand that switching costs affect customer loyalty not solely through the contractual cost component of switching costs, but also through the learning and transaction cost components. A customer's, or potential customer's, belief that he or she will end up with a 'bad deal' financially in switching to a new provider—and that assessment will include all of the implicit as well as explicit costs—is the most important issue in the churn decision. This highlights the point that managing customer relationships, so that they remain positive, acts to keep the customer attached, whether this is a result of satisfaction outweighing perceived benefits or simply of customer inertia (Burnham et al, 2003; Caruana, 2004). It also highlights the need for poaching strategies to emphasize not only the financial benefits, but the relational benefits as well (Hu & Hwang, 2006). It should be noted that existing studies point out that one of the primary issues affecting the learning cost component has been the lack of time to undertake a complete comparison of the many offerings in the market. Additionally, providers have tended in the past to couch their offerings in terms that vary widely from their competitors', thus introducing a level of uncertainty and confusion in the minds of the analyzing consumer (Hu & Hwang, 2006). These factors are becoming less and less viable as consumers turn to the Internet for their purchasing information and guidance, and as consumers demand—and get—a certain level of standardization in the offerings of providers in the market, whether that standardization comes from the providers themselves or from organizations that perform such analyses and offer them to the consuming (Internet- or magazine-based) consumer. Additionally, the increasing homogeneity of pricing strategies and service packages will lead to a lessening of the impact of explicit (transaction and contractual) switching costs on the churn decision (Hu & Hwang, 2006). Thus, management needs to concentrate on customer relationships. Swartz (2000) quotes two senior executives in the industry:

If service is poor, then customers will pay any cancellation fees to get rid of the service and choose another provider....

You have to look at your reasons for churn... You can't use a contract to make up for poor service. If your service is poor, you can lock them in for a year... but they're gone the minute month 13 rolls around.

More research needs to be done on the notion of overall loyalty as a multidimensional construct. Are there positive barriers, such as interpersonal relationships, as well as negative? What relative influence on customer satisfaction do core and non-core services have? How sensitive are costs as barriers? Research into whether or not there are services that are perceived as having low barriers as opposed to services that are perceived as having high barriers within the market offerings would help refine our understanding of the role of various costs.

REFERENCES

Aydin, S., Özer, G., & Arasil, Ö. (2005). Customer loyalty and the effect of switching costs as a modifier variable: A case in the Turkish mobile phone market. *Marketing Intelligence and Planning, 23*(1), 89-103.

Bateson, J. E. G., & Hoffman, K. D. (1999). *Managing services marketing, text and readings* (4th ed.). Fort Worth, TX: Dryden Press.

Burnham, T. A., Frels, J. K., & Mahajan, V. (2003). Consumer switching costs: A typology, antecedents and consequences. *Journal of the Academy of Marketing Science, 31*(2), 109-126.

Caruana, A. (2004). The impact of switching costs on customer loyalty: A study among corporate customers of mobile telephony. *Journal of Targeting, Measurement and Analysis for Marketing, 12*(3), 256-268.

Day, G. S. (1969). A two dimensional concept of brand loyalty. *Journal of Advertising Research, 9*(3), 29-36.

Fornell, C. (1992). A national consumer satisfaction barometer: The Swedish experience. *Journal of Marketing, 56*(1), 6-21.

Fornell, C., Johnson, M. D., Anderson, E. W., Cha, J., & Bryant, B. E. (1996). The American Customer Satisfaction Index: Nature, purpose, and findings. *Journal of Marketing, 60*(4), 7-18.

Guiltnan, J. P. (1989). A classification of switching costs with implications for relationship marketing. In T. L. Childers & R. P. Bagozzi (Eds.), *Proceedings of the Winter Educators' Conference: Marketing Theory and Practice* (pp. 216-220), Chicago.

Hu, A. W.-L., & Hwang, I.-S. (2006). Measuring the effects of consumer switching costs on switching intention in Taiwan mobile telecommunications services. *Journal of American Academy of Business, 9*(1), 75-85.

Jones, M. A., Mothersbaugh, D. L., & Beatty, S. E. (2000). Switching barriers and repurchase intentions in services. *Journal of Retailing, 76*(2), 259-274.

Jones, T. O, & Sasser, W. E. (1995). Why satisfied customers defect. *Harvard Business Review, 73*(1), 88-99.

Kim, H.-S., & Yoon, C.-H. (2004). Determinants of subscriber churn and customer loyalty in the Korean mobile telephony market. *Telecommunications Policy, 28,* 751-756.

Klemperer, P. (1987). Markets with consumer switching costs. *The Quarterly Journal of Economics, 102,* 375-394.

Lee, J., Lee, J., & Feick, L. (2001). The impact of switching costs on customer satisfaction-loyalty link: Mobile phone service in France. *Journal of Services Marketing, 15*(1), 35-48.

Newman, J. W., & Werbel, R. A. (1973). Multivariate analysis of brand loyalty for major household appliances. *Journal of Marketing Research, 10*(4), 404-409.

Nunnally, J.C., & Bernstein, I.H. (1994). *Psychometric theory* (3rd ed.). New York: McGraw-Hill.

Oliver, R. L. (1996). *Satisfaction: A behavioral perspective on the consumer.* New York: McGraw-Hill.

Porter, M. E. (2003). *Competitive strategy: Techniques for analyzing industries and competitors.* New York: MacMillan.

Serenko, A., Turel, O., & Yol, S. (2006). Moderating roles of user demographics in the American customer satisfaction model within the context of mobile services. *Journal of Information Technology Management, 17*(4): in-press.

Swartz, N. (2000). Reconsidering contracts. *Wireless Review, 17*(4), 48-52.

Thibault, J. W., & Kelley, H. H. (1959). *The social psychology of groups.* New York: John Wiley & Sons.

Turel, O., & Serenko, A. (2006). Satisfaction with mobile services in Canada: An empirical investigation. *Telecommunications Policy, 30*(5-6), 314-331.

Turel, O., Serenko, A., & Bontis, N. (2007). User acceptance of wireless short messaging services: Deconstructing perceived value. *Information & Management, 44*(1), 63-73.

Turel, O., Serenko, A., Detlor, B., Collan, M., Nam, I., & Puhakainen, J. (2006). Investigating the determinants of satisfaction and usage of mobile IT services in four countries. *Journal of Global Information Technology Management, 9*(4), 6-27.

Weiss, A.M., & Anderson, E. (1992). Converting from independent to employee sales forces: The role of perceived switching costs. *Journal of Marketing Research, 29*(1), 101-115.

KEY TERMS

American Customer Satisfaction Model: The original model suggests that satisfaction affects the overall customer loyalty, where loyalty is a unidimensional construct that consists of price tolerance (i.e., the probability of staying with the current provider if it increases prices or if a competitor decreases prices) and repurchase likelihood (i.e., the probability of purchasing the same service again). If the customer's expectations of product quality, service quality, and price are exceeded, a firm will achieve high levels of customer satisfaction and will create *customer delight*. If the customer's expectations are not met, customer dissatisfaction will result. And the lower the satisfaction level, the more likely the customer is to stop buying from the firm.

Churn: This refers to the notion that a company will, over any given period of time, lose existing customers and gain new customers. Churn is, currently, mostly created by the luring away of competitors' customers.

Customer Loyalty: The notion that a customer will continue to use a particular brand or product; the behavior customers exhibit when they make frequent repeat purchases of a brand or product.

Price Tolerance: The extent to which price is an important criterion in the customer's decision-making process; thus a price-sensitive customer is likely to notice a price rise and switch to a cheaper brand or supplier.

Repurchase Likelihood: The probability of buying new services from the current provider when these purchases are not affected by prior contractual obligations, for example, when a contract has expired.

Switching Cost: One-time cost facing the consumer/buyer of switching from one supplier to another. Switching costs are composed of transaction costs (costs incurred when a consumer begins a relationship with a provider, and includes the costs associated with ending that relationship or terminating an existing relationship), learning costs (costs associated with the effort required by the consumer to achieve the same level of knowledge and comfort acquired using a particular supplier's product, but which may not be transferable to the same/similar products of other suppliers), and contractual costs (costs that are directly provider induced in order to penalize churn and which are intended to prevent poaching of customers by other suppliers).

Convergence Technology for Enabling Technologies

G. Sivaradje
Pondicherry Engineering College, India

I. Saravanan
Pondicherry Engineering College, India

P. Dananjayan
Pondicherry Engineering College, India

INTRODUCTION

Today, we find a large number of wireless networks based on different radio access technologies (RATs). Every existing RAT has its own merits. Now the focus is turned towards the next-generation communication networks (Akyildiz, Mohanty, & Xie, 2005), which will seamlessly integrate various existing wireless communication networks, such as wireless local area networks (WLANs, e.g., IEEE 802.11 a/b/g and HIPERLAN/2), wireless wide area networks (WWANs, e.g., 1G, 2G, 3G, IEEE 802.20), wireless personal area networks (WPANs, e.g., Bluetooth, IEEE 802.15.1/3/4), and wireless metropolitan area networks (WMANs, e.g., IEEE 802.16) to form a converged heterogeneous architecture (Cavalcanti, Agrawal, Cordeiro, Xie, & Kumar, 2005). Seamless integration does not mean that the RATs are converged into a single network. Instead the services offered by the existing RATs are integrated as shown in Figure 1.

Convergence technology is a technology that combines different existing access technologies such as cellular, cordless, WLAN-type systems, short-range wireless connectivity, and wired systems on a common platform to complement each other in an optimum way and to provide a multiplicity of possibilities for current and future services and applications to users in a single terminal. After creating a converged heterogeneous architecture, the next step is to perform a common radio resource management (RRM) (Magnusson, Lundsjo, Sachs, & Wallentin, 2004). RRM helps to maximize the use of available spectrum resources, support mixed traffic types with different QoS requirements, increase trunking capacity and grade of service (GoS), improve spectrum usage by selecting the best RAT based on radio conditions (e.g., path loss), minimize inter-system handover latency, preserve QoS across multiple RATs, and reduce signaling delay. A typical converged heterogeneous architecture (Song, Jiang, Zhuang, & Shen, 2005) is shown in Figure 2.

Figure 1. Convergence of services

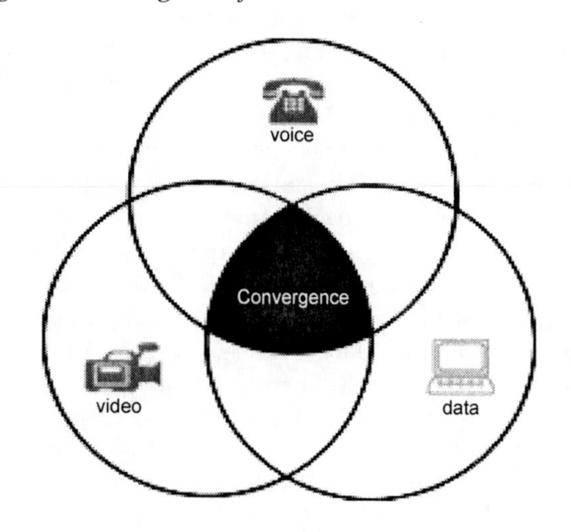

CHALLENGES

The integration of different networks to provide services as a single interworking network requires many difficult challenges to be addressed. Because existing networks do not have fair RRM, the major challenge that needs to be addressed has to be mobility management. The heterogeneous network architecture will be based on IP protocol that will enhance the interoperability and flexibility. IETF Mobile IP protocol is used to support macro mobility management. But both IP protocol and mobile IP protocol (Pack & Choi, 2004; Montavont & Noel, 2002) was not basically designed to support the real-time applications. So, during the handoff between systems, users will experience the service discontinuity, such as long service time gap or network disconnection. Besides this service discontinuity, the different service characteristics of these interworked networks may degrade the quality of service (QoS).

Some of the other challenges include topology and routing, vertical handoff management, load balancing, unified

Figure 2. Converged heterogeneous network architecture

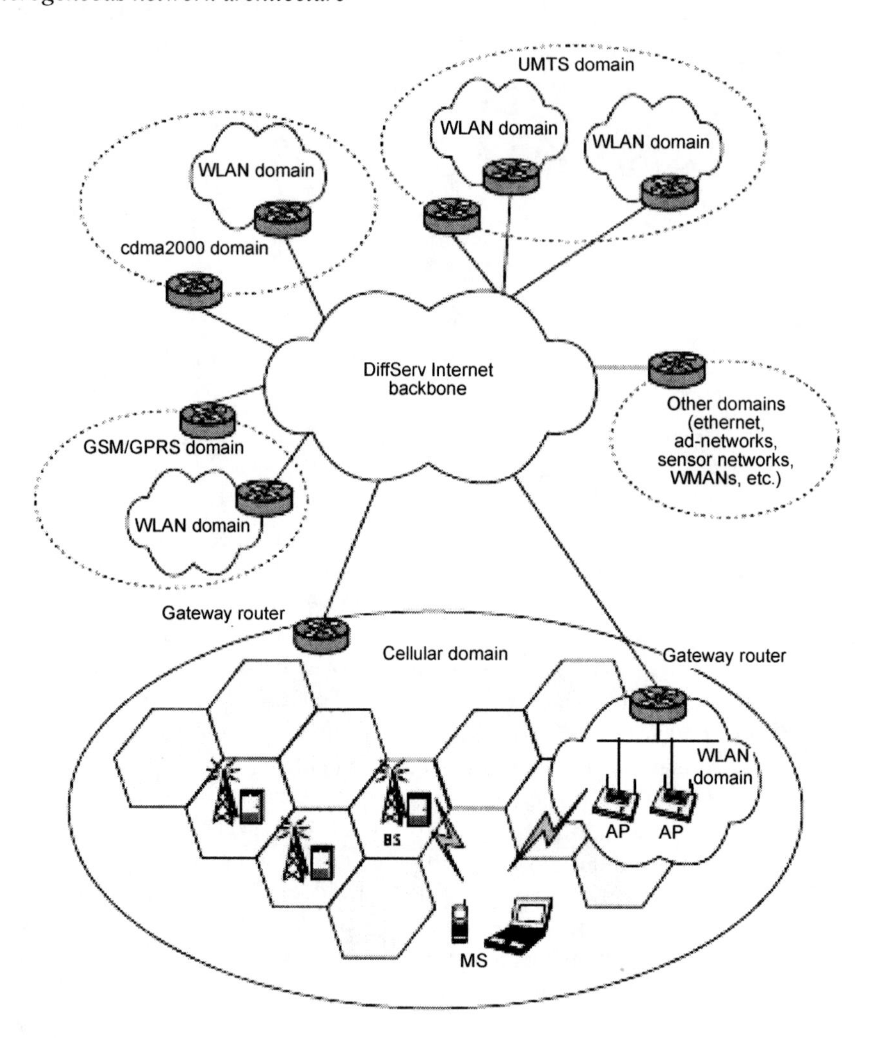

accounting and billing, and last but not least the protocol stack of mobile station (MS), which should contain various wireless air-interfaces integrated into one wireless open terminal so that same end equipment can flexibly work in the wireless access domain as well as in the mobile cellular networks.

PROTOCOL STACK

In a homogeneous network, all network entities run the same protocol stack, where each layer has a particular goal and provides services to the upper layers. The integration of different technologies with different capabilities and functionalities is an extremely complex task and involves issues at all the layers of the protocol stack. So in a heterogeneous environment, different mobile devices can execute different protocols for a given layer. For example, the protocol stack of a dual-mode MS is given in Figure 3.

This protocol stack consists of multiple physical, data link, and medium access control (MAC) layers, and network, transport, and application layers. Therefore, it is critical to select the most appropriate combination of lower layers (link, MAC, and physical) that could provide the best service to the upper layers. Furthermore, some control planes such as mobility management and connection management can be added. These control planes can eventually use information from several layers to implement their functionalities. The network layer has a fundamental role in this process, since it is the interface between available communications interfaces (or access technologies) that operate in a point-to-point fashion, and the end-to-end (transport and application) layers. In other words, the task of the network layer is to provide a uniform substrate over which transport (e.g., TCP and UDP), and application protocols can efficiently run, independent of the access technologies used in each of the point-to-point links in an end-to-end connection. Although there are issues in all layers, the network layer has received more attention than

Figure 3. Protocol stack of a dual-mode MS

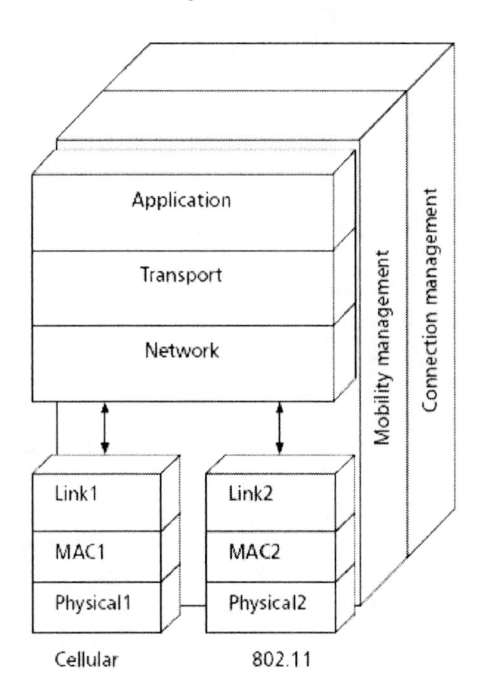

any other layer, and little integration-related work has been done at the lower layers. Indeed, integrated architectures are expected not to require modifications at the lower layers so that different wireless technologies can operate independently. However, this integration task is extremely complex, and it requires the support of integration architecture in terms of mobility and connection management. Seamless handoffs for "out of coverage" terminals and resource management can be provided by the two control planes.

ROUTING ISSUES

All RATS in the integrated architecture is considered as IPv6-based networks, and each element in the internetworking networks has a distinct ID number corresponding to the network routing address (Liu & Zhou, 2004). The infrastructure of a network is mapped into IPv6 addresses as shown in Figure 4. For example, the mapping of infrastructure of cellular network and IEEE 802.11 WLAN are shown in Figures 5 and 6. WLAN is given some reservation IDs, so that they can be utilized by mobile nodes under MANET mode.

VERTICAL HANDOFF MANAGEMENT

Vertical handoff is the handoff between different RATs. The major challenge in vertical handoff is that it is difficult to support a seamless service during inter-access network handoff (Wu, Banerjee, Basu, & Das, 2005; Ma, Yu, Leung, & Randhawa, 2004). The service interworking architecture and procedures, the way to provide the network and user securities, the control scheme for minimizing performance decrease caused by different service data rates, and the interworking network detection and selection methods are typical problems and to be addressed to provide stable and continuous services to users.

Unlike in the homogeneous wired networks, providing QoS for integrated architecture has some fundamental bottlenecks. This is because each radio access technology has different transmission-rate capacity over the radio interfaces, therefore the handoff between the two systems makes the maintenance of QoS connection very hard. For example, WLAN can provide a transmission speed from 11Mb/s up to 54Mb/s theoretically, while UMTS has only 144kb/s at vehicular speed, 384 kb/s at pedestrian speed, and 2 Mb/s when used indoors. If we keep the QoS resource assigned by UMTS to a connection that is actually in a WAN hotspot, the advantage of the high speed of the WLAN is not fully taken. On the other hand, if we use a WLAN parameter for a station in the UMTS network, the connection may not be admitted at all (Zhang et al., 2003). Therefore, to maintain a sensible QoS framework, one has to consider the significant difference transmission capacity between two systems especially when user handover takes place.

Figure 4. Mapping infrastructure into IPv6 format

IP Header	ID Mapping	Data Packets

Figure 5. Mapping infrastructure of IEEE 802.11 WLAN into IPv6 address

Router ID	Access Point ID	Mobile Node ID

Figure 6. Mapping infrastructure of cellular network into IPv6 address

Network ID	RNC ID	Base Station ID	Mobile Node ID

APPLICATIONS

Convergence technology gives the possibility to combine audio and video, data, graphics, slides and documents, and Internet services in any way you like, so as to maximize the effectiveness of the communication. Integrating all traffic types enables more versatile and efficient ways of working, not just internally to the organization, but externally to customers, partners, and suppliers. It also creates a multi-system environment where a single service could be offered at different speeds at different locations/times via separate systems. The flexibility of convergence technology provides many applications and services to the user community. Some of the applications are:

- **Find-Me-Follow-Me:** This is a customizable service that makes it easy for callers to 'find' a user. Using a Web portal customers can choose how incoming calls should be handled. Options include ringing multiple phones simultaneously, or picking the order of phones to ring sequentially. Ubiquity's SIP A/S is used to dial out, in parallel or sequentially, to the user's contact numbers. Using IVR, the user can then accept the call or forward it to voicemail.
- **InfoChannels:** This is a multimedia content subscription application that pushes information and entertainment to users in real time. Users subscribe to content services through a Web portal, and new content is delivered to their designated device (mobile phone, PDA, PC browser) as soon as it is available.
- **Rich Media Conferencing:** Speak conference director is a highly scalable, carrier-class, IP conferencing application that enables conferencing service providers (CSPs) to offer hosted audio and Web conferencing services. This easy-to-use, browser-based solution offers a complete conferencing application feature set, as well as a Web portal for scheduling, initiating, managing, and terminating multi-party conferences.

Some of the services that the convergence provides to the user community are:

- **Unified Messaging:** Same inbox handling data, voice and fax.
- **Hosted IP Voice:** A complete, outsourced telephone service offering all PBX-type features.
- **IP Fax:** Delivery of e-mail to fax and fax to e-mail in a large number of countries.
- **IP Telephony:** A combination of quality transmission globally across the WAN and the LAN, with tailored consulting and end-to-end support.
- **Voice for IP VPN:** Integrated voice and data transmission, using a specific voice.
- **Video for IP VPN:** Point-to-point video transmission over the IP VPN network, using a specific class of service, called RT Vi.
- **Virtual Contact Center Services:** Optimization of agent resources while reducing costs, by allowing the routing of calls based on the agent's skills.
- **Voiceover Wi-Fi:** Full corporate mobility with a converged voice and data wireless solution.

CONCLUSION

This article provides features about convergence technology. The convergence of all existing networks will provide access to all available services using a single-user terminal. But there are many challenges to be addressed in converging the networks. In spite of converging the networks, management of the converged network is more challengeable. This article illustrates some of the challenges, and many are still open issues. Considering all the factors discussed, convergence technology is going to provide future flexibility to the wireless communication world. The complexity of this interesting technology must be addressed in the near future.

REFERENCES

Akyildiz, I. F., Mohanty, S., & Xie, J. (2005). A ubiquitous mobile communication architecture for next-generation heterogeneous wireless systems. *IEEE Radio Communications, 43*(6), S29-S36.

Cavalcanti, D., Agrawal, D., Cordeiro, C., Xie, B., & Kumar, A. (2005). Issues in integrating cellular networks, WLANS, and MANETs: A futuristic heterogeneous wireless network. *IEEE Wireless Communications, 12*(3), 30-41.

Liu, C., & Zhou, C. (2004). HCRAS: A novel hybrid internetworking architecture between WLAN and UMTS cellular networks. In *Proceedings of IEEE 2004* (pp. 374-379).

Ma, L., Yu, F., & Leung, V. C. M., & Randhawa, T. (2004). A new method to support UMTS/WLAN vertical handover using SCTP. *IEEE Wireless Communication, 11*(4), 44-51.

Magnusson, P., Lundsjo, J., Sachs, J., & Wallentin, P. (2004). Radio resource management distribution in a Beyond 3G Multi-Radio Access architecture. In *Proceedings of the IEEE Communications Society, Globecom* (pp. 3372-3477).

Montavont, N., & Noel, T. (2002). Handover management for mobile nodes in IPv6 networks. *IEEE Communications Magazine, 40*(8), 38-43.

Pack, S., & Choi, Y. (2004). A study on performance of hierarchical mobile IPv6 in IP-based cellular networks. *IEICE Transactions on Communication, E87-B*(3), 546-551.

Song, W., Jiang, H., Zhuang, W., & Shen, X. (2005). Resource management for QoS support in cellular/WLAN interworking. *IEEE Network, 19*(5), 12-18.

Wu, W., Banerjee, N., Basu, K., & Das, S. K. (2005). SIP-based vertical handoff between WWANS and WLANS. *IEEE Wireless Communications, 12*(3), 66-72.

Zhang, Q., Guo, C., Guo, Z., & Zhu, W. (2003). Efficient mobility management for vertical handoff between WWAN and WLAN. *IEEE Communication Magazine, 41*(11), 102-108.

KEY TERMS

Communication Network: Network of telecommunications links arranged so that messages may be passed from one part of the network to another over multiple links.

Grade of Service (GoS): A measurement of the quality of communications service in terms of the availability of circuits when calls are to be made. Grade of service is based on the busiest hour of the day and is measured as either the percentage of calls blocked in dial access situations or average delay in manual situations.

Heterogeneous Network: A network that consists of workstations, servers, network interface cards, operating systems, and applications from many vendors, all working together as a single unit.

Radio Access Technology (RAT): Technology or system used for the cellular system (e.g., GSM, UMTS, etc.).

Wireless Local Area Network (WLAN): Wireless network that uses radio frequency technology to transmit network messages through the air for relatively short distances, like across an office building or college campus.

Wireless Metropolitan Area Network (WMAN): A regional wireless computer or communication network spanning the area covered by an average to large city.

Wireless Personal Area Network (WPAN): Personal, short-distance area wireless network for interconnecting devices centered around an individual person's workspace.

Wireless Wide Area Network (WWAN): Wireless network that enables users to establish wireless connections over remote private or public networks using radio, satellite, and mobile phone technologies instead of traditional cable networking solutions like telephone systems or cable modems over large geographical areas.

Cooperative Caching in a Mobile Environment

Say Ying Lim
Monash University, Australia

INTRODUCTION

There has been a rapid recent development in the usage of mobile devices, such as personal digital assistant (PDA), mobile notebooks and other mobile electronic devices. This has been increasingly in demand due to new smaller designs and easy to carry around features. The mobile computing technology has pushed the uses of mobile device even more by providing the ability for mobile users to access information anytime, anywhere. However, mobile computing environments have certain limitations, such as short battery power and limited storage and involving communication cost and bandwidth limitations (Kossman et al., 2001). Hence, it is of great interest to provide efficient query processing with quick response rate and with low transfer cost. Due to the inherent factors like low bandwidth and low reliability of wireless channels in the mobile computing environment, it is therefore important for a mobile client to cache its frequently accessed database items into its local storage (Chan, Si, & Leong, 2001).

Caching is a key strategy in improving data retrieval performance of mobile clients (Barbara & Imielinkski, 1994; Chow, Leong, & Chan, 2005). In order to retain the frequently accessed database items in the client's local memory, a caching mechanism is needed. By having the caching mechanisms, it allows a client to serve database queries at least partially during connection, which is an inherent constraint of the mobile environment. Thus, by having an effective caching mechanism in keeping the frequently accessed items, the more queries could be served in case of disconnection (Chan, Si, & Leong, 2001). The aim of caching is beneficial to the mobile environment, because having the data cached into the local memory can help future queries to be answered more quickly or to access the data faster, with low latency time and reduced start up delays that may be caused on the client side (Kara & Edwards, 2003). In addition to improving the access latency, it also helps to save power due to the ability to allow lower data transmission, as well as improvement in terms of data availability in situations of disconnection (Wu, Yu, & Chen, 1996; Lee, Xu, & Zheng, 2002).

In this article, we concentrate particularly on cooperative caching, which is basically a type of caching strategy that not only allows mobile clients to retrieve database items from the servers, but also from the cache in their peers.

BACKGROUND

The effect of having the ability to cache data is of great importance, particularly in the mobile computing environment. This is due to the reason that contacting the remote servers for data is expensive in the wireless environment and the vulnerability to frequent disconnection can further increase the communication costs (Leong & Si, 1997). Also, caching the frequently accessed data in the mobile devices will help reduce tune-in time and power consumption because requested data can be fetched for the cache without tuning into the communication channel for retrieval (Lee & Lee, 1999). Caching has also been proven as a significant technique in improving the performance of data retrieval in peer to peer network in helping to save bandwidth for each data retrieval that are made by the mobile clients (Joseph et al., 2005). The cached data are meant to support disconnected or intermitted connected operations. There are many different types of caching strategies that serve the purpose to improve query response time and to reduce contention on narrow bandwidth (Zheng, Lee, & Lee, 2004).

Regardless of which caching strategy one chooses, they have their own advantages and disadvantages in some way. In principle, caching can improve system performance in two ways: (a) It can eliminate multiple requests for the same data and (b) improve performance by off-loading the work. The first way can be achieved by allowing mobile clients to share data among each other by allowing them to access each other's cache within a reasonably boundary between them. The second way, however, can be demonstrated as in an example of a mobile user who is interested in keeping stock prices and caches them into his mobile device. By having copies in his own mobile device, he can perform his own data analysis based on the cached data without communicating directly to the server over the wireless channel.

COOPERATIVE CACHING

The cooperative caching is a kind of information sharing that was developed by the heavy influence and emergence of the robust yet reliable peer-to-peer (P2P) technologies, which allows mobile clients (Kortuem et al., 2001). This type of information sharing among clients in a mobile

Figure 1. An overview of cooperative caching architecture

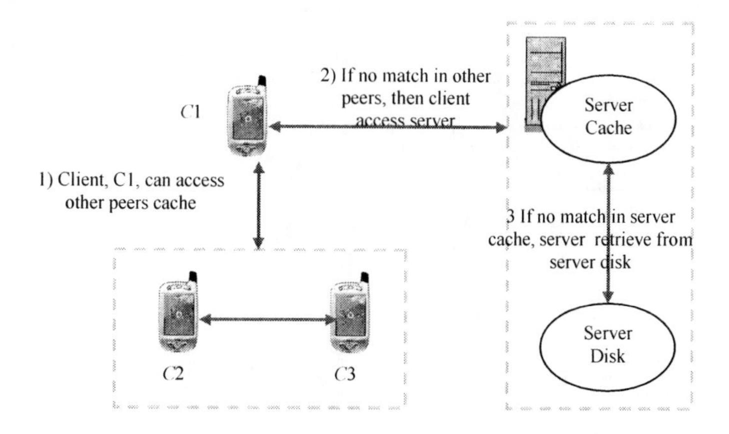

environment has generally allowed the clients to directly communicate among themselves by being able to share cached information through accessing data items from the cache in their neighboring peers rather than having to rely on their communication to the server for each query request (Chow, Leong, & Chan, 2004).

There are several distinctive and significant benefits that cooperative caching brings to a mobile computing environment. These include improving access latency, reducing servers' workloads and alleviating point–to-point channel congestion. Though the benefits outweigh the drawbacks, there is still a main concern that cooperative caching may produce. This refers to the possibility of the increase in the communication overhead among mobile clients (Ku, Zimmermann, & Wan, 2005)

Framework Design of Cooperative Caching in a Mobile Environment

Several clients and servers are connected together within a wireless channel in a mobile environment. Basically, the clients which denote the mobile hosts are connected to each other wirelessly within a certain boundary among each other, and they can exchange information by allowing each other to access the other peers' cache (Cao, Yin, & Das, 2004).

Figure 1 illustrates an example of the framework architecture design of the cooperative caching, which provides the ability for mobile clients to communicate among each other. If the client encounters a *local cache miss*, it will send a query to request, from its neighboring peers, to obtain a communication and the desired data from its peer's cache. Otherwise, it will be known as a *local cache hit* if the desired data exists in its local cache. As for trying to obtain data from its peers, if the desired data is available from its neighboring peers or if the other peers can turn in the requested data before it is broadcast on the channel, then it is known as a *global cache hit*; otherwise it is called a *global cache miss* and the client would have to wait for the desired data to arrive in the broadcast channel or access the server cache and, if that fails, then the server would retrieve the desired data from the disk (Sarkar & Hartman, 2001; Hara, 2002; Chow, Leong, & Chan, 2005).

As a summary, a mobile client can choose to either (a) retrieve data from the server directly by having a direct communication through issuing a query (Hu & Johnson, 2000) or (b) capture the data from a scalable broadcast channel (Su, Tassiulas, & Tsotras, 1999). These are known as a pull-base and push-base mechanism respectively. Further investigation on pull and push-based environments are made in the subsequent subsections.

Using Cooperative Caching in an On-Demand (Pull-Based) Environment

A pull-based environment refers to relating the use of traditional point–to-point scenario similar to client-server communication directly. It can also be known as an on demand query or server strategy whereby processing can be done on the server upon request sent by the mobile clients. Figure 2 illustrates an example of a pull-based architecture. It can be seen that the mobile client issues a direct query to the server over a dedicated channel to be processed. Processing takes

Figure 2. Example of an on demand environment system (pull-based system)

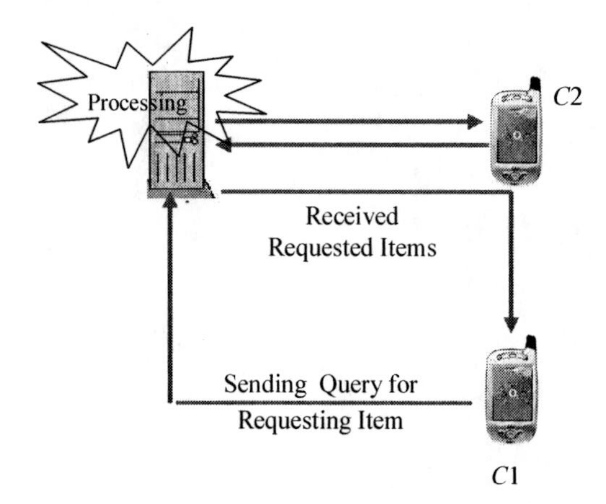

place in the server and, once the requested data items have been obtained, it will returned back to the client directly.

The main advantage of this system is that a client can issue query directly to the server and wait for the server to process and return the results according to the query being issued. Also, it is appropriate in situations where privacy is a major concern. On the other hand, the limitation is that this is not desirable in a mobile environment where there are limited resources to satisfy each individual client directly. Thus, this shows limitation when it comes to large-scale systems.

Example 1: Looking at Figure 3, suppose a shopper in a shopping complex wants to know which shops to visit by obtaining information from the store directories. Imagine that this client is denoted as *C*2 in Figure 3. If this shopper, *C*2, finds the target shop in its local cache, as she has previously visited this shopping complex and has cached the store directories in her mobile device. If this is the case, then she can just merely obtain the information from the cache. If not, then it will send request to its neighboring peers, which in this example are *C*1 and *C*3 since the boundary of the wireless transmission for *C*2 covers clients *C*1 and *C*3. So *C*2 can obtain the desired data from either *C*1 or *C*3. Assuming *C*3 has the data that *C*2 wanted, this means *C*2 can obtain the data from the cache in his peer, in this example *C*3. Otherwise, *C*2 would have to obtain it from the server directly.

Using Cooperative Caching in a Broadcast (Push-Based) Environment

In this section, we focus on using cooperating caching in a broadcast environment, which can sometimes be referred to as a push-based system or also known as on-air strategy. In a broadcast environment, a mobile client is able to tune into the broadcast channel to retrieve the data that they want by having the server broadcast the data items into the air for the user to tune into (Waluyo, Srinivasan, & Taniar, 2005). Thus, each client can access a piece of data by waiting for its data to arrive. Figure 4 depicts an example of a broadcast environment. The main advantages of this broadcast environment in terms of its delivery mechanism is its higher throughput for data access for a larger number of clients because, with the absence of the communication contention between the clients requesting data, they are able to share the bandwidth more efficiently (Hara, 2002). Information broadcast appears to be an essential method for information dissemination for mobiles users, since it is able to broadcast to an arbitrary number of mobile users (Lee & Lee, 1999).

There are advantages and disadvantages of the broadcast environment. The strength would be its ability to disseminate

Figure 3. Using cooperative caching in an on demand environment

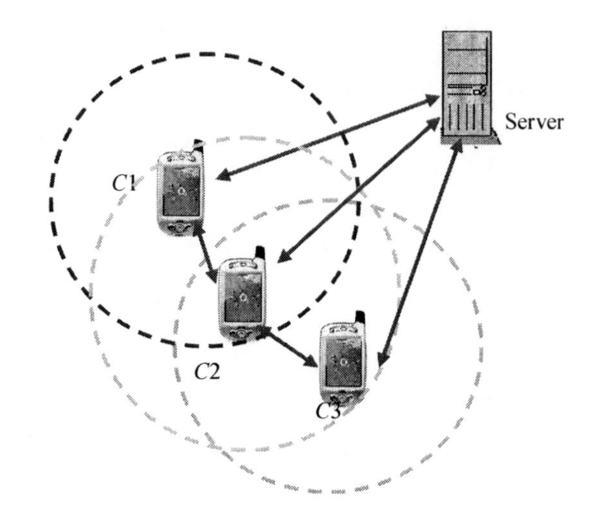

Figure 4. Example of a broadcast environment system (push-based system)

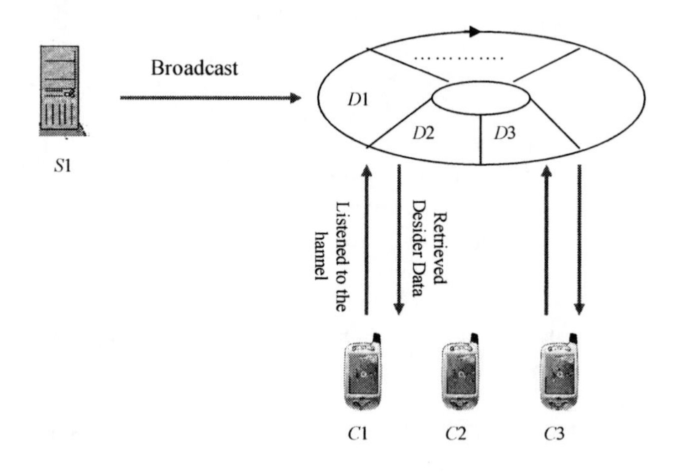

data to an immense number of mobile clients. However, the greatest disadvantage lies in the way in which the data item are broadcast in a sequential way. This will lead to longer access latency if there is a substantial increase in the number of data items being broadcast. Also, by broadcasting the data item on the air, this means the mobile clients have to consume considerable amounts of power to listen to the broadcast channel until the target data items reaches its turn and appears in the broadcast channel.

In this section, we would like to illustrate the use of cooperative caching in the broadcast environment. It is always a benefit to allow the client to have access to more data in the mobile environment by improving the data availability and data accessibility (Cao, Yin, & Das, 2004). By having the clients share bandwidth together in the wireless channels; it is believed that cooperative caching is able to further save bandwidth for each data retrieval (Joseph et al., 2005). An example of an application utilizing this push-based mechanism can be a situation where, in the airport, up-to-date schedules could be broadcast and passengers with their mobile devices are able to receive and store the information.

Generally, there are a series of steps involved in cooperative caching in a broadcast environment. Basically, when a client issues a request to a particular data item on the broadcast channel, the issued request client would first check whether the desired data has been previously cached. If it has, then the answer would be immediately returned and succeed at once. However, if it has not, then it will secondly check to see if the response time in accessing the desired item that is cached by the neighboring peers is shorter than to wait for the item to appear on the broadcast channel. If it is, then the client will obtain it from its peers; otherwise it may be easier just to obtain it from the broadcast channel if it appears to be faster. Otherwise, it received a reply from other neighboring peers that cached the desired data item and the request would be completed and succeeded when the transmission of the item is completed. However, in case of a situation where no neighboring peers obtained the desired data, then the client would have to wait for the next broadcast period for the desired data item to appear (Hara, 2002).

FUTURE TRENDS

There have been several researches done in the area of cooperative caching in a mobile environment. Problems and limitations incurred in the usage of cooperative caching in the mobile environment have generated a lot of attention from researches in finding a good cache strategy that is specifically designed for use only in the mobile computing environment.

In the future, it is desirable to design techniques that can further improve the efficiency of caching in reducing the overhead of maintaining and identifying the cached items. Also exploring the semantics of the data items cached by the mobile clients is beneficial. Practical evaluations from various factors, such as communication and computational overhead, can be essential to determine which strategy appears most appropriate for a real mobile environment. It is also valuable to look into cache replacement and cache invalidation and addressing them in an environment where updates features takes place.

It is also advantageous to look further into incorporation of an effective replication scheme that may increase data availability and accessibility, as well as improve the average query latency. It would be useful to always take into account reducing the number of server requests and power consumption, as well as shortening the access latency as more neighboring peers increase. As far as most situations explain in this article, the focus does not take into account location dependency queries. Thus, it can be a good idea to consider location dependent queries, as well as accessing multiple non-collaborative servers instead of just obtaining data from a single server.

CONCLUSION

Caching appears to be a key factor in helping to improve the performance of answering queries in mobile environment. A new state-of-the-art information sharing known as cooperative caching allows mobile clients an alternative way to obtain desired data items. Clients can now have the ability to access data items from the cache in their neighboring peers' devices with the implementation of cooperative caching.

In this article, we have described issues of caching in a mobile environment by focusing mainly on cooperative caching in the mobile environment. Our discussion assumed a broadcast environment. This article serves as a helpful foundation for those who wish to gain preliminary knowledge about the usefulness and benefits of caching in the mobile environment, particularly cooperative caching.

REFERENCES

Barbara, D., & Imielinski, T. (1994, November). Sleepers and workaholics: Caching strategies in mobile environments. *MOBIDATA: An Interactive Journal of Mobile Computing, 1*(1).

Chan, B. Y., Si, A., & Leong H. V. (1998). Cache management for mobile databases: Design and evaluation. In *Proceedings of the International Conference on Data Engineering (ICDE)* (pp.54-63).

Chow, C. Y., Leong, H. V., & Chan, A. T. S. (2005). Distributed group-based cooperative caching in a mobile environment. In *Proceedings of the 6th international conference on Mobile Data Management (MDM)* (pp. 97-106).

Chow, C. Y., Leong, H. V., & Chan, A. T. S. (2004). Group-based cooperative cache management for mobile clients in a mobile environment. In *Proceedings of the 33rd International Conference on Parallel Processing (ICPP)* (pp. 83-90).

Chow, C. Y., Leong, H. V., & Chan, A. T. S. (in press). Utilizing the cache space of low-activity clients a mobile cooperative caching environment. *International Journal of Wireless and Mobile Computing (IJWMC).*

Chow, C. Y., Leong, H. V., & Chan, A. T. S, "Peer-to-Peer Cooperative Caching in a Hybrid Data Delivery Environment", In *Proceedings of the 2004 International Symposium on Parallel Architectures, Algorithms and Networks (ISPAN2004),* 2004.

Chow, C.Y., Leong, H.V., & Chan A.T.S. (2004). Cache signatures for peer-to-peer cooperative caching in mobile environments. In *Proceedings of 2004 International Conference on Advanced Information Networking and Applications.*

Cortés, T., Girona, S., & Labarta, J. (1997). *Design issues of a cooperative cache with no coherence problems.* In *Fifth Workshop on I/O in Parallel and Distributed Systems (IOPADS'97)* (pp. 37-46).

Hara, T. (2002). Cooperative caching by mobile clients in push-based information systems. In *Proceedings of ACM International Conference on Information and Knowledge Management (ACM CIKM'02)* (pp.186-193).

Huron A.R., & Jiao, Y. (2005). Data broadcasting in mobile environment. In D. Katsaros, A. Nanopoulos, & Y. Manolopaulos (Eds.), *Wireless information highways* (Chapter 4). Hershey, PA: IRM Press.

Imielinski, T., & Badrinath, B. (1994). Mobile wireless computing: Challenges in data management. *Communications of the ACM, 37*(10), 18-28.

Imielinski, T., Viswanathan, S., & Badrinath, B. R. (1997). Data on air: Organisation and access. *IEEE Transactions on Knowledge and Data Engineering, 9*(3), 353-371.

Joseph, M. S., Kumar, M., Shen, H., & Das, S. K. (2005). Energy efficient data retrieval and caching in mobile peer-to-peer networks. In *Proceedings of 2nd Workshop on Mobile Peer-to-Peer Networks (MP2P)* (pp. 50-54).

Kara, H., & Edwards, C. (2003). A caching architecture for content delivery to mobile devices. In *Proceedings of the 29th EUROMICRO Conference: New Waves in System Architecture (EUROMICRO'03).*

Kortuem, G., Schneider, J., Preuitt, D., Thompson, C., Fickas, S., & Segall, Z. (2001). When peer-to-peer comes face to face: Collaborative peer to peer computing in mobile ad hoc networks. In *Proceedings of the First International Conference on Peer to Peer Computing.*

Lee, W. C., & Lee, D. L. (1996). Using signature techniques for information filtering in wireless and mobile environments. *Journal on Distributed and Parallel Databases, 4*(3), 205-227.

Papadopouli, M., & Issarny, V. (2001). Effects of power conservation, wireless coverage and cooperation on data dissemination among mobile devices. In *Proceedings. of the 2nd ACM International Symposium on Mobile Ad HOC Networking and Computing (MobiHoc)* (pp. 117-127).

Prabhajara, K., Hua, K.A., & Oh, J.H. (2000). Multi-level, multi-channel air cache designs for broadcasting in a mobile environment. In *Proceedings of the 16th International Conference on Data Engineering* (pp. 167-186).

Roy, N., Roy, A., Basu, K., & Das, S. K. (2005). A cooperative learning framework for mobility-aware resource management in multi-inhabitant smart homes. In *Proceedings of 2nd Annual IEEE International Conference on Mobile and Ubiquitous Systems: Networking and Services* (pp. 393-403).

Sailhan, F., & Issarny, V. (2003). Cooperative caching in ad hoc network. In *Proceedings. of the 4th International Conference on Mobile Data Management (MDM)* (pp.13-28).

Sarkar, P., & Hartman, J. H. (2000). Hint-based cooperative caching. *ACM Transactions on Computer Systems, 18*(4), 387-419.

Shen, H., Joseph, M. S., Kumar, M., & Das, S. K. (2005). PReCinCt: An energy efficient data retrieval scheme for ,mobile peer-to-peer networks. In *Proceedings of 19th IEEE International Parallel and Distributed Processing Symposium (IPDPS).*

Shen, H., Joseph, M. S., Kumar, M., & Das, S. K. (2004). Cooperative caching with optimal radius in hybrid wireless networks. In *Proceedings of Third IFIP-TC6 Networking Conference* (LNCS 3042, pp. 841-853).

Waluyo, A. B., Srinivasan, B., & Taniar, D. (2005). Indexing schemes for multi channel data broadcasting in mobile databases. *International Journal of Wireless and Mobile Computing, 1*(6).

Waluyo, A.B., Srinivasan, B., & Taniar, D. (2005). Research in mobile database query optimization and processing. *Mobile Information Systems*, 1(4).

Xu, J., Hu, Q., Lee, D. L., & Lee, W.-C. (2000). SAIU: An efficient cache replacement policy for wireless on-demand broadcasts. In *Proceedings of the 9th International Conference on Information and Knowledge Management* (pp.46-53).

Xu, J., Hu, Q., Lee, W.-C., & Lee D. L. (2004). Performance evaluation of an optimal cache replacement policy for wireless data dissemination. *IEEE Transaction on Knowledge and Data Engineering (TKDE)*, *16*(1), 125-139.

Yajima, E., Hara, T., Tsukamoto, M., & Nishio, S. (2001). Scheduling and caching strategies for correlated data in push-based information systems. *ACM SIGAPP Applied Computing Review*, *9*(1). 22-28.

Zheng, B., Xu, J., & Lee, D. L. (2002, October). Cache invalidation and replacement strategies for location-dependent data in mobile environments. *IEEE Transactions on Computers*, *51*(10), 1141-1153.

KEY TERMS

Broadcast Environment: also known as push-based sy stem where the server would broadcast a set of data to the air for a population of mobile users to tune in for their required data.

Caching: Techniques of temporarily storing frequently accessed data designed to reduce network transfers and therefore increase speed of download

Caching Management Strategy: A strategy that relates to how a client manipulates the data that has been cached in an efficient and effective way by maintaining the data items in the client's local storage.

Cooperative Caching: A type of caching strategy that not only allows mobile clients to retrieve database items from the servers, but also from the caches of their peers.

Mobile Environment: Refers to a set of database servers that may or not be collaborative with one another that disseminate data via wireless channels to multiple mobile users.

On Demand Environment: also known as a pull–based environment, which relates to techniques that enable the server to process requests that are sent from mobile users.

Peer-to-Peer: Facilitate the features of data sharing among groups of people.

CORBA on Mobile Devices

Markus Aleksy
University of Mannheim, Germany

Axel Korthaus
University of Mannheim, Germany

Martin Schader
University of Mannheim, Germany

INTRODUCTION

Since its introduction in 1991, the Common Object Request Broker Architecture (CORBA) standard, defined by the Object Management Group (OMG, 2004b), has undergone several major revisions and has spread far throughout the domain of object-oriented and distributed systems. It not only brings about independence of computer architectures, operating systems, and programming languages, but also ensures freedom of choice with respect to Object Request Broker (ORB) product vendors. The latter benefit was the result of the introduction of a globally unique object reference, the Interoperable Object Reference (IOR), and a standard transmission protocol, the Internet Inter-ORB Protocol (IIOP), in CORBA 2.0.

CORBA uses an Interface Definition Language (IDL) to specify the interfaces that objects present to clients in order to offer their services. IDL is a purely declarative language—that is, it is used to describe the data types and interfaces in terms of their attributes, operations, and exceptions, but not to define implementation algorithms for their operations. IDL forms the foundation of CORBA's programming language independence, and language-specific IDL compilers must be used to translate IDL interface definitions into concrete programming languages. Besides the language mappings defined in the OMG standard (i.e., mappings from IDL to Ada, C, C++, COBOL, Java, Lisp, PL/1, Python, and Smalltalk), there are also non-standard language mappings to programming languages like Eiffel, Objective C, and Perl, which are exclusively implemented in certain ORB products.

While CORBA has been very successful in the domain of enterprise computing, its adoption for mobile devices is obstructed by a central problem: the limited resources of such devices. If standard-compliant CORBA-based applications are to be executed on mobile devices, storage requirements, for example, represent a major bottleneck. But for all that, several research groups have made an effort over the past few years to establish the CORBA standard in the domain of mobile devices. The existing approaches can be divided into three categories:

1. approaches that are restricted in the sense that they use an implementation of the IIOP protocol only,
2. approaches that build on the minumumCORBA specification, and
3. approaches that rely on other ways to reduce the memory footprint of a CORBA implementation.

In the following sections of this chapter, these approaches will be discussed in detail.

THE PROTOCOL APPROACH

The first alternative to realize a CORBA infrastructure on mobile devices can be described as the "protocol approach." Instead of providing an implementation of the complete CORBA specification, only the IIOP protocol is implemented in this solution.

As already mentioned, the CORBA 2.0 standard for the first time introduced the definition of a protocol for the communication between different ORBs. On an abstract level, the General Inter-ORB Protocol (GIOP) was defined, which specifies a standardized transmission syntax, the common data representation (CDR), and several message formats. Among the characteristics of CDR is a complete mapping of all data types defined in IDL and the support of different byte orders. With CORBA 2.1, the set of possible GIOP message formats was extended by "Fragment" messages. The bi-directional variant of GIOP permits both the client and the server to act as the initiator of a message for all possible kinds of messages. Since GIOP is an abstract protocol, the actual communication is routed via the Internet Inter-ORB Protocol, which provides a mapping between GIOP messages and the Transmission Control Protocol/Internet Protocol (TCP/IP) layer used in the Internet. Apart from IIOP, so-called Environment-specific Inter-ORB Protocols (ESIOPs) can be used. Currently, four different IIOP versions (1.0, 1.1, 1.2, and 1.3) can be encountered. To avoid interoperability problems that might occur when ORBs implementing different protocol versions need to communicate, the OMG specifies requirements as to which protocol has to be supported in which context (cf. OMG, 2004b, pp. 15-51).

In the context of mobile devices, a challenge lies in the realization of GIOP over wireless networks. Current CORBA implementations typically use IIOP to guarantee interoperability between CORBA-based applications. However, TCP/IP is not a suitable transport layer for wireless communication (Amir, Balakrishnan, Seshan, & Katz, 1995), so better alternatives like the Mobility Layer (Haahr, Cunningham, & Cahill, 1999, 2000) or WAP (Ruggaber, Schiller, & Seitz, 1999; Ruggaber & Seitz, 2000) were developed for that purpose. In order to provide a standardized solution for this aspect too, the OMG has adopted the wireless access and terminal mobility in CORBA specification (OMG, 2005). Furthermore, to account for the fact that the Bluetooth protocol has gained increasing popularity in the area of mobile devices, the OMG has issued the GIOP Tunneling Over Bluetooth specification (OMG, 2003).

The protocol approach was implemented in the context of several projects. For example, the work described by Haahr et al. (1999) is one of the first solutions belonging to that category. Moreover, BASE (Becker, Schiele, Gubbels, & Rothermel, 2003) and LegORB (Roman, Mickunas, Kon, & Campbell, 2000) are representatives of the protocol approach. Among the defining characteristics of BASE are, according to Becker et al. (2003), the uniform access to remote services and device-specific capabilities, the decoupling of the application communication model and the underlying interoperability protocols, and its dynamic extensibility supporting the whole range of devices from simple sensors to high-capacity workstations. There are two ways to generate invocations in BASE: they can be either generated by proxies representing services, or they are encoded analogous to a CORBA DII invocation by the application developers. The "micro-broker" coming with BASE only necessitates a plug-in to transport (marshal and send) an operation invocation. The return values an invocation might possibly construct may be accepted by an additional transport plug-in.

LegORB implements a microkernel-type architecture. Its core only contains components for low-level services. Application developers have to implement specific policies or simply select them suitably, whenever they are packaged with the ORB. LegORB's core consists of three customized components: the LegORB configurator, the client-side configurator, and the server-side configurator. They provide the glue necessary to put all the components together. LegORB itself is an assembly of components with different functional scopes with duties concerning network, marshaling, demarshaling, and so forth. The actual service capability as well as the size of LegORB is determined by the number and type of components that are composed in a concrete development project.

On the one hand, the protocol approach is sufficient to enable communication between different mobile as well as stationary CORBA-based applications, but on the other hand, it has considerable limitations. The first restriction pertains to a lack of source code portability. Since the presented solutions are specifically designed to address the communication aspect, this approach requires conceptual rethinking with respect to the way the mobile CORBA applications have to be developed. This not only holds for the migration of existing CORBA-based applications to mobile devices. Moreover, modifications to the "traditional" development process used for conventional CORBA-based applications are needed to meet the changed conditions in a mobile applications setting. The conventional development process usually starts with the specification of the IDL interfaces required in the application, if static operation invocations are intended, which is the normal case. These IDL definitions are then translated using an IDL compiler. Subsequently, the developers use different standardized CORBA classes and interfaces for ORB initialization or object activation. However, the protocol-based approaches often do not provide IDL support and do not necessarily allow for the use of the stipulated classes and interfaces for routine CORBA tasks, so that they often lead to considerable learning curves, even for experienced CORBA developers, in order to adapt to the changed programming conditions.

APPROACHES BASED ON THE MINIMUMCORBA STANDARD

The first dedicated OMG specification targeted on the reduction of the footprint of CORBA-based solutions was the minimumCORBA specification (OMG, 2002). In this specification, the OMG identified parts of the full CORBA standard that might be dispensable under certain circumstances like in the case of the limited resources available on mobile devices:

The features of CORBA omitted by this profile clearly have value in mainstream CORBA applications. However, they are provided at some cost, in terms of resources, and there is a significant class of applications for which that cost cannot be justified.

The omissions and reductions adopted by the OMG mainly concern the following points:

- **Omission of the Dynamic Invocation Interface:** The Dynamic Invocation Interface (DII) provides functionality that enables a client application to invoke operations of objects and to receive the returned results at runtime without the need to have knowledge about the corresponding signatures and types at compile time.

- **Omission of the Dynamic Skeleton Interface:** The Dynamic Skeleton Interface (DSI) became part of the CORBA standard in version 2.0. It represents a runtime mechanism to integrate components that do not have any IDL-based compiled skeletons at compile time. The DSI provides an interface that is able to accept a specific type of invocation. Incoming messages are analyzed with regard to their intended receiver object and the operation to be performed. The DSI can process both static and dynamic operation invocations.
- **Omission of the Interface Repository:** The interface repository (IR) is a runtime repository containing machine-readable versions of IDL definitions. It provides the type information needed by the DII. With CORBA 2.0, globally unique identifiers for components and their interfaces were introduced, which are system generated, so that the ORB is able to deal with entries from different IRs without the risk of ambiguities.
- **Omission of the DynAny Functionality:** The functionality of this interface allows for runtime generation of new data types that are unknown at compile time.
- **Reduction of the ORB Interface:** The operations omitted from the ORB interface specification provided functionality that was required for DII operation and could therefore be disposed of.
- **Reduction of the Object Interface:** The operations omitted from the Object interface specification mainly provided functionality that was required for DII and IR operation, were relevant for older CORBA versions, or required types that are omitted in the profile.
- **Reduction of the Portable Object Adapter Interface:** By introducing the Portable Object Adapter (POA) interface as replacement of the underspecified Basic Object Adapter (BOA), portability of CORBA-based applications could be increased. The POA serves as a link between the ORB and the servants. In the context of the minimumCORBA specification, its functionality has been reduced.

An example of an approach that belongs to the category of minimumCORBA-based solutions is K-ORB, which was created by the Distributed Systems Group in Dublin (cf. http://www.dsg.cs.tcd.ie).

A central advantage of approaches relying on minimumCORBA is the fact that the application development process remains largely unchanged. Although the use of well-known classes and interfaces like ORB and POA is restricted, it is still possible. This approach therefore requires considerably less reorientation on the part of developers already familiar with the construction of CORBA-based applications. Relying on this approach, it is much easier to port existing applications or to realize new projects than with the protocol approach.

OTHER APPROACHES

Other endeavors of research groups to achieve a reduction of the scale and scope of CORBA-based applications for their deployment on mobile devices can be summarized as being based on the following general ideas:

- using a microkernel architecture,
- following a component-based approach,
- employing a reflection-based technique, or
- providing capabilities for reconfiguration and adaptation.

The projects *MICO on Palm Pilot* and *Mico on iPAQ* (Puder, 2002) were among the first attempts to establish MICO ORB (Puder, 2000), an open source implementation of CORBA, in the domain of mobile applications. They used MICO's microkernel architecture to reduce the required memory. The use of a microkernel had already been tested successfully in several architectures and was generalized in the description of the microkernel design pattern in Buschmann, Meunier, Rohnert, Sommerlad, & Stal (1996). The developers of ZEN ORB start from the basic premise that the simultaneous support of all protocol versions at the same time is not generally required. Thus, the introduction of a "plug-and-play" interface, which can dynamically reload components on demand—such as protocol classes of a specific version—is seen as a possible solution to reduce memory requirements. Klefstad, Rao, and Schmidt (2003) describe how such an ORB could be designed and identify additional optimization factors, such as object adapters, protocol transports, object resolvers, IOR parsers, any data type handlers, buffer allocators, and CDR streams.

Most of the documented approaches, however, rely on a mix of concepts. K-ORB and ZEN, for example, use component technology together with reconfiguration capabilities. LegORB and Universally Interoperable Core (Roman, Kon, & Campbell, 2001) implement all of the architectural concepts discussed so far.

USING CORBASERVICES ON MOBILE DEVICES

The ORB represents the fundamental communication component in distributed CORBA applications. However, to support the application developer further, the OMG has standardized a number of system-level services, called CORBAservices. These services extend the basic functionality provided by the ORB by offering frequently required functionality (e.g., concerning service lookup, event management, transactions, etc.).

Although the use of CORBAservices promotes aspects like portability and reusability and enables experienced developers to produce their applications more quickly, the back side is that they generate additional memory requirements. Even though it is not necessary to run the various services directly on the mobile devices (they are normally hosted on PCs or workstations), mobile applications that need to make use of those services still need to have the corresponding stub files generated by an IDL compiler at their disposal locally, that is, on the mobile device. Typically, a CORBA-based application does not require all of the CORBAservices available. On the contrary, most applications do without CORBAservices completely, employ only the Naming Service, or use a relatively small amount of additional services, depending on their application purpose.

For PDAs, which often do not have any storage facilities apart from their Random Access Memory (RAM), stubs of CORBAservices that are never needed by a client application running on that PDA would undesirably lock memory resources that are urgently needed for other purposes.

One way to reduce the amount of memory required by those stub files is to use OMG's Lightweight Services specification (OMG, 2004a). It only contains the definition of three functionally reduced CORBAservices: the Lightweight Naming service, the Lightweight Event service, and the Lightweight Time service. In Aleksy, Korthaus, and Schader (2003), we use the so-called "Janus" approach to make the memory footprint significantly smaller. The basic idea of the Janus approach is to adapt the appearance of a component to its specific use. In our case, we have simplified and reduced the management interfaces of the event service as perceived by the mobile application. To this end, we developed a dedicated component, called Event Service Proxy, which offers a drastically simplified version of the event service interfaces and is responsible for the main part of the registration process. The actual event transmission occurs as always—that is, without utilizing the Event Service Proxy. In this way, the approach does not affect the performance of the application, except for the registration process.

FUTURE TRENDS

An important challenge for the future is the extension of the spectrum of lightweight CORBAservices to be used by mobile applications. For example, the design and realization of a "mobile" variant of the transaction service (OMG, 2000b) will be of crucial importance for the advancement of m-business applications. Similarly, a "mobile" version of the trading service (OMG, 2000a) is needed in order to provide CORBA applications running on mobile devices with extended functionality for the detection of new services.

CONCLUSION

In this article, we have presented different research efforts aiming at the goal of establishing CORBA in the domain of mobile devices. While the protocol-based approach is easier to implement and enjoys great popularity despite its limitations, the minimumCORBA-based approach does not yet meet the expectations. The ongoing advancement of mobile devices and corresponding communication infrastructures, however, raise hope that this approach will increasingly attract attention in the near future. Another future challenge will be the use of CORBAservices on mobile devices. By adopting the Lightweight Services specification, the OMG has made a first step in that direction. Nevertheless, this aspect will continue to necessitate the development of ideas for new solutions in the years to come.

REFERENCES

Aleksy, M., Korthaus, A., & Schader, M. (2003). CARLA—a CORBA-based architecture for lightweight agents. *Proceedings of the International Conference on Intelligent Agent Technology* (IAT'03) (pp. 111-118), Halifax, Canada.

Amir, E., Balakrishnan, H., Seshan, S., & Katz, R.H. (1995). Efficient TCP over networks with wireless links. *Proceedings of the 5th IEEE Workshop of Hot Topics in Operating Systems,* Orcas Island, WA.

Becker, C., Schiele, G., Gubbels, H., & Rothermel, K. (2003). BASE—a micro-broker-based middleware for pervasive computing. *Proceedings of the 1st IEEE International Conference on Pervasive Computing and Communications* (PerCom'03), Fort Worth, TX.

Buschmann, F., Meunier, R., Rohnert, H., Sommerlad, P., & Stal, M. (1996). *Pattern-oriented software architecture—a system of patterns.* Chichester: John Wiley & Sons.

Haahr, M., Cunningham R., & Cahill, V. (1999). Supporting CORBA applications in a mobile environment. *Proceedings of the 5th International Conference on Mobile Computing and Networking* (MobiCom'99), Seattle, WA.

Haahr, M., Cunningham, R., & Cahill, V. (2000). Towards a generic architecture for mobile object-oriented applications. *Proceedings of the Workshop on Service Portability and Virtual Customer Environments (SerP 2000)*, San Francisco, CA.

Klefstad, R., Rao, S., & Schmidt, D. (2003). Design and performance of a dynamically configurable messaging protocols framework for real-time CORBA. *Proceedings of the 36th Hawaii International Conference on System Sciences* (HICSS-36), Big Island, HI.

OMG (Object Management Group). (2003). *GIOP tunneling over Bluetooth.* OMG Technical Document Number dtc/2003-10-03. Retrieved from http://www.omg.org/cgi-bin/doc?dtc/2003-10-03

OMG. (2004a). *Lightweight services specification, version 1.0.* OMG Technical Document Number formal/04-10-01. Retrieved from ftp://ftp.omg.org/pub/docs/formal/04-10-01.pdf

OMG. (2002). *Minimum CORBA specification.* OMG Technical Document formal/02-08-01. Retrieved from http://www.omg.org/cgi-bin/ doc?formal/2002-08-01.pdf

OMG. (2004b). *The Common Object Request Broker: Architecture and specification. Version 3.0.3.* OMG Technical Document Number formal/04-03-01. Retrieved from ftp://ftp.omg.org/pub/docs/formal/04-03-01.pdf

OMG. (2000a). *Trading object service specification. Version 1.0.* OMG Technical Document Number formal/00-06-27. Retrieved from http://www.omg.org/cgi-bin/doc?formal/2000-06-27

OMG. (2000b). *Transaction service specification. Version 1.4.* OMG Technical Document Number formal/03-09-02. Retrieved from http://www.omg.org/cgi-bin/doc?formal/2003-09-02

OMG. (2005). *Wireless access and terminal mobility in CORBA. Version 1.2.* OMG Technical Document Number formal/05-05-02. Retrieved from http://www.omg.org/cgi-bin/doc?formal/2005-05-02

Puder, A. (2002). Middleware for handheld devices. *Proceedings of the AT&T Software Symposium,* Middletown, NJ.

Puder, A., & Römer, K. (2000). *MICO—an open source CORBA implementation.* San Francisco: Morgan Kaufmann.

Roman, M., Mickunas, D., Kon, F., & Campbell, R.H. (2000). LegORB and ubiquitous CORBA. *Proceedings of the Workshop on Reflective Middleware (IFIP/ACM Middlewar 2000),* IBM Palisades Executive Conference Center, NY.

Roman, M., Kon, F., & Campbell, R.H. (2001). Reflective middleware: From your desk to your hand. *Distributed Systems Online, 2*(5).

Ruggaber, R., Schiller, J., & Seitz, J. (1999). Using WAP as the enabling technology for CORBA in mobile and wireless environments. *Proceedings of the 7th IEEE Workshop on Future Trends of Distributed Computing Systems* (pp. 69-74), Cape Town, South Africa.

Ruggaber, R., & Seitz, J. (2000). Using CORBA applications in nomadic environments. *Proceedings of the 3rd IEEE Workshop on Mobile Computing Systems and Applications* (pp. 161-170), Monterey, CA.

KEY TERMS

CORBAservices (a.k.a. CORBA Services): General-purpose, system-related extensions of the core functionality of an ORB that are relevant to the basic operation of a distributed application.

General Inter-ORB Protocol (GIOP): Protocol specifying a standardized transmission syntax, together with several message formats.

Interface Definition Language (IDL): A purely declarative language used for the description of a CORBA application's data types and interfaces—with their attributes, operation signatures, and exceptions—independently of a concrete implementation language.

Internet Inter-ORB Protocol (IIOP): Protocol specifying how GIOP messages can be exchanged over Transmission control protocol/Internet protocol (TCP/IP) connections.

Object Adapter: Technically, the connecting link between the ORB and the proper object implementations. From a logical perspective, it connects CORBA objects that are specified by means of IDL to their implementations, which were written in a concrete programming language.

Object Request Broker (ORB): Constitutes the architecture's communication component and sometimes denoted as the "object bus."

Reflection: Means that a program is able to gain insight into its own structure. Reflection makes it possible to query meta-information about classes and their instances at runtime. Also called Introspection.

Cross-Layer RRM in Wireless Data Networks

Amoakoh Gyasi-Agyei
Central Queensland University, Australia

INTRODUCTION

A wireless data network is the infrastructure for mobile computing, which is the act of communicating while on-the-move via portable computers. Hence, wireless data networks (WDNs) and associated issues are enabling technologies for mobile computing. Usually, a WDN does not stand alone; it is connected to the fixed Internet. Hence, it is also referred to as *wireless Internet.* The unit of information transfer and processing in a packet-switched WDN is packet, which is a bunch of bits with the identifying fields needed for efficient forwarding. Advances in digitization enable a packet's content to be of varying nature, such as conversational voice samples, streaming video scenes, or non-real-time data.

Every WDN allowing multiple users to share its services requires a radio resource management (RRM) to coordinate the efficient use of the wireless transmission medium or channel. The wireless channel used by WDNs has time-, frequency-, and environment-dependent quality due to multi-path signal propagation, shadowing, path loss, and user mobility. There are two main RRM philosophies: link/rate adaptation, and opportunistic communications or opportunistic RRM. The link adaptation philosophy views the inherent channel variability as bad and hence mitigates it via complex mechanisms such as interleaving, power control, equalization, and spatial diversity (Gyasi-Agyei, 2005).

Opportunistic communications is the recent RRM philosophy, which exploits the dynamics in wireless channel quality to improve system throughput. In fact, measures have been proposed to enhance or even induce channel variability if necessary to further improve throughput (Viswanath, Tse, & Laroia, 2002). RRM has several functionalities, such as traffic admission control, power control, and scheduling. This chapter focuses on opportunistic communications or scheduling (OS). In fact, link/rate adaptation-based RRM is also somewhat cross-layer protocol engineering, as its uses physical layer information, but for a different purpose than OS. In principle, OS can be designed for single-carrier or multi-carrier, single-antenna or multi-antenna, single-hop or multi-hop, centralized or distributed wireless networks. It can also serve both real-time and non-real-time network traffic. However, it is much easier to design an OS scheme for single-hop, centralized wireless networks serving non-real-time data traffic. The achievable throughput gains are also maximized when no traffic timing constraints are embedded in the OS policy.

Scheduling is the dynamic process of allocating a shared resource to multiple parallel users in order to optimize some desirable performance metrics. Metrics of interest include: maximization of system throughput, minimization of packet delay and jitter, and the provision of fairness. Scheduling is a key mechanism in RRM and operates in the medium access control (MAC) layer. Only three things can happen to the transmission medium of a multiuser network: resource hogging, resource clogging, or equitable resource sharing. Without a MAC protocol, the desirable third option can hardly occur. In the following we discuss some general aspects of OS, propose a generalized OS design framework, discuss future trends of OS, and list some open issues in OS design.

BACKGROUND

This section reviews some background material on OS.

Why Cross-Layer Protocol Engineering?

A protocol is a set of agreed-upon rules by which two entities communicate efficiently. This includes both semantics and syntax. The telecommunications industry has been sustained so far by the open systems interconnect (OSI) reference architecture designed by the International Standards Organization (ISO). Hence this architecture is referred to as the ISO/OSI model. The ISO/OSI model is a brick-wall protocol architecture whereby the functions of a complete workable communications system is broken into a set of functionalities (not without duplications) and each set is called a layer. Each layer provides a service to the overlying layer. However, the service user is not privy to how the service is provided, and neither does it know the features of its service provider. Non-adjacent layers on the same machine have no interaction, but layers at the same level on two machines communicating interact logically to exchange data and signaling. This horizontal interaction is referred to as *peer-to-peer communication.* Indeed, no practical system is strictly designed according to the ISO/OSI model. However, the popular protocol used on the Internet, TCP/IP model, is designed based on the ISO/OSI philosophy.

The ISO/OSI philosophy has been embraced so far as great, as its modularity enables upgrading of one layer with minimal impact on other layers, until the recent drive towards system performance optimization via cross-layer engineering.

Table 1. Protocol layering and modularity vs. protocol efficiency

Layering Protocol Engineering	Cross-Layer Protocol Engineering
Traditional protocol design approach	Modern protocol design approach
Modularity oriented	Efficiency oriented
Strict layering abstraction, no interactions between non-adjacent protocol layers on the same machine	Exploits inter-layer interactions to optimize overall system performance
Partitioning of protocol functionalities reduces protocol complexity	High computational complexity
Allows protocol specialization	Requires interdisciplinary knowledge
	Incompatible with existing protocols
Easy to manage and maintain	Complicated system management and maintenance
	Revolutionary approach to system design
Only lower protocol layers are network specific	Entire protocol may be network specific
	Can reduce device energy consumption

Figure 1. An OS algorithm showing cross-layer signaling exchanges

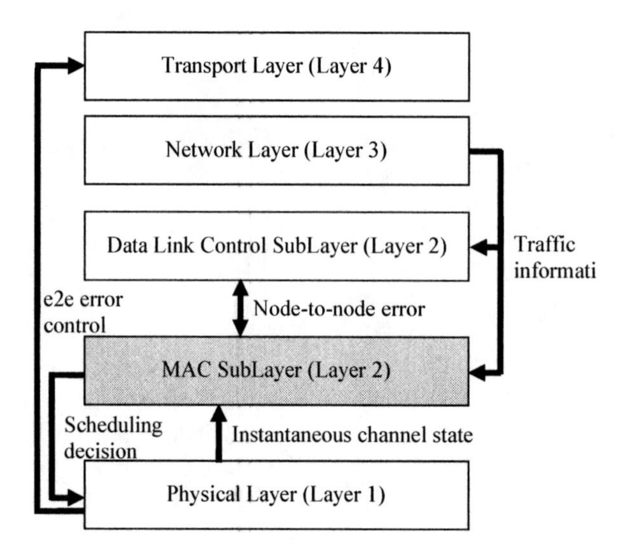

This new era of *cross-layer protocol engineering* has opened both opportunities and challenges for protocol designers. Fourth-generation networks and beyond are expected to exploit the benefits in cross-layer protocol engineering (Shakkottai, Rappaport, & Karlsson, 2003). *Cross-layer protocol design leverages runtime information across different layers to enhance the performance of the entire wireless system.* It can also be used to reduce functionality duplications during protocol design. The idea originates from the notion that layers can adapt to the instantaneous condition of other layers to improve the overall service provided to network applications. Table 1 summarizes the basic features of both protocol design philosophies.

Figure 1 illustrates the design of a cross-layer wireless scheduler. We can observe interactions in several directions between layers 1-4. These inter-layer interactions can allow higher layers to adapt to the random variability in physical layer properties to optimize system performance. Specifically, Figure 1 shows channel state information (CSI) from Layer 1 and traffic information from Layer 3 communicated to Layer 2, where the scheduler operates. These parameters enable the MAC layer to make informative decisions. For example, the

CSI can be in the form of instantaneous channel signal-to-noise ratio, the received signal strength, mobile user information, user speed of motion, or channel impulse response (Verikouskis, Alonso, & Giamalis). The traffic information can be the service history, required packet timing constraints, minimum throughput requirement, or minimum reliability requirements. The traffic delay information and the CSI enable the link layer to use the appropriate error control scheme depending on channel error pattern and traffic delay requirements. That is, whether backward-error or forward-error, correction should be used. The delay information also influences the scheduling decision. The CSI feedback to the connection-oriented transport layer allows the latter's end-to-end (e2e) error control mechanism to distinguish congestion-related from channel-state-related packet losses.

Why Opportunistic Scheduling?

As discussed above, the wireless channel quality varies with time, space, and operating frequency—that is, specto-spatio-temporally varying. Hence, multiple users sharing a wireless resource experience independently varying channel qualities at the same time: some users experience poor channels, while others experience good channels. This is referred to as *multiuser diversity* (Knopp & Humblet, 1995), the basis of opportunistic scheduling. The origin of opportunistic communications is attributed to the works of Knopp and Humblet (1995) and Viswanath et al. (2002). Opportunistic scheduling (Shakkottai & Stolyar, 2001; Gyasi-Agyei, 2003; Liu, Chong, & Shroff, 2003; Liu, Grul, & Knightly, 2003; Hu, Zhang, & Sadowsky, 2004; Gyasi-Agyei & Kim, 2006) is a wireless scheduling which exploits multiuser diversity to maximize the total system throughput. The per-user throughput is also maximized if users have symmetric fading statistics. The throughput gain is achieved by picking a user among all active users at a given time, which has the relatively best channel condition and hence the highest data transfer speed. The achieved gain in OS over a comparable non-OS scheme is referred to as *multiuser diversity gain*. Multiuser diversity gain increases with the rate of channel variations, the dynamic range of channel variability and randomization, and hence the size of user population sharing a wireless resource. The larger the user population, the higher the chance of picking a user at a high channel quality at any time, and hence the higher the multiuser diversity gain. This is the motivation behind proposals to enhance channel scattering if necessary (Viswanath et al., 2002). Besides the above limitations, the wireless spectrum is a finite resource. Hence, novel efforts such as opportunistic communications are required to maximize its utility. While conventional diversity techniques combat channel slow fading to achieve error-free communications, multiuser diversity exploits channel fast fading to boost system throughput.

Figure 2. Time-slotted, single-cell system using OS to serve multiple heterogeneous users

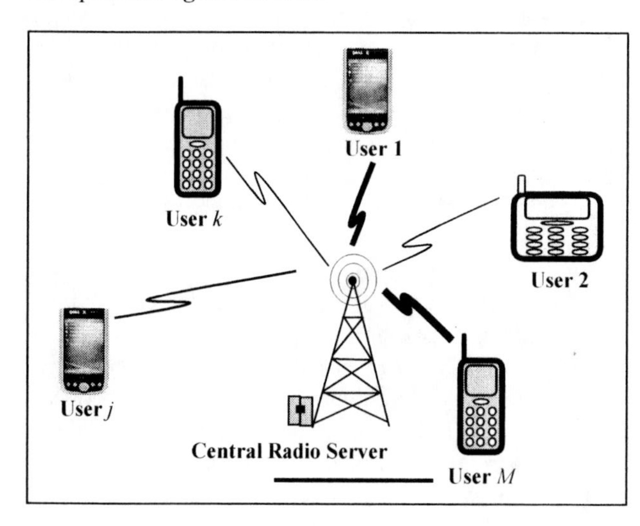

Consider Figure 2 where M users share a wireless server offering time-slotted service. The feasible data transmission speed of each user varies over time in accordance with its channel quality variations. The maximum speed that user m can send data per Hertz of bandwidth at time t lies in the range:

$$0 \leq R_m(t) < \min\left[\log_2\left(1 + SNR_m(t)\right),\ 2\log_2 K\right] \quad (1)$$

where $K \geq 1$ is the number of signaling levels used in the system and $SNR_m(t)$ is user m's instantaneous signal-to-noise ratio. By scheduling a user with the highest instantaneous rate, a greedy OS is able to achieve the throughput bound at the cost of all other system performance metrics. However, guarantee on bounded delay is necessary if OS has to support real-time traffic. This can be achieved by embedding traffic temporal parameters into OS and making optimum trade-off between delay, throughput, and fairness.

Design Issues and Performance Metrics of Cross-Layer RRM Algorithms

The design of efficient cross-layer RRM mechanisms requires the consideration of several issues. Some of the crucial design metrics are: throughput, energy efficiency, equity, algorithmic complexity, scalability and optimality, feasibility, stability, timeliness, algorithmic convergence, near-far, hidden-terminal, and exposed-terminal problems.

The wireless medium used by WDNs is a finite resource which is getting crowded, requiring an economized usage. Energy efficiency is an interesting issue due to the power constraints of battery-powered wireless terminals. Also, in some network architectures, notably wireless sensor networks operating in a harsh environment, the power of the wireless nodes cannot be easily replenished. Hence, power failure of a single node can destabilize an entire network. RRM schemes must be fair to prevent hogging and/or starvation. However, fairness does not necessary mean that all queues receive the same level of service. For this reason several fairness models are designed; examples are utilitarian fairness and proportional fairness.

Algorithms with low complexity reduce energy dissipation in wireless terminals and prolong their battery lifetime. Low-complexity algorithms can also reduce the processing time of traffic and hence reduce the end-to-end packet delay. Scalable and stable algorithms are able to support network growth under all traffic patterns. Hence, RRM algorithms should answer questions like: Can the algorithm work efficiently in networks of all sizes? Is the algorithm stable under all possible traffic arrival patterns and user population? Stability ensures that the length of every queue in an *admissible system* is bounded at steady state—that is:

$$\lim_{x\to\infty} \sup E\left[\|\mathbf{q}_t\|\right] \le \alpha$$

where

$$\mathbf{q}_t = (q_t^1, q_{t,}^2 \cdots q_t^n) \quad (2)$$

where q_t^k is the length of the kth queue at time t, n is the number of simultaneous queues in the system at a given time, and α is a small non-negative number. Algorithm convergence is necessary to maintain a stable system. A queuing system is said to be admissible if there is a service process which is able to maintain stable queues under a given traffic arrival process.

The near-far problem (Figure 3c) occurs when two transmitters send signals to the same receiver about the same time and one has a much stronger signal than the other at the receiver. Thus the stronger signal prevents the receiver from detecting the weaker signal. This term is coined, as the transmitter of the stronger signal is usually closer to the receiver than that of the weaker signal. This issue is more troublesome in CDMA systems. One popular solution to this problem is the use of transmitter power control.

Figure 3. Illustration of some wireless communication issues

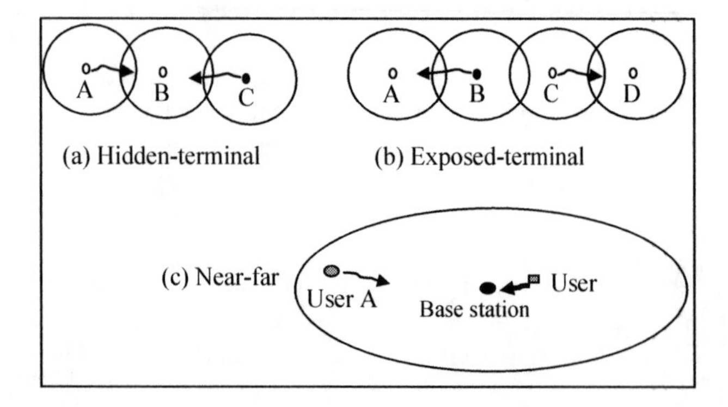

Figure 3 illustrates the *hidden-terminal* and the *exposed-terminal* problems in wireless networks. The former occurs when two nodes (here A and C) that are out of range with each other transmit signals to the same node (here B) and collide at the receiver B. The exposed-terminal problem occurs when an ongoing communication between two users unnecessarily prevents communication between another pair of users. Combining power control with scheduling prevents all the issues in Figure 3.

A Framework for Utility-Based Opportunistic Schedulers

Utility functions are widely applied in economics to quantify the benefit in using finite resources. Maximizing a utility function results in maximizing the benefits in the corresponding resources. In terms of wireless communications, the resources are power, transmission channels, bandwidth, and so forth. Let x be a vector of network parameters or their indices. We can thus define parametric utility functions $U(x,t)$ at any time t which consider the RRM design issues discussed above which are of interest in a given situation. Such a generalized parametric model enables the classification of several OS schemes in the literature. Consider the three-part utility function:

$$U(x,t) = s(x,t)\cdot f(x,t)\cdot d(x,t) \quad (3)$$

For example, maximizing $f(x,t)$ means enhancing fairness and maximizing $s(x,t)$ enhances throughput. $U(x,t)$ in (3) is simple and covers three key OS design metrics.

As a simple design example, consider the time-slotted multiuser system shown in Figure 2, and let $s(x,t) = R_m(t)$, $f(x,t) = 1/B_i(t)$, and $d(x, t) = \exp[\{\kappa_x W_x(t) - \eta(t)\}/q_x(t)]$.

This results in the delay-aware BLOT (D-BLOT) policy, a variation of the Best Link LOwest Throughput (BLOT) first scheduling (Gyasi-Agyei & Kim, 2006). The resulting utility function is:

$$U(x,t) = \frac{R_m(t)}{B_i(t)}\exp\left[\frac{\kappa_x W_x(t) - \eta(t)}{q_x(t)}\right]$$

Here, $x = (B_i(t), R_m(t))^T$ or $x = (i,m)^T$. For simplicity we can require that $\eta(t) = \sum_x (t)/\|x\|$ and $q_x(t) = \eta(t)$. $W_x(t)$ is the waiting time of connection x at time t, and κ_x is a constant factor reflecting a connection's timing constraints. We note that any of the three functionals in equation (3) can be set to unity. For example, all non-real-time OS schemes use $d(x,t) = 1$. It is worthy to remark that both BLOT and D-BLOT guarantee a minimum service to multiple queues that are concurrently active at a given user, as detailed in Gyasi-Agyei and Kim (2006) and Gyasi-Agyei (2005).

FUTURE TRENDS AND OPEN ISSUES

Gyasi-Agyei and Kim (2006) classify OS schemes into eight types under four subgroups and recognize that Type F OS is a field yet to be explored. Li and Niu (2004) attempt this problem. Below, we discuss some possible scenarios for such OS. Figure 4 illustrates the downlink of a multi-carrier (OFDM) wireless system using multiple antennas, OS, and dynamic subcarrier assignment. In this architecture, each OFDM symbol carries the data for a single connection, and disjointed subsets of OFDM subcarriers are allocated to different connections in a given time slot. This provides mutual orthogonality across the signals received at the receiver, as the spatial signatures are usually correlated. Elements of the antenna array at the transmitter transmit data of independent connections. However, multiple active connections may terminate at the same receiver. Assume that there are X concurrent connections and L multiple antennas per radio server. If $L=X$, then X mutually orthogonal sets of random beams can serve the X connections simultaneously. If $L>X$, then spatial multiplexing gain can be exploited on the remaining $L-X$ beams.

The architecture in Figure 4, whose details are left for future research, may operate as follows. Channel qualities of OFDM subcarriers from Layer 1 and traffic information from the application layer through Layer 3 are fed into the OS. The OS then uses this

information to build a utility function. In each scheduling epoch, the queue that has the highest utility on a subcarrier

Figure 4. A downlink architecture for Type F opportunistic scheduler

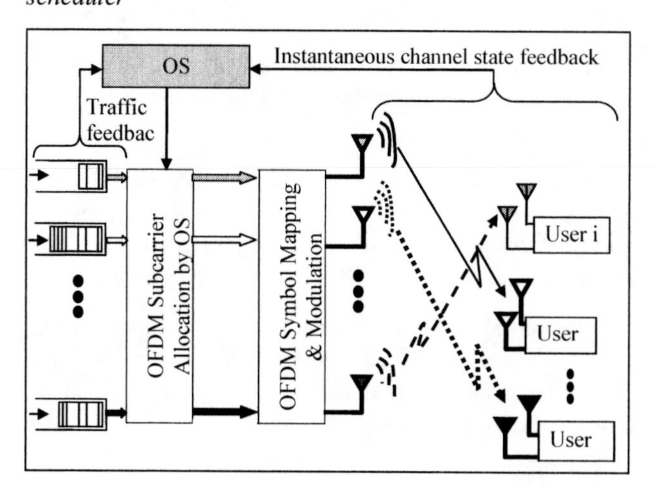

is allowed to transmit/receive information on that subcarrier. The set of subcarriers allocated to each connection in each time slot is used by the OFDM transmitter to construct an OFDM modulating symbol and then transmitted to the corresponding user.

OFDM enables broadband communications over otherwise frequency-selective fading channels without remarkable inter-symbol interference. Time-slotted OFDM-based networks exploit the synergies between frequency and time multiplexing. Another advantage in OFDM transmissions is that different power levels can be allocated to different subcarriers (i.e., adaptive subcarrier allocation) based on their runtime qualities to meet a given reliability requirement. This is referred to as the multiuser *water-filling principle*. One issue with this approach is the signaling load involved in estimating the CSI on all subcarriers and communicating them to the scheduling engine. A subcarrier clustering approach bundling a set of subcarriers into a subchannel and assigning a single CSI has been proposed to reduce this issue.

Mobile computing and its enabling infrastructure are faced with several challenges. For example, as user terminals must be carried around, they impose ergonomic constraints. These constraints in turn penalize power supply, device memory size, disk capacity, and processor speeds. Although these issues have been improved in recent years, more remains to be done to boost the technology's uptake. These end-user terminal constraints affect the issues discussed under 'Design Issues of RRM Algorithms'. Some of the open issues and challenges in OS include the need to:

- Develop efficient mechanisms using CSI to help a connection-oriented transport layer to distinguish channel-dependent packet losses from network congestion-dependent packet losses so as to minimize throughput degradation due to end-to-end congestion control mechanisms. Current proposals to solve this problem include the use of the explicit congestion notification flag in packet header, snoop transport protocol, and explicit loss notification (Jiang, Zhuang, & Shen, 2005).
- Develop channel-state-dependent data compression schemes to reduce data transfer times.
- Investigate implications of cross-layer design on the overall networking architecture.
- Develop CSI feedback without errors and with minimal delay. Also, the effect of errors and delays in CSI on scheduling performance is an interesting task. Estimation and communication of CSI from physical to higher layers can be quite signaling intensive and hence impact transmission efficiency. This issue is exacerbated in networks with high mobility and/or changing topology. Hence, novel techniques with optimum trade-off between currency of CSI and frequency of CSI estimation are needed.

- Develop OS for multi-hop and distributed wireless networks, for example, sensor networks.
- Design OS schemes to serve traffic of varying characteristics over varying interfaces.
- Develop OS for multi-carrier wireless networks using spatial diversity.
- Design OS for variable-size time slots to serve variable-length packets.
- Develop simple but efficient online OS algorithms for wireless networks.

CONCLUSION

We have presented a handy introduction to opportunistic communications, an idea whose time has come. It is hoped that the open issues underscored become a basis for further R&D on the topic. Opportunistic communications is inherently cross-layer in nature and a disruptive technology, as it makes fading-combating techniques counter-productive. This is the reality facing the well-established RRM based on ISO/OSI model and link adaptation.

REFERENCES

Gyasi-Agyei, A. (2003, September). BL^{2}xF–channel state-dependent scheduling algorithms for wireless IP networks. *Proceedings of the IEEE International Conference on Networks* (pp. 623-628). Sydney.

Gyasi-Agyei, A. (2005). Multiuser diversity based opportunistic scheduling for wireless data networks. *IEEE Communications Letters*, *9*(7), 670-672.

Gyasi-Agyei, A., & Kim, S.-L. (2006, January). Comparison of opportunistic scheduling policies in time-slotted AMC wireless networks. *Proceedings of the IEEE International Symposium on Wireless Pervasive Computing* (pp. 1-6). Phuket, Thailand.

Gyasi-Agyei, A., & Kim, S.-L. (2006). Cross-layer multiservice opportunistic scheduling for wireless networks. *IEEE Communications Magazine, 44*(6).

Hu, M., Zhang, J., & Sadowsky, J. (2004). Traffic aided opportunistic scheduling for wireless networks: Algorithms and performance bounds. *Elsevier Computer Network,* (November), 505-518.

Jian, H., Zhuang, W., & Shen, H. S. (2005). Cross-layer design for resource allocation in 3G wireless networks and beyond. *IEEE Communications Magazine,* (December), 120-126.

Knopp, R., & Humblet, P.A. (1995, June). Information capacity and power control in single-cell multiuser communications. *Proceedings of the IEEE International Conference on Communications* (pp. 331-335). Seattle, WA.

Li, L., & Niu, Z. (2004, September). A multi-dimensional radio resource scheduling scheme for MIMO-OFDM systems with channel dependent parallel weighted fair queuing (CD-PWFQ). *Proceedings of the IEEE International Symposium on PIMRC* (pp. 2367-2371).

Liu, X., Chong, E. K. P., & Shroff, N. B. (2003, October). A framework for opportunistic scheduling in wireless networks. *Computer Networks, 41*(4), 451-474.

Liu, Y., Grul, S., & Knightly, E. W. (2003). WVFQ: An opportunistic wireless scheduler with statistical fairness bounds. *IEEE Transactions on Wireless Communications, 2*(5), 1017-1028.

Shakkottai, S., Rappaport, T. S., & Karlsson, P. C. (2003). Cross-layer design for wireless networks. *IEEE Communications Magazine, 41*(10), 74-80.

Verikouskis, C., Alonso, L., & Giamalis, T. (2005). Cross-layer optimization for wireless systems: A European research key challenge. *IEEE Communications Magazine, 43*(7).

Viswanath, P., Tse, D. N. C., & Laroia, R. (2002). Opportunistic beamforming using dumb antennas. *IEEE Transactions on Information Theory, 48*(6), 1277-1294.

KEY TERMS

Cross-Layer Protocol (CLP): Any communications protocol that interacts with a protocol operating on any other layer of the ISO/OSI protocol model.

Fading: Dynamically changing attenuation, usually experienced on wireless channels.

Mobile Computing: Communication via portable computer over a wireless data network while on-the-move.

Multiuser Diversity (MUD): The statistically independent variability of channel qualities (states) across multiple users in a multiuser system at a given time.

Multiuser Diversity Gain (MDG): The gain achieved (compared to a non-channel-aware version of the same algorithm) when a radio resource manager (RRM), such as a packet scheduler, exploits MUD on wireless connectivity.

Opportunistic MAC (OMAC): A medium access control protocol that exploits MUD in its operation. An example of OMAC is an opportunistic scheduling. OMAC is also referred to as opportunistic communications.

Opportunistic Scheduler (OS): A traffic scheduler that utilizes MUD to enhance desirable system performance such as system throughput or spectral efficiency. OS is also referred to as channel-aware or channel-state-dependent scheduling.

Radio Resource Management: The process of allocating wireless network resources (e.g., transmission channels, power, and spectrum) to radio nodes in an optimum manner—optimum in the sense that the service requirements of most of the users are met and system throughput is maximized.

Scheduling: The dynamic process of allocating a shared resource to multiple competing users in order to optimize some desirable performance metrics.

Data Caching in Mobile Ad-Hoc Networks

Narottam Chand
Indian Institute of Technology Roorkee, India

R. C. Joshi
Indian Institute of Technology Roorkee, India

Manoj Misra
Indian Institute of Technology Roorkee, India

INTRODUCTION

Mobile wireless networks allow a more flexible communication structure than traditional networks. Wireless communication enables information transfer among a network of disconnected, and often mobile, users. Popular wireless networks such as mobile phone networks and wireless local area networks (LANs), are traditionally infrastructure based—that is, base stations (BSs), access points (APs), and servers are deployed before the network can be used. A mobile ad hoc network (MANET) consists of a group of mobile hosts that may communicate with each other without fixed wireless infrastructure. In contrast to conventional cellular systems, there is no master-slave relationship between nodes, such as base station to mobile users in ad-hoc networks. Communication between nodes can be supported by direct connection or multi-hop relays. The nodes have the responsibility of self-organizing so that the network is robust to the variations in network topology due to node mobility as well as the fluctuations of the signal quality in the wireless environment. All of these guarantee anywhere and anytime communication. Recently, mobile ad-hoc networks have been receiving increasing attention in both commercial and military applications.

The dynamic and self-organizing nature of ad-hoc networks makes them particularly useful in situations where rapid network deployments are required or it is prohibitively costly to deploy and manage network infrastructure. Some example applications include:

- attendees in a conference room sharing documents and other information via their laptops and PDAs (personal digital assistants);
- armed forces creating a tactical network in unfamiliar territory for communications and distribution of situational awareness information;
- small sensor devices located in animals and other strategic locations that collectively monitor habitats and environmental conditions; and
- emergency services communicating in a disaster area and sharing video updates of specific locations among workers in the field, and back to headquarters.

Unfortunately, the ad-hoc nature that makes these networks attractive also introduces many complex communication problems. From a communications perspective, the main characteristics of ad-hoc networks include:

1. lack of pre-configuration, meaning network configuration and management must be automatic and dynamic;
2. node mobility, resulting in constantly changing network topologies;
3. multi-hop routing;
4. resource-limited devices, for example, laptops, PDAs, and mobile phones have power and CPU processing constraints;
5. resource-limited wireless communications, for example, a few kilobits per second per node; and
6. potentially large networks, for example, a network of sensors may comprise thousands or even tens of thousands of mobile nodes.

A key research challenge in ad-hoc networks is to increase the efficiency of data transfer, while handling the harsh environmental conditions such as energy constraints and highly mobile devices. Presently, most of the researches in ad-hoc networks focus on the development of dynamic routing protocols that can improve the connectivity among mobile nodes which are connected to each other by one-hop/multi-hop links. Although routing is an important issue in ad-hoc networks, other issues such as information/data access are also very important since the ultimate goal of using such networks is to provide information access to mobile nodes. One of the most attractive techniques that improves data availability is caching. In general, caching results in:

1. enhanced QoS at the nodes—lower jitter, latency, and packet loss;

2. reduced network bandwidth consumption; and
3. reduced data server/source workload.

In addition, reduction in bandwidth consumption infers that a properly implemented caching architecture for ad-hoc network can potentially improve battery life in mobile nodes.

BACKGROUND

Caching has been proved to be an important technique for improving the data retrieval performance in mobile environments (Chand, Joshi, & Misra, 2004, 2005; Cao, 2002, 2003). With caching, the data access delay is reduced since data access requests can be served from the local cache, thereby obviating the need for data transmission over the scarce wireless links. However, caching techniques used in one-hop mobile environments (i.e., cellular networks) may not be applicable to multi-hop mobile environments since the data or request may need to go through multiple hops. As mobile clients in ad-hoc networks may have similar tasks and share common interest, cooperative caching, which allows the sharing and coordination of cached data among multiple clients can be used to reduce the bandwidth and power consumption.

To date there are some works in literature on cooperative caching in ad-hoc networks, such as consistency (Yin & Cao, 2004; Cao, Yin, & Das, 2004), placement (Zhang, Yin, & Cao, 2004; Nuggehalli, Srinivasan, & Chiasserini, 2003; Papadopouli & Schulzrinne, 2001), discovery (Takaaki & Aida, 2003), and proxy caching (Lau, Kumar, & Venkatesh, 2002; Friedman, Gradinariu, & Simon, 2004; Sailhan & Issarny, 2003; Lim, Lee, Cao, & Das, 2004, 2006). However, efficient cache replacement is not considered yet.

Cache management in mobile ad-hoc networks, in general, includes the following issues to be addressed:

1. The cache discovery algorithm that is used to efficiently discover, select, and deliver the requested data item(s) from neighboring nodes. In a cooperative architecture, the order of looking for an item follows local cache to neighboring nodes, and then to the original server.
2. The design of cache replacement algorithm—when the cache space is sufficient for storing one new item, the node places the item in the cache. Otherwise, the possibility of replacing other cached item(s) with the new item is considered.
3. Cache admission control—this is to decide what data items can be cached to improve the performance of the caching system.
4. The cache consistency algorithm, which ensures that updates are propagated to the copies elsewhere, and no stale data items are present.

In this article, a *Zone Cooperative (ZC)* cache is proposed for mobile ad-hoc networks. Mobile nodes belonging to the neighborhood (zone) of a given node form a cooperative cache system for this node since the cost for communication with them is low both in terms of energy consumption and message exchanges. In ZC caching, each mobile node has a cache to store the frequently accessed data items. The cache at a node is a nonvolatile memory such as hard disk. The data items in the cache satisfy not only the node's own requests, but also the data requests passing through it from other nodes. For a data miss in the local cache, the node first searches the data in its cooperation zone before forwarding the request to the next node that lies on a path towards server. The caching scheme includes a discovery process and a cache management technique. The proposed cache discovery algorithm ensures that requested data are returned from the nearest node or server. As a part of cache management, cache admission control, Least Utility Value (LUV)-based replacement policy, and cache consistency technique are developed. The admission control prevents high data replication by enforcing a minimum distance between the same data item, while the replacement policy helps in improving the cache hit ratio and accessibility. Cache consistency ensures that clients only access valid states of the data.

SYSTEM ENVIRONMENT

The system environment is assumed to be an ad-hoc network where a mobile host accesses data items held as originals by other mobile hosts. A mobile host that holds the original value of a data item is called data server. A data source may be connected to the wired network. A data request initiated by a host is forwarded hop-by-hop along the routing path until it reaches the data source and then the data source sends back the requested data. Each mobile host maintains local cache in its hard disk. To reduce the bandwidth consumption and query latency, the number of hops between the data source/cache and the requester should be as small as possible. Most mobile hosts, however, do not have sufficient cache storage, and hence the caching strategy is to be devised efficiently. In this system environment, we also make the following assumptions:

- Assign a unique host identifier to each mobile host in the system. The system has a total of M hosts, and MH_i $(1 \leq i \leq M)$ is a host identifier. Each host moves freely.
- We assign a unique data identifier to each data item located in the system. The set of all data items is denoted by $D = \{d_1, d_2, ..., d_N\}$, where N is the total number of data items and d_j $(1 \leq j \leq N)$ is a data identifier. D_i denotes the actual data of the item with id d_i. Size of

Figure 1. Service of a client request in ZC caching strategy

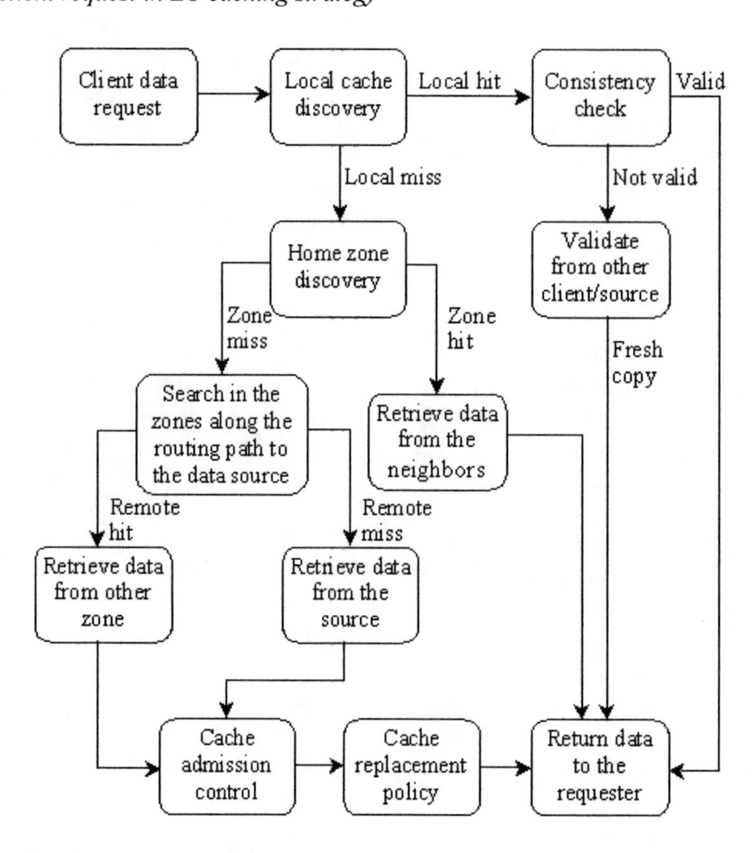

data item d_i is s_i (in bytes)—that is, $s_i = |D_i|$. The origin of each data item is held by a particular data source.

- Each mobile host has a cache space of C bytes.
- Each data item is periodically updated at data source. After a data item is updated, its cached copy (maintained on one or more hosts) may become invalid.

ZONE COOPERATIVE CACHING

The design rationale of Zone Cooperative (ZC) caching is that it is considered advantageous for a client to share cache with its neighbors lying in the zone (i.e., clients that are accessible in one-hop). Mobile clients belonging to the cooperation zone of a given client then form a cooperative cache system for this client since the cost for communicating with them is low both in terms of energy consumption and message exchange. Figure 1 shows the behavior of ZC caching strategy for a client request.

When a data request is initiated at a client, it first looks for the item in its own cache. If there is a local cache miss, the client checks if the data item is cached in other clients within its home zone. When a client receives the request and has the data item in its local cache (i.e., a zone cache hit), it will send a reply to the requester to acknowledge that it has the data item. In case of a zone cache miss, the request is forwarded to the neighbor node along the routing path. Before forwarding a request, each client along the path searches the item in its local cache or zone as described above. If the data item is not found on the zones along the routing path (i.e., a remote cache miss), the request finally reaches the data source and the data source sends back the requested data.

When a client receives the requested data, a cache admission control is triggered to decide whether it should be brought into the cache. Inserting a data item into the cache might not always be favorable, because incorrect decision can lower the probability of cache hits. In ZC, the cache admission control allows a client to cache a data item based on the distance of data source or other client that has the requested data. If the original of the data resides in the same zone of the requesting client, then the item is not cached, because it is unnecessary to replicate a data item in the same zone since cached data can be used by closely located clients. In general, same data items are cached at least two hops away.

The ZC caching uses a simple weak consistency/invalidation model based on Time-To-Live (TTL), in which a client considers a cached copy up-to-date if its TTL has not expired.

The client removes the cached data when the TTL expires. A client refreshes a cached item and its TTL if a fresh copy of the item passes by.

UTILITY-BASED CACHE REPLACEMENT

The traditional cache replacement algorithms (e.g., LRU) might be unsuitable for ad-hoc networks due to non-uniform data size, varying distance between requester and data source, and frequent data updates.

We have developed a Least Utility Value-based cache replacement policy, where data items with the lowest utility are those that are removed from the cache. Four factors are considered while computing utility value of a data item at a client.

- **Popularity:** The access probability reflects the popularity of a data item for a host. An item with lower access probability should be chosen for replacement. At a client, the access probability A_i for data item d_i is given as

$$A_i = a_i \Big/ \sum_{k=1}^{N} a_k,$$

where a_i is the mean access rate to data item d_i. a_i can be estimated by employing *sliding window* method of last K access times. We keep a sliding window of K most recent access timestamps $(t_i^1, t_i^2, \ldots, t_i^K)$ for data item d_i in the cache. The access rate is updated using the formula

$$a_i = \frac{K}{t^c - t_i^K},$$

where t^c is the current time and t_i^K is the timestamp of oldest access to item d_i in the sliding window. When fewer than K samples are available, all the samples are used to estimate a_i. To reduce the computational complexity, the access rates for all cached items are not updated during each replacement; rather the access rate for an item is updated only when the item is accessed. K can be as small as 2 or 3 to achieve the best performance. Thus the spatial overhead to store recent access timestamps is relatively small.

- **Distance:** Distance (δ) is measured as the number of hops between the requesting client and the responding client (source or cache). The greater the distance, the greater is the *utility* value of the data item. This is because caching data items that are further away saves bandwidth and reduces latency for subsequent requests.
- **Coherency:** A data item d_i is valid for a limited lifetime, which is known using the TTL_i field. An item that is valid for a shorter period should be preferred for replacement.
- **Size (s):** A data item with larger size should be chosen for replacement, because the cache can accommodate more items and satisfy more access requests.

Based on the above factors, the *utility$_i$* function for an item d_i is computed as:

$$utility_i = \frac{A_i . \delta_i . TTL_i}{s_i}.$$

The idea is to maximize the total utility value for the data items kept in the cache. For a cache of size C such that the size s_i of each data item d_i is very much less than C, the principle of optimality implies that the mobile client MH_x should always retain a set C_x of data items in its cache such that

$$\sum_{d_i \in C_x} utility_i$$

is maximized subject to

$$\sum_{d_i \in C_x} s_i \le C.$$

Maximizing the above objective function implies a minimization of the response time per reference. The task of LUV is to make this optimal decision for every replacement. A binary min-heap data structure is used to implement the LUV policy. The key field for the heap is the $utility_i$ value for each cached data item d_i. When the events of cache replacement occur, the root item of the heap is deleted. This operation is repeated until sufficient space is obtained for the incoming data item. Let N_c denote the number of cached items and S the victim set size. Every deletion operation has a complexity of $O(\log N_c)$. An insertion operation also has an $O(\log N_c)$ complexity. Thus the time complexity for every cache replacement operation is $O(S \log N_c)$.

CONCLUSION

Spurred by the progress of technologies and deployment at low cost, the use of ad-hoc networks is expected to be largely exploited for mobile computing, and no longer be restricted to specific applications (e.g., crisis applications as in military and emergency/rescue operations or disaster recovery). In particular, ad-hoc networks effectively support ubiquitous networking, providing users with network access in most situations. Data access in ad-hoc networks is an important issue where mobile nodes communicate with each other via short-range transmissions to share information. Caching of frequently accessed data in such an environment is a potential technique that can improve the data access performance and availability. In this article, we propose a novel scheme called *Zone Cooperative (ZC)* for caching in mobile ad-hoc networks. The scheme enables nodes to share their data which helps alleviate the longer average query latency and limited data accessibility problems in ad-hoc networks.

REFERENCES

Cao, G. (2002). On improving the performance of cache invalidation in mobile environments. *Mobile Networks and Applications, 7*(4), 291-303.

Cao, G. (2003). A scalable low-latency cache invalidation strategy for mobile environments. *IEEE Transactions on Knowledge and Data Engineering, 15*(5), 1251-1265.

Cao, G., Yin, L., & Das, C. (2004). Cooperative cache based data access framework for ad hoc networks. *IEEE Computer Magazine,* 32-39.

Chand, N., Joshi, R. C., & Misra, M. (2004). Broadcast based cache invalidation and prefetching in mobile environment. *Proceedings of HiPC* (LNCS 3296, pp. 410-419), Bangalore, India. Berlin: Springer-Verlag.

Chand, N., Joshi, R. C., & Misra, M. (2005). Energy efficient cache invalidation in a mobile environment. *International Journal of Digital Information Management (Special Issue on Distributed Data Management), 3*(2), 119-125.

Friedman, R., Gradinariu, M., & Simon, G. (2004). Locating cache proxies in MANETs. *Proceedings of the ACM International Symposium on Mobile Ad Hoc Networking and Computing* (pp. 175-186), Tokyo, Japan.

Lau, W. H. O., Kumar, M., & Venkatesh, S. (2002). Cooperative cache architecture in support of caching multimedia objects in MANETs. *ACM International Workshop on Wireless Mobile Multimedia* (pp. 56-63), Atlanta, GA.

Lim, S., Lee, W.-C., Cao, G., & Das, C. R. (2004). Performance comparison of cache invalidation strategies for Internet-based mobile ad hoc networks. *Proceedings of the IEEE International Conference on Mobile Ad Hoc and Sensor Systems (MASS)* (pp. 104-113), FL.

Lim, S., Lee, W.-C., Cao, G., & Das, C. R. (2006). A novel caching scheme for improving Internet-based mobile ad hoc networks performance. *Ad Hoc Networks Journal, 4*(2), 225-239.

Nuggehalli, P., Srinivasan, V., & Chiasserini, C.-F. (2003). Energy-efficient caching strategies in ad hoc wireless networks. *Proceedings of the ACM Symposium on Mobile Ad Hoc Networking and Computing (MobiHoc)* (pp. 25-34), MD.

Papadopouli, M., & Schulzrinne, H. (2001). Effects of power conservation, wireless coverage and cooperation on data dissemination among mobile devices. *Proceedings of the ACM Symposium on Mobile Ad Hoc Networking and Computing (MobiHoc)* (pp. 117-127), CA.

Sailhan, F., & Issarny, V. (2003). Cooperative caching in ad hoc networks. *Proceedings of the International Conference on Mobile Data Management (MDM)* (pp. 13-28), Melbourne, Australia.

Takaaki, M., & Aida, H. (2003). Cache data access system in ad hoc networks. *Proceedings of the Vehicular Technology Conference (VTC)* (pp. 1228-1232), FL.

Yin, L., & Cao, G. (2004). Supporting cooperative caching in ad hoc networks. *Proceedings of IEEE INFOCOM* (pp. 2537-2547), Hong Kong.

Zhang, W., Yin, L., & Cao, G. (2004). Secure cooperative cache based data access in ad hoc networks. *Proceedings of the NSF International Workshop on Theoretical and Algorithmic Aspects of Wireless Ad Hoc, Sensor, and Peer-to-Peer Networks,* Chicago.

KEY TERMS

Access Point (AP): A transceiver in a wireless LAN that can connect a network to one or many wireless devices. APs can also bridge to one another.

Cache Consistency: A technique to ensure that the data at a client cache has same value as on the original server.

Cache Discovery: Searching the requested data in a MANET.

Cache Replacement: The process of eviction of an item from the cache when a new item is to be stored in the cache.

Caching: A technique where a copy of the remote data is stored locally to improve data availability and reduce access delay.

Cooperation Zone: One-hop neighbors of a mobile client form the cooperation zone for the client.

Data Server: A client that holds the original value of data.

Mobile Ad-Hoc Network: An autonomous network of mobile clients connected by wireless links where network topology may change rapidly and unpredictably.

Decision Analysis for Business to Adopt RFID

Koong Lin
Tainan National University of the Arts, Taiwan

Chad Lin
Edith Cowan University, Australia

Huei Leu
Industrial Technology Research Institute, Taiwan

INTRODUCTION

The spending for RFID (radio frequency identification) has been increasing rapidly in recent years. According to Gartner, global spending on RFID is likely to reach US$3 billion by 2010 (CNET, 2005). In addition, interests continue to grow for the adoption of this mobile computing and commerce device in many different types of applications (ABI, 2006). In 2005, Wal-Mart asked its top 100 suppliers to use RFID tags, and this had a profound effect on the projected growth of RFID technology as well as potential applications in the industrial, defense, and retails sectors (Albertsons, 2004).

However, very few studies have examined and evaluated the adoption of RFID options by the organizations. Organizations face various risks and uncertainties when assessing the adopted mobile technologies. Different organizations are likely to encounter different challenges and problems. This research aims to develop a mechanism that can help organizations to specify their risks and choose a suitable adoption alternative. This research has adopted the AHP (analytic hierarchy process) methodology to analyze the data, as it is useful for analyzing different RFID adoption alternatives and can assist organizations in predicting the possible issues and challenges when adopting RFID.

The objectives of this article are to: (1) describe basic components of a mobile computing and commerce device, RFID; and (2) explore the current practices, issues, and applications in this mobile technology.

BACKGROUND

RFID is a built-in wireless technology that incorporates a smart IC (integrated circuit) tag. It allows organizations to capture accurate information about the location and status of products, and track them as they move from the assembly line to the retail store (Albertsons, 2004). The three major components of RFID are: tags, readers, and software systems. RFID tags consist of silicon chips and antennas. Each tag uses an ID coding system and contains a unique serial number of a product. This enables the tag to store some information of the product. At present, the most well-known ID coding system is called EPC (electronic product code), which is formulated by MIT and used by Wal-Mart. The RFID EPC Network is constructed from the ONS (object name service), Savant (a middleware specific to RFID), and PML (Physical Markup Language) (AutoID, 2006; Lin et al., 2004).

An RFID reader is used to communicate with RFID tags. In reading, the signal is sent out continually by the active tags. In interrogating, the reader sends a signal to the tags and listens. It can also send radio waves to energize the passive tags in order to receive their data. On the other hand, RFID software systems are the glue that integrates the RFID systems. The software systems manage the basic functions of the RFID reader and other components that route information to servers.

Organizations are using RFID in a number of data collection applications specific to their own industry, ranging from retail environments to hospitals, as well as what is currently being leveraged in warehouses to keep track of inventory and shipping. RFID has many advantages and can be deployed to assist organizations in improving global integration, as well as used as an effective tool in the areas of, for example, retail inventory tracking, customer relationship management, supply chain management, or any other situation where the tracking of the movement of goods or people is critical (Finkenzeller, 2003).

However, there are some business and technical problems and issues with the use of RFID technology (such as data sharing, data usefulness, accuracy, costs and benefits, security and privacy, and RFID standards) and this calls for further research. In essence, RFID requires collective and collaborative actions by stakeholders and organizations as a whole to ensure successful adoption and functioning of this technology. This is often affected by the divergent factors and perceptions of the internal and external stakeholders within an organization in the process of adopting RFIDs. For example, an organization must consider the potential costs in mastering collaborative planning and implementa-

Figure1. The research framework for evaluating RFID adoption

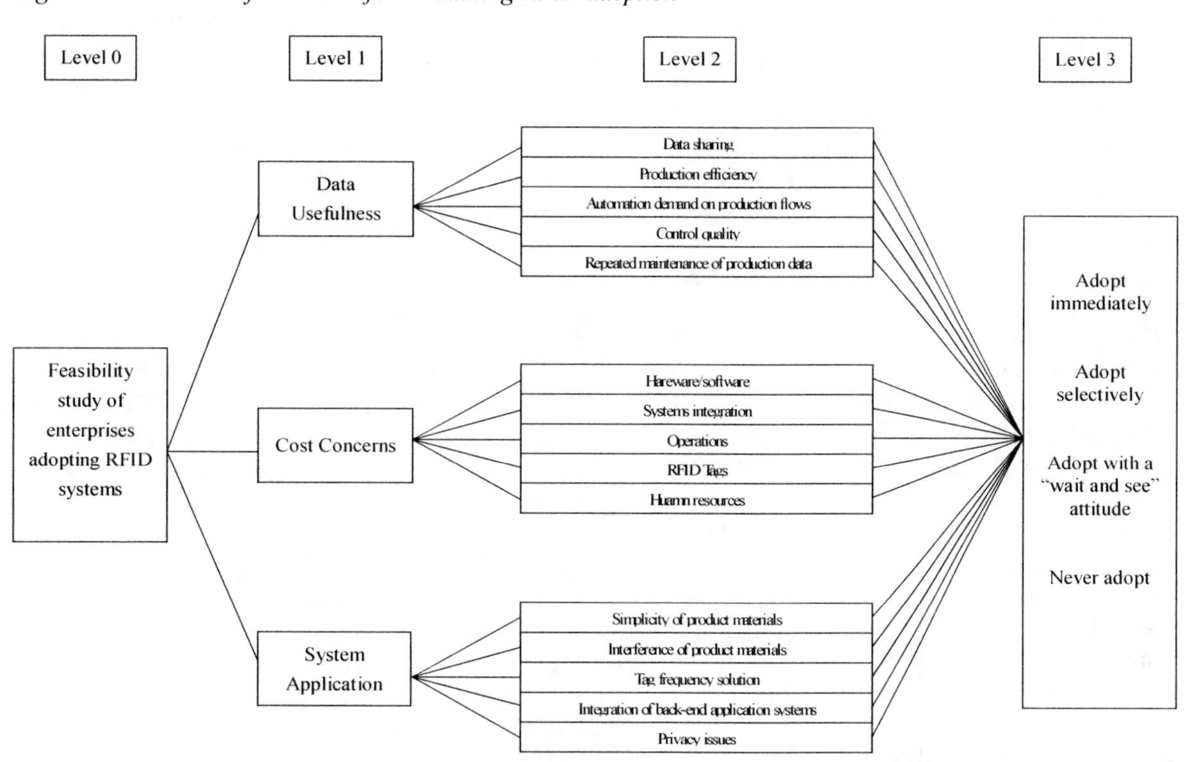

tion with its partners before attempting to share and use the RFID data (EPCglobal, 2006).

RESEARCH METHODOLOGY AND DESIGN

The AHP methodology is deployed to analyze the data collected and to build a decision support system. AHP was developed by Satty (1980) to reflect the way people actually think, and it continues to be the most highly regarded and widely used decision-making theory. In essence, AHP is a process that transforms a complicated problem into a hierarchical structure (Lin & Liu, 2005). By reducing complex decisions to a series of one-on-one comparisons and then synthesizing the results, AHP not only assists decision makers in arriving at the best decision, but also provides a clear rationale that it is the best (Lin et al., 2005).

The AHP methodology is useful for analyzing the RFID adoption alternatives as it can assist organizations in developing an integrated assessment of the entire organizational structure. AHP can also help to assess the inter-organizational issues among different divisions within an organization. Moreover, AHP can help to predict possible risks and challenges when adopting RFID so that the organizations are able to formulate appropriate strategies in order to minimize them (Satty, 1980; Wang & Yuan, 2001).

The following steps and considerations need to be taken into account when analyzing RFID adoption using AHP (Hair, Anderson, Tatham, & Black, 1998):

1. Issues may arise at the divisional level within an organization when adopting RFID.
2. The hierarchical structure for the organizations studied needs to be built using the collected data.
3. The questionnaire needs to be designed appropriately in order to identify and assess these issues.
4. Suggestions for improvement and/or alternatives need to be put forward in order to minimize the RFID adoption risks for the participating organizations.
5. Suggestions for improvement and/or alternatives also need to be communicated to all divisions within an organization, and alternatives need to be adjusted and revised accordingly.

Figure 1 depicts the AHP analysis hierarchical research framework for the evaluation of RFID adoption. Two types of questionnaires were designed for this research: (1) the expert questionnaire: to be completed by the RFID researchers and experts (expert evaluators); and (2) the industry questionnaires: to be completed by the RFID decision makers in those organizations that had a financial capital of at least US$1billion (industry evaluators).

- **Level 0:** The goals for this level were to conduct feasibility study of the industries involved with the adoption of RFID systems, as well as to identify and evaluate the importance of all major adoption factors (Lin et al., 2005).
- **Level 1:** After confirming the scope of the feasibility study for adoption of RFID in industries, three major factors were identified: data usefulness, cost concerns, and system application.
- **Level 2:** The three major factors identified in Level 1 were then decomposed into several criteria (in Level 2) which were evaluated according to their relative importance. These criteria were identified via interviews with the respondents, literature review, and surveys of industry characteristics.
- **Level 3:** Four alternatives were proposed for the adoption of RFID systems. The four proposed alternatives were: adopt immediately, adopt selectively, adopt with a "wait and see" attitude, and never adopt.

Data Analysis and Results

A total of 53 questionnaires were returned and the responses were analyzed using the AHP software. Significant differences were found between the responses from the RFID expert evaluators and industry evaluators. For example, "data sharing" was viewed by the expert evaluators as the most important issue for the "data usefulness" factor, whereas the "quality control" was the most important criterion for the industry evaluators.

Following this, the RFID experts were invited to provide their viewpoints on RFID techniques and theories. Using the AHP method, we found that both industry and expert evaluators had ranked the "cost concerns" of RFID as their number one factor for the adoption of RFID. The "application system" factor was ranked as the second most important concern by these evaluators, while they were least concerned about the "data usefulness" of RFID. In particular, costs of RFID tags and hardware were considered as the most important cost factor.

Weighting Analysis

The software, ExpertChoice, was used to compute the weights from the responses. The consistency test was used to calculate the inconsistency ratio (IR) of the adoption criteria. All criteria have received consistent responses as their IR value is less than 0.1 (Satty, 1980). The AHP analysis indicated that both expert and industry evaluators ranked the adoption factors as follows: cost concerns (0.486 & 0.633), system applications (0.377 & 0.249), and data usefulness (0.137 & 0.118). The results are shown in Tables 1 and 2. Some research findings from Tables 1 and 2 are presented as follows.

Table 1. Adoption factor weightings of expert evaluators

Adoption Factors	Criteria	Weights
Data Usefulness (0.137)	Data sharing	0.034
	Production efficiency	0.023
	Automation demand on production flows	0.027
	Control quality	0.027
	Repeated maintenance of production data	0.025
Cost Concerns (0.486)	Hardware/software	0.126
	System integration	0.074
	Operations	0.085
	RFID tags	0.168
	Human resources	0.035
System Applications (0.377)	Simplicity of product materials	0.026
	Interference of product materials	0.064
	Tag frequency solution	0.037
	Integration of back-end application systems	0.082
	Privacy issues	0.167

Table 2. Adoption factor weightings of industry evaluators

Adoption Factors	Criteria	Weights
Data Usefulness (0.118)	Data sharing	0.017
	Production efficiency	0.029
	Automation demand on production flows	0.019
	Control quality	0.033
	Repeated maintenance of production data	0.020
Cost Concerns (0.633)	Hardware/software	0.155
	System integration	0.151
	Operations	0.116
	RFID tags	0.127
	Human resources	0.085
System Applications (0.249)	Simplicity of product materials	0.035
	Interference of product materials	0.032
	Tag frequency solution	0.030
	Integration of back-end application systems	0.058
	Privacy issues	0.095

Table 3. A comparison of weights by expert evaluators and industry evaluators

Criteria	Expert Weights	Industry Weights	Differences
RFID tag costs	0.168	0.127	0.041
Privacy issues	0.167	0.095	0.072
Hardware/software costs	0.126	0.155	0.029
Development/operation costs	0.085	0.116	0.031
Integration of back-end application systems	0.082	0.058	0.024
System integration costs	0.074	0.151	0.077
Interference of product materials	0.064	0.032	0.032
Tag frequency selection	0.037	0.03	0.007
Human resource costs	0.035	0.085	0.05
Data sharing	0.034	0.017	0.017
Automation of production flows	0.027	0.019	0.008
Quality control	0.027	0.033	0.006
Simplicity of product materials	0.026	0.035	0.009
Repeated maintenance of product data	0.025	0.02	0.005
Production efficiency	0.023	0.029	0.006

Data Usefulness

The responses from expert evaluators indicated that "data sharing" is far more important than the other adoption factors. "Data sharing" in RFID is generally defined as using a standardized data format to communicate between RFID supply chain suppliers. These respondents considered data format standardization as the most important issue in the process of adopting RFID. On the other hand, the responses from industry evaluators stated that "quality control" factor is their number one concern. This is not surprising given that businesses place great emphasis on business quality and service (RFID Journal, 2004).

Cost Concerns

As mentioned earlier, costs of RFID tags and hardware were considered as the most important cost factors from the expert evaluators' point of view, whereas the industry evaluators were more concerned about hardware and system integration costs. The high cost of RFID tags was viewed by the expert evaluators as the main obstacle for the adoption of RFID. On the other hand, industry evaluators were extremely concerned about the integration between the existing hardware and the new RFID systems. Most organizations would prefer to retain their current systems in order to avoid the system compatibility problem and increase the overall system performance (Traub, 2005).

Application System

Privacy was viewed as the most significant criterion for the "application system" factor by both the expert and industry evaluators. This was followed by the integration of a back-end system. This had reflected the fact that the privacy concern would likely affect the degree of trust during the future implementation of RFID. In addition, the problem with the integration of various back-end systems will also present security problems and other challenges for organizations to handle (Floerkemeier, 2003).

Table 3 depicts a comparison of weights by expert evaluators and industry evaluators.

In Table 4, criteria for RFID adoption were ranked according to their weights by both the expert and industry evaluators.

FUTURE TRENDS

Despite the fact that RFID has been widely adopted in many different fields at an increasing rate in recent years, the issues of security and privacy remain the key challenges

Table 4. Ranking of adoption criteria by expert evaluators and industry evaluators

Ranking	Expert		Industry	
	Criteria	Weights	Criteria	Weights
1	RFID tag costs	0.168	Hardware/software costs	0.155
2	Privacy issues	0.167	System integration costs	0.151
3	Hardware/software costs	0.126	RFID tag costs	0.127
4	Development/operation costs	0.085	Development/operation costs	0.116
5	Integration of back-end applications systems	0.082	Privacy issues	0.095
6	System integration costs	0.074	Human resource costs	0.085
7	Interference of product materials	0.064	Integration of back-end application systems	0.058
8	Tag frequency selection	0.037	Simplicity of product materials	0.035
9	Human resource costs	0.035	Quality control	0.033
10	Data sharing	0.034	Interference of product materials	0.032
11	Automation of production flows	0.027	Tag frequency selection	0.030
12	Quality control	0.027	Production efficiency	0.029
13	Simplicity of product materials	0.026	Repeated maintenance of product data	0.020
14	Repeated maintenance of product data	0.025	Automation of production flows	0.019
15	Production efficiency	0.023	Data sharing	0.017

in promoting the technology. Many industry experts have pointed out that the RFID-included objects should be targeted more efficiently by real-time tracking and instant management. However, the transmission of data is very vulnerable to eavesdropping because of the contact-less type of RFID remote retrieval. A primary security concern surrounding the RFID technology is the unsolicited tracking of consumer location and analyzing of their shopping habits or behavior. This is one important issue that needs to be addressed by RFID vendors.

In addition, the RFID technology can be further promoted by: (1) reducing the total costs of RFID; (2) resolving the interference problems; (3) improving the identification accuracy; (4) protecting the intellectual property rights; (5) establishing international standards for encoding systems, reader protocols, and programming environments; and (6) developing better software supports, including middleware, EPC information systems, and ONS design. Finally, it is expected that the RFID system will have a great impact on the way we work and live.

DISCUSSION AND CONCLUSION

The last section in the questionnaire asked the respondents to give scores to all four alternatives. The results have shown that both expert and industry evaluators preferred the second RFID adoption alternative (that is, to adopt RFID selectively).

The evaluators were also asked to weigh the relative importance of the three main factors as well as the criteria in each factor by a pair-wise comparison. The ranking of the adoption alternatives were obtained by multiplying the weighting given by expert evaluators and the weighting given by industry evaluators, by the scores obtained from the identified criteria in a matrix format. Some of the key findings from the AHP analysis are presented below:

1. When asked about the "data usefulness" factor, both expert and industry evaluators preferred the "adopt selectively" alternative. These evaluators indicated that the usefulness of data sharing and quick turn-around time in RFID could significantly improve the production efficiency and lower the costs of inventory if it was adopted in SCM (supply chain management) environments.
2. There were significant differences of opinion from both the expert and industry evaluators on the "cost concerns" factor. Expert evaluators preferred the "adopt selectively" alternative despite the high cost of implementing RFID. This was due to the fact that expert evaluators believed that the benefits of RFID would exceed the costs. On the other hand, industry experts were more conservative and preferred the

"adopt with a 'wait and see' attitude" due to the high costs of implementing RFID.

3. There were also significant differences of opinion from both the expert and industry evaluators on the "system application" factor. Expert evaluators were more concerned about the privacy issues than the industry evaluators.
4. Overall, the ranking for these four alternatives were: (1) adopt selectively, (2) adopt with a 'wait and see' attitude; (3) never adopt, and (4) adopt immediately. There was no difference in terms of the preference for these four alternatives by both the expert and industry evaluators.

In conclusion, organizations with the ability to quickly identify benefits, costs, and risks associated with RFID technologies and to effectively deploy them are more likely to gain competitive advantages from RFID. Regular evaluation of RFID technologies allows organizations to benefit from their implementation. Those organizations that regularly maintain strategic evaluation of RFID technologies can help to ensure that they will achieve RFID's true benefits.

REFERENCES

ABI. (2006). *RFID realism: Improving our ability to execute RFID projects with holistic solutions.* Retrieved from http://ww.abiresearch.com

Albertsons. (2004). The nation's second largest food and drug retailer, has launched its first RFID pilot project and announced that it will require its top 100 suppliers to tag pallets and cartons by April 2005. *RFID Journal.*

AutoID Labs. (2006). Retrieved from http://www.autoidlabs.org/

CNET. (2005). *Gartner sees RFID as $3 billion business by 2010.* Retrieved from http://www.news.com

EPCglobal. (2006). Retrieved from http://www.epcglobal.org.tw

Finkenzeller K. (2003). *RFID handbook: Fundamentals and applications in contactless smart cards and identification* (Trans. Rachel Waddington). Chichester, UK/New York: John Wiley & Sons.

Floerkemeier, C. (2003). *PML specification 1.0.* Retrieved from http//www.epcglobalinc.org/standards_technology/Secure/v1.0/PML_Core_Specification_v1.0.pdf

Hair, J. Jr., Anderson, R., Tatham, R., & Black, W. (1998). *Multivariate data analysis* (5th ed., p. 447). Upper Saddle River, NJ: Prentice Hall.

Lin, K. H., & Liu, L. (2005, December 23). On the development of digital TV commerce using the AHP mechanism. *Proceedings of the Technology & Business Forum 2005,* Hualien, Taiwan.

Lin, K. H., Chen, C. T., Ke, J. C., Leu, H., Yen, Y. C., & Yang, S. H. (2005, March 25-26). Using AHP technology to design a RFID adopting decision analysis system. *Proceedings of the 2005 Conference of Electronic Commerce and Digital Life (ECDL2005).*

Lin, K. H., Chen, P., Juang, W., Dai, J., Jeng, W., Kuo, T., et al. (2004, June 10). Exploring the EPC network architecture for RFID technology application. *Proceedings of the 2004 Information Management Application and Development.*

RFID Journal. (2004). Target expects top vendor partners to apply tags to all pallets and cases and start shipping to select regional distribution facilities beginning late spring 2005. *RFID Journal,* (February 20).

Satty, T. L. (1980). *The Analystic Hierarchy Process.* New York: McGraw-Hill.

Traub, K. (2005). *ALE: A new standard for data access.* Retrieved from http://www.rfidjournal.com/article/articleview/1493/1/82/, 4/18

Wang, M., & Yuan, B. J. C. (2001). Application of Analytic Hierarchy Process (AHP) on the evaluation of alternative technologies—A case of digital TV receivers. *Tamsui Oxford Journal of Management Sciences, 17,* 29-42.

KEY TERMS

Analytic Hierarchy Process (AHP): Methodology developed by Satty (1980) to reflect the way people actually think; it continues to be the most highly regarded and widely used decision-making theory.

Electronic Product Code (EPC): It contains digits to identify the manufacturer, the product category, and the individual item.

Object Name Service (ONS): It looks up unique electronic product codes and points computers to information about the item associated with the code.

Physical Markup Language (PML): A method of describing products in a way computers can understand. PML, based on the widely accepted eXtensible Markup Language, is used to share information via the Internet in a format all computers can understand and use.

Radio Frequency Identification (RFID): A built-in wireless technology that incorporates a smart IC (integrated circuit) tag. The three major components of RFID are: tags, readers, and software systems.

RFID Reader: Used to communicate with RFID tags. In reading, the signal is sent out continually by the active tags. In interrogating, the reader sends a signal to the tags and listens.

RFID Software: RFID software systems are the glue that integrates the RFID systems. The software systems manage the basic functions of the RFID reader and other components that route information to servers.

RFID Tag: RFID tags consist of silicon chips and antennas. Each tag uses an ID coding system and contains a unique serial number of a product. This enables the tag to store some information of the product.

Definitions, Key Characteristics, and Generations of Mobile Games

Eui Jun Jeong
Michigan State University, USA

Dan J. Kim
University of Houston Clear Lake, USA

INTRODUCTION

In the emerging wireless environment of digital media communications represented as *ubiquitous* and *convergence,* rapid distribution of handheld mobile devices has brought the explosive growth of the mobile content market. Along with the development of the mobile content industry, mobile games supported by mobile features such as portability (mobility), accessibility (generality), and convenience (simplicity) have shown the highest growth rate in the world game market these days.

In-Stat/MDR (2004) and Ovum (2004) expect that the mobile games' annual growth rate between 2005 and 2009 will be around 50% in the United States and 30% in the world. According to KGDI (2005) and CESA (2005), compared to the rate of the whole game market (5%) of the world, it is about six times higher, and it exceeds the rate of video console (10%) and online games (25%). Mobile games thus are predicted to be one of the leading platforms in the world game market in 10 years' time. In addition, as the competition among game companies has been enhanced with the convergence of game platforms, mobile games are being regarded as a breakthrough for the presently stagnant game market, which has focused on heavy users.

However, due to the relative novelty of mobile games, there are a few visible barriers in the mobile game industry. First, definitions and terminologies and key characteristics related to mobile games are not clearly arranged as yet. Second, there is little research on the classification and development trends of mobile games. Therefore, this article is designed to contribute insights into these barriers in three ways. Firstly, the article provides narrow and broad definitions of mobile games. Secondly, key characteristics, platforms, and service types of mobile games are discussed. Finally, following the broad definition of mobile games, this article classifies mobile games as one to fourth generations and one pre-generation. Characteristics and examples of each generation are also presented.

DEFINITIONS OF MOBILE GAMES

Each country and each game research institution has different definitions and terminologies. The definition of mobile games is important because the functions of mobile devices are being converged with those of other devices. Mobile games—more precisely, mobile network games—are narrowly defined as *games conducted in handheld devices with network functionality.* The two key elements of this definition are *portability* and *networkability.* In this definition, mobile games are generally referred to as the games played in handheld mobile devices such as cell phones and PDAs with wireless communication functionality. In terms of portability and networkability, the characteristics of mobile games are different from other device platforms such as PC and console games, which do not have both portability and wireless capability. For example, Game Boy (GB) with no communication functionality was only regarded as a portable console device. However, this concept has lost some of its ground in the market since the advent of new mobile game devices from portable consoles such as Play Station Portable (PSP) and Nintendo Dual Screens (NDS), as wireless networked games began to be serviced through the new mobile game devices.

Mobile games can be broadly defined as *embedded, downloaded, or networked games conducted in handheld devices such as mobile phones, portable consoles, and PDAs.* The key element of this concept is portability: all games in portable devices can be thought of as mobile games without regard to wireless functions. Therefore, this concept expands mobile games by including video games in portable consoles and embedded games in general portable devices such as PDAs, calculators, and dictionaries. As most game devices have been adopted with wireless networking functions, this definition becomes more powerful in game markets.

Recently, the narrow definition of mobile games has been generally used. However, since the meaning of *mobile* includes that of *portable and network (either wired or wireless function is embedded),* the broad definition of mobile games including portable game-dedicated devices such as GBs and PSPs should be used. This definition is more persuasive in the present and future game market. For instance, the competition between Nokia's N-gage (i.e., a cell phone integrating the functions of MP3 and games) and Sony's PSP (i.e., a portable game machine including functions of MP3 and networking) is for the preoccupation of a future mobile platform.

KEY CHARACTERISTICS, PLATFORMS, AND SERVICE TYPES

Characteristics and Limitations of Mobile Games

Mobile games are differentiated from other platform games such as console, PC, and arcade games in terms of their portability, accessibility, networkability, and simplicity. Owing to the *portability* (i.e., mobility), users can play games anytime. This characteristic has attracted many light users, who play simple games such as puzzle, card, or word games, because these games can be played in one's spare time in a short amount of time. Compared to players in other genres such as role playing games (RPGs) and simulation games that require a long time to play, light users vary broadly in terms of age, and many women players also belong to this group. This is one of the strongest potentials of mobile games. The second characteristic of mobile games is *accessibility.* This can be defined as to the extent one can use a mobile device to play games at anytime and at anyplace. Console games are restricted to owners who have console machines and who want to enjoy games for a long time in a particular place. Likewise, most PC games and arcade games need to be somewhere in front of game devices with network facilities. However, mobile games—especially using mobile devices—are easy to access, because people almost always bring those devices anywhere and can download games anywhere as long as wireless networks are available. The third characteristic is *networkability.* Through wired or wireless connections, online games and console games are transplanted into mobile games to facilitate game usages. For example, some online games are linked to mobile games, so those games can be used both in PCs and mobile devices: game users can play the games with no limits in terms of location, machine, and time. Furthermore, mobile game users can play multi-user real-time games such as MMORPG (massively multiplayer online role-playing game) and real-time strategy (RTS) games. The final characteristic is their *simplicity* to use: mobile devices are simpler to handle than other platform machines. In addition, it is much easier to acquire the skills of the games and use them than those of other platforms.

Because of these characteristics, mobile games develop faster than other platform games. According to W2F (2003) and KGDI (2005), the development of a PC or console game usually takes at least two years to develop with more than 20 trained people and about $3 million. But in mobile games, about three to six months are spent with five people and less than $150,000. This is why the initial market entry barrier of mobile games is lower than that of other platform game markets. However, the average lifecycle of mobile games is less than six months, and value chains are more complex than those of other platform games.

Despite the major advantages of mobile games, there are drawbacks in some points. The most essential point is from not-unified platforms. With Internet and console games, converting of original games is not necessary, because the original games can be available in any PC via the Internet. However, mobile games should be converted to make them fit to other platforms, even in the same area. In other words, the conversion is necessary for service to be available in other mobile devices. The second is small screens and low capable devices. Although 3D networked games are being serviced, small screens and monotonous sounds are not sufficient to maximize the feelings of presence for users, and mobile game devices still do not have enough capacity to download high-capacity games through mobile networks.

Mobile Game Platforms

Mobile platforms function as game engines by running applications: a game engine is the core code handling the basic functionality of a game. Each mobile device has its own platform, so developers make games based on the formats of those platforms. With the development of platforms, downloaded, 3D games, and more advanced games are now serviced. These platforms are either freely opened or purchased with license fees. Platform holders have tried to expand their platforms, because the prevalence of their platforms implies a strong influence in mobile markets. These days, Java is the most influential platform both in mobile phone games and in handset manufacture. The Java 2 Micro Edition (J2ME) is a freeware version of Java; Execution Engine (ExEn) and Mophun are also freeware platforms distributed mainly in Europe. Brew is the licensed platform mainly used in the United States, Japan, and Korea. Different from mobile phone games, portable console games such as GB, N-Gage, PSP, and NDS have their own development tools for the platforms. Developers who want to make mobile games in portable consoles should use such development kits with the charge of license fees. Since developers adopt more prevailing game kits for the better benefit of their games, the market prevalence of console platforms is parallel with the amount of license fees for portable console manufacturers.

Mobile Game Service Types

With the development of mobile service technologies, mobile game services have evolved from single/embedded to multiplayer networked games. Single/embedded are games with which just one player can play without network services. These embedded games are still used in many mobile devices as a service for device customers. Message-based are games using short message service (SMS). These types are played in wireless network environments through WAP (Wireless Application Protocol) browsing environments, but these games are shifting into multi-media message service (MMS) with high capabilities providing enhanced messaging services such as graphics, sound, animation, cartoons, and texts. Downloaded games have been developed with the advent of download platforms such as Java, ExEn, and Mophun. These games have been taken usually from mobile portals managed by mobile network operators, with charges based on both content and traffic fees. Networked games are the newest type of mobile games which are activated with the advent of the flat sum systems.

GENERATIONS OF MOBILE GAMES

From the broad definition of mobile games perspective, portable console machines were the first mobile devices that emerged in the 1970s. These games have been categorized as console games because most hit games were published by console game companies such as Nintendo, Sega, and Sony, and game users were the same as those of console games. However, they have expanded their ranges into color graphic games in the 1990s and mobile network games in the 2000s, so users of such games are no longer limited to young boys not yet in their teens. Following the broad definition of mobile games, this article includes portable console games as a part of mobile games. Portable console games, made before the advent of network portable console games around 2003, are regarded as "portable embedded games," which are categorized as the pre-generation of mobile games.

The first wireless mobile phone game, *Snake,* was serviced as a text-based (or early-graphic) game in 1997. However, today's state-of-the-art games are 3D, fully networked multi-user games with high definitions in wide color screens. As the development of mobile interfaces and network functionalities continues, mobile games can be divided into four generations and one pre-generation. These generations are categorized by stand-alone (off-line) or networked, text-based or graphic-based, and 2D or 3D graphics.

Pre-generation (Pre-G) refers to portable console games that are played in standalone portable devices. In the 1970s, these games were all embedded in only-one-game-use portable game machines such as *Auto Race, Merlin,* and *Missile Attack* by American vendors such as Mattel, Entex, and Tomy. However, in the 1980s, both embedded and cartridge usable games were pervaded with the initiatives of Japanese game companies such as Nintendo and Bandai. From 1989, portable console games were converged into the Nintendo Game Boy era with cartridge games. In the mid-1990s, these games were serviced with color graphic games. With various games usually transplanted from console games, these portable console games flourished with the development of console games.

The *first generation (1G)* refers to text-based mobile phone games like puzzle games. They had been usually serviced by wireless application protocol (WAP) from 1998, and most of them are single-player embedded games. Some early-graphic games were embodied by white and black dots. These games spread until around 2001 when mobile platforms such as Java, Brew, and ExEn began to spread for the development of mobile graphic games. The *second generation (2G)* refers to graphic games. Developers transported popular games in PC or console games into mobile devices. At first, all graphic services were 2D white and black, but from around 2002, color phones rapidly spread with color graphic games, and some functions such as chatting and reviewing were added. With the prevalence of download platforms, downloaded games generally began to be provided by mobile portals. Traditional board games such as card and chess games were also translated into mobile graphic games in this generation with the concept of licensed games.

The *third generation (3G)* refers to networked games with simple network functionalities. Around 2003, most games were 2D graphic: 3D games were just a state of experiment. Network functionality was not fully serviced, because of the high cost of network use and low capabilities of mobile devices. However, owing to network capability, new games such as various simulations, multi-user role-playing games (MRPG), and location-based service (LBS) games were firstly developed in this period. Additionally, new mobile devices such as N-gage, PSP, and NDS had changed the traditional concepts of mobile games with the mixture of wireless and networked game services. These devices are estimated to have promoted the degree of mobile games as much as that of console games. With the prevalence of device convergences between mobile and console devices, from this generation there is no accurate difference among game genres. The *fourth generation (4G)* games refer to 3D and full networked games such as massively multi-user online role playing games (MMORPG). In addition, around 2004, full 3D mobile game services began to be serviced, and many developers joined the development of 3D network games. With the spread of new 3D graphic mobile phones and flat sum systems, 3D networked games are steadily gaining their shares in game markets. Table 1 illustrates the mobile game generations, key characteristics of each generation, and examples.

Table 1. Generations of mobile games

	Outset	Characteristic Game Genres	Examples
Pre-G	1970s	Portable console games Portable console color games (Embedded or cartridge games)	*Auto Race, Football* *Bomberman* *Pac Man, Tetris (Console)*
1G	1997/1998	Text-based games (early-graphic) WAP games	*Snake* *Dataclash, Gladiator*
2G	2001/2002	Downloaded games (Java, Brew) 2D color graphic games	*Tetris, Chess Mobile* *Samurai Romanesque*
3G	2003/2004	Portable console network games Half network games 3D graphic games	*Pokemon Ruby* *Badlands, Samgukji* *3D Pool*
4G	2004/2005	Full 3D graphic games Full 3D network games	*3D Golf, 3D Bass Fishing* *Homerun King Mobile*

CONCLUSION

With the expansion of a convergence and ubiquitous environment, the range of mobile games has grown to include all games available in handheld devices with portability. At first, mobile games were regarded as embedded single-user games. However, through the development of network and graphic technology, mobile games have been played as both full network games with multi-users and 3D graphic games with high-definition devices. So, many high-capability games such as MMORPG and multi-user simulation games have been adopted in mobile devices with high-speed network capability. In addition, with the convergence of game devices, the boundary between mobile phones and console devices has been eroded, while games in PC and console machines have been transformed into mobile games. Due to the accessibility, portability, and ease of use, mobile games have a wide range of users and do not impose limitations in age, sex, and social status. Traditionally game users were usually young males, but when it comes to mobile games, the game users are more diversified: not only young boys, but also elderly people and women are joining in on the new mobile gaming era. With the development of mobile technology, diversification of mobile content services, and generalization of mobile game users, mobile games will continue to gain more power within game markets.

Mobile games are summarized along with taxonomies. In addition, recent trends with game application areas will be discussed in the next article, "Mobile Games Part II: Taxonomies, Applications, and Trends."

REFERENCES

CESA. (2005). *2005 CESA game white paper.* Tokyo: Computer Entertainment Suppliers' Association.

DeMaria, R., & Wilson, J. L. (2002). *High score: The illustrated history of electronic games.* Berkeley: Osborne Media Group.

Entertainment Software Association. (2002). *Top ten industry facts.* Retrieved from www.theesa.com/pressro om.html

Ermi, L., & Mäyrä, F. (2005). Challenges for pervasive mobile game design: Examining players' emotional responses. *Proceedings of the ACM SIGCHI International Conference on Advances in Computer Entertainment Technology* (pp. 47-55), Valencia, Spain.

Hall, J. (2005). Future of games: Mobile gaming. In J. Raessens & J. Goldstein (Eds.) *Handbook of computer game studies* (pp. 47-55). Cambridge, MA: MIT Press.

In-Stat/MDR. (2004). Mobile gaming services in the U.S., 2004-2009. *In-Stat/MDR,* (August).

KGDI. (2005). *2005 game white paper.* Seoul: Korea Game Development & Promotion Institute.

Newman, J. (2004). *Videogames.* London: Routledge.

Nokia. (2003). *Introduction to mobile game development, Nokia Corporation.* Retrieved from www.forum.nokia.com/html_reader/main/1,,2768,00.html

Ovum. (2004, December). *Ovum forecasts global wireless market.* Ovum.

Ring, L. (2004). *The mobile connection: The cell phone's impact on society.* San Francisco: Morgan Kaufmann.

Taylor Nelson Sofres. (2002). *Wireless and Internet technology adoption by consumers around the world.* Retrieved from www.tnsofres.com/IndustryExpertise/IT/WirelessandInternetAdoption byConsumersAroudtheWorldA4.pdf

W2F. (2003, October). *Winning and losing in mobile games.* W2F Limited.

KEY TERMS

Device Platform: A device such as a cell phone, PDA, PC, or console machine through which games can be played.

Local-Based Service (LBS) Game: A mobile network game played within a local place around a telecommunication base with the information of a user's position.

Massively Multi-player Online Game (MMOG): A game where a huge number of users can play simultaneously based on their roles or missions.

Mobile Game Platform: Core code handling of the basic functionality of a game such as downloading, networking, or activating 3D graphics.

Mobile Game: An embedded, downloaded, or networked game conducted in a handheld device such as a mobile phone, portable console, or PDA.

Portable Console (Device): Handheld console machine such as PSP, NDS, and GBA with portable capabilities.

Role Playing Game (RPG): A game where a gamer takes a role and uses items in accomplishing missions or quests.

3D Network Game: A game played in connection with other users, using 3D graphics.

Design Methodology for Mobile Information Systems

Zakaria Maamar
Zayed University, UAE

Qusay H. Mahmoud
University of Guelph, Canada

INTRODUCTION

Mobile information systems (MISs) are having a major impact on businesses and individuals. No longer confined to the office or home, people can use devices that they carry with them, along with wireless communication networks, to access the systems and data that they need. In many cases MISs do not just replace traditional wired information systems or even provide similar functionality. Instead, they are planned, designed, and implemented with the unique characteristics of wireless communication and mobile client use in mind. These unique characteristics feature the need for specific design and development methodologies for MISs. Design methods allow considering systems independently of the existing information technologies, and thus enable the development of lasting solutions. Among the characteristics that a MIS design method needs to consider, we cite: unrestricted mobility of persons, scarcity of mobile devices' power-source, and frequent disconnections of these devices.

The field of MISs is the result of the convergence of high-speed wireless networks and personal mobile devices. The aim of MISs is to provide the ability to compute, communicate, and collaborate anywhere, anytime. Wireless technologies for communication are the link between mobile clients and other system components. Mobile client devices include various types, for example, mobile phones, personal digital assistants, and laptops. Samples of MIS applications are mobile commerce (Andreou et al., 2002), inventory systems in which stock clerks use special-purpose mobile devices to check inventory, police systems that allow officers to access criminal databases from laptops in their patrol cars, and tracking information systems with which truck drivers can check information on their loads, destinations, and revenues using mobile phones. MISs can be used in different domains and target different categories of people.

In this article, we report on the rationale of having a method for designing and developing mobile information systems. This method includes a conceptual model, a set of requirements, and different steps for developing the system. The development of a method for MISs is an appropriate response to the need of professionals in the field of MISs. Indeed, this need is motivated by the increased demand that is emerging from multiple bodies: wireless service providers, wireless equipment manufacturers, companies developing applications over wireless systems, and businesses for which MISs are offered. Besides all these bodies, high-speed wireless data services are emerging (e.g., GPRS, UMTS), requiring some sort of new expertise. A design and development method for MISs should support professionals in their work.

MOBILE COMPUTING MODEL

The general mobile computing model in a wireless environment consists of two distinct sets of entities (Figure 1): mobile clients (MCs) and fixed hosts. Some of the fixed hosts, called mobile support stations (MSSs), are enhanced with wireless interfaces. An MSS can communicate with the MCs within its radio coverage area called wireless cell. An MC can communicate with a fixed host/server via an MSS over a wireless channel. The wireless channel is logically separated into two sub-channels: an uplink channel and a downlink channel. The uplink channel is used by MCs to submit queries to the server via an MSS, whereas the downlink channel is used by MSSs to disseminate information or to forward the responses from the server to a target client. Each cell has an identifier (CID) for identification purposes. A CID is periodically broadcasted to all the MCs residing in a corresponding cell.

The wireless application protocol (WAP) is a technology that plays a major role in the field deployment of the mobile computing model (Open Mobile Alliance). WAP is an open, global specification that empowers users with mobile devices to easily access and interact with information and services instantly. It describes how to send requests and responses over a wireless connection, using the wireless session protocol (WSP), which is an extended and byte-coded version of HTPP 1.1. A WSP request is sent from a mobile device to a WAP gateway/proxy to establish an HTTP session with the target Web server. Over this session, the WSP request, converted into HTTP, is sent. The content, typically presented

Figure 1. Representation of the mobile computing model

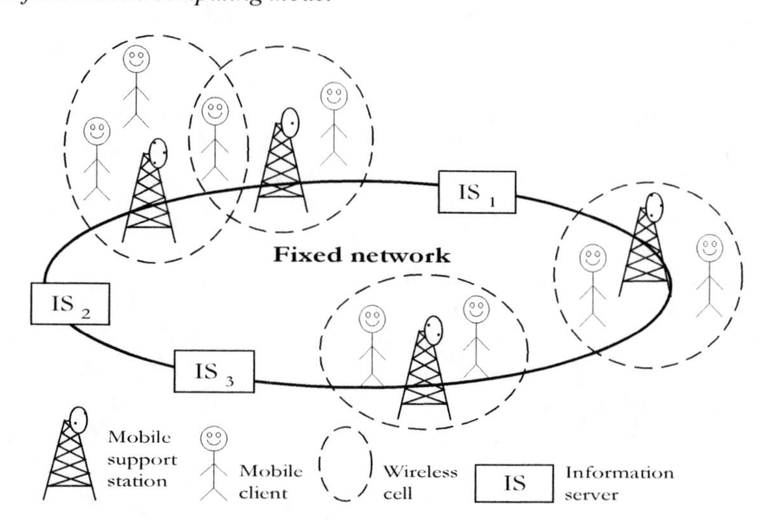

in the Wireless Markup Language (WML), is sent back to the WAP gateway, where it is byte-coded and sent to the device over the WSP session.

REQUIREMENTS FOR MISs

The role of an MIS is to provide information to mobile users through wireless communication networks. Two aspects are highlighted here: information and network. Information has to be available, taking into account terrain and propagation techniques. Plus, the information exchange has to be secured. A security problem inherent to all wireless communication networks consists of third parties being able to easily capture the radio signals while in the air. Thus, appropriate data protection and privacy safeguards must be ensured. Regarding the network element, this latter needs to consider failure cases and recover from them.

1. **Information Availability Requirement:** This illustrates the need for a user to have uninterrupted and secure access to information on the network. Aspects to consider are: survivability and fault tolerance, ability to recover from security breaches and failures, network design for fault tolerance, and design of protocols for automatic reconfiguration of information flow after failure or security breach.
2. **Network Survivability Requirement:** This illustrates the need to maintain the communication network "alive" despite of potential failures. Aspects to consider are: understand system functionality in the case of failures, minimize the impact of failures on users, and provide means to overcome failures.
3. **Information Security Requirement:** This illustrates the importance of providing reliable and unaltered information. Aspects to consider are: confidentiality to protect information from unauthorized disclosure, and integrity to protect information from unauthorized modification and ensure that information is accurate, complete, and can be relied upon.
4. **Network Security Requirement:** This illustrates the information security using network security. Aspects to consider are: confidentiality, sender authentication, access control, and identification.
5. **Additional Requirements of MIS Have Been Put Forward:** Indeed, the increasing reliance and growth in information-based wireless services impose three requirements—availability, scalability, and cost efficiency—on the services to be provided. Availability means that users can count on accessing any wireless service from anywhere, anytime, regardless of the site, network load, or device type. Availability also means that the site provides services meeting some measures of quality such as short, acceptable, and predictable response time. Scalability means that service providers should be able to serve a fast-growing number of customers with minimal performance degradation. Finally, cost effectiveness means that the quality of wireless services (e.g., availability, response time) should come with adequate expenditures in IT infrastructure and personnel.

CHALLENGES AND POSSIBLE SOLUTIONS IN MISs

The requirements discussed above pose several crucial challenges, which must be faced in order for MIS applications to function correctly in the target environment.

- **Transmission Errors:** Messages sent over wireless links are exposed to interference (and varying delays) that can alter the content received by the user, the target device, or the serer. Applications must be prepared to handle these problems. Transmission errors may occur at any point in a wireless transaction and at any point during the sending or receiving of a message. They can occur after a request has been initiated, in the middle of the transaction, or after a reply has been sent.
- **Message Latency:** Message latency, or the time it takes to deliver a message, is primarily affected by the nature of each system that handles the message, and by the processing time needed and delays that may occur at each node from origin to destination. Message latency should be handled, and users of wireless applications should be kept informed of processing delays. It is especially important to remember that a message may be delivered to a user long after the time it is sent. A long delay might be due to coverage problems or transmission errors, or the user's device might be switched off or have a dead battery.
- **Security:** Any information transmitted over wireless links is subject to interception. Some of that information could be sensitive, like credit card numbers and other personal information. The solution needed really depends on the level of sensitivity.

Here are some practical hints useful to consider when developing mobile applications. These hints back the development of the proposed method for designing mobile information systems.

- **Understand the Environment and Do Some Research Up Front:** As with developing any other software application, we must understand the needs of the potential users and the requirements imposed by all networks and systems the service will rely on.
- **Choose an Appropriate Architecture:** The architecture of the mobile application is very important. No optimization techniques will make up for an ill-considered architecture. The two most important design goals should be to minimize the amount of data transmitted over the wireless link, and to anticipate errors and handle them intelligently.
- **Partition the Application:** Think carefully when deciding which operations should be performed on the server and which on the handheld device. Downloadable wireless applications allow locating much of an application's functionality of the device; it can retrieve data from the server efficiently, then perform calculations and display information locally. This approach can dramatically reduce costly interaction over the wireless link, but it is feasible only if the device can handle the processing the application needs to perform.
- **Use Compact Data Representation:** Data can be represented in many forms, some more compact than others. Consider the available representations and select the one that requires fewer bits to be transmitted. For example, numbers will usually be much more compact if transmitted in binary rather than string forms.
- **Manage Message Latency:** In some applications, it may be possible to do other work while a message is being processed. If the delay is appreciable—and especially if the information is likely to go stale—it is important to keep the user informed of progress. Design the user interface of your applications to handle message latency appropriately.
- **Simplify the Interface:** Keep the application's interface simple enough that the user seldom needs to refer to a user manual to perform a task. To do so, reduce the amount of information displayed on the device, and make input sequences concise so the user can accomplish tasks with the minimum number of button clicks.

PROPOSED METHOD

The first step towards a successful wireless implementation project is a thorough business analysis, which serves as the backbone of any project bearing a fruitful return of investment. The analysis ensures that the project's requirements result in a wireless implementation that will successfully meet users' expectations and needs. Next, determinations about development features, approach, and constraints are made. This ensures that the wireless implementation is a good fit with the planned usage and infrastructure of the company. Finally, a choice needs to be made with regard to the software and hardware systems.

Business Analysis

When considering a wireless implementation, several questions have to be considered:

- What are the overall goals for implementing wireless services?
- What are the new markets to be targeted?
- What are the goals for giving mobile customer/staff wireless remote access?
- What technologies are currently in place towards supporting a wireless enterprise?
- Is interactivity important to the company?
- What current functions are suitable for wireless use?

- How prepared is the infrastructure to develop and host wireless applications?
- Are resources available to develop, implement, and support the wireless project?
- Is it more economical to have a wireless solution compared to a wired one?

Development

There are several factors to take into account when determining the wireless development solution. Indeed, MISs are expanding rapidly and changing from largely voice-oriented to increasingly data and multimedia systems.

- **Online vs. Off-Line:** On one hand, online applications include functions that require continuous connectivity, for example, looking for an inventory status or checking for available flights. Online wireless applications require real-time connectivity to be effective and useful. On the other hand, offline applications do not require real-time connectivity. Instead, they reside locally on a particular wireless device and are always available for use, but not always in real time. In addition, their use is limited to that particular device.
- **Screen Size:** With a much smaller display area than traditional desktops/laptops, it is important to fit in as much user-required functionality as possible, while trying to format the information in such a way that it appears attractive and appealing to end users.
- **Color:** Not all wireless devices support color, and some of them support a broader palette than others. Therefore, it is important to consider each device's color support, especially if multicolor content is to be provided, such as maps or advertisement banners.
- **Ergonomics:** Wireless devices vary widely in their standards and capabilities from one to the other. Which devices should be supported, which ones are best suited for the application's needs, and can all these devices be supported at the same time?

The above-listed questions have to be associated with a development lifecycle of the MIS. We advocate the consideration of four stages to constitute that lifecycle:

1. Requirements Stage
   - Identify key information that users need when mobile.
   - Establish use-case scenarios for such information.
   - Illustrate these scenarios to users for validation purposes.
2. Analysis Stage
   - Analyze and compare similar systems (wired or mobile) to the future mobile system.
   - Identify the elements that are directly linked to wireless aspects.
   - Highlight features of different wireless devices that the future wireless system will support.
   - Identify the needed wireless communication technologies as well as the network topologies on which the future wireless system will be built.
   - Analyze the various technologies to get users' queries and return responses (e-mails, SMS, WML, etc.). Dempsey and Donnelly (2002) listed some of the key features of an m-interface: usability, intelligent and personalized services, security, consultation capabilities, and pervasive and flexible payment mechanisms.
   - Analyze security and scalability problems.
3. Design Stage
   - Analyze security and scalability problems.
   - Use existing information resources or tailor them for mobile use.
   - Develop the architecture of the future application at data and process levels.
   - Discuss the location of data and processes, and who is in charge of maintaining these data and implementing these processes.
   - Provide solutions to potential security and scalability problems.
4. Implementation Stage
   - Develop and test the new application using for example Java 2 Micro Edition (J2ME) platform (http://java.sun.com).
   - Deploy the new application on the field.

CONCLUSION

In this article, we overviewed our vision of the importance of having a dedicated design method for mobile information systems. This importance is motivated by the continuous pressure on the professionals of MISs, who are demanded to put new solutions according to the latest advances in the mobile field. For instance, it is no longer accepted to postpone operations just because there is no connection to a fixed computing desktop. Mobile devices are permitting new opportunities when it comes to banking, messaging, and shopping, just to cite a few.

REFERENCES

Andreou, A. S., Chrysostomou, C., Leonidou, C., Mavromoustakos, S., Pitsillides, A., Samaras, G., Samaras, C., & Schizas, C. (2002). Mobile commerce applications and services: A design and development approach. *Proceedings of the 1st International Conference on Mobile Business (MBusiness 2002)*, Athens, Greece.

Bellavista, P., Corradi, A., & Stefanelli, C. (2002). The ubiquitous provisioning of Internet services to portable devices. *IEEE Pervasive Computing, 1*(3).

Campo, C. (2002). Service discovery in pervasive multi-agent systems. *Proceedings of the 1st International Workshop on Ubiquitous Agents on Embedded, Wearable, and Mobile Devices* (in conjunction with *AAMAS'2002*), Bologna, Italy.

Castano, A., Ferrara, S., Montanelli, S., Pagani, E., & Rossi, G. P. (2003). Ontology-addressable contents in P2P networks. *Proceedings of the 1st Workshop on Semantics in Peer-to-Peer and Grid Computing* (in conjunction with *WWW'2003*), Budapest, Hungary.

Dempsey, S., & Donnelly, W. (2002). Identifying the building blocks of mobile commerce. *Proceedings of the 1st International Conference on Mobile Business (MBusiness'2002)*, Athens, Greece.

Elsen, I., Hartung, F., Horn, U., Kampmann, M., & Peters, L. (2001). Streaming technology in 3G mobile communication systems. *IEEE Computer, 34*(9).

Jose, R., Moreira, A., & Rodrigues, H. (2003). The AROUND architecture for dynamic location-based services. *Mobile Networks and Applications, 8*(4).

Karakasidis, A., & Pitoura, E. (2002). DBGlobe: A data-centric approach to global computing. *Proceedings of the 22nd International Conference on Distributed Computing Systems Workshops (ICDCSW 2002)*, Vienna, Austria.

Konig-Ries, B., & Klein, M. (2002). Information services to support e-learning in ad-hoc networks. *Proceedings of the 1st International Workshop on Wireless Information Systems* (in conjunction with *ICEIS 2002*), Ciudad Real, Spain.

Maamar, Z., Ben-Younes, K., & Al-Khatib, G. (2003). Scenarios of supporting mobile users in wireless networks. *Proceedings of the 2nd International Workshop on Wireless Information Systems* (in conjunction with *ICEIS 2003*), Angers, France.

Open Mobile Alliance. (2005). Retrieved June 2005 from http://www.wapforum.org

Raghu, T. S., Ramesh, R., & Whinston, A. B. (2002). Next steps for mobile entertainment portals. *IEEE Computer, 35*(5).

Ratsimor, O., Chakraborty, D., Tolia, S., Kushraj, D., Kunjithapatham, A., Gupta, G., Joshi, A., & Finin, T. (2002). Allia: Alliance-based service discovery for ad-hoc environments. *Proceedings of the 2nd ACM Mobile Commerce Workshop* (in conjunction with *MOBICOM 2002*), Atlanta, GA.

KEY TERMS

Cell: A geographic area defining the range in which a mobile support station supports a mobile client. Each cell has a cell identifier (CID) that uniquely describes it.

General Packet Radio Service (GPRS): A mobile data service available to users of global system for mobile communications (GSM) users.

Java 2 Micro Edition (J2ME): An edition of the Java platform for developing applications that can run on consumer wireless devices such as mobile phones.

Mobile Client (MC): A user with a handheld wireless device that is able to move while maintaining its connection to the network.

Mobile Information System (MIS): A computing information system designed to support users of handheld wireless devices.

Mobile Support Station (MSS): A static host that facilitates the communication with mobile clients. An MSS supports mobile clients within a geographic area known as a cell.

Short Message Service (SMS): A service for sending text messages, up to 160 characters each, to mobile phones.

Universal Mobile Telecommunications System (UMTS): A third-generation (3G) mobile phone technology.

Wireless Application Protocol (WAP): A standard specification for enabling mobile users to access information through their handheld wireless devices.

Wireless Markup Language (WML): A scripting language that is part of the WAP specification.

Distributed Approach for QoS Guarantee to Wireless Multimedia

Kumar S. Chetan
NetDevices India Pvt Ltd, India

P. Venkataram
Indian Institute of Science, India

Ranapratap Sircar
Wipro Technologies, India

INTRODUCTION

Providing support for QoS at the MAC layer in the IEEE 802.11 is one of the very active research areas. There are various methods that are being worked out to achieve QoS at MAC level. In this article we describe a proposed enhancement to the DCF (distributed coordination function) access method to provide QoS guarantee for wireless multimedia applications.

Wireless Multimedia Applications

With the advancement in wireless communication networks and portable computing technologies, the transport of real-time multimedia traffic over the wireless channel provides new services to the users. Transport is challenging due to the severe resource constraints of the wireless link and mobility. Key characteristics of multimedia-type application service are that they require different quality of service (QoS) guarantees.

The following characteristics of WLAN add to the design challenge:

- Low bandwidth of a few Mbps compared to wired LANs bandwidth of tens or hundreds Mbps.
- Communication range is limited to a few hundred feet.
- Noisy environment that leads to high probability of message loss.
- Co-existence with other potential WLANs competing on the same communication channel.

Successful launching of multimedia applications requires satisfying the application's QoS requirements.

The main metrics (or constraints) mentioned in such guidelines and that eventually influence the MAC design are: time delay, time delay variation and data rate. We develop a scheme to provide guaranteed data rate for different applications in WLAN environment.

PROPOSED ENHANCEMENT OF DCF TO PROVIDE QoS

The proposed enhancement is developed as a modular system, which integrates with DCF MAC of 802.11b wireless LAN.

Salient Features of the Modular System

- Provides throughput guarantee for traffic flow between a pair of mobile stations.
- Works in distributed mode.
- Provides MAC level admission control for traffic flow.
- Applications on the mobile stations can send resource reservations request for each call (session).
- Works with backward compatibility, on hosts that do not support QoS enhancements.

Based on the basic principle of DCF access mode, each mobile station transmits data independent of other mobile station. Also, the AP (access point) has no role to play during the data transmission. Under this scenario, the throughput control (and guarantee) has to be achieved in a distributed manner.

One has to restrain a station from accessing the medium if there are other stations in the BSS that has requested for higher resource. If there is no such other station, the station is allowed to access the medium. We propose to use eight different priority flows. The queue manager at mobile stations maintains queues for these flows. Also, the state of these queues (if there are applications that are using this flow) is synchronized across all the mobile station via the beacon messages sent by AP. The scheduler transmits the

Figure 1. Block schematic of proposed system at AP

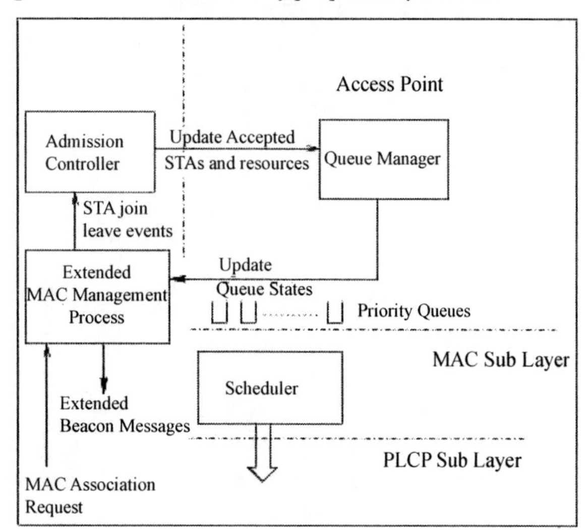

Figure 2. Block schematic of proposed system at mobile station

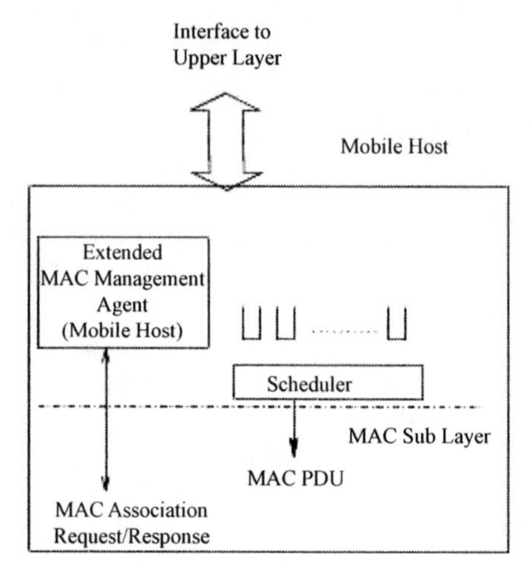

packets from these priority queues using a priority algorithm. The admission controller admits a call to particular flow, if the acceptance of call to the flow do not over-shoot the throughput for that flow.

ARCHITECTURE OF THE MODULAR SYSTEM

The block schematics of the proposed system are given in Figures 1 and 2, for AP and mobile station respectively. The system has four major components:

1. Extended MAC management process
2. Admission controller
3. Queue manager
4. Scheduler

The detailed functioning of each of the components is explained in the following subsections.

Enhanced MAC Management Process

To signal the QoS messages, we propose an extension to MAC layer management messages to carry the resource request and responses.

To signal mobile host resource requirements to AP, we propose extension to the existing MAC management frames. This approach does not need any changes in the core MAC layer. The SME (station management entity), which is normally residing in a separate management plane, needs modifications, which can be easily incorporated.

We incorporate enhanced MAC management with two segments.

MAC Management Process at AP

Apart from receiving and transmitting extended management frames to signal QoS, here the AP also broadcasts the queue states as an extension to beacon message.

MAC Management agent at Mobile Host

Apart from normal functionalities, the agent also receives the extended beacon message with queue states and passes this information to QM.

Management Message Modified

The beacon message is extended to include the queue states in a bit mask format. The queue state is an eight-bit field with each field representing a priority queue state. The queue is considered active if the bit is set, else inactive.

Admission Controller

The admission controller gets triggered by the extended MAC management process. The admission controller is a parameter based admission controller. However the decision process is modified to suit the distributed nature of DCF functionalities. While working in DCF mode each mobile station transmits/receives PDUs independent of AP and independent of other mobile stations. However the queue

Figure 3. Queue state field

Priority Queue 1	Priority Queue 2	Priority Queue 3	Priority Queue 4	Priority Queue 5	Priority Queue 6	Priority Queue 7	Priority Queue 8
bit 1							bit 8

Bit = 1 Queue is active
Bit = 0 Queue is inactive

Algorithm 1. Call admission controller algorithm

```
Begin
        for ( i = Min_Priority;i <= Max_Priority; i++ )
            do
                if ( Requested_Tput + Current_Tput[i] <= Max_Tput[i] )
                    then
                        goto accept_call
                    else
                        continue
                endif
            done
        Call RejectCall() /* No priority flow could fit this call */
        goto end

        accept_call:
            Current_Tput[i] = Current_Tput[i] + Request_Tput
            Call AcceptCall()
End
```

manager and scheduler as designed such that each station (while being completely independent) will transmit the PDUs at certain priority class. In order for the admission controller to admit a call, it needs to identify the maximum available throughput for a particular flow.

Based on [3] and [4] maximum achievable throughput per priority class is estimated as follows.

$$\text{Max Tput}[i] = \frac{P(\text{successful tx} \mid \text{flow} = i) * \text{Data pay load size}}{P(\text{collision}) * Dur_{collision} + P(\text{slot is idle}) * \text{aSlotTime} + P(\text{successful tx}) * Dur_{success}} \quad (1)$$

where $Dur_{success}$ is time duration for successful transmission of PDU, $Dur_{collision}$ is collision duration and aSlotTime is duration of one slot interval.

When a new call request arrives, the call is tried to fit in a flow of highest priority. By doing so, if the call requested throughput is achieved and existing calls are not disturbed, the call is accepted, or else the priority is decreased till the call is acceptable, if the call is not acceptable beyond least priority, the call is rejected. The admission control algorithm is described as follows.

Queue Manager

The main functionality of the queue manager is to synchronize the queue states across all mobile stations. Due to lack of any centralized coordinator, the queue manager synchronizes the states using broadcast messages. The queue manager at the AP broadcasts the states of queues to all the mobile stations in the BSS using extension to Beacon frames.

A queue for a particular flow is active if there are any calls in that flow, otherwise the queue is inactive. The queue states are broadcasted using bit masks in the extended beacon message, with each bit representing a flow. If the bit is set to 1, the queue for that flow is active; otherwise the queue is inactive. The queue manager module on station receives these broadcast messages and updates the queue state.

When all the stations have exchanged the states of the queue, each of the mobile stations would have synchronized queue states. Thus all the stations in the BSS will have a queue at particular priority level as active.

Scheduler

The scheduler's job to pick a packet from the priority flow queue and schedule them for transmission. The scheduler

Algorithm 2. Scheduling packets

```
begin
        r = rand()   /* r[0,1] */
        CurrentProb  0
        for(i=1 to N) /* N is number of active flows */
                do
                        if (r < CurrentProb)
                                Call TransmitPacket(i)
                        else
                                CurrentProb  = CurrentProb+NormPriority[i]
                        endif
                done
end
```

picks the packet from active queues. Irrespective of the fact the queue has data or not, the PacketTransmit function is called by the scheduler. The scheduler picks up the packet randomly from any of the active queues only. Now the probability to choose a queue is equal to normalized value of the priority of the queue. We define the normalized priority P_n as

$$P_{ni} = \frac{P_i}{\sum_K P_k} \qquad (2)$$

where P_i is the priority of ith queue.

We have assumed equal sized packets in all queues. For non-equal packet sizes the normalized priority would become:

$$P_{ni} = \frac{\frac{P_i}{S_i}}{\sum_k \frac{P_k}{S_k}} \qquad (3)$$

where S_k is the size of k_{th} packet.

Each of the active queues is arranged in the ascending order of the priority. The scheduler selects the queue to be scheduled, and send the packet for transmission. The scheduling algorithm is described in Algorithm 2.

Packet Transmission Process

The PacketTransmit function checks for the packet and hands over the packet to MAC layer if the queue has data.

If the queue is empty, no packet is passed to the MAC function; however the packet transmission will back-off for other stations to transmit. A timer call back for a duration that represents the time taken by an average-sized packet is registered. This will provide opportunity for other stations with the packet in the same priority level queue to transmit the packet on to the medium.

The MAC will perform DCF access method to access the medium and make transmission. If the packet could not be transmitted due to wireless medium loss or collusion, the packet is retransmitted by the MAC layer, the scheduling is not changed due to retransmission. The packet transmission algorithm is described in Algorithm 3.

FUNCTIONING OF THE MODULAR SYSTEM

The functions of modular system are to gather the QoS requirements from the mobile stations and schedule their applications as per the specified QoS.

The functioning of the system is explained by considering the following cases.

Case 1. When New Mobile Station Enters the BSS with New Application

When a new station enters the BSS, the station joins the BSS using normal association procedure (as per IEEE802.11 standard). The AP instantiates a MAC management agent. The agent is migrated to the mobile host. The station uses the "Extended MAC Management Agent" to signal the re-

Algorithm 3. Packet transmission algorithm

```
Begin
        if( i == active) /* Is the queue non-empty */
                Call SendPacket /* Call the MAC function */
                Call TimerCallback(packet_size)
        else
                Call TimerCallback(Ave_packet_size)
        endif
end
```

source requirements for new application. The AP receives the resource request, and passes on the QoS parameters to admission controller. The admission controller examines the resource requests, applies the admission control algorithm and accepts/rejects the request. The accept/reject response is conveyed back to the mobile station via the "Extended MAC Management process" module and updates the queue states in the QM. The QM updates the queue states for each of the flows, and sends them to the mobile stations as an extension to beacon message.

Upon reception of the response for the call request, the mobile station can start sending traffic. The QM at the mobile station receives the queue states in the extended beacon message and updates local queue state. The scheduler the mobile station schedules the traffic using the queue state information.

Case 2. When a Mobile Station Leaves the BSS

When the mobile station is about leave the BSS, the station uses the enhanced MAC management process to signal the AP. The AP receives the station ID, and updates the QM and admission control to relinquish the resource used all application under this station and changes the queue state if required.

Case 3. When New Application Has to be Admitted

A mobile station is already associated with the AP, and may be running some applications or otherwise. Under this condition, the station and AP follow the normal steps (as in Case 1), as this is no different from Case 1.

Case 4. When Applications are Terminated at the Mobile Station

A mobile station stops the current application, but the station is still in the BSS. When the application is stopped, the station uses the enhanced MAC management process to signal the AP that the application has stopped. The AP follows steps in Case 2 to relinquish the resource and update queue state.

IMPLEMENTATION CONSIDERATION

The system has two components, that is to say, AP component and mobile station component.

- The AC is implemented at the centralized location, normally co-located with the AP.
- Extension to MAC management layer, QM and scheduler are implemented in both AP and mobile station.

AP Components Implementation

It is easier to enhance the AP to support functionalities such as CAC, QM and scheduler since there is relatively very few numbers of APs and also the APs are normally controlled by the service provider, who wishes to provide QoS in the BSS. The AC, QM and scheduler are implemented in high-level system language (C) and integrated into the AP software. The MLME functions on the AP are modified to understand the extensions in the MAC frames and pass the QoS parameters to the CAC. The receive (RX) and transmit (TX) functions on the AP are modified to call the QM and scheduler functions respectively.

Mobile Station Components Implementation

The station components are implemented as Java byte-code, and the agent is instantiated at the AP and migrated to the mobile station, as the mobile station gets associated with the AP. The extended MAC management agent, QM and scheduler are implemented in Java.

SIMULATION AND RESULTS

We have used ns-2 [7] as simulator to validate the proposed system. Ns are a discrete event simulator targeted at networking research. Ns provides substantial support for simulation of TCP, routing, and multicast protocols over wired and wireless (local and satellite) networks.

Simulation Setup and Procedure

Using ns-2, a sixteen-node network working in BSS and iBSS modes is setup. We have modified the MAC layer functionalities in ns-2 to support the proposed system.

In the simulated system, the mobile nodes communicate with each other and external world over the 802.11b 11MBPS channel. The transmission power of the mobile is chosen such that stations are not hidden from one another. If not stated otherwise, during the simulation neither RTS/CTS nor the fragmentation is used. For most of our simulation we have considered on-off traffic with exponential distribution, and log-normal distribution for the radio channel errors. We have done several experiments by considering traffic in both upstream and downstream direction. The allowed traffic has been classified as VBR (variable bit-rate) and

Table 1. Simulation parameters

Data Rate	11 Mbps
RTS Threshold	3000
Physical Layer Frequency	2457e+6
Transmission Power	31 mW
Receiver Threshold	1.15209e-10
Radio Propagation Model	TwoRayGround
LongRetryLimit	2

Figure 4. Throughput plot for mobile station with CBR traffic under DCF mode

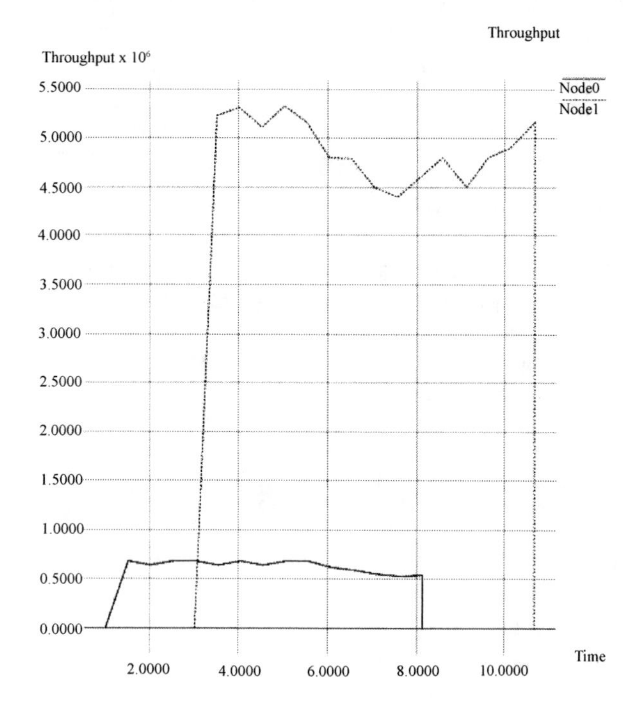

CBR (constant bit-rate) with random arrivals. The duration of each experiment varies from 100 seconds to 1,000 seconds. The simulation parameters that are significant are listed in Table 1.

We have simulated the proposed systems as three individual units. We explain the simulation for each case with CBR and VBR traffics.

The DCF MAC functions at each station are modified according to the proposed scheme. Each of the mobile station generates traffic destined to other nodes with data packets of 1,500bytes, 1,000 bytes, 512 bytes and 64 bytes.

In our first experiment, we considered two nodes generating CBR traffic. Each node requested a throughput of 10\% and 90\% of the net throughput. In Figure 4, we have plotted the throughput vs. time for two nodes. It can be observed that both the stations get the allocated bandwidth.

Figure 5. Throughput plot for mobile station with VBR traffic under DCF mode

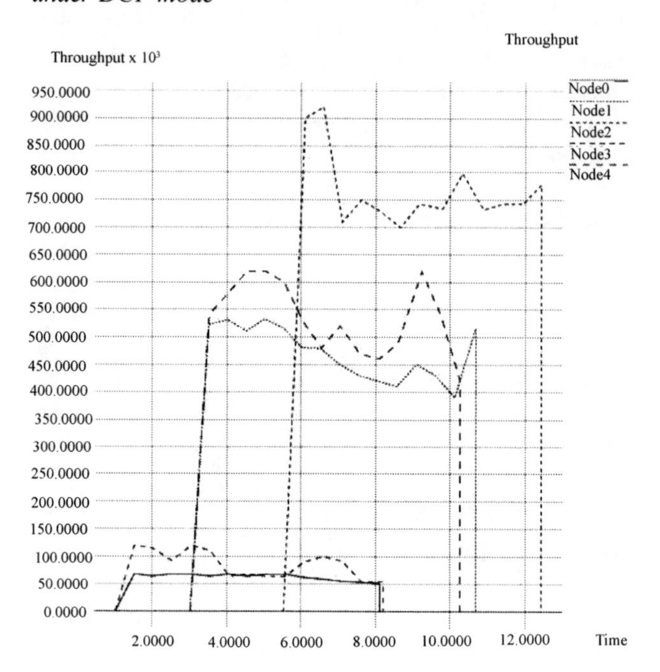

Figure 6. Normalized throughput for proposed scheme and standard DCF mod

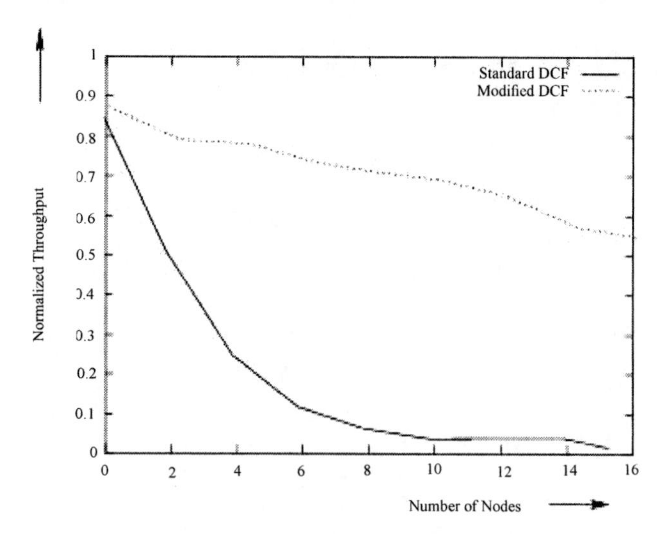

Also when the node0 application is terminated, the node1 uses up the excess bandwidth.

In the next experiment, we have considered BSS with five nodes admitted for scheduling. The nodes requested throughput requirements are at 50, 100, 500, 550, and 750 Kbps. In Figure 5, we have plotted the throughput versus time for all the five nodes. It can be observed from the plot, each of the nodes is scheduled to get the allocated bandwidth.

Comparison with the Standard DCF Scheme

Again we use the normalized throughput as a measure to compare the proposed DCF scheme against the standard DCF scheme.

We have plotted the normalized throughput for a node working proposed DCF mode (with CBR traffic) against standard DCF mode in Figure 6. With the proposed scheme, the mobile 6station is admitted to the BSS via admission controller and the scheduler schedules the allocated bandwidth for the station. Hence the normalized throughput stays close to one. With the standard DCF, the mobile stations share the bandwidth in uncontrolled manner. Thus, as the number of stations increases in the BSS, there is no control over the bandwidth usage.

SUMMARY

In this article we have proposed method of providing QoS in wireless LAN operating in DCF mode. Providing QoS guarantee in a distributed environment has to be a distributed approach due to the distributed nature of the DCF. The proposed method requires small changes (extensions) to the MAC management entity and scheduler function operating at each station above the MAC layer.

REFERENCES

Anker, T., Cohen, R., Dolev, D., & Singer, Y. (2001). Probabilistic fair queuing. In *IEEE Workshop on High Performance Switching and Routing.* Dallas, TX.

Bianchi, G. (2000). Performance analysis of the IEEE 802.11 distributed coordination function. *IEEE JSAC,* 3, 535-47.

Chetan Kumar, S., Venkataram, P., & Pratap Singh, R. (2005). Distributed approach for QoS guarantee to wireless multimedia. In *International Conference on Advances in Mobile Multimedia,* Kuala Lumpur, Malaysia.

Network Simulator NS-2. (n.d.). Retrieved from http://www.isi.edu/nsnam/ns/.

Ni, Q., Romdhani, L., Turletti, T., & Aad, I. (2002). QoS issues and enhancements for IEEE 802.11 wireless LAN. In *Rapport de recherche de l INRIA.* Sophia Antipolis, Equipe: PLANETE.

Pong, D., & Moors, T. (2003). Call admission control for IEEE 802.11 contention access mechanism. *Globecom 2003,* San Francisco USA.

Tay, Y. C., & Chua, K.C. (2001). A capacity analysis for the IEEE 802.11 MAC protocol. *Journal of Wireless Networks, 7.*

KEY TERMS

Access Point (AP): An entity in the wireless LAN that is normally connected to a wired backbone and coordinates the operation of wireless mobile stations that are operating in infrastructure mode.

Call Admission Controller (CAC): A set of functions and procedures that control the admission of new calls into the system predictable based on predefined set of rules.

Media Access Controller (MAC): A set of procedures that governs the access to the media in a multiple access system. DCF (distributed coordination function) and PCF (point coordination function) are two MAC functions defined in IEEE 802.11 standard for wireless LAN.

Quality of Service (QoS): The term quality of service defines of quantitative representation of network resources that affect the application performance.

Distributed Computing in Wireless Sensor Networks

Hong Huang
New Mexico State University, USA

INTRODUCTION

Wireless sensor networks (WSNs) recently have attracted a great amount of attention because of their potential to dramatically change how humans interact with the physical world (Estrin, Culler, Pister, & Sukhatme, 2002; Akyildiz, Su, Sankarasubramaniam, & Cayirci, 2002). A wireless sensor network is composed of many tiny, wirelessly connected devices, which observe and perhaps interact with the physical world. The applications of WSN are many and wide-ranging, including wildlife habitat monitoring, smart home and building, quality monitoring in manufacturing, target tracking in battlefields, detection of biochemical agents, and so forth.

The emerging WSN technology promises to fundamentally change the way humans observe and interact with the physical world. To realize such a vision, *distributed computing* is necessary for at least two reasons. First, sending all the raw data to a base station for centralized processing is very costly in terms of energy consumption and often impractical for large networks because of the scalability problem of wireless networks' transport capacity (Gupta & Kumar, 2000). Second, merely using a large number of inexpensive devices to collect data hardly fundamentally changes the way humans interact with the physical world; and it is the intelligence embedded inside the network (i.e., distributed computing) that can have a profound impact.

WSN presents a very difficult environment for distributed computing. Sensors have severe limitations in processing power and memory size, and being battery-powered, they are particularly energy constrained. As a reference, the hardware capabilities of a typical sensor node are listed in Table 1. In WSN, device failures can be frequent, sensory data may be corrupted by error, and the wireless communications exhibit complex and unpredictable behavior. In such an environment, traditional methods for distributed computing face fundamental difficulties. Now, communications links are neither reliable nor predictable: they can come and go at any time. Packet routing is difficult since maintaining and storing routing tables for a massive number of nodes is out of the question. Routing to a single destination seems to have a solution (Intanagonwiwat, Govindan, & Estrin, 2000), complex message routing for distributed computing remains difficult. Also, distributed organizing and grouping of sensory data using traditional methods is costly in terms of protocol message overhead.

This article is organized as follows. We first describe WSN infrastructures required to support distributed computing, followed by a description of typical, important distributed computing applications in WSN; we the conclude the article.

INFRASTRUCTURE SUPPORT FOR DISTRIBUTED COMPUTING IN WSN

In order for WSN to effectively perform distributed computing, some necessary infrastructure needs to be established. The type of infrastructure required varies according to the specific application in question, but the common ones include neighbor discovery and management, synchronization, localization, clustering and grouping, and data collection infrastructure. We elaborate on each item as follows.

Table 1. Hardware capabilities of typical sensor nodes (Crossbow Technology —www.xbow.com)

	CPU	Nonvolatile Memory	Radio Transceiver	Power
MICA2	ATMega 128L 8 MHz, 8 bit	512 KB	869/915, 434, 315 MHz, FSK ~40 Kbps	2 AA 2850 mAh
MICA2DOT	ATMega 128L 8 MHz, 8 bit	512 KB	869/915, 434, 315 MHz, FSK ~40 Kbps	Coin cell 1000 mAh

Neighbor discovery and management refers to the process in which sensors discover their neighbors, learn their properties, and control which neighbors to communicate with. Discovery is typically done through sensors exchanging hello messages within radio range. In the process, sensors discover not only neighbors' presence, but also optionally their node type, node identifier, power level, location/coordinates, and so forth. Frequently, sensors can also control how many neighbors to communicate with through the use of power control—that is, a sensor can increase or decrease the scope of its immediate neighborhood by increasing or decreasing its transmitting power, respectively. This is also called topology control (Li & Hou, 2004), and its purpose is to allow sensors to use just enough, but no more power to ensure adequate connectivity.

Synchronization refers to the process in which sensors synchronize their clocks. Synchronization is necessary because sensory data is often not useful if not put in a proper temporal reference frame. Traditional methods for synchronization in a network, such as NTP (Mills, 1994), do not apply very well in WSN. This is because the assumptions on which tradition network synchronization methods are based, such as availability of high-precision clocks, stable connections, and consistent delays, are no longer true in WSN, causing considerable difficulties. The approach to deal with such difficulties in WSN is to relax the requirements. For example, only local, not global, synchronization is maintained, or only event ordering, not precise timing, is kept (Romer, 2001).

Localization refers to the process in which sensors obtain their position/coordinates information. Similar to synchronization, localization is necessary because sensory data needs to be put in a spatial reference frame. Sensors with global positioning system (GPS) capability are currently commercially available; they obtain their coordinates from satellites with a few meters' accuracy. The downside with using GPS is the cost, and the unavailability indoors or under dense foliage. In a WSN without GPS capability, it is still possible to localize relatively to a few reference points in the network (Bulusu, Heidemann, & Estrin, 2000).

Clustering and grouping refers to the process in which sensors organize themselves into clusters or groups for some specific function. A cluster or a group typically consists of a leader and a few members. The leader represents the cluster or group and maintains external communication, while the members report data to the leader and do not communicate with the outside. Such organization is advantageous for scalability, since a large network can now be reduced to a set of cluster or groups. Task-specific clusters or groups can be formed. For example, sensors around a moving target form a tracking group, which moves with target, while sensors not in the tracking group can be put to sleep to save energy (Liu, Reich, Cheung, & Zhao, 2003).

Data collecting infrastructure ensures that sensory data is transported correctly and efficiently to one or a few collection points, sometimes called data sinks. A typical approach is publish and subscribe with attribute-based naming, where a sink broadcasts its interest for some data attributes, and sensors send their data if it matches the interests. An example of such an approach is directed diffusion, in which an infrastructure based on the hop count to the sink is established and refreshed periodically (Intanagonwiwat et al., 2000).

TYPICAL DISTRIBUTED COMPUTING APPLICATIONS IN WSN

In this section, we describe typical distributed computing applications in WSN which include distributed query and search, collaborative signal processing, distributed detection and estimation, and distributed target tracking.

Distributed query and search refers to the process in which a user query or search for an event or events inside the network in a distributed fashion. There are two major types of such applications: blind and structured. In a blind search, no prior information about the target exists. In a structured search, some kind of infrastructure exists which points to the target location in a distributed manner. We elaborate on these two types of searches below.

There are three major approaches to perform blind search. The first one is flooding, in which the query message is flooded to the entire network and the target responds with a reply. The advantages of flooding are simplicity and low response latency. The disadvantage is the high communications cost in terms of number of messages transmitted. To mitigate the high communication cost of flooding, a second approach, iterative, limited flooding, can be used (Chang & Liu, 2004). In such an approach, a sequence of limited broadcasts of increasing hop-count limits is tried until the target is found, in the hope that the target will be found during a low-cost, limited broadcast. The expected communications cost reduction of this approach comes at the expense of higher search latency. In the third approach, a query packet carries out a random walk in the network, which continues until the search target is encountered (Avin & Brito, 2004). This approach can further reduce communications cost but at the expense of even higher latency.

In a structured search, indices or pointers for targets are distributed in the network. A typical approach uses a distributed hash table, where the name of a target is randomly and uniformly hashed to a number that identifies a node, or a location, where the target information is stored (Ratnasamy et al., 2002). The search becomes a simple matter of evaluating the hash function of the target name which points to a node that stores the target information. This simplification comes at the cost of maintaining an infrastructure that stores target information in a distributed manner, which can be costly if the network is highly dynamic, for example, nodes join or leave the network frequently.

Collaborative signal processing (CSP) refers to the process in which sensors process sensory data collaboratively rather than individually. CSP has three main advantages. First, CSP promotes robustness because, while an individual sensor can be faulty, consensus from many sensors is reliable. Second, CSP promotes accuracy since the results fuse the observations from many sensors, canceling noise from individual sensors. Third, CSP promotes efficiency since only a fraction of sensors needs to be activated—those which have highest potential to eliminate uncertainty (Zhao, Shin, & Reich, 2002). In addition, sophisticated multi-resolution analysis techniques, such as wavelet transform, can be used to reduce the amount of data transported. CSP is a common method used in different distributed applications, and its specifics vary according to the particular application in question and are described in the following sections.

Distributed detection and estimation refers to the process in which WSN makes a decision about the occurrence of an event (detection) or about the value of a physical variable (estimation) in a distributed manner. In distributed detection, under both Bayes and Neyman-Peason criterion, the optimum test is a likelihood ratio test. In the case of independent observations, it suffices to send likelihood ratios from individual sensors rather than the raw data, leading to a large savings in communications cost. In distributed estimation, the principal of sufficient statistics can be appealed to drastically reduce the amount of data transported. For example, to estimate the mean of a Gaussian variable, the sufficient statistic is simply the sum of sampled values and the count of number of samples, both of which can be collected using a running sum or count rather than the entirety of the sampled values, resulting in large savings in communications cost.

Research issues related to distributed detection and estimation include sensor data censoring to exclude spurious data (Patwari & Hero, 2003), distributed, iterative sensor fusion based on consensus (Xiao, Boyd, & Lall, 2005), and finding the optimal tradeoff between decision fidelity or estimation error, and sensor density, sensitivity, quantization level, communications cost, and power consumption (Aldosari & Moura, 2004).

Distributed target tracking refers to the process in which sensors collaboratively track one or more discrete, moving targets. In a sense, distributed target tracking is a subclass of distributed detection and estimation, but faces the severe challenge caused by target movement. To deal with such difficulty, sensors typically form a tracking group (Li, Wong, Hu, & Sayeed, 2002). The group membership can be determined in a number of ways. A sensor can declare itself a member if it receives a signal from the target that exceeds a certain threshold. Or sensors can be added to the group sequentially in the order of its capability to reduce uncertainty (Zhao et al., 2002). The sensory data from members of the tracking group is collected at the group leader, which can be selected as the sensor with the highest signal level. For the fusion of sensory data, a number of standard methods are available, such as Kalman filtering (Spanos, Olfati-Saber, & Murray, 2005; Mokashi, Huang, Kuppireddy, & Varghese, 2005), maximum likelihood estimation, and so forth.

Research issues related to distributed target tracking include the distribution of state (target) information among individual sensors (Liu, Chu, Reich, & Zhao, 2004), managing tracking groups, and trading off power consumption and surveillance quality (Gui & Mohapatra, 2004).

CONCLUSION

In this article we have provided an introduction to WSN and its characteristics, and the challenges WSN poses for distributed computing. We have described some infrastructure support for distributed computing, including neighbor discovery and management, synchronization, localization, clustering and grouping, and data collection. We have discussed typical, important distributed computing applications in WSN, including distributed query and search, collaborative signal processing, distributed detection and estimation, and distributed target tracking.

WSN presents new, exciting challenges to distributed computing, such as how to achieve complex, high-level goals from a large number of simple devices, and how to obtain robustness and certainty through coordinating unreliable devices and processing uncertain data. Successfully addressing such challenges will significantly advance the state of the art of distributed computing.

REFERENCES

Akyildiz, I. F., Su, W., Sankarasubramaniam, Y., & Cayirci, E. (2002). A survey on sensor networks. *Computer Networks, 38*(4), 393-422.

Aldosari, S. A., & Moura, J. M. F. (2004). Fusion in sensor networks with communication constraints. *Proceedings of the Conference on Information Processing in Sensor Networks (IPSN'04)*, Berkeley, CA.

Avin, C., & Brito, C. (2004). Efficient and robust query processing in dynamic environments using random walk techniques. *Proceedings of the Conference on Information Processing in Sensor Networks (IPSN'04)*, Berkeley, CA.

Bulusu, N., Heidemann, J., & Estrin, D. (2000). GPS-less low-cost outdoor localization for very small devices. *IEEE Personal Communications, 7*(4), 28-34.

Chang, N., & Liu, M. (2004). Revisiting the TTL-based controlled flooding search: Optimality and randomization. *Proceedings of the International Conference on Mobile Computing and Networks (MobiCom'04)*, Philadelphia, PA.

Estrin, D., Culler, D., Pister, K., & Sukhatme, G. (2002). Connecting the physical world with pervasive networks. *IEEE Pervasive Computing, 1*(1), 59-69.

Gui, C., & Mohapatra, P. (2004). Power conservation and quality of surveillance in target tracking sensor networks. *Proceedings of the International Conference on Mobile Computing and Networks (MobiCom'04)*, Philadelphia, PA.

Gupta, P., & Kumar, P. R. (2000). The capacity of wireless networks. *IEEE Transactions on Information Theory, 46*(2), 388-404.

Intanagonwiwat, C., Govindan, R., & Estrin, D. (2000). Directed diffusion: A scalable and robust communication paradigm for sensor networks. *Proceedings of the International Conference on Mobile Computing and Networks (MobiCom'00)*, Boston.

Li, D., Wong, K., Hu, Y., & Sayeed, A. (2002). Detection, classification, tracking of targets in micro-sensor networks. *IEEE Signal Processing Magazine, 19*(2), 17-30.

Li, N., & Hou, J. C. (2004). Topology control in heterogeneous wireless networks: Problems and solutions. *Proceedings of the IEEE Conference on Computer Communications (INFOCOM'04)*, Hong Kong, China.

Liu, J., Reich, J., Cheung, P., & Zhao, F. (2003). Distributed group management for track initiation and maintenance in target localization applications. *Proceedings of the Conference on Information Processing in Sensor Networks (IPSN'03)*, Palo Alto, CA.

Liu, J., Chu, M., Reich, J., & Zhao, F. (2004). Distributed state representation for tracking problems in sensor networks. *Proceedings of the Conference on Information Processing in Sensor Networks (IPSN'04)*, Berkeley, CA.

Mills, D. L. (1994). Precision synchronization of computer network clocks. *ACM Computer Communication Review, 24*(2), 28-43.

Mokashi, G., Huang, H., Kuppireddy, B., & Varghese, S. (2005). A robust scheme to track moving targets in sensor nets using amorphous clustering and Kalman filtering. *Proceedings of the IEEE Military Communications Conference (Milcom'05)*, Atlantic City, NJ.

Patwari, N., & Hero, A. O. (2003). Hierarchical censoring for distributed detection in wireless sensor networks. *Proceedings of the IEEE International Conference on Acoustics, Speech, and Signal Processing (ICASSP'03)*, Hong Kong, China.

Ratnasamy, S., Karp, B., Yin, L., Yu, F., Estrin, D., Govindan, R., & Shenker, S. (2002). GHT: A geographic hash table for data-centric storage. *Proceedings of the ACM International Workshop on Wireless Sensor Networks and Applications (WSNA'02)*, Atlanta, GA.

Romer, K. (2001). Time synchronization in ad hoc networks. *Proceedings of the ACM International Symposium on Mobile Ad Hoc Networking and Computing (MobiHoc'01)*, Long Beach, CA.

Spanos, D., Olfati-Saber, R., & Murray, R. (2005). Distributed Kalman filtering in sensor networks with quantifiable performance. *Proceedings of the Conference on Information Processing in Sensor Networks (IPSN'05)*, Los Angeles, CA.

Xiao, L., Boyd, S., & Lall, S. (2005). A scheme for asynchronous distributed sensor fusion based on average consensus. *Proceedings of the Conference on Information Processing in Sensor Networks (IPSN'05)*, Los Angeles, CA.

Zhao, F., Shin, J., & Reich, J. (2002). Information-driven dynamic sensor collaboration for tracking applications. *IEEE Signal Processing Magazine, 19*(2), 61-72.

KEY TERMS

Clustering and Grouping: The process in which sensors organize themselves into clusters or groups for some specific function.

Collaborative Signal Processing (CSP): The process in which sensors process sensory data collaboratively rather than individually.

Data Collecting Infrastructure: Ensures that sensory data is transported correctly and efficiently to one or a few collection points.

Distributed Computing: The kind of computing realized through, and distributed among, many discrete devices.

Distributed Detection and Estimation: The process in which a WSN makes a decision about the occurrence of an event (detection) or about the value of a physical variable (estimation) in a distributed manner.

Distributed Query and Search: The process in which a user queries or searches for an event or events inside the network in a distributed fashion.

Distributed Target Tracking: The process in which sensors collaboratively track one or more discrete, moving targets.

Localization: The process in which sensors obtain their position/coordinates information.

Neighbor Discovery and Management: The process in which sensors discover their neighbors, learn their properties, and control which neighbors to communicate with.

Synchronization: The process in which sensors synchronize their clocks.

Wireless Sensor Network (WSN): A type of network composed of many tiny, wirelessly connected sensing devices.

Distributed Heterogeneous Tracking for Augmented Reality

Mihran Tuceryan
Indiana University Purdue University Indianapolis, USA

Rajeev R. Raje
Indiana University Purdue University Indianapolis, USA

INTRODUCTION

Augmented reality (AR) is a technique in which a user's view of the real world is enhanced or augmented with additional information generated from a computer model (Azuma et al., 2001). The enhancement may consist of virtual artifacts to be fitted into the environment or a display of non-geometric information about existing real objects. Mobile AR (MAR) systems implement this interaction paradigm in an environment in which the user moves, possibly over wide areas (Feiner, MacIntyre, Hoellerer, & Webster, 1997). This is in contrast to non-mobile AR systems that are utilized in limited spaces such as a computer-aided surgery or by a technician's aid in a repair shop. There are a number of challenges to implementing successful AR systems. These include a proper calibration of the optical properties of cameras and display systems (Tuceryan et al., 1995; Tuceryan, Genc, & Navab, 2002), and an accurate registration of three-dimensional objects with their physical counterparts and environments (Breen, Whitaker, Rose, & Tuceryan, 1996; Whitaker, Crampton, Breen, Tuceryan, & Rose, 1995). In particular, as the observer (or an object of interest) moves over time, the 3D graphics need to be properly updated so that the realism of the resulting scene and/or alignment of necessary objects and graphics are maintained. Furthermore, this has to be done in real time and with high accuracy. The technology that allows this real-time update of the graphics as users and objects move is a *tracking system* that measures the position and orientation of the tracked objects (Koller et al., 1997). The ability to track objects, therefore, is one of the big challenges in MAR systems. This article describes a software framework for realizing such a distributed tracking environment by discovering independently deployed, possibly heterogeneous trackers and fusing the data from them while roaming over a wide area. In addition to the MAR domain, this kind of a tracking capability would also be useful in other domains such as robotics and location-aware applications. The novelty of this research lies in the amalgamation of the theoretical principles from the domains of AR/VR, data fusion, and the distributed software systems to create a sensor-based, wide-area tracking environment.

BACKGROUND

Although a few approaches for tracking have been proposed (e.g., Hightower & Borriello, 2001; Koller et al., 1997; Neumann & Cho, 1996; State, Hirota, Chen, Garrett, & Livingston, 1996), the ability to track objects accurately and in real time over a *wide area* does not yet have a satisfactory solution. Moreover, many of these approaches require a highly engineered environment with a uniform set of trackers whose architecture is known in advance (Welch, & Foxlin, 2002; Ubisense, 2006). Assuming that trackers have been deployed and are operating and exist in the environment, this research deals with questions of how to discover what trackers exist in a local area, what quality-of-service (QoS) properties they have, and how to make the best use of their measurements in a mobile and dynamic environment.

The wide-area, ubiquitous tracking that is the focus of this article has been addressed mainly in the pervasive/ubiquitous computing community. An early tracking system was HiBall that utilized a ceiling instrumented by LED lights (Welch, 1999). The HiBall tracker covered a wider area than a typical magnetic tracking system, and in the implementation its range covered a room or a lab. The scalability of such a system was limited because of the increased cost of extending beyond the size of a lab. The BAT system, which used ultrasound as the core technology (Harter, Hopper, Steggles, Ward, & Webster, 2002; Newman, Ingram, & Hopper, 2001), had a limited resolution. The location sensing system, by Ubisense (2006) uses the ultra-wide-band technology and has a better resolution (6 inches positional accuracy, according to company Web sites).

Researchers at Intel Research studied the use of existing wireless hotspots and cell phone towers to compute location information over wide-areas (Schilit et al., 2003; Borriello, Chalmers, LaMarca, & Nixon, 2005). Bahl and others studied localization techniques using existing Wi-Fi wireless hubs (Bahl & Padmanabhan, 2000; Balachandran, Voelker, & Bahl, 2003). Their methods assume a ubiquitous infrastructure that exists for other purposes (networking) that can be tapped into for localization of users. Typically, their resolution tends to be low and is not sufficient for typical AR applications.

Ubiquitous tracking systems specifically for AR systems have been also studied by Bauer et al. (2002). Newman et al. (2004), and Reitmayr (2001), and have resulted in prototypical systems, some of which are component based (e.g., DWARF by [Bauer, Bruegge, Klinker, MacWilliams, Reicher, Riss, et al., 2001]).

THE UNIFRAME-BASED MOBILE TRACKING SERVICE FOR AR

As indicated earlier, the distributed tracking system is an example of a heterogeneous, distributed computing system. The overall architecture and various components of the distributed tracker subsystem that is the focus of this article are shown in Figure 1.

The software realization of this tracking system is based on the principles of uniframe (Olson, Raje, Bryant, Burt, & Auguston, 2005). Uniframe provides an environment for an interoperation of heterogeneous and distributed software components, and uses the principles a meta-component model, service-oriented architectures, generative programming, and two-level (TLG) and event grammars (EG). The realization of the distributed tracking system, using the UniFrame, begins with a generative domain model (GDM) (Czarnecki & Eisenecker, 2000) created by experts from the tracking domain. This GDM contains various details, such as the software architecture of the tracking system expressed in terms of underlying components, their interactions, the rules for generating middleware, and the rules for the prediction and monitoring of the quality of the integrated system. Each component is defined by a Unified Meta-component Model (UMM) specification (Raje, 2000). The UMM has three parts: (a) components; (b) services, offered by components, and associated guarantees; and (c) infrastructure for deploying and discovering components. A developer who wishes to create specific components for the tracking system consults the GDM and creates implementations using the UMM specifications encoded there. After a component is developed, it is validated against the quality requirements, both functional and QoS. The developer also creates an associated UMM specification for that component. This specification and the component are deployed on the network. These components are also registered with the uniframe resource discovery system (URDS) (Siram, 2002).

A system integrator planning to create the tracking system, from independently developed and deployed components, issues a query consisting of the requirements the tracking system must meet. The query processing consults the GDM, divides the query into many sub-queries, each corresponding to a single component UMM specification. These sub-queries are passed to the URDS, which searches for appropriate matching components. If such components are found, these are returned to the system integrator, who then selects a subset of these results, provides any proprietary components, and requests the process to assemble the integrated system conforming to the design. The uniframe system generator (Huang, 2003) carries out the generation of the integrated system. The key challenges in creating the tracking system are: (a) designing the GDM, (b) the discovery of components, and (c) the generation of a prototypical tracking system. These are briefly discussed below.

Designing the GDM

The GDM is developed by the domain experts and contains the software architecture of the family of systems, along with many associated details. For a tracking system, it can be either handcrafted or generated via the uniframe system generator (Huang, 2003). One important piece of the GDM relates to the specification of components that make the software architecture of the tracking system. The specification provides

Figure 1. Architecture and components of the distributed tracking subsystem for mobile AR applications

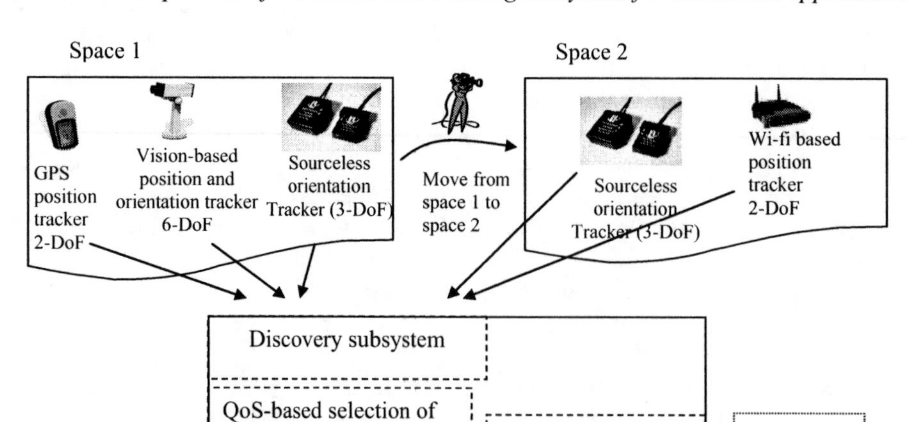

the necessary details during the discovery process and the system generation process. The approach for specifying the components is indicated as follows.

UMM-Specifications

Each sensing device used in the tracking system is represented by a corresponding software component that encapsulates its behavior. For example, a GPS sensor in the tracking system is encapsulated as a component offering a service that provides 3DOF position information with certain accuracy. As indicated earlier, each component in uniframe has an associated UMM specification. This specification contains many attributes (Raje, 2000) that reflect various details related to that component. In the context of the tracking system, the functional and the QoS attributes of a component are the most important ones. For example, a partial specification (for the sake of brevity) for an Inertia Cube Tracker component is:

Component Name: InertiaCubeTracker
Domain Name: Distributed Tracking
Informal Description: Provides the orientation information.
Computational Attributes
Inherent Attributes:
Id: cs.iupui.edu/InertiaCubeTracker;
...
Validity: 12/1/07 Registration: pegasus.cs.iupui.edu/HH1
Functional Attributes:
Functional Description: Provides the orientation of a tracked object.
Algorithm: Kalman Filter;
Complexity: O (n^3)
Syntactical Contract:
Vector getOrientation();
Semantic Contract:
Pre-condition: {calibrated (InertiaCubeTracker)== true}
Post-condition: {sizeof (orientationVector) == 3)}
Synchronization Contract:
Nature: Multi-threaded Synchronization Policy
Implementation Mechanism: semaphore
Technology: CORBA
....
Quality of Service Attributes
QoS Metrics: resolution, drift, lag-time,
resolution: 0.1 degrees; drift: 0.01 degrees/sec; lag-time: 1 ms
....

The preceding partial specification shows many important characteristics of the UMM-specification. These are: (a) the specification is comprehensive and highlights many aspects of a component; (b) the specification is an enhanced realization of the concept of multi-level contracts (Beugnard, Jezequel, Plouzeau, & Watkins, 1999), thus the specification not only describes the functional aspects, but also emphasizes the QoS attributes of a component; and (c) the specification is consistent with the concepts of a service-oriented view of software components. The detailed nature of the specification provides sufficient information for the discovery of components to create the tracking system.

Discovery Process

In a distributed tracking environment, it is conceivable that there would be many different software instances offering similar types of services. For example, there may be multiple trackers (each encapsulating a different inertia tracker) implemented in a variety of ways and possibly offering different qualities of the tracking results. Hence, to realize such a tracking system, it is necessary to discover various alternatives available over the network. That is the role of the discovery service. The discovery process is realized using the principles of the URDS (Siram, 2002). The URDS is a hierarchical, proactive, and interoperable discovery service. The components are selected based on their type and the QoS (e.g., resolution) values for a specific tracking system.

Generation of the Prototypical System

Uniframe uses the principles of glues of wrappers that are generated using the concepts of TLG and EG. The purposes of the glue and wrapper code are to allow an interoperation between heterogeneous components and also to insert any instrumentation code that can collect event traces to observe the actual QoS values during the execution of the system. A system generator (Huang, 2003) accepts the selected components as the input and uses the information in the GDM to semi-automatically create the distributed system. The rules for the generation are a part of the GDM and are developed during the formalization of the GDM. Another important facet of the generation process is the ability to make static predictions (based on the QoS properties of the selected components) and compare them against the actual execution results. The process of prediction is based on the principles of sensor fusion, as described in the following section.

DATA FUSION

Once the individual tracker components are selected through the discovery process described earlier, their results need to be combined (fused) in order to get the best estimate of the position and orientation of the tracked object. The usual

Figure 2. Federated Kalman filter architecture (Carlson, 1990)

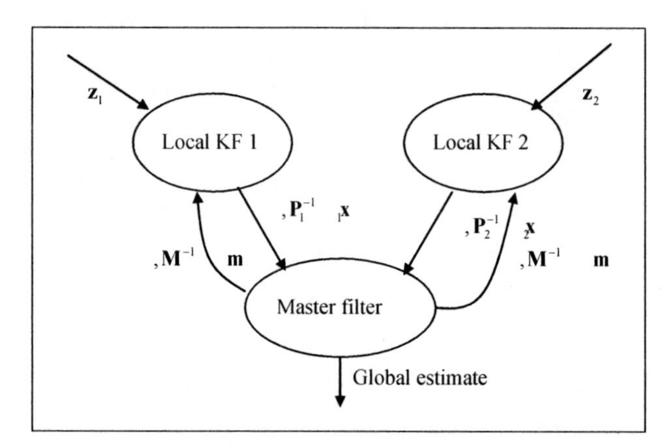

*Note: Here the $\hat{x}_i$ and M_i are the state vector and covariance matrices for the local filters. The z_is are the measurements for each sensor i. The quantities **m** and **M** are the global state vector and covariance matrix estimated by the master filter, respectively. In this figure, the global estimates are fed back to the local filters.*

framework for fusing multi-modal sensor data for tracking is the various modifications and extensions of the Kalman filter (Brown & Hwang, 1997; Kalman, 1960; Welch & Bishop, 2001). The Kalman filter is a recursive estimation method that tries to estimate the state of a discrete-time controlled process (i.e., the pose of a tracked object, possibly with velocity and acceleration information) using observable measurements (i.e., data from tracker sensors). The state is estimated in each time step by a set of update equations in the form of "predict" and "correct" cycle. The typical Kalman filter formulation requires that all the state variables for the available devices and the relevant equations be set up globally at the beginning. Given the dynamic and distributed nature of the framework described in this article, this approach is not practical. Instead, the federated Kalman filter first described by Carlson (1990) is to be used. In this framework, the sensors have their own local Kalman filter running that estimates the local state and covariance. Then for each object that is tracked, there is a master Kalman filter that uses the estimates coming from the local Kalman filters and computes a global estimate of the state and covariance. The results of the master filter can be fed back to the local filters to improve their local state estimates also. The rough architecture of such a federated Kalman filter is shown in Figure 2.

The federated Kalman filter allows for the assembling of a master filter depending on the new set of sensors found through the discovery process. The covariance matrix is a measure of the error in the state information and can be used as part of the QoS information.

FUTURE TRENDS

The future plans include testing the methods by an integrated system that consists of multiple trackers distributed over a sufficiently large area. A variety of trackers such as vision-based trackers (e.g., AR Toolkit), Wi-Fi based trackers, magnetic trackers, and inertial trackers will be utilized in testing this prototype system. Application prototypes will be created to show the effectiveness of the proposed research. These prototypes will act as the test-beds and will provide feedback to the principles of the proposed discovery service.

CONCLUSION

The software framework outlined in this article is a promising approach to developing practical wide-area tracking systems that can utilize existing tracker infrastructures. The limitations of the framework are mainly due to the hardware. For example, one cannot foresee all the possible trackers and thus equip the tracked object accordingly ahead of time. In order to accommodate this, the framework assumes multiple, heterogeneous *classes* of tracking systems rather than instances of trackers. An example of tracker class might be a vision-based tracker. Thus, there may be many instances of such trackers, but one needs to equip the object with a standard fiducial marker.

REFERENCES

Azuma, R., Baillot, Y., Behringer, R., Feiner, S., Julier, S., & MacIntyre, B. (2001). Recent advances in augmented reality. *IEEE Computer Graphics & Applications, 21*(6), 34-47.

Bahl, P., & Padmanabhan, V. (2000). RADAR: An in-building RF-based user location and tracking system. *Proceedings of the IEEE Infocom 2000* (pp. 775-784). IEEE CS Press.

Balachandran, A., Voelker, G., & Bahl, P. (2003). Wireless hotspots: Current challenges and future directions. *Proceedings of the 1st ACM International Workshop on Wireless Mobile Applications and Services on WLAN Hotspots* (pp. 1-9). ACM Press.

Bauer, M., Bruegge, B., Klinker, G., MacWilliams, A., Reicher, T., Riss, S., Sandor, C., & Wagner, M. (2001) Design of a component-based augmented reality framework, In *Proceedings of the IEEE and ACM International Symposium on Augmented Reality (ISAR 2001)* (pp.45-54).

Bauer, M., Bruegge, B., Klinker, G., MacWilliams, A., Reicher, T., Sandor, C., & Wagner, M. (2002). An architecture concept for ubiquitous computing aware wearable computers. *Proceedings of the International Workshop on Smart Appliances and Wearable Computing.*

Beugnard, A., Jezequel, J., Plouzeau, N., & Watkins, D. (1999). Making components contract aware. *IEEE Computer, 32*(7), 38-45.

Borriello, G., Chalmers, M., LaMarca, A., & Nixon, P. (2005). Delivering real-world ubiquitous location systems. *Communications of the ACM, 48,* 36-41.

Breen, D., Whitaker, R., Rose, E., & Tuceryan, M. (1996). Interactive occlusion and automatic object placement for augmented reality. *Proceedings of the Eurographics '96 Conference* (pp. 11-22).

Brown, R., & Hwang, P. (1997). *Introduction to random signals and applied Kalman filtering* (3rd ed.). New York: John Wiley & Sons.

Carlson, N. A. (1990). Federated square root filter for decentralized parallel processes. *IEEE Transactions on Aerospace and Electronic Systems, 26*(3), 517-525.

Czarnecki, K., & Eisenecker, U. (2000). *Generative programming: Methods, tools, and applications.* Boston: Addison-Wesley.

Feiner, S., MacIntyre, B., Hoellerer, T., & Webster, T. (1997). A touring machine: Prototyping 3D mobile augmented reality systems for exploring the urban environment. *Proceedings of the International Symposium on Wearable Computers (ISWC'97)* (pp. 74-81).

Foxlin, E., & Durlach. N. (1994). An inertial head-orientation tracker with automatic drift compensation for use with HMDs. *Proceedings of the VRST '94 Conference* (pp. 159-173), Singapore.

Harter, A., Hopper, A., Steggles, P., Ward, A., & Webster, P. (2002). The anatomy of a context-aware application. *Wireless Networks, 8*(2-3), 187-197.

Hightower, J., & Borriello, G. (2001). Location systems for ubiquitous computing. *Computer, 34*(8), 57-66.

Huang, Z. (2003). *The uniframe system-level generative programming framework.* Unpublished MS thesis, Department of Computer and Information Science, Indiana University Purdue University Indianapolis, USA. Retrieved May 11, 2006, from http://www.cs.iupui.edu/uniFrame/

Kalman, R. E. (1960). A new approach to linear filtering and prediction problems. *Transactions of the ASME—Journal of Basic Engineering,* 35-45.

Koller, D., Klinker, G., Rose, E., Breen, D., Whitaker, R., & Tuceryan, M. (1997). Real-time vision-based camera tracking for augmented reality applications. *Proceedings of the Symposium on Virtual Reality Software and Technology* (pp. 87-94).

Neumann, U., & Cho, Y. (1996). A self-tracking augmented reality system. *Proceedings of the ACM Symposium on Virtual Reality and Applications* (pp. 109-115).

Newman, J., Ingram, D., & Hopper, A. (2001). Augmented reality in a wide area sentient environment. *Proceedings of the International Symposium on Augmented Reality* (pp. 77-86), New York. IEEE Press.

Newman, J., Wagner, M., Bauer, M., MacWilliams, A., Pintaric, T., Beyer, D., et al. (2004). Ubiquitous tracking for augmented reality. *Proceedings of the 3rd IEEE and ACM International Symposium on Mixed and Augmented Reality* (pp. 192-201).

Olson, A., Raje, R., Bryant, B., Burt, C., & Auguston, M. (2005). UniFrame: A unified framework for developing service-oriented, component-based, distributed software systems. In Z. Stojanovic & A. Dahanayake (Eds.), *Service oriented software system engineering: Challenges and practices* (Chapter IV, pp. 68-87). Hershey, PA: Idea Group Inc.

Raje, R. (2000). UMM: Unified Meta-object Model for open distributed systems. *Proceedings of the 4th IEEE International Conference on Algorithms and Architecture for Parallel Processing* (pp. 454-465). IEEE Press.

Reitmayr, G., & Schmalstieg, D. (2001). An open software architecture for virtual reality interaction. *Proceedings of the ACM Symposium on Virtual Reality Software and Technology* (pp. 47-54). ACM Press.

Schilit, B.N., LaMarca, A., Borriello, G., Griswold, W.G., McDonald, D., Lazowska, E., et al. (2003). Challenge: Ubiquitous location-aware computing and the "Place Lab" initiative. *Proceedings of the 1st ACM International Workshop on Wireless Mobile Applications and Services on WLAN Hotspots* (pp. 29-35). ACM Press.

Siram, N. (2002). *An architecture for discovery of heterogeneous software components.* Unpublished MS thesis, Department of Computer and Information Science, Indiana University Purdue University Indianapolis, USA. Retrieved May 11, 2006, from http://www.cs.iupui.edu/uniFrame/

State, A., Hirota, G., Chen, D., Garrett, W., & Livingston, M. (1996). Superior augmented reality registration by integrating landmark tracking and magnetic tracking. *Proceedings of SIGGRAPH '96* (pp. 429-438).

Tuceryan, M., Greer, D., Whitaker, R., Breen, D., Crampton, C., Rose, E., & Ahlers, K. (1995). Calibration requirements and procedures for a monitor-based augmented reality system. *IEEE Transactions on Visualization and Computer Graphics, 1*(3), 255-273.

Tuceryan, M., Genc, Y., & Navab, N. (2002). Single Point Active Alignment Method (SPAAM) for optical see-through (HMD) calibration for augmented reality. *Presence: Teleoperators and Virtual Environments, 11*(3), 259-276.

Ubisense. (2006). Retrieved May 3, 2006, from http://www.ubisense.net/

Welch, G., & Bishop, G. (2001). An introduction to the Kalman filter. *Proceedings of the Siggraph Course,* Los Angeles.

Welch, G., & Foxlin, E. (2002). Motion tracking: No silver bullet, but a respectable arsenal. *IEEE Computer Graphics and Applications, 22*(6), 24-38.

Whitaker, R., Crampton, C., Breen, D., Tuceryan, M., & Rose, E. (1995). Object calibration for augmented reality. *Proceedings of Eurographics '95* (pp. 15-27).

KEY TERMS

Augmented Reality: Superimposing information on the view of the physical world for the purposes of providing information.

Degrees-of-Freedom (DOF): For a tracker of a particular type, the number of independent dimensions of information obtained from the sensor hardware.

Kalman Filter: A linear estimation technique first proposed by Rudolph Kalman in 1960 that is extensively used in tracking and navigation applications.

Tracking: A hardware/software system that can provide the position and/or orientation of an object being tracked in real time.

UniFrame: A unifying framework that supports a seamless integration of distributed and heterogeneous components.

UniFrame Resource Discovery System (URDS): A system that provides an infrastructure for proactively discovering components deployed over a network.

Distributed Web GIS

Jihong Guan
Tongji University, China

Shuigeng Zhou
Fudan University, China

Jiaogen Zhou
Wuhan University, China

Fubao Zhu
Wuhan University, China

INTRODUCTION

The popularity of World Wide Web and the diversity of GISs on the Internet have led to an increasing number of geo-referenced information (GRI) sources that spread over the Internet. How to integrate the heterogeneous and autonomous GISs to facilitate GRI accessing, data sharing, and interoperability is still a big challenge. Furthermore, the rapidly emerging mobile Internet and constantly increasing number of wireless subscribers bring new opportunities to geographic information services. Putting the Internet GIS in the palm will enable us to access geographic information with personal devices anytime and anywhere.

In the past decade, a lot of research has been done on designing interoperable systems in which collections of autonomous and heterogeneous GISs can cooperate to carry out query tasks. However, as far as system architecture is concerned, current solutions for integration of distributed GIS applications are mainly based on either C/S or B/S mode. The inherent limitations of these modes—for example, requiring a proper bandwidth, high-quality and stable network connection, less supporting of group awareness, and high-level cooperation—make them incompetent to fulfill various requirements of a dynamic, complicated, and distributed network computing environment, especially the mobile network environment, where the wireless communication networks have low bandwidth, frequent disconnections, and long latency, and the mobile devices (PDAs or mobile phones) have limited power, memory, computational power, and displaying capability. Such a situation calls for a new framework to support globally geographic information accessing and sharing in the (mobile) Internet environment.

The mobile agent is a recently developed computing paradigm that offers a full-featured infrastructure for development and management of network-efficient applications. Mobile agents are processes dispatched from one host to another during its execution on behalf of its owner or creator to accomplish a specified task. Agent-based computing can benefit Internet (especially mobile Internet) applications by *providing asynchronous task execution and more dynamics, supporting flexible and extensible cooperation, reducing communication bandwidth, enhancing real-time abilities and a higher degree of robustness, enabling off-line processing and disconnected operation.* Thus it is natural to introduce mobile agents into accessing and sharing distributed geographic information in a (mobile) Internet environment.

This article presents the MADGIS (Mobile Agent-based Distributed Geographic Information System) project, which aims at integrating distributed Web GIS applications by using mobile agent technologies to overcome the limitations of traditional distributed computing paradigms in a (mobile) Internet context.

MADGIS FRAMEWORK

The MADGIS system consists of client sites (or clients), sever sites (or servers), a (mobile) Internet or intranet connecting these sites, and mobile agents roaming on the Internet/intranet for retrieving information on behalf of the clients. Figure 1 is an overview of MADGIS.

In MADGIS, a client site refers to a client machine, which can be a desktop, a laptop personal computer, a PDA, or a mobile phone used for query submission and results presentation. A server site is also a MADGIS server that provides spatial information services for local or remote requests. A user submits a query from a client machine to a server *via* Web browser. The query is analyzed and optimized by the server, from which one or multiple mobile agents are created and dispatched to accomplish the query task cooperatively. Each mobile agent along with its sub-task travels from one remote server to another to gather the related information. Retrieved information is then taken back to the original site after the mobile agent finishes its mission. All returned

Figure 1. MADGIS overview

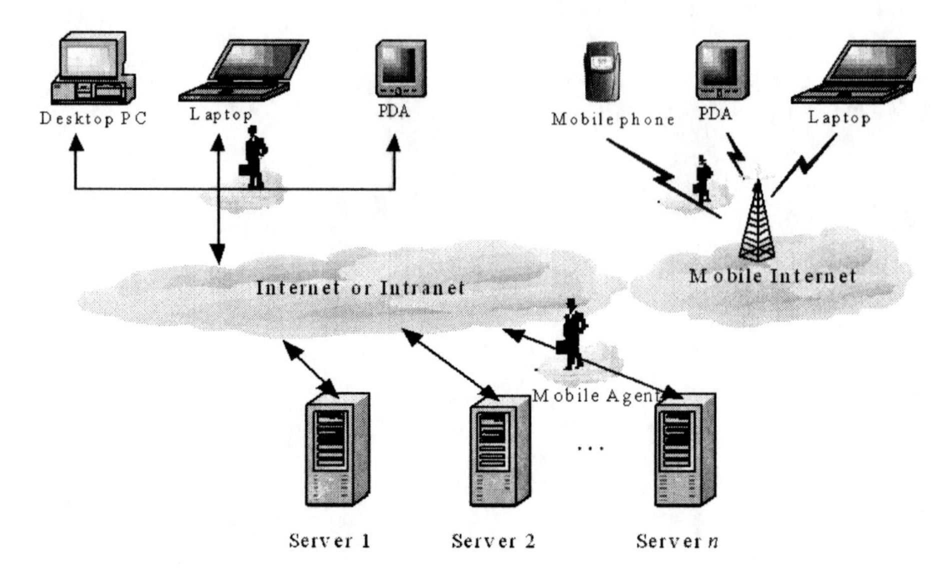

information is further merged there and presented to the user. The servers also provide a docking facility for mobile agents in case they cannot travel back to the destinations promptly due to network problems.

The Client Site of MADGIS

A user can access any GISs within the MADGIS system via a client or a local server. First, a user should log into one server in the system. Then the server returns a Web (HTML) page to the client, in which there is a Java Applet termed as Client-Applet composed of one mobile agent environment (MAE), one stationary agent, and one mobile agent. The client-applet is executed at the client site to establish the MAE for the client and to start the stationary agent encoded in the Client-Applet. We call this stationary agent "client-agent", which is responsible for two tasks:

1. To obtain the data sources description information (DSDI) from the visited server, which includes all data sources' metadata (e.g., names, URLs, and schema of each data source). Users submit queries to or browse the MADGIS system according to the DSDI. The client-agent gets DSDI from the stationary agent of the visited server. At each server, DSDI is maintained by the local stationary agent. Besides responsibility for maintaining local DSDI, the stationary agent should also send messages to other servers in the system to notify of updates of DSDI, so as to keep the global DSDI updating simultaneously.
2. To create the query interface (QI) in the Web browser with which the user submits queries and gets retrieved data.

Thus when the query environment is set up at the client site, what a user can see is only the QI, while client-agent, mobile agent, and its execution environment are at the back-end. The user starts his or her query operations via QI, and the server accessed will take charge of query processing and mobile agent manipulation. Typically, a whole query session consists of the following steps:

1. At a client site, a user visits one server *via* Web browser by specifying the server's URL.
2. The accessed server returns a Web page including a client-applet.
3. The client-applet is executed at the client to establish MAE and to start the stationary client-agent.
4. The client-agent obtains the DSDI from the server and creates the QI for the user.
5. The user constructs his or her query and submits it via the QI to the server.
6. When the client-agent gets the user's query, it initiates a mobile agent to take over the query task.
7. The mobile agent with user's query task migrates to the server to which the client first visited for further query processing.
8. After the query task is completed at the server, the mobile agent moves back to the client and returns the results to the user via the browser.

Above, steps 1 to 4 are necessary for a client to access MADGIS. After that, the client can submit queries that are answered by following steps 5 to 8 repeatedly. The process described above and the interaction among client, server, and mobile agent are demonstrated in Figure 2.

Figure 2. The client site of MADGIS

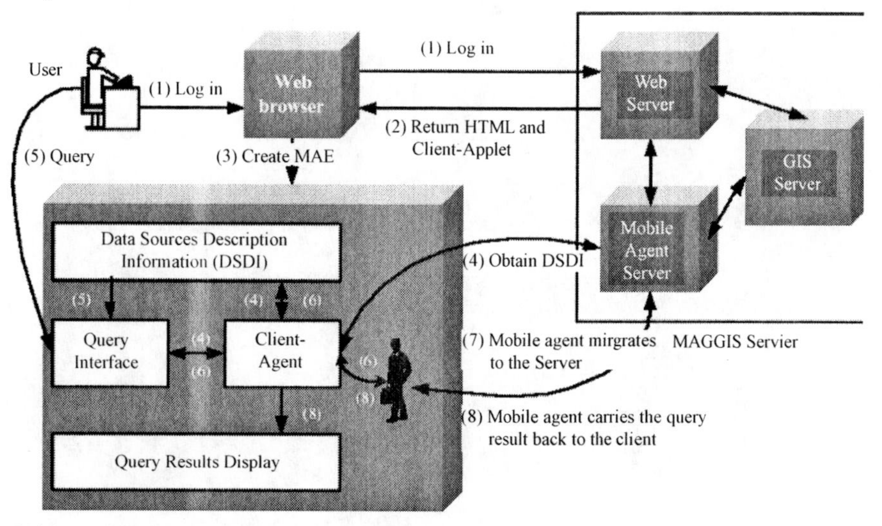

The Server Site of MADGIS

A server in MADGIS consists of at least three main components: Web server (or simply WSever), GIS server (GSever), and mobile agent server (MAServer). WServer is the interface for a user to connect to a MADGIS server and responsible for providing client-applet to the user. GServer is composed of a GIS database and a spatial query processing engine that provides support for local query processing. MAServer is the key component of a MADGIS server; it provides the facilities to support mobile agents to carry out query processing. Figure 3 illustrates the architecture of a typical MADGIS server. In what follows, we discuss MAServer in detail.

A MAServer contains four stationary agents: local services agent (LSA), query optimization agent (QOA), querying and wrapping agent (QWA), and mediation and transformation agent (MTA).

LSA has the following duties: (a) providing the system's DSDI for clients who log into the server; (b) maintaining local DSDI and notifying local DSDI updates to remote servers; (c) maintaining the statistical information of local data resources; and (d) sensing and detecting the status of network traffic between the local server and remote servers or clients.

QOA is responsible for analyzing and optimizing the user's query. A user's query first performs grammar-check and is parsed by the Client-Agent; it is then taken to the server site by a mobile agent (termed main query agent). The QOA at the server site takes charge of the query's optimization and determines query strategy accordingly. A user query may be split into several sub-queries, each of which is related to one server's data source.

QWA is responsible for fulfilling the sub-query task related to local data and wrapping the query results into a standard GML document. When a user query involves data of other data sources or sites, data retrieved from multiple sites has to be merged, which is done by MTA. Then the merged query results will be further transformed into a SVG document and taken back to the client by mobile agent.

Figure 3 also illustrates the procedure of how a query task is completed by using mobile agent. The procedure includes the following steps:

1. The query task is shifted to QOA for optimization to reduce computational cost and network transmission volume incurred by query processing. After optimization, a query plan is created, which includes a set of sub-queries, sites on which the sub-queries are executed respectively, and the execution sequences of these sub-queries. QOA returns the result of query optimization to the main query agent. The main query agent then decides whether additional mobile agents are requested to carry out sub-queries processing.
2. If the query involves only local data, the main query agent will go on to finish the query task itself without necessity of spawning other mobile agents. The main query agent assigns the query task to QWA, who is in charge of retrieving data and wrapping the results.
3. If the query involves data of multiple server sites, and the sub-queries are requested to evaluate in sequence, then the main query agent or a mobile agent created by the main query agent will take over the query task. The mobile agent will be dispatched out according to its itinerary arranged previously. Since each sub-query involves one remote site, the sub-queries will

Figure 3. The server site of MADGIS

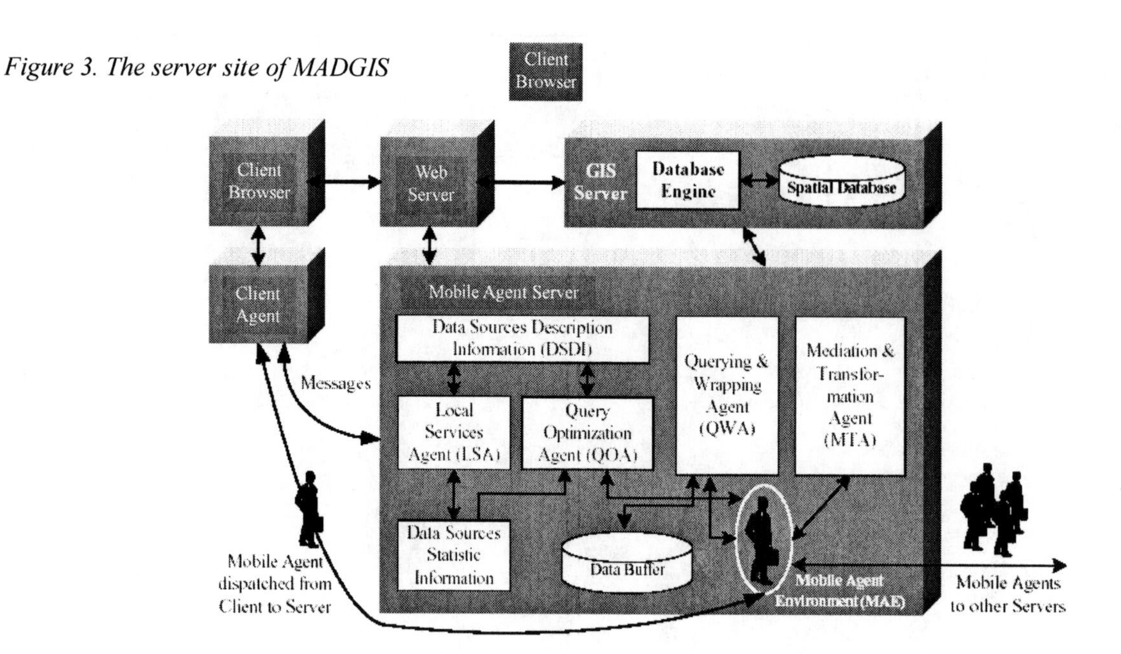

be accomplished one by one. After the last sub-query is finished, the dispatched mobile agent will go back to the client site directly or to its original server site.

4. Otherwise, if the query involves data of multiple sites, and the sub-queries are requested to evaluate in parallel, and the local site and other remote sites corresponding to the sub-queries are connected, then multiple mobile agents will be cloned by the main query agent to execute the sub-queries in parallel so as to gain better efficiency. The main query agent may join the mobile agents group to finish the query task or just take the role of coordinator of the multiple agents.
5. Query results obtained from all related sites are brought back to the original server, then MTA at the server site will do data integration and transformation, and one SVG document as the final result of the query will be created.
6. The main query agent carries the final query result in SVG format to the client, which is presented to the user via browser.

Another function of the MADGIS server is to provide the docking mechanism for mobile agents when connection between the current site and the destination site of agent migration is disrupted. LSA will take the role of deactivation and activation of mobile agents when such a situation happens.

Mobile Agent Environment

The mobile agent environment (MAE) exists at both the client and the server sites. It provides an environment for mobile agents to create, execute, dispatch, and migrate. Besides the mobile agents, MAE is composed of the following functional modules: mobile agent manager (MAM), mobile agent transportation (MAT), mobile agent naming (MAN), mobile agent communication (MAC), and mobile agent security (MAS).

MAM, the heart of MAE, is responsible for all kinds of management of mobile agents. It mainly provides a full-fledged environment for agent creating and executing, basic functions to make the mobile agent migrate precisely to its destination, functions for agent scheduling locally, and support for an agent's remote management. MAT controls the transferring of mobile agents—that is, sending and receiving mobile agents to and from remote sites. MAN manages a mobile agents' naming service, which provides the mechanism of tracing mobile agents. MAC serves mobile agent communication, which serves as the protocol of communication, collaboration, and events transmission among agents. MAS provides a two-facet security mechanism. On the one hand, it is responsible for distinguishing users and authenticating their mobile agents in order to protect server resources from being illegally accessed or even maliciously attacked. On the other hand, it ensures mobile agents not be tampered by malicious hosts or other agents.

MADGIS PROTOTYPE

We use Aglet Software Development Kit (ASDK) 2.0 (http://www.trl.ibm.co.jp/aglets/), JDK 1.3, and Geotools 0.8.0 for implementing the MADGIS prototype. Geographic information of states, cities, rivers, roads, and lakes of the United States, Canada, and Mexico from ArcView GIS 3.2 samples and Guangzhou (a city in China) Agricultural Information Systems are processed and stored on five computers, which take the roles of servers. Users can access MADGIS transpar-

Figure 4. Query result 1 of the prototype

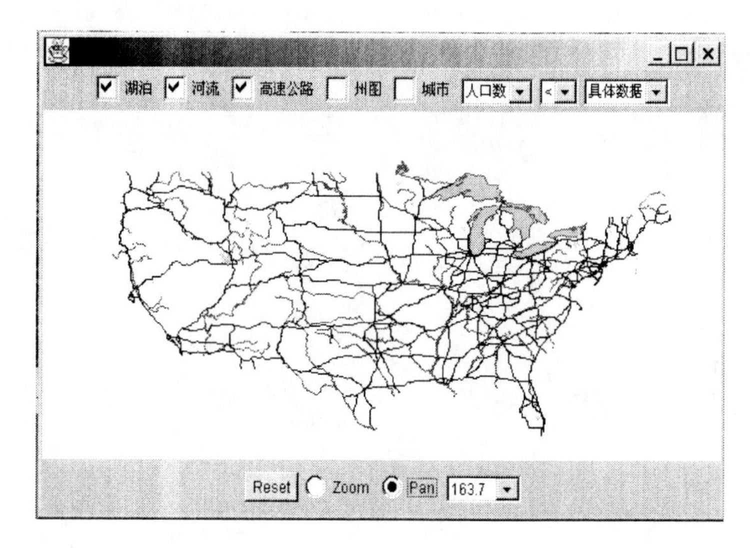

Figure 5. Query result 2 of the prototype

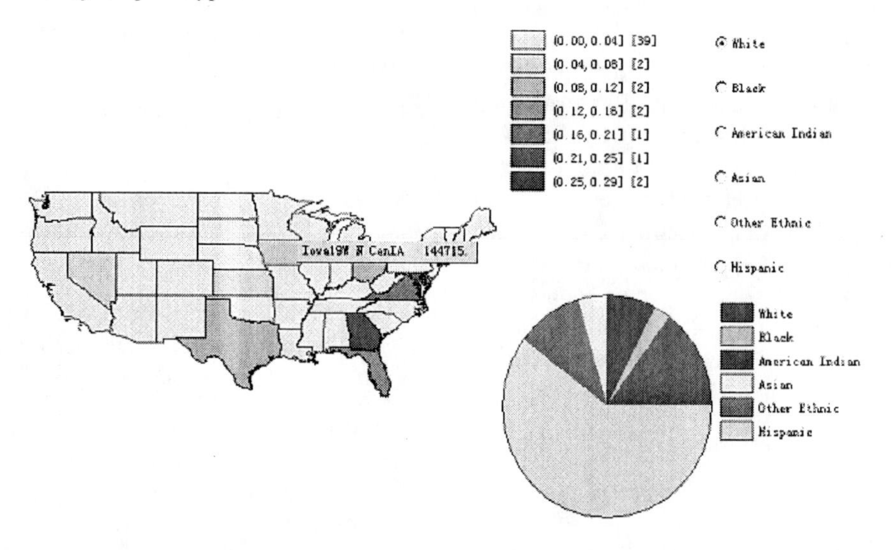

ently from any client machine connected to the Internet for geographic information stored in the system. Figures 4 and 5 show the interface of the prototype and the query results for roads, rivers, and lakes, and population distribution.

RELATED WORK

We give a brief survey on related work of distributed GISs and mobile agent areas in this part.

For distributed GISs, quite a lot research has been done in the past decade to design distributed interoperable systems (Leclercq, Benslimane, & Yetongnon, 1997; Fonseca & Egenhofer, 1999; Abel, Ooi, Tan, & Tan, 1998; Wang, 2000), which can be classified into three levels of integration: platform level, syntactical level, and application level (Fonseca & Egenhofer, 1999). Platform-level integration is concerned with hardware, operating systems, and network protocols, providing support for transferring of flat structure files between systems. Syntactic-level integration provides functionalities and tools for defining persistent and uniform views over multiple heterogeneous spatial data sources. Application-level integration aims at defining seamless system integration. However, these existing approaches and systems are mainly based on a connection-oriented mechanism (C/S and B/S modes) not suitable for efficient accessing, sharing, and integration of distributed systems in a (mobile) Internet environment.

For mobile agent-based applications, as a new and promising distributed computing paradigm, mobile agents have been employed in many applications, which include information retrieval (Brewington et al., 1999), Web search (Kato, 1999), electronic commerce (Dasgupta, 1999), and personalization Web services (Samaras & Panayioyou, 2002). Tsou and Buttenfield (2000) proposed to use agents to provide distributed GIS services; however, agents referred to in this article are *static* agents, which are quite different from the *mobile* agents employed in this article. A static agent is just a program with a certain autonomy or even intelligence, but it cannot migrate from one site to another to carry out missions on behalf of its own.

CONCLUSION

The Internet has greatly changed the ways of geographic data accessing, sharing, and disseminating. The emergence of mobile agent technologies brings new opportunities and challenges to Web-based geographic information services. To fulfill the requirements of distributed GIS accessing and sharing under a (mobile) Internet environment, we propose the MADGIS framework for accessing distributed Web-based GISs by using mobile agent technologies. By introducing mobile agents to carry out query tasks, MADGIS can save significant bandwidth by moving locally to the resources needed; carry the code to retrieve remote information, without needing the remote availability of a specific server; proceed without continuous network connections, for interacting entities can be moved to the same site when connections are available, and can then interact without requiring further network connections; and work with mobile computing systems (e.g., laptop, palmtop).

ACKNOWLEDGMENTS

This work was supported by grants numbered 60573183 and 60373019 from NSFC, grant No. 20045006071-16 from the Chenguang Program of Wuhan Municipality, grant No. WKL(04)0303 from the Open Researches Fund Program of LIESMARS, and the Shuguang Scholar Program of Shanghai Education Development Foundation.

REFERENCES

Abel, D., Ooi, B., Tan, K., & Tan, S.H. (1998). Towards integrated geographical information geoprocessing. *International Journal of Geographical Information Science,* 353-371.

Brewington, B., Gray, R., Moizumi, K., Kotz, D., Cybenko, G., & Rus, D. (1999). Mobile agents in distributed information retrieval. In M. Klusch (Ed.), *Intelligent Information Agents* (pp. 355-395). Berlin: Springer-Verlag.

Dasgupta, P. (1999). MagNET: Mobile agents for networked electronic trading. *IEEE Transactions on Knowledge and Data Engineering, 11*(4), 509-525.

Fonseca, F., & Egenhofer, M. (1999). Ontology-driven geographic information systems. *Proceedings of the 7th ACM Symposium on Advances in Geographic Information Systems* (pp. 14-19), Kansas City.

Geo-Agent. (n.d.). Retrieved from http://map.sdsu.edu/geo-agent/

Kato, K. (1999). An approach to mobile software robots for the WWW. *IEEE Transactions on Knowledge and Data Engineering, 11*(4), 526-548.

Leclercq, E., Benslimane, D., & Yetongnon, K. (1997). Amun: An object-oriented model for cooperative spatial information systems. *Proceedings of the IEEE Knowledge and Data Engineering Exchange Workshop* (pp. 73-80), Newport Beach, CA.

Open GIS Consortium (OGC). (n.d.). Retrieved from http://www.opengis.org/

Sahai, A. (n.d.). *Intelligent agents for a mobile network manager (MNM).* Retrieved from http://www.irisa.fr/solidor/doc/ps97/

Samaras, G., & Panayioyou, C. (2002). A flexible personalization architecture for wireless Internet based on mobile agents. *Proceedings of ADBIS 2002* (LNCS 2435, pp. 120-134).

Tsou, M.H., & Buttenfield, B.P. (2000). Agent-based mechanisms for distributing geographic information services on the Internet. *Proceedings of the 1st International Conference on Geographic Information Science,* Savannah, GA.

Wang, F. (2000). A distributed geographic information system on the Common Object Request Broker Architecture (CORBA). *Geoinformatica, 4*(1), 89-115.

World Wide Web Consortium. (1999, August). *Scalable Vector Graphics (SVG) 1.0 specification* (W3C work draft). Retrieved from http://www.w3.org/tr/svg

XSLT. (n.d.). Retrieved from http://www.w3.org/XSLT

KEY TERMS

Agent: Takes the roles of a representative and acts on behalf of other persons or organizations; is often a software routine that waits in the background and performs an action when a specified event occurs.

Client: A computer or program that can download files for manipulation, run applications, or request application-based services from a file server.

Data Sharing: The ability to share the same data resource with multiple applications or users. Implies that the data are stored in one or more servers in the network and that there is some software locking mechanism that prevents the same set of data from being changed by two people at the same time.

Geographic Information System (GIS): A computer application used to store, view, and analyze geographical information, especially maps. Often called "mapping software," it links attributes and characteristics of an area to its geographic location.

MADGIS: Mobile Agent-based Geographic Information System.

Mobile Agent: A composition of computer software and data which is able to migrate (move) from one computer to another autonomously and continue its execution on the destination computer with the features of autonomy, social ability, learning, and most important, mobility.

Mobile Agent Environment (MAE): Provides an environment for mobile agents to create, execute, dispatch, and migrate.

Server: A computer that processes requests for HTML and other documents that are components of Web pages. The term "server" may refer to both the hardware and software (the entire computer system) or just the software that performs the service.

Web GIS: Short for Web-based GIS; a geographic information system that deals with spatial information and provides it Web users via Web browsers.

Dynamic Pricing Based on Net Cost for Mobile Content Services

Nopparat Srikhutkhao
Kasetsart University, Thailand

Sukumal Kitisin
Kasetsart University, Thailand

INTRODUCTION

In the past few years, the mobile phone's performance has increased rapidly. According to IDC's Worldwide Mobile Phone 2004-2008 Forecast and Analysis, sales of 2.5G mobile phones will drive market growth for the next several years, with sales of 3G mobile phones finally surpassing the 100 million annual unit mark in 2007. Future mobile phones can support more than 20,000 colors. With the advancements in functionality and performance of mobile phones, users will use them for all sorts of activities, and that will increase mobile content service requests. Currently, the pricing of mobile content service is up to each provider; typically they implement a fixed price called a market price because the providers do not have a formula to estimate the price according to the actual cost of their services. This article proposes a dynamic pricing model based on net cost for mobile content services.

BACKGROUND

A mobile phone today can support various format data causing mobile content service popularity among all mobile phone users. They can request a music VDO clip, a song, or a mobile phone game program. The price of each mobile content service differs for each different format of data. For example, the price of a true-tone ring tone is 35 baht (Sanook.com, 2005), while a Java game download costs 40 baht (Siam2you, 2005).

Conventionally, an operator set a fixed market price for each mobile content service. The prices can vary from operator to operator. The pricing has not been calculated based on the net cost for the requested service. Therefore, the set price can be lower or much higher than the actual cost. To come up with a way for a provider to be able to set a mobile content service price based on its actual cost, the provider must be able to quantify its actual cost for service. This article presents mobile content service interaction models and formulas for estimating the actual cost of a mobile content service; a provider can refer to these models and formulas when pricing its services.

Data Formats

The previous section discusses improving the performance of mobile devices and the variety of content available. We can classify mobile content service into four types: audio, image, video, and application (ClearSky Mobile Media, 2005). Users can request an audio clip and use it as their ring tone. They can leave voice messages for each other or download an mp3 song for their entertainment (Nokia, 2005; Sony Ericsson, 2005; Samsung, 2005). The audio content can be of three sub-types: monophonic (Sonic Spot, 2005), polyphonic (Cakewalk, 2005), and true tone. Image format can be either static or dynamic/animation. Users can request a music VDO clip and play it on their mobile phones, and apparently, a few companies have started to provide NetTV on mobile phones as well. Lastly, an example of application content users widely request could be a Java game application.

Parties in Mobile Content Services

Providing mobile content services involves many parties. We consider the following five participants (Bratsberg & Wasenden, 2004; Andreas, 2001). The first party is a user requesting mobile content services. The second party is a mobile operator (MO), which is the owner of a mobile phone service frequency. When a user requests mobile content service, the user will send a request to his or her content provider through the MO's network. The third party is a content provider (CP), which is an organization to serve mobile contents. The MO may or may not have license on the contents. The fourth party is the content owner (CO), which could be a person or an organization that has authorization for legal distribution of the mobile contents. And the last party is a content aggregator (CA), a middleman between a user and a content provider. The CA can help increase the channel to serve mobile content services.

Figure 1. Mobile content service model 1

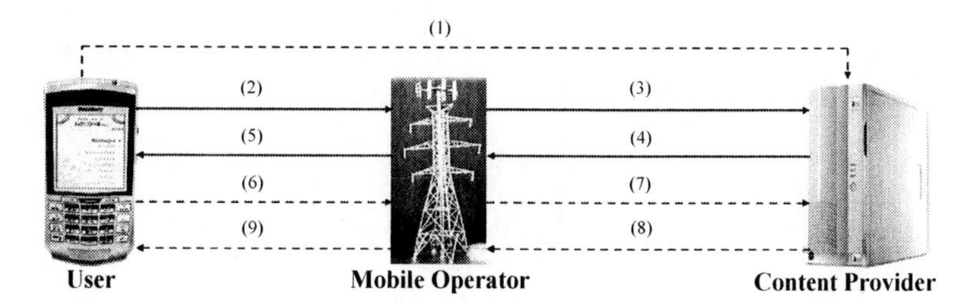

Figure 2. Mobile content service model 2

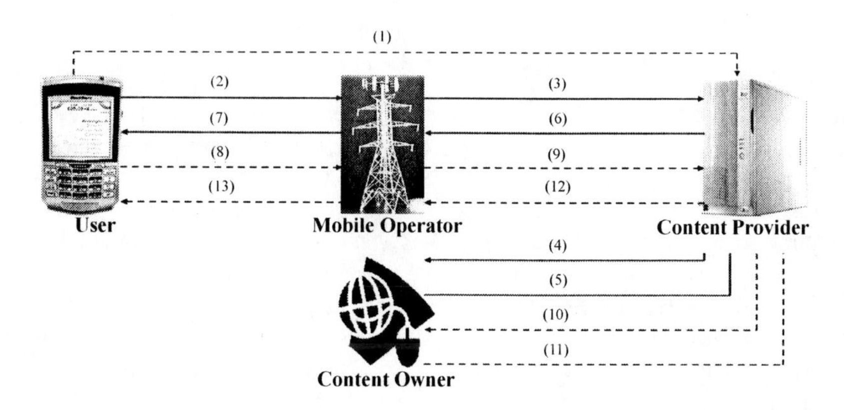

Request and Response Formats

When a user wants to use a mobile content service, he or she makes a request for the desired content from a CA or CP. Then the CA or CP responds to the user with the requested content. Requests and responses can be of the following four types: a Web request through a Web page, a short message service (SMS), an interactive voice responder (IVR), or a WAP request via a mobile internet WAP page. When the CA or CP responds successfully, the response can be sent using one of these four formats: a short message service (SMS), a smart message, a WAP push format, and a multimedia message service (MMS).

Mobile Content Service Models

We categorized all mobile content services into the following interaction models based on involved parties and content providing methods.

Model 1: Parties involved are user, MO, and CP. A user requests content from a CP by using an SMS, IVR, WAP, or Web request. For any request format except a Web request, the request is sent to the CP through the MO's network. For a Web request format, the request is transferred to the CP directly. After the CP processes the request, the CP will reply to the user with the requested content information in the form of a WAP push, a WAP URL, or a bookmark. For a monophonic ring tone request, the CP will send content information in a smart message format. Other content formats can be replied to with an SMS, a WAP push, or an MMS. The workflow of model 1, as shown in Figure 1, is:

1. User requests content via a Web page.
2. User requests content via SMS, IVR, or WAP page.
3. MO forwards the request to CP.
4. CP sends the content information to user through MO.
5. MO forwards the content information from CP to user.
6. User connects to WAP page or open WAP push for retrieving the content file through MO.
7. MO redirects the file request to CP.
8. CP sends the content file to the user through network of MO.
9. MO transfers content file to the user.

Model 2: Parties involved are user, MO, CP, and CO, with the CP as the content file sender. A user sends a request to a CP in SMS, IVR, WAP or Web format. The CP, after receiving the request, sends this request to a CO for content information. The CO sends content information to the user via the CP in smart message format, SMS (URL for retrieving

Figure 3. Mobile content service model 3

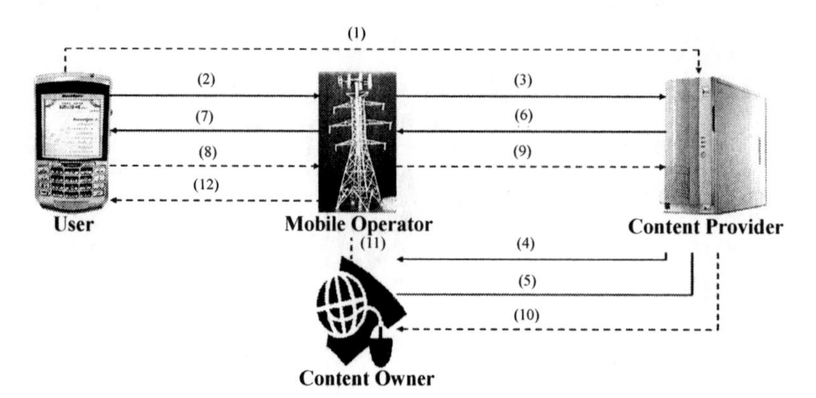

content file), or a WAP push. After that, the CP forwards the information to user. The workflow of model 2, as shown in Figure 2, is as follows. Steps 1-3 are the same as in model 1. In step 4, the CP forwards the request to the CO. In step 5, the CO sends the content information to the user via the CP. Steps 6-9 of this model are the same as steps 4-7 of model 1. In step 10, the CP forwards the request to the CO. In step 11, the CO sends the content file to the user via the CP. In step 12, the CP sends the content file to the user through the network of the MO. In step 13, the MO transfers the content file to the user.

Model 3: Parties involved are user, MO, CP, and CO, with the CO as the content file sender. The CO has permission to distribute the content files. The CP is a middleman between the CO and the user. The user requests content from the CP, which then sends the request to the CO. The CO processes the request and sends the content file directly to the user. The workflow for model 3 is: steps 1-10 are the same as steps 1-10 of model 2. In step 11, the CO sends the content file to the user through the network of the MO. In step 12, the MO transfers the content file to the user.

Model 4: Parties are user, MO, CA, CP, and CO, with the CA as the content file sender. For model 4, there is a middleman between the user and the CP. When the user requests a service, the user sends a request to the CA. Then the request is forwarded to the CP and the CO. After the CO processes the request, the content file will be sent to the user via the CP and the CA. The workflow of model 4 is as follows:

1. User requests content via a Web page.
2. User requests content via SMS, IVR, or WAP page.
3. MO forwards request to CA.
4. CA forwards request to CP.
5. CP forwards request to CO.

6-9. CO sends content information to user via CP, CA, and MO.

10-13. User connects to WAP page or open WAP push for retrieving the content file through MO, CA, and CP.

14-17. CO sends the content file to user via CP, CA, and MO.

Model 5: Parties involved are user, MO, CA, CP, and CO, with the CO as the content file sender. Methods for requesting content in this model are the same as those of model 4. They can be via an SMS, IVR, or WAP request. Requests will be sent via the network of the MO. Another

Figure 4. Mobile content service model 4

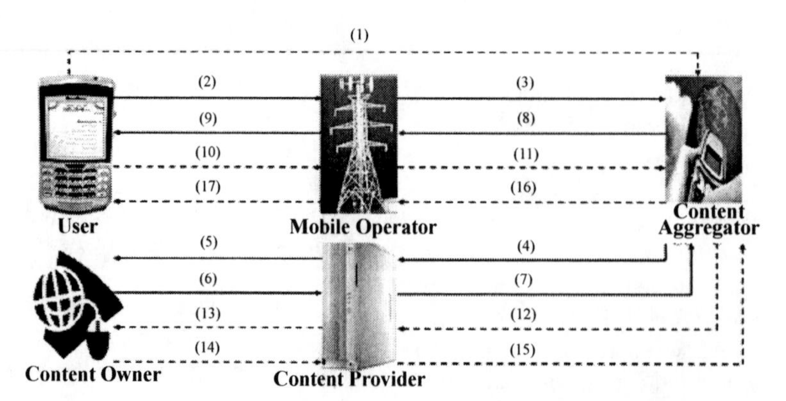

Figure 5. Mobile content service model 5

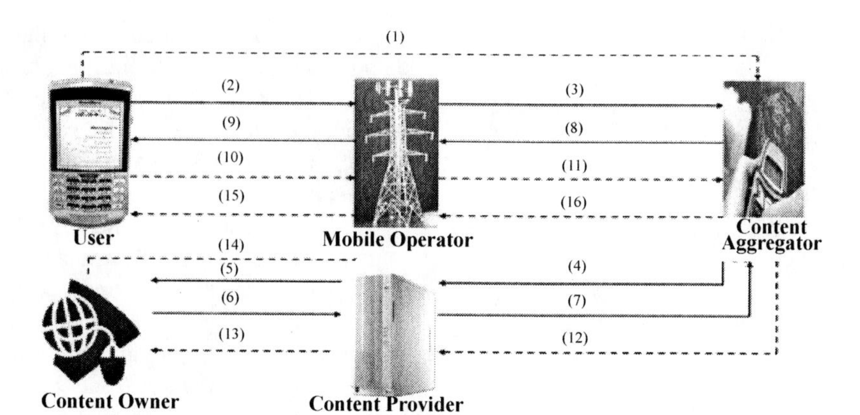

option is that a request can be sent to the CA directly. When the CO finishes processing the request, the CP then sends the content file directly to the user. The workflow of model 5 is as follows. Step 1-13 are the same as those in model 4. In step 14, the CO sends the content file to the user through the network of the MO. In step 15, the MO transfers the content file to the user.

Cost of Mobile Content Service

To be able to calculate the actual cost of a service, the providers must know the actual cost of providing the service. Two factors contributing to the cost of providing a mobile content service are the cost of software or content for the content service and the operational cost. Each party has a different operational cost and pays a different content fee or has a different revenue sharing model for a mobile content service (Smorodinsky, 2002; Kivisaari & Luukkainen, 2003; Stiller, Reichl, & Leinen, 2001). Operating costs for the MO are mainly the bandwidth cost for sending the content to the user. For the CP, the operating costs can be the cost of the bandwidth, the operation cost, the revenue sharing or fee for the MO, and the revenue sharing or fee for the CO. For the CO, its operating costs are from the cost of the bandwidth and the operation cost. And the CA's operating costs come from the cost of the bandwidth, the operation cost, the revenue sharing or fee for the MO, and the revenue sharing or fee for the CP.

Formula for Calculating an Actual Cost

Formulas depend on the mobile content service interaction models, the format of the mobile content, and its transfer venues. Parameters used in the formulas are as follows:

- S refers to the software value or value of a content file.
- I_A refers to the operation cost of CA.
- I_P refers to the operation cost of CP.
- I_O refers to the operation cost of CO.
- I refers to the total operation cost.
- B refers to content file size (Bit).
- D_M refers to bandwidth cost per bit for MO.
- D_A refers to bandwidth cost per bit for CA.
- D_p refers to bandwidth cost per bit of CP.
- D_O refers to bandwidth cost per bit of CO.
- A refers to bandwidth cost for CA.
- P refers to bandwidth cost of CP.
- O refers to bandwidth cost of CO.
- M refers to bandwidth cost of MO.
- C refers to the sum of bandwidth cost for CP and MO.
- R_1 refers to revenue sharing for MO.
- R_{2_1} refers to content fee for CO.
- R_{2_2} refers to revenue sharing for CO.
- R_{3_1} refers to fee for CP.
- R_{3_2} refers to revenue sharing for CP.
- W refers to the sum of dynamic costs before calculation revenue sharing and content fee for CO.
- E refers to the sum of dynamic cost before calculation revenue sharing and content fee for CP.
- T refers to dynamic cost before calculate revenue sharing for MO.
- N refers to the actual cost.

Formula 1: for model 1:

$I_P = I$, $B^* D_p = P$, $B^* D_M = M$, $P+M = C$, $S+I+C = T$, $T/(1-R_1) = N$

Formula 2: for model 2:

$I_P + I_O = I$, $B^* D_O = O$, $B^* D_p = P$, $B^* D_M = M$, $(O+P+M) = C$,

$W + R_{2\_1} = T$ *OR* $W / (1 - R_{2\_2}) = T$, $T/(1- R_1) = N$

Formula 3: for model 3:
$I_P + I_O = I$, $B^* D_O = O$, $B^* D_M = M$, $O+M = C$, $S+I+C = W$,

$W + R_{2\_1} = T$ *OR* $W / (1 - R_{2\_2}) = T$, $T/(1- R_1) = N$

Formula 4: for model 4:

$I_A + I_P + I_O = I$, $B^* D_O = O$, $B^* D_p = P$, $B^* D_A = A$, $B^* D_M = M$, $(O+P+A+M) = C$

$S+I+C = W$, $W + R_{2\_1} = E$ *OR* $W /(1- R_{2\_2}) = E$, $E + R_{3\_1} = T$ *OR* $E / (1- R_{3\_2}) = T$

$T / (1- R_1) = N$

Formula 5: for model 5:

$I_A + I_P + I_O = I$, $B^* D_O = O$, $B^* D_M = M$, $O+M = C$, $S+I+C = W$,

$W + R_{2\_1} = E$ *OR* $W /(1- R_{2\_2}) = E$, $E + R_{3\_1} = T$ *OR* $E / (1- R_{3\_2}) = T$

$T / (1- R_1) = N$

RESULTS AND ANALYSIS

We analyzed our actual cost formulas presented above by doing experiments based on three different types of transmitting channels: ADSL, leased line, and satellite. Cost for transmitting file content is determined by an average rate from ISPs in Thailand (True Internet, 2005; LOXINFO, 2005; Internet KSC, 2005; Ji-NET, 2005; INET, 2005). Each party uses the same sending channel. For the operation cost, we randomly selected a cost. The random method is Gaussians, with a base cost of 0.8262 and deviation of 0.14711. The software/content cost is a randomly selected value ranging from 0.1058 to 2.1160 baht. The sending channel of our results is satellite. Figures 6 and 7 show the average actual costs of true tone. The average operation costs for CA, CP, and CO is 0.829624 baht. Figures 8 and 9 shows the monophonic actual cost. And the average operation cost for CA, CP, and CO is 0.821213 baht.

Figure 6 shows average costs of true tone content for all service models and the market price. The average costs of models 1-5 are 1.66, 6.66, 6.64, 20.02, and 19.95 respectively. The maximum actual costs are less than market about 0.6 times. Thus, we found the market price for true tone content overpriced.

Figure 7 shows the probability of the customer being willing to pay the cost price of true tone content. For Figure 7, we found customers almost willing to pay the cost price of model 1 and customers willing to pay about 60% of the market price.

Figure 8 shows the average costs for a monophonic through satellite compared with the market price. From this figure, we found the market price is more than the actual

Figure 6. Average actual cost for a true tone content service

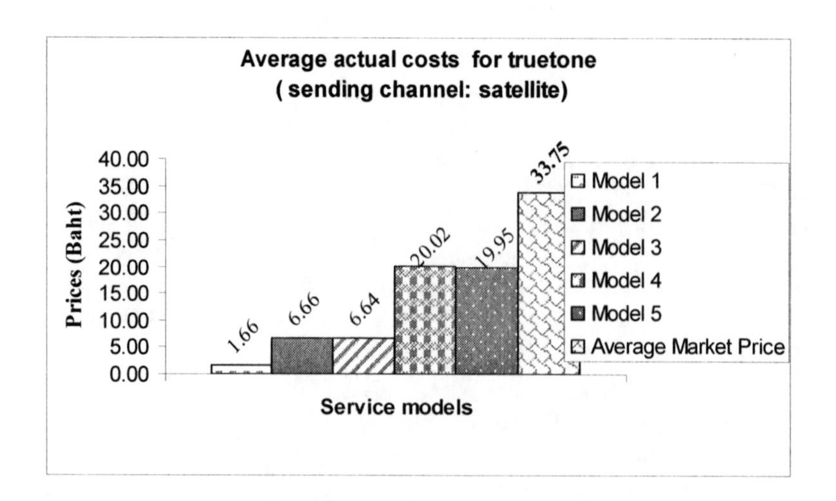

Figure 7. Probability of customer willing to pay the cost price of true tone content

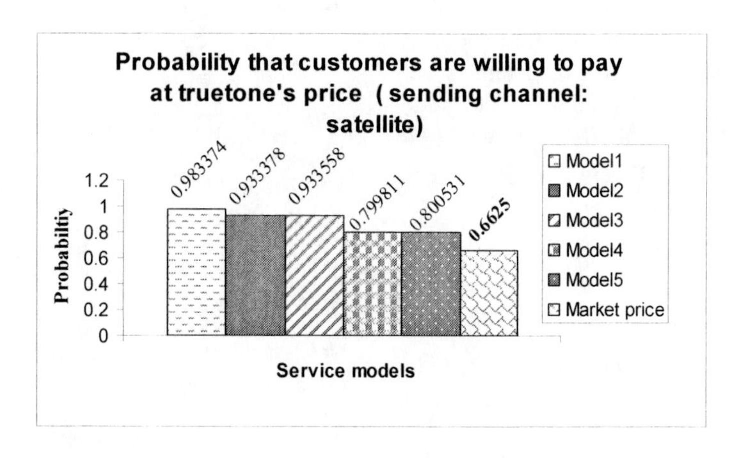

Figure 8. Average actual cost for a monophonic

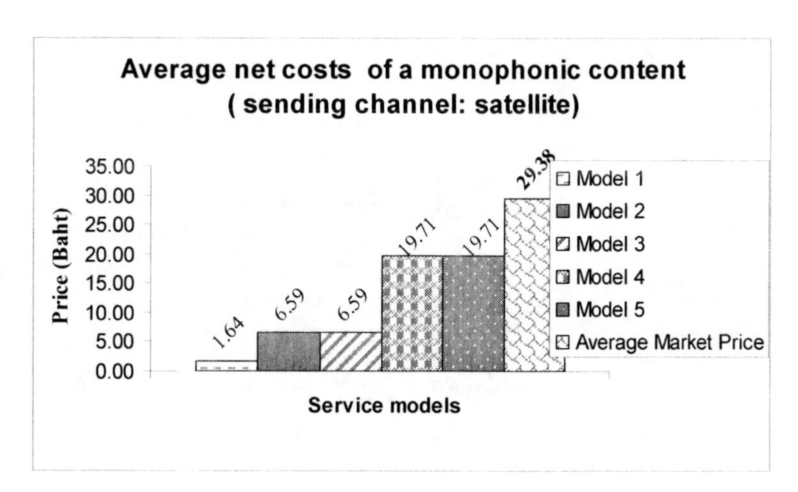

Figure 9. A probability that customers are willing to pay for monophonic content

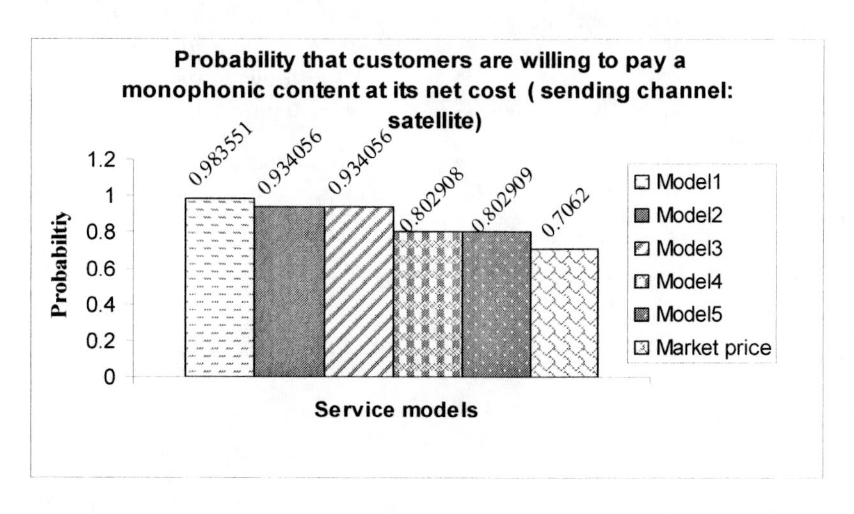

costs for all models. Market prices are very expensive. It is about 18 times more than model 1's actual cost.

Figure 9 shows the probability that the customers are willing to pay for actual costs. We found customers will gladly pay the cost prices for model 1, model 2, and model 3. Also, we found the probability of customers willing to pay market price is less than the probability of model 1 by nearly 30%.

SUMMARY

From our preliminary results using our formulas, the size of file, number of parties, and the sender affect the calculation of the actual cost. We presented an alternative method in pricing the mobile content services based on the actual cost. The method allows a provider to dynamically set its service price accordingly.

REFERENCES

Andreas S. (2001). *The digital content network receiver service market.* White Paper, HTRC Group, Canada. Retrieved March 17, 2005, from http://www.htrcgroup.com/pdffiles/dcnr.pdf

Bratsberg, H., & Wasenden, O. (2004, September). *Changing regulation—Impacts on mobile content distribution.* Retrieved March 16, 2005, from http://web.si.umich.edu/tprc/papers/2004/373/bratsberg_wasenden_tprc04_mobile_content_distribution_final.pdf

Cakewalk. (2005). *Desktop music handbook: Glossary of MIDI and digital audio terms.* Retrieved August 28, 2005, from http://www.cakewalk.com/tips/desktop-glossary.asp

ClearSky Mobile Media. (2005). Retrieved July 13, 2005, from http://www.clearskymobilemedia.com/carriersol/en-cont.asp

INET. (2005). *Always by your side.* Retrieved August 25, 2005, from http://www.inet.co.th

Internet KSC. (2005). Retrieved August 25, 2005, from http://www.ksc.net

Ji-NET. (2005). Retrieved August 25, 2005, from http://www.ji-net.com

Kivisaari, E., & Luukkainen, S. (2003, March). Content-based pricing of services in the mobile Internet. *Proceedings of the 7th IASTED International Conference on Internet and Multimedia Systems and Applications.* Retrieved March 15, 2003, from http://www.tml.tkk.fi/~sakaril/Content-based_pricing.pdf

LOXINFO. (2005). Retrieved August 25, 2005, from http://www.csloxinfo.co.th/

Nokia. (2005). Retrieved August 20, 2005 from http://www.nokia.co.th/nokia/0,,51297,00.html

Samsung. (2005). *Digital world.* Retrieved August 23, 2005, from http://product.samsung.com/cgi-bin/nabc/product/b2c_product_type.jsp?eUser=&prod_path=/Phones+and+Fax+Machines%2fWireless+Phones

Sanook.com. (2005). Retrieved June 10, 2005, from http://mobilemagic.sanook.com

Siam2you. (2005). Retrieved June 10, 2005, from http://www.siam2you.com

Smorodinsky. R. (2002). *Mobile entertainment—A value chain analysis and reference business scenario.* Retrieved from http://www.fing.org/ref/upload/GlobalCommunicationsrevised.pdf

Sonic Spot. (2005). *Glossary.* Retrieved September 1, 2005, from http://www.sonicspot.com/guide/glossary.html

Sony Ericsson. (2005). *Products.* Retrieved August 18, 2005, from http://www.sonyericsson.com/spg.jsp?cc=global&lc=en&ver=4001&template=pg1_1&zone=pp

Stiller, B., Reichl, P., & Leinen, S. (2001, March). Pricing and cost recovery for Internet services: Practical review, classification and application of relevant models. *NETNOMICS: Economic Research and Electronic Networking, 3*(1). Retrieved January 12, 2005, from http://userver.ftw.at/~reichl/publications/NETNOMICS00.pdf

True Internet. (2005). Retrieved October 7, 2005, from http://www.asianet.co.th/home.htm

Efficient and Scalable Group Key Management in Wireless Networks

Yiling Wang
Monash University, Australia

Phu Dung Le
Monash University, Australia

INTRODUCTION

Multicast is an efficient paradigm to support group communications, as it reduces the traffic by simultaneously delivering a single stream of information to multiple receivers on a large scale. Along with widespread deployment of wireless networks and fast improving capabilities of mobile devices, it is reasonable to believe that the integration of wireless and multicast will result in enormous benefits. Before users can enjoy the flexibility and efficiency of wireless multicast, security must be achieved. The core issue of wireless multicast security is access control, which means that only authorized users can participate in the group communications. Access control can be achieved by encrypting the communication data with a cryptographic key, known as group key. Group key is shared by all the registered users, so that only authorized members can gain access to the group communication contents. Several group key management approaches (Challal & Seba, 2005; Sherman & McGrew, 2003; Kostas, Kiwior, Rajappan, & Dalal, 2003; Wong, Gouda, & Lam, 2000; Wallner, Harder, & Agee, 1999; Harney & Muckenhirn, 1997; Steiner, Tsudik, & Waidner, 1996; Mittra, 1997; Hardjono, Cain, & Monga, 2000; Kim, Perrig, & Tsudik, 2004; Perrig, Song, & Tygar, 2001) have been proposed in the literature, most of them directed towards wired networks. Although some approaches can be employed in the wireless environment, they cannot achieve the same efficiency as in the wired networks. The complexity of group key management in wireless networks cannot be confined only to the limitation of wireless networks such as higher data error rate and limited bandwidth, but also from the properties of wireless devices, such as insufficient computation power, limited power supply, and inadequate storage space.

BACKGROUND

Over the last decade, a large number of group key management approaches have been proposed. Among them, the most prominent is the logical key hierarchy (LKH) (Wong et al., 2000; Wallner et al., 1999). In LKH, a key tree is formed comprising group and other auxiliary keys (key encryption key, KEK) that are used to distribute the group key to the users. Figure 1 depicts a typical LKH key tree. In the LKH key tree, users are associated with the leaf nodes, and each user must store a set of keys along the path from leaf node up to the root. When membership changes such as join or departure, the rekeying procedure is invoked to update the keys along the path, thereby ensuring security. This update affects all the members in the tree. LKH algorithm has some drawbacks which prevent its application in the wireless environment:

- **1-Affects-n:** As mentioned above, one membership change affects all the group members. But some changes are unnecessary to the members, especially in cellular networks, because the users in the cell are not only logical neighbors in the key tree but also physical (Sun, Trappe, & Liu, 2002).
- **Storage Inefficiency:** In the LKH algorithm, users have to store a set of keys. As the size of group increases, so does the number of keys stored by each user. This results in storage inefficiency of the lightweight mobile devices due to the limitation of storage space.

Figure1. Typical LKH key tree

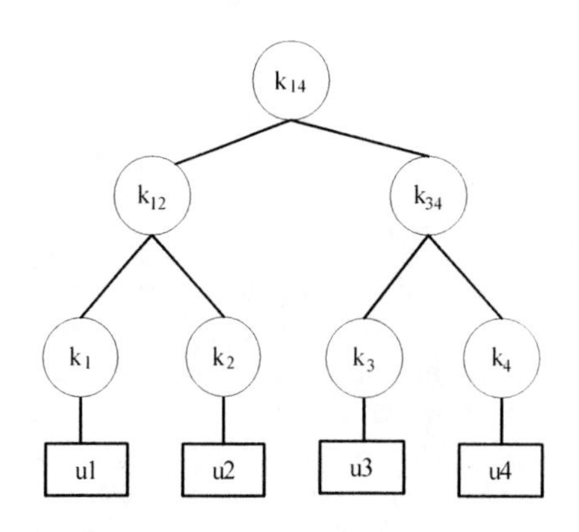

Figure 2. Group key management structure

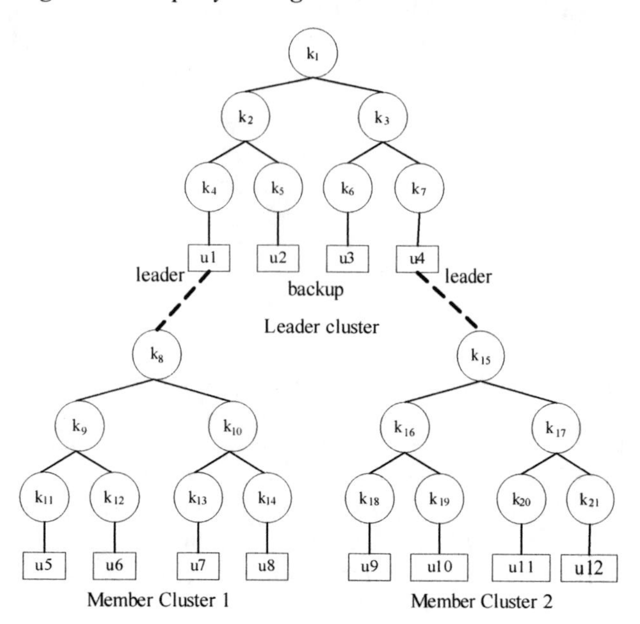

Group Key Management Algorithm

Group key management algorithm is a core part of multicast security. It maintains the logical key structure and performs the procedures to assign, distribute, and update the group key and other KEKs.

Notation

In this section, we depict the notations that we will use in the following sections:

- bs: base station
- BS: a set of the base stations
- {x}k: message x is encrypted by the key k
- A → B {message}: A sends message to B via unicast
- A⇒B {message}: A sends message to B via broadcast or multicast

The Proposed Logical Key Structure

In each cell of wireless network, the base station is responsible for managing the group key. In our proposed work the base station categorizes the authorized members into several clusters to form a two-level key management structure. Figure 2 depicts this logical structure within the cell. The group key management area is divided into smaller areas called clusters. Each cluster has its own cluster key for communication. The members of the leader cluster are assigned as leaders of the lower level member cluster—that is, one leader for one member cluster. The leader is responsible for distributing the rekeying messages to the lower level cluster users, thereby reducing the communication and computation overhead of the key server. The leftover users in the leader cluster serve as leadership backups.

The steps to build our proposed logical key management structure are as follows:

- **Step 1:** The base station groups users into several clusters based on the cluster policy, which defines the size of clusters and the ratio of leaders and backups. One cluster is assigned as the leader cluster, and others are member clusters.
- **Step 2:** Separate key trees are built for each cluster. The members in the leader cluster are assigned as leaders of member clusters—that is, one leader for one member cluster.
- **Step 3:** The base station assigns a local multicast address to each cluster for cluster communications.

The Proposed Group Key Management Algorithm

There are three main operations in the wireless group key management (multiple subgroups): member join, member leaving, and handoff. The rekeying procedures of these operations occur independently in each wireless cell. We illustrate our algorithm in each of these operations in the following subsections.

Member Join

When a user wants to join a group, backward secrecy must be maintained to prevent the new member from accessing the previous group communication details. The join procedure starts with the group join request sent by the user to the group key server (GKS).

u → GKS: {join request}

After authentication, GKS updates the group key and distributes to the base stations.

GKS ⇒ BS: {new group key}

There are two types of join: leader cluster join and member cluster join. The base station assigns the new member into a cluster according to the cluster policy, where leader cluster is given priority over member clusters. When the cluster is decided, the base station invokes the join procedure to rekey the cluster key tree. For example, in Figure 2, if u2 wants to join the group, then the rekeying procedure is invoked at the leader cluster. The base station needs to send the following two messages:

bs → u1: {new k_1, k_2, group key}k_4

bs ⇒ u3, u4: {new k_1, group key}k_3

u1 ⇒ cluster 1: {new group key}$k_{(cluster-1)}$

u4 ⇒ cluster 2: {new group key}$k_{(cluster-2)}$

The join procedure for the member cluster is similar to the leader cluster join. In Figure 2, if u6 wants to join the group, the rekeying is as follows:

bs → u5: { new k_8, k_9, cluster-1 key, group key}k_{11}

bs ⇒ u7, u8: {new k_8, cluster-1 key, group key}k_{10}

bs → u1: {new cluster-1 key, group key}k_4

bs ⇒ u2, u3, u4: {new group key}k_1

u4 ⇒ cluster 2: {new group key}$k_{(cluster-2)}$

Member Leaving

When a member leaves the group, forward secrecy has to be guaranteed to keep the departing user from accessing the future group communication content. The leaving procedure can either be invoked by the user or initiated by the key server. In our proposal, there are two types of leaving: member cluster departure and leader cluster departure. First, we describe the user departure from member cluster, which is the most frequently occurring event. The rekeying procedure for this leaving is as follows:

GKS generates a new group key, and sends to base stations.

GKS ⇒ BS: {new group key}

The base station updates the affected keys in the member cluster key tree. For instance, in Figure 2, if u5 leaves the group, the base station sends four rekeying messages:

bs → u6: {new k_8, k_9, cluster-1 key, group key}k_{12}

bs ⇒ u7,u8: {new k_8, cluster-1 key, group key}k_{10}

bs → u1: {new cluster-1 key, group key}k_4

bs ⇒ u2, u3,u4: {new group key}k_1

u4 ⇒ cluster 2: {new group key}$k_{(cluster-2)}$

As for the user departure from leader cluster, the situation becomes much more complex because of the need to select the next leader of the member cluster. This scenario can be classified into three categories:

1. The first scenario is when the leadership backup leaves the group. Since the backup has no association with the member cluster, the rekeying is limited to the leader cluster, followed by the rekeying procedure for the leaving operation. For example, in Figure 2, when u2 leaves the group:
   GKS generates a new group key and delivers it to the base stations.

   GKS ⇒ BS: {new group key}

   The base station updates the key tree of leader cluster.

   bs → u1: {new k_1, k_2, group key}k_4

   bs ⇒ u3, u4: {new k_1, group key}k_3

   u1 ⇒ cluster 1: {new group key}$k_{(cluster-1)}$

   u4 ⇒ cluster 2: {new group key}$k_{(cluster-2)}$

2. The second scenario is when the leader departs and the backup is to be elected as the next leader. Given a situation, the base station invokes the rekeying procedure in the leader cluster and assigns a backup to be the new leader of the affected cluster. For instance, as shown in the Figure 3, if u1 leaves the group, the base station assigns u3 to be the new leader of member cluster 1.

   bs → u2: {new k_1, k_2, group key}k_5

   bs ⇒ u3, u4: {new k_1, group key}k_3

   bs→ u3: {new cluster-1 key}k_6

   bs ⇒ cluster 1: {new cluster-1 key, group key}k_8

   u4 ⇒ cluster 2: {new group key}$k_{(cluster-2)}$

3. The last scenario is the worst case where a leader leaves the group and there is no backup available. The base station needs to select a user from the affected member cluster and assign her/him as the new leader. Instead of rekeying the affected member cluster immediately, the KEK update is delayed until join, leave, or eventual departure of the chosen user takes place. The base station invokes the rekeying only in the leader cluster in order to update the group key. For instance, as shown in the Figure 2, when u1, u2, and

u3 leave the group, u5 is selected and assigned as the new leader of member cluster 1.

bs $\rightarrow$ u5: {new k_1, k_2, cluster-1 key, group key}k_{11}

bs $\rightarrow$ u4: {new k_1, k_3, group key}k_7

bs $\Rightarrow$ cluster 1: {new cluster-1 key, group key}k_8

u4 $\Rightarrow$ cluster 2: {new group key}$k_{(cluster-2)}$

Handoff

Handoff is a unique operation in the wireless group communications. Several approaches (DeCleene et al., 2001; Sun et al., 2002) have been proposed in literature to address the group key management during the handoff. DeCleene et al. (2001) proposed a delayed rekeying scheme to postpone local rekeying until a particular criterion is satisfied, such as join, leave, or eventual departure of handoff users. We enhance this approach with the handoff authentication to further reduce the rekeying cost.

We assume that the mobile devices can detect the signals from two base stations on the edge of wireless cells. When a user moves from one cell to another at the edge of the cell, s/he switches to the new base station, sends a handoff join message, and then switches back to the old base station. The user authentication is performed by the two base stations (Wang & Le, 2005). The new base station sends a user authentication request to the old one:

$bs_{new} \rightarrow bs_{old}$: {AUTHENTICATION_REQUEST}

The old base station replies to the new base station with the authentication information of the moving user:

$bs_{old} \rightarrow bs_{new}$: {AUTHENTICATION_REPLY}

After a predefined time, the user switches to the new base station. When the user moves into the new cell, the new base station assigns the handoff user into a suitable cluster according to the join procedure, and sends a set of new keys to the user.

Instead of immediately rekeying, the old base station records the handoff user into a *Handoff User List* and postpones the rekeying until join or leaving happens in the cell or eventual departure of the moving user. When the local rekeying affects the handoff user, the base station deletes the user from the *Handoff User List.*

Performance Discussion

To measure the performance of a system, many parameters can be taken into consideration. However, we believe that for the group key management system, efficiency is one of the most significant parameters. In this section, we describe the efficiency of our proposed work in comparison with LKH. We explain the efficiency from three perspectives: communication cost, computation cost, and storage cost.

Communication Cost

Communication overhead is recorded as a measure of the number of rekeying messages transmitted per operation by the key server. We evaluate the communication cost for the join and leave operations. Without any loss of generality, we employ a binary tree to build the key tree.

Communication Cost During Join

In our structure, when a user joins the group, the new member can be either assigned into a leader cluster or member cluster. For the leader cluster join, the key update is restricted only to the leader cluster. Therefore, the communication cost is $\log_2 n_l$; n_l is the number of users in the leader cluster. As for the member cluster join, the communication cost is $\log_2 n_m + 2$; n_m is the number of users in the member cluster. According to LKH, the communication cost of join is $\log_2 n$, where n is the number of group users in the cell. In order to better explain the efficiency of our work, we calculate the join cost in one cell comprising 1,023 members. According to the cluster policy, the ratio of leaders and backups is 1:1, the members are divided into 17 clusters, 15 clusters having 64 members each and 2 clusters with 32 users. One 32-user cluster is assigned as the leader cluster. Now, we consider the join cost of a new member, namely 1024th user. In our approach, when the user joins the leader cluster, the cost is $\log_2 32 = 5$. If the user joins the member cluster, the overhead is $\log_2 64 + 2 = 8$. As for the LKH, the cost is $\log_2 1024 = 10$. So by comparison, we can see that our proposal achieves the significant improvement in the communication cost by 20% for member cluster join, and even better by 50% for leader cluster join.

Communication Cost During Leave

There are two types of leaving cost in our proposal: (1) user departure from member cluster, and (2) user departure from leader cluster. As for the user leaving from the member cluster, the communication cost is $\log_2 n_m + 2$, the summation of the rekeying cost of the member cluster key tree, and two extra messages sent to the leader cluster.

As for the user departing from the leader cluster, there are three situations: (1) backup user leaving, (2) cluster leader leaving with backup available, and (3) cluster leader leaving with no backup available. We calculate the communication cost of these three scenarios as follows:

Table 1. Comparison of communication cost

	Member Join	Member Leaving
Our Proposal	$\log_2 n_l$ / $\log_2 n_m + 2$	$\log_2 n_m + 2$ / $\log_2 n_l$ / $\log_2 n_l + 2$ / $\log_2 n_l + 2$
LKH	$\log_2 n$	$\log_2 n$

n: the number of group members in the cell
n_m: the number of users in the member cluster
n_l: the number of users in the leader cluster

When backup user leaves the group, the rekeying procedure is confined in the leader cluster. The communication cost is $\log_2 n_l$, where n_l is the number of users in the leader cluster. For the second situation, when the current cluster leader leaves the group, the backup can take over the leadership, so the communication cost is $\log_2 n_l + 2$. As for the third scenario where the leader and the backups move from the group, due to the delayed rekeying in the affected member cluster, the rekeying procedure is invoked only in the leader cluster. Hence, the communication cost is $\log_2 n_l + 2$.

In LKH, the leaving communication cost is $\log_2 n$, where n is the number of group members in the cell.

We consider the same example as described above to explain the situation further. In our proposal, when the user of the member cluster leaves, the communication cost is $\log_2 64 + 2 = 8$. When the user in the leader cluster leaves the group, the communication cost of three scenarios as described is $\log_2 32 = 5$, $\log_2 32 + 2 = 7$ and $\log_2 32 + 2 = 7$ respectively, whereas the rekeying cost for LKH is $\log_2 1024 = 10$. From this example, it can be observed that our proposal improves the communication efficiency by 30% and 50%.

We tabulate the comparison of our approach and LKH in Table 1.

Computation Cost

Computation cost is to measure the overhead of encryption and decryption during the rekeying procedure. This cost is firmly associated with the two factors: (1) the communication cost, and (2) the length of the message. In our approach, we group users into clusters and for each cluster a smaller key tree is built. Hence, during the rekeying procedure, the number of keys that needs to be updated is much less when compared to that of LKH, which means that the length of our rekeying message is much shorter than the message used in LKH. Additionally, from the above analysis, we can see that our proposal has a greater advantage over LKH with respect to the communication cost. By combining these two factors, it can be inferred that the proposed approach is much more computationally efficient than LKH.

Storage Cost

The storage efficiency is to measure the number of keys stored in the key server and at the user side. In our proposal, users are associated with the cluster key tree whose size is much smaller than the LKH key tree. Therefore our approach has less number of keys stored at the user side than that in LKH. For the users in the member cluster, the number of stored keys is $\log_2 n_m + 3$, and for the users in the leader cluster, the number of stored keys is $\log_2 n_l + 3$, whereas in LKH each user needs to store $\log_2 n + 1$ keys. Considering the same example, the user in the leader cluster needs to store 8 keys, and the number of keys stored by the user in the member cluster is 9. When compared to the 11 keys stored by the user in LKH, our approach proves to offer better efficiency. On the key server side, the number of keys stored in our proposal and LKH is $n_c \times 2n_m + 2n_l \approx 2n$ and $2n - 1$ respectively. We tabulate the comparison in Table 2.

FUTURE TRENDS

Along with the fast development in wireless technology and mobile devices, more and more multicast group applications and services will be emerging on wireless networks. The future research will focus on combining the proposed

Table 2. Comparison of key storage cost

	User	Key Server
Our Proposal	$\log_2 n_m + 3$ / $\log_2 n_l + 3$	$n_c \times 2n_m + 2n_l$
LKH	$\log_2 n + 1$	$2n - 1$

n_m: the number of users in the member cluster
n_l: the number of users in the leader cluster
n_c: the number of member clusters
n: the number of group members in the cell

work with other key management structures to provide a solution to secure and efficient group key management in wireless networks.

CONCLUSION

Here, we proposed a new, efficient group key management approach for wireless networks. The proposed scheme has a two-tier logical key structure where the users are grouped into clusters, which help in significantly improving the communication, computation, and storage efficiency on the client side.

REFERENCES

Chen, J., & Chao, T. (2004). *IP-based next-generation wireless networks.* New York: John Wiley & Sons.

Challal, Y., & Seba, H. (2005). Group key management protocols: A novel taxonomy. *International Journal of Information Technology, 2*(1), 105-118.

DeCleene, B., Dondeti, L. R., Griffin, S., Hardjono, T., Kiwior, D., Kurose, J., Towsley, D., Vasudevan, S., & Zhang, C. (2001, June). Secure group communications for wireless networks. *Proceedings of MILCOM.*

Hardjono, T., Cain, B., & Monga, I. (2000). *Intra-domain group key management protocol.* Retrieved from draft-ietf-ipsec-intragkm-00.txt

Harney, H., & Muckenhirn, C. (1997). Group key management protocol (GKMP) architecture. *RFC, 2094.*

Kim, Y., Perrig, A., & Tsudik, G. (2004). Tree-based group key agreement. *ACM Transactions on Information and System Security, 7,* 60-96.

Kostas, T., Kiwior, D., Rajappan, G., & Dalal, M. (2003). Key management for secure multicast group communication in mobile networks. *Proceedings of the DARPA Information Survivability Conference and Exposition* (Vol. 2, pp. 41-43).

Mittra, S. (1997). Iolus: A framework for scalable secure multicasting. *ACM SIGCOMM, 27*(4).

Perrig, A., Song, D., & Tygar, D. (2001). ELK, a new protocol for efficient large-group key distribution. *Proceedings of the IEEE Symposium on Security and Privacy* (pp. 247-262).

Sherman, A. T., & McGrew, D. A. (2003). Key establishment in large dynamic groups using one-way function trees. *IEEE Transactions on Software Engineering, 29,* 444-458.

Steiner, M., Tsudik, G., & Waidner, M. (1996). Diffie-Hellman key distribution extended to group communication. *Proceedings of the 3rd ACM Conference on Computer and Communications Security (CCS '96)* (pp. 31-37).

Sun, Y., Trappe, W., & Liu, K. J. R. (2002). An efficient key management scheme for secure wireless multicast. *Proceedings of the IEEE International Conference on Communications* (Vol. 2, pp. 1236-1240).

Wallner, D., Harder, E., & Agee, R. (1999). Key management for multicast: Issues and Architectures. *RFC, 2627.*

Wang, YL., & Le, P.D. (2005, September). Secure group communications in wireless networks. *Proceedings of the 3rd International Conference on Advances in Mobile Multimedia,* Malaysia.

Wong, C. K., Gouda, M., & Lam, S. S. (2000). Secure group communications using key graphs. *IEEE/ACM Transactions on Networking, 8,* 16-30.

KEY TERMS

Backward Secrecy: To prevent new group members from accessing previous group communications, which they may have recorded.

Forward Secrecy: To prevent departing members from decoding future group data traffic.

Handoff: In a cellular wireless network, the transition of signal for any given user from one base station to a geographically adjacent base station as the user moves around.

Key Encryption Key (KEK): A key used to encrypt the other keys for distribution in the multicast group.

Key Management Algorithm: In the group key management system, an algorithm is applied to maintain the logical key structure held by the group members and other entities.

Logical Key Hierarchy (LKH): This type of algorithm is a tree structure for efficient group rekeying. Each node of the tree represents a key, with the root node representing the group key. Each leaf node represents a group member, and each member knows all the keys in its path to the root.

Multicast: A communication mechanism to delivery a single message to multiple receivers on a network. The message will be duplicated automatically by routers when multiple copies are needed.

1-Affects-n: When one group membership changes, the rekeying procedure will affect all the remaining members.

E

Efficient Replication Management Techniques for Mobile Databases

Ziyad Tariq Abdul-Mehdi
Multimedia University, Malaysia

Ali Bin Mamat
Universiti Putra Malaysia, Malaysia

Hamidah Ibrahim
Universiti Putra Malaysia, Malaysia

Mustafa M. Dirs
College University Technology Tun Hussein Onn, Malaysia

INTRODUCTION

Mobile databases permeate everywhere into today's computing and communication environment. One envisions application infrastructures that will increasingly rely on mobile technology. Current mobility applications tend to have a large central server and use mobile platforms only as caching devices. We want to elevate the role of mobile computers to first class entities in the sense that they will allow the mobile user to work/update capabilities independent of a central server. In such an environment, several mobile computers may collectively form the entire distributed system of interest. These mobile computers may communicate to each other in an ad hoc manner by communicating through networks that are formed on demand. Such communication may occur through wired (fixed) or wireless (ad hoc) networks. At any given time, a subset of the computer collection may connect and would require reliable and dependable access to relevant data of interest. Peer-to-peer (P2P) computing, basically, is an ad hoc network and it can be built on the fixed or along a wireless network. With P2P, computers can communicate directly and share both data and resources. So far, many applications such as ICQ (where users exchange personal messages), similar to Napster and Freenet (where users exchange music files), have taken the advantage of P2P technology. However, data management is an outstanding issue and leads directly to the problem of low data availability. Data availability is the central issue in P2P data management. The most important characteristic that affects data availability in P2P environment is the nature of the network. In the case of an ad hoc network, hosts are connected to the network only temporarily. Furthermore, hosts play the role of router, and they communicate with each other directly without any dedicated hosts. If there are no dedicated hosts that act as a router, obviously the network connections are prone to get disconnected and/or become unreliable. Thus, it is difficult to guarantee one-copy "serializability," since one relies on the mobile hosts, not the fixed hosts, in order to communicate with other hosts not reachable directly (Faiz & Zaslavsky, 1995). When hosts disconnect more often, due to the applications that have high transaction rates, the deadlock and reconciliation rate will experience a cubic growth (Faiz & Zaslavsky, 1995) and, the database is in an inconsistent state and there is no obvious way to repair this problem or allow for this eventuality. In the case of fixed network, the network connection is relatively stable, but the availability of sufficient computing resource depends on the strategies of replication.

Walborn and Chrysanthis (1997) describe the use of mobile computers in the trucking industry. Each truck has a computer with a satellite/ radio link—this is able to interact with the corporate database. Other applications include involving avoiding remote or disaster areas and for military applications with mobile computers forming ad hoc networks without communications and/or with stationary computers. Faiz and Zaslavsky (1995) discuss the impact of wireless technologies and mobile hosts on a variety of replication strategies. Distributed replicated file systems such as Ficus and Coda (Reiher, Heidemann, Ratner, Skinner, & Popek, 1994) have extensive experience with disconnected operations.

In this article, we consider the distributed database that can make up mobile nodes as well as peer-to-peer concepts. These nodes and peers may be replicated both for fault tolerance (dependability), and to compensate for nodes that are currently disconnected. Thus we have a distributed replicated database, where several sites must participate in the synchronization of a transaction. The capabilities of the distributed replicated database are extended to allow mobile nodes to plan for a disconnection, with the capability of update, and for the database—on behalf of mobile node by using fixed proxy server—to make these updates during the mobile disconnection. Once a mobile reconnects, it automatically synchronizes and integrates into database.

By using the notion of planned disconnection (*sign-off and check-out modes*), we present a framework, which allows the replicated data of mobile nodes to be available to access and update for low costs in reading and writing.

This article is organized as follows; in the second section, we review the *read one write all* (ROWA) technique; in the third section, the model of the *diagonal replication on grid* (DRG) technique is presented, and we also present an algorithm to allow disconnection nodes to update using a sign-off and check–out idea adapted in the system. In the fourth section, the correctness and the performance of the proposed technique is analyzed in terms of *communication cost and availability* comparing ROWA techniques and DRG techniques. In the final section, the conclusion is given.

VIEW OF ROWA STRUCTURE TECHNIQUE

The simplest technique to maintain replicated data is when a read-only operation is allowed to read any copy, and write operation is required to write all copies. This is called a read one write all (ROWA) protocol. This protocol only works correctly when a transaction process from one correct state to another correct state is carried out. The ROWA has the lowest read cost because only one replica is accessed by a read operation. The weakness of this method is the low write availability, because a write operation cannot be done in a failure of any replica.

The available copies technique proposed by Bernstein, Hadzilacos, and Goodman (1987) is an enhanced version of the ROWA approach in terms of the availability of write operations. Every read is translated into a "read" of any replica of the data object and every "write" is translated into write of all available copies of that data object. This technique can handle each site either when it is operational or down—and that all operational sites can communicate with them. If a site does not respond to a message within the timeout period, then it is assumed to be down. However, writing is very expensive when all copies are available: forcing read-write transactions to write all replicas.

Lazy replication protocol does not attempt to perform the write operation on all copies of the data object within the context of the transaction that updates that data object. Instead, it performs the update on one or more copies of the data object and later propagates the changes to all the other copies in all other sites. A lazy replication scheme can be characterized using four basic parameters(Bernstein, Hadzilacos, & Goodman, 1987; Borowski, 1996; Goldring, 1995; Ozsu & Valduriez, 1999). The ownership parameter defines the permissions for updating copies. If a copy is updateable it is called primary copy, otherwise it is called a secondary copy. The site that stores the primary copy of a data object is called a master for this data object, while the sites that store its secondary copies are called slaves. The propagation parameter defines when the updates copy must be propagated towards the sites storing the other copies of the same data object. Generally, lazy replication protocols can be classified into two groups. (1) The first group consists of lazy replication methods where all copies are updateable. In this case, there is group ownership on the copies. A conflict happens if two or more sites update the same replica. There are several policies for conflict detection and resolution that can be based on timestamp ordering, node priority, and others (Buretta, 1997; Helal, Heddaya, & Bhargava, 1996). The problem with conflict resolution is that during a certain period of time the database may be in an inconsistent state. (2) The second group consists of protocols where there is a single master that is updated, and each time a query is submitted for execution, the secondary copies that are read by the query are refreshed by executing all received refresh transaction. Therefore, a delay may be caused by that query.

DATABASE MODEL

In considering this environment, it has two types of networks; that is, a fixed network or an ad hoc network. For the fixed network, all sites are logically organized in the form of two-dimensional grid structure (i.e., 5*5) and in a nodes (ad hoc network), this consists of N nodes labeled $N_1, N_2, \ldots, N_n$ (as shown in Figure 1). The data will only be partially or fully replicated in that data items are stored redundantly at multiple sites. Information about the location of all copies of a data item may be stored at each site or kept in directories at several of the sites. Users interact with the database by invoking transactions at any one of the database sites. A

Figure 1. Database model

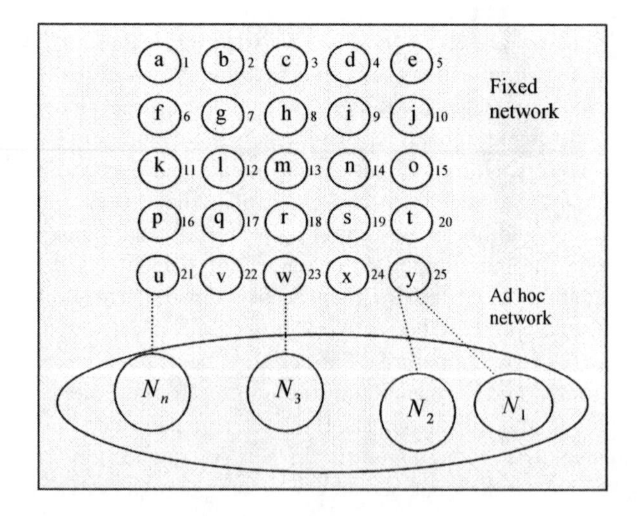

transaction is a sequence of read and writes operations on the data items that are executed automatically. The criterion for correctness in databases is the *serializable* execution of transactions (Wolfson, Jojodia, & Huang, 1997). Serializable executions are guaranteed by using a *concurrency control* mechanism, such as two-phase locking, timestamp ordering, or optimistic concurrency control (Agrawal & Abbadi, 1996). Since two-phase locking is widely used, we assume that each site in the distributed system enforces two-phase locking locally. Multiple copies of a data object must appear as a single logical data object to the transactions. This is termed as "one-copy equivalence" and is enforced by the replica control technique; as used in this study of diagonal replication grid (DRG) technique. The correctness criterion for replicated database is one-copy serializability (Berstein & Goodman, 1994), which ensures both one-copy equivalence and the serializable execution of transactions. In order to ensure one-copy serializability, a replicated data object may be read by reading a quorum of copies, and it may be written by writing a quorum of copies, The selection of a quorum is restricted by the quorum intersection property to ensure one-copy equivalence: for any two operations o [x] and o' [x] on a data object x, where at least one of them is a write, the quorum must have a non-empty intersection. The quorum for an operation is defined as a set of copies whose number is sufficient to execute that operation. We assume that a mechanism to enforce one-copy serializability is used. This could be a "synchronous control protocol" using diagonal replication on grid (DRG) technique, that is, a site *initiates* a DRG transaction to update its data object. For all accessible data objects, a DRG transaction attempts to access a DRG quorum. If a DRG transaction gets a DRG write quorum without non-empty intersection, it is accepted for execution and completion, otherwise it is rejected. We assume that for a read quorum, that if two transactions attempt to read common data objects, read operations do not change the values of the data object. Since read and write quorums must intersect and any two DRG quorums must also intersect, then all transaction executions are one-copy serializable.

In this article, we consider two models—*sign-off model and check- out model*. They work under these conditions:

1. The distributed system must be able to process database updates even though some of the nodes are not available. It is to designate one of the members as the fixed proxy site at fixed network, which can present mobile node when mobile disconnection.
2. To optimize the communication costs and the availability of the system of replicated data in the distributed systems under the fixed and ad-hoc network in P2P environment.
3. The fixed network, we are describing in the diagonal replication on grid (DRG) technique, considers only that the diagonal sides will be replicated—that of the data at a fixed network based on quorum intersection property.
4. The members of mobile nodes are fixed and databases at ad-hoc network (mobile nodes) will be replicated at the node that is the most commonly visited node.
5. The fixed proxy should be selected from the nearest one to the mobile database.
6. Transfer any data between two mobiles nodes should be through fixed proxy.

Basic Sign-off

In a simple planned disconnection or *basic sign-off* (Holliday, Agrawa, & Abbadi, 2000), the database of the disconnected node becomes read-only. The connected sites continue to read and update the data item or objects. DRG technique adjusts the replication at diagonal sites at fixed network so that the transaction can complete in spite of mobile node disconnects. The planned disconnection will be accomplished with the help of a fixed *proxy*. When a node disconnects, it appoints another site as a fixed network to vote on its behalf to ensure that replicas can be updated, and in any other actions of the distributed system that require consensus. The power to vote on behalf of a mobile node at fixed network is called a fixed proxy and the site with that power is also referred to as a fixed proxy.

If node N_1 wants to disconnect from a distributed database, it carries out a disconnect dialog so that the system is aware that it has not failed, but will merely disconnect for a period of time. The node N_1 contacts the nearest fixed network Y to be N_1's fixed proxy. During the disconnection, N_1 can only read its local copy of the data. When the fixed proxy Y sees a message for N_1, it answers on behalf of N_1 while N_1 is disconnected. The fixed proxy Y also keeps track of the updates to the database that N_1 has missed because of the disconnection. (We assume that all updates are sent to all copies in a diagonal set, and so, fixed proxy Y can respond on behalf on N_1).

Assuming that N_1 is a mobile computer and wishes to disconnect using sign-off procedure:

1. Node N_1 selects the nearest fixed proxy Y from fixed network, the fixed proxy Y should be peer to N_1. The fixed proxy Y should inform N_1 which will be the fixed proxy to N_1 in case of fixed proxy Y does not work (failure).
2. Node N_1 informs fixed proxy Y the new data that is not replicated at fixed proxy Y, N_1 transfer a new data.
3. The fixed proxy Y will replicate the N_1 data to all diagonal set on the request to it.
4. The data items at N_1 that are replicated at fixed proxy Y, the fixed proxy Y has the right to vote for N_1 in matters concerning writes to the data items.

When node N_1 wants to reconnect, it should also go through sign-on:

1. Node N_1 reconnects and contacts the fixed proxy Y.
2. Fixed proxy Y transfers all new data that has been missed during disconnection to node N_1.
3. If the fixed proxy Y itself disconnects during system failure, the other site that is appointed from fixed proxy Y can work as fixed proxy to node N_1.
4. During connection of node N_1, the fixed proxy Y (or the one on behalf of the fixed proxy Y) allows services to transfer data anew to node N_1—all which has been missed during the disconnection of node N_1.

In case of node N_1 being disconnected forever, the node N_1 will inform fixed proxy Y whether to delete all N_1's data or keep it with fixed proxy Y forever, or node N_1 will leave their data at fixed proxy Y when planned disconnect and forget it, this will depend on node N_1.

Check-Out Model

If the node N_1 wants to disconnect and still be able to update a particular data object, it declares its intention to do so, but before disconnection—and "check-out" or "takes" the object for writing. This is accomplished by obtaining a lock on the item before disconnection. In order to maintain serializability in *check-out* mode, the fixed proxy Y and its diagonal set including the mobile nodes are prevented from accessing the object which N_1 has checked-out (as if N_1 had a write-lock). An object can only be checked-out to one mobile node at a time. Since many database systems use two-phase locking, it makes sense to implement *check-out* mode using the existing locking mechanisms. The mobile node that wishes to disconnect, for example, N_1, acquires a write-lock on the item or object it wants to update while disconnected. This write-lock is like an ordinary write-lock except that the "transaction" that holds it should not be aborted due to a deadlock with ordinary transactions. The mechanism for obtaining the lock might be via a transaction or through some other means. In order to distinguish these "transactions" from ordinary user transactions, we will call them *pseudo-transactions*.

To preserve correctness, it must be possible to serialize all of the transactions executed by node N_1 during disconnection and at the point in time of disconnection. This can be done if:

1. Only those items write-locked by pseudo-transactions at disconnect time can be modified by node N_1 during disconnect.
2. Items write-locked by pseudo-transactions at disconnect time can neither be read or written by other sites (a consequence of maintaining the write-lock) and the pseudo-transaction cannot abort in order to release the lock.
3. Items not write-locked by pseudo-transactions at disconnect time are treated as read-only by node N_1 during disconnect (unless they were currently locked by other transactions at the time of disconnection).

Assuming that node N_1 wishes to disconnect and "check-out" a set of items—X. The disconnect procedure is as follows:

1. Node N_1 selects a fixed proxy—Y, as in basic sign-off mode—and follows all the same steps to handle voting rights for replicated and non-replicated items.
2. At the same time, node N_1 initiates a pseudo-transaction to obtain write-lock on the items in X.
3. If the pseudo-transaction is successful, N_1 disconnects with update privileges on all the items in X. If the pseudo-transaction is not successful, N_1 will try again or disconnect without update rights to X.

When node N_1 wants to reconnect, it should also go through check-out:

1. Node N_1 reconnects and contacts the fixed proxy—Y.
2. Node N_1 will transfer any new data from X to fixed proxy Y.
3. Fixed proxy Y will commit the value of item X and release the lock from item X.
4. Fixed proxy Y transfers any new data that has been missed during disconnection to node N_1.
5. If the fixed proxy Y itself disconnects (e.g., due system failure), the other site that has been appointed from fixed proxy Y can work as fixed proxy to node N_1.

DRG Technique

This environment has two types of networks, (1) the fixed network and (2) the ad-hoc network. For the fixed network, all sites are logically organized in the form of a two-dimensional grid structure. For example, if a DRG consists of twenty-five sites, it will logically organized in the form of 5 x 5 grid (as shown in Figure 1), each site having a master data file. In the remainder of this study, we are assuming that all replica copies are data files. A site is either operational or failed, and the state (operational or failed) of each site is statistically independent to the others. When a site is operational, the copy at the site is available; otherwise it is unavailable. In the fixed network, the data file will replicate to *diagonal sites*. While in the ad-hoc network, the data file will replicate asynchronously at only one node based on the most frequently visited site (when the node reconnects the fixed proxy will transfer to it any new and recent updated data). The logical structure for fixed and ad hoc network is shown as in Figure 1. The circles in the grid represent the sites under the fixed network environment and $a, b, ..., y$ represent the master data files located at site 1,2,...,25 respectively. The circles S1,S2,...., Sn represent the master file at mobile node located at node 26,27,28,...,n; as shown in the oblong shape are nodes under the ad-hoc network.

The commonly visited site is defined as the most frequented node request for the same data at a fixed network (the commonly visited sites can be given either by a user or selected automatically from a log file/database at each center). This site will replicate the data asynchronously (by/once mobile node reconnects to a fixed proxy), until then it will not be considered for read and write quorums on fixed network, but mobile nodes can read their own data during disconnection without any update of the data. Since the data file is replicated to only the diagonal sites at the fixed network, therefore it minimizes the number of database update operations, misrouted (and dropped out calls). Also, sites are autonomous in processing different queries or update operations; this consequently reduces the query response time. The number of data replication, d, can be calculated using Property 1, described as follows:

Property 1. One assumes the number of data replication from each site, $d = n$

Proof: Let N = n x n be a set of all sites that are logically organized in a two-dimensional grid structure form as shown in Figure 1. Based on definition 1:

The number of diagonal sites = number of sites in a diagonal set

$= |D(s)| = n.$
$\because n=5$
$\because n=|D(s)|$
$\therefore |D(s)|=5$

Definition 1. Assuming that the fixed network environment consists of n x n sites that they are logically organized in the form of a two-dimensional grid structure. These sites are labeled $s(i,j)$, $1 \le i \le n$, $1 \le j \le n$. The *diagonal site* to $s(i,j)$ is $\{s(k,l) | k=i+1, l=j+1$; and $\{k,l \le n$, if i=n, initialized i=0, if j=n, initialized j=0$\}$. A diagonal set, D(s), is a set of diagonal sites. As an example, Assume that n=5, then the diagonal site to s(1,1) is s(2,2), the diagonal site to s(2,2) is s(3,3), the diagonal site to s(2,1) is (3,2), and so forth.

Thus, based on this technique, sites in the diagonal set will have the replica copies in common. From Figure 1, one of the diagonal sets is {s(1,1), s(2,2), s(3,3), s(4,4),s(5,5)}, and each site will have the same replica copies, that is, {a,g,m,s,y}.

The number of diagonal set equals to n, and the mth diagonal set is noted as $D^m(s)$, for m=1,2,...n.

For example, from *Figure 1*, if n=5, then the diagonal sets are:

$D^1(s) = \{s(1,1), s(2,2), s(3,3), s(4,4), s(5,5)\}$,
$D^2(s) = \{s(2,1), s(3,2), s(4,3), s(5,4), s(1,5)\}$,
$D^3(s) = \{s(3,1), s(4,2), s(5,3), s(1,4), s(2,5)\}$,
$D^4(s) = \{s(4,1), s(5,2), s(1,3), s(2,4), s(3,5)\}$, and
$D^5(s) = \{s(5,1), s(1,2), s(2,3), s(3,4), s(4,5)\}$,

The primary site of any data file and, for simplicity, its diagonal sites, are assigned with vote one and vote zero, which is analogous to binary vote assignment proposed in (Mat Deris, Evans, Saman, & Noraziah, 2000). A vote assignment on grid, B, is a function such that

$$B(s(i,j)) \in \{0,1\}, 1 \le i \le n, 1 \le j \le n$$

where $B(s(i,j))$ is the vote assigned to site $s(i,j)$. This assignment is treated as an allocation of replicated copies and a vote assigned to the site results in a copy allocated at the diagonal site. That is, 1 vote ≡ 1 copy.

Let

$$L_B = \sum_{s(i,j) \in D(s)} B(s(i,j))$$

where, L_B is the total number of votes assigned to the primary site and its diagonal sites. Thus, $L_B = d$.

Let r and w denote the read quorum and write quorum, respectively. To ensure that the read operation always gets up-to-date values, r + w must be greater than the total number of copies (votes) assigned to all sites. The following conditions are used to ensure consistency:

$1 \le r \le L_B$, $1 \le w \le L_B$,

$r + w = L_B + 1.$

Conditions (1) and (2) ensure that there is a non-empty intersection of copies between every pair of read and write operations. Thus, the conditions ensure that a read operation can access the most recently updated copy of the replicated data. Timestamps can be used to determine which copies are most recently updated.

Let S(B) be the set of sites at which replicated copies are stored corresponding to the assignment B. Then

$$S(B) = \{s(i,j) | B(s(i,j)) = 1, 1 \le i \le n, 1 \le j \le n\}.$$

Definition 2. For a quorum q, a *quorum group* is any subset of S(B) whose size is greater than or equal to q. The collection of quorum group is defined as the *quorum set*.

Let Q(B,q) be the quorum set with respect to the assignment B and quorum q, then

$Q(B,q) = \{G|\ G \subseteq S(B) \text{ and } |G| \geq q\}$

For example, from *Figure 1*, let site s(1,1) be the primary site of the master data file *a*. Its diagonal sites are s(2,2),s(3,3)),s(4,4), and s(5,5). Consider an assignment B for the data file *a*, such that

$B_a(s(1,1))=B_a(s(2,2))=B_a(s(3,3))=B_a(s(4,4))=B_a(s(5,5)) = 1$

and

$L_{B_a} = B_a(s(1,1))+B_a(s(2,2))+B_a(s(3,3))+ B_a(s(4,4)) + B_a(s(5,5)) = 5.$

Therefore, $S(B_a)$ = {s(1,1),s(2,2),s(3,3), s(4,4),s(5,5)}.

If a read quorum for data file *a*, r =2 and a write quorum w = L_{Ba}-r+1 = 4, then the quorum sets for read and write operations are $Q(B_a,2)$ and $Q(B_a,4)$, respectively, where

$Q(B_a,2)$={s(1,1),s(2,2)},{s(1,1),s(3,3)},{s(1,1),s(4,4)},{s(1,1),s(5,5)},{s(2,2),s(3,3)},{s(2,2),s(4,4)},{s(2,2),s(5,5)},{s(3,3),s(4,4)},{s(4,4),s(5,5)},{s(1,1),s(2,2),s(3,3)},

{s(1,1),s(2,2),s(4,4)},{s(1,1),s(2,2),s(5,5)},{s(1,1),s(3,3),s(4,4)},{s(1,1),s(3,3),s(5,5)},

[s(1,1),s(4,4),s(5,5)},{s(2,2),s(3,3),s(4,4)},{s(2,2),s(3,3),s(5,5)},{s(2,2),s(4,4),s(5,5)},

{s(3,3),s(4,4),s(5,5)},{s(1,1),s(2,2),s(3,3),s(4,4)},{s(1,1),s(2,2),s(3,3),s(5,5)}, {s(1,1),s(2,2),s(4,4),s(5,5)},{s(1,1),s(3,3),s(4,4),s(5,5)},{s(2,2),s(3,3),s(4,4),

s(5,5)},{s(1,1),s(2,2),s(3,3),s(4,4),s(5,5)}And

$Q(B_a,4)$={s(1,1),s(2,2),s(3,3),s(4,4)},{s(1,1),s(2,2),s(3,3),s(5,5)},{s(1,1),s(2,2),s(4,4),s(5,5)},

{s(1,1),s(3,3),s(4,4),s(5,5)},{s(2,2),s(3,3),s(4,4),s(5,5)},{s(1,1),s(2,2),s(3,3),s(4,4),s(5,5)}

The Correctness of DRG

In this section, the study will show that the DRG protocol is one-copy serializable. The sets of groups (coterie) (Maekawa, 1985) will be defined, and to avoid confusion we refer to sets of copies as groups. Thus, a set of groups are/is a set of sets of copies.

Definition 3. Coterie. Let U be a set of groups that compose the system. A set of groups T is a coterie under U if and only if:

1. *G Î T implies that G ¹ Æ and G Í U*
2. *G, H Î T then G Ç H ¹ Æ* (inter*section property*).
3. *There are no G, H Î T such that G Ì H* (min*imality*)

By the definition of coterie and definition from 3.3.2, then Q (B, w) is a coterie, because it satisfies all coteries' properties. The correct criterion for replicated database is one-copy serializable. The next theorem provides us with a mechanism to check whether DRG is correct.

Assertion 3. The history H is one-copy serializable if Ti Î H, i=1,2,...n satisfy quorum intersection properties.

Proof: Suppose history H satisfies quorum intersection properties. Assume that Ti, Tj Î H, then at least one of Ti's operations precedes and conflicts one of Tj's operations. Then Ti→Tj. Thus H is an cyclic RDSG. By theorem (4) H is one- copy serializable.

Theorem 3. The DRG protocol is one-copy serializable.

Proof: as in Assertion 1. The theorem holds on condition that the DRG protocol satisfies the quorum intersection properties. Since read operations do not change the value of the accessed data object, a read quorum does not need to satisfy the intersection property. To ensure that a read operation can access the most recently updated copy of the replicated data, that means the two conditions as follow must be conformed.

1. $1 \leq r \leq L_B$, $1 \leq w \leq L_B$
2. $r + w = L_B + 1$

While a write quorum needs to satisfy read-write and write-write intersection properties. For case of write-write intersection, since W is coterie then it satisfies write-write intersection. However, for the case of read-write intersection, it can be easily shown that *∀ G Î R and ∀ H Î W, then G Ç H ¹ Æ*.

PERFORMANCE ANALYSIS AND COMPARISON

In this section, we analyze and compare the performance of the DRG technique with the ROWA technique on the communication cost and the data availability.

Communication Costs and Availability Analysis

The communication cost of an operation is directly proportional to the size of the quorum required to execute the operation. Therefore, one represents the communication cost in terms of the quorum size. In estimating the availability of this, all copies are assumed to have the same availability *p*.

$C_{X,Y}$ denotes the communication cost with X technique for Y operation, which is R(read) or W(write).

The ROWA Technique

Let N be the number of copies which are organized as a dimension n x n. Read operation needs only one copy, while a write operation needs to access n copies (a copy in each replica) in the system. Thus, the communication cost of a read operation is:

$$C_{ROWA,R} = 1$$

and the communication cost of write operation is:

$$C_{ROWA,W} = n$$

In the case of quorum, ROWA requires a read on any one of the copies, therefore the availability for read operation in the ROWA technique is

$$A_{ROWA,R} = \sum_{i=1}^{n} \binom{n}{i} P^i (1-P)^{n-i} = 1-(1-P)^n. \qquad (1)$$

While in write operation it needs to writes in all copies. Let $A_{X,Y}$ be the availability with X technique for Y operation, then the write availability in the ROWA technique is:

$$A_{ROWA,W} = \sum_{i=1}^{n} \binom{n}{i} P^i (1-P)^{n-i} = P^n. \qquad \text{.....(2)}$$

The DRG Technique

Let p_i denote the availability of site i. Read operations on the replicated data are executed by acquiring a read quorum and write operations are executed by acquiring a write quorum. For simplicity, one chooses the read quorum equal to the write quorum. Thus, the communication cost for read and write operations equals to $\lfloor L_{Ba}/2 \rfloor$, that is,

$$C_{DRG,R} = C_{DRG,W} = \lfloor L_{Ba}/2 \rfloor.$$

For example, if the primary site has four neighbors, each of which has vote one, then

$$C_{DRG,R} = C_{DRG,W} = \lfloor 5/2 \rfloor = 3.$$

For any assignment B and quorum q for the data file x, define $\varphi(B_a,q)$ to be the probability that at least q sites in $S(B_a)$ are available, then:

$$\varphi(B_a,q) = \Pr\{\text{at least q sites in } S(B_a) \text{ are available}\}$$

$$= \sum_{q}^{n} \binom{|S(B_x|}{q} p^q (1-p)^{|S(B_x|-q} \qquad \text{..... (3)}$$

Thus, the availability of read and write operations for the data file *a*, are $\varphi(B_a,r)$ and $\varphi(B_a,w)$, respectively. Let $Av(B_a,r,w)$ denote the read/write availability corresponding to the assignment B_a, read quorum r and write quorum w. If the probability that an arriving operation of read and write for data file *a* are *f* and (1-*f*), respectively, then:

$$Av(B_a,r,w) = f\,\varphi(B_a,r) + (1-f)\,\varphi(B_a,w). \qquad \text{...... (4)}$$

Definition 4. Let Av(x) be the availability function with respect to x. Av(x) is in the closed form if $0 \le x \le 1$ then $0 \le Av(x) \le 1$.

Theorem 4. The read/write availability under DRG technique are in the closed form.

***Proof*:** For the case of read availability, from equation (3) and by definition 3.1.2, as $0 \le p_i \le 1$, $i=1,2,\ldots,L_{Ba}$ then $0 \le \varphi(B_a,r) \le 1$. Similarly, for the case of write availability where $0 \le \varphi(B_a,w) \le 1$ as $0 \le p_i \le 1$

Performance Comparison

Comparison of Costs

The communication cost of an operation is directly proportional to the size of the quorum required to execute the operation. Therefore, one represents the communication cost in terms of the quorum size. Table 1 shows the read and write costs of the two techniques between DRG and ROWA for different total number of copies, n = 16, 25, 36, 49, 64, and 81. The compared data of ROWA is derived from MUSTAFA. From Table 1, it is apparent that DRG has the lowest cost for write operations in spite of having a bigger

Table 1. Comparison of the read and write cost between GS and DRG under the different copies

	Number of copies in the system					
	16	25	36	49	64	81
ROWA (R)	1	1	1	1	1	1
ROWA (W)	16	25	36	49	64	81
DRG(R)	2	3	3	4	4	5
DRG (W)	3	3	4	4	5	5

Table 2. Comparison of the read availability between ROWA and DRG

Techniques	Read Availability								
	0.1	**0.2**	**0.3**	**0.4**	**0.5**	**0.6**	**0.7**	**0.8**	**0.9**
ROWA, N=25	1	1	1	1	1	1	1	1	1
ROWA, N=36	1	1	1	1	1	1	1	1	1
ROWA, N=64	1	1	1	1	1	1	1	1	1
DRG, N=25	0.837	0.837	0.837	0.837	0.837	0.837	0.837	0.837	0.837
DRG, N=36	0.763	0.781	0.8	0.818	0.837	0.855	0.874	0.892	0.911
DRG, N=64	0.82	0.833	0.847	0.86	0.874	0.888	0.901	0.915	0.928

Table 3. Comparison of the write availability between ROWA and DRG

Techniques	Write Availability								
	0.1	**0.2**	**0.3**	**0.4**	**0.5**	**0.6**	**0.7**	**0.8**	**0.9**
ROWA, N=25	0	0	0	0	0	0	0	0	0
ROWA, N=36	0	0	0	0	0	0	0	0	0
ROWA, N=64	0	0	0	0	0	0	0	0	0
DRG, N=25	0.837	0.837	0.837	0.837	0.837	0.837	0.837	0.837	0.837
DRG, N=36	0.763	0.781	0.8	0.818	0.837	0.855	0.874	0.892	0.911
DRG, N=64	0.82	0.833	0.847	0.86	0.874	0.888	0.901	0.915	0.928

number of copies when compared with ROWA quorums. It can be seen that DRG needs only 3 copies for the write quorums with 25 copies. On the opposite, the write cost is 25 for the ROWA with 25 copies. It increases more than DRG as a number of copies increases. For example, it increases to 81 for the write cost for ROWA with 81 copies, while for DRG; it increases to 5 for the write cost. While in the read operation it is apparent that ROWA has the lowest cost for read operations in spite of having a bigger number of copies when compared with DRG quorums because the data object will replicate at all the sites and (in this case) will increase the accessibility and availability of the data and make the cost of reading equal to 1 compared with a DRG cost between 2 and 5 for copies between 16 and 81. Looking at the lowest cost or reading in the ROWA it is still not efficient because the copy of the database is not replicated at all the sites causing a disadvantage in that this take large space for storage with an increase in the response time. So still the read at DRG technique is better than ROWA because the data item will replicated to diagonal sites only and this case will decrease the response time and updating.

Figure 2. Comparison of the read availability between ROWA and DRG

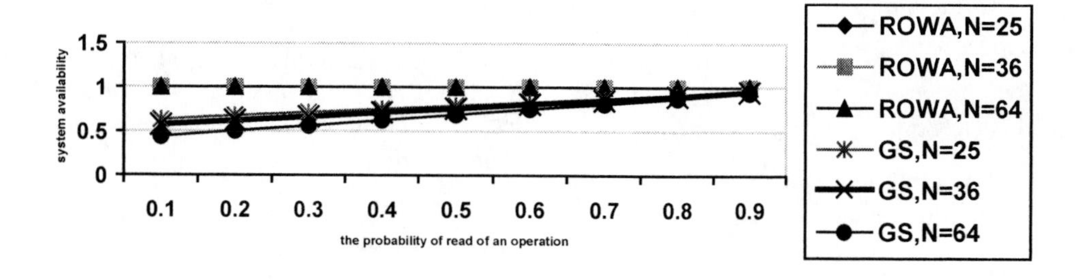

Figure 3. Comparison of the read availability between ROWA and DRG

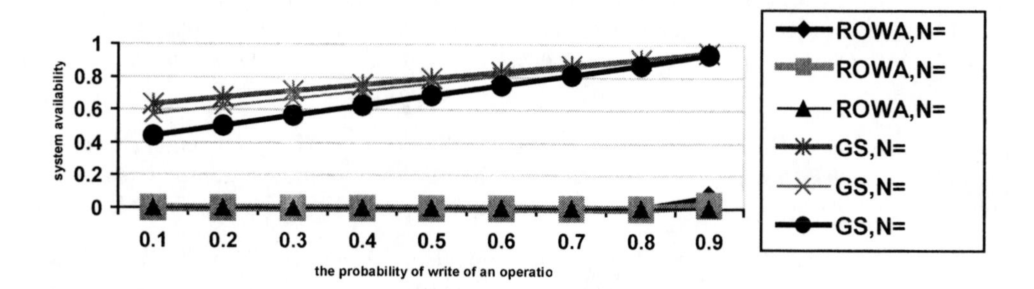

Comparisons of Read/Write Availabilities

In this section, one will compare the performance on the read/write availability of the ROWA technique based on equations (1) and (2), and our DRG technique based on equations (3) and (4) for the case of n= 25, 36, and 64. In estimating the availability of operations, all copies are assumed to have the same availability.

Figures 2 and 3 and Tables 2 and 3 show the results obtained from the analysis for read and write availabilities between those two protocols when n=25,36 and 64.We assume that all data copies have the same availability, p, and varies from 0.1 to 0.9. From Tables 2 and 3, note that read availability for ROWA outperformed the DRG technique. This is due to the fact that ROWA needs only one copy for the communication cost. However those copies have the availability of more than 90% when individual copy has the availability of 70% and above. On opposite, the write availability for DRG outperformed from ROWA. This is due to fact that the number of copies needed to construct.

CONCLUSION

In this article we show the typical two-planned disconnection models to be suitable with P2P system, but we have presented a new technique, called the "Hybrid Replication Technique." This has been proposed to manage data replication in the fixed and ad hoc P2P network environment. The replication technique for the fixed network is proposed and based on diagonal replication technique (DRG), while the replication for the ad hoc network is done asynchronously only to the most frequently visited site by using one of the planned disconnection models. The analysis of the DRG has been presented in terms of read/write availability and communication costs. This has showed that, the DRG technique provides a convenient approach to high availability for update-frequent operations. This is due to the minimum number of quorum size required. In comparison to the ROWA, DRG requires significantly lower communication cost for an operation, while providing higher system availability, which is preferred for P2P environment.

REFERENCES

Agrawal, D., & El Abbadi, A. (1996). Using reconfiguration for efficient management of replicated data. *IEEE Transactions on Knowledge and Data Engineering*, *8*(5), 786-801.

Bernstein, P. A., Hadzilacos, V., & Goodman, N. (1987). Concurrency control and recovery in database systems. Reading, MA: Addison Wesley.

Bernstein, P. A., & Goodman, N. (1994). An algorithm for concurrency control and recovery in replicated distributed databases. *ACM Transactions on Database Systems, 9*(4), 596-615.

Borowski, S. (1996). *Oracle 7 Concepts Release 7.3*. Redwood City, CA: Oracle Corp.

Buretta, M. (1997). Data replication: Tools and techniques for managing distributed information. New York: John Wiley.

Faiz, M., & Zaslavsky, A. (1995, March). Database replica management strategies in multidatabase systems with mobile hosts. In *Proceedings of the 6th International Hong Kong Computer Society Database Workshop.*

Goldring, R. (1995, May). Things very update replication customer should know. In *Proceedings of ACM SIGMOD International Conference On Management of Data* (pp. 439-440).

Helal, A. A., Heddaya, A. A., & Bhargava, B. B. (1996). *Replication techniques in distributed systems*. MA: Kluwer Academic Publishers.

Holliday, J., Agrawal, D., & Abbadi, A. E. (2000, February). Exploiting planned disconnections in mobile environments. In *Proceedings of the 10th IEEE Workshop on Research Issues in Data Engineering (RIDE2000)* (pp. 25–29).

Maekawa, M. (1985). A $\sqrt{n}$ algorithm for mutual exclusion in decentralized systems. *ACM Transactions on Computer Systems, 3*(2), 145-159.

Mat Deris, M., Evans, D. J., Saman, M. Y., & Noraziah, A. (2003). Binary vote assignment on a grid for efficient access of replicated data. *International Journal of Computer Mathematics, 80.*

MAT DERIS, M. (2001). *Efficient access of replication data in distributed database systems*. PhD thesis, University Putra Malaysia.

Ozsu, M. T., & Valduriez, P. (1999). *Principles of distributed database system* (2nd ed.). Prentice Hall.

Reiher, P. P., Heidemann, J., Ratner, D., Skinner, G., & Popek, G. (1994, June). Resolving file conflicts in the Ficus file system. In *Proceedings of the Summer USENIX Conference* (pp. 183-195).

Walborn, S., & Chrysanthis, P. K. (1997). Pro-motion: Management of mobile transactions. In *Proceedings of the 11th ACM Symposium on Applied Computing.*

Wolfson, O., Jajodia, S., & Huang, Y. (1997). An adaptive data replication algorithm. *ACM Transactions on Database Systems, 22*(2), 255-314.

Embedded Agents for Mobile Services

John F. Bradley
University College Dublin, Ireland

Conor Muldoon
University College Dublin, Ireland

Gregory M. P. O'Hare
University College Dublin, Ireland

Michael J. O'Grady
University College Dublin, Ireland

INTRODUCTION

A significant rise in the use of mobile computing technologies has been witnessed in recent years. Various interpretations of the mobile computing paradigm, for example, ubiquitous and pervasive computing (Weiser, 1991) and more recently, ambient intelligence (Aarts & Marzano, 2003)—have been the subject of much research. The vision of mobile computing is often held as one of "smart" devices operating seamlessly and dynamically, forming ad-hoc networks with other related devices, and presenting the user with a truly ubiquitous intelligent environment. This vision offers many similarities with the concept of distributed artificial intelligence where autonomous entities, known as agents, interact with one another forming ad-hoc alliances, and working both reactively and proactively to achieve individual and common objectives.

This article will focus on the current state of the art in the deployment of multi-agent systems on mobile devices and smart phones. A number of platforms will be described, along with some practical issues concerning the deployment of agents in mobile applications.

BACKGROUND

In the most general terms, an agent is one entity that acts, or has the authority to act, on behalf of another. In terms of information technology, an agent is a computational entity that acts on behalf of a human user, software entity, or another agent. Agents have a number of attributes that distinguish them from other software (Bradshaw, 1997; Etzioni & Weld, 1995; Franklin & Graesser, 1996; Wooldridge & Jennings, 1995):

- **Autonomy:** The ability to operate without the direct intervention from any entity, and possess control over their own actions and internal state.
- **Reactivity:** The ability to perceive their environment and react to changes in an appropriate fashion.
- **Proactivity:** The ability to exhibit goal-directed behavior by taking the initiative.
- **Inferential Capability:** The ability to make decisions based on current knowledge of self, environment, and general goals.
- **Social Ability:** The ability to collaborate and communicate with other entities.
- **Temporal Persistence:** The ability to have attributes like identity and internal state to continue over time.
- **Personality:** The ability to demonstrate the attributes of a believable character.
- **Mobility:** The ability to migrate self, either proactively or reactively, from one host device to another.
- **Adaptivity:** The ability to change based on experience.

An agent requires some space where it can exist and function, and this is provided for by an agent platform (AP). An AP comprises "the machine(s), operating system, agent support software,...agent management components...and agents" (FIPA, 2000, p. 6). The AP allows for agent creation, execution, and communication.

The majority of computer systems currently in operation use algorithms that are based on the concept of perfect information. The problem is that in the real world, businesses often require software functionality that is much more complex than this (Georgeff, Pell, Pollack, Tambe, & Wooldridge, 1999). Typically, computational entities within these systems should have an innate ability to deal with partial information and uncertainty within their environment. These types of systems are highly complex and are intractable using traditional approaches to software development. The rate at which business systems must change, due to market pressures and new information coming to light, requires software architectures and languages that more efficiently manage the complexity that results from alterations being made to the code and the specifications.

Agent architectures, and in particular belief-desire-intention (BDI) (Rao & Geogeff, 1995) agent architectures, are specifically designed to deal with these types of issues and thus contain mechanisms for dealing with uncertainty and change. A problem with traditional systems is that they assume that they exist within a static or constant world that contains perfect information. The types of mobile systems that we are concerned with are dynamic and perhaps even chaotic, embedded with agents that have a partial view of the world and which are resource bounded.

Agents rarely exist in isolation, but usually form a coalition of agents in what is termed a multi-agent system (MAS). Though endowed with particular responsibilities, each individual agent collaborates with other agents to fulfill the objectives of the MAS. Fundamental to this collaboration is the existence of an Agent Communications Language (ACL), which is shared and understood by all agents. The necessity to support inter-agent communication has led to the development of an international ACL standard, which has been ratified by the Foundation for Intelligent Physical Agents (FIPA).

JAVA 2 MICRO EDITION (J2ME)

Most agent platforms developed for mobile devices have been written in the Java programming language—on mobile devices that usually means Java 2 Micro Edition (J2ME). This edition of Java contains a cut down API, a reduced footprint Java Virtual Machine, and a slightly different syntax (e.g., parameterized classes in Java 5). Java applications that contain dependencies on the idiosyncrasies of the different editions cannot be ported to a different range of devices without making alterations to the code. Their performance, however, is improved because the code is no longer developed to the lowest common denominator. Different algorithms and coding styles are now used for desktop machines and embedded devices rather than adopting comprised or overarching approaches that do not maximize the performance or maintainability of either.

A NUMBER OF AGENT PLATFORMS EXISTS FOR MOBILE DEVICES

3APL-M

3APL-M (Koch, 2005) is a platform that enables the fabrication of agents using the Artificial Autonomous Agents Programming Language (3APL) (Dastani, Riemsdijk, Dignum, & Meye, 2003) for internal knowledge representation. Its binary version is distributed in J2ME and J2SE compilations. 3APL provides programming constructs for implementing agents' beliefs, goals, basic capabilities, and a set of practical reasoning rules. The framework comprises an API that allows a Java application to call 3APL logic and deliberation structures.

Agent Factory Micro Edition

Agent Factory Micro Edition (AFME) (Muldoon, O'Hare, Collier, & O'Grady, 2006) is an agent platform developed for the construction of lightweight intelligent agents for cellular digital mobile phones and other compatible mobile devices. AFME is broadly based on Agent Factory (Collier, 2001), a pre-existing J2SE framework for the fabrication and deployment of agents. AFME differs from the original version of the system in that it has been designed to operate on top of the Constrained Limited Device Configuration (CLDC) Java platform augmented with the Mobile Information Device Profile (MIDP). CLDC and MIDP form a subset of the J2ME platform specifications. Though sharing the same broad objectives of the other projects mentioned in this section, AFME differs in a number of ways. With a jar size of 85k, it is probably the smallest footprint FIPA-compliant deliberative agent platform in the world. The platform supports the development of a type of software agent that is: autonomous, situated, socially able, intentional, rational, and mobile. An agent-oriented programming language and interpreter facilitate the expression of an agent's behavior through the formal notions of belief and commitment. This approach is consistent with a BDI agent model.

LEAP

Probably the most widely known agent platform for resource-constrained devices is the Light Extensible Agent Platform (LEAP) (Berger, Rusitschka, Toropov, Watzke, & Schichte, 2002). LEAP is a FIPA-compliant agent platform developed to be capable of operating on both fixed and mobile devices with various operating systems in wired or wireless networks. Since version 3.0, LEAP extends the Java Agent DEvelopment Framework (JADE) (Bellifemine, Caire, Poggi, & Rimassa, 2003) by using a set of profiles that allow it to be configured for various Java Virtual Machines (JVMs). The architecture of the platform is modular and contains components for managing the lifecycle of the agents and controlling the array of communication protocols. The platform is split into several agent containers, one for every device used. These containers are responsible for passing messages between agents and choosing the appropriate communication protocol. One of these containers, known as the main container, includes agents that fulfill the white and yellow pages services as necessitated by the FIPA specification.

MAE

The MAE (mobile agent environment) (Mihailescu, Binder, & Kendall, 2002) agent platform has been designed to be independent of device and language implementations. To accomplish this, the platform is divided into a reference API specification, reference implementation, and non-standard implementation additions. The reference API specification is not dependent on programming language or hardware, and it contains the core platform components. The Reference Implementation contains all the device-dependent code required by the reference API specification. The third part, non-standard implementation additions, is used for application-specific components.

While this approach gives a high degree of platform independence, unless it is being deployed in an environment of homogeneous devices, it means a lot of work as each platform may require its own implementation.

MicroFIPA-OS

MicroFIPA-OS is an agent toolkit based on the standard FIPA-OS but optimized for resource-constrained mobile devices (Tarkoma & Laukkanen, 2002). It targets the personal Java platform and thus operates on personal data assistants. The system can run in minimal mode whereby agents do not use task and conversation managers. Yellow and white page services are provided in compliance with the FIPA specification. The platform is entirely embedded, however it is recommended that only one agent operate on low-specification devices.

NON-EMBEDDED AGENTS FOR MOBILE SERVICES

There are other types of agent platforms suitable for mobile services that do not embed the agents in the mobile device:

- platforms that use the mobile device as just an interface while the agents are executed on more capable hosts, for example MobiAgent; and
- platforms that do part of the execution on the mobile device, while simultaneously executing the remainder the task on other hosts such as KSACI (Hübner, 2000a, 2000b).

MobiAgent

A MobiAgent (Mahmoud, 2001) platform comprises a handheld mobile wireless device and an agent gateway, which are networked and communicate through hypertext transfer protocol (HTTP). The agent gateway executes the agent and its associated apparatus. The user interacts with the agent through an interface on the mobile device, which connects to the agent gateway and configures the agent. After the agent carries out a task, it reports back through the interface. This approach requires the minimum amount of processing and memory resources on the mobile device, but it makes the connectivity essential.

KSACI

Simple agent communication infrastructure (SACI) is a framework for creating agents that communicate using the Knowledge, Query, and Manipulation Language (KQML) (Finin, 1997). Each SACI agent has a mailbox to communicate with other agents. Infrastructure support is provided for white and yellow pages, but the platform is not FIPA compliant. KSACI is a smaller version of SACI suitable for running on the kVM (Albuquerque, Hübner, de Paula, Sichman, & Ramalho, 2001). The platform is not entirely situated on the constrained device and only supports the running of a single agent, which communicates via HTTP with a proxy running on a desktop machine.

DISCUSSION

A mobile computing environment is typified by resource constraints. Issues like processing power, memory, battery life, connectivity, and input/ouput (I/O) all require careful consideration.

It is often reported that intelligent agent platforms are unsuited for mobile applications because of their excessive computational overhead. This problem is usually due to particular agent platform implementations rather than an innate problem with the agent paradigm itself. Improving the efficiency of the reasoning algorithms within these systems can often lead to significant gains in efficiency. Additionally, the programming style adopted by the developer can have a considerable impact on performance. Developing in a style that conforms to the Law of Demeter (Lieberherr, Holland, & Riel, 1988) can reduce the footprint of the software by minimizing duplicated code while also improving maintainability in that internal implementation details of the object model are hidden. Further performance gains may be obtained through the use of autonomic procedures. An example of such a procedure, termed Collaborative Agent Tuning, may be found in Muldoon, O'Hare, and O'Grady (2005). Tuning enables agents collectively to alter their response times and computational overhead so as to maximize system performance.

The communication infrastructure is another fundamental resource that must be managed astutely when developing multi-agent systems. It is particularly important when working with lightweight devices that have limited battery power since sending messages consumes significantly more battery resources than normal processing. Mobile devices often have limited bandwidth and must make intelligent decisions as to what information to download and when to download it.

Additionally there is the issue of human-agent communication. Consideration must be made for the I/O capabilities of the devices. Most would have some form of keyboard input in the form of a touch screen or keypad. How the agent would convey information would be a bigger modality issue—is there a screen, does it allow for graphics or just text, how big is the screen, and how much of it is available to the agent?

FUTURE TRENDS

In the future, agents will emerge that are endowed with autonomy, mobility, and human-computer interaction facilities (Bradley, Duffy, O'Hare, Martin, & Schön, 2004). Such agents will opportunistically migrate, based on their tasks at hand, to different platforms (each offering varying capabilities and prospects), which would usually be for the benefit of an associated user. The presence of the agent moving through cyberspace as the user moves through physical space allows the associated user to be contactable at anytime through the agent.

A clear application of such nomadic agents is that of an autonomous "intelligent" digital assistant that is independent of any one physical device. These entities will effectively give any user his or her own personal assistant that will help with the information overload in daily life, assisting with personal communications and offer a generic interface to any number of devices. These devices will have the ability to react to the current needs of their user, and beyond this, grow and learn to anticipate future needs and requirements. Perhaps our vision can be best summed up by Luc Steels' metaphor for what the robots of the future will be like:

[It] is related to the age-old mythological concept of angels. Almost every culture has imagined persistent beings which help humans through their life. These beings are ascribed cognitive powers, often beyond those of humans, and are supposed to be able to perceive and act in the real world by materialising themselves in a bodily form at will. (Steels, 1999)

He goes on to detail how angels may "project the idea of someone protecting you, preventing you from making bad decisions or actions, empowering you, and defending you in places of influence."

CONCLUSION

Agents encapsulate a number of features that make them an attractive and viable option for realizing mobile services. At a basic level, their autonomous nature, ability to react to external events, as well as an inherent capability to be proactive in fulfilling their objectives make them particularly suitable for operating in complex and dynamic environments. Should an agent be endowed with a mobility capability, its ability to adapt and respond to unexpected events is further enhanced. However, there are a few negative aspects to using agents. These systems can be more complex and require more device resources than the equivalent application-specific programs. Having no native support, agents require their own agent platforms for creation, execution, and communication. These problems will be reduced with advancements in mobile computing technologies, however in order to optimize system performance, agents will still have to manage their resources in a prudent and intelligent manner.

REFERENCES

Aarts, E., & Marzano, S. (Eds.). (2003). *The new everyday: Views on ambient intelligence.* Rotterdam, The Netherlands: 010.

Albuquerque, R. L., Hübner, J. F., de Paula, G. E., Sichman, J. S., & Ramalho, G. L. (2001, August 1-3). KSACI: A handheld device infrastructure for agents communication. *Pre-proceedings of the 8th International Workshop on Agent Theories, Architectures, and Languages (ATAL-2001)*, Seattle, WA.

Bellifemine, F., Caire, G., Poggi, A., & Rimassa, G. (2003, September). *JADE.* White Paper.

Berger, M., Rusitschka, S., Toropov, D., Watzke, M., & Schichte, M. (2002). Porting distributed agent-middleware to small mobile devices. *Proceedings of the Workshop on Ubiquitous Agents on Embedded, Wearable, and Mobile Devices held in conjunction with the Joint Conference on Autonomous Agents and Multi-Agent Systems (AAMAS)*, Bologna, Italy.

Bradley, J. F., Duffy, B. R., O'Hare, G. M. P., Martin, A. N., & Schön, B. (2004, September 7-8). Virtual personal assistants in a pervasive computing world. *Proceedings of IEEE Systems, Man and Cybernetics, the UK-RI 3rd Workshop on Intelligent Cybernetic Systems (ICS'04)*, Derry, Northern Ireland.

Bradshaw, J. M. (1997). An introduction to software agents. In J. M. Bradshaw (Ed.), *Software agents* (pp. 3-46). Boston: MIT Press.

Collier, R. W. (2001, March). *Agent factory: A framework for the engineering of agent-oriented applications.* PhD thesis, Department of Computer Science, University College Dublin, National University of Ireland.

Dastani, M., Riemsdijk, B., Dignum, F., & Meye, J. J. (2003). A programming language for cognitive agents: Goal directed 3APL. *Proceedings of the 1st Workshop on Programming Multiagent Systems: Languages, Frameworks, Techniques, and Tools* (ProMAS), Melbourne.

Etzioni, O., & Weld, D. S. (1995). Intelligent agents on the Internet: Fact, fiction, and forecast. *IEEE Expert, 10*(4), 44-49.

Finin, T., & Labrou, Y. (1997). KQML as an agent communication language. In J. M. Bradshaw (Ed.), *Software agents* (pp. 291-316). Boston: The MIT Press.

FIPA (Foundation for Intelligent Physical Agents). (2000). *FIPA agent management specification.* Retrieved from http://www.fipa.org

Franklin, S., & Graesser, A. (1996). Is it an agent or just a program? A taxonomy for autonomous agents. *Proceedings of the 3rd International Workshop on Agent Theories, Architectures, and Languages.* New York: Springer-Verlag.

Georgeff, M., Pell, B., Pollack, M., Tambe, M., & Wooldridge, M. (1999). The belief-desire-intention model of agency. *Proceedings of the 5th International Workshop on Intelligent Agents V: Agent Theories, Architectures, and Languages (ATAL-98)*, Paris, France.

Hübner, J. F., & Sichman, J. S. (2000a). SACI: Uma ferramenta para implementação e monitoração da comunicação entre agents. *Proceedings of IBERAMIA.*

Hübner, J. F., & Sichman, J. S. (2000b). *SACI programming guide.*

Koch, F. (2005, July 25-29). 3APL-M platform for deliberative agents in mobile devices. *Proceedings of the 4th International Joint Conference on Autonomous Agents and Multiagent Systems (AAMAS"05)* (pp. 153-154), The Netherlands. New York: ACM Press.

Lieberherr, K. J., Holland, I., & Riel, A. J. (1988). Object-oriented programming: An objective sense of style. Object oriented programming systems, languages and applications conference. *SIGPLAN Notices (Special Issue),* (11), 323-334.

Mahmoud, Q. H. (2001). MobiAgent: An agent-based approach to wireless information systems. *Proceedings of the 3rd International Bi-Conference Workshop on Agent-Oriented Information Systems,* Montreal, Canada.

Mihailescu, P., Binder, W., & Kendall, E. (2002). MAE: A mobile agent platform for building wireless m-commerce applications. *Proceedings of the 8th ECOOP Workshop on Mobile Object Systems: Agent Applications and New Frontiers,* Málaga, Spain.

Muldoon, C., O'Hare, G. M. P., Collier, R. W., & O'Grady, M. J. (2006, May 28-31). Agent factory micro edition: A framework for ambient applications. *Proceedings of Intelligent Agents in Computing Systems, a Workshop of the International Conference on Computational Science (ICCS 2006)*, Reading.

Muldoon, C., O'Hare, G. M. P., & O'Grady, M. J. (2005). Collaborative agent tuning. *Proceedings of the 6th International Workshop on Engineering Societies in the Agents' World (ESAW 2005)*, Kusadasi, Turkey.

Rao, A. S., & Georgeff, M. P. (1995, June). BDI agents: From theory to practice. *Proceedings of the 1st International Conference on Multi-Agent Systems (ICMAS'95)* (pp. 312-319), San Francisco.

Steels, L. (1999). *Digital angels.* Retrieved from http://arti.vub.ac.be/steels/sued-deutsche.pdf

Tarkoma, S., & Laukkanen, M. (2002). Supporting software agents on small devices. *Proceedings of the 1st International Joint Conference on Autonomous Agents and Multi-Agent Systems (AAMAS)*, Bologna, Italy.

Weiser, M. (1991). The computer for the twenty-first century. *Scientific American,* (September), 94-100.

Wooldridge, M., & Jennings, N. R. (1995). Intelligent agents: Theory and practice. *Knowledge Engineering Review, 10*(2), 115-152.

KEY TERMS

Agent: A computational entity that acts or has the authority to act on behalf of a human user, software entity, or another agent.

Agent Communication Language: A formal language used for communication between agents.

Agent Platform: Provides the necessary infrastructure on which an agent operates.

Ambient Intelligence: Computing and networking technology that is unobtrusively embedded in the environment.

Embedded Agent: An agent that is contained wholly, along with its platform, on a particular device.

Mobile Service: One of several services provided through devices in a mobile computing environment (i.e., mobile phones, personal data assistants, wearable computers, etc.).

Multi-Agent System: A system comprising several agents—on the same platform or across multiple platforms—with a common goal.

Pervasive Computing: Computing involving computers, usually mobile devices, in all aspects of daily life.

Ubiquitous Computing: Computing in which the computers are embedded in everyday objects and all computing is done in the background.

Enabling Mobile Chat Using Bluetooth

Ádrian Lívio Vasconcelos Guedes
Federal University of Campina Grande, Brazil

Jerônimo Silva Rocha
Federal University of Campina Grande, Brazil

Hyggo Almeida
Federal University of Campina Grande, Brazil

Angelo Perkusich
Federal University of Campina Grande, Brazil

INTRODUCTION

Mobile chat applications can be seen as an alternative and effective way of communicating for people without the need of using the mobile telephony system. Based on the new generation of cellular phones with support for communication technologies, such as Bluetooth and Wi-Fi, it is possible to develop applications to enable mobile chats. Such applications can provide mechanisms to discover and communicate with other devices in a shorter range, but with low or no communication costs.

This article introduces Let's Talk, a mobile chat and relationship application. It allows a Symbian OS Series 60 mobile phone user to create a profile and share it with other users. Also, it is possible to invite other users to a chat in a session. The profile sharing and the chat communication data are transferred over a Bluetooth connection. After creating your profile, a user can search for other profiles in the range of the Bluetooth connection and make your profile available to other users.

In this article, we discuss design and implementation issues related to the application development using a Symbian-based cellular phone and the C++ programming language. The remainder of this article is organized as follows. We first present the technologies used to develop the application: Symbian OS, the Series 60 platform, the Bluetooth wireless technology, and the Cobain Framework. We then present the Let's Talk software and the use of the technologies presented in the Background section. Possible improvements for the application and trends related to the theme are then offered, followed by final remarks in the Conclusion section.

BACKGROUND

Symbian OS

The Symbian OS (http://www.symbian.com/) is an operating system designed for mobile devices; it is an industry standard, used in smart phones of many manufacturers, such as Nokia, Siemens, Motorola, Sansung, and others.

Symbian is optimized for mobile devices that have low memory and processing power, with low runtime memory requirements. It is designed to optimize the device performance and the battery life. It is a multi-tasking operating system, allowing many applications to run concurrently. To reduce recourse consumption, Symbian provides multi-thread support to the programmer through the concept of active objects, which are a lightweight alternative to threads.

The Symbian OS development model is based on an object-oriented architecture using the C++ programming language with optimized memory management for embedded software.

Series 60

The Series 60 platform (http://www.s60.com/) was developed by Nokia, but it is also licensed to other manufacturers. It was built over the Symbian Operating System, providing a configurable graphical user interface library and a set of applications and other general-purpose implementations.

The set of applications includes personal information management (PIM) and multimedia applications, such as calendars, contacts, text and multimedia messaging (SMS,

Figure 1. Series 60 user interface

MMS), e-mail, browsing using WAP or others, and so forth.

Some of the main features of Series 60 are the large color screen with a minimum specification of 172 by 208 pixels, and at least 4,096 colors (64K colors in Series 60 2.x) and many interaction models, such as two soft keys, five-way navigator, and other dedicated keys (Edwards, Barker, & Staff of EMCC Software, 2004). The Series 60 User interface is illustrated in Figure 1.

Bluetooth

Bluetooth (IEEE 802.15.1) is a wireless specification for personal area networks (PANs). It provides a way to connect and exchange information between devices such as personal digital assistants (PDAs), mobile phones, laptops, PCs, printers, and digital cameras via a secure, low-cost, globally available short-range radio frequency (http://www.bluetooth.com/). Bluetooth is available in most Series 60 devices providing connectivity to these devices.

The Cobain Framework

Cobain is an API (application programming interface) that permits the development of Bluetooth applications, simplifying the development process for this kind of application in the Symbian OS (Dahlbom & Kokkola, 2004). It consists of a lightweight ad-hoc networking framework, providing a Unix-like API socket, hiding details of implementation, such as Active Objects handling.

Figure. 2 Profile form

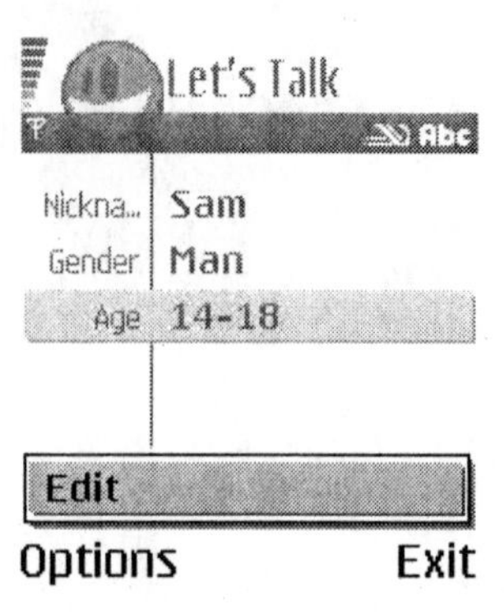

LET'S TALK

Let's Talk is a chat application for Symbian OS Series 60 mobile phones, allowing users to contact another people and establish a conversation. The profile sharing mechanism allows a user to create a profile and share it with other users. The profile contains personal information, such as name, age, and gender. This information is made available to other Let's Talk users that can invite this user to chat after viewing his/her profile. This feature enables the establishment of relationships between users based on the level of interest in their profiles.

The application may be running in two modes: waiting or searching. The searching application can discover all devices running the application in the waiting mode and allows the user to request the profile of any discovered user. The profile request contains the searching user profile data, which will be evaluated by the waiting user.

The waiting mode informs the user that there is an incoming profile request and shows the profile of the requesting user in a form. In response to this request, the user sends the waiting user profile data to the searching application, which shows this profile to the searching user.

In each device, Let's Talk creates a form that contains the profile of the other user using the data sent via Bluetooth. In the Series 60 platform, the forms provide a way for the user quickly and easily to enter or edit many items of data in the application. The form could also be used in *view mode* to display information about the user profile. If a form has a view mode, like the profile creation form, the form focus appears as a solid block as illustrated in Figure 2. Switching to edit mode is achieved by selection of Edit from the Options menu (Edwards et al., 2004).

The profile sharing and the chat communication data are transferred over a Bluetooth connection. The connection is established using the Cobain API that is responsible for: (1) discovering devices available for Bluetooth connection, (2) discovering services available at the selected device, (3) connecting to the given service, (4) sending and receiving

Figure 3. Chat user interface

the chat and profile data to and from the remote service, (5) and closing the connection (Dahlbom & Kokkola, 2004). The chat user interface is depicted in Figure 3.

The application can be used in several places like restaurants, shopping centers, subways, coffee shops, and other places with great people concentration. For these various scenarios the application behavior is similar, but improvements could be performed for specific contexts. For example, the profile of chat applications for restaurants could include preferences related to gastronomy.

FUTURE TRENDS

Application Improvements

In a future version of the application, some new functionalities may be implemented, such as search by profile information, context-sensitive profile, and automatic profile comparison.

- **Search by Profile Information:** The application can provide a search field for specific profile information. A user can search for all users that have a specific age, for example.
- **Context-Sensitive Profile:** The profile form fields may change depending on the application scenario. In a professional ambient such as an academic lab, the user profile may contain information about his/her skills in a specific area. Another user with a problem in a specific knowledge area can find someone to help him/her.
- **Automatic Profile Comparison:** The application may automatically share the profile and compare the contents. In a party, a user may be advertised that there is another user in this party that likes the same music style as him/her.
- **Server Connection:** Using a Bluetooth access point, the application may have access to a central service that can provide functionalities like profile database access, or improve the range of application over than a Bluetooth connection.

Future Trends in Mobile Chat

In the context of mobile chat, future trends include the usage of new technologies related to multimedia, content, and connection mechanisms. Considering connection mechanisms, Bluetooth architecture could work together with wired infrastructure to enable long-range chat.

A multimedia-enabled mobile chat application can provide creation, exchange, and presentation of videos, sounds, and images during a chat session. Video and audio chats could also be possible, without using the telephony infrastructure. For these applications, though, mechanisms of lower battery consumption are fundamental.

Access to content and services of wired infrastructures during chat sessions is also an interesting feature to be considered. A chat user could obtain content related to his/her conversation and services to support it, like dictionaries and translators.

CONCLUSION

Let's Talk is an interesting application for personal relationships, providing costless, wireless communication. The profile sharing mechanism enables users to choose persons to chat with based on profile descriptions.

With the improvements proposed earlier in this article, the application can become a powerful tool enabling the user to find and communicate with persons of interest.

The enhancement of new technologies enables the applications to have access to Web and multimedia contents. These contents also may be used in mobile chat applications like Let's Talk.

REFERENCES

Dahlbom, M., & Kokkola, M. (2004). *Cobain architectural specification.* Retrieved from http://irssibot.777-team.org/cobain/index.html

Edwards, L., Barker, R., & Staff of EMCC Software. (2004). *Developing Series 60 applications: A guide for Symbian OS C++ developers.* Boston: Addison-Wesley.

KEY TERMS

API: Application programming interface.

IEEE: Institute of Electrical and Electronics Engineers.

MMS: Multimedia message service.

PAN: Personal area network.

PDA: Personal digital assistant.

PIM: Personal information management.

SMS: Short message service.

WAP: Wireless application protocol.

Enabling Mobility in IPv6 Networks

Saidal Razalli Bin Azzuhri
Malaysia University of Science and Technology, Malaysia

K. Daniel Wong
Malaysia University of Science and Technology, Malaysia

INTRODUCTION

With the explosive growth in Internet usage over the last decade, the need for a larger address space is unavoidable, since all the addresses in IPv4 are nearly fully occupied. IPv6 (Deering & Hinden, 1998), with 128-bit addresses compared to IPv4 with 32-bit addresses and other advantages (like auto-configuration and IP mobility), can overcome many of the problems that IPv4 had before.

One of the requirements for the modern Internet is IP mobility support. In IPv4, a special router is needed to act as a foreign agent in the visited/foreign network and the need of a network element in the home network known as a home agent for a mobile host. IPv6 does away with the need for the foreign agent and operates in any location without any special support from a local router. Route optimization is inherent in IPv6, and this feature eliminates the triangle-routing (routing through the home agent) problem that exists in IPv4. IPv6 enjoys many network optimizations that are already built in within IPv6.

IP MOBILITY

IP mobility can be defined as referring to situations where there is a change in a node's IP address due to a change of its attachment point within the Internet topology (Soliman, 2004). This change may be caused by physical moment, such as someone moving her computer from one room to another or someone sitting in a moving vehicle that traverses different links. IP mobility can also occur due to change in the topology, which causes a node to change its address. Mobile IPv6 is a suite of protocols for IPv6 nodes to handle IP mobility.

Mobile IPv6 allows an IPv6 host to leave its home subnet while transparently maintaining all its connections and remaining reachable to the rest of the Internet. The use of IP in wireless technologies, such as local area networks (LANs; e.g., IEEE 802.11a, b, and g) to wide area networks (WANs; e.g., 3G), makes mobility in wireless devices an interesting research field. The popularity of wireless technologies allows users (hosts) to move freely within large geographical areas, but requires good support for mobility. Mobile IPv6 is the more prominent solution for mobility for IP wireless devices (Samad & Ishak, 2004). We will first review some relevant features of IPv6 before explaining how Mobile IPv6 works.

RELEVANT FEATURES OF IPV6

IPv6 specification was already defined in RFC 2460 (Deering & Hinden, 1998). Figures 1 and 2 show the header comparison between IPv4 and IPv6 headers. These include the header format, extension headers, and their processing fields. As can be seen in the figures, only the IPv6 header contains the minimum amounts of information necessary for IPv6 hosts to communicates with each other. The IPv6 packet header is much simpler than the IPv4 one. It is now a fixed size with no optional fields. Options in IPv4 are replaced by an IPv6

Figure 1. IPv4 header

8 bits		8 bits	8 bits	8 bits
version	IHL	type of service	total length	
identification			flags	fragment offset
TTL		protocol	header checksum	
source address (32 bits)				
destination address (32 bits)				
options				padding

Figure 2. IPv6 header

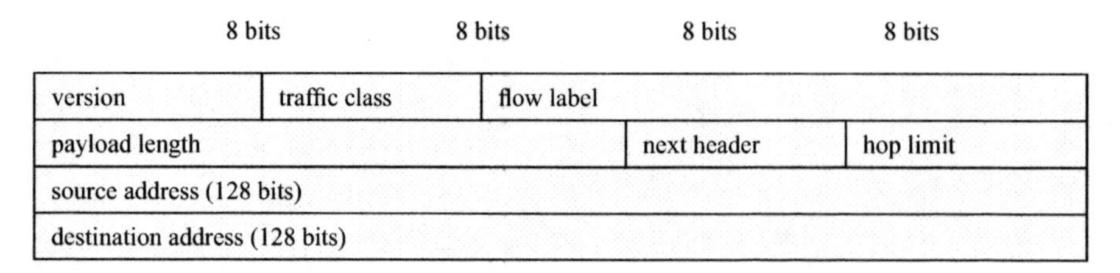

Figure 3. Extension header

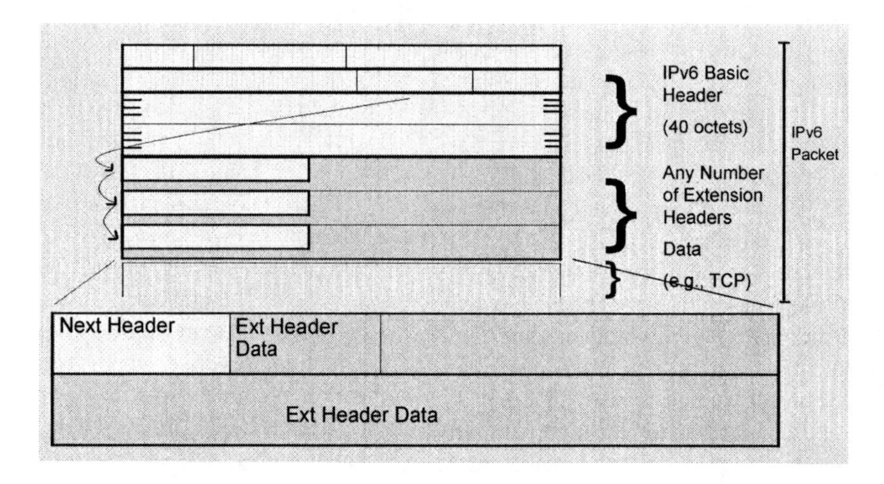

extension header (will be explained later), which includes additional parameters for hosts or routers to receive IPv6 packets. An extension header in IPv6 may contain one or more extension headers when necessary for the processing of such a packet (Deering & Hinden, 1998).

The IPv6 header removes some of the fields that were previously included in the IPv4, and added new fields. It has been slimmed down to the necessary minimum header compression, which now has 8 fields, compare to the previous 13 fields in IPv4. The following sections explain further some of these design choices.

Extension Header

Mobile IPv6 has an optional header called *extension header.* IPv6 extension headers are defined to encode certain options that are needed for processing of the IPv6 packet and its subsequent packets. Encoding options must minimize the amount of time needed in order to classify the header and forward the packet on the correct route. The benefits of extension headers can be best explained when comparing them to option fields in IPv4 headers. Consider the router receiving an IPv4 packet including one or more options. The router would first determine that the packet is carrying IP options. The next step is the router must parse or classify the IP header to find out which options require processing by the router itself, as opposed to processing by the ultimate receiver of the packet. The process of parsing this header and its options takes some time and can reduce efficiency.

Routing Header

IPv6 defines a fixed size 40-bytes header and extension header for additional options. The routing header includes addresses of nodes that must be in the path taken by a packet on its way to its ultimate destination. Thus, the routing header is a form of source routing and can be used to make sure the packet goes to certain nodes/addresses on its way to its ultimate destination. It also allows routing to certain special-purpose routers for special reasons (e.g., mobility support).

Hop-by-Hop Options Header

As the name implies, this header includes options that need to be processed by every node (routers) along the packet's delivery path. It specifies delivery parameters at each hop on the way to the destination. Some of the fields in this type of header are used to alert a router to things like multicast listener discovery, that is, that this packet is part of a multicast and requires special processing.

Destination Options Header

The destination option is used to specify a process that needs to be performed by the destination node, whose address is the destination address in the IPv6 header. It is useful for Mobile IPv6 as the destination options header used to exchange the registration messages between mobile nodes and the home agent. Delivery parameters are either intermediate hops or the final destination, similar to the hop-by-hop options. The difference between destination and hop-by-hop options is that the former is processed by nodes that the packets are destined to, while the latter is processed by every node along the network path until the last receiver.

Authentication Header

The authentication header is a mechanism to protect a packet's integrity following the establishment of a security association. It is used to provide data authentication and integrity checking information, but not encryption. In addition, the authentication header protects against replay attacks by including a sequence number field, which is incremented each time the packets are sent. It is mechanism of security in IPv6, not fully bullet-proof, but it provides first level of data security (Faccin & Le, 2003).

Fragment Header

Fragment header is used similarly as in IPv4. It indicates that this packet is part of a fragmented stream, but fragmentation is only allowed on the part of the sender. Routers are not allowed to fragment payloads, which makes for better quality-of-service overall. In IPv6, only the sending host can perform this function.

Tunneling

Tunneling can be defined as a process whose node (host or router) encapsulates an IPv6 packet in another IPv6 header, which can be two or more packets (if encapsulation is done more than once) (Jeong, Park, & Kim, 2004). There are several terms associated with tunneling:

- **Tunnel Entry Point:** Originating tunnel node.
- **Tunnel Exit Point:** Terminating tunnel mode.

The tunnel will act as a virtual point-to-point link when seen by the original IPv6 header, starting at the tunnel entry point and ending at the tunnel exit point. The header of the new IPv6 packet is shown in Figure 4. Note that the tunnel exit point decapsulates the packet, and the decapsulated packet will be sent to the host that its destined to.

Tunneled IPv6 Packet

When an IPv6 packet is tunneled inside a new IPv6 packet, the router along the tunnel point will only recognize the outer header, which contains a new source and destination address. Tunneling is a very important mechanism in mobility, which will be discussed later. When packets arrive at a tunnel exit point, the outer header is thrown away by the tunnel exit router, and the original header will be processed (decapsulation process). Mobile IPv6 makes use of tunnels where the source address of tunneled packets will be the home agent address and the destination address will be the mobile node (MN) care-of-address (COA).

Router Discovery

Router Solicitations Mechanism

A router solicitation is used by hosts to discover one or more default routers neighboring the solicitations host. It is sent to an all-routers multicast address (link-local multicast address). This link-local multiple address is hardcoded in all router implementations.

Router Advertisement Mechanism

A router advertisement can be described as a response to router solicitations. When a router is solicited, the advertisement is sent to a unicast address of soliciting node. By contrast,

Figure 4. IPv6 tunneling

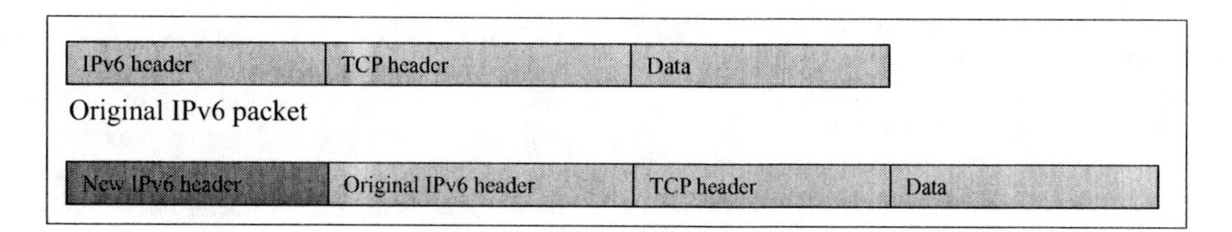

an unsolicited router advertisement is sent to 'all-nodes multicast addresses'.

Stateless Address Autoconfiguration

Using the prefix information obtained from a router advertisement (RA), nodes can append an EUI-64 bit interface identifier to the advertised prefix to form a unique IPv6 address. Thus, it can obtain a global address in a stateless way. The term stateless refers to the fact that the address is configured without the need for keeping a record for such address allocation in any node (no state), except for the node that assigned the address to one of its interfaces. This eliminates the need for a stateful server like a DHCP server that keeps track of addresses allocated on the link.

HOW IPV6 MOBILITY WORKS

Mobile IPv6 allows IPv6 nodes to be mobile. It also allows nodes to be reachable and maintain ongoing connections while changing their location within the network topology (Silva, Camilo, Costa, Matos, & Boavida, 2004). Connection maintenance is done by the IP layer using Mobile IPv6 messages, options, and processes that ensure correct delivery of data regardless of the mobile node's location. This operation is transparent to upper layers, in order to maintain sessions as the mobile node changes its location. It is important to understand the components of Mobile IPv6 first before going on to details on mobility operation.

Since the Internet protocol itself does not address node mobility, modification is needed to enable nodes to continue to receive packets when they change their points of attachment to the Internet. This raises the need for a mobile Internet protocol. The mobility support protocol for IPv6 developed by IETF is known as Mobile IPv6 (RFC3775) (Johnson, Perkins, & Arkko, 2004). Just like the relationship between IPv6 and IPv4, Mobile IPv6 evolved from its counterpart known as Mobile IPv4 (or just Mobile IP for short). However, several new features in IPv6 make it more accommodating to mobility than IPv4.

Mobile IPv6 Components

Mobile Node (MN)

Mobile node is an IPv6 node that can change links, and therefore addresses, while maintaining reachability using its home address. A mobile node has awareness of its home address and the global address for the link to which it is attached (known as the care-of address), and indicates its home address/care-of address mapping to the home agent and Mobile IPv6-capable nodes with which it is communicating.

Correspondent Node (CN)

Correspondent node is an IPv6 node that communicates with a mobile node. A correspondent node does not have to be Mobile IPv6-capable. However, if the correspondent node is Mobile IPv6-capable, it can also be a mobile node that is away from home.

Home Address

An address is assigned to the mobile node when it is attached at the home link and through which the mobile node is always reachable, regardless of its location on an IPv6 network. If the mobile node is attached to the home link, Mobile IPv6 processes are not used, and communication occurs normally. If the mobile node is away from home (not attached to the home link), packets addressed to the mobile node's home address are intercepted by the home agent and tunneled to the mobile node's current location on an IPv6 network. Because the mobile node is always assigned the home address, it is always logically connected to the home link, even if it is physically somewhere else.

Home Link (HL)

This is a link to which a home address prefix is assigned, from which the mobile node obtains its home address. The home agent resides on the home link.

Foreign Link (FL)

This is a link that is not the mobile node's home link, but one that is visited by mobile node.

Home Agent (HA)

A home agent is a router on the home link that maintains registrations of mobile nodes that are away from home and the different addresses that they are currently using. If the mobile node is away from home, it registers its current address with the home agent, which tunnels data sent to the mobile node's home address to the mobile node's current address on an IPv6 network and forwards tunneled data sent by the mobile node.

Care-of-Address (COA)

A COA is an address used by a mobile node while it is attached to a foreign link. For stateless address configuration, the care-of address is a combination of the foreign subnet prefix and an interface ID determined by the mobile node. The association of a home address with a care-of address for a mobile node is known as a *binding*. Correspondent nodes and home agents keep information on bindings in a *binding cache.*

Overall components of Mobile IPv6 and their placement in a typical network can be seen in Figure 5. Please note that the mobile node itself can be in the home link, in which case it will be using its home IPv6 address and normal IPv6 packet forwarding. Mobile IPv6 allows an IPv6 hosts to leave its home, while transparently maintaining all its connections and remaining unreachable to the rest of the Internet.

Mobile IPv6 Procedures

Figure 5 shows the Mobile IPv6 architecture. A mobile node (MN) first needs to determine whether it is currently connected to its home link or a foreign link. If it detects it has moved to a foreign link, it will obtain a care-of-address; it also reports its COA to the correspondent node (CN). These two procedures are called *binding update,* and once acknowledged (same as binding update, but in reverse direction, known as 'binding acknowledgement'), binding update with the correspondent node will be known as *route optimization.* Once the CN knows the mobile node's COA, it will be able to send further packets directly to mobile node's COA, without going through the triangle route via the mobile node's home agent as shown in Figure 5. The above brief overview of Mobile IPv6 operation contains three key components of the protocol, namely, *router discovery, address notification,* and *packet routing,* which will be further illustrated in the following sections.

Mobile IPv6 Router Discovery

Router discovery has three main functions for a mobile node. Firstly, router discovery determines whether the mobile node is currently connected to its home link or a foreign link. Two types of messages, *router advertisements* and *router solicitation,* are involved. Router advertisements are used by the routers and home agents to announce their capabilities to mobile nodes. Specifically, a router advertisement is periodically transmitted as broadcast on each attached network where the node is configured to perform as a home agent or a Mobile IPv6 router, or both. Router solicitations are sent by mobile nodes that do not have the patience to wait for the next periodic transmission of a router advertisement. So the only purpose of a router solicitation is to force any routers or home agents on the network to immediately transmit a router advertisement. This is useful when the frequency at which routers and home agents are transmitting is too low compared with the moving frequency of the mobile node.

Secondly, router discovery can detect whether the mobile node has moved from one network to another. The mobile node may perform location and movement detection by examining the network prefixes contained in a received advertisement. If any of these prefixes matches the network-prefix of the mobile node *home address,* then the mobile node is connected to its home link; otherwise, if none of the prefixes matches the network prefixes of the mobile node's home address, then the mobile node decides that it is connected to a foreign link.

Figure 5. Components of Mobile IPv6

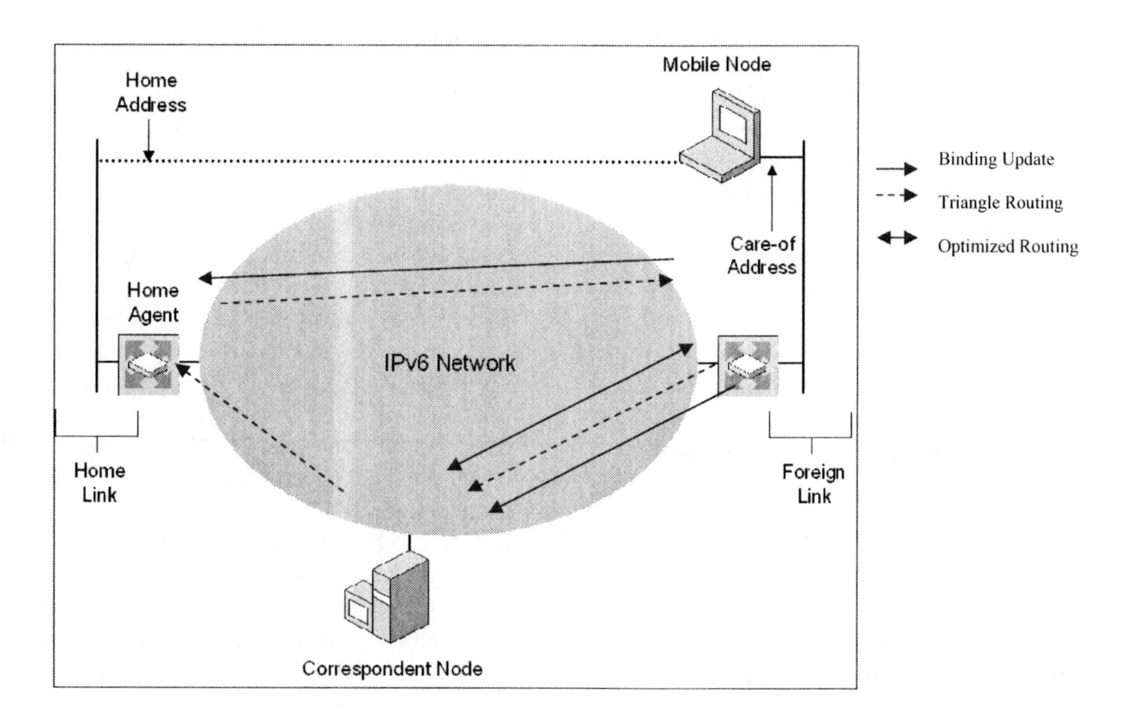

Thirdly, router discovery helps the mobile node obtain a COA in the foreign link. There are two methods by which a mobile node can acquire a COA in IPv6. The first method is *stateful address autoconfiguration,* in which the mobile node simply asks a server for an address and uses it as a care-of-address. The second method of acquiring a COA is *stateless address autoconfiguration,* in which a mobile node automatically forms a COA by concatenating a network prefix with an interface token. The first method is very similar to DHCP in IPv4 (the name is converted to DHCPv6), while the second is new to IPv6. The choice of specific method depends on the information contained in the received router advertisements.

Mobile IPv6 Address Notification

The Mobile IPv6 address notification is the process by which a mobile node informs both its home agent and various correspondent nodes of its current COA (Chirovolu, Agrawal, & Vandenhoute, 1999). The home agent uses the COA as the tunnel destination to forward a packet to the mobile node when it is away from home. The correspondent node uses the COA to route packets directly to the mobile node, without going through the mobile node's home agent. Thus, Mobile IPv6 has a built-in route optimization.

The messages used for notification include *binding request, binding update,* and *binding acknowledgment* (Qi, 2001). A binding request sent by a correspondent node to a mobile node can also initiate the binding update from it. Otherwise, the mobile node can also initiate the binding update by itself to both the home agent and correspondent node to inform them of its COA. The binding acknowledgement is sent by the receiver to tell the mobile node whether the received binding update is accepted or rejected. The mobile node can specify in the binding update whether a binding acknowledgement from the receiver is required or not.

Unlike in Mobile IPv4 where the address updating messages are carried as payloads of UDP/IP packets, all three types of Mobile IPv6 address notification messages are encoded as options to be carried within an IPv6 *destinations options header.* Therefore, these messages are only examined by the ultimate destination and not by any intervening routers along the path.

Mobile IPv6 Packet Routing

When connected to their home network or link, the mobile nodes send and receive packets just as any other stationary node. When a mobile node is connected to a foreign network:

1. *For packets route from the mobile node,* the mobile node must be able to determine a router that can forward packets generated by the mobile node and then just uses standard IP routing to deliver each packet to its destination. The search for a router is easier in IPv6 than in IPv4 because all IPv6 routers are required to implement router discovery, which is not the case in IPv4. As a result, a mobile node can select any router on the foreign link from which it has received router advertisement and configures its routing table to send all packets generated by itself to that router.
2. *For packets routed to the mobile node,* if the correspondent node is aware of mobile node's current COA, it will use the mobile node's COA in the IPv6 destination address field and put the mobile node's home address in a routing header. When the mobile node receives the packets at its COA, it looks within the routing header and finds its own home address as the ultimate destination of the IPv6 packet. Thus, the mobile node consumes the packet by sending it to the higher-layer protocols.

If the correspondent node is ignorant of mobile node's COA, it puts the mobile node's home address in the IPv6 destination address field and places its own address in the IPv6 Source address field and sends the packet out. The packet will then be received by the mobile node's home agent and tunneled to the mobile node COA. Furthermore, the mobile node interprets the presence of the tunnel to mean that the correspondent node does not know the mobile node's current COA and thus will send a binding update to inform the correspondent node of its address. Once this occurs, the correspondent node will then send packets directly to the mobile node as described above (route optimization).

DISCUSSION AND CONCLUSION

Our analysis of Mobile IPv6 finds it to be better than Mobile IPv4 in several ways. It is better integrated with IPv6 than Mobile IPv4 was with IPv4, perhaps because IPv6 was designed with mobility support as one of the requirements. Route optimization allows correspondent nodes to be directly informed about the current location of the MN, so to avoid triangular routing.

However, Mobile IPv6, like all complex protocols, has its weaknesses. Researchers have found various problems. For example, Mobile IPv6 is vulnerable to certain security attacks, such as on the address notification process. An attacker that eavesdrops on the right packets could send a falsely authenticated binding update message supposedly from the MN, and thus cause traffic meant for the MN to be diverted to an arbitrary address. Preliminary works on solutions to such problems have been reported (Lim & Wong, 2005). Another example of a problem with Mobile IPv6 is its vulnerability to the "simultaneous mobility" problem. When both MN and CN are mobile, and they move simultaneously, the address notification procedure may fail. Details and proposed solutions are given in Wong and Dutta (2005).

It is expected that most of the major "holes" in Mobile IPv6 should be fixed soon, as researchers report their findings and ideas.

REFERENCES

Chirovolu, G., Agrawal, A., & Vandenhoute, M. (1999). Mobility and QoS support for IPv6-based real-time wireless Internet traffic. *IEEE Communications Magazine.*

Deering, S., & Hinden, R. (1998). *Internet protocol, version 6.* RFC 2460, IETF.

Faccin, S. M., & Le, F. (2003). A secure and efficient solution to the IPv6 address ownership problem. *Proceedings of the International Conference on Communication Technology.*

Jeong, J., Park, J., & Kim, H. (2004). Dynamic tunnel management protocol for IPv4 traversal of IPv6 mobile network. *IEEE Personal Communications Magazine.*

Johnson, D., Perkins, C.E., & Arkko, J. (2004). *Mobility support in IPv6.* RFC 3775, IETF.

Lim, E., & Wong, K. D. (2005). Binding update alternatives for Mobile IP version 6. *Proceedings of the 4th International Conference on Information Technology in Asia (CITA 2005)*, Kuching, Malaysia.

Qi, S (2001). *On providing flow transparent mobility support for IPv6-based wireless real-time services.* MEng thesis, Department of Electrical and Computer Engineering, National University of Singapore.

Samad, M., & Ishak, R. (2004). Deployment of wireless Mobile IPv6 in Malaysia. *Proceedings of the RF and Microwave Conference.*

Silva, J., Camilo, T., Costa, A., Matos, C., & Boavida, F. (2004). Exploring IPv6 mobility in IPv6 environments—issues and lessons learnt. *Proceedings of the IEEE International Conference on System, Man and Cybernatics.*

Soliman, H. (2004). *MobileIPv6—mobility on wireless Internet.* Boston: Addison-Wesley.

Wong, K. D., & Dutta, A. (2005). Simultaneous mobility in MIPv6. *Proceedings of the IEEE Electro/Information Technology Conference (EIT)*, Lincoln, NE.

KEY TERMS

Binding Acknowledgement: Same as *binding update,* but in reverse direction; acknowledgement sent by a home agent to mobile nodes, as a response to binding update.

Binding Update: Notification of a mobile node's care-of-address to its home agent, sent by mobile nodes to a home agent.

Binding: The association between a mobile node's home address and care-of-address.

Duplicate Address Detection (DAD): The way of checking that no node on the link is already using that address.

Neighbor Unreachability Detection (NUD): The way of checking that another host is still available and reachable on the link.

Route Optimization: Direct communication between correspondent node to any mobile node, without needing to pass through the mobile node's home network and be forwarded by its home agent; thus eliminates the problem of triangle routing and improves routing efficiency.

Triangle Routing: Indirect communication between correspondent node and mobile node. the packet from the correspondent node will travel to its home agent first, before going to the mobile node, which is not efficient.

Enabling Multimedia Applications in Memory-Limited Mobile Devices

Raul Fernandes Herbster
Federal University of Campina Grande, Brazil

Hyggo Almeida
Federal University of Campina Grande, Brazil

Angelo Perkusich
Federal University of Campina Grande, Brazil

Marcos Morais
Federal University of Campina Grande, Brazil

INTRODUCTION

Embedded systems have several constraints which make the development of applications for such platforms a difficult task: memory, cost, power consumption, user interface, and much more. These characteristics restrict the variety of applications that can be developed for embedded systems. For example, storing and playing large videos with good resolution in a limited memory and processing power mobile device is not viable.

Usually, a client-server application is developed to share tasks: clients show results while servers process data. In such a context, another hard task for limited memory/processing devices could be delegated to the server: storage of large data. If the client needs data, it can be sent piece by piece from the server to the client.

In this article we propose a layered architecture that makes possible the visualization of large videos, and even other multimedia documents, in memory/processing limited devices. Storage of videos is performed at the server side, and the client plays the video without worrying about storage space in the device. Data available in the server is divided into small pieces of readable data for mobile devices, generally JPEG files. For example, when the client requests videos from the server, the videos are sent as JPEG files and shown at an ideal rate for users. The video frames are sent through a wireless connection.

The remainder of this article is organized as follows. We begin by describing background concepts on embedded systems and client-server applications, and then present our solution to enable multimedia applications in memory-limited mobile devices. We next discuss some future trends in mobile multimedia systems, and finally, present concluding remarks.

BACKGROUND

Embedded Systems

An embedded system is not intended to be a general-purpose computer. It is a device designed to perform specific tasks, including a programmable computer. A considerable number of products use embedded systems in their design: automobiles, personal digital assistants, and even household appliances (Wayne, 2005). These limited systems have some constraints that must be carefully analyzed while designing the applications for them: size, time constraints, power consumption, memory usage and disposal, and much more (Yaghmour, 2003).

These constraints restrict the variety of software for embedded systems. The development of applications which demand a large amount of memory, for example, is not viable for embedded systems, because the memory of such devices is limited. Extra memory can also be provided, but the total cost of application is very high. Another example is multimedia applications, such as video players: storing and playing large videos with good resolution in a limited memory and processing power mobile device is a very hard task.

There are specific platforms that were developed to perform multimedia tasks: embedded video decoders and embedded digital cameras, for example. However, other considerable parts of embedded systems, like personal digital assistants (PDAs) and cell phones, are not designed to play videos with good quality, store large amount of data, and encode/decode videos. Thus, it is important to design solutions enabling multimedia environments in this variety of memory/processing-limited devices.

Figure 1. Client/server architecture

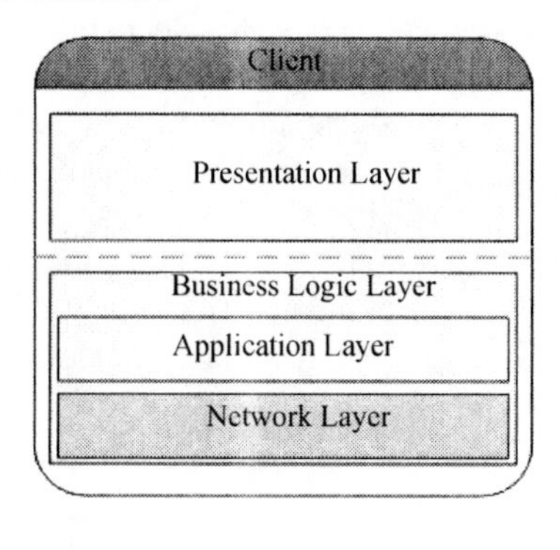

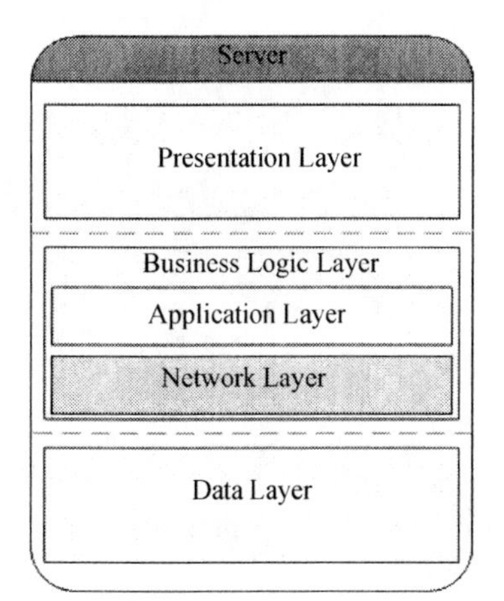

Layered and Client-Server Architectures

Layered architectures share services through a hierarchical organization: each layer provides specific services to the layers above it and also acts as a client to the layer below (Shaw & Garlan, 1996). This characteristic increases the level of abstraction, allowing the partition of complex problems into a set of tasks easier to perform. Layered architectures also decouple modules of the software, so reuse is also more easily supported. As communication of layers is made through contracts specified as interfaces, the implementation of each module can be modified interchangeably (Bass & Kazman, 1998).

Most of the applications have three major layers with different functionalities: presentation, which handles inputs from devices and outputs to screen display; application or business logic, which has the main functionalities of the application; and data, which provides services for storing the data of the application (Fastie, 1999).

The client-server architecture has two elements that establish communication with each other: the front-end or client portion, which makes a service request to another program, called server; and the back-end or server portion, which provides service to the request. The client-server architecture allows an efficient way to interconnect programs that are distributed at different places (Jorwekar, 2005). However, the client-server architecture is more than just a separation of a user from a server computer (Fastie, 1999). Each portion has also its own modules: presentation, application, and data.

ENABLING MULTIMEDIA APPLICATIONS

Multimedia applications demand a considerable amount of resources from the environment in order to guarantee quality of service, which can be defined in terms of security, availability, or efficiency (Banâtre, 2001). Embedded systems have several constraints, like limited memory (Yaghmour, 2003), which make it very difficult to implement multimedia applications in an embedded platform.

Today, the growing interest for mobile devices and multimedia products requires the development of multimedia applications for embedded systems (Banâtre, 2001). There are approaches (Grun, Balasa, & Dutt, 1998; Leeman et al., 2005) that try to enhance embedded systems memory and other system aspects, such as processing, to provide better results in multimedia applications. However, most of the solutions available focus on hardware architecture, and a large number of programmers are not used to programming at the hardware level.

A solution based on client-server architecture is a good proposal for limited-memory/processing mobile devices because harder tasks can be performed by the server side whereas the client just displays results. By designing applications based on an architecture that shares tasks, constraints like limited memory and low computing power are partially solved. In this article, we propose a layered, client/server architecture that allows playing and storing large videos on limited-memory/processing mobile devices. The data is sent through a wireless intranet.

Client/Server Architecture

In Figure 1, both client and server modules are illustrated. Each module of both elements can be changed at any time, except the application layer because the rules of application are defined on it: if business logic changes, so does the application.

The server architecture is a standard three-tier architecture:

Figure 2. Logic model of video descriptions information

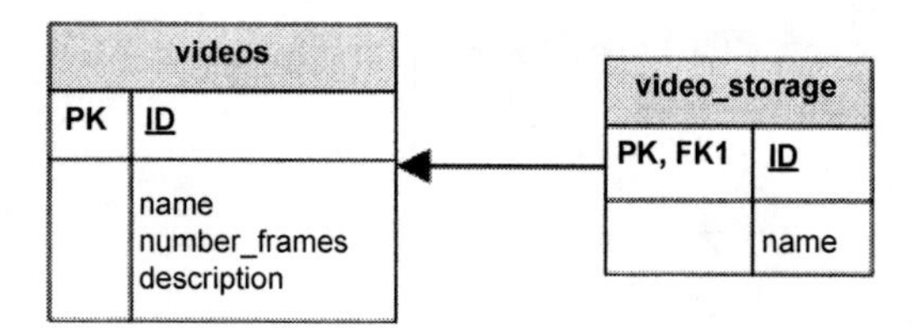

- **Presentation Layer:** This layer interacts directly with the client. Its functionalities are related to display forms so that the user adds multimedia content to the server repository.
- **Business Logic Layer:** This has two sub-layers: the network layer, which manages the connection of server and client, receiving requests and sending responses; and the application layer, which requests services for the data layer, such as document storage and reports.
- **Data Layer:** This layer stores the multimedia documents as JPEG files. Each document has information about management files, including ID, number of frames (JPEG files), and specific description elements, which depends on the multimedia document type. Figure 2 illustrates the logic model of the tables that contain such information. The server stores in a table (video_storage) the titles and ID of all videos. All the information about a video is stored in another table (videos).

The client is a single two-tier application; the data layer is not defined.

- **Presentation Layer:** This consists of a video player with buttons to select the video and a screen to display the video.
- **Business Logic Layer:** This has two sub-layers: the network layer, which manages the connection with the client, sending requests to the server and receiving data from it; and the application layer, which gets frames from the network layer and also controls the tax rate of displaying the frames.

Execution Scenario

In what follows, we present an execution scenario of the mobile multimedia architecture. For this, consider that, at server side, a video was divided into small pieces of readable data (JPEG files). The quantity of pieces depends on the desired quality of the video that will be played at the client side.

The communication process between client and server is illustrated in Figure 3 and is described as follows. Whenever the server receives a connection request from a client, it sends to the client the list of available videos (steps 1 and 2). The client receives data and displays this information whenever required (step 3). Then, the client sends to the server the ID of the requested video (step 4). The server receives the request and starts the transmission of the video, piece by piece (steps 5 and 6).

At the client side, the pieces of data (JPEG files) are received in a tax rate that depends on the network (traffic, band, etc.). However, to display frames to the client in a constant tax rate and guarantee quality of service, it is necessary to maintain a buffer, which is controlled by the video player.

Figure 3. Client/server communication

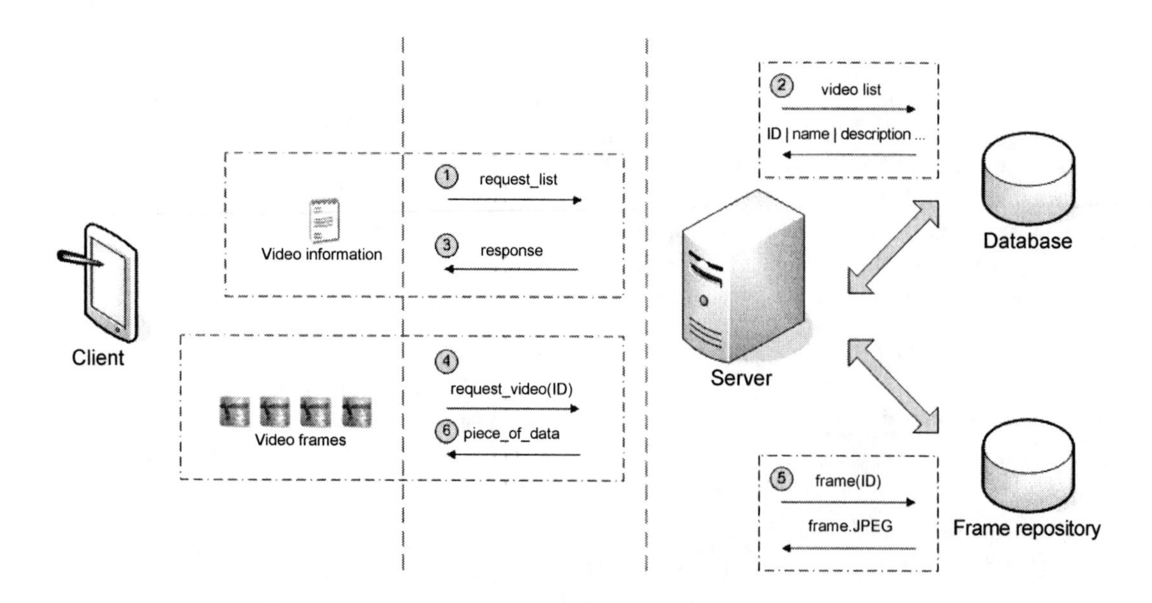

In the architecture described, the data is sent through a wireless intranet. The pieces of data are JPEG files, but could also be in Motion JPEG (MJPEG) format. The tax of frames depends on the quality of the video player on the client side: generally, a high video quality rate is 30 frames per second, whereas a low video quality rate is 10 frames per second.

In a wireless network, this solution needs a large part of the network bandwidth. Therefore, there is a tradeoff between memory/processing capacity and network bandwidth. Nevertheless, considering home entertainment environments, such a tradeoff is worth the cost mainly because wireless networks in home environments have enough bandwidth to be used in such a context.

The architecture can also be used for other kinds of multimedia documents. For example, large PDF documents cannot be visualized with good quality on memory/processing-limited devices. The PDF documents can be also shared as JPEG files and sent to the client.

FUTURE TRENDS

A protocol defines communication between client and server. It is an important element to guarantee QoS. As for future work, we suggest a deeper study of protocols enabling a good service for a given situation. It is important to focus on protocols that do not demand a lot of resources from the network, such as bandwidth.

There are some protocols that are implemented over UDP, for example, trivial file transport protocol (TFTP) (RFC 783, 1981) and real-time transfer protocol (RTP) (RFC, 1996). These protocols were implemented to demand few resources of the network, and to transfer a considerable number of files through the network.

Another interesting research approach is to measure variables of the network while using an application based on the architecture described, for example, to define how many devices running such applications the network supports.

CONCLUSION

Multimedia applications demand a considerable amount of resources from systems. To develop multimedia applications for embedded systems, it is necessary to tackle constraints inherent to such platforms, such as limited memory and processing power.

In this article, we described a general architecture used for enabling multimedia applications in memory/processing systems. The architecture has two parts: a server, which receives requests and sends responses to clients; and clients, which make requests. Both parts have a layered architecture.

The solution proposed is relatively simple to implement and is easy to maintain, because the modules are decoupled and can be modified interchangeably. However, the architecture demands a considerable amount of network bandwidth, because the number of packages sent by the server to the client is large. Nevertheless, considering that the application is implemented over a wireless network in home environments, the bandwidth tradeoff is worth the cost.

REFERENCES

Banâtre, M. (2001). Ubiquitous computing and embedded operating systems design. *ERCIM News,* (47).

Bass, C., & Kazman. (1998). *Software architecture in practice.* Boston: Addison Wesley Longman.

Fastie, W. (1999). Understanding client/server computing. *PC Magazine,* 229-230.

Grun, P., Balasa, F., & Dutt, N. (1998). Memory size estimation for multimedia applications. *International Conference on Hardware Software Codesign, Proceedings of the 6th International Workshop on Hardware/Software Codesign* (pp. 145-149), Seattle, WA.

Yahgmour, K. (2003). *Building embedded Linux systems.* CA: O'Reilly.

Jorwekar, S. (2005). *Client server software architecture.*

Leeman, M., Atienza, D., Deconinck, G., De Florio, V., Mendías, J.M., Ykman-Couvreur, C., Catthoor, F., & Lauwereins, R. (2005). Methodology for refinement and optimisation of dynamic memory management for embedded systems in multimedia applications. *Journal of VLSI Signal Processing, 40*(3), 383-396.

RFC 783. (1981). *The TFTP protocol (revision 2).* Retrieved April 6, 2006, from http://www.ietf.org/rfc/rfc0783.txt?number=0783

RFC 1889. (1996). *RTP: A transport protocol for real-time applications.* Retrieved April 6, 2006, from http://www.ietf.org/rfc/rfc1889.txt?number=1889

Shaw, M., & Garlan, D. (1996). *Software architecture: Perspectives on an emerging discipline.* Englewood Cliffs, NJ: Prentice Hall.

Wolf, W. (2005). *Computer as components: Principles of embedded computing system design.* San Francisco: Morgan Kaufmann.

KEY TERMS

Client-Server Architecture: A basic concept used in computer networking, wherein servers retrieve information requested by clients, and clients desplay that information to the user.

Embedded Systems: An embedded system is a special-purpose computer system, which is completely encapsulated by the device it controls. An embedded system has specific requirements and performs pre-defined tasks, unlike a general purpose personal computer.

Embedded Software: Software designed for embedded systems.

Layered Architecture: The division of a network model into multiple discrete layers, or levels, through which messages pass as they are prepared for transmission.

Mobile Devices: Any portable device used to access a network (Internet, for example).

Multimedia Application: Applications that support the interactive use of text, audio, still images, video, and graphics.

Network Protocols: A set of rules and procedures governing communication between entities connected by the network.

Wireless Network: Networks without connecting cables, that rely on radio waves for transmission of data.

Enabling Technologies for Mobile Multimedia

Kevin Curran
University of Ulster, Northern Ireland

INTRODUCTION

Mobile communications is a continually growing sector in industry, and a wide variety of visual services such as video-on-demand have been created that are limited by low-bandwidth network infrastructures. The distinction between mobile phones and personal device assistants (PDAs) has already become blurred, with pervasive computing being the term coined to describe the tendency to integrate computing and communication into everyday life. Audio quality is highly sensitive to jitter, and video is sensitive to available bandwidth. For lip synchronization, audio and video streams need to be synchronized to within 80-100 milliseconds for skew to be imperceptible (Tannenbaum, 2005). Packets are effectively passed automatically through to the presentation device. Interpretation of the delivered information is left to human perception; because humans are far more tolerant than computers, lost packets are likely to be perceived merely as a temporary quality reduction. Nevertheless packet loss is still a significant problem for isochronous interactions. For example, since a typical packet size is generally above the threshold for audible loss (approximately 20 milliseconds), the loss of a single audio packet can be noticeable to the receiver. Resource reservation protocols are an attempt to resolve these difficulties by allocating resources prior to communication. Uncompressed multimedia data require a lot of storage capacity and very high bandwidth. Thus the use of multimedia compression is very essential. Since the source should encode the streams and the destination should decode them, multimedia compression imposes substantial loads on processing resources, such as CPU power (Yan & Mabo, 2004). New technologies for connecting devices like wireless communication and high bandwidth networks make the network connections even more heterogeneous. Additionally, the network topology is no longer static, due to the increasing mobility of users. Ubiquitous computing is a term often associated with this type of networking.

BACKGROUND

The creation of low bit rate standards such as H.263 (Harrysson, 2002) allows reasonable quality video through the existing Internet and is an important step in paving the way forward. As these new media services become available, the demand for multimedia through mobile devices will invariably increase. Corporations such as Intel do not plan to be left behind. Intel has created a new breed of mobile chip code named Banias. Intel's president and chief operating officer Paul Otellino states that "eventually every single chip that Intel produces will contain a radio transmitter that handles wireless protocols, which will allow users to move seamlessly among networks. Among our employees this initiative is affectionately referred to as 'radio free Intel'."

Figure 1. PDAs

Products such as Real Audio (www.realaudio.com) and IPCast (www.ipcast.com) for streaming media are also becoming increasingly common; however, multimedia, due to its timely nature, requires guarantees different in nature with regards to delivery of data from TCP traffic such as HTTP requests. In addition, multimedia applications increase the set of requirements in terms of throughput, end-to-end delay, delay jitter, and clock synchronization. These requirements may not all be directly met by the networks, therefore end-system protocols enrich network services to provide the quality of service (QoS) required by applications. In ubiquitous computing, software is used by roaming users interacting with the electronic world through a collection of devices ranging from handhelds such as PDAs (Figure 1) and mobile phones (Figure 2) to personal computers (Figure 3) and laptops (Figure 4).

The Java language, thanks to its portability and support for code mobility, is seen as the best candidate for such settings (Román et al., 2002; Kochnev & Terekhov, 2003). The heterogeneity added by modern smart devices is also characterized by an additional property, which is that many of these devices are typically tailored to distinct purposes. Therefore, not only memory and storage capabilities differ widely, but local device capabilities, in addition to the availability of resources changing over time (e.g., a global

Figure 2. Mobiles

Figure 3. Desktops

Figure 4. Laptops

positioning satellite (GPS) system cannot work indoors unless one uses specialized repeaters—see Jee, Boo, Choi, & Kim, 2003), thus a need exists for middleware to be aware of these pervasive computing properties. With regards to multimedia, applications that use group communication (e.g., videoconferencing) mechanisms must be able to scale from small groups with few members, up to groups with thousands of receivers (Tojo, Enokido, & Takizawa, 2003).

The protocols underlying the Internet were not designed for the latest cellular type networks with their low bandwidth, high error losses, and roaming users, thus many 'fixes' have arisen to solve the problem of efficient data delivery to mobile resource-constrained devices (Saber & Mirenkov, 2003). Mobility requires adaptability, meaning that systems must be location-aware and situation-aware, taking advantage of this information in order to dynamically reconfigure in a distributed fashion (Solon, McKevitt, & Curran, 2003; Matthur & Mundur, 2003). However, situations in which a user moves an end-device and uses information services can be challenging. In these situations the placement of different cooperating parts is a research challenge.

ENABLING TECHNOLOGIES FOR MOBILE MULTIMEDIA

In 1946, the first car-based telephone was set up in St. Louis in the United States. The system used a single radio transmitter on top of a tall building. A single channel was used, and therefore a button was pushed to talk and released to listen (Tanenbaum, 2005). This half-duplex system is still used by modern-day CB-radio systems used by police and taxi operators. In the 1960s the system was improved to a two-channel system, called improved mobile telephone system (IMTS). The system could not support many users, as frequencies were limited. The problem was solved by the idea of using cells to facilitate the re-use of frequencies. More users can be supported in such a cellular radio system. It was implemented for the first time in the advanced mobile phone system (AMPS). Wide-area wireless data services have been more of a promise than a reality. It can be argued that success for wireless data depends on the development of a digital communications architecture that integrates and interoperates across regional-area, wide-area, metropolitan-area, campus-area, in-building, and in-room wireless networks.

The convergence of two technological developments has made mobile computing a reality. In the last few years, the UK and other developed countries have spent large amounts of money to install and deploy wireless communication facilities. Originally aimed at telephone services (which still account for the majority of usage), the same infrastructure is increasingly used to transfer data. The second development is the continuing reduction in the size of computer hardware, leading to portable computation devices such as laptops, palmtops, or functionally enhanced cell phones. Unlike second-generation cellular networks, future cellular systems will cover an area with a variety of non-homogeneous cells that may overlap. This allows the network operators to tune the system layout to subscriber density and subscribed services. Cells of different sizes will offer widely varying bandwidths: very high bandwidths with low error rates in pico-cells, and very low bandwidths with higher error rates in macro-cells as illustrated in Table 1. Again, depending on the current location, the sets of available services might also differ.

Unlike traditional computer systems characterized by short-lived connections that are bursty in nature, Streaming Audio/Video sessions are typically long lived (the length of a presentation) and require continuous transfer of data. Streaming services will require, by today's standards, the delivery of enormous volumes of data to customer homes. For example, entertainment NTSC video compressed using

Table 1. Characteristics of various wireless networks

Type of Network	Bandwidth Latency	Latency	Mobility	Typical Video Performance	Typical Audio Performance
In-Room/Building (Radio Frequency Infrared)	>> 1 Mbps RF: 2-20 Mbps IR: 1-50 Mbps	<< 10 ms	Pedestrian	2-Way, Interactive, Full Frame Rate (Compressed)	High Quality, 16 bit samples, 22 KHz rate
Campus-Area Packet Relay	Approx. 64 kbps	Approx. 100 ms	Pedestrian	Medium Quality Slow Scan	Medium Quality Reduced Rate
Wide-Area (Cellular, PCS)	19.2 kbps	> 100 ms	Pedestrian/ Vehicular	Video Phone or Freeze Frame	Asynchronous "Voicemail"
Regional-Area (LEO/VSAT DBS)	Asymmetric Up/Dn 100 bps to 4.8 kbps 12 Mbps	>> 100 ms	Pedestrian/ Vehicular Stationary	Asynchronous Video Playback	Asynchronous "Voice Mail"

the MPEG standards requires bandwidths between 1.5 and 6 Mb/s. Many signaling schemes have been developed that can deliver data at this rate to homes over existing communications links (Forouzan, 2002). Some signaling schemes suitable for high-speed video delivery are:

- **Asymmetrical Digital Subscriber Loop:** ADSL (Bingham, 2000) takes advantage of the advances in coding to provide a customer with a downstream wideband signal, an upstream control. The cost to the end user is quite low in this scheme, as it requires little change to the existing equipment.
- **Cable TV:** CATV (Forouzan, 2002) uses a broadband coaxial cable system and can support multiple MPEG-compressed video streams. CATV has enormous bandwidth capability and can support hundreds of simultaneous connections. Furthermore, as cable is quite widely deployed, the cost of supporting video-on-demand and other services is significantly lower. However, it requires adaptation to allow bi-directional signaling in the support of interactive services.

A cellular wireless network consists of fixed based stations connecting mobile devices through a wired backbone network where each mobile device establishes contact through their local base stations. The available bandwidth on a wireless link is limited, and channels are more prone to errors. It is argued that future evolution of network services will be driven by the ability of network elements to provide enhanced multimedia services to any client anywhere (Harrysson, 2002). Future network elements must be capable of transparently accommodating and adjusting to client and content heterogeneity. There are benefits to filtering IP packets in the wireless network so that minimal application data is carried to the mobile hosts to preserve radio resources and prevent the overloading of mobile hosts with unnecessary information and ultimately wasteful processing. A proxy is an intermediary component between a source and a sink, which transforms the data in some manner. In the case of mobile hosts, a proxy is often an application that executes in the wired network to support the host. This location is frequently the base station, the machine in the wired network that provides the radio interface. As the user moves, the proxy may also move to remain on the communication path from the mobile device to the fixed network. The proxy hides the mobile from the server, which thinks that it communicates with a standard client (i.e., a PC directly connected to the wired network) (Kammann & Blachnitzky, 2002).

Wireless links are characterized by relatively low bandwidth and high transmission error rates (Chakravorty & Pratt, 2002). Furthermore, mobile devices often have computational constraints that preclude the use of standard Internet video formats on them; thus by placing a mobile transcoding proxy at the base station (BS), the incoming video stream can be transcoded to a lower bandwidth stream, perhaps to a format more suitable to the nature of the device, and control the rate of output transmission over the wireless link (Joshi, 2000).

Figure 5 illustrates a scenario where a transcoding gateway is configured to transcode MPEG streams to H.261. In the architecture, the transcoding gateway may also simply forward MPEG or H.261 packets to an alternate session (in both directions) without performing transcoding. Figure 6 illustrates locations in which intelligence about available network services may be placed. Client may utilize this network knowledge to select the most appropriate server and mechanism in order to obtain appropriate content. As an alternative, this knowledge (and the associated burden)

Figure 5. A transcoding proxy

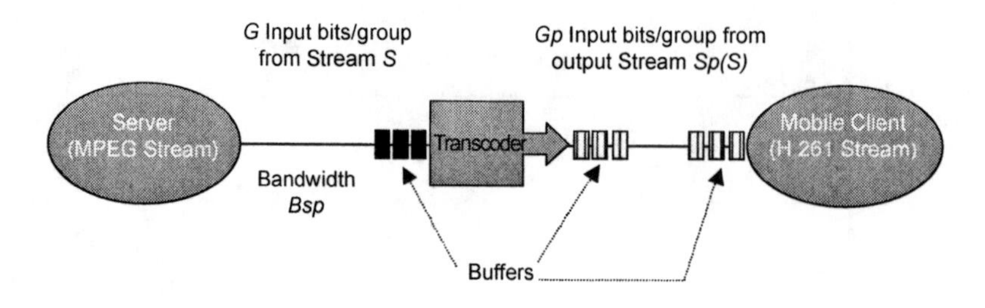

Figure 6. Variations in client-server connectivity

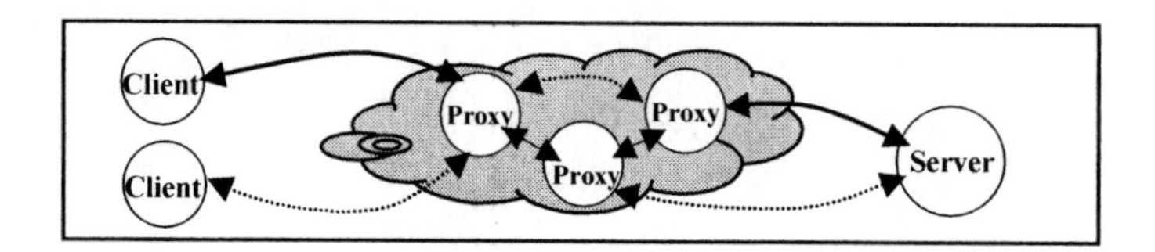

could be entirely or partially transferred to the individual servers or could reside inside the network.

Image transcoding is where an image is converted from one format to another (Vetro, Sun, & Wang, 2001). This may be performed by altering the Qscale (basically applying compression to reduce quality). This is sometimes known as simply resolution reduction. Another method is to scale down the dimensions of the image (spatial transcoding) (Chandra, Gehani, Schlatter Ellis, & Vahdat, 2001) so to reduce the overall byte size (e.g., scaling a 160Kb frame by 50% to 32KB). Another method known as temporal transcoding is where frames are simply dropped (this can sometimes be known as simply rate reduction), while another method may be simply to transcode the image to grayscale, which may be useful for monochrome PDAs (again this transcoding process results in reduced byte size of the image or video frame). Recently there has been increased research into intelligent intermediaries). Support for streaming media in the form of media filters has also been proposed for programmable heterogeneous networking. Canfora, Di Santo, Venturi, Zimeo, and Zito (2005) propose multiple proxy caches serving as intelligent intermediaries, improving content delivery performance by caching content. A key feature of these proxies is that they can be moved and re-configured to exploit geographic locality and content access patterns, thus reducing network server load. Proxies may also perform content translation on static multimedia in addition to distillation functions in order to support content and client heterogeneity (Yu, Katz, & Laksham, 2005). Another example is fast forward networks broadcast overlay architecture (Fast, 2005), where there are media bridges in the network which can be used in combination with RealAudio or other multimedia streams to provide an application-layer multicast overlay network. One could adopt the view at this time that "boxes" are being placed in the network to aid applications.

Padmanabhan, Wang, and Chou (2003) consider the problem of distributing "live" streaming media content to a potentially large and highly dynamic population of mobile hosts. Peer-to-peer content distribution is attractive in this setting because the bandwidth available to serve content scales with demand. A key challenge, however, is making content distribution robust to peer transience. Others (Topic, 2002) have proposed a hierarchy of retransmission servers positioned around expensive or over-utilized links. The servers operate a negative acknowledgement (NACK)-based reliable protocol between them, and receivers use a similar scheme for requesting lost packets. Their proposal significantly improves reception quality, but requires manual configuration of the retransmission servers. Streaming of stored data makes little sense unless browsing and selective playback is a requirement. For totally non-real-time scenarios, a normal transport protocol and pre-fetch can be used to achieve perfect audio quality. Media service frameworks are middleware aimed at integrating multimedia services with mobile service platforms. One such framework is PARLAY, which is the service framework of the 3rd Generation Partnership Project (3GPP, 2003). The use of the PARLAY APIs is proposed for the control of multimedia services (Parley, 2003). PARLAY is a forum established by key vendors and carriers with the goal to define object-oriented APIs that allow third-party application developers to access network resources in a generic and technology-independent way.

FUTURE TRENDS

Mobile phone technologies have evolved in several major phases denoted by "Generations" or "G" for short. Three generations of mobile phones have evolved so far, each successive generation more reliable and flexible than the previous. The first of these is referred to as the first generation or 1G. This generation was developed during the 1980s and early 1990s, and only provided an analog voice service with no data services available (Bates, 2002). The second generation or 2G of mobile technologies used circuit-based digital networks. Since 2G networks are digital, they are capable of carrying data transmissions, with an average speed of around 9.6K bps (bits per second). Because 2G networks can support the transfer of data, they are able to support Java-enabled phones. Some manufacturers are providing Java 2 Micro Edition (J2ME) (Knudsen & Li, 2005) phones for 2G networks, though the majority are designing their Java-enabled phones for the 2.5G and 3G networks, where the increased bandwidth and data transmission speed will make these applications more usable (Hoffman, 2002). These are packet based and allow for "always on" connectivity. The third generation of mobile communications (3G) (http://www.3gnewsroom.com) is digital mobile multimedia offering broadband mobile communications with voice, video, graphics, audio, and other forms of information. 3G builds upon the knowledge and experience derived from the preceding generations of mobile communication, namely 2G and 2.5G, although 3G networks use different transmission frequencies from these previous generations and therefore require a different infrastructure (Camarillo & Garcia-Martin, 2005). These networks will improve data transmission speed up to 144K bps in a high-speed moving environment, 384K bps in a low-speed moving environment, and 2Mbps in a stationary environment. 3G services see the logical convergence of two of the biggest technology trends of recent times, the Internet and mobile telephony.

Some of the services that will be enabled by the broadband bandwidth of the 3G networks include downloadable and streaming audio and video, voice-over Internet protocol (VoIP), sending and receiving high-quality color images, electronic agents that roam communications networks delivering/receiving messages or looking for information or services, and the capability to determine geographic position of a mobile device using the global positioning system (Barnes et al., 2003). 3G will also facilitate many other new services that have not previously been available over mobile networks due to the limitations in data transmission speeds. These new wireless applications will provide solutions to companies with distributed workforces, where employees need access to a wide range of information and services via their corporate intranets when they are working off-site with no access to a desktop (Camarillo & Garcia-Martin, 2005).

CONCLUSION

Flexible and adaptive frameworks are necessary in order to develop distributed multimedia applications in such heterogeneous end-systems and network environments. The processing capability differs substantially for many of these devices, with PDAs being severely resource constrained in comparison to leading desktop computers. The networks connecting these devices and machines range from GSM, Ethernet LAN, and Ethernet 802.11 to Gigabit Ethernet. Networking has been examined at a low-level micro-protocol level and again from a high-level middleware framework viewpoint. Transcoding proxies were introduced as a promising way to achieving dynamic configuration, especially because of the resulting openness, which enables the programmer to customize the structure of the system; other issues regarding mobility were also discussed.

REFERENCES

Barnes, J., Rizos, C., Wang, J., Small, D., Voigt, G., & Gambale, N. (2003, September 9-12). LocataNet: A new positioning technology for high precision indoor and outdoor positioning. *Proceedings of ION (Institute of Navigation) GPS/GNSS 2003,* Portland, OR, (pp. 1779-1789).

Bates, J. (2002, May). *Optimizing voice transmission in ATM/IP mobile networks.* New York: McGraw-Hill Telecom Engineering.

Bingham, J. (2000, January). *ADSL, VDSL, and multicarrier modulation (1st ed.).* Wiley-Interscience.

Camarillo, G., & Garcia-Martin, M. (2005, December). *The 3G IP Multimedia Subsystem (IMS): Merging the Internet and the cellular worlds* (2nd ed.). New York: John Wiley & Sons.

Canfora, G., Di Santo, G., Venturi, G., Zimeo, E., & Zito, M. (2005). Migrating Web application sessions in mobile computing. *Proceedings of the International World Wide Web Conference* (pp. 56-62).

Chandra, S., Gehani, A., Schlatter Ellis, C., & Vahdat, A. (2001, January). Transcoding characteristics of Web images. *Proceedings of the SPIE Multimedia Computing and Networking Conference* (pp. 135-149).

Chakravorty, R., & Pratt, I. (2002, September 9-11). WWW performance over GPRS. *Proceedings of the 4th IEEE Conference on Mobile and Wireless Communications Networks (MWCN 2002),* Stockholm, Sweden, (pp. 191-195).

Feng, Y., & Zhu, J. (2001). *Wireless Java programming with J2ME* (1st ed.). Sams Publishing.

Forouzan, B. (2002). *Data communications and networking* (2nd ed.). New York: McGraw-Hill.

Harrysson, A. (2002, September 9-11). Industry challenges for mobile services. *Proceedings of the 4th IEEE Conference on Mobile and Wireless Communications Networks (MWCN 2002),* Stockholm, Sweden, (pp. 42-48).

Hoffman, J. (2002, September). *GPRS demystified* (1st ed.). New York: McGraw-Hill Professional.

Jee, G., Boo, S., Choi, J., & Kim, H. (2003, September 9-12). An indoor positioning using GPS repeater. *Proceedings of ION (Institute of Navigation) GPS/GNSS 2003,* Portland, OR, (pp. 42-48).

Kammann, J., & Blachnitzky, T. (2002, September 9-11). Split-proxy concept for application layer handover in mobile communication systems. *Proceedings of the 4th IEEE Conference on Mobile and Wireless Communications Networks (MWCN 2002),* Stockholm, Sweden, (pp. 532-536).

Knudsen, J., & Li, S. (2005, May). *Beginning J2ME: From novice to professional.* APress.

Kochnev, D., & Terekhov, A.(2003). Surviving Java for mobiles. *IEEE Pervasive Computing, 2*(2), 90-95.

Matthur, A., & Mundur, P. (2003, September 24-26). Congestion adaptive streaming: An integrated approach. *Proceedings of DMS'2003, the 9th International Conference on Distributed Multimedia Systems,* Miami, FL (pp. 109-113).

Padmanabhan, V. N., Wang, H. J., & Chou, P. A. (2003, March). *Resilient peer-to-peer streaming.* Technical Report MSR-TR-2003-11, Microsoft Research, Redmond, WA.

Parlay Group. (2003). *PARLAY specification 2.1.* Retrieved from http://www.parlay.org

Román, M., Hess, C., Cerqueira, R., Ranganathan, A., Campbell, R., & Nahrstedt, K. (2002). A middleware infrastructure for active spaces. *IEEE Pervasive Computing, 1*(4), 74-83.

Saber, M., & Mirenkov. N. (2003, September 24-26). A multimedia programming environment for cellular automata systems. *Proceedings of DMS'2003, the 9th International Conference on Distributed Multimedia Systems,* Miami, FL (pp. 84-89).

Solon, T., McKevitt, P., & Curran, K. (2003, October 22-23). Telemorph—Bandwidth determined mobile multimodal presentation. *Proceedings of IT&T 2003,* Donegal, Ireland, (pp. 90-101).

Tanenbaum, A. (2005, May). *Computer networks* (5th ed.). Englewood Cliffs, NJ: Prentice Hall.

Tojo, T., Enokido, T., & Takizawa, M. (2003, September 24-26). Notification-based QoS control protocol for group communication. *Proceedings of DMS'2003, the 9th International Conference on Distributed Multimedia Systems,* Miami, FL.

Topic, M. (2002). *Streaming media demystified.* New York: McGraw-Hill Education.

Vetro, A., Sun, H., & Wang, Y. (2001). Object-based transcoding for adaptable video content delivery. *IEEE Transactions on Circuits and Systems for Video Technology, 11*(3), 387-397.

Yan, B., & Mabo, R. (2004). QoS control for video and audio communication in conventional and active networks: Approaches and comparison. *IEEE Communications Surveys & Tutorials, 2004* (pp. 42-49). Retrieved from http://www.comsoc.org/livepubs/surveys/public/2004/jan/bai.html

Yu, F., Katz, R., & Laksham, T. (2005). Efficient multi-match packet classification and lookup with TCAM. *IEEE Micro Magazine, 25*(1), 50-59.

KEY TERMS

Bandwidth: The amount of data that can be transferred from one point to another, usually between a server and client; it is a measure of the range of frequencies a transmitted signal occupies. Bandwidth is the data speed in bits per second. In analog systems, bandwidth is measured in terms of the difference between the highest-frequency signal component and the lowest-frequency signal component.

Broadband: The telecommunication that provides multiple channels of data over a single communications medium.

Cellular Network: A cellular wireless network consists of fixed based stations connecting mobile devices through a wired backbone network where each mobile device establishes contact through their local base stations. The available bandwidth on a wireless link is limited, and channels are more prone to errors.

Encoding: Accomplishes two main objectives: (1) it reduces the size of video and audio files, by means of compression, making Internet delivery feasible; and (2) it saves files in a format that can be read and played back on the desktops of the targeted audience. Encoding may be handled by a software application, or by specialized hardware with encoding software built in.

Media: A term with many different meanings; in the context of *streaming media,* it refers to video, animation, and audio. The term "media" may also refer to something used for storage or transmission, such as tapes, diskettes, CD-ROMs, DVDs, or networks such as the Internet.

Multiple Bit Rate Video: The support of multiple encoded video streams within one media stream. By using multiple bit rate video in an encoder, you can create media-based content that has a variety of video streams at variable bandwidths. After receiving this multiple encoded stream, the server determines which bandwidth to stream based on the network bandwidth available. Multiple bit rate video is not supported on generic HTTP servers.

Streaming Video: A sequence of moving images that are transmitted in compressed form over the Internet and displayed by a viewer as they arrive; it is usually sent from pre-recorded video files, but can be distributed as part of a live broadcast feed.

Third-Generation (3G) Mobile Communications: Digital mobile multimedia offering broadband mobile communications with voice, video, graphics, audio, and other forms of information.

Enabling Technologies for Pervasive Computing

João Henrique Kleinschmidt
State University of Campinas, Brazil

Walter da Cunha Borelli
State University of Campinas, Brazil

INTRODUCTION

Bluetooth (Bluetooth SIG, 2004) and ZigBee (ZigBee Alliance, 2004) are short-range radio technologies designed for wireless personal area networks (WPANs), where the devices must have low power consumption and require little infrastructure to operate, or none at all. These devices will enable many applications of mobile and pervasive computing. Bluetooth is the IEEE 802.15.1 (2002) standard and focuses on cable replacement for consumer devices and voice applications for medium data rate networks. ZigBee is the IEEE 802.15.4 (2003) standard for low data rate networks for sensors and control devices. The IEEE defines only the physical (PHY) and medium access control (MAC) layers of the standards (Baker, 2005). Both standards have alliances formed by different companies that develop the specifications for the other layers, such as network, link, security, and application. Although designed for different applications, there exists some overlap among these technologies, which are both competitive and complementary. This article makes a comparison of the two standards, addressing the differences, similarities, and coexistence issues. Some research challenges are described, such as quality of service, security, energy-saving methods and protocols for network formation, routing, and scheduling.

BLUETOOTH

Bluetooth originated in 1994 when Ericsson started to develop a technology for cable replacement between mobile phones and accessories. Some years later Ericsson and other companies joined together to form the Bluetooth Special Interest Group (SIG), and in 1998 the specification 1.0 was released. The IEEE published the 802.15.1 standard in 2002, adopting the lower layers of Bluetooth. The specification Bluetooth 2.0+EDR (Enhanced Data Rate) was released in 2004 (Bluetooth SIG, 2004).

Bluetooth is a low-cost wireless radio technology designed to eliminate wires and cables between mobile and fixed devices over short distances, allowing the formation of ad hoc networks. The core protocols of Bluetooth are the radio, baseband, link manager protocol (LMP), logical link control and adaptation protocol (L2CAP), and service discovery protocol (SDP). The radio specifies details of the air interface, including frequency, modulation scheme, and transmit power. The baseband is responsible for connection establishment, addressing, packet format, timing, and power control. The LMP is used for link setup between devices and link management, while the L2CAP adapts upper-layer protocols to the baseband layer. The SDP is concerned with device information and services offered by Bluetooth devices.

Bluetooth operates on the 2.4 GHz ISM (Industrial, Scientific, and Medical) band employing a frequency-hopping spread spectrum (FHSS) technique. There are 79 hopping frequencies, each having a bandwidth of 1MHz. The transmission rate is up to 1 Mbps in version 1.2 (Bluetooth SIG, 2003) using GFSK (Gaussian frequency shift keying) modulation. In version 2.0+EDR new modes of 2 Mbps and 3 Mbps were introduced. These modes use GFSK modulation for the header and access code of the packets, but employ different modulation for data. The $\pi/4$ differential quadrature phase-shift keying (DQPSK) modulation and 8 differential phase-shift keying (DPSK) modulation are employed in 2 Mbps and 3 Mbps mode, respectively.

The communication channel can support both data (asynchronous) and voice (synchronous) communications. The synchronous voice channels are provided using circuit switching with a slot reservation at fixed intervals. The asynchronous data channels are provided using packet switching utilizing a polling access scheme. The channel is divided in time slots of 625 μs. A time-division duplex (TDD) scheme is used for full-duplex operation.

Each Bluetooth data packet has three fields: the access code (72 bits), header (54 bits), and payload. The access code is used for synchronization and the header has information such as packet type, flow control, and acknowledgement. Three error correction schemes are defined for Bluetooth. A 1/3 rate FEC (forward error correction) is used for packet header; for data, 2/3 rate FEC and ARQ (automatic retransmission request). The ARQ scheme asks for a retransmission

Figure 1. Piconet and scatternet

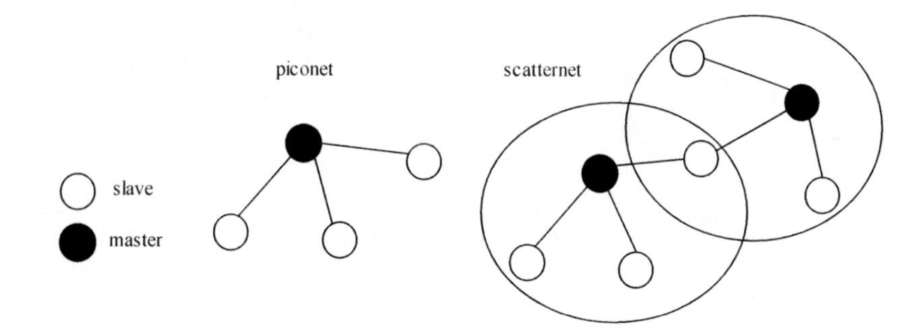

of the packet any time the CRC (cyclic redundancy check) code detects errors. The 2/3 rate FEC is a (15,10) Hamming code used in some packets. The ARQ scheme is not used for synchronous packets such as voice.

The devices can communicate with each other forming a network, called piconet, with up to eight nodes. Within a piconet, one device is assigned as a master node and the others act as slave nodes. In the case of multiple slaves, the channel (and bandwidth) is shared among all the devices in the piconet. Devices in different piconets can communicate using a structure called scatternet, as shown in Figure 1. A slave in one piconet can participate in another piconet as either a master or slave. In a scatternet, two or more piconets are not synchronized in either time or frequency. Each of the them operates in its own frequency-hopping channel while the bridge nodes in multiple piconets participate at the appropriate time via TDD. The range of Bluetooth devices depends on the class power, ranging from 10 to 100 meters.

ZIGBEE

ZigBee has its origins in 1998, when Motorola started to develop a wireless technology for low-power mesh networking (Baker, 2005). The IEEE 802.15.4 standard was ratified in May 2003 based on Motorola's proposal. Other companies joined together and formed the ZigBee Alliance in 2002. The ZigBee specification was ratified in December 2004, covering the network, security, and application layers (Baker, 2005).

ZigBee has been designed for low power consumption, low cost, and low data rates for monitoring, control, and sensor applications (Akyildiz, Su, Sankarasubramaniam, & Cayirci, 2002). The lifetime of the networks are expected to be of many months to years with non-rechargeable batteries. The devices operate in unlicensed bands: 2.4 GHz (global), 902-928 MHz (Americas), and 868 MHz (Europe). At 2.4 GHz (16 channels), the raw data rates can achieve up to 250 Kbps, with offset-quadrature phase-shift keying (OQPSK) modulation and direct sequence spread spectrum (DSSS). The 868 MHz (1 channel) and 915 MHz (10 channels) bands also use DSSS, but with binary-phase-shift keying (BPSK) modulation, achieving data rates up to 20 Kbps and 40 Kbps, respectively. The expected range is from 10-100m, depending on environment characteristics.

Each packet, called PHY protocol data unit (PPDU), contains a preamble sequence, a start of frame delimiter, the frame length, and a payload field, the PHY service data unit (PSDU). The 32-bit preamble is designed for acquisition of symbol and chip timing. The payload length can vary from 2 to 127 bytes. A frame check sequence improves the reliability of a packet in difficult conditions. There are four basic frame types: data, acknowledgement (ACK), MAC command, and beacon. The ACK frame confirms to the transmitter that the packet was received without error. The MAC command frame can be used for remote control and nodes configuration.

In 802.15.4 two channel-access mechanisms are implemented, for non-beacon and beacon network. A non-beacon network uses carrier-sense medium access with collision avoidance (CSMA-CA) with positive acknowledgements for successfully received packets. For a beacon-enabled network, a structure called superframe controls the channel access to guarantee dedicated bandwidth and low latency. The network coordinator is responsible for set up of the superframe to transmit beacons at predetermined intervals and to provide 16 equal-width time slots between beacons for contention-free channel access in each time slot (IEEE Std. 802.15.4, 2003; Gutierrez, Callaway, & Barret, 2003).

A ZigBee network can support up to 65.535 nodes, which can be a network coordinator, a full function device (FFD), or a reduced function device (RFD). The network coordinator has general network information and requires the most memory and computing capabilities of the three types. An FFD supports all 802.15.4 functions, and an RFD has limited functionalities to reduce cost and complexity. Two topologies are supported by the standard: star and peer-to-peer, as shown in Figure 2. In the star topology, the communication is performed between network devices and a single central controller, called the PAN coordinator, responsible

Figure 2. ZigBee topologies

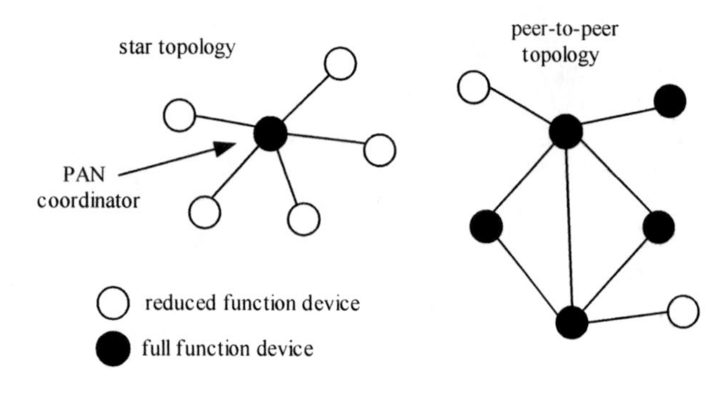

for managing all the star functionality. In the peer-to-peer topology, every network device can communicate with any other within its range. This topology also contains a PAN coordinator, which acts as the root of the network. Peer-to-peer topology allows more complex network formations to be implemented, such as the cluster-tree. The cluster-tree network is a special case of a peer-to-peer network in which most devices are FFDs.

COMPARING ZIGBEE AND BLUETOOTH

Bluetooth and ZigBee have been designed for different applications, and this section makes a comparison between some features of both technologies, such as data rate, power consumption, network latency, complexity, topology, and scalability (Baker, 2005).

In applications where higher data rates are required, Bluetooth always has advantages, especially the 2.0+EDR version (Bluetooth SIG, 2004). While ZigBee mainly supports applications as periodic or intermittent data, achieving rates up to 250 Kbps, Bluetooth can support different traffic types, including not only periodical data, but also multimedia and voice traffic.

ZigBee devices are able to sleep frequently for extended periods to conserve power. This feature works well for energy savings, but increases the network latency because the node will have to awake in order to transmit or receive data. In Bluetooth, the devices do not sleep very often because the nodes are frequently waiting for new nodes or to join other networks. Consequently, data transmission and networks access is fast. Bluetooth devices in sleep mode have to synchronize with the network for communication, while in ZigBee this is not necessary.

The Bluetooth protocol stack is relatively complex when compared to ZigBee. The protocol stack size for ZigBee is about 28 Kbytes and for Bluetooth approximately 100 Kbytes (Geer, 2005). Bluetooth is also more complex if we consider the number of devices. A piconet can have only eight nodes, and a scatternet structure has to be formed to accommodate more nodes (Persson, Manivannan, & Singhal, 2005; Whitaker, Hodge, & Chlamtac, 2005). The Bluetooth SIG does not specify the protocols for scatternet formation. This task is easier in ZigBee networks, since no additional protocols have to be used. In terms of scalability the ZigBee also has some advantages, because network growth is easier to be implemented with flexible topologies. A Bluetooth network growth requires a flexible scatternet formation and routing protocol.

The applications have to consider these characteristics of both protocols when deciding which is the most advantageous for that specific implementation. Bluetooth will fit better in short-range cable replacement, extending LANs to Bluetooth devices and in industries for communication between fixed equipment and mobile devices or machine-to-machine communication (Baker, 2005). ZigBee is most likely to be applied in wireless sensor networks and industries wireless networks, or any other application where battery replacement is difficult and the networks have to live for months to years without human intervention. Many networks may also implement both protocols in complementary roles using the more suitable characteristic of each for that application. Table 1 shows some features of both technologies.

RESEARCH CHALLENGES

In the Bluetooth specification there is no information on how a scatternet topology should be formed, maintained, or operated (Persson et al., 2005; Whitaker et al., 2005). Two scatternet topologies that are created from separate approaches can have different characteristics. The complexity of these tasks significantly increases when moving from single piconets to multiple connected piconets.

Some research challenges in Bluetooth scatternets are formation, device status, routing, and intra and inter-piconet scheduling schemes. Each device needs to determine its role with respect to (possibly) multiple piconets, whether master and/or slave. Whitaker et al. (2005) state:

There is a large degree of freedom in the number of feasible alternative scatternets, which defines a significant combinatorial optimization problem. This is made more difficult by the decentralized nature of the problem, characterized by a lack of a centralized entity with global knowledge.

The task of packet routing in a scatternet also is not so easy because the packet may have to be transmitted in multiple piconets until it reaches its destination. In a Bluetooth piconet, the master controls the channel access. A slave can send a packet only if it receives a polling packet from the master. Some slaves may participate in multiple piconets,

Table 1. Comparison between Bluetooth and ZigBee

Characteristic	Bluetooth	ZigBee
Data rate	1 Mbps (version 1.2) 3 Mbps (version 2.0)	20-250 Kbps
Expected battery duration	Days	Years
Operating frequency	2.4 GHz ISM	868 MHz, 902-928 MHz, 2.4 GHz ISM
Security	64 bit, 128 bit	128 bit AES
Network topology	Piconet and scatternet	Star, peer-to-peer, cluster tree
Protocol stack size	~100 KB	~28 KB
Transmission range	10-100 meters (depending on power class)	10-100 meters (depending on the environment)
Network latency (typical) New device enumeration Changing from sleep to active Active device channel access	 12 s 3 s 2 ms	 30 ms 15 ms 15 ms
Applications	Cable replacement, wireless USB, handset, headset	Remote control, sensors, battery-operated products

so they become more important than others and the scheduling scheme may give priority for these slaves. The nodes involved in many piconets can only be active one at a time, and the scheduling strategy has to consider this characteristic. As stated in Whitaker et al. (2005), many factors influence the design of scatternet protocols, such as distribution of devices, scalability, device differentiation, environmental dynamism, integration between coordination issues, and level of centralization. The design of efficient protocols could make Bluetooth fit for a wider range of applications.

Although in ZigBee the formation of networks with many nodes is not a great problem, the management of a network with thousands of nodes has not been addressed and may be very difficult (Geer, 2005; Zheng & Lee, 2004). Since ZigBee specification was released after Bluetooth, many important issues have not been addressed, and some distributed protocols will have to be designed for these networks. Both standards have security features, including algorithms for authentication, key exchange, and encryption, but its efficiency still has to be analyzed in networks with many nodes.

Other important issue concerning ZigBee and Bluetooth is the coexistence of both devices, as they use the same 2.4 GHz band, and channel allocation conflicts are inevitable between these WPAN technologies (Chen, Sun, & Gerla, 2006; Howitt & Gutierrez, 2003). This band is also used by wireless LANs based on IEEE 802.11 standard cordless phones and microwave ovens. Interference between near devices may be very common, so coexistence strategies have to be implemented. It is important to study the characteristics of each channel allocation scheme and how each channel allocation scheme interacts with the others. The discussion on the coexistence issue between IEEE 802.11 and the IEEE 802.15-based WPAN technologies has been included in the IEEE 802.15.2 standard.

CONCLUSION

Bluetooth and ZigBee are wireless technologies that may enable many applications of ubiquitous and pervasive computing envisioned by Weiser (1991). Millions of devices are expected to be equipped with one or both technologies in the next few years. This work addressed some of the main features and made some comparisons between them. Some research challenges were described. These issues must be properly studied for the widespread use of ZigBee and Bluetooth technologies.

REFERENCES

Akyildiz, I. F., Su, W., Sankarasubramaniam, Y. & Cayirci, E. (2002). A survey on sensor networks. *IEEE Communications Magazine, 40*(8), 102-114.

Baker, N. (2005). ZigBee and Bluetooth: Strengths and weakness for industrial applications. *IEEE Computing and Control Engineering Journal, 16*(2), 20-25.

Bluetooth SIG. (2004). *Specification of the Bluetooth system. Core, version 2.0 + EDR.* Retrieved from http://www.bluetooth.com

Bluetooth SIG. (2003). *Specification of the Bluetooth system. Core, version 1.2.* Retrieved from http://www.bluetooth.com

Chen, L., Sun, T., & Gerla, M. (2006). Modeling channel conflict probabilities between IEEE 802.15 based wireless personal area networks. *Proceedings of the IEEE International Conference on Communications,* Istanbul, Turkey.

Geer, D. (2005). Users make a beeline for ZigBee sensor technology. *IEEE Computer, 38*(12), 16-19.

Gutierrez, J., Callaway, E., & Barret, R. (2003). *IEEE 802.15.4 low-rate wireless personal area networks: Enabling wireless sensor networks.* Institute of Electrical & Electronics Engineer (IEEE).

Howitt, I., & Gutierrez, J. (2003). IEEE 802.15 low rate-wireless personal area network coexistence issues. *Proceedings of the IEEE Wireless Communication and Networking Conference* (pp. 1481-1486), New Orleans, LA.

IEEE Std. 802.15.1. (2002). *IEEE standard for wireless medium access control (MAC) and physical layer (PHY) specifications for wireless personal area networks.*

IEEE Std. 802.15.4. (2003). *IEEE standard for wireless medium access control (MAC) and physical layer (PHY) specifications for low-rate wireless personal area networks (LR-WPANs).*

Persson, K. E., Manivannan, D., & Singhal, M. (2005). Bluetooth scatternet formation: Criteria, models and classification. *Elsevier Ad Hoc Networks, 3*(6), 777-794.

Weiser, M. (1991). The computer for the 21st century. *Scientific American, 265*(3), 94-104.

Whitaker, R. M., Hodge, L. E., & Chlamtac, I. (2005). Bluetooth scatternet formation: A survey. *Elsevier Ad Hoc Networks, 3*(4), 403-450.

Zheng, J., & Lee, M. J. (2004). Will IEEE 802.15.4 make ubiquitous networking a reality? A discussion on a potential low power, low bit rate standard. *IEEE Communications Magazine, 42,* 140-146.

ZigBee Alliance. (2004). *ZigBee specification version 1.0.* Retrieved from http://www.zigbee.com

KEY TERMS

Carrier-Sense Medium Access with Collision Avoidance (CSMA-CA): A network contention protocol that listens to a network in order to avoid collisions.

Direct Sequence Spread Spectrum (DSSS): A technique that spreads the data into a large coded stream that takes the full bandwidth of the channel.

Frequency Hopping Spread Spectrum (FHSS): A method of transmitting signals by rapidly switching a carrier among many frequency channels using a pseudorandom sequence known to both transmitter and receiver.

Medium Access Control (MAC): A network layer that determines who is allowed to access the physical media at any one time.

Modulation: The process in which information signals are impressed on an radio frequency carrier wave by varying the amplitude, frequency, or phase.

Pervasive Computing: An environment where devices are always available and communicate with each other over wireless networks without any interaction required by the user.

Scatternet: A group of independent and non-synchronized piconets that share at least one common Bluetooth device.

Sensor Network: A network of spatially distributed devices using sensors to monitor conditions at different locations, such as temperature, sound, pressure, and so forth.

Wireless Personal Area Network (WPAN): A logical grouping of wireless devices that is typically limited to a small cell radius.

Extreme Programming for Mobile Applications

Pankaj Kamthan
Concordia University, Canada

INTRODUCTION

The liberty, expediency, and flexibility that come with mobile access have led to proliferation of mobile applications. At the same time, these applications face constant challenges posed by new implementation languages, variations in user agents, and demands for new services from user classes of different cultural backgrounds, age groups, and capabilities.

To address that, we require a methodical approach towards the development lifecycle and maintenance of mobile applications that can adequately respond to this constantly changing environment. In other words, it needs to be *agile* (Highsmith, 2002). In this article, we propose the use of an agile methodology, Extreme Programming (XP) (Beck & Andres, 2005), for a systematic development of mobile applications.

The organization of the article is as follows. We first outline the background necessary for the discussion that follows and state our position. This is followed by a discussion of the applicability and feasibility of XP practices as they pertain to mobile applications. Then the limitations of XP towards mobile applications, particularly those that are developed in an open source setting, are highlighted, and suggestions for improvement are presented. Next, challenges and directions for future research are outlined. Finally, concluding remarks are given.

BACKGROUND

In recent years, ongoing efforts towards affordability of mobile devices by public-at-large and increasing contact points (service providers) have opened new vistas in the arena of mobile applications. This has also resulted in increased expectations, including sophisticated services, from mobile users. As a consequence, mobile applications continue to become increasingly large and complex. This growth, however, needs to be carefully controlled and sustained. For that, it is important that the lessons learned from the successes and failures (Nguyen, Johnson, & Hackett, 2003) in the evolution of Web applications not be ignored. Specifically, a systematic approach for creating mobile applications is desirable. We call this *Mobile Web Engineering*, inspired by traditional software engineering and Web engineering (Ginige & Murugesan, 2001).

The focus in the literature (Hjelm, 2000), however, has primarily been on implementation languages rather than the *process*. In Salmre (2005), a systematic approach to developing mobile applications is advocated, but the discussion is within the technology-specific context of Microsoft .Net Framework and Visual Basic. It is unclear how these can scale to the changing technological environment. One of the purposes of this article is to fill this gap.

We adopt the most broadly used and well-tested agile methodology, namely XP, for the development of mobile applications. XP is a test-driven "lightweight" methodology designed for small teams which emphasizes customer satisfaction and promotes teamwork. XP was created to tackle uncertainties in development environment, and in doing so, put more emphasis on the social (people) component (engineer, customer, and end user). The XP practices are set up to mitigate project risks (dynamically changing requirements, new system due by a specific timeline, and so on) and increase the likelihood of success. The use of XP has been suggested for a "rapid application development" of Web applications (Wallace, Raggett, & Aufgang, 2002; Maurer & Martel, 2002).

It is not the purpose of this article to evaluate the merit of XP on its own or with respect to other agile methodologies; such assessments have been carried out elsewhere (Turk, France, & Rumpe, 2002; Mnkandla & Dwolatzky, 2004).

ENGINEERING MOBILE APPLICATIONS USING EXTREME PROGRAMMING

In this section, we discuss in detail how the practices put forth by XP manifest themselves in the development of mobile applications (see Table 1). The 12 XP practices are: *The Planning Game, Small Releases, Metaphor Guide, Simple Design, Testing, Refactoring, Pair Programming, Collective Ownership, Continuous Integration, 40-Hour Week, On-Site Customer,* and *Coding Standards.*

We note that some of these practices such as *Testing, Refactoring*, or *Pair Programming* are not native to XP and were discovered in other contexts previously. In this sense, by aggregating them in a coherent manner, XP bases itself on "best practices." These practices are also not necessarily mutually exclusive, and we point out the relationships among them where necessary. We also draw attention to the obstacles

Table 1. XP practices corresponding to process workflows in a mobile application

Process Workflow	XP Practices
Planning	40-Hour Week, The Planning Game (Project Velocity)
Analysis (Domain Modeling, Requirements)	On-Site Customer, The Planning Game (User Stories)
Design	Metaphor Guide (Natural Naming), Simple Design, Refactoring
Implementation	Collective Ownership, On-Site Customer, Metaphor Guide, Coding Standards, Pair Programming, Continuous Integration
Verification and Validation	On-Site Customer, Testing (Unit Tests, Acceptance Tests)
Delivery	Small Releases

in the realization of these practices that pose challenges to the deployment of XP for mobile applications.

The Planning Game

The purpose of *The Planning Game* is to determine the scope of the project and future releases by combining business priorities and technical estimates. For that, it solicits input from the "customer" to define the business value of desired features and uses cost estimates provided by the programmers. This input comes in form of *user stories* (Alexander & Maiden, 2004). A user story is a user experience informally expressed in a few lines with a mobile application such as navigating or using a search engine. The estimation is limited to the assessment of *project velocity*, a tangible metric that determines the pace at which the team can produce deliverables. The plan is prone to modifications based on the current reality.

Small Releases

The idea behind *Small Releases* is to have a simple system (an evolutionary prototype) into production early, and then via short cycles, iteratively and/or incrementally, reach the final system. To have a concrete proof-of-concept up and running can be used to solicit feedback for future versions and can help convince customers and managers of the viability of the project. This is useful for mobile applications that are highly interactive. However, there is cost associated with prototypes and therefore their number should be kept under control.

Metaphor Guide

The use of metaphors (Boyd, 1999) is prevalent in all aspects of software development. A *Metaphor Guide* is an effort to streamline and standardize efforts for naming software objects and is available for team-wide use. Natural naming (Keller, 1990) is a technique initially used in source code contexts that encourages the use of names that consist of one or more full words of the natural language for program elements in preference to acronyms or abbreviations. Indeed, natural naming strengthens the link between the underlying conceptual entity and its given name. For example, MobileProfile is a combination of two real-world metaphors placed into a natural naming scheme. The two main concerns in naming are: (1) length due to the constrained interfaces on mobile devices, and (2) user familiarity, as user background is often non-technical. For example, it is preferable to use EnterSearchWords as an indicator inside the form interface (to save space) rather than RegularExpressionForQueryString outside the form, although the latter may be a more accurate description.

Simple Design

The motivation behind a *Simple Design* is that in XP's view, requirements are *not* complete when the design commences. This is in line with the reality of mobile applications, which have to respond to the market pressures and the competition that are beyond their control, or other unavoidable circumstances such as variations in implementation technology. Therefore, the design is minimal based on *current* (not future) requirements. It aims for simplicity, and to ensure "good" design its quality (specifically, structural complexity) is improved by frequent revisitations, that is, *Refactoring*.

Testing

There is a strong emphasis in XP on validation and verification of the software at all times. By being test driven, there is transition from one phase to another only if the tests succeed. The tests range from unit tests (using tools such as HTMLUnit, HTTPUnit, XMLUnit, XSLTUnit, and JUnit) written by programmers to acceptance tests involving customers (to satisfy customer requirements). There are variations in user agents (browsers) with respect to their support for mobile markup languages, and often a single mobile device does not have the capabilities for multiple installations. Therefore, interface testing is uniquely critical to mobile applications. As part of that activity, syntactical validation of documents being served is critical. A detailed treatment of tools, techniques, and methods for testing mobile applications as well as for test plans is given in Nguyen et al. (2003).

Refactoring

The artifacts created during analysis or design may need to evolve for reasons such as discovery of "impurities" (or code "smells") or obsolescence. The refactoring (Fowler, Beck, Brant, Opdyke, & Roberts, 1999) methods are structural transformations that help eradicate the undesirables without changing the functionality of the application. Examples of such smells include inconsistent names of classes, operations, or attributes that hinder communication, redundancy (duplication), classes with unnecessary responsibility (non-cohesivity), and so on. The goal of *Refactoring* is to improve the design of the system throughout the entire development.

Pair Programming

This is one of the practices of XP that highlights the social aspects of engineering. The idea behind *Pair Programming* is to encourage collaborative work. In some controlled experiments (Williams & Kessler, 2003), Pair Programming has been shown to produce better code at similar or lower cost than programmers working alone. Empirical studies (Katira, 2004) have shown that some level of compatibility among partners in Pair Programming is necessary for it to be effective. The notion of Pair Programming can be extended to artifacts created during early stages (namely, analysis and design phases) of the development process that focus on modeling (Kamthan, 2005). For example, the use of the Unified Modeling Language (UML) (Booch, Jacobson, & Rumbaugh, 2005) for modeling mobile applications has been suggested (Grassi, Mirandola, & Sabetta, 2004). A pair can be responsible for several other practices such as using *Refactoring* to obtain a *Simple Design, Continuous Integration,* and *Testing.* The *On-Site Customer* can be a partner in a pair, but only as a co-pilot.

Collective Ownership

According to the XP philosophy, one of reasons for inertia in modifications to software is that when change is warranted, the team has to wait for specific personnel to carry it out. Therefore, in XP, all the code belongs to all the programmers and anyone can change code anywhere in the system at any time. However, for such an arrangement to be effective, configuration management that provides trace of person, date/time, and of nature and location of the change carried out is needed.

Continuous Integration

In an incremental and iterative approach of XP, standalone units (such as a corporate logo, navigation icons, and so on) are created and then integrated. However, it is not automatic that if the individual parts work, then their sum would also work. For example, a graphical navigation bar may work well individually, but may not when included in an XHTML Basic document displayed on a personal digital assistant (PDA) due to, say, incorrect encoding or link syntax. By "continuous," XP means integrating and building the software system multiple times a day. The advantage of *Continuous Integration* is minimal propagation of errors (limited to the last addition).

40-Hour Week

The term *40-Hour Week* is to be taken figuratively rather than literally. It simply implies that, due to the emphasis on the social aspect in XP, "overwork" is not recommended. XP believes that excessive overtime leads to low productivity in the long term. For example, tired programmers are prone to more mistakes, which in turn may slow down progress of the project.

On-Site Customer

The availability of a full-time *On-Site Customer* helps in understanding the application domain, determining requirements, setting priorities, and answering questions as the programmers have them. In XP, every contributor to the project, including the customer, is an integral part of the *entire* team. This has two major implications for the team: its structure is not hierarchical, and it requires physical proximity of the participants to function.

Coding Standards

There are a variety of languages for expressing information in mobile applications. The Extensible HyperText Markup Language (XHTML) is a recast of the HyperText Markup Language (HTML) in XML. XHTML Basic, a successor of Compact HTML (cHTML) and of the Wireless Markup Language (WML) 1.0, has native support for elementary constructs for structuring information like paragraphs, lists, tables, and so on. XHTML Mobile Profile (XHTML-MP) from Openwave Systems and WML 2.0 from Open Mobile Alliance (OMA) extend the functionality of XHTML Basic by adding modules as defined by the Modularization of XHTML. SVG Tiny and SVG Basic are scalable vector graphics (SVG) profiles targeted towards cellular phones and PDAs, respectively. They support two-dimensional vector graphics that work across output resolutions, across color spaces, and across a range of available bandwidths. SMIL Basic is a Synchronized Multimedia Integration Language (SMIL) profile that meets the needs of resource-constrained devices such as mobile phones and portable disc players. It allows description of temporal behavior of a multimedia presentation, associates hyperlinks with media objects, and describes the layout of the presentation on a screen. The CSS Mobile Profile is a subset of the Cascading Style Sheets (CSSs) tailored to the needs and constraints of mobile devices, and provides the presentation semantics on the client.

Information based on these languages can be served using any general or special-purpose programming language. It is critical that instances based on these languages be communicable. For example, for *Pair Programming* and for *Collective Ownership* to be effective, there needs to be a common understanding. *Coding Standards* provide means for doing that. It is known (Schneidewind & Fenton, 1996) that, when applied judiciously, standards can contribute towards quality improvement.

CHALLENGES TO THE DEPLOYMENT OF EXTREME PROGRAMMING TO MOBILE APPLICATIONS

In this section, we highlight certain caveats of applying XP practices as-is, as well as certain aspects that are essential to mobile applications but are not covered by these practices per se.

- XP does not mandate a rigorous feasibility study, including a formal means for cost estimation, as part of *The Planning Game.* A feasibility study could for instance determine if one could take advantage of reuse. For example, the functionality of a mobile application for different classes of cellular phones should not be all that different.
- The notions of *Pair Programming, 40-Hour Week,* and *On-Site Customer* make sense for a development in a non-distributed environment only. This would present a coordination obstacle if a mobile application were being developed in different natural languages, each in different parts of the world.
- Testing for usability in general and in case of mobile applications in particular can be prohibitive for small-to-medium-size enterprises, particularly if it involves specialized rooms, dedicated infrastructure with video monitoring, and recordings for subsequent analysis. Also, testing cannot always detect all errors in systems. For example, that user supplied correct address or that internal documentation corresponds to source code are beyond the scope of testing. Furthermore, testing is only one approach to verification and defect removal. XP does not include any support for formal inspections (Wiegers, 2002), although that is somewhat ameliorated by support for *Pair Programming,* which can be viewed as "informal" inspections.

XP does not explicitly take into account the licensing conditions under which the software is developed. For example, mobile applications that are open source (Kamthan, 2006) or outsourced will not be able to comply with some of the practices mandated by XP.

Finally, we point out that agility is not a panacea (Boehm & Turner, 2004). In fact, there are several issues associated with agile methodologies in general and XP in particular (Turk et al., 2002). For example, XP is not applicable to large (greater than 15) team sizes, distributed development, or for very large projects. However, XP provides a feasible first step from an ad-hoc approach to an organized view of developing mobile applications.

FUTURE TRENDS

COCOMO II (Boehm et al., 2001) provides a rigorous approach to cost (time, effort) estimation and could assist in *The Planning Game.* But that would require some adjustments in measures and calibrations of data, as mobile applications are different from traditional applications for which the COCOMO II cost estimation model is defined.

Since XP is driven by testing, there is an urgent need for unit testing frameworks (Hamill, 2004) beyond what are currently available and tailored to mobile markup languages.

For large-scale mobile applications, a "heterogeneous" process environment approach that mixes agility with discipline (Boehm & Turner, 2004) could be useful. A natural extension of the previous discussion would be to deploy a simplified version of the Unified Process (UP) (Jacobson, Booch, & Rumbaugh, 1999), which is a process *framework* that can be tailored to produce a process model for mobile applications. Indeed, such a MobileUP would on one hand be model driven, iterative, customer centric, and would on the other hand still emphasize a top-down team hierarchy and document-based communication.

CONCLUSION

Mobile applications continue to increase in size and complexity, and to sustain and manage this growth, require a systematic approach towards their development. For that, there is a need to move away from thinking at the implementation language level and focus on abstractions created *earlier* in the process. At the same time, we wish to avoid the bureaucracy in development processes that have plagued software engineering in the past.

XP provides one such viable option for development of small-to-medium-size mobile applications. The aforementioned shortcomings inherent to XP are resolvable, and pave the way towards improvements as well as considerations for other process models tailored to mobile applications.

REFERENCES

Alexander, I., & Maiden, N. (2004). *Scenarios, stories, use cases through the systems development life-cycle.* New York: John Wiley & Sons.

Beck, K., & Andres, C. (2005). *Extreme programming explained: Embrace change* (2nd ed.). Boston: Addison-Wesley.

Boehm, B. W., Abts, C., Brown, A. W., Chulani, S., Clark, B. K., Horowitz, E., et al. (2001). *Software cost estimation with COCOMO II.* Englewood Cliffs, NJ: Prentice-Hall.

Boehm, B., & Turner, R. (2004). *Balancing agility and discipline: A guide for the perplexed.* Boston: Addison-Wesley.

Booch, G., Jacobson, I., & Rumbaugh, J. (2005). *The Unified Modeling Language reference manual* (2nd ed.). Boston: Addison-Wesley.

Boyd, N. S. (1999). Using natural language in software development. *Journal of Object-Oriented Programming, 11*(9).

Fowler, M., Beck, K., Brant, J., Opdyke, W., & Roberts, D. (1999). *Refactoring: Improving the design of existing code.* Boston: Addison-Wesley.

Grassi, V., Mirandola, R., & Sabetta, A. (2004, October 11-15). A UML profile to model mobile systems. *Proceedings of the 7th International Conference on the Unified Modeling Language (UML 2004)*, Lisbon, Portugal.

Hamill, P. (2004). *Unit test frameworks: Tools for high-quality software development.* O'Reilly Media.

Hjelm, J. (2000). *Designing wireless information services.* New York: John Wiley & Sons.

Highsmith, J. (2002). *Agile software development ecosystems.* Boston: Addison-Wesley.

Jacobson, I., Booch, G., & Rumbaugh, J. (1999). *The Unified Software Development process.* Boston: Addison-Wesley.

Kamthan, P. (2005, January 14-16). Pair modeling. *Proceedings of the 2005 Canadian University Software Engineering Conference (CUSEC 2005)*, Ottawa, Canada.

Kamthan, P. (2006, January 19-21). Open source software in software engineering education: No free lunch. *Proceedings of the 2006 Canadian University Software Engineering Conference* (CUSEC 2006), Montreal, Canada.

Katira, N. (2004). *Understanding the compatibility of pair programmers.* MSc Thesis, North Carolina State University, USA.

Keller, D. (1990). A guide to natural naming. *ACM SIGPLAN Notices, 25*(5), 95-102.

Maurer, F., & Martel, S. (2002). Extreme programming. Rapid development for Web-based applications. *IEEE Internet Computing, 6*(1), 86-90.

Mnkandla, E., & Dwolatzky, B. (2004). A survey of agile methodologies. *Transactions of the South African Institute of Electrical Engineers, 95*(4), 236-247.

Nguyen, H.Q., Johnson, R., & Hackett, M. (2003). *Testing applications on the Web: Test planning for mobile and Internet-based systems* (2nd ed.). New York: John Wiley & Sons.

Salmre, I. (2005). *Writing mobile code: Essential software engineering for building mobile applications.* Boston: Addison-Wesley.

Schneidewind, N. F., & Fenton, N. E. (1996). Do standards improve product quality? *IEEE Software, 13*(1), 22-24.

Turk, D., France, R., & Rumpe, B. (2002, May 26-29). Limitations of agile software processes. *Proceedings of the 3rd International Conference on eXtreme Programming and Agile Processes in Software Engineering* (XP 2002) (pp. 43-46), Sardinia, Italy.

Wallace, D., Raggett, I., & Aufgang, J. (2002). *Extreme programming for Web projects.* Boston: Addison-Wesley.

Wiegers, K. (2002). *Peer reviews in software: A practical guide.* Boston: Addison-Wesley.

Williams, L., & Kessler, R. (2003). *Pair programming illuminated.* Boston: Addison-Wesley.

KEY TERMS

Agile Development: A philosophy that embraces uncertainty, encourages team communication, values customer satisfaction, vies for early delivery, and promotes sustainable development.

Coding Standard: A documented agreement that addresses the use of a formal (such as markup or programming) language.

Mobile Web Engineering: A discipline concerned with the establishment and use of sound scientific, engineering, and management principles, and disciplined and systematic approaches to the successful development, deployment, and maintenance of high-quality mobile Web applications.

Natural Naming: A technique for using full names based on the terminology of the application domain that consist of one or more words of the natural language instead of acronyms or abbreviations for elements in a software representation.

Pair Programming: A practice that involves two people such that one person (the primary person or the pilot) works on the artifact while the other (the secondary person or the co-pilot) provides support in decision making and provides input and critical feedback on all aspects of the artifact as it evolves.

Refactoring: A structureal transformation that provides a systematic way of eradicating the undesirable(s) from an artifact while preserving its behavioral semantics.

Software Engineering: A discipline that advocates a systematic approach of developing high-quality software on a large scale while taking into account the factors of sustainability and longevity, as well as organizational constraints of time and resources.

Factors Affecting Mobile Commerce and Level of Involvement

Frederick Hong Kit Yim
Drexel University, USA

Alan ching Biu Tse
The Chinese University of Hong Kong, Hong Kong

King Yin Wong
The Chinese University of Hong Kong, Hong Kong

INTRODUCTION

Driven by the accelerating advancement in information technology (IT), the penetration of the Internet and other communications services has increased substantially. Hoffman (2000), one of the most renowned scholars in the realm of Internet research, considers the Internet as "the most important innovation since the development of the printing press." Indeed, the omnipresent nature of the Internet and the World Wide Web (WWW) has been a defining characteristic of the "new world" of electronic commerce (Dutta, Kwan, & Segev, 1998). There are a good number of academics and practitioners who predict that the Internet and the WWW will be the central focus of all commercial activities in the coming decades (e.g., Dholakia, 1998). In particular, Jarvenpaa & Todd (1996) argue that the Internet is alive with the potential to act as a commercial medium and market. Figuratively, discussing the business prospects of the Internet and the WWW is somehow analogous to discussing the Gold Rush of the 19th century (Dholakia, 1995).

Admittedly, the close down of a lot of dot.coms since 2000 has been a concern for many people. However, the statistical figures we have up to now show that the growth pattern continues to be exponential. For example, the latest Forrester Online Retail Index released in January 2002 indicates that consumers spent $5.7 billion online in December, compared to $4.9 billion in November (Forrester Research, 2002a). There is yet another sign of optimism for online shopping: The Internet Confidence Index (as released in September 2002), jointly developed by Yahoo and ACNielsen, rose 13 points over the inaugural survey released in June 2001, indicating a strengthening in consumers' attitudes and confidence in e-commerce services (Yahoo Media Relations, 2002). Hence, we believe that the setback is only temporary and is part of a normal business adjustment. The future trend is very clear to us. Everybody, be it multinationals or small firms, should be convinced of the need to be on the Web.

While researchers like Sheth and Sisodia (1999) have described the growth of the Internet as astonishing, an even more startling growth is projected in the area of wireless Internet access via mobile devices. The general consensus is that mobile commerce, a variant of Internet commerce (Lucas, 2001) that lets users "surf" their phones (Wolfinbarger & Gilly, 2001), will become part of the next evolutionary stage of e-commerce (e.g., Keen, 2001; Leung & Antypas, 2001; Tausz, 2001). Mobile commerce involves the different processes of content delivery (notification and reporting) and transactions (purchasing and data entry) on mobile devices, and its current landscape resembles the Internet in its first generation in the early 1990s (Leung & Antypas, 2001). According to a study by Strategy Analytics, the rise in demand for mobile commerce services will lead to a market value of $230 billion by 2006 (Patel, 2001). Also a cause for optimism in mobile commerce services is the estimates made by the Yankee Group that the value of goods and services purchased via mobile devices will exceed $50 billion by 2005, up from $100 million in 2000 (Yankee Group, 2001). According to Yankee, the number of wireless consumers using financial services in North America alone will reach more than 35 million in 2005, a leap from the current 500,000.

Research on consumers' online behavior has so far been centered on the World Wide Web. Very few, if any, have specifically focused on mobile access despite the fact that mobile handsets are becoming increasingly popular. This is an important area of study, as the mobile phone is quickly bypassing the PC as the means of Internet access and online shopping. According to the Computer Industry Almanac, there will be an estimated 1.46 billion Internet users by 2007, compared to the 533 million today. Currently, wireless access constitutes a significant, yet limited user share of 16.0%, but by 2007, this number would have increased dramatically to 56.8% (Computer Industry Almanac, 2002). These optimistic projections are further supported by the prediction of Forrester Research that, within 5 years, up to 2.3 million wired phone subscribers in the U.S. would make the switch to wireless access, making an average of 2.2 wireless phones per household by 2007 (Forrester Research, 2002b).

Aided by staggering advances in information technology, mobile devices are now capable of offering a number of Internet-based and Internet-centric services, fueling the growth of mobile commerce. The ascendancy of mobile commerce as a marketing channel warrants researchers' and practitioners' alert even in its current rudimentary stage, not only because of the huge market potential projected, but also because mobile commerce can offer new channels through which enterprises can interact with customers (Leung & Antypas, 2001). In a bid to fill the research void in the realm of mobile commerce, and to afford some insights to firms battling over the electronic commerce arena, this research was conducted with the following two objectives in mind. The first objective is to scrutinize what constitutes the weighty factors as far as transacting through mobile devices is concerned. The second one is to find out how the importance of these factors would vary when consumers are confronted with two different transactions, each with a varying degree of involvement (Celsi & Olson, 1988). The first type of transaction is a low involvement one that involves buying movie tickets with little financial commitment, while the second one is undertaking stock transactions where the stake is high.

In the following, we would briefly summarize what the literature says about important factors that affect online shopping, which forms the basis for us to speculate on factors that may be important for consumers shopping via their mobile phones, the latter being one kind of online shopping, which should resemble to some degree other forms of shopping on the Internet as far as important factors affecting consumer behavior is concerned. Hypotheses are then formulated, which is followed by the methodology. After presenting the results, we discussed the implications and conclusions of this study.

CONCEPTUALIZATION

Regardless of the mode of access, the popularity of online shopping can be partially attributed to the effectiveness and efficiency to acquire information about vendor prices and product offerings (Alba et al., 1997; Bakos, 1997; Cook & Coupey, 1998; Klein, 1998; Peterson, Balasubramanian, & Bronnenberg, 1997; Sheth, Sisodia, & Sharma, 2000; Wolfinbarger & Gilly, 2001), and convenience in overcoming geographical and time barriers (Peterson, Balasubramanian, & Bronnenberg, 1997; Sheth & Sisodia, 1999). In sum, previous literature has found that convenience, site design and financial security are dominant in determining e-satisfaction and likelihood of using the Internet as a shopping channel (Eighmey & McCord, 1998; Szymanski & Hise, 2000; Tse & Yim, 2001).

Given that mobile commerce is also one kind of online shopping, we posit that "convenience," "site design" and "financial security" are the three crucial factors affecting consumers' propensity to transact through mobile phones:

Convenience

One of the widely held perceptions that drives consumers to go online is convenience (e.g., Donthu, 1999; Wind & Mahajan, 2002). The information superhighway has been promoted as a convenient avenue for shopping (Szymanski & Hise, 2000). Driven by the growth of mobile commerce, the convenience of online shopping is further enhanced (Lucas, 2001). Li et al. (1999) find that convenience is a robust predictor of users' online buying status. Similarly, Becker-Olsen (2000) expounds that one of the most important factors that determines whether consumers buy online is the extent to which they perceive the Internet as convenient. The convenience instilled in the electronic marketplace is manifested in time savings, effort economization and accessibility, as perceived by online consumers (Wolfinbarger & Gilly, 2001). Like shopping using a PC, consumers buying movie tickets or completing stock transactions via their mobile phones would be able to save a lot of time and effort that would otherwise be wasted in dealing with agents or ticket offices.

As buying movie tickets is a transaction of low involvement and that undertaking stock transactions is of high involvement, it can be logically reckoned that the convenience factor is different in significance depending upon the situation. Convenience may have a more significant impact on consumers' propensity to transact online in the context of a ticket transaction, as compared to a stock transaction. Consumers should experience greater satisfaction when they can buy movie tickets anytime and anywhere breaking the time- and location-bound facets of traditional "gravitational" commerce (Sheth & Sisodia, 1999). On the other hand, for stock investment, consumers' major concern is security, as the consequence of any mistake can result in a great loss (Rosenbloom, 2000). Hence, we speculate that if an online stock trading system is too convenient, online investors may actually refrain from using it. For example, if, for the sake of convenience, a user is not required to enter a second password to confirm a transaction, the user may end up feeling highly insecure and less satisfied. To test our assertions, we put forward the following hypotheses:

- **H_{1a}:** Convenience significantly affects willingness to transact online for both movie ticket and stock transactions.
- **H_{1b}:** The importance of convenience in determining whether consumers transact online is greater for a ticket transaction than a stock transaction.

- **H_{1c}:** The greater the level of convenience is in an online stock transaction, the lower the intention is in using the system.

Site Design

Site design is considered important in the realm of electronic commerce (e.g., Eighmey & McCord, 1998; Lohse & Spiller, 1999; Wolfinbarger & Gilly, 2001) and mobile commerce (Lucas, 2001). Like any diffusion of innovation, there is a learning curve for most consumers to utilize electronic commerce in a way they feel most comfortable (Li, Kuo, & Russell, 1999). The success akin to the adoption of mobile transactions hinges on, at least in part, the complexity of the utilization of this innovation (Childers et al., 2001; Rogers, 1983). It can thus be understood that meticulously crafted Web sites tend to be more successful in terms of ushering in first-time online shoppers and gaining repeat visits. Empirically, Novak et al. (2000) find that a compelling online experience, which can be engendered by deliberate contrivance of Web sites, is positively associated with expected use in the future and the amount of time consumers spend online. As far as a stock transaction is concerned, we speculate that site design plays a crucial role in facilitating consumers to carry out the transaction. Users would undoubtedly require an effective design that allows them to orchestrate their portfolio without committing any mistake. Regarding the importance of site design in the use of mobile phones for ticket transactions, we undertook ten in-depth interviews with buyers who had bought tickets with their mobile phone before, and found that since most of the ticket transactions currently completed via mobile devices do not afford consumers the latitude to choose the specific seats they like because of the small screen size, they do not expect the same level of good site design as for stock transactions. Hence, our second hypothesis is formulated as below:

- **H_{2a}:** Site design significantly affects willingness to transact online for both movie ticket and stock transactions.
- **H_{2b}:** The importance of site design in determining whether consumers transact online is greater for a stock transaction than a ticket transaction.

Financial Security

It has been argued for long that the issue of financial security is pivotal in electronic transactions (e.g., Guglielmo, 1998; Kluger, 2000; Rosenbloom, 2000; Stateman, 1997), specifically in those conducted via wireless devices (Gair, 2001; Goldman, 2001; Hurley, 2001; Laughlin, 2001; Teerikorpi, 2001). As stock transactions entail a significantly larger amount of financial risk than ticket transactions, it can be logically reasoned that financial security plays a more salient role in affecting consumers' propensity to transact online when consumers are confronted with a stock transaction, as compared to a ticket transaction. Our third hypothesis is arrived at as follows:

- **H_{3a}:** Financial security significantly affects willingness to transact online for both movie ticket and stock transactions.
- **H_{3b}:** The importance of financial security in determining whether consumers transact online is greater for a stock transaction than a ticket transaction.

METHOD

Using a survey design, 192 respondents were selected by convenience sampling and interviewed in three different locations in Hong Kong representing a good cross-section of three different strata of socio-economic groups. The Hong Kong sample is deemed appropriate, by virtue of the fact that Hong Kong is the perfect location for a mobile commerce solutions player and is alive with opportunities fueled by its more than 5.77 million mobile phone subscribers, which represents a mobile penetration rate of 88.88% (Leung, 2002). The research was conducted in shopping malls, where a high customer flow can be found. Respondents were asked to make their evaluations when confronted first with a movie ticket buying scenario, and subsequently with a stock transaction using a mobile phone. For each scenario, before respondents indicate their likelihood of using the channel for the transaction, they were given a detailed description and explanation of the features and functionalities of the phone and the Web site that provide the service. Their likelihood of using the mobile service is captured by a seven-point scale ranging from "–3" to "3," where "–3" means "highly unlikely," "0" means "neither unlikely nor likely," and "3" represents "highly likely."

Each factor described above—convenience, site design, and financial security—is measured by at least four items so that the consistency/reliability of respondents' replies can be assessed. The items for each factor are based on those suggested by Keeney (1999). The factors are presented in bold type below followed by the items used for the factor. For each of the statements below, respondents were asked to indicate the extent to which they agree or disagree with it using again a seven-point scale ranging from "–3" to "3," where "–3" means "strongly disagree," "0" means "neither disagree nor agree," and "3" represents "strongly agree."

Convenience

- The mobile service is convenient.
- The mobile service can maximize transactional speed.

Table 1. Cronbach alphas for all the subscales

Factors	Ticket Transaction	Stock Transaction
Convenience	0.6236	0.8752
Site Design	0.5325	0.7122
Financial Security	0.7552	0.8870

- The mobile service can minimize waiting time.
- The mobile service can minimize personal travel.

Site Design

- The interface is easy to use.
- The interface is designed in such a way that I can contact a service staff easily if needed.
- The interface allows me to get a variety of services easily.
- The design of the interface is of high quality.
- I enjoy using the interface.

Financial Security

- The system is secure.
- Transaction conducted through the system is accurate.
- The system allows me to keep track of my previous transactions without error.
- The system provides me with clear information of my previous transactions.
- The system protects my personal financial information and privacy.

RESULTS

Table 1 reports the reliabilities of the items for each of the categories for both types of transactions: ticket and stock. All Cronbach's alpha values are greater than 0.7, except those of Convenience and Site Design for Ticket Transactions, which are marginal. However, for exploratory studies like this one, a value of Cronbach's alpha that exceeds 0.5 would be considered acceptable (Nunnally, 1978), and so the summated score of the items under each factor would be used as predictor variables for the statistical analyses described as follows:

Two regression analyses were conducted using the summated scores of the factors as the independent variables and the likelihood of transacting online (buying tickets or consummating stock transactions) as the dependent variable. The results of the regression analyses are depicted in Table 2.

Table 2. Regression coefficients for predictors of transacting online

Panel A: Regression findings for ticket transaction

Predictor Variable	Standardized Coefficient (SE)	t-value (p-level)
Convenience	0.079 (0.046)	0.787 (0.433)
Site Design	-0.122 (0.047)	-1.096 (0.275)
Financial Security	0.097 (0.034)	0.854 (0.395)

Panel B: Regression findings for stock transaction

Predictor Variable	Standardized Coefficient (SE)	t-value (p-level)
Convenience	-0.059 (0.074)	-0.274 (0.785)
Site Design	-0.328 (0.081)	-1.373 (0.175)
Financial Security	0.557 (0.061)	2.216 (0.030)

As evinced in Table 2, none of the factors significantly affect the intention to use mobile phone for buying movie tickets online. For online stock transactions, financial security exhibits the greatest impact on the likelihood of using the system, with the other two factors playing a non-significant role. Hence, our results do not lend support to $H_{1a,}$ H_{1c} and H_{2a}, and only H_{3a} is supported in so far as high involvement products are concerned.

To test the remaining hypotheses, we need to compute the difference in effect sizes for each factor in the two buying situations. The required differences, computed using the method suggested by Cohen (1977), are shown in Table 3.

The group difference effect sizes shown in Table 3 are free of the original measurement units. They measure the difference in effects of each of the factors under consideration—namely convenience, site design and financial security—on willingness to use a mobile phone to undertake the two different types of transactions. As seen in Table 3, we find that the difference in effect sizes associated with convenience is negative, which is in the expected direction. Tallmadge (1977) provides rough guidelines of difference = 0.25 indicating small effect, and difference = 0.33 for medium effect. Using this guideline, the effect size difference for convenience, is very small, so H_{1b} cannot be said to be supported although the negative direction is consistent with H_{1b}. On the other hand, the differences for site design and financial security are medium in magnitude, thus supporting H_{2b} and H_{3b}.

Table 3. Difference in effect sizes between ticket and stock transactions

	Convenience			
	Mean[1]	Standard Deviation	Pooled Within Group Standard Deviation	Difference in Effect Sizes[2]
Ticket	6.260	3.965	4.444	-0.058
Stock	6.000	5.206		
	Site Design			
	Mean	Standard deviation	Pooled within group standard deviation	Difference in Effect Sizes
Ticket	1.349	4.255	4.687	0.355
Stock	3.014	5.383		
	Security			
	Mean	Standard deviation	Pooled within group standard deviation	Difference in Effect Sizes
Ticket	1.945	6.065	6.561	0.328
Stock	4.100	7.380		

Note: [1] *Mean level of willingness to undertake respective online transactions.*
[2] *positive effect size difference is analogous to a positive treatment effect size using stock as the experimental group and ticket as the control group.*

DISCUSSION AND IMPLICATIONS

The next profound shift in the use of IT will obviously be toward wireless and mobile commerce (Keen, 2001), an emerging discipline (Varshney, 2002). The whole world of mobile commerce is about to explode (Martin, 2002). In many ways, m-commerce is, per se, the continuation of e-commerce with the Palm handheld, wireless laptops and a new generation of Web-enabled digital phones already on the market. It is even believed that portable devices such as phones, pagers and computers with mobile modems will quickly surpass desktop PCs as the Internet access devices of choice (Lindquist, 2001). The race for dominance in mobile commerce has begun (Nohria & Leestma, 2001). As addressed by Hoffman (2000), scholarly research on the Internet cannot keep abreast with business practice, let alone the scanty, if any, research on the newly emerging mobile commerce. To the very best of our knowledge, the survey we have conducted serves as a pioneer study in the realm of mobile commerce.

Aligning with our initial surmise that the salience of convenience in determining whether consumers transact online is larger for a ticket transaction than a stock transaction, our results, though only marginally supporting the hypothesis, shed light on what is deemed weighty in providing mobile commerce transactions. Enterprises providing electronic ticketing services of recreational activities, which are of low involvement, should pay heightened attention to how the convenience of their services can be enhanced. For example, the waiting time for consummating a ticket transaction as well as the transaction time required should be minimized.

Another noteworthy issue is that the coefficient associated with convenience as a predictor variable of the likelihood of consummating a stock transaction is negative. This may mean that respondents may associate increased level of convenience with increased level of inherent risks, thus hampering their propensity to transact online when the transactions in question are of high involvement. Firms facilitating online stock transactions, or other high involvement transactions, should thus be alert to this issue—they should promulgate their commitment to reduce their clients' risks in line with providing convenience.

Financial security is of paramount concern to online consumers seeking to consummate stock transactions, lending support to our third hypothesis. Indeed, the coefficient associated with financial security for stock transaction is of both practical and statistical significance (0.557; p-level = 0.030), signifying the colossal effects exerting from financial security on the likelihood to transact online. Meanwhile, the effect size difference for security is medium. The implication for our results is pronounced: financial security should be given overwhelming priority to high involvement transactions. Practically, clients should be continually and periodically informed of their online transactions, expressed in unequivocal terms. Building trust with clients is a proper and effective way to alleviate their worry about financial security (Shneiderman, 2000). This can be accomplished by nurturing and fostering a firm's relationship with its clients (Price & Arnould, 1999).

The insignificance of site design serving as a predictor of the likelihood of online transaction is contrary to what is addressed in the extant literature pertaining to traditional electronic commerce. Given the very nature of mobile commerce, we are yet afforded with some novel insights: as screens of mobile phones are miniatures of desktop monitors (Lucas, 2001), the possible designs for a site are constrained—overly fancy site designs cannot be demonstrated in the realm of mobile commerce, rendering site design insignificant in predicting transacting online. Although our study shows that site design does not play a significant role in affecting willingness to transact online, the difference in effect sizes are medium for the two different types of transactions. The latter is an indication that online firms intending to sell high involvement products should spend more on interface design when compared with their counterparts selling low involvement items.

DIRECTIONS FOR FUTURE RESEARCH

As the costs pertaining to the two online transactions delineated previously are somehow kept constant, we have not yet examined the effects of costs on propensity to transact online. The issue should be addressed in future research, since minimizing costs is identified as an objective in online transactions (Keeney, 1999; Leavy, 1999).

The intriguing result of the negative coefficient associated with convenience in the realm of high involvement transactions also warrants further research. Not until forthcoming research renders our intuitive interpretation to empirical scrutiny will a more comprehensive understanding of mobile commerce ensue.

Furthermore, other types of products should be chosen in addition to the ones we have used in this research to improve the generalizability of our findings. Meanwhile, in addition to classifying products based on level of involvement, we may categorize products in other ways. For example, we may classify products as search, experience and credence goods (Klein, 1998), and the relative importance of the three factors studied may be different for these three categories of products.

REFERENCES

Alba, J., Lynch, J., Weitz, B., Janiszewski, C., Lutz, R., Sawyer, A., & Wood, S. (1997, July). Interactive home shopping: Consumer, retailer, and manufacturer incentives to participate in electronic marketplaces. *Journal of Marketing, 61,* 38-53.

Bakos, J. Y. (1997). Reducing buyer search costs: Implications for electronics marketplaces. *Management Science, 43*(12), 1676-1692.

Becker-Olsen, K. L. (2000). *Point, click and shop: An exploratory investigation of consumer perceptions of online shopping.* Paper presented at AMA summer conference.

Celsi, R. L., & Olson, J. C. (1988). The role of involvement in attention and comprehension processes. *Journal of Consumer Research, 15,* 210-224.

Childers, T. L., Carr, C. L., Peck, J., & Carson, S. (2001). Hedonic and utilitarian motivations for online retail shopping behavior. *Journal of Retailing, 77*(4), 511-535.

Cohen, J. (1977). *Statistical power analysis for the behavioral sciences.* New York: Academic Press.

Computer Industry Almanac. (2002). *Internet users will top 1 billion in 2005. Wireless Internet users will reach 48% in 2005.* Retrived December 31, 2005, from http://www.c-i-a.com/pr032102.htm

Cook, D. L., & Coupey, E. (1998). Consumer behavior and unresolved regulatory issues in electronic marketing. *Journal of Business Research, 41,* 231-238.

Dholakia, R. R. (1995). *Connecting to the Net: Marketing actions and market responses.* Paper presented at the International Seminar on Impact of Information Technology hosted by CIET-SENAI, Rio de Janeiro, Brazil, December 6, 1995.

Dholakia, R. R. (1998). Special issue on conducting business in the new electronic environment: Prospects and problems. *Journal of Business Research, 41,* 175-177.

Donthu, N. (1999). The Internet shopper. *Journal of Advertising Research, 39*(3), 52-58.

Dutta, S., Kwan, S., & Segev, A. (1998). Business transformation in electronic commerce: A study of sectoral and regional trends. *European Management Journal, 16*(5), 540-551.

Eighmey, J., & McCord, L. (1998). Adding value in the information age: Uses and gratifications of sites on the World Wide Web. *Journal of Business Research, 41,* 187-194.

Forrester Research. (2002a). *December shopping up from last year in spite of rough economy, according to the Forrester Research Online Retail Index.* Retrieved December 31, 2005, from http://www.forrester.nl/ER/Press/Release/0,1769,678,00.html

Forrester Research. (2002b). Retrieved from http://www.forrester.com

Gair, C. (2001). The next big thing? *Black Enterprise, 31*(10), 62.

Goldman, C. (2001). Banking on Security. *Wireless Review, 18*(7), 22-24.

Guglielmo, C. (1998). Security fears still dog Web sales. *Inter@ctive Week, 5*(273), 44-47.

Hoffman, D. L. (2000). The revolution will not be televised: Introduction to the special issue on marketing science and the Internet. *Marketing Science, 19*(1), 1-3.

Hurley, H. (2001). Pocket-sized security. *Telephony, 240*(18), 42-50.

Jarvenpaa, S. L., & Todd, P. A. (1996). Consumer reactions to electronic shopping on the World Wide Web. *International Journal of Electronic Commerce, 1*(2), 59-88.

Keen, P. G. W. (2001). Go mobile—now! *Computerworld, 35*(24), 36.

Keeney, R. L. (1999). The value of Internet commerce to the customer. *Journal of the Institute for Operations Research and the Management Sciences, 45*(4), 533-542.

Klein, L. R. (1998). Evaluating the potential of interactive media through a new lens: Search versus experience goods. *Journal of Business Research, 41,* 195-203.

Kluger, J. (2000). Extortion on the Internet. *Time, 155*(3), 56-58.

Laughlin, K. (2001). Banking on wireless. *America's Network, 105*(1), 56-60.

Leavy, B. (1999). Organization and competitiveness—Towards a new perspective. *Journal of General Management, 24*(3), 33-52.

Leung, K., & Antypas, J. (2001). Improving returns on m-commerce investments. *Journal of Business Strategy, 22*(5), 12-13.

Leung, T. (2002, May 27). HK trails in mobile data. *Asia Computer Weekly,* 1.

Li, H., Kuo, C., & Russell, M. G. (1999). The impact of perceived channel utilities, shopping orientations, and demographics on the consumer's online buying behavior. *Journal of Computer Mediated Communication, 5*(2).

Lindquist, C. (2001). Mobile Internet access exploding. *CIO, 14*(13), 138.

Lohse, G. L., & Spiller, P. (1999). Internet retail store design: How the user interface influences traffic and sales? *Journal of Computer Mediated Communication, 5*(2).

Lucas, P. (2001). M-commerce gets personal. *Credit Card Management, 14*(1), 24-27.

Martin, N. (2002). Content a la Wmode: Serving up solutions for wireless content. *EContent, 25*(1), 48-49.

Nohria, N., & Leestma, M. (2001). A moving target: The mobile-commerce customer. *MIT Sloan Management Review, 42*(3), 104.

Novak, T. P., Hoffman, D. L., & Yung, Y. (2000). Measuring the customer experience in online environments: A structural modeling approach. *Marketing Science, 19*(1), 22-42.

Nunnally, J. C. (1978). *Psychometric Theory* (2nd ed.). New York: McGraw-Hill.

Patel, N. (2001). *Mobile commerce market update.* Retrieved December 31, 2005, from http://www.strategyanalytics.net/default.aspx?mod=ReportAbstractViewer&a0=839

Peterson, R. A., Balasubramanian, S., & Bronnenberg, B. J. (1997). Exploring the implications of the Internet for consumer marketing. *Journal of the Academy of Marketing Science, 25*(4), 329-346.

Price, L. L., & Arnould, E. J. (1999). Commercial friendships: service provider-client relationships in context. *Journal of Marketing, 63,* 38-56.

Rogers, E. M. (1983). *Diffusion of innovations* (3rd ed.). New York: Free Press.

Rosenbloom, A. (2000). Trusting technology. *Communications of the ACM, 43*(12), 31-32.

Sheth, J. N., & Sisodia, R. S. (1999). Revisiting marketing's lawlike generalizations. *Journal of the Academy of Marketing Science, 27*(1), 71-87.

Sheth, J. N., Sisodia, R. S., & Sharma, A. (2000). The antecedents and consequences of customer-centric marketing. *Journal of the Academy of Marketing Science, 28*(1), 55-66.

Shneiderman, B. (2000). Designing trust into online experiences. *Communications of the ACM, 43*(12), 57-59.

Stateman, A. (1997). Security issues impact online buying habits. *Public Relations Tactics, 4*(10), 8-16.

Szymanski, D. M., & Hise, R. T. (2000). E-satisfaction: An initial examination. *Journal of Retailing, 76*(3), 309-322.

Tallmadge, G. K. (1977). *The Joint Dissemination Review Panel ideabook.* Washington, DC: National Institute of Education and U.S. Office of Education.

Tausz, A. (2001). Customizing your world. *CMA Management, 75*(2), 48-51.

Teerikorpi, E. (2001). How secure is the wireless Internet. *Telecommunications, 35*(5), 46-47.

Tse, A. C. B., & Yim, F. (2001). Factors affecting the choice of channels: Online vs. conventional. *Journal of International Consumer Marketing, 14*(2/3), 137-152.

Varshney, U. (2002). Multicast support in mobile commerce applications. *Computer, 35*(2), 115-117.

Wind, Y., & Mahajan, V. (2002). Convergence marketing. *Journal of Interactive Marketing, 16*(2), 64-79.

Wolfinbarger, M., & Gilly, M. C. (2001). Shopping online for freedom, control, and fun. *California Management Review, 43*(2), 34-55.

Yahoo Media Relations. (2001). *Internet confidence index.* Retrieved December 31, 2005, from http://docs.yahoo.com/docs/info/yici/06-02.html

Yankee Group. (2001). Retrieved from http://www.yankeegroup.com

KEY TERMS

Convenience: One of the determining factors for e-satisfaction and likelihood of using the Internet as a shopping channel. It is manifested in time savings, effort economization and accessibility, as perceived by online consumers.

Financial Security: One of the determining factors for e-satisfaction and likelihood of using the Internet as a shopping channel. It refers to the personal financial information protection for the consumers who make online transactions.

Involvement: A consumer's overall subjective feeling of personal relevance.

Mobile Commerce: A variant of Internet commerce that lets users "surf" their mobile devices, for example, mobile phone, PDA.

Online Shopping: Transactions made via Internet rather than at a physical location by consumers.

Site Design: One of the determining factors for e-satisfaction and likelihood of using the Internet as a shopping channel. It refers to the interface quality that a company provides for its consumers to do online transactions.

A Game-Based Methodology for Collaborative Mobile Applications

Michael Massimi
University of Toronto, Canada

Craig H. Ganoe
The Pennsylvania State University, USA

John M. Carroll
The Pennsylvania State University, USA

INTRODUCTION

Mobile computing, perhaps more so than traditional desktop computing, requires methods for allowing application designers to try ideas, create prototypes, and explore the problem space. This need can be met with rapid prototyping. Rapid prototyping is a technique that permits members of a design team to iterate through several versions of their low-level designs (Thompson & Wishbow, 1992). During each cycle of each prototype, the design team identifies critical use cases, verifies requirements are being met, and gathers both subjective and objective data regarding usability. Because "shallow" or low-fidelity prototypes can be quickly created, used, and thrown away (Sefelin, Tscheligi, & Giller, 2003), the team can explore many options and designs with less effort than it would take to create "deep" or high-fidelity versions of each prototype (Rudd, Stern, & Isensee, 1996).

Rapid prototyping techniques are especially valuable when the application is intended for a mobile user. This is for three primary reasons. First, the mobile user is likely to be simultaneously attending to a dynamic or unpredictable environment. This environment taxes the user's cognitive abilities. Users must navigate to their destinations, avoiding obstacles and responding to changing conditions. Non-technical aspects can change, like weather or available routes. Many times, the user must "make place" in order to use the system, stopping to seek out an area to use the software (Kristoffersen & Ljunberg, 1999). Technical aspects of the system, such as network availability and power levels, can also be difficult to accurately predict and may require complex adaptation algorithms (Noble et al., 1997; de Lara, Kumar, Wallach, & Zwaenepoel, 2003; Welch, 1995). Compared to a stationary environment, the number of things that can go wrong seems to skyrocket.

Second, interpersonal communication changes when a dimension of mobility is introduced. When working collaboratively on a task, users require awareness of the tasks their collaborators are performing (Ganoe et al., 2003) in order to prevent redundancy and achieve an equitable distribution of work. When users are mobile, however, awareness is no longer simply *what* other people are doing, but also *where* they are doing it. This introduces a need for additional application support for mobile collaborative systems.

Third, heterogeneity of devices results in different interaction styles. Mobile phones provide an excellent example of this problem. Each manufacturer repositions buttons based on hardware and space constraints. Even within a manufacturer's own product line, multiple key configurations occur. This is to say nothing of the variety of mobile devices available—PDAs, tablet PCs, wearable computers, and so on. Some of the large manufacturers, like Palm, provide human interface guidelines to third-party developers (Ostrem, 2003). Most do not.

In terms of evaluating systems, Abowd and Mynatt (2000) argue that our current methods are not sufficient. The traditional task-based evaluation methods no longer apply in a world where we cannot always experimentally control the environment, and where there is not a clear, single indicator of task performance. There are not established tests that can be performed to determine the effectiveness of deployed systems, mainly because there are not many of them in the world yet. Because we do not have a base of knowledge regarding how to design for mobile interaction, early affirmations of whether the application will serve a human need are critical, and Abowd and Mynatt state that we should "understand how a new system is used by its intended population before performing more quantitative studies on its impact" (p. 47).

Mobile systems need fast, inexpensive ways of prototyping and gathering usability results. This entry describes previous work in rapid prototyping for mobile systems. We then contribute a novel rapid prototyping methodology for mobile systems, which we call "Scavenger Hunt." It is anticipated that this methodology will be useful not only for those interested in rapid prototyping and design methodologies, but also for design teams with real deadlines to meet. Finally, we identify future trends in prototype evaluation of mobile systems.

BACKGROUND

Games

Our prototype evaluation methodology is based on a game—specifically, a Scavenger Hunt. The basis for this choice stems from success with using games as a tool for design and testing for non-mobile applications.

Twidale and Marty (2005) used a "game show" format during a conference, wherein contestants found usability problems in software, cheered on by an audience. They argue that "it is worth exploring the power of rapid, lightweight methods to catch relatively uncontroversial and easily fixed usability flaws." Scavenger Hunt does this as well, although the focus of the participant is not on the actual discovery of the flaw, but on completing a higher-level task.

Spool, Snyder, Ballman, and Schroeder (1994) created a game where designers are placed onto teams and are given a time limit to create a UI. Then, test users move from design to design and must complete the same task on each one. The design with the quickest task completion time is the winner. Here, the goal is to teach designers how to create usable software by rewarding them in a game. In this study, the game is used educationally. The goal of the game is to teach the player how to create good designs, or how to use a particular evaluation method (e.g., heuristic evaluation). Instead, we use a game itself to *evaluate* the prototype. This game-based evaluation is designed to compliment other lightweight usability evaluation metrics like heuristic evaluations (Nielsen & Molich, 1994).

Pedersen and Buur (2000) created a board game to help participatory design teams conceptualize their sessions. The board, modeled after the industrial plant where the users worked, was populated with foam pieces representing artifacts and people. The design partners took turns moving the pieces to explain processes in the plant, and this opened the door to discussion about what should and should not occur during a particular process. The notion of turn-taking is especially noteworthy, as it allows design partners to offer their thoughts and obtain equal footing in the design process. We move from a board game to a "real-life" game in the SH process. In addition, we are interested in using a game as an evaluation tool rather than a design tool. Despite these differences, the past successes with games as parts of the design lifecycle are very encouraging.

Mobile Design and Usability

In experiments conducted by Virzi, Sokolov, and Karis (1996), it was found that testing with low-fidelity prototypes found almost as many usability problems as their high-fidelity counterparts. We argue, however, that paper prototypes will not be suitable for mobile interaction, and that low-fidelity computer-based versions of prototypes should be used instead.

SCAVENGER HUNT

Motivation

To gather usability metrics about mobile collaboration systems, we have developed a methodology we call "Scavenger Hunt" (SH). SH emulates the children's game where players are given a list of items that they must collect and bring to a pre-ordained location. In our methodology, the "players" are in fact target users, and each is equipped with the appropriate mobile device and prototype software under scrutiny.

By basing the rapid prototyping technique on a well-known game, the users can quickly be brought up to speed on how to complete the usability test. Further, they are motivated to "win" the game by completing all the tasks to the best of their ability. This combats the ennui that might otherwise set in when a user is simply asked to perform a series of artificial tasks. In fact, a savvy usability tester might pit two teams against one another to see who wins first and by what methods. Extreme use cases are more likely to emerge when users push the system to its boundaries to win.

Study Details

We conducted a pilot study wherein we used the SH method to evaluate a collaboration tool prototype. The specific details we have used to conduct this SH session follow and are meant to serve as an early model for future applications of this method. These details and parameters can, of course, be tailored to meet the needs of a particular design team, product, or schedule.

Software

In order to pilot the Scavenger Hunt method, we developed a Weblog prototype as the software under scrutiny. The Weblog (which we call SH Blog) allowed multiple people to add posts to it, edit each others' posts, and reorganize the ordering of the posts. We purposefully did not create a "polished" version of the software. The prototype was representative of a first pass through coding the system and was written in approximately five hours.

The prototype was written in PHP and HTML. Clients ran Microsoft Internet Explorer for Pocket PC and rendered pages from an Apache Web server running on Linux. Data was stored server-side in a MySQL relational database.

Participants

Eleven participants were recruited from a summer school program for gifted youth. They were divided into groups of three (one participant failed to arrive). Each team was self-selected and worked together on a project during the summer school program, so the participants were comfortable working with each other. Overall they reported high levels of comfort with technology, but did not use mobile computers very often.

Pre-Session Setup

Before the SH session, we distributed 24 clues throughout the building. These clues were evenly divided among the three floors of the building. All clues were printed on brightly colored paper and were hung on walls or placed on tables. We attempted to disperse the clues throughout the building evenly so that a participant would have a chance to find a clue in consistent time intervals (e.g., after about 45 seconds of walking). We ensured that all clues were in public areas so that participants would have access to them and would not feel awkward entering private offices.

The SH Blog was engineered to capture data about user interactions before the session. The time of posting and user who posted were logged. A software engineer monitored the MySQL database that stored SH blog posts and noted the progress of the team. This monitoring was essential to the evaluation of the prototype from the software engineering perspective, as it allowed us to look "under the hood" of the software during the session.

Finally, we ensured IEEE 802.11b wireless networking was available in all areas where there were clues. Some areas, such as stairwells or elevators, could not receive a signal; this is characteristic of most mobile computing environments, however.

Starting an SH Session

We gathered each team individually at the beginning of the session and had them complete a questionnaire that asked questions about their comfort level with mobile computers, their experience with working while mobile, and their preferences for group work. Each team was then given an overview of the game that explained the following:

- There are clues throughout the building. They are all in plain view and are in public spaces.
- You need to collect as many clues as you can in order to answer a riddle.
- You must use the SH Blog to share the clues and to work on solving the riddle. You may not use software on your mobile computer besides the SH Blog.
- You will have one hour to complete the task and return here with the answer to the riddle.

Participants were then given the riddle and asked to begin. They immediately began the task and started to walk around the building, entering clues into the SH Blog.

Collecting Data During an SH Session

During the study, participants were videotaped by researchers with camcorders in order to later analyze comments and note salient themes. Participants were asked to think aloud in order to capture the cognition accompanying the interaction and problem solving. At the end of the session, users completed a questionnaire about their experience with the SH Blog prototype. Finally, based on observations during the task and questionnaire responses, we conducted a brief semi-structured interview wherein we asked questions about problems, ideas for changes, and experiential preferences. By using three different research instruments, we are able to collect a wide range of data and make design suggestions based on both explicit and implicit behaviors.

At the system level, we captured information about the number of posts made by each team member, the total number of posts, the movement of posts, the deletion of posts, and so on. By charting these over time and comparing the different groups, we can note differences in usage and system support. For example, one group in our study generated a new post for each clue they discovered; another chose to accumulate all clues in a single post. These varying styles indicate, for example, the need for the SH Blog to accommodate both large numbers of posts and large, monolithic posts.

Evaluation and Results

Some of the salient results of our trial run are presented. It is important to note the different types of problems that the method identified—they run the gamut from usability issues, to systems issues, to social issues. Many of these insights may not have been found by traditional task-based user tests.

We noted that no group collected the entire set of clues. This was for a variety of reasons. A team member might believe that a different team member already collected the clue. A team might miss the clue completely. A team might have found it and subsequently deleted it, reasoning that it was redundant or useless. Without a sufficient subset of the clues collected, teams could not solve the problem.

Different teams approached the game differently, even though they were all given the same starting conditions. One team chose to divide the building into floors and then assigned a floor to each member. Another team chose to have one member act as an analyst back at the "base," while the other two members walked around and focused solely on the collection aspect. This indicates that high-level "gaming" strategies must be supported in the software.

In every trial, the members initially split up and went separate ways to find clues. Again, in every trial, the members eventually met face-to-face once they thought they had collected all of the clues. We noticed a shift in work styles from an individual, mobile worker to a stationary, group worker. This indicates that the software must accommodate individual work and group work separately, and provide a transition between the two work styles.

Of the four trials, only one team actually solved the problem. Even this team had boiled it down to an educated guess. Based on this outcome and the ones given above, we determined that the SH Blog did not allow the users to accomplish the task and needed revisions. The ability to reorder posts more easily and the ability to draw free-form tables were the primary revisions that users identified in the questionnaires and interviews. In one exceptional case, a user accidentally deleted the aggregate post that contained all of their collected clues! It was only then that we realized there was no undo function.

As these four themes suggest, users identified numerous flaws and areas for improvement in the software during interviews, questionnaires, and observations. Because the goal of this study was for feasibility purposes, we have not evaluated our method against other methods on similar tasks. We do, however, feel that the insights gained from using SH were well worth the setup costs. The major contribution of using SH is to show that rapid, low-cost usability and systems testing can be conducted early in the design process. This method may be of use to design teams in both research and industry settings.

FUTURE TRENDS

As mobile applications are developed more frequently in order to suit the needs of the third-wave information worker, we believe that prototyping, iterative design, and usability testing will become more and more important. Cell phones demonstrate that users prefer mobile devices when they are used for interpersonal contact, and methods for evaluating collaboration on-the-go are essential for this task. Techniques like SH are useful for software engineers and usability engineers alike. We believe more tools like it should be developed.

Further, the evaluation of these tools is an open research problem. How do we demonstrate that one mobile system suits its users' requirements more effectively than another? What are the outcomes to be measured? In the absence of a long history of software deployment and use, these questions remain to be answered.

CONCLUSION

As Abowd and Mynatt (2000) observed, it is extremely difficult to conduct evaluations of mobile computing systems because of the always-on, dynamic environment. For this reason, it is imperative that we have tools for early, non-trivial user testing, and we have presented a novel method for doing so. Our method is lightweight and can be applied repeatedly in the design process to ensure that requirements are met before an expensive deployment begins. Although it does not replace actual field trials, it can identify systems-level and interface-level flaws by simulating a representative task in the problem domain. Continued work in identifying the critical components of evaluating mobile systems is important, as is the need for early prototyping and validation.

REFERENCES

Abowd, G., & Mynatt, E. (2000). Charting past, present, and future research in ubiquitous computing. *ACM Transactions on Computer-Human Interaction, 7*(1), 29-58.

de Lara, E., Kumar, R., Wallach, D. S., & Zwaenepoel, W. (2003). Collaboration and multimedia authoring on mobile devices. *Proceedings of the 1st International Conference on Mobile Systems, Applications, and Services (MobiSyS)* (pp. 287-301).

Ganoe, C., Somervell, J., Neale, D., Isenhour, P., Carroll, J., Rosson, M. B., et al. (2003). Classroom BRIDGE: Using collaborative public and desktop timelines to support activity awareness. *Proceedings of the 16th ACM Symposium on User Interface Software and Technology (UIST)* (pp. 21-30).

Kristoffersen, S., & Ljunberg, F. (1999). Making place to make it work: Empirical exploration of HCI for mobile CSCW. *Proceedings of the International ACM SIGGROUP Conference on Supporting Group Work 1999* (pp. 276-285).

Nielsen, J., & Molich, R. (1990). Heuristic evaluation of user interfaces. *Proceedings of the ACM SIGCHI Conference on Human Factors in Computing Systems 1990* (pp. 373-380).

Noble, B., Satyanarayanan, M., Narayanan, D., Tilton, J. E., Flinn, J., & Walker, K. (1997). Agile application-aware adaptation for mobility. *Proceedings of the 16th ACM Symposium on Operating Systems Principles* (pp. 276-287).

Ostrem, J. (2003). *Palm OS user interface guidelines.* Retrieved January 6, 2006, from http://www.palmos.com/dev/support/docs/ui/UIGuidelinesTOC.html

Pedersen, J., & Buur, J. (2000). Games and movies—Towards innovative co-design with users. *Proceedings of the CoDesigning Conference,* Coventry, UK.

Rudd, J., Stern, K., & Isensee, S. (1996). Low vs. high-fidelity prototyping debate. *ACM Interactions, 3*(1), 76-85.

Sefelin, R., Tscheligi, M., & Giller, V. (2003). Paper prototyping—What is it good for? A comparison of paper- and computer-based low-fidelity prototyping. *CHI 2003 Extended Abstracts,* 778-779.

Spool, J. M., Snyder, C., Ballman, D., & Schroeder, W. (1994). Using a game to teach a design process. *Proceedings of the ACM SIGCHI Conference on Human Factors in Computing Systems* (pp. 117-118).

Thompson, M., & Wishbow, N. (1992). Prototyping: Tools and techniques: Improving software and documentation quality through rapid prototyping. *Proceedings of the 10th Annual International Conference on Systems Documentation* (pp. 191-199).

Twidale, M., & Marty, P. (2005). Come on down! A game show approach to illustration usability evaluation methods. *IEEE Interactions, 12*(6), 24-27.

Virzi, R. A., Sokolov, J. L., & Karis, D. (1996). Usability problem identification using both low- and high-fidelity prototypes. *Proceedings of the ACM SIGCHI Conference on Human Factors in Computing Systems* (pp. 236-243).

Welch, G. (1995). A survey of power management techniques in mobile computing operating systems. *ACM SIGOPS Operating Systems Review, 29*(4), 47-56.

KEY TERMS

Evaluation Methodology: A procedure for determining the quality of a system in relation to how it satisfies user needs. Scavenger Hunt is an example of an evaluation methodology for rapidly prototyped mobile collaboration systems.

Extreme Use Case: Unlike a "critical use case" where a task is identified that is essential to the operation of the system, an "extreme use case" is the situation that arises when users interact with the software under stressful (i.e., timed) conditions and push the software to its limits, both in terms of system-level support and usability.

Mobile Collaboration: The situation that arises when two or more people must work together on a problem, while one or more of them is in the process of changing location or is in the field.

Participatory Design: The process of designing *with* users instead of designing *for* users, by actually including end users on the design team and mutually learning from one another.

Rapid Prototyping: The process of creating, evaluating, and refining low-cost, easily fabricated prototypes in order to quickly identify and fix flaws.

Scavenger Hunt: A lightweight method for evaluating prototypes early in the design of a mobile or ubiquitous computing system, wherein participants play a game while the design team identifies systems- and interface-level flaws.

SH Blog: A collaborative mobile Weblog that is shared among a group of people working on the same task. The people who are involved are also posting from mobile devices.

Gender Difference in the Motivations of Mobile Internet Usage

Shintaro Okazaki
Autonomous University of Madrid, Spain

INTRODUCTION

The rapid pace of adoption of Web-enabled mobile handsets in worldwide markets has become an increasingly important issue for information systems professionals. A recent survey indicates that the number of global mobile Internet adopters is expected to reach nearly 600 million by 2008 (Ipsos-Insight, 2004; Probe Group, 2004), while the number of Internet-connected mobile phones will exceed the number of Internet-connected PCs by 2005 (*The Economist*, 2001). Such drastic convergence of the Internet and the mobile handset has been led by Asian and Scandinavian countries, where penetration has been especially meteoric. For example, roughly 70 million people in Japan, or 55% of the population, have signed up for mobile Internet access, in comparison to 12% in the United States (Faiola, 2004; Greenspan, 2003). Consequently, mobile phones or *Keitai* have been converted into devices for surfing the Internet, and by 2004 monthly mobile spending per consumer exceeded 35 euro.

Much of this success can be traced back to 1999, when NTT DoCoMo introduced the "i-mode" service. i-mode is a mobile service offering continuous Internet access based on packet-switching technology (Barnes & Huff, 2003). Through an i-mode handset, users can access a main micro-browser, which offers such typical services as e-mail, data search, instant messaging, Internet, and "i-menu." The "i-menu" acts as a mobile portal that leads to approximately 4,100 official and 50,000 unofficial sites (NTT DoCoMo 2003). Many such mobile portal sites can thus be considered as a pull-type advertising platform, where consumers can satisfy diverse information needs.

Several researchers have attempted to conceptualize the success of i-mode in comparison to WAP (Baldi & Thaung 2002) and in the light of the technology acceptance model (TAM) (Barnes & Huff 2003). Okazaki (2004) examined factors influencing consumer adoption of the i-mode pull-type advertising platform. However, there is a dearth of empirical research in this area, and especially in developing a model that captures the specific dimensions of mobile Internet adoption. In this respect, this study aims to propose a measurement scale of consumer perceptions of mobile portal sites.

The present study adopts, as its principal framework, the attitudinal model suggested by Dabholkar (1994). This includes "ease of use," "fun," and "performance" as important determinants of attitude. These are often referred to as "ease of use," "usefulness," and "enjoyment" in, for example, the TAM proposed by Davis (1986; Davis, Bagozzi, & Warshaw, 1989, 1992). The relevant literature suggests that dimensions similar to "ease of use" and "fun" are important antecedents of new technology adoption. For example, Shih (2004) and Szymanski and Hise (2000) found "perceived ease of use" and "convenient," respectively, as important antecedents of online behavior. Likewise, Moon and Kim (2001) found "perceived playfulness" to be a factor influencing WWW usage behavior, similar to the "fun" dimension. However, unlike earlier studies of m-commerce adoption, this study drops the third dimension of the TAM, "usefulness," in favor of "performance," because the former is appropriate only for tangible products, but not relevant for technology-based services (Dabholkar & Bagozzi, 2002). In contrast, "performance" represents a dimension that encompasses the reliability and accuracy of the technology-based service, as perceived by the consumer (Dabholkar, 1994). These three dimensions capture customer perceptions, which would initiate the attitude-intention-behavior causal chain (Davis, 1986).

BACKGROUND

Prior Theories on Technology Adoption

The technology acceptance model has been used to explain online user behavior (Featherman & Pavlous, 2002; Moon & Kim, 2001). Originally, TAM was based on Ajzen and Fishbein's (1980) theory of reasoned action (TRA), which is concerned with the determinants of consciously intended behaviors. TRA has been described as one of the most widely studied models in social psychology (Ajzen & Fishbein, 1980; Fishbein & Ajzen, 1975). According to TRA, "a person's performance of a specified behavior is determined by his or her behavioral intention (BI) to perform the behavior, and BI is jointly determined by the person's attitude (A) and subjective norm (SN) concerning the behavior in question (Figure 1), with relative weights typically estimated by regression: BI = A + SN" (Davis et al., 1989). Here, BI refers to the degree of strength of one's intention to perform a specified behavior, while A is defined as an evaluative effect regarding performing the target behavior. SN is meant to be "the person's perception that most people who are important to

Figure 1. Theory of reasoned action (TRA)

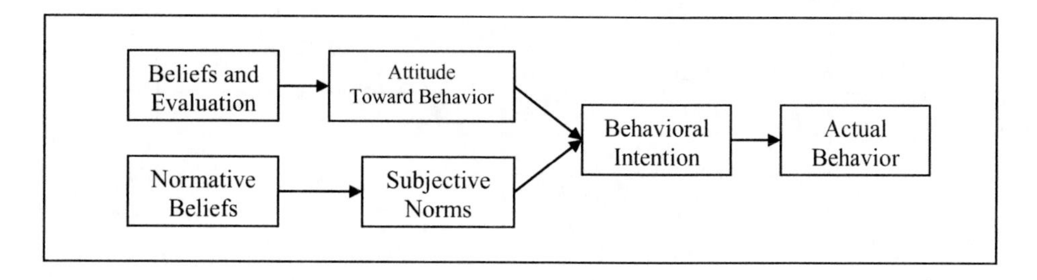

Figure 2. Technology acceptance model (TAM)

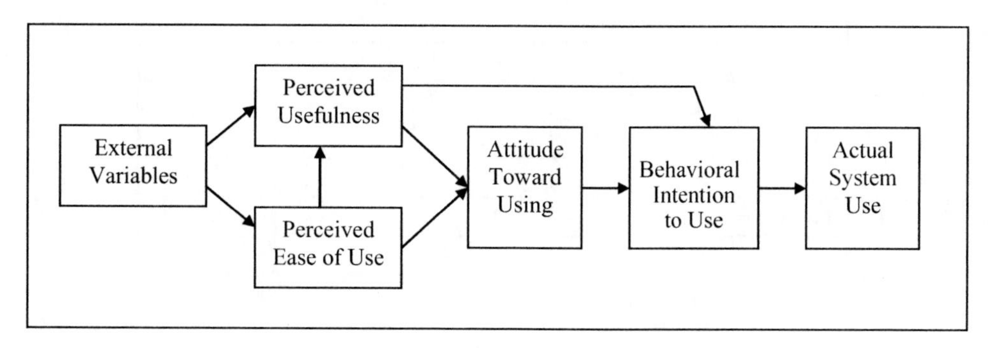

him think he should or should not perform the behavior in question" (Fishbein & Ajzen, 1975).

TAM extends TRA with attempts to explain the antecedents of computer-usage behavior. TAM comprises five fundamental salient beliefs: perceived ease of use, perceived usefulness, attitudes toward use, intention to use, and actual use. Perceived usefulness is defined as "the degree to which a person believes that using a particular system would enhance his or her job performance," while perceived ease of use is "the degree to which a person believes that using a particular system would be free of effort" (Davis et al., 1989). Although they are not the only variables of interest in explaining user behavior, perceived ease of use and perceived usefulness have been proven empirically to be key determinants of behavior in a wide range of academic disciplines, such as the learning process of a computer language, evaluation of information reports, and adoption of alternative communication technologies, among others. However, TAM excludes the "influence of social and personal control factors on behavior" (Taylor & Todd, 1995). Consequently, Ajzen and Fishbein (1980) proposed another extension of TRA, the theory of planned behavior (TPB), to account for those conditions in which individuals may not have complete control over their own behavior (Taylor & Todd, 1995).

A key objective of TAM is "to assess the value of IT to an organization and to understand the determinants of that value" (Taylor & Todd, 1995). Hence, much IT research has aimed to enhance companies' effective IT resource management. However, TAM has been expanded to emerging new forms of IT, such as the wired as well as the wireless Internet.

In a pioneering study, Moon and Kim (2001) conducted empirical research on extending TAM to the World Wide Web context. They constructed an extension of TAM based on an individual's intrinsic motivation theory, and found that "perceived playfulness" had a positive effect on individuals' attitudes toward using the WWW and their behavioral intentions to use the WWW. Furthermore, TAM has been applied to explain the adoption of telecommunication technology, such as telework (Hun, Ku, & Chang, 2003), mobile devices (Kwon & Chidambaram, 2000), and m-commerce services (Pedersen & Ling, 2003). Generally, these studies suggest certain modifications of the original TAM in order to include social influence and behavioral control variables (Pedersen & Ling, 2003). In the following section, we explore the possible extension of TAM to m-commerce adoption.

Gender Differences in Mobile Internet Adoption

Gender has frequently been used as part of the social and the cultural meanings associated with developing marketing strategy via advertising messages, and in market segmentation strategy in particular, because it is easily: (1) identifiable, (2) accessible, (3) measurable, (4) responsive to marketing mix, (5) sufficiently large, and (6) profitable (Darley & Smith, 1995). However, although there has been much research on

Table 1. Rotated component matrix

		Component 1	Component 2	Component 3
Performance	Detailed	.695		
	Updated	.639		
	Intelligible	.619		
	Reliable	.449		
Ease of Use	Easy		.644	
	Killing Time		.589	
	Interesting		.536	
	Free		.440	
	Educational		.410	
Fun	Appealing			.642
	Helpful			.629
	Practical			.437
Total Variance		21.1	30.8	40.0
Eigenvalue		2.53	1.17	1.08

new technology adoption, little attention has been paid to gender differences in electronic communication. Yang and Lester (2005) argue that "research on gender and CMC has consistently demonstrated that gender inequalities define professional and scholarly electronic communication and that men are over-represented in electronic communities". This is considered a serious lacuna, since evidence has been found of important gender differences in human communication, including advertising (Wolin, 1999).

Our literature review found only one study that examined gender differences in online purchasing behavior. Yang and Lester (2005) conducted a series of studies on purchasing textbooks online at universities, and found that female students at an urban university tended to demonstrate fewer computer/Internet skills than male students, and that their level of skill was a more consistent predictor of purchasing textbooks online: the higher their level of skill, the more likely female students were to buy books online, and the effect of level of skill was greater for female than for male students.

To date, no gender studies of mobile Internet adoption have been reported. However, following Yang and Lester (2005), we may assume that, in learning and accessing wireless Internet with mobile handsets, female users may be less skillful than their male counterparts. For example, in terms of TAM, females may perceive more negatively ease of use, which is one of the essential determinants of attitude toward new technology adoption.

PROPOSED MODEL OF CONSUMER MOTIVATIONS

Although the specific motivations to use *wired* and *wireless* Internet must differ between individuals, the *overall* motivations of online information search may be similar for the two media. Thus, we adopted three primary motivations from prior research on wired Internet adoption: (1) performance, (2) ease of use, and (3) fun. First, Shih (2004) empirically examined online purchasing behavior, and found perceived usefulness to be the major determinant of behavioral intentions to use the Internet, while perceived ease of use is a secondary determinant. We adopt these concepts as performance and ease of use. It has been pointed out that the term *performance* is preferred to *usefulness*, in the case of intangible technology adoption. Second, Moon and Kim (2001) introduced an additional determinant of attitude formation, perceived playfulness or fun, to capture WWW usage behavior. Hence, we propose these three constructs as the principal drivers or motivations of enhanced mobile Internet usage. These constructs are essentially in line with Davis et al.'s (1989) TAM, which has frequently been used to explain and predict user adoption of a new information technology. Hence, our aim in this study is to examine whether there are any important differences between male and female mobile Internet users in terms of these constructs.

Table 2. Logistic regression results

Theoretical Constructs	Variables	Mobile Site		Internet		Satellite TV		Newspaper		WOM	
Performance	Detailed	-.117		-.005		.209		-.070		.005	
	Updated	.070		-.053		-.117		.202	*	-.107	
	Intelligible	.117		.073		.036		-.025		.042	
	Reliable	-.832	**	-.419	**	-.206		-.022		-.057	
Ease of Use	Easy	.253	**	.294	**	.241		.134		.297	***
	Killing Time	.311	**	.131		.255	*	-.285	**	-.345	**
	Interesting	.126		-.023		-.210		.098		.206	*
	Free	-.490	*	-.531	***	-.712	**	-.218		-.413	***
	Educational	-.042		-.021		.043		.137		-.318	**
Fun	Appealing	-.566		-.101		.021		-.226		.017	
	Helpful	-.019		.133		-.505		-.138		.518	**
	Practical	.103		.197		-.364	*	.189		.835	***

SURVEY METHOD

The survey was conducted via an online questionnaire that was made available in a popular commercial Web site in Japan. There were no restrictions on access, and the survey was open to the public audience. The questionnaire consists of a variety of questions, on general demographics, media usage, habits and spending, motives to use mobile Internet site, and so forth. As an incentive to complete the questionnaire, respondents were given an e-coupon as a reward for their participation. In total, 1,637 responses were obtained.

We assigned four adjectives for each of the three constructs: detailed, reliable, educational, and updated for performance; interesting, appealing, helpful, and killing time for fun; and easy, free, intelligible, and practical for ease of use. In order to identify the importance of each item, we used a dichotomous measure, asking whether respondents perceived a given adjective as describing his or her own perception of the mobile Internet site. For example, if they accessed a mobile Internet site because it seemed "reliable," they marked the answer "yes." In order to conduct statistical analysis, these dichotomous variables were converted into fictitious variables by assigning "1" to "yes" and "0" to "no."

RESULTS

With regard to the demographic composition by gender, the distribution of age and marital status differ little across gender; important differences can be observed in education and occupation. The proportion of people with bachelor or higher degrees is much greater in males than in females. On the other hand, females dominate junior college graduates. With regard to occupation, administrative, managerial, and professional workers are primarily male. A similar tendency can be seen in self-employed and skilled labor, although the magnitude is much less. There are more female workers in services.

To examine the dimensionality of the variables, we first conducted an exploratory factor analysis (EFA) with a principal component method. Although dichotomous variables are not ideal in EFA, fictitious variables are considered acceptable in this usage (Hair, Anderson, Tatham, & Black, 1998). The Varimax rotation was used, while a scree plot was carefully examined. Only variables with eigenvalue greater than 1 were retained. After several attempts using trial and error, we determined a three-factor solution to be the best, in which 12 proposed items were converged. However, as Table 1 shows, some of the items were classified into different constructs. Because of the exploratory nature of the study, we deemed this convergence to be reasonable and acceptable for the subsequent analysis.

Next, a logistic regression was performed with gender as a dependent variable and the importance (existence or absence) of adjective items as independent variables. It was possible to use binary data for both dependent and independent variables, because logistic regression does not require the normality assumption, as multiple regression does (Hair et al., 1998). However, because multicollinearity can seriously distort the results, a diagnostic was carried out via VIF and Tolerance values. Both values for each independent variable ranged from .80 to 1.23, showing no serious presence of multicollinearity.

The results of logistic regression are shown in Table 2. As clearly shown, ease of use plays an important role in separating male and female mobile Internet users. Chi-square tests reveal significant differences between male and female users in terms of easy, killing time, and free. Interestingly, female users are likely to perceive mobile Internet sites as an easy medium for killing time significantly more than their male counterparts. The opposite is true for free: male users essentially appreciate a mobile Internet site as a free information source. With regard to reliability in performance, male users tend to perceive this item more strongly than female users. Finally, logistic regression was also performed for different media, such as (wired) Internet, satellite TV, newspapers, and word of mouth (WOM). Despite the dangers of oversimplification, it seems that a mobile Internet site exhibits the combined effects of a wired Internet and word of mouth.

IMPLICATIONS

Our findings clearly show that there are important differences between male and female mobile users in terms of motivations to access mobile Internet sites. Specifically, female users are more prone to access a mobile Internet site for spare-time leisure and ease of use, while male users do so for free information. It should be noted that both genders perceive a mobile Internet site as a reliable, updated, and intelligible information source. In comparison with other media, a mobile Internet site is considered to be a combination of Internet and word of mouth. This makes sense because a mobile device is essentially and uniquely characterized as a personalized telecommunication medium, which is accessible only via a mobile telephone.

Given that Japanese mobile Internet services focus on information and entertainment (Okazaki, 2004), the findings of this study may provide useful implications for IT managers. That is, female users are more likely to appreciate a mobile Internet site for its entertainment value, while male users may seek more pragmatic results or outcomes, that is, reliable daily information. For example, typical male white-collar workers may seek daily financial news and replace newspapers with a mobile device as an information source. On the other hand, female users are attracted by more enjoyable usage, which provides an easy escape from daily routine. In this regard, IT managers should be aware of the importance of tailoring the content of mobile Internet according to gender-specific needs and wants.

Limitations

A few limitations should be recognized to make our findings more objective. First, our study examined only 12 adjective items with three proposed constructs. Future research should include a larger variety of items that may be related to consumers' perceptions associated with mobile Internet sites. Second, this study did not specify a type of "mobile Internet site." That is, our findings should be interpreted at most as general evaluations of mobile Internet adoption. Given a rapid proliferation of mobile Internet services, future research should specify the type of mobile Internet site and its benefits as a unit of analysis. Finally, the binary nature of data means that the construct reliability and validity were not established. Any future study should use a semantic differential scale, instead of a dichotomous scale, as a measure, and much effort should be made to improve the reliability indices.

REFERENCES

Ajzen, I. (1985). From intentions to actions: A theory of planned behaviour. In *Action control: From cognition to behaviour* (pp. 11-39). New York: Springer-Verlag.

Darley, W., & Smith, R. (1995). Gender differences in information processing strategies: An empirical test of the selectivity model in advertising response. *Journal of Advertising, 24*(1), 41-56.

Davis, F., Bagozzi, R., & Warshaw, P. (1989). User acceptance of computer technology: A comparison of two theoretical models. *Management Science, 35*(8), 982-1003.

Durlacher. (1999, November). *Mobile commerce report.* Retrieved from http://www.durlacher.com/fr-research-reps.htm

Featherman, M., & Pavlos, P. (2002). Predicting e-services adoption: A perceived risk facets perspective. *Proceedings of AMCIS 2002,* Dallas, TX.

Fishbein, M., & Ajzen, I. (1975). *Belief, attitude, intention, behaviour: An introduction to theory and research.* Reading, MA: Addison-Wesley.

Hair, J. Jr., Anderson, R., Tatham, R., & Black, W. (1998). *Multivariate data analysis* (5th ed.). Upper Saddle River, NJ: Prentice Hall.

Höflich, J., & Rössler, P. (2001). Mobile schriftliche Kommunikation oder: E-mail fürdas handy. *Medien & Kommunikationswissenschaft, 49,* 437-461. Cited by Pedersen & Ling (2003).

Hun, S., Ku, C., & Chang, C. (2003). Critical factors of WAP services adoption: An empirical study. *Electronic Commerce Research and Applications, 2*(1), 42-60.

Juniperresearch.com. (2004). Mobile data sales predicted to bolster operator revenues. *New Media Age,* (October 21), 6.

Kleijnen, M., Wetzels, M., & Ruyter, K. (2004). Consumer acceptance of wireless finance. *Journal of Financial Service Marketing, 8*(3), 206-217.

Lin, C. (1996). Looking back: The contribution of Blumler and Katz's 'Uses of mass communication' to communication research. *Journal of Broadcasting & Electronic Media, 40*(4), 574-581.

Moon, J., & Kim, Y. (2001). Extending the TAM for a World-Wide-Web context. *Information & Management, 38,* 217-230.

NTT DoCoMo. (2003, October 30). *I-mode subscribers surpass 40 million.* Retrieved January 2004 from http://www.nttdocomo.com/

Okazaki, S. (2004). How do Japanese consumers perceive wireless advertising? A multivariate analysis. *International Journal of Advertising, 23*(4), 429-454.

Pagani, M. (2004). Determinants of adoption of third generation mobile multimedia services. *Journal of Interactive Marketing, 18*(3), 46-59.

Pedersen, P., & Ling, R. (2003). Modifying adoption research for mobile Internet service adoption: Cross-disciplinary interactions. *Proceedings of the 36th IEEE Hawaii International Conference on System Sciences 2003,* (pp. 90-91).

Sadeh, N. (2002). *M-commerce: Technologies, services, and business models.* New York: John Wiley & Sons.

Shih, H. (2004). Extended technology acceptance model of Internet utilization behaviour. *Information & Management, 41,* 719-729.

Taylor, S., & Todd, P. (1995). Understanding information technology usage: A test of competing models. *Information Systems Research, 6*(2), 144-176.

Wolin, L. (2003). Gender issues in advertising—An oversight synthesis of research: 1970-2002. *Journal of Advertising Research, 43,* 111-129.

Yang, B., & Lester, D. (2005). Sex differences in purchasing textbooks online. *Computers in Human Behaviour, 21,* 147-152.

KEY TERMS

i-mode: A broad range of Internet services for a monthly fee of approximately 3 Euro, including e-mail, transaction services (e.g., banking, trading, shopping, ticket reservations, etc.), infotainment services (e.g., news, weather, sports, games, music download, karaoke, etc.), and directory services (e.g., telephone directory, restaurant guide, city information, etc.), which offers more than 3,000 official sites accessible through the i-mode menu.

Mobile Commerce (M-Commerce): Any transaction with a monetary value that is conducted via a mobile telecommunications network. In a broader sense, it can be defined as the emerging set of applications and services people can access from their Internet-enabled mobile devices.

Mobile Portal: Typically, m-commerce takes place in a strategic platform called a "mobile portal," where third-generation (3G) mobile communication systems offer a high degree of commonality of worldwide roaming capability, support for a wide range of Internet and multimedia applications and services, and data rates in excess of 144 kbps. Examples of such mobile portals take many forms: NTT DoCoMo's i-mode portal, Nordea's WAP Solo portal, Webraska's SmartZone platform, among others. So far, Japan's i-mode portal has been asserted to be "the most successful and most comprehensive example of m-commerce today."

Technology Acceptance Model (TAM): Extends TRA with attempts to explain the antecedents of computer-usage behavior. TAM comprises five fundamental salient beliefs: perceived ease of use, perceived usefulness, attitudes toward use, intention to use, and actual use.

Theory of Reasoned Action (TRA): This model explains that a person's performance of a specified behavior is determined by his or her behavioral intention (BI) to perform the behavior, and BI is jointly determined by the person's attitude (A) and subjective norm (SN) concerning the behavior in question, with relative weights typically estimated by regression: BI = A + SN.

Uses and Gratifications Theory: A theory derived from media communication studies that focuses on individual users' needs or motivations of a particular medium. According to a recent study of mobile phone users, seven gratifications were identified: fashion/status, affection/sociability, relaxation, mobility, immediate access, instrumentality, and reassurance.

Handheld Computing and J2ME for Internet-Enabled Mobile Handheld Devices

Wen-Chen Hu
University of North Dakota, USA

Jyh-haw Yeh
Boise State University, USA

I-Lung Kao
IBM, USA

Yapin Zhong
Shandong Institute of Physical Education and Sport, China

INTRODUCTION

Mobile commerce or m-commerce is defined as the exchange or buying and selling of commodities, services, or information on the Internet through the use of Internet-enabled mobile handheld devices (Hu, Lee, & Yeh, 2004). It is expected to be the next milestone after electronic commerce blossoming in the late-1990s. Internet-enabled mobile handheld devices are one of the core components of a mobile commerce system, making it possible for mobile users to directly interact with mobile commerce applications. Much of a mobile user's first impression of the application will be formed by his or her interaction with the device, therefore the success of mobile commerce applications is greatly dependent on how easy they are to use. However, programming for handheld devices is never an easy task not only because the programming languages and environments are significantly different from the traditional ones, but also because various languages and operating systems are used by handheld devices and none of them dominates.

This article gives a study of handheld computing, especially J2ME (Java 2 Platform, Micro Edition) programming, for mobile commerce. Various environments/languages are available for client-side handheld programming. Five of the most popular are (1) BREW, (2) J2ME, (3) Palm OS, (4) Symbian OS, and (v) Windows Mobile. They apply different approaches to accomplishing the development of mobile applications. Three themes of this article are:

1. Introduction of handheld computing, which includes server- and client- side computing.
2. Brief introductions of four kinds of client-side computing.
3. Detailed discussion of J2ME and J2ME programming.

Other important issues such as a handheld computing development cycle will also be discussed.

BACKGROUND

Handheld computing is a fairly new computing area and a formal definition of it is not found yet. Nevertheless, the authors define it as follows: Handheld computing is the programming for handheld devices such as smart cellular phones and PDAs (personal digital assistants). It consists of two kinds of programming: client- and server- side programming.

The definitions of client- and server- side computing are given as follows:

- **Client-Side Handheld Computing:** It is the programming for handheld devices and it does not need the support from server-side programs. Typical applications created by it include (1) address books, (2) video games, (3) note pads, and (4) to-do list.
- **Server-Side Handheld Computing:** It is the programming for wireless mobile handheld devices and it needs the support from server-side programs. Typical applications created by it include (1) instant messages, (2) mobile Web contents, (3) online video games, and (4) wireless telephony.

This article will focus on the client-side computing. The server-side computing is briefly given next.

Server-Side Handheld Computing

Most applications created by this kind of programming, such as instant messaging, require network programming such as TCP/IP programming, which will not be covered in this chapter. The most popular application of server-side handheld computing is database-driven mobile Web sites, whose structure is shown in Figure 1. A database-driven

Table 1. A comparison among five handheld-computing languages/environments

	BREW	J2ME	Palm OS	Symbian OS	Windows Mobile
Creator	Qualcomm Inc.	Sun Microsystems Inc.	PalmSource Inc.	Symbian Ltd.	Microsoft Corp.
Language/ Environment	Environment	Language	Environment	Environment	Environment
Market Share (PDA) as of 2004	N/A	N/A	2nd	N/A	1st
Market Share (Smartphone) as of 2005	?	N/A	3rd	1st	2nd
Primary Host Language	C/C++	Java	C/C++	C/C++	C/C++
Target Devices	Phones	PDAs & phones	PDAs	Phones	PDAs & phones

Figure 1. A generalized system structure of a database-driven mobile Web site

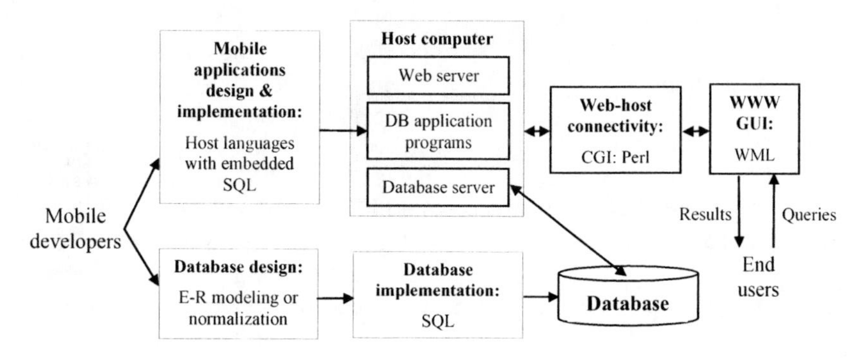

mobile Web site is often implemented by using a three-tiered client/server architecture consisting of three layers:

1. **User Interface:** It runs on a handheld device (the client) and uses a standard graphical user interface (GUI).
2. **Functional Module:** This level actually processes data. It may consist of one or more separate modules running on a workstation or application server. This tier may be multi-tiered itself.
3. **Database Management System (DBMS):** A DBMS on a host computer stores the data required by the middle tier.

The three-tier design has many advantages over traditional two- or single- tier design, the chief one being: the added modularity makes it easier to modify or replace one tier without affecting the other tiers.

CLIENT-SIDE HANDHELD COMPUTING

Various environments/languages are available for client-side handheld programming. Five of the most popular are (1) BREW, (2) J2ME, (3) Palm OS, (4) Symbian OS, and (5) Windows Mobile. They apply different approaches to accomplishing the development of mobile applications. Figure 2 shows a generalized development cycle applied by them and Table 1 gives the comparison among the five languages/environments. The second half of this article is devoted to J2ME details and brief introductions of the other four are given in this section.

BREW (Binary Runtime Environment for Wireless)

BREW is an application development platform created by Qualcomm Inc. for CDMA-based mobile phones (Qualcomm Inc., 2003). CDMA is a digital wireless telephony transmission technique and its standards used for 2G mobile

Figure 2. A generalized client-side handheld computing development cycle

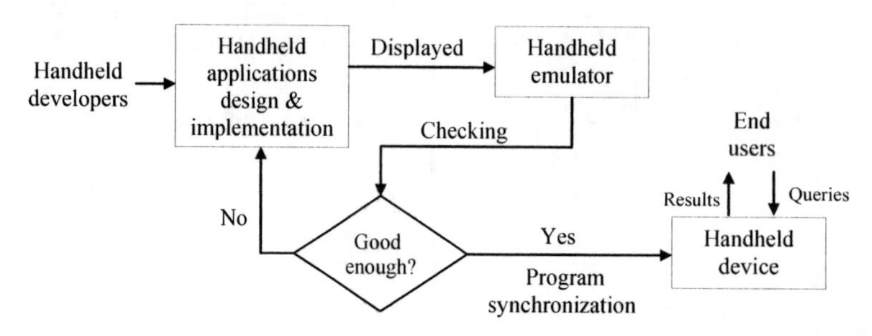

telephony are the IS-95 standards championed by Qualcomm. BREW is a complete, end-to-end solution for wireless applications development, device configuration, application distribution, and billing and payment. The complete BREW solution includes

- BREW SDK (software development kit) for application developers,
- BREW client software and porting tools for device manufacturers, and
- BREW distribution system (BDS) that is controlled and managed by operators—enabling them to easily get applications from developers to market and coordinate the billing and payment process.

Palm OS

Palm OS, developed by Palm Source Inc., is a fully ARM-native, 32-bit operating system running on handheld devices (PalmSource Inc., 2002). Palm OS runs on almost two out of every three PDAs. Its popularity can be attributed to its many advantages, such as its long battery life, support for a wide variety of wireless standards, and the abundant software available. The plain design of the Palm OS has resulted in a long battery life, approximately twice that of its rivals. It supports many important wireless standards, including Bluetooth and 802.11b local wireless and GSM, Mo-bitex, and CDMA wide-area wireless networks. Two major versions of Palm OS are currently under development:

- **Palm OS Garnet:** It is an enhanced version of Palm OS 5 and provides features such as dynamic input area, improved network communication, and support for a broad range of screen resolutions including QVGA.
- **Palm OS Cobalt:** It is Palm OS 6, which focuses on enabling faster and more efficient development of smartphones and integrated wireless (WiFi/Bluetooth) handhelds.

As of August 2005, no hardware products run Palm OS Cobalt and all devices use Palm OS Garnet. Likely as a result of Palm OS Cobalt's lack of adoption, PalmSource has shifted to developing Palm OS Cobalt's APIs on top of a Linux kernel.

Symbian OS

Symbian Ltd. is a software licensing company that develops and supplies the advanced, open, standard operating system—Symbian OS—for data-enabled mobile phones (Symbian Ltd., 2005). It is an independent, for-profit company whose mission is to establish Symbian OS as the world standard for mobile digital data systems, primarily for use in cellular telecoms. Symbian OS includes a multi-tasking multithreaded core, a user interface framework, data services enablers, application engines, integrated PIM functionality, and wireless communications. It is a descendant of EPOC, which is a range of operating systems developed by Psion for handheld devices.

Windows Mobile

Windows Mobile is a compact operating system for mobile devices based on the Microsoft Win32 API (Microsoft Corp., 2005). It is designed to be similar to desktop versions of Windows. In 1996, Microsoft launched Windows CE, a version of the Microsoft Windows operating system designed specially for a variety of embedded products, including handheld devices. However, it was not well received primarily because of battery-hungry hardware and limited functionality, possibly due to the way that Windows CE was adapted for handheld devices from other Microsoft 32-bit desktop operating systems. Windows Mobile includes three major kinds of software:

- **Pocket PCs:** Pocket PC enables you to store and retrieve e-mail, contacts, appointments, games, exchange text messages with MSN Messenger, browse the Web, and so on.
- **Smartphones:** Smartphone supplies functions of a mobile phone, but also integrates PDA-type functional-

Figure 3. A screenshot of KToolbar after launching

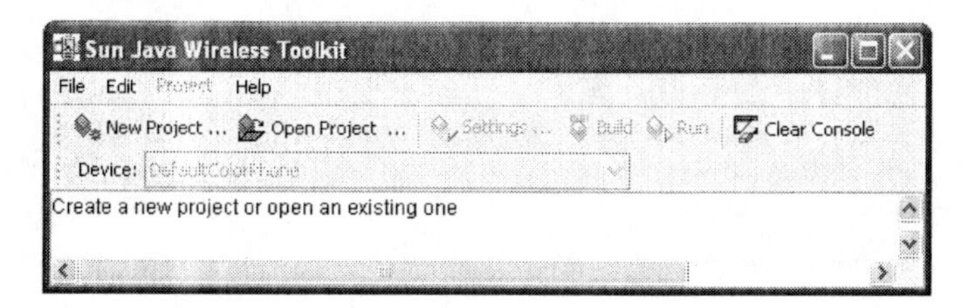

Figure 4. A screenshot of a pop-up window after clicking on the button New Project of KToolbar

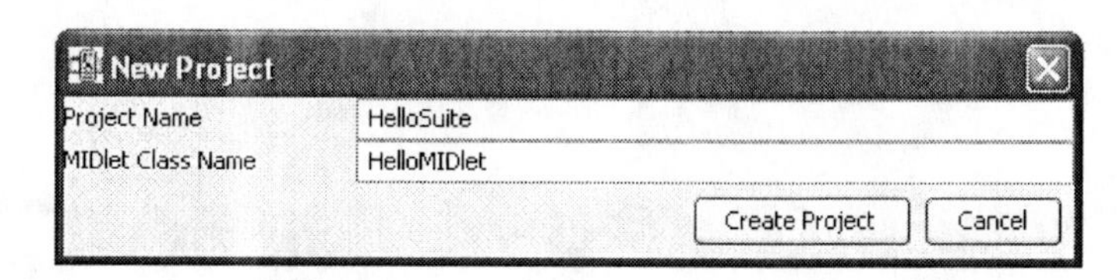

ity, such as e-mails, instant messages, music, and Web surfing, into a voice-centric handset.

- **Portable Media Centers:** Portable media centers let users take recorded TV programs, movies, home videos, music, and photos transferred from Microsoft Windows XP-based PC anywhere.

Windows Mobile-Based Pocket PCs

Pocket PCs were designed with better service for mobile users in mind and offers far more computing power than Windows CE. It provides scaled-down versions of many popular desktop applications, including Microsoft Outlook, Internet Explorer, Word, Excel, Windows Media Player, and others. It also includes three major kinds of software:

- **Pocket PC:** It puts the power of Windows software into a Pocket PC, giving you time to do more with the people and things that matter.
- **Pocket PC Phone Edition:** It combines all the standard functionality of a Windows Mobile-based Pocket PC with that of a feature-rich mobile phone.
- **Ruggedized Pocket PC:** It lets you do more of what matters to you even in the toughest user environments.

Windows Mobile-Based Smartphones

Windows Mobile-based smartphone integrates PDA-type functionality into a voice-centric handset comparable in size to today's mobile phones. It is designed for one-handed operation with keypad access to both voice and data features. The Smartphone is a Windows CE-based cellular phone. Like the Pocket PC, all Smartphones regardless of manufacturer share the same configuration of Windows CE. Also, Smartphones come bundled with a set of applications such as an address book, calendar, and e-mail program.

J2ME (JAVA 2 PLATFORM, MICRO EDITION)

J2ME provides an environment for applications running on consumer devices, such as mobile phones, PDAs, and TV set-top boxes, as well as a broad range of embedded devices (Sun Microsystem Inc., 2002a). Like its counterparts for the enterprise (J2EE), desktop (J2SE) and smart card (Java Card) environments, J2ME includes Java virtual machines and a set of standard Java APIs defined through the Java Community Process, by expert groups whose members include device manufacturers, software vendors, and service providers.

J2ME Architecture

The J2ME architecture comprises a variety of configurations, profiles, and optional packages that implementers and developers can choose from, and combine to construct a complete Java runtime environment that closely fits the requirements of a particular range of devices and a target market. There are two sets of J2ME packages, which target different devices:

- **High-End Devices:** They include connected device configuration (CDC), foundation and personal profile.
- **Entry-Level Devices and Smart Phones:** They include connected limited device configuration (CLDC) and mobile information device profile (MIDP).

Figure 5. A screenshot of KToolbar after a project HelloSuite created

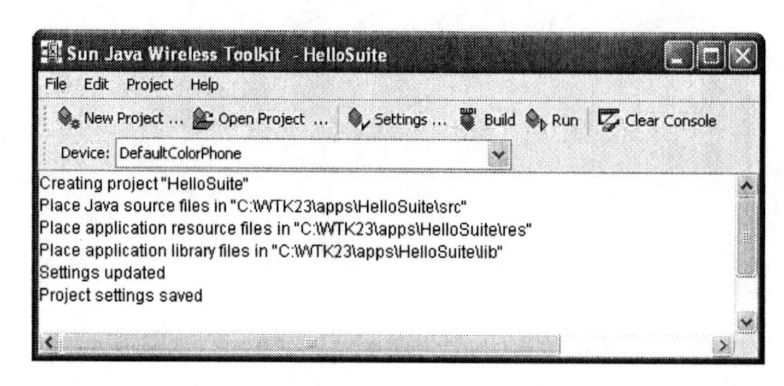

Figure 6. An example of an MIDlet program HelloMIDlet.java

C:\WTK23\apps\HelloSuite\src\HelloMIDlet.java

```
// This package defines MIDP applications and the interactions between
// the application and the environment in which the application runs.
import  javax.microedition.midlet.*;

// This package provides a set of features for user interfaces.
import  javax.microedition.lcdui.*;

public class HelloMIDlet  extends MIDlet  implements CommandListener {

  public void  startApp( ) {
    Display  display     = Display.getDisplay( this );
    Form     mainForm    = new  Form    ( "HelloMIDlet" );
    Ticker   ticker      = new  Ticker  ( "Greeting, World" );
    Command  exitCommand = new  Command( "Exit", Command.EXIT, 0 );

    mainForm.append            ( "\n\n          Hello, World!" );
    mainForm.setTicker         ( ticker );
    mainForm.addCommand        ( exitCommand );
    mainForm.setCommandListener( this );
    display.setCurrent         ( mainForm );
  }

  public void  pauseApp  ( ) { }

  public void  destroyApp( boolean unconditional ) {
    notifyDestroyed( );
  }

  public void  commandAction( Command c, Displayable s ) {
    if ( c.getCommandType( ) == Command.EXIT )
      notifyDestroyed( );
  }
}
```

Configurations comprise a virtual machine and a minimal set of class libraries and they provide the base functionality for a particular range of devices that share similar characteristics, such as network connectivity and memory footprint. Profiles provide a complete runtime environment for a specific device category.

J2ME Programming

This sub-section gives an example of J2ME programming (Sun Microsystem Inc., 2004). Other client-side handheld programming is similar to this. Figure 3 shows the Sun Java Wireless Toolkit©, which is a toolbox for developing wireless applications that are based on J2ME's CLDC and MIDP. The

Table 2. Mobile Information Device Profile (MIDP) package list

Package	Classes and Descriptions
User Interface	`javax.microedition.lcdui`: The UI API provides a set of features for implementation of user interfaces for MIDP applications.
	`javax.microedition.lcdui.game`: The Game API package provides a series of classes that enable the development of rich gaming content for wireless devices.
Persistence	`javax.microedition.rms`: It provides a mechanism for MIDlets to persistently store data and later retrieve it.
Application Lifecycle	`javax.microedition.midlet`: The MIDlet package defines MIDP applications and the interactions between the application and the environment in which the application runs.
Networking	`javax.microedition.io`: The MID Profile includes networking support based on the `Generic Connection` framework from the *Connected, Limited Device Configuration*.
Audio	`javax.microedition.media`: The MIDP 2.0 Media API is a directly compatible building block of the Mobile Media API (JSR-135) specification.
	`javax.microedition.media.control`: This package defines the specific Control types that can be used with a `Player`.
Public Key	`javax.microedition.pki`: Certificates are used to authenticate information for secure Connections.
Core	`java.io`: Provides classes for input and output through data streams.
	`java.lang`: MID Profile Language Classes included from Java 2 Standard Edition.
	`java.util`: MID Profile Utility Classes included from Java 2 Standard Edition.

Figure 7. A screenshot of an emulator displaying the execution results of HelloSuite

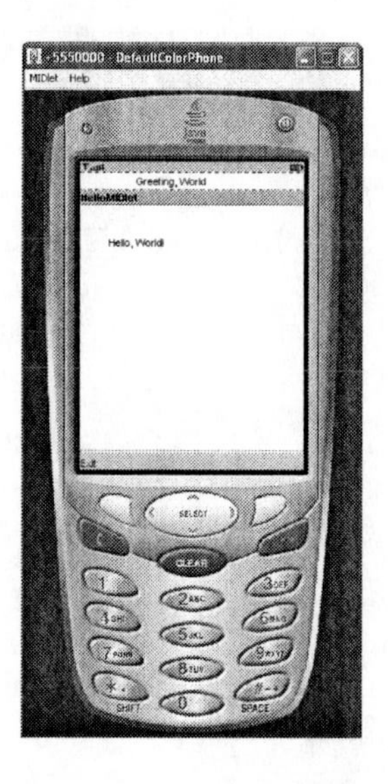

toolkit includes the emulation environments, performance optimization and tuning features, documentation, and examples that developers need to bring efficient and successful wireless applications to market quickly. The following steps show how to develop an MIDP application, a simple "Hello, World!" program, under Microsoft Windows XP:

1. Download Sun Java Wireless Toolkit 2.3 Beta, which includes a set of tools and utilities and an emulator for creating Java applications that run on handheld devices, at http://java.sun.com/products/sjwtoolkit/download-2_3.html .
2. Run MIDlet, an MIDP application, development environment KToolbar as shown in Figure 3 by selecting the following Windows commands:

   Start ▶ All Programs ▶ Sun Java Wireless Toolkit 2.3 Beta ▶ KToolbar

3. Create a new project by giving a project name such as HelloSuite and a class name such as HelloMIDlet as shown in Figure 4. After the project HelloSuite is created, the KToolbar will display the message shown in Figure 5, which tells where to put the Java source files, application resource files, and application library files.
4. Create a J2ME source program and put it in the directory C:\WTK23\apps\HelloSuite\src\. Figure 6 gives a J2ME example, which displays the text "Hello, World!" and a ticker with a message "Greeting, world."
5. Build the project by clicking on the `Build` button. The `Build` includes compilation and pre-verifying.
6. Run the project by clicking on the `Run` button. An emulator will be popped up and displays the execution results of the built project. For example, Figure 7 shows an emulator displays the execution results of HelloSuite.
7. Upload the application to handheld devices by using USB cables, infrared ports, or Bluetooth wireless technology.

Mobile Information Device Profile (MIDP) Packages

Table 2 shows the packages provided by the MIDP (Sun Microsystem Inc., 2002b). The packages javax.* are the extensions to standard Java packages. They are not included in the JDK or JRE. They must be downloaded separately.

FUTURE TRENDS

A number of mobile operating systems with small footprints and reduced storage capacity have emerged to support the computing-related functions of mobile devices. For example, Rearch In Motion Ltd.'s BlackBerry 8700 smartphone uses RIM OS and provides Web access, as well as wireless voice, address book, and appointment applications (Research In Motion Ltd., 2005). Because the handheld device is small and has limited power and memory, the mobile OSs' requirements are significantly less than those of desktop OSs. Although a wide range of mobile handheld devices are available in the market, the operating systems, the hub of the devices, are dominated by just few major organizations. The following two lists show the operating systems used in the top brands of smart cellular phones and PDAs in descending order of market share:

- **Smart Cellular Phones:** Symbian OS, Microsoft Smartphone, Palm OS, Linux, and RIM OS (Symbian Ltd., n.d.).
- **PDAs:** Microsoft Pocket PC, Palm OS, RIM OS, and Linux (WindowsForDevices, 2004).

The market share is changing frequently and claims concerning the share vary enormously. It is almost impossible to predict which will be the ultimate winner in the battle of mobile operating systems.

CONCLUSION

Mobile commerce is a coming milestone after electronic commerce blossoming in the late-1990s. The success of mobile commerce applications is greatly dependent on handheld devices, by which mobile users perform the mobile transactions. Handheld computing is defined as the programming for handheld devices such as smart cellular phones and PDAs. It consists of two kinds of programming: client- and server- side programming. Various environments/languages are available for client-side handheld programming. Five of the most popular are

1. **BREW:** It is created by Qualcomm Inc. for CDMA-based smartphones.
2. **J2ME:** J2ME is an edition of the Java platform that is targeted at small, standalone or connectable consumer and embedded devices.
3. **Palm OS:** It is a fully ARM-native, 32-bit operating system running on handheld devices.
4. **Symbian OS:** Symbian OS is an industry standard operating system for smartphones, a joint venture originally set up by Ericsson, Nokia, and Psion.
5. **Windows Mobile:** Windows Mobile is a compact operating system for handheld devices based on the Microsoft Win32 API. It is a small version of Windows, and features ma ny "pocket" versions of popular Microsoft applications, such as Pocket Word, Excel, Access, PowerPoint, and Internet Explorer.

They apply different approaches to accomplishing the development of handheld applications and it is almost impossible to predict which approaches will dominate the client-side handheld computing in the future, as the Windows to desktop PCs.

REFERENCES

Hu, W.-C., Lee, C.-W., & Yeh, J.-H. (2004). Mobile commerce systems. In Shi Nansi (Ed.), *Mobile Commerce Applications* (pp. 1-23). Hershey, PA: Idea Group Publishing.

Microsoft Corp. (2005). *What's new for developers in Windows Mobile 5.0?* Retrieved August 29, 2005, from http://msdn.microsoft.com/mobility/windowsmobile/howto/documentation/default.aspx?pull=/library/en-us/dnppcgen/html/whatsnew_wm5.asp

PalmSource Inc. (2002). *Why PalmOS?* Retrieved June 23, 2005, from http://www.palmsource.com/palmos/Advantage/index_files/v3_document.htm

Qualcomm Inc. (2003). *BREW and J2ME: A complete wireless solution for operators committed to Java.* Retrieved February 12, 2005, from http://brew.qualcomm.com/brew/en/img/about/pdf/brew_j2me.pdf

Research In Motion Ltd. (2005). *BlackBerry application control: An overview for application developers.* Retrieved January 05, 2006, from http://www.blackberry.com/knowledgecenterpublic/livelink.exe/fetch/2000/7979/1181821/832210/BlackBerry_Application_Control_Overview_for_Developers.pdf?nodeid=1106734&vernum=0

Sun Microsystem Inc. (2002a). *Java 2 Platform, Micro Edition.* Retrieved January 12, 2006, from http://java.sun.com/j2me/docs/j2me-ds.pdf

Sun Microsystem Inc. (2002b). *Mobile information device profile specification 2.0*. Retrieved October 25, 2005, from http://jcp.org/aboutJava/communityprocess/final/jsr118/

Sun Microsystem Inc. (2004). *J2ME Wireless Toolkit 2.2: User's guide*. Retrieved October 21, 2005, from http://java.sun.com/j2me/docs/wtk2.2/docs/UserGuide.pdf

Symbian Ltd. (2005). *Symbain OS Version 9.2*. Retrieved December 20, 2005, from http://www.symbian.com/technology/symbianOSv9.2_ds_0905.pdf

Symbian Ltd. (n.d.). *Symbian fast facts*. Retrieved January 26, 2005, from http://www.symbian.com/about/fastfaqs.html

Wilson, J. (2005). *What's new for developers in Windows Mobile 5.0*. Retrieved January 14, 2006, from http://msdn.microsoft.com/smartclient/default.aspx?pull=/library/en-us/dnppcgen/html/whatsnew_wm5.asp&print=true

WindowsForDevices.com. (2004). *Windows CE zooms past Palm*. Retrieved August 23, 2005, form http://www.windowsfordevices.com/news/NS6887329036.html

KEY TERMS

Binary Runtime Environment for Wireless (BREW): BREW is an application development platform created by Qualcomm Inc. for CDMA-based mobile phones.

Client-Side Handheld Programming: It is the programming for handheld devices and it does not need the supports from server-side programs. Typical applications created by it include (1) address books, (2) video games, (3) note pads, and (4) to-do list.

Handheld Computing: It is the programming for handheld devices such as smart cellular phones and PDAs (Personal Digital Assistants). It consists of two kinds of programming: client- and server-side programming.

Java 2 Platform, Micro Edition (J2ME): J2ME provides an environment for applications running on consumer devices, such as mobile phones, PDAs, and TV set-top boxes, as well as a broad range of embedded devices.

Palm OS: Palm OS, developed by Palm Source Inc., is a fully ARM-native, 32-bit operating system running on handheld devices.

Server-Side Handheld Programming: It is the programming for wireless mobile handheld devices and it needs the supports from server-side programs. Typical applications created by it include (1) instant messages, (2) mobile Web contents, (3) online video games, and (4) wireless telephony.

Symbian OS: Symbian Ltd. is a software licensing company that develops and supplies the advanced, open, standard operating system—Symbian OS—for data-enabled mobile phones.

Windows Mobile: Windows Mobile is a compact operating system for mobile devices based on the Microsoft Win32 API. It is designed to be similar to desktop versions of Windows.

An Infrastructural Perspective on U-Commerce

Stephen Keegan
University College Dublin, Ireland

Caroline Byrne
Institute of Technology Carlow, Ireland

Peter O'Hare
University College Dublin, Ireland

Gregory M. P. O'Hare
University College Dublin, Ireland

INTRODUCTION

In modern mobile-equipped businesses, the scales of economics sway between increasing economic returns and flawlessly decreasing expenditures while providing a worthwhile service for their customer base. Early mobile computing adopters realized that the scales of economic solvency weighed in favor of businesses that seamlessly delivered and managed customer expectations. This is only feasible if all frontline staff are endowed with relevant technological advances and educated appropriately in their usage. Timely, adequate responses to customer requests results in retaining satisfied customers and an expanding customer base. Efficient use of mobile advances can reduce mundane office tasks by preventing replication of work through data transfer between mobile devices and workstations. These streamlined tasks can often tilt the scales favorably for a struggling company.

Mobile computing encourages technological advances at the company's cutting edge while supporting its employees' daily duties by optimizing tasks. This is achieved via various handheld devices, each operating daily as unique satellite data stations, wirelessly updating the central company computer system. Another recent phenomenon is that of astute consumers comparing and contrasting products and prices prior to purchase via the internet. Mobile computing allows us the luxury of comparison from our closest physical retail outlet. As we physically view the product desired, our mobile enabled handheld device can navigate the Internet for comparable products at more competitive prices, thus allowing us the power to purchase under the canopy of an informed choice.

We define u-commerce as "the use of ubiquitous networks to support personalized and uninterrupted communications and transactions between an organization and its various stakeholders to provide a level of value over, above, and beyond traditional commerce" (Junglas & Watson 2003). U-commerce encompasses concepts that are ubiquitous, universal, unique, and unison. We take this opportunity to explore each of these in some depth.

Ubiquitous

Computers are already ubiquitous in our society. With continually decreasing hardware costs, relentless miniaturization, and the adoption of high-speed networks, this trend is likely to continue. Modern automobiles already contain dozens of microprocessors, while the unabated popularity of third-generation mobile phones means that mobile computing is now within reach of people in their daily lives.

Universal

The utility factor of u-commerce-enabling accessories like laptops, mobile phones, and PDAs has been limited by the fact that they are often not universally usable. Perhaps the most well-known instance of this type of incompatibility lies within the domain of mobile phones. People traveling between Europe and the United States often find that their European (GSM) phones operating at the 900 MHz and/or 1800 MHz frequencies are incompatible with those in the United States (CDMA), which typically operate at a frequency of 1900 MHz.

Unique

Many current retail delivery systems fail to exploit the unique characteristics of each individual user. Within u-commerce we envisage a model whereby users interact with information and services based upon the context at that point in time. Here context can entail such factors as temporal informa-

Figure 1. The development of commerce

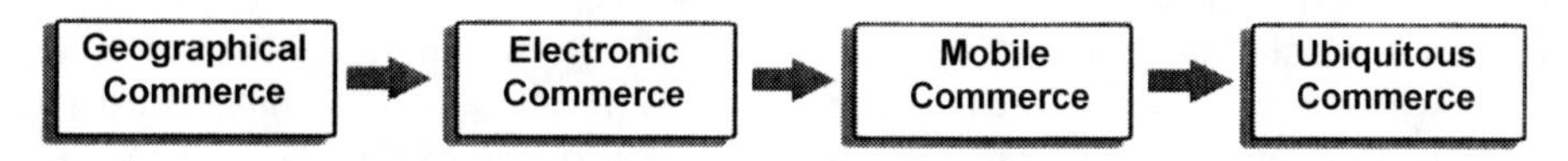

tion (e.g., What time of day is it?), user preferences (Does the user like ice-cream?), location data (e.g., How far is the user from our shop?), or user profile data (e.g., Is this user a female tourist?).

Unison

U-commerce relies on unison between all electronic data and equipment relevant in the user's life. Appropriate data such as profile information, product preferences, and financial data is securely shared in a distributed fashion and is readily retrievable at the appropriate time. Unison delivers the integration of various communication systems so there is a single interface or connection point.

BACKGROUND

Over the past decade the emergence of new electronic mobile and communications technologies has driven the way we conduct our business. Traditionally, commerce was geographic, with consumers seeking out and physically purchasing a product or service. The rapid deployment and ready accessibility of the Internet led to the dawn of *electronic commerce* (e-commerce). E-commerce enabled consumers to purchase products and service electronically via the Internet (http://www.ebay.com, http://www.amazon.com).

The development and widespread deployment of wireless technologies has ensured that mobile computing is spawning a dominant new culture (Rheingold, 2002). The mobile culture has gripped modern society with people regularly using their cellular phones, PDAs, MP3 players, and digital cameras. The development of new wireless hardware, software, and services is now occurring at an exponential rate. As a result m-commerce and u- commerce applications and services must be developed if they wish to evolve with the available technology.

M-commerce and u-commerce have significant differences from the geographical and electronic commerce which preceded them. Mobile devices impose a number of constraints upon business and service providers, including: smaller screen sizes, reduced interface interactivity, shorter battery life, and a restricted computational power. These restrictions have direct implications upon the mobile consumers, with users being less tolerant of irrelevant information and as a consequence having a shorter attention span. M-commerce and u-commerce business and service providers must address these restrictions and resolve them in creative and intelligent ways.

STATE OF THE ART

Shoppers today face a bewildering array of choices, whether they are shopping online or in the real world. To help shoppers cope with all of these choices, online merchants have deployed recommender systems that guide people toward products they are more likely to find interesting (Sarwar, Karypis, Konstan, & Reidl, 2001). Many of these online recommender systems operate by suggesting products that complement products you have purchased in the past. Others suggest products that complement those you have in your shopping cart at checkout time. If you have ever bought a book at Amazon.com or browsed musical listings at yahoo.com, you may have used a recommender system. Some of these systems, though ingenious, can prove to be of limited utility when applied to a mobile scenario. Dynamic pricing, mobile users, and limited hardware capabilities mean that new approaches are imperative.

Movielens Unplugged (Miller, Albert, Lam, Konstan, & Riedl, 2003) attempted to transpose the usability of the Movielens project to a selection of mobile devices. Particular emphasis was placed on developing a user interface that was capable of supporting multiple front ends and multiple devices. A set of generalized design principles was derived during a user trial. MobyRek (Ricci, Nguyen, & Cavada, 2004) is an on-tour recommendation system that becomes operational when a mobile traveler requests MobyRek to find some interesting travel products and ends when the traveler either selects a product or quits the session. It evolves in cycles, and in each cycle a set of recommended products is shown to the user. The recommendation process that it employs consists of four logical components: initialization, interaction, adaptation, and retainment.

Mobitip (Rudström, Svensson, Cöster, & Höök, 2004) is a mobile recommender system that allows people to create, rate, and share information using short-range Bluetooth communication, while occasionally synchronizing with a central server. It is argued that a real-time distribution schema of user profile data in impractical. The proposed solution involves storing a user's profile on the mobile device together, with a ranked list of predictions from a central server computed the last time the user docked. These predictions are recalculated as and when new data becomes available.

Of particular interest to us are the several companies who are experimenting with using Bluetooth to deliver the personal shopping assistant vision. WideRay (http://www.wideray.com/) has placed several of its BlueRay kiosks in selected music outlets, video shops, and theatres at a number of locations in Europe and the United States. When customers come within 10 meters of the kiosks, they receive a text message asking if they are interested in getting more information about various items the store is selling, perhaps music, ring tones, videos, or games. If customers are interested, they can go to the kiosk and choose what to download. Moonstorm (http://www.moonstorm.com/) recently released software called Cellfire that can automatically download coupons to mobile phones for stores in the customer's area. Customers do not receive any intrusive text messages, but must click on the Cellfire icon on their phones to examine and use the coupons.

A recent project, eNcentive, (Ratsimor, Finin, Joshi, & Yesha, 2003), is an infrastructure that facilitates peer-to-peer electronic commerce in the mobile environment. The system functions by aggressively broadcasting coupons, advertisements, and promotions through a geographical region populated by businesses (i.e., restaurants, dry cleaners, etc.) onto users' PDAs.

iGrocer (Shekar, Nair, & Helal, 2003) is a smart grocery shopping assistant that is hosted on a smart phone that comes with a bar code scanner. iGrocer helps users create and maintain weekly grocery shopping lists. Another feature of iGrocer is the nutrition indicator, which recommends foods based on a compatibility check between the user's health profile and the nutritional value of the food. The device can also act as a 'trusted checkout' and thereby act on behalf of the store and the customer.

PARTICIPANTS

There are some prominent challenges implicit in delivering the u-commerce vision. Some of these may be familiar to the reader in that they are equally applicable to the e-commerce domain, while others are of particular pertinence in the domain of u-commerce. The ability to provide users with simple, convenient, and trusted means of purchasing goods and services is paramount to the realization of u-commerce. There are a number of different participants upon whom this depends.

The Consumer

The consumer embodies the demand side of all u-commerce product and service acquisition. It is the consumer who identifies a need at a particular point and seeks to consolidate that need in a convenient fashion as specified in the consumer buying behavior model (Guttman, Moukas, & Maes, 1998). We assume that the u-commerce consumer is equipped with a networked device or set of devices to support the timely acquisition of products and services. The terms *user* and *consumer* are interchangeable within the scope of the remainder of this article.

The Provisor

The provisor is analogous to a real-world retailer—that is, a *bricks-and-mortar store*. In the context of u-commerce, however, the provisor may or may not have a real-world presence. Each provisor competes for consumers on the basis of a dynamic set of information garnered from a range of sources. The consumer's personal profile may be stored on the consumer's device, while an annotated record of that consumer's past purchasing behavior may be retrieved from a different location at the behest of the provisor (assuming authorization from the consumer). This composite set of data is utilized by the provisor in making appropriate product offerings to the consumer.

The Mediator

The mediator acts as a conduit between buyer (consumer) and seller (provisor). Its principal role is to ensure that trade occurs between all parties in an efficient and fair manner. A secondary role is to maintain and coordinate access to additional infrastructure required for u-commerce and not directly supplied by the consumer or the provisor. The most obvious example of this kind of infrastructure is a data network; this class of participant includes fixed and mobile network operators.

CHALLENGES

After considering the requirements of these three participants, a series of challenges can be identified with relation to the implementation of a supportive u-commerce architecture. We can classify these challenges in accordance with these participants as follows.

Infrastructural Considerations

Since commercial transitions within the system are completely data-flow based: the rate at which information can be transferred around the system in an efficient and timely manner is of prime importance. This rate is dependent upon the capabilities of the underlying infrastructure. It is envisaged that u-commerce data transfer requirements will accelerate, with a growing number of consumers demanding an increasingly *immersive experience* with multimedia aspects. The continued evolution of mobile data network infrastructures can support this growth.

Hardware Constraints

Challenges such as limited memory and processing power, although of particular concern to the consumer, are considerable design factors for both the provisor and infrastructural operator. A multitude of (often heterogeneous) hardware modules are required to operate in tandem to deliver commercial services. Bluetooth hotspots, GPS units, and electronic compasses are examples of this type of module.

Human-Computer Interface Issues

As with hardware considerations, *human-computer interface* (HCI) issues are most relevant when discussing the interface requirements and capabilities of the consumer's portable (and therefore diminutive) hardware—for example, mobile phone, PDA, personal GPS unit. In a mobile device environment, more than any other platform, each layer of the application architecture must be carefully considered and prioritized to maximize the device's physical capacity.

Standards and Interoperability

The ability of different, sometimes competing, parties to buy and sell goods is dependent upon a set of well-defined interoperability metrics. A number of initiatives have emerged to counter this problem. The incorporation of XML (http://www.w3.org/XML/) and markup-derived ontologies is seen as a suitable mechanism to deliver fair and standardized trading channels. Foremost of these is the UNSPSC (http://www.unspsc.org/), an extendible system of 18,000 terms to classify both products and services jointly developed by the United Nations Development Program (UNDP) and Dun & Bradstreet Corporation (D&B) in 1998.

Cultural Aspects

The intrepidation sometimes observed in consumers when adopting new commerce channels is perhaps understandable. The level of this reluctance is not universal. The modest uptake in the recent and much-heralded introduction of *i-mode* services on European networks, when contrasted with the blistering growth of i-mode in recent years in its home market of Japan only, serves to reinforce the importance of cultural factors. The role of cultural acceptance is often underestimated by proponents of u-commerce who may be predisposed with a technical bias.

Security Concerns

In the late 1990s the potential of online shopping was seen as being underdeveloped. A primary factor was the well-founded concerns of potential online shoppers over security. Security on the seemingly anarchic structure of the World Wide Web is only as strong as its weakest link. Some important lessons can be drawn from this. Firstly, all facets of a u-commerce architecture must be secure, and perhaps more importantly the system must not be just secure, but must be seen as being secure.

Legal/Regulatory Issues

A major challenge for u-commerce is that legal and regulatory stipulations in some markets around the world sometimes prevent network operators from brokering the sale of non-communication goods or services without the legal status of a bank or financial institution. To overcome these restrictions, there needs to be close cooperation between:

- banks, as they are the trusted intermediaries between consumers and provisor;
- credit card issuers, with their global coverage and extensive experience;
- network operators, who maintain an established subscriber base and who hold a central position in the communication value chain; and
- provisors.

Viable and Fair Business Model

The rise to prominence of eBay and Amazon, among others, is partly due to the utilization of appropriate business models. Successful models recognize the need to ensure that all participants are rewarded. This need must be balanced with an assurance that the cost of airtime must not be prohibitive compared with the amount of the transaction. Revenue-sharing models must provide win/win resolutions for all parties, with each participant receiving a reward, however small, for all transactions in which they played a part.

Suitable Payment Solutions

A u-commerce payment system must be capable of integrating different provisors, financial institutions, as well as other payment systems. The payment solution must comply with existing legal regulations and be flexible enough to be *localized* in accordance with different practices around the world, and must be capable of meeting any future legal requirements.

Figure 2. Easishop system overview

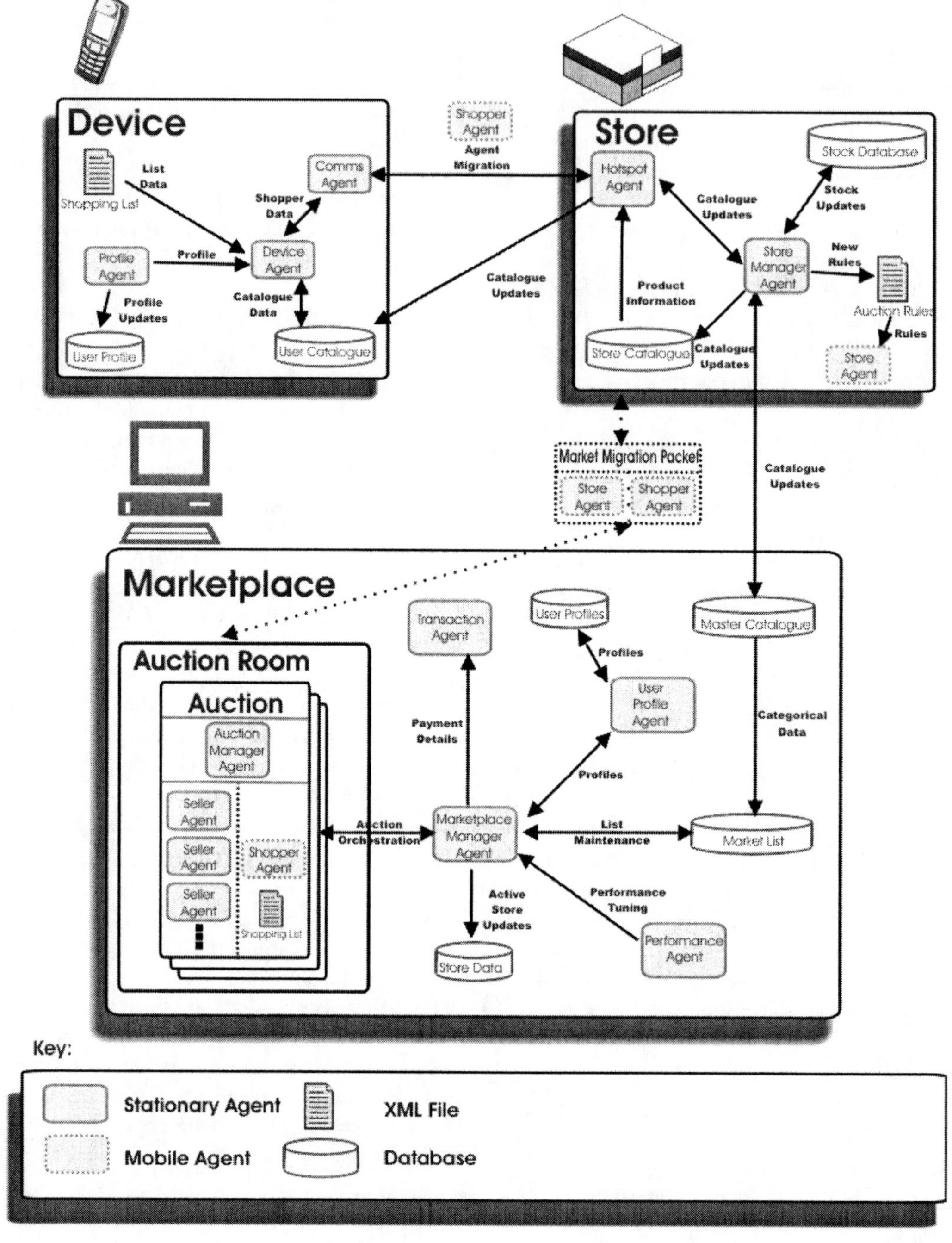

IMPLEMENTATION

An example of a typical u-commerce framework is Easishop (Keegan & O'Hare, 2004). A three-tiered architecture, the system is composed of a *marketplace* (mediator), a set of competing, independent *stores* (provisors), as well as a set of mobile *device* users (consumers). This architecture is represented in Figure 2.

Easishop has been implemented both on an archetypical mobile phone and PDA, namely the SonyEricsson P910i and HP IPAQ 3870 respectively. All software has been implemented in Java, with the J2ME variant being deployed on the mobile devices and the standard edition J2SE V.5 used on the Easishop network nodes. Over the Bluetooth connection, the serial port profile (SPP) is used while standard IP is employed between shop nodes. The shop nodes themselves have been implemented on standard workstations connected via Ethernet. All GUI elements have been realized using the Thinlet (http://www.thinlet.org/) toolkit.

The notion of agency is fundamental to Easishop. All agents have been designed and implemented in Agent Factory (Collier, O'Hare, Lowen, & Rooney, 2003). The resultant agents enable an effective mechanism for delivering u-commerce interoperation. Specifically, Agent Factory supports the creation of a type of mobile agent that is autonomous, situated, socially able, intentional, rational, and mobile. The reasoning mechanism used by such agents conforms to a belief-desire-intention (BDI) (Rao & Georgeff, 1996) architecture.

CONCLUSION

U-commerce virtually envelops consumers with a bewildering array of telecommunication extravaganzas, empowering all equally with opportunities to have their business/social/entertainment desires instantly fulfilled. The enhanced security features that are penetrating the mobile industry provide society with a panacea regarding any previous security concerns. The mobile moguls are constantly diversifying in order to retain their market share, and offering their customers the facility to buy/sell/gamble as and when they desire is an opportunity too lucrative to ignore.

Several hardware/software manufacturers are responding to market and consumer demands with a variety of devices and applications that facilitate consumers' frenetic lifestyles by assisting with grocery/retail shopping. Consumers will continue to benefit from technological advances as the costs of mobile hardware and per second billing steadily decrease, and our suppliers dazzle us with applications that enhance our daily interactions.

This wave of telecommunications forces us to re-engineer our beliefs and perceptions on what constitutes mobile usage in today's culture and embrace the possibility of having what we want, when we want it. Undoubtedly, providers have numerous hardware or memory challenges ahead and will fervently endeavor to absorb these issues and provide pertinent solutions. Competition increases productivity, and ultimately it will be consumers who benefit from the current economic race to the u-commerce summit.

REFERENCES

Brody, A. B., & Gottsman, E. J. (1999). Pocket BargainFinder: A handheld device for augmented commerce. *Proceedings of the 1st International Symposium on Handheld and Ubiquitous Computing (HUC '99)*, Karlsruhe, Germany.

Collier, R. W., O'Hare, G. M. P., Lowen, T. D., & Rooney, C. F. B. (2003). Beyond prototyping in the factory of agents. *Proceedings of the 3rd International Central & Eastern European Conference on Multi-Agent Systems (CEEMAS 2003)*, Prague, Czech Republic.

Guttman, R. H., Moukas, A. G., & Maes, P. (1998). Agents as mediators in electronic commerce. *Electronic Markets, 8*(1), 22-27.

Junglas, I. A., & Watson, R. T. (2003). U-commerce: A conceptual extension of e- and m-commerce. *Proceedings of the International Conference on Information Systems*, Seattle, WA.

Keegan, S., & O'Hare, G. M. P. (2004). Easishop—Agent-based cross merchant product comparison shopping for the mobile user. *Proceedings of the 1st International Conference on Information & Communication Technologies: From Theory to Applications (ICTTA '04)*, Damascus, Syria.

Miller, B. N., Albert, I., Lam, S. K., Konstan, J. A., & Riedl, J. (2003, January). MovieLens unplugged: Experiences with an occasionally connected recommender system. *Proceedings of the 2003 International Conference on Intelligent User Interfaces (IUI '03)*, Miami Beach, FL.

Rao, A. S., & Georgeff, M. P. (1996). BDI agents: From theory to practice. *Proceedings of the 1st International Conference on Multi-Agent Systems (ICMAS-96)*, San Francisco.

Ratsimor, O.V., Finin, T., Joshi, A., & Yesha, Y. (2003). eNcentive: A framework for intelligent marketing in mobile peer-to-peer environments. *Proceedings of the 5th International Conference on Electronic Commerce (ICEC 2003)*, Pittsburgh, PA.

Rheingold. H. (2002). *Smart mobs: The next social revolution.* Perseus.

Ricci, F., Nguyen, Q., & Cavada, D. (2004). On-tour interactive travel recommendations. *Proceedings of the 11th International Conference on Information and Communication Technologies in Travel and Tourism (ENTER 2004)*, Cairo, Egypt.

Rudström, Å., Svensson, M., Cöster, R., & Höök, K. (2004). MobiTip: Using Bluetooth as a mediator of social context. *Adjunct Proceedings of Ubicomp 2004,* Nottingham, UK.

Sarwar, B., Karypis G., Konstan, J., & Reidl, J. (2001). Item-based collaborative filtering recommendation algorithms. *Proceedings of the 10th International Conference on the World Wide Web (WWW '01)*, Hong Kong.

Shekar, S., Nair, P., & Helal, A. (2003). iGrocer—A ubiquitous and pervasive smart grocer shopping system. *Proceed-*

ings of the ACM Symposium on Applied Computing (SAC), Melbourne, FL.

KEY TERMS

CDMA: Code division multiple access.

GPS: Global positioning system.

J2ME: Java Platform, Micro Edition.

PDA: Personal digital assistant.

UNSPSC: United Nations Standard Products and Services Code.

XML: Extensible Markup Language.

Integrating Pedagogy, Infrastructure, and Tools for Mobile Learning

David M. Kennedy
Hong Kong University, Hong Kong

Doug Vogel
City University of Hong Kong, Hong Kong

INTRODUCTION

With the advent of the Web, students are empowered with environments that support a wide variety of interactions. These include engagement with authentic tasks, using a range of learning resources, and engaging with teachers and/or other students in knowledge-building communicative interactions. However, the concept of the fully wired world where students can learn anytime/anywhere is still unrealized. Instead, the growth of wireless networks has been substantial, with some countries limiting the construction of wired environments in preference to wireless connectivity. Thus, student learning environments and student expectations for convenience and flexibility are evolving to include wireless solutions along with wired Internet access at home or university.

A key issue associated with the growth of wireless services is the corresponding trade-off of service quality compared to wired computing (Associated Press, 2005). The availability of services is perceived as more important than high bandwidth and high security. The growth of wireless networks in the past 10 years has been spectacular, with a raft of technologies and standards arising to provide connectivity (Fenn & Linden, 2005). There is one note of caution: one of the leaders of research into wireless technologies, Cornel University in the USA, believes that due to competing technologies, even a fully wireless campus is still some time away (Vernon, 2006). This is due to:

- limitations in the interoperability of different wireless systems;
- high power requirements of the 802.11 wireless standard necessitating powerful (heavy) batteries for PDAs and smart phones;
- lower security than wired links; and
- potential interference, resulting in frustrated users.

Therefore, the concept of the fully wired or wireless connected world is still unrealized and will remain so for some time to come. Instead, the creation of local wireless hotspots has been suggested as a more cost-effective method (in the long term) for providing greater connectivity *and flexibility* to students (Boerner, 2002). Local wireless networking is already providing wireless links for students at cafés, shopping centers, airports, schools, and universities.

American students now have very high expectations that wireless will be available in all locations on a campus (Green, 2004). In Hong Kong, government statistics list the ownership of mobile devices as having reached an extraordinary level of 122.6% of market penetration (Office of the Telecommunications Authority of Hong Kong, 2005). Students may have a multitude of mobile devices, from mobile phones, iPods, digital cameras, and personal digital assistants (PDAs), to laptop computers. It would be remiss of teachers if they did not attempt to make the effort to utilize such pervasive technologies for teaching and learning as students increasingly try to "cram learning into the interstices of daily life" (Sharples, Taylor, & Vavoula, 2005, p. 58).

In this article the approach adopted for the design of m-learning tools and infrastructure is predicated on the idea that there will be intermittent wireless connectivity with limited bandwidth (e.g., grainy video at best). Students in Hong Kong (like most places) lead busy lives, and access to always-on Internet connections may not be possible or desired. Instead, the concept of flexibility of learning—learning at a time most suitable to the student—is seen as a primary driving factor for our work. What follows is a description of a framework for development of learning tools and institutional infrastructure designed to take advantage mobile devices and the flexibility offered by m-learning.

MOBILE DEVICES AS SEMIOTIC TOOLS

Tools were once seen as some form of some physical object (e.g., a screwdriver, the pulley, a hammer, or the cogs on a bicycle). The purpose of tools was to enhance human strength and/or human capabilities. Traditional learning included the humble pen-and-paper or an abacus. However, humans have also created semiotic tools (Vygotsky, 1978), which are intangible tools to mediate cognition. These semiotic tools include language, numbers, algebraic notation, mnemonic

Figure 1. Classification of mobile technologies (Adapted from Naismith, Lonsdale, Vavoula, & Sharples, 2005, p. 7)

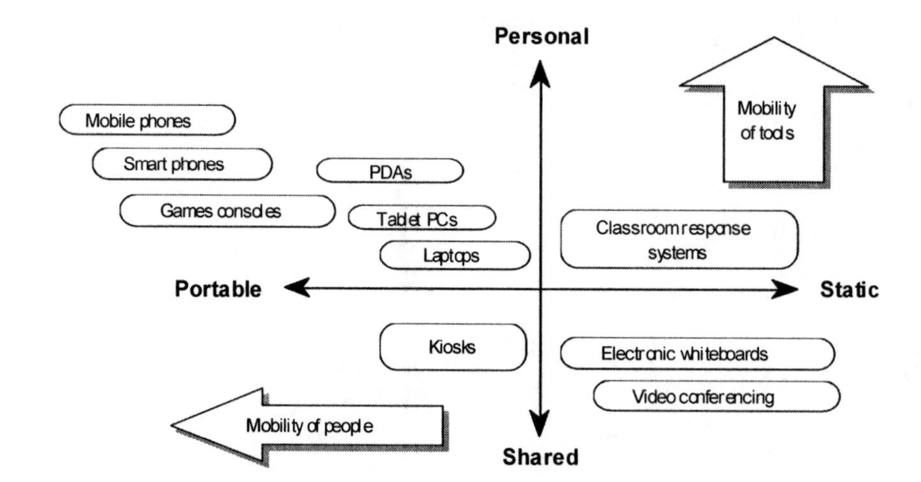

techniques, graphs, and diagrams—most of which may be expressed in the form of media elements that are easily stored, retrieved, and manipulated by computers (Kennedy, 2001). Since Vygotsky's time, technical tools (computers, PDAs, mobile phones, smart phones) have come to encompass devices that can utilize and manipulate signs (intangible tools) to enhance human cognitive processes (Duffy & Cunningham, 1996; Jonassen & Reeves, 1996). Current mobile devices function as computer-based cognitive tools, helping people to store, organize, structure, communicate, annotate, capture information, play, and engage in increasingly complex tasks, blurring the distinction between tangible (hardware) and intangible (software and signs) tools—one without the other is meaningless.

The feature set of mobile devices is improving rapidly as the power of the central processing unit (CPU) increases, following Moore's Law as desktop computers have for past decades (Zheng & Ni, 2006a). The future looks bright with the convergence of personal digital assistants, mobile telephones, and digital imaging into devices described as smart phones (Zheng & Ni, 2006b). The growth of computing power in such devices offers many opportunities for learning. Already such devices are endowed with features and facilities in the realm of science fiction just a few years ago, running a variety of operating systems with support for the .NET framework from Microsoft, Java, multimedia capability, and storage capacity in the multi-gigabyte realm, rapidly overcoming limitations described only a few years ago by Csete, Wong, and Vogel (2004). For example, connection speeds have risen dramatically.

However, if the potential of mobile tools for learning is not to be wasted, there is a pressing need to develop appropriate learning tools that can provide structure to the student experience. Such learning tools need to be pedagogically sound, offer high levels of interactivity, and be compliant with the available infrastructure. It has been shown that placing content on the Web or storing it in a learning management system (LMS) is not sufficient for learning to occur (Ehrmann, 1995; Reeves, 2003; Rehak & Mason, 2003). It is even more disadvantageous to do so in a mobile environment with limitations on screen size, battery life, and processing power, notwithstanding the rapid development of functionalities and features. Some examples are the virtual keyboard (http://www.virtual-laser-keyboard.com/) and more powerful batteries that enable faster, power-demanding CPUs and hard drives to be used for longer periods of time (http://www.medistechnologies.com/).

DEVELOPING FOR THE MOBILE LEARNER

Vavoula and Sharples (2002) suggested that mobility is an intrinsic property of learning. They argue that learning has spatial (workplace, university, home), temporal (days, evenings, weekends), and developmental components (the learning needs/life skills of individuals which change depending upon age, interest, or employment). Figure 1 is a diagrammatic representation of this view.

In Figure 1 there are two arrows. The horizontal arrow indicates increasing mobility of people (right to left), while the vertical arrow indicates increasing mobility of the device. In the work described in this article, the focus is on the top left quadrant, with high mobility for people and devices. Applications (mobile learning tools or m-learning applications) that support mobility of devices and people have a number of criteria that differ widely from the desktop environment. In Table 1 the basic design elements suggested by Zheng

Table 1. Design elements for user interface design for mobile learning (Adapted from Zheng & Ni, 2006a, p. 473)

Design Element	Description	Guide to UID for Mobile Learning
Context	Functional design that accounts for screen size, processor speed, educational needs	Set of icons that simplify the UI, legible font size, and adaptation of the desktop environment to suit the limitations of the mobile platform
Content	Resources that can be presented, annotated, queried, and answered	Limits on text length, image size, length of input required, use of menu-driven options for data input
Community	Information sharing (Bluetooth and WiFi)	Text, instant messaging, image messaging
Customization	Tailoring the device to the personal needs of the student	Linking to on-campus resources
Communication	Human-computer interaction	A variety of input methods, wizard-like dialogues
Connection	A variety of connection methods need to be supported	Customization of connection settings

and Ni (2006a) for user interface design (UID) of mobile devices are shown.

However, the elements shown in Table 1 neglect the all important pedagogical perspective needed to create flexible learning environments that incorporate m-learning effectively. Richards (2004) has provided a way of simplifying the paradigm wars that often plague discussions about the use of information and communication technologies (ICTs) in education by coining the expressions "new learning" and "old learning". Effectively, new learning (student-centered) is based around a social constructivist view, in which ICTs (including m-learning) are intended to:

- engage students actively in the construction of knowledge;
- involve problem solving rather than solving problems (algorithms);
- encourage and support collaboration, articulation, and discussion;
- enhance understanding rather than rote learning;
- anchor skill development and learning tasks in meaningful, authentic, and information-rich environments;
- promote motivation by interactivity;
- promote learning through cooperative, group-based activity; and
- focus on engaging activities that require higher-level skills by integrating lower-level skills (Roblyer, Edwards, & Havriluk, 2002, p. 51).

The change from old learning to new learning also has a technological equivalent. The development of a raft of mobile devices that are both personal and portable (see Figure 1) is exemplified by the relationships articulated in Table 2. In Table 2, Sharples et al. (in press) link the changes in learning methodology and current thinking to the developments in technology that support the current focus.

In summary, mobile devices support a synergy of tools, both tangible and intangible, providing ever-increasing functionality and modes by which knowledge can be constructed, annotated, stored, shared, and created.

FOUR EXAMPLES OF MOBILE TOOLS

At the City University of Hong Kong, the development of mobile tools has focused on providing more flexible possibilities for learning by students. Students in Hong Kong are like most others, juggling busy lives and moving in and out of wired and wireless environments. Zheng and Ni (2006a) have suggested a model for the manner in which one may

Table 2. New learning and technology (Adapted from Sharples et al., in press)

New Learning	New Technology
Personalized	Personal
Learner-Centered	User-Centered
Situated in Time and Place	Mobile
Collaborative	Collaborative and Networked
Ubiquitous	Ubiquitous
Lifelong and Life-Wide	Connected
Activity-Based	Greater Functionality

Figure 2. Some notions of mobile computing (Adapted from Zheng & Ni, 2006a, p. 471)

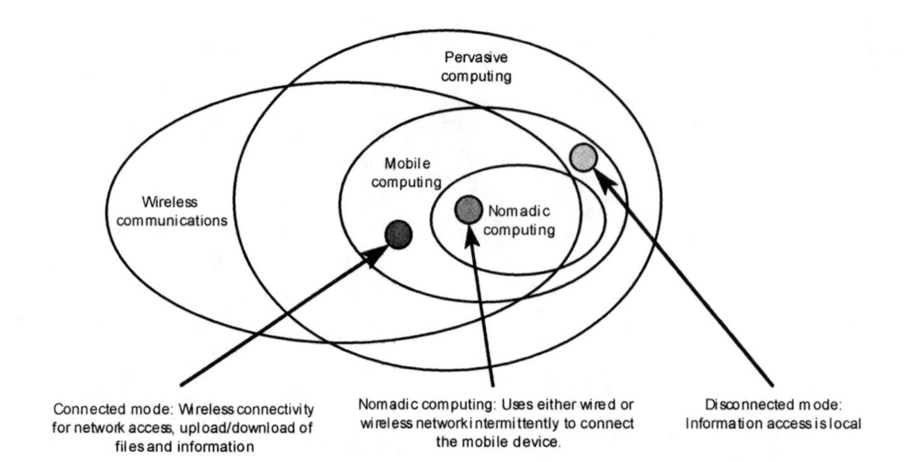

consider the elements of mobile computing (see Figure 2) which is more congruent with current practice and infrastructure. The Hong Kong students move between the:

- *connected mode* (at the campus),
- *nomadic mode* (at home or connect to a desktop computer on campus or at home), and
- *disconnected mode* (on public transport, away from wired or wireless connections).

Current technology does not support *pervasive computing,* which will be reached when the hardware and the software become deeply embedded in the user's physical environment, so much so that the user may not even notice that he or she is interacting with a computing environment. The current environment, particularly in education, is still some way from this goal (Zheng & Ni, 2006a).

The current project is a holistic approach that seeks to:

1. develop and research the use of a range of tools that are designed to more readily support academic teachers in their quest to match the student learning outcomes with appropriate activities, assessment, and feedback (Kennedy, Vogel, & Xu, 2004);
2. develop the technical infrastructure that enables academics to publish activities to the Web or personal digital assistants (PDAs) or smart phones;
3. develop the technical infrastructure to allow lecturers to monitor student activity and record student learning outcomes about student interactions on their PDAs to the lecturers from within the university LMS, BlackBoard; and
4. provide advice (mainly pedagogical) and support (with examples) for developing content suitable for m-learning.

The project has been operating for nearly 18 months and is providing a degree of the flexibility demanded by students. Figure 3 provides an overview of the three components: learning tools, authoring environment, and e-token server. The concept of an e-token was suggested by tutors at City University as a method of stimulating interest and motivation among students. Students are given a number of e-tokens at the commencement of their course of study. They can use these e-tokens to buy questions. When they upload their answers and responses back to the e-token server, the log of their success or failure to solve the tasks determines how many e-tokens they receive for their efforts. An honor board of the top three students is kept on the e-token Web site.

The e-token server provides a secure connection between the university LMS (BlackBoard) and the institutional wireless network. The e-token server synchronizes with the main BlackBoard server once a day. At the time of writing, work was being undertaken to integrate this function into BlackBoard so that:

1. students can access any of the mobile resources (software and questions) directly from their BlackBoard course area (a single sign-in);
2. students upload their responses to either a personal e-portfolio or an area that is monitored by the lecturer;
3. lecturers can publish resources directly to students in either a Web-based or mobile form; and
4. lecturers can receive an instantaneous update of student downloads, uploads, and responses (rather than having to wait until the next day).

The four tools currently in use are a quiz tool based on 'Hot Potatoes' (freely available to non-profit educational institutions; http://hotpot.uvic.ca), an interactive graphing tutorial tool (IGO), a Crossword tool, and a simulation called

Figure 3. E-token server, mobile content creation, and publishing

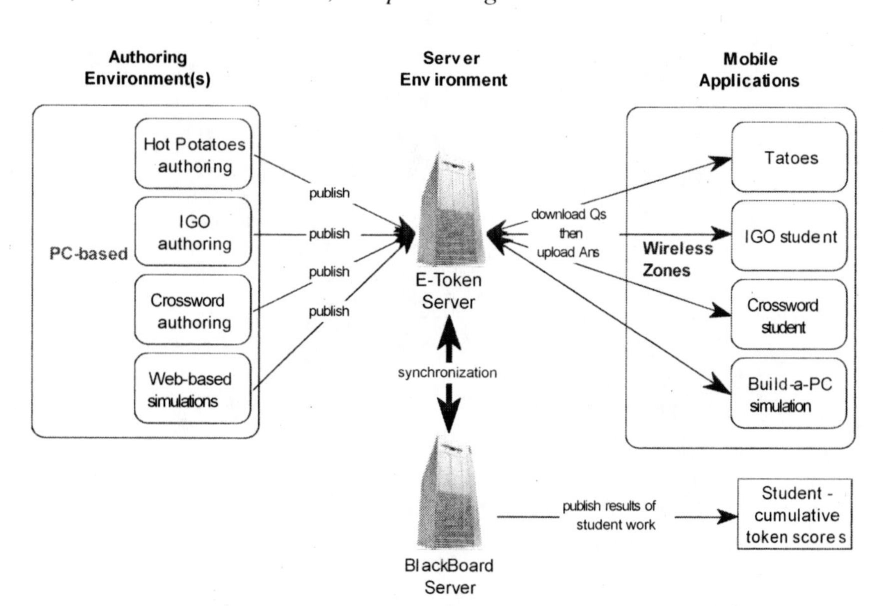

'Build a PC'. The project team has created a set of icons to provide a consistent experience for students. All of the mobile tools use the same or application-specific icons (see Figures 4, 5, and 6). The development of a set of common icons for common tasks (e.g., save, upload, download, log-in, check answer, hints, information) are consistent in each mobile application.

Tatoes

Tatoes authoring is done on a PC. Tatoes adapts quiz-based content (multiple-choice, fill-in-the-gap, ordered lists, matching exercises) created with the software Hot Potatoes to be exported to any device using the Windows .NET mobile platform. A lecturer may create a series of questions using Hot Potatoes and either publish them to the Web or in a form that students may download from the e-token server (see Figure 3). In this way instructors are only required to create one set of questions for students but have the ability to save the questions for Web or mobile delivery. One of the key pedagogical factors was to provide detailed feedback for incorrect choices in the multiple-choice questions. Charman (1999) suggests that one of the most important determinants of the success of multiple-choice questions for learning is to:

- indicate clearly if the response is correct;
- provide sufficient time for the feedback to be understood;
- consider using graphics to enhance the feedback;
- be positive, friendly, and use a neutral tone;
- provide feedback for each distractor; and
- provide references or links to other sources to improve understanding.

In the Tatoes questions, lecturers have been encouraged to adopt these principles. A sample question *Which one of the following statements best describes global IT platforms?* is shown in Table 3. Feedback regarding student use of this resource has been collected, and it will be discussed in the next section.

Table 3. An example of an MCQ item with feedback

Choice	Statement	Feedback
A:	Hardware choices are difficult in some countries because of high tariffs and import restrictions.	Yes, you are correct.
B:	Software packages are compatible in all countries when you buy from the same hardware vendor.	No, software packages developed in Europe may be incompatible with American or Asian versions. Try again.
C:	Increasing hardware costs is one of the chief reasons for the trend toward the use of global IT.	There is no evidence that hardware costs are increasing. Try again.
D:	Hardware choices are easier in global markets than in the U.S. because of short lead times for government approvals.	No, lead times for government approvals vary from country to country.

Figure 4. Mobile version of the IGO

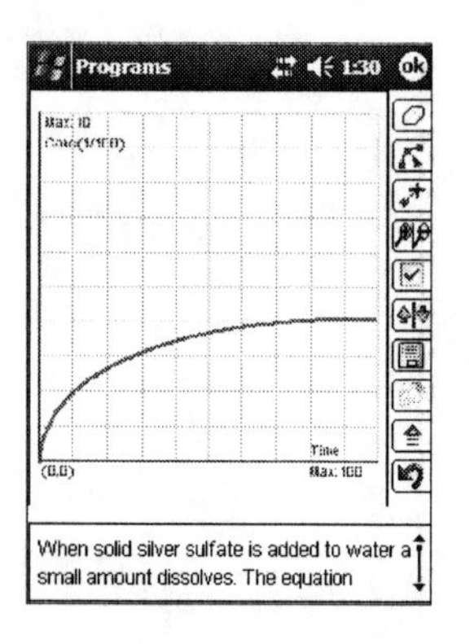

An initial curve has been drawn as part of a question in chemistry. The icons represent, in order:

1. Erase
2. Show handles for adjusting the shape of the curve
3. Change the format to a graph where the origin is placed centrally in the image palette (allowing graphs with positive values in the x and y axis to be displayed)
4. Shrink or grow the image
5. Next question in the sequence
6. Save a question
7. Open a question
8. Upload an answer to the e-token server
9. Go back

Figure 5. The Crossword puzzle

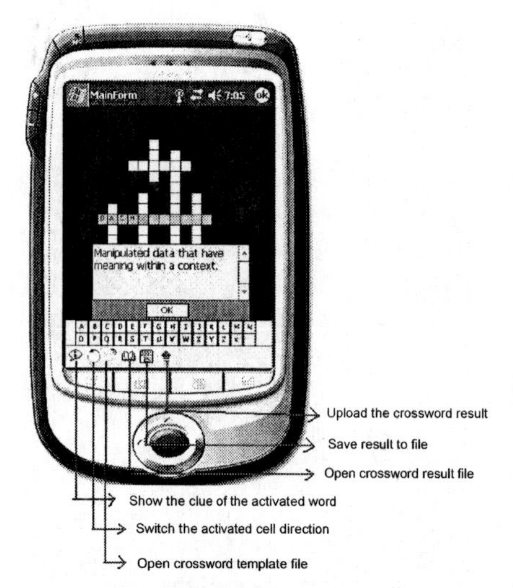

The Crossword puzzle uses the same basic icons set as all other applications being developed for this project. It has proved to be very popular with students. However, there are application-specific icons including "show a clue for the current selection" and "switch the active cell direction." The use of icons and minimal descriptions and hints has provided an interface that students do not find difficult to engage with.

Interactive Graphing Object (IGO)

Graphs are used for a large number of subjects to provide a visual representation between two or three variables. Moreover, many concepts are best represented by graphs. Achieving understanding of key concepts in science, business, and medical sciences is frequently more efficacious using graphs or other visual representations (Kremer, 1998; Kennedy, 2002; Kozma, 2000). The interactive graphing object (IGO) was originally developed for the Web (Kennedy, 2004). The IGO enabled students to sketch a graph onscreen in order to articulate their knowledge directly. This is in contrast to watching an animation or choosing the correct graph from a set of static images. The current project has created a Java-based version of the IGO that operates on either the Palm PDA or a PDA with the Windows Mobile operating system (Kennedy et al., 2004). A screen capture of the mobile version is shown in Figure 4. The IGO authoring environment enables complex questions and feedback involving graphical representations to be created for either the Web or the PDA. At this time there is only limited evaluation data available for this highly interactive tool.

Figure 6. The Build-a-PC mobile interface; (a) the introduction, (b) the role interface of Build-a-PC, (c) interactive exercise, and (d) expert comparison

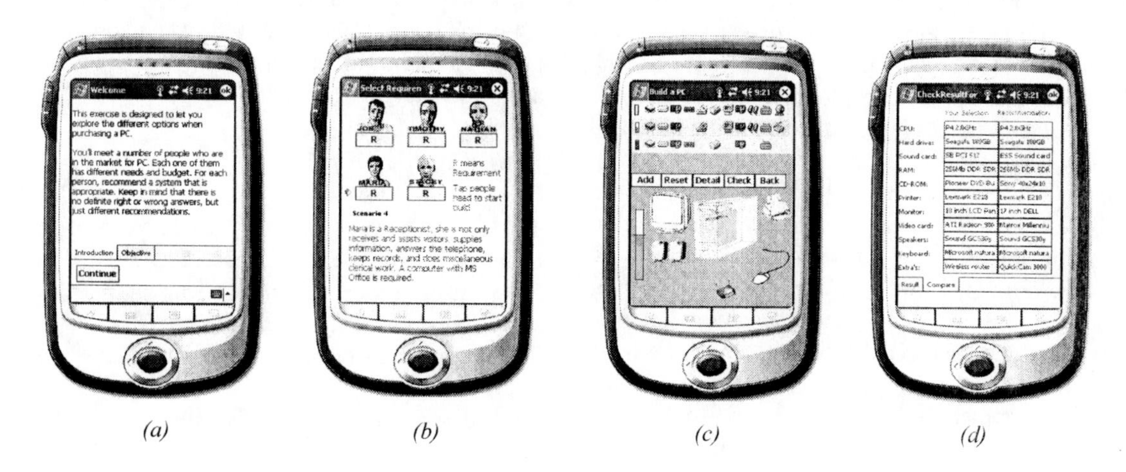

Crossword Puzzle

The Crossword tool creates crossword puzzles with hints and clues for the PDA. It is an application that has been designed to operate specifically on the PDA and does not have a Web version. Two screen captures of the interface are shown in Figure 5.

Build-a-PC Tool

The Build-a-PC has both a Web form and a mobile form (similar to the IGO). The four elements of the user interface are shown in Figures 6A to 6D. The task involves the student in an authentic task: to select the appropriate desktop system computer system for a range of people in an organization. Students must make decisions based upon the position held, responsibilities of the individual in the organization, and the budget available for the purchase.

At this stage the tool is available and has been used in two courses. However, the programming to collect log data is not yet in place.

In Figures 6(a), 6(b), and 6(c)), the interface for the activity is shown. Figure 6(a) shows the introduction to the task where a student is told that she or he is the purchasing officer for a company. Figure 6(b) provides the list of company personnel and a description of their positions in the company for whom the student must decide which configuration is most cost efficient and effective. Figure 6(c) is the interactive portion is which students identify components on the top lines and insert into the computer configuration. Feedback is available in text form relative to task completion as well as a 'budget bar'. At the conclusion of the task, the student can compare his or her choices against the expert view (Figure 6(d)).

CURRENT EVALUATION DATA OF TATOES, IGO, AND CROSSWORD

Both qualitative and quantitative evaluations are underway with the tools that have been developed. To date, perceptual data has been gathered from two courses of study. The first is from 416 students in multiple sections of an introductory business course (of approximately 800 students) in Semester 1, 2005-2006 year, and the second an introductory business class in Semester 2 of 2006. The use of the PDA was not mandated in either course, but was a voluntary activity. Current feedback from the tutors indicates that students who engaged with the use of mobile devices for learning found the experience of using a PDA to be a significantly better learning experience. Currently 186 students have used the e-token system from a total of 812 enrolled in the Semester 2 course. The use of the e-token system is entirely voluntary. Results are encrypted to discourage inappropriate sharing of results. Evaluation of each of the tools will be discussed separately.

Tatoes

A total of 87 students downloaded Tatoes-based exercises. However, only 28 students uploaded results back to the server. This evaluation is based upon analysis of this data. Only one student (of those who uploaded data) actively explored alternative answers after arriving at the correct answer—looking at the final distracter of one question after making three attempts to arrive at the correct answer. All other students stopped exploring alternatives once they discerned the correct answer (sometimes after one, two, or even three wrong answers first). This is in contrast to a study undertaken by Fritze (1994) with undergraduate chemistry students. In that study, there was evidence that students will take the opportunity to explore explanations provided as feedback to alternatives in problems. Fritze (1994) provided a method of visually mapping problem-solving strategies from which he identified a number of problem exploration strategies, some of which involved students repeatedly examining hints and explanations. However, this strategy was not observed in this pilot study. There is some evidence that the approach adopted by students in Hong Kong depends to a significant degree on how the assessment task is perceived (Tang & Biggs, 1996). If a novel situation or task is given without sufficient guidance (e.g., a set of new strategies or a framework of undertaking the task), the students tend to revert to established patterns of coping "in a highly surface-oriented assessment environment" (Tang & Biggs, 1996, p. 179). So while the students may wish to adopt a deep approach to learning, the perception of an assessment task may prevent them from doing so.

The current data bears this out. The pedagogical approach adopted when the questions were written was intended to encourage a deep approach to learning. In particular, high-quality feedback was provided to the students for all distracters (see Table 3). It is clear from the log data that students ceased exploring alternatives once they arrived at the right answer (except for the single student indicated above). Some of the other 59 students who downloaded the Tatoes exercises may have used a more exploratory approach, but this data is not available at this time (interviews were to be scheduled in the fall semester of 2006).

In the case of the students who uploaded their responses back to the server to receive e-tokens, there may have been some confusion with respect to trying to achieve the best possible mark in light of the assessment-based culture typical of Hong Kong education. The Tatoes log record does not penalize students for looking at alternatives once the correct answer has been selected. However, students may not have realized this, or having looked at the alternatives, then decided there was little point to uploading their scores back to the e-token server. Penalties are only incurred for incorrect selections leading up to the correct answer. Therefore, encouraging active use of the distracters (after achieving the correct response) to confirm or query why these are not the correct answer is a challenge for the next iteration of the project.

The Crossword Puzzle

At the time of writing, 47 students have downloaded on average eight crosswords each. This resource has proved very popular for students revising for their examinations, especially for an activity that does not specifically carry any grades. The nature of the crossword puzzle is focused on keywords and concepts. Students may compare answers with the crossword, but may not beam the answer to each other since that feature has been disabled by the Crossword software. It is expected that students will use the crossword puzzles as points of discussion and debate, rather than as an assessment target to be met.

The IGO

The IGO has been evaluated by an international panel of educational technology experts (Jones, 2004; Jones & McNaught, 2005). These evaluations have been reported elsewhere but some elements are shown here (Kennedy, 2004). Examples of key comments made by the expert reviewers are:

This project epitomizes a very important kind of learning object. It can be the basis of an unlimited number of applications across many fields in mathematics, science and social science.

The learning object supports core learning processes that are rarely dealt with in Web-based materials.

The IGO is similar to Data Works and Language Works in that it calls on students to simultaneously deal with textual/conceptual and graphical representations. It is significantly more sophisticated in that it allows for seven different graph forms and these graphs are based on student-supplied data. Students create a graph and receive feedback on several key aspects of their work. Equally important, they are able to redo their graph in light of this feedback. The reiterative process in which the student acts, receives feedback, and acts on that feedback, all in the context of conceptual understanding, is potentially an extraordinarily powerful one. It is also different in that it is a template that allows lecturers to create their own graphs, questions and feedback without the need for a programmer or Web developer.

However, this is one of the issues of developing new technology-based tools: gaining the cooperation of lecturers who have control of the curriculum and resources. While expert evaluators were very positive, it is often difficult to implement unfamiliar tools into a curriculum, especially if the development of content is seen by lecturers as too time consuming. An earlier iteration of the IGO based upon Shockwave is still in use at The University of Melbourne (Kemm et al., 2000), and while the current version of the IGO is a more robust and user-friendly learning object to author and deploy, there remain issues of developing questions in academic disciplines: the designers need the active participation of the academic teachers to develop content-specific questions and establish a community of practice.

Table 4. Current inhibitions and future solutions to the development of mobile tools for education (Adapted from Csete et al., 2004)

Inhibitors	Relationship to Current M-Learning Functionality	Future Solution
Small screen size	Current screen size creates overlapping text and/or graphics, especially for the feedback to questions.	Flexible film display
Non-ergonomic input method	The stylus or onscreen text recognition (e.g., the graffiti script for Palm devices) limits the speed and flexibility of input an annotation.	Voice recognition Projection keyboard Cursive hand-writing recognition
Slow CPU speed	Applications run more slowly than on a desktop computer.	New breed of architecture for faster CPU
Limited memory	The size and power requirements of more powerful CPUs limit what can be placed in mobile devices.	Expansion memory card Increase internal RAM capacity
Limited battery span	Extended use is limited by current technology.	New breeds of lithium batteries or fuel cells
Ever-changing OS	This is a current and likely future problem not resolvable in the short term with many competing systems (e.g., Symbian, Windows mobile and Palm).	Open-source OS for mobile devices (e.g., Linux has been run successfully on a mobile phone)
Infrastructure compatibility	There is a plethora of wireless standards, some of which require large amounts of power.	Standards are still being developed to bridge the mobile platform.
Connectivity bandwidth	Current wireless networks have sacrificed speed for access.	3G mobile capacity and Bluetooth v.1.2 More efficient wireless protocols than currently available

THE FUTURE

The future is bright for mobile devices for education as CPU speed, battery life, and number of applications increase (see Table 4). However, what is more important is the establishment of sound pedagogical practice based upon experience and research into how students use the tools in practice, what the specific learning needs are, and how more effective feedback, communication, and collaboration can be enhanced. The three tools, Crossword, IGO, and Build-a-PC, have been designed with the facility for lecturers to provide high-quality feedback to students. The use of the Tatoes environment provides limited evidence that writing multiple-choice questions with detailed feedback may encourage some students to not only look for the correct answers to a question, but to spend time examining what makes other distracters wrong.

What this project has also made clear is the need for more support for individual academics in the development of questions for formative evaluation for students with high-quality feedback, as well as more encouragement for the students to actively explore alternatives to broaden their own understanding. In an assessment-driven educational culture, students need to be convinced that exploring explanations and feedback will not harm their overall scores, and may even help them understand the content more deeply—thus satisfying a recognized characteristic of students in Hong Kong for deep learning *and* improved exam results (Tang & Biggs, 1996).

REFERENCES

Associated Press. (2005). *Google offers free Wi-Fi for San Francisco.* Retrieved May 31, 2006, from http://www.msnbc.msn.com/id/9551548/

Boerner, G. L. (2002). The brave new world of wireless technologies: A primer for educators. *Campus Technology,*

16(3). Retrieved May 31, 2006, from http://www.campus-technology.com/mag.asp?month=10&year=2002

Charman, D. (1999). Issues and impacts of using computer-based assessments (CBAs) for formative assessment. In S. Brown, P. Race, & J. Bull (Eds.), *Computer-assisted assessment in higher education* (pp. 85-94). London: Kogan Page.

Csete, J., Wong, Y.-H., & Vogel, D. (2004). Mobile devices in and out of the classroom. In L. Cantoni & C. McLoughlin (Eds.), *ED-MEDIA 2004* (pp. 4729-4736). *Proceedings of the 16th World Conference on Educational Multimedia and Hypermedia & World Conference on Educational Telecommunications,* Lugano, Switzerland. Norfolk VA: Association for the Advancement of Computing in Education.

Ehrmann, S. C. (1995). New technology: Old trap. *The Educom Review, 30*(5), 41-43.

Fenn, J., & Linden, A. (2005). *Gartner's hype cycle special report for 2005.* Retrieved May 31, 2006, from http://www.gartner.com/DisplayDocument?id=484424

Fritze, P. (1994). A visual approach to the evaluation of computer-based learning materials. In K. Beattie, C. McNaught, & S. Wills (Eds.), *Interactive multimedia in university education: Designing for change in teaching and learning* (pp. 273-285). Amsterdam: Elsevier.

Green, K. (2004). *The 2003 national survey of information technology in US higher education.* Retrieved May 31, 2006, from http://www.campuscomputing.net/

Jones, J. (2004). *The Interactive Graphing Object—Compiled evaluation report.* Hong Kong: The University of Hong Kong. Retrieved May 31, 2006, from http://learnet.hku.hk/production/evaluation/IGO%20-%20Display.pdf

Jones, J., & McNaught, C. (2005). Using learning object evaluation: Challenges and lessons learned in the Hong Kong context. In G. Richards & P. Kommers (Eds.), *ED-MEDIA 2005. Proceedings of the 17th Annual World Conference on Educational Multimedia, Hypermedia & Telecommunications,* Montreal, Canada (pp. 3580-3585). Norfolk VA: Association for the Advancement of Computers in Education.

Kemm, R. E., Kavnoudias, H., Weaver, D. A., Fritze, P. A., Stone, N., & Williams, N. T. (2000). Collaborative learning: An effective and enjoyable experience! A successful computer-facilitated environment for tertiary students. In J. Bourdeau & R. Heller (Eds.), *ED-MEDIA 2000* (pp. 9-20). *Proceedings of the 12th Annual World Conference on Educational Multimedia, Hypermedia & Telecommunications,* Montreal, Canada. Norfolk, VA: Association for the Advancement of Computers in Education.

Kennedy, D.M. (2001). *The design, development and evaluation of generic interactive computer-based learning tools in higher education.* Unpublished doctoral dissertation, The University of Melbourne, Australia.

Kennedy, D. M. (2002). Visual mapping: A tool for design, development and communication in the development of IT-rich learning environments. In A. Williamson, C. Gunn, A. Young, & T. Clear (Eds.), *Winds of change in the sea of learning: Charting the course digital education* (Vol. 1, pp. 339-348). *Proceedings of the 19th Annual Conference of the Australasian Society for Computers in Learning in Tertiary Education.* Auckland, New Zealand: UNITEC Institute of Technology. Retrieved May 31, 2006, from http://www.ascilite.org.au/conferences/auckland02/proceedings/papers/150.pdf

Kennedy, D. M. (2004). Continuous refinement of reusable learning objects: The case of the Interactive Graphing Object. In L. Cantoni & C. McLoughlin (Eds.), *ED-MEDIA 2004* (pp. 1398-1404). *Proceedings of the 16th World Conference on Educational Multimedia and Hypermedia & World Conference on Educational Telecommunications,* Lugano, Switzerland. Norfolk, VA: Association for the Advancement of Computing in Education.

Kennedy, D. M., Vogel, D. R., & Xu, T. (2004). Increasing opportunities for learning: Mobile graphing. In R. Atkinson, C. McBeath, D. Jonas-Dwyer, & R. Phillips (Eds.), *ASCILITE 2004: Beyond the comfort zone* (pp. 493-502). *Proceedings of the 21st Annual Conference of the Australian Society for Computers in Learning in Tertiary Education,* Perth, Western Australia. Retrieved May 31, 2006, from http://www.ascilite.org.au/conferences/perth04/procs/kennedy.html

Kozma, R. B. (2000). The use of multiple representations and the social construction of understanding in chemistry. In M. Jacobson & R. Kozma (Eds.), *Innovations in science and mathematics education: Advanced designs for technologies of learning* (pp. 11-46). Mahwah, NJ: Lawrence Erlbaum.

Kremer, R. (1998). *Visual languages for knowledge representation.* Retrieved May 31, 2006, from http://pages.cpsc.ucalgary.ca/~kremer/papers/KAW98/visual/kremer-visual.html

Naismith, L., Lonsdale, P., Vavoula, G., & Sharples, M. (2005). *Literature review in mobile technologies and learning* (11). Bristol, UK: FutureLab. Retrieved May 31, 2006, from http://www.futurelab.org.uk/download/pdfs/research/lit_reviews/futurelab_review_11.pdf

Office of the Telecommunications Authority of Hong Kong. (2004). *Key telecommunications statistics: December 2005.* Retrieved May 31, 2006, from http://www.ofta.gov.hk/en/datastat/key_stat.html

Reeves, T.C. (2003). Storm clouds on the digital education horizon. *Journal of Computing in Higher Education, 15*(1), 3-26.

Richards, C. (2004). From old to new learning: Global dilemmas, exemplary Asian contexts, and ICT as a key to cultural change in education. *Globalisation, Societies and Education, 2*(3), 399-414.

Rehak, D. R., & Mason, R. (2003). Keeping the learning in learning objects. In A. Littlejohn (Ed.), *Reusing online resources: A sustainable approach to e-learning* (pp. 20-34). London: Kogan Page.

Roblyer, M. D., Edwards, J., & Havriluk, M. A. (2002). *Integrating educational technology into teaching.* Columbus, OH: Prentice-Hall.

Sharples, M., Taylor, J., & Vavoula, G. (2005). Towards a theory of mobile learning. Mobile technology: The future of learning in your hands. *Proceedings of the 4th World Conference on M-Learning (M-Learn2005)*, Cape Town, South Africa. Retrieved May 31, 2006, from http://www.mlearn.org.za/CD/papers/Sharples-%20Theory%20of%20Mobile.pdf

Sharples, M., Taylor, J., & Vavoula, G. (in press). A theory of learning for the mobile age. In R. Andrews & C. Haythornwaite (Eds.), *Handbook of e-learning research.* London: Sage.

Tang, C., & Biggs, J. (1996). How Hong Kong students cope with assessment. In D. A. Watkins & J. B. Biggs (Eds.), *The Chinese learner: Cultural, psychological and contextual influences* (pp. 159-182). Hong Kong: Comparative Education Research Centre & Camberwell: Australian Council for Educational Research.

Trinder, J. J., Magill, J. V., & Roy, S. (2005). Portable assessment: Towards ubiquitous education. *International Journal of Electrical Engineering Education, 42,* 73-78. Retrieved May 31, 2006, from http://www.findarticles.com/p/articles/mi_qa3792/is_200501/ai_n13507163

Vavoula, G. (2005). *A study of mobile learning practices (D4.4).* Birmingham, UK: MOBIlearn. Retrieved May 31, 2006, from http://www.mobilearn.org/download/results/public_deliverables/MOBIlearn_D4.4_Final.pdf

Vavoula, G., & Sharples, M. (2002). kLeOS: A personal, mobile, knowledge and learning organisation system. In M. Milrad, U. Hoppe, & Kinshuk (Eds.), *Proceedings of the IEEE International Workshop on Mobile and Wireless Technologies in Education (WMTE2002)* (pp. 152-156). Vaxjo, Sweden: Institute of Electrical and Electronics Engineers (IEEE).

Vernon, R.D. (2006). *Cornell data networking: Wired vs. wireless?* Retrieved May 1, 2006, from http://www.cit.cornell.edu/oit/Arch-Init/WIRELESS.pdf

Zheng, P., & Ni, L. M. (2006a). *Smart phone & next generation mobile computing.* San Francisco: Elsevier.

Zheng, P., & Ni, L. M. (2006b). The rise of the smart phone. *IEEE Distributed Systems Online 1541-4922, 7*(3). Retrieved May 31, 2006, from http://csdl2.computer.org/comp/mags/ds/2006/03/o3003.pdf

KEY TERMS

Computer-Based Cognitive Tool: One of a set of software tools that support user cognition or thinking to construct knowledge by the manipulation of signs (e.g., graphs, language, and/or mathematics) using computers.

Intangible: Not able to be physically grasped. For example, language is an intangible tool, one that supports the mediation of cognition/thinking.

M-Learning: The use of electronic mobile devices to support learning, either wholly or partially, formal or informal. M-learning is a part of e-learning that makes use of the convenience of mobile devices to provide flexibility when and where the student learns.

New Learning: Student-centered learning based upon a social constructivist view of teaching and learning, involving authentic, meaningful activities, problem solving, cooperation, collaboration, articulation, and discussion.

Semiotic Tool: One of a set of intangible tools that enable meaning to be conveyed (e.g., language, mathematics, or computer software).

Smart Phone: A device that represents the convergence of a raft of technologies, including personal digital assistants (PDAs), global positioning satellite (GPS) systems, digital cameras, mp3 and video players, and mobile telephony.

Wireless: A raft of technologies that enable the transfer of data to and from appropriately equipped devices without the aid of wires. Wireless technologies include infrared, Bluetooth, and the wireless protocols in mobile telephony.

Intelligent Medium Access Control Protocol for WSN

Haroon Malik
Acadia University, Canada

Elhadi Shakshuki
Acadia University, Canada

Mieso Kabeto Denko
University of Guelph, Canada

INTRODUCTION

This article reports an ongoing research that proposes an approach to the expansion of sensor-MAC (S-MAC), a cluster-based contention protocol to intelligent medium access control (I-MAC) protocol. I-MAC protocol is designed especially for wireless sensor networks (WSNs). A sensor network uses battery-operated computing and sensing devices. A network of these devices are used in many applications, such as agriculture and environmental monitoring.

The S-MAC protocol is based on a unique feature: it conserves battery power by powering off nodes that are not actively transmitting or receiving packets. In doing so, nodes also turn off their radios. The manner in which nodes power themselves off does not influence any delay or throughput characteristics of the protocol. Inspired by the energy conservation mechanism of the S-MAC, we unmitigated our efforts to augment the node lifetime in a sensor network. In such a network, border nodes act as shared nodes between virtual clusters. Virtual clusters are formed on the basis of sleep and wake schedule of nodes. To prolong the lifetime of the network, nodes are allowed to intelligently switch to cluster where they experience minimum energy drain. Towards this end, we propose a multi-agent system at each node. This system includes two types of agents: stationary monitoring agent (SMA) and mobile mote agent (MMA). SMA is a static agent and has the functionality to monitor the node events. MMA is a mobile agent with the ability to roam in WSN. This article focuses on the architecture of the proposed system.

BACKGROUND

Recently, there has been development and adoption of many commercial and industry wireless communication standards. WSNs are a new class of wireless networks that have appeared in the last few years. Sensor networks consist of individual nodes that are able to interact with their environment by sensing or controlling physical parameters. They perform local computation based on the data gathered and transmit the results to their neighbors. Sensor nodes need to collaborate with each other to fulfill their tasks when a single node is incapable of doing so. The range of applications of WSN is rapidly growing. Some envisaged areas of applications include monitoring the environment, assisting healthcare professionals, military communication, and precision agriculture (Akyildiz, Su, Sankarasubramaniam, & Cayirci, 2002; Stojmenovic, 2006).

WSNs differ from traditional wireless networks in several ways (Akyildiz et al., 2002). First, sensor networks consist of a number of nodes and have high network density that competes for the same channel. Second, most nodes in sensor networks are battery powered, and it is often very difficult to change batteries for those nodes. Third, nodes are often deployed in an ad-hoc fashion rather than with careful pre-planning; they must then organize themselves into a communication network. Fourth, sensor networks are prone to node, network failures and self-organization. Fifth, broadcasting is the main mode of communication in sensor networks and this may cause channel contentions. Finally, most traffic in the network is triggered by sensing events, and it can be extreme at times.

These characteristics of WSNs suggest that traditional MAC protocols are not suitable for wireless sensor networks without modifications. To this end, a number of MAC protocols have been proposed for WSNs (Frazer et al., 1999; Katayoun & Gregory, 1999; Heinzelman, Chandrakasan, & Balakrishnan, 2000).

These protocols are based on different design principles, including the number of physical channels used, the way a node is notified of an incoming message, and the degree of organization. Among these protocols, sensor MAC (S-MAC) uses in-channel signaling, slot structure, and a collective listening approach per slot to reduce idle-listening problem. We believe that the design principles of S-MAC enhance the energy performance for WSNs. This motivated us to propose an intelligent medium access protocol, which is an extension to sensor-MAC (S-MAC) integrating a multi-agent approach. This approach allows the border nodes to intelligently select a cluster that will experience minimum energy drain.

Many research efforts are being made to propose MAC protocol in the area of WSN that is energy efficient with minimum collision and increase of throughput (Katayoun & Gregory, 1999; Heinzelman, Chandrakasan, & Balakrishnan, 2000). The MAC layer is considered as a sub-layer of the data link layer in the network protocol stack. MAC protocols have been extensively studied in traditional areas of wireless voice and data communications. Time division multiple access (TDMA), frequency division multiple access (FDMA), and code decision multiple access (CDMA) are MAC protocols that are widely used in modern cellular communication systems (Rappaport, 1996). The basic idea of these protocols is to avoid interference by scheduling nodes onto different sub-channels that are divided either by time, frequency, or orthogonal codes. Since these sub-channels do not interfere with each other, MAC protocols in this group are largely collision-free. These protocols are called scheduled MAC protocols. Other types of MAC protocols are based on channel contention. Rather than pre-allocating transmissions, nodes compete for a shared channel, resulting in probabilistic coordination. Collision happens during the contention procedure in such systems. Classical examples of contention-based MAC protocols include ALOHA (Leonard & Fouad, 1975) and carrier sensor multiple access (CSMA) (Norman, 1985). In pure ALOHA (Norman, 1985), a node transmits a packet when it is generated, while in slotted ALOHA a node transmits at the next available slot. Packets that collide are discarded and will retransmit again. In CSMA, a node listens to the channel before transmitting. If it detects a busy channel, it delays access and retries to transmit later. The CSMA protocol has been widely studied and extended. Today, it is the basis of several widely used standards, including IEEE 802.11 (LAN/MAN, 1999).

TDMA-based protocols are effective at avoiding collisions and have a built-in duty cycle extenuating idle listening. Contention-based protocols in contrast to TDMA simplify the activities and do not require any dedicated access point in a cluster. MACAW (Mark & Randy, 1997) is an example of contention-based protocol. It is widely used in wireless sensor networks and in ad-hoc networks, because of its simplicity and robustness to hidden terminal problems. The standardized IEEE 802.11 distributed coordination function (DCF) (LAN MAN, 1999) is also an example of the contention-based protocol and is mainly built on a MACAW protocol. Most of the research work (Wei & John, 2004) showed that energy consumption of the MAC protocol is very high when nodes are in idle mode. This is due to idle listening. PAMAS (Singh & Raghavendra, 1998) provided an improvement by avoiding the overhearing among neighboring nodes.

S-MAC (Wei & John, 2004) is a slot-based MAC protocol specifically designed for wireless sensor networks. Built on contention-based protocols like IEEE 802.11, S-MAC retains the flexibility of contention-based protocols while improving energy efficiency in multi-hop networks. S-MAC implements approaches to reduce energy consumption from all the major sources of energy as idle listening, collision, overhearing, and control overhead.

S-MAC: Highlights

The smooth operation of any wireless network depends, to a large extent, on the effectiveness of the low-level medium access control (MAC) sub-layer. MAC in WSN aims to ensure that no two nodes are interfering with each other's transmissions, and deals with the situation when they do. S-MAC contention-based MAC protocol not only addresses the transmission interfering issues, but also extends its efforts in minimizing the protocol-overhead, overhearing, and idle-listening. S-MAC principle is based on locally managed synchronizations and periodic sleep-listen schedules. Neighboring nodes form virtual clusters to set up common sleep schedules. If two neighboring nodes reside in two different virtual clusters, they should wakeup at listen periods of both clusters. Schedule exchange is accomplished by periodical SYNC packet broadcasts to immediate neighbors. The period for each node to send a SYNC packet is called the synchronization period.

Figure 1 shows an example scenario for sender-receiver communication. Collision avoidance is achieved by carrier sense (CS). Furthermore, Ready to send and clear to send (RTS/CTS) packet exchanges are used or unicast type data packets. An important feature of S-MAC is the concept of message-passing where long messages are divided into frames and sent in a burst. With this technique, one may achieve energy savings by minimizing communication overhead at the expense of unfairness in medium access. Periodic sleep may result in high latency, especially in multi-hop routing

Figure 1. Sender-receiver communication

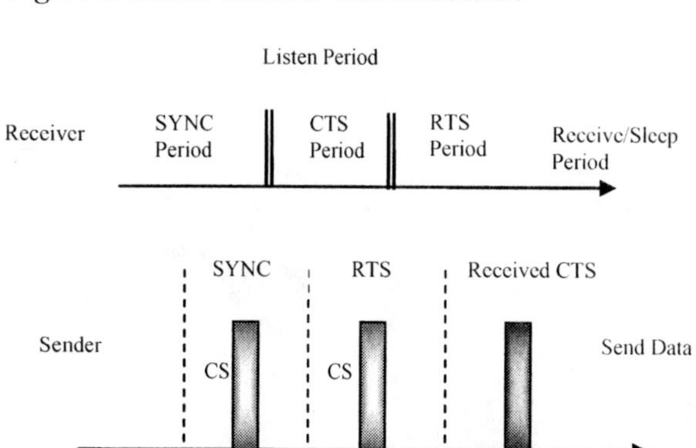

Figure 2. Border nodes

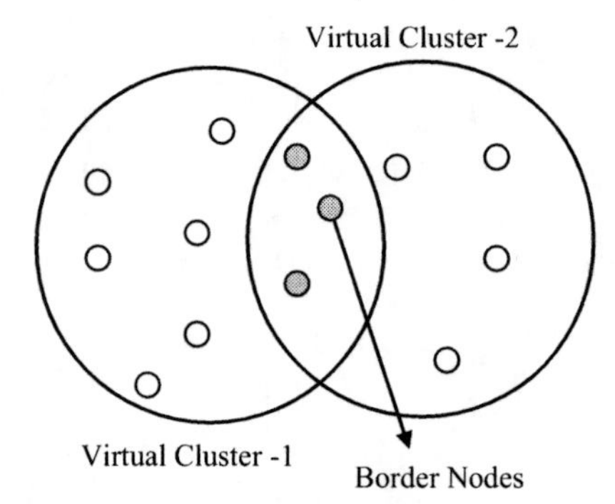

algorithm, since all immediate nodes have their own sleep schedules. The latency caused by periodic sleeping is called sleep delay (Wei & John, 2004). An adaptive listening technique is proposed to improve the sleep delay and thus the overall latency. In adaptive listening technique, the node who overhears its neighbor's transmission wakes up for a short time at the end of the transmission.

Hence, if the node is the next-hop node, its neighbor could pass data immediately. The end of the transmissions is known by the duration field of RTS/CTS packets. The energy waste caused by idle listening is reduced by sleep schedules. Broadcast data packets do not use RTS/CTS which increases collision probability. It may incur overhearing if the packets are not destined to the listening node. Sleep and listen periods are predefined and constant, which decreases the efficiency of the algorithm under variable traffic load.

One of the major problems of S-MAC protocol is the possibility of following two different schedules of neighboring nodes. This results in more energy consumptions via idle listening and overhearing. In this article, we address these problems by extending S-MAC to our proposed I-MAC protocol.

SYSTEM ARCHITECTURE

This section discusses our proposed agent-based medium access protocol for wireless sensor network, which addresses the problem of border nodes in S-MAC. This resulted when nodes try to adopt different schedules. If the radio frequency (RF) signal of a node in a cluster overlaps with the RF signal of a node in another cluster, they should have the ability to communicate with each other. If the nodes adapt the schedule of the two or more neighboring nodes, it is called border node (BN), as shown in Figure 2. Border nodes lose more energy, as they have to wake up longer than the regular nodes in a cluster. They wake up with the wake up schedule of nodes in both clusters, causing more energy drain. This raises the threat of minimizing their lifetime in the sensor network.

A node can be restricted to adopt only one schedule of its neighbors. This will stop any node from becoming border node. However, using this approach a node will remain in only one cluster for all its lifetime until it expires. In our work, a multi-agent systems approach is proposed to deal with the problem of border node, as well to provide better energy-efficient cluster. This enables the BN to join a cluster that best fits it to provide more energy efficiency. The proposed multi-agent system architecture consists of two types of agents, including stationary monitoring agent and mobile mote agent, as shown in Figure 3.

The proposed system architecture shown in Figure 3 is designed for BN to find the optimal energy-efficient cluster at given times, prolonging their lives in WSN.

The stationary monitoring agent (SMA) closely monitors the mote activity and correspondingly updates its energy model. After the border node has stationed itself to the most energy-efficient cluster, the mobile mote agent comes in to action. MMA makes use of its mobile capability by visiting and querying the energy model of BN's neighboring nodes in other virtual clusters. Thus, it periodically updates the BN about the most energy-efficient cluster it can switch.

Figure 3. System architecture

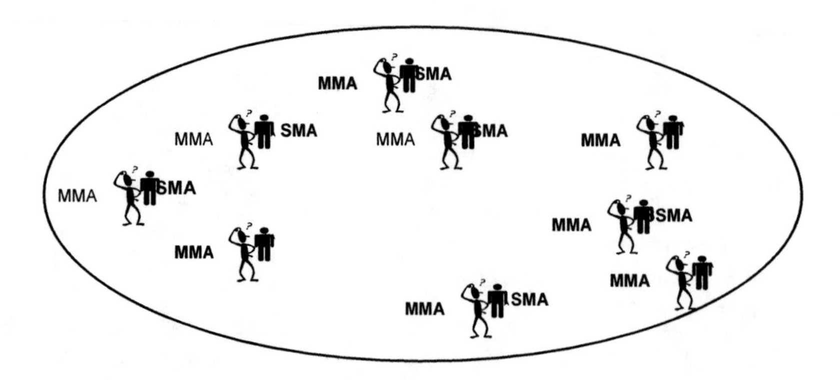

Figure 4. Agent's architecture

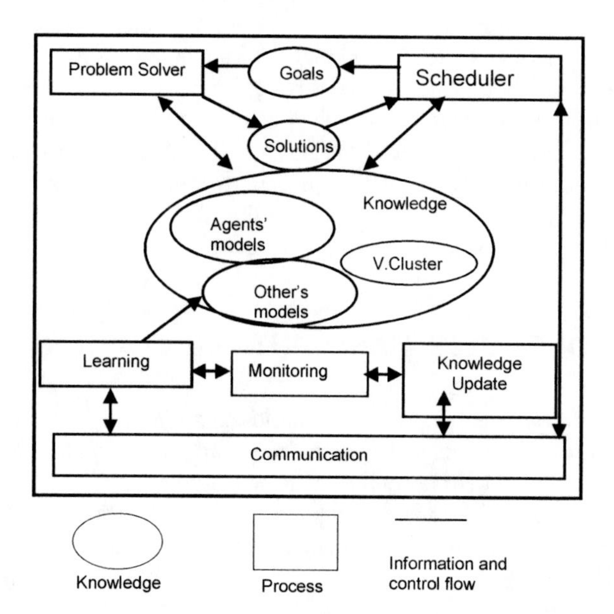

Figure 5. SMA's monitoring process

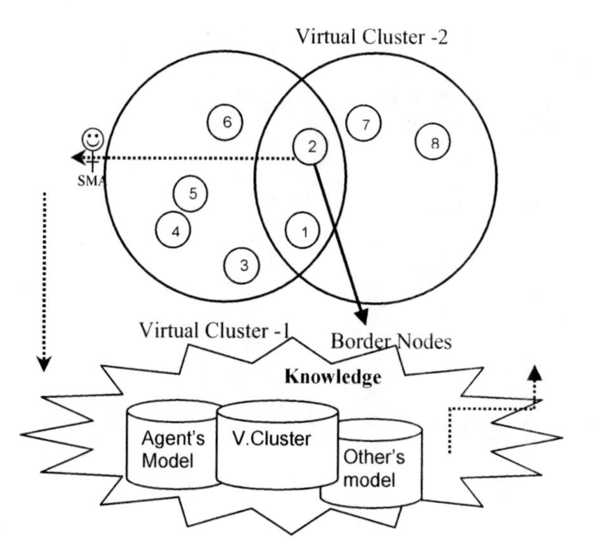

AGENT'S ARCHITECTURE

The architecture of all the proposed agents is based on the agent model proposed in Shakshuki, Ghenniwa, and Kamel (2003). Each agent poses the basic structure as shown in Figure 4, with the addition of more components based on functionality.

As shown in Figure 4, the agent architecture consists of knowledge components and executable components. The knowledge component contains the information about the WSN environment such as the number of cluster nodes it belongs to, its neighbor node, goals that need to be satisfied, and possible solutions generated to satisfy a goal. The learning component provides the agent with the capability to learn; it uses the monitored observations stored in its knowledge and runs machine-learning techniques, such as genetic algorithms, to know the energy-efficient cluster. The scheduler component provides the agent with a time agenda to start and stop certain activities such as monitoring and mobility. The communication component allows the agent to exchange messages with another agent and with an event occurring in a node. The two proposed agents (SMA and MMA) are the subject of the following two sections.

STATIONARY MONITORING AGENT

Each wireless sensor node running I-MAC protocol is equipped with a stationary monitoring agent. The SMA monitors the node and records each activity that results in loss of un-negligible energy drop. This includes transmitting data and receiving messages as data burst. The SMA closely monitors the activity of the node and records its observation into its knowledge. This includes the neighboring node ID with which the node is communicating, the virtual cluster to which the neighbor node belongs, and the amount of energy exhausted for an activity. SMA periodically updates the node's energy model.

MOBILE MOTE AGENT

The mobile mote agent resides on every node in the wireless sensor network and works closely with the SMA. The MMA has the ability to travel from one node to another and benefits from knowledge acquired by its SMA that resides in the same node. In addition to moving ability, it is also able to learn using learning techniques. Use of learning techniques makes MMA scalable. As energy is a vital issue in WSN, MMA helps to learn the best energy-efficient cluster for the border node. Indeed after the border node switch to the energy-efficient cluster, the MMA still continues visiting its neighbor nodes in other virtual clusters to find the most energy-efficient cluster.

Taking advantage of the knowledge acquired by SMA shown in Figure 6, the MMA discovers virtual cluster 1 to be more energy efficient for border node 2 than virtual cluster 1. It switches the border node 2 to virtual cluster 2. Node 2 no longer remains a border node, as shown in Figure 6.

Meanwhile the SMA updates others models by marking the neighbor of node 2 in a virtual cluster as inactive. Now, node 2 will only adopt the sleep schedule of virtual cluster 2,

Figure 6. MMA on move

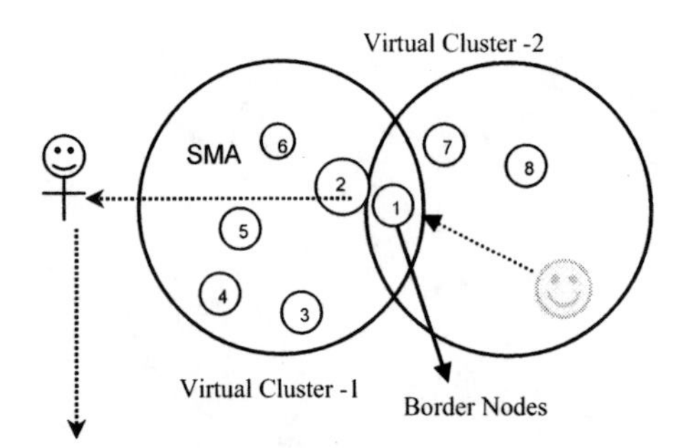

Id	cluster	Trans	Energy
1	1	RTS	0.00029J
2	1	CTS	0.00007J
2	1	ACK	0.0010 J

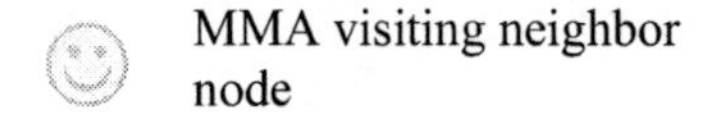

MMA not available at host

hence reducing its duty cycle. At this point all the monitoring events of the SMA on node 2 will belong to cluster 1.

Sensing for a particular region may increase or decrease by demand, due to the nature of the commercial application of sensor networks (Estrin, Govindan, Heidemann, & Kumar, 1999). Hence, it is vital to highlight that the node should always try to join the most energy-efficient virtual cluster at all times. Mobility of MMA now comes handy. It will visit node 2 neighbors in virtual cluster 2. It will determine the energy efficiency based on the knowledge acquired by the neighbor's SMA. In time, if the MMA finds virtual cluster 2 outperforming virtual cluster 1, it will switch node 2 to cluster 2. In doing so, the SMA will toggle the others models—that is, update by marking the neighbor in cluster 1 as inactive and in cluster 2 as active.

FUTURE TENDS

Although a number of MAC protocols have been proposed for sensor networks, several aspects still require further research. Some of the aspects and areas for further research are described as follows:

- Most existing MAC protocols for sensor networks do not support mobile sensors. Further research is needed to adapt existing MAC protocols to support mobility or to design new MAC protocols with mobility in mind.
- MAC protocols for sensor networks so far consider energy efficiency as a primary design goal. Although this is indisputably the main challenge in WSNs, future implementations and deployment need to consider performance issues such as reliable data delivery, reduced latency, and higher throughput.
- Although there are research efforts on real implementation testbeds, such work is relatively low compared to simulation-based performance evaluation of MAC protocols. In the future, performance evaluation should involve both simulation and experiments in implementation testbeds.

CONCLUSION AND FUTURE WORK

In this article, we presented a multi-agent-based system architecture to reduce the energy consumption among border nodes in S-MAC protocol. Our proposed system consists of two types of agents, including SMA and MMA. SMA monitors the events associated with a mote during communication and builds its knowledge. MMA benefits from the knowledge of SMA to predict the most energy-efficient cluster for border node at all times.

In the future, we plan to develop a simulator using Java. The environment of the simulator will include a testbed of Mica-2 motes running on Tiny OS. We also plan to deploy our agents that occupy minimal memory in motes and can easily transverse in a sensor network with fewer transmissions.

REFERENCES

Abramson, N. (1985). Development of the ALOHANET. *IEEE Transactions on Information Theory, 31,* 119-123.

Akyildiz, I. F., Su, W., Sankarasubramaniam, Y., & Cayirci, E. (2002). A survey on sensor networks. *IEEE Communications Magazine, 40*(8), 102-116.

Bennett, F., Clarke, D., Evans, B. J., Hopper, A., Jones, A., & Leask, D. (1999). Piconet: Embedded mobile networking. *IEEE Personal Communications Magazine, 4*(5), 8-15.

Estrin, D., Govindan, R., Heidemann, J., & Kumar, S. (1999). Next century challenges: Scalable coordination in sensor networks. *Proceedings of ACM MobiCom '99* (pp. 263-270), Washington, DC.

Heinzelman, W. R., Chandrakasan, A., & Balakrishnan, H. (2000, January). Energy-efficient communication protocols for wireless microsensor networks. In *Proceedings of the Hawaii International Conference on Systems Sciences* (vol. 8, pp. 8020-8030), Maui, HI.

Kleinrock, L., & Tobagi, F. (1975). Packet switching in radio channels: Part I—carrier sense multiple access modes and their throughput delay characteristics. *IEEE Transactions on Communications, 23*(12), 1400-1416.

LAN/MAN. (1999). *Wireless LAN medium access control (MAC) and physical layer (PHY) specification.* Standards Committee, IEEE Computer Society.

Rappaport, T. S. (1996). *Wireless communications: Principles and practice.* Englewood Cliffs, NJ: Prentice Hall.

Shakshuki, E., Ghenniwa, H., & Kamel, M. (2003). Agent-based system architecture for dynamic and open environments. *International Journal of Information Technology and Decision Making, 2*(1), 105-133.

Singh, S., & Raghavendra, C. S. (1998). PAMAS: Power aware multi-access protocol with signalling for ad hoc networks. *ACM Computer Communication, 28*(3), 5-26.

Sohrabi, K., & Pottie, J. G. (1999). Performance of a novel self organization protocol for wireless ad hoc sensor networks. *Proceedings of the IEEE 50th Vehicular Technology Conference* (pp. 1222-1226).

Stemm, M., & Katz, K. H. (1997). Measuring and reducing energy consumption of network interfaces in hand-held devices. *IEICE Transactions on Communications, 80*(8) 1125-1131.

Stojmenovic, I. (2006). Localized network layer protocols in sensor networks based on optimizing cost over progress ratio. *IEEE Network, 20*(1), 21-27.

Woo, A., & Culler, D. (2001). A transmission control scheme for media access in sensor networks. *Proceedings of the ACM/IEEE International Conference on Mobile Computing and Networking* (pp. 221-235), Rome.

Ye, W., & Heidemann, J. (2004). Medium access control in wireless sensor networks. In C. S. Raghavendra, K. Sivalingam, & T. Znati (Eds.), *Wireless sensor networks* (pp. 73-92). Kluwer Academic.

KEY TERMS

Energy Model: A theoretical construct that represents the energy consumption of a sensor mote by set of variables.

Mobile Mote Agent (MMA): A mobile agent that can roam around WSN. It can query the energy model of the sensor nodes.

Stationary Monitoring Agent (SMA): A static agent designed for monitoring the node activity and energy model in the WSN.

Virtual Cluster: Refers to the logical relation existing between a set of nodes based on their wake and sleep schedules. The virtual cluster is not dependant upon geographical boundaries.

Intelligent User Preference Detection for Product Brokering

Sheng-Uei Guan
Brunel University, UK

INTRODUCTION

A good business-to-consumer environment can be developed through the creation of intelligent software agents (Maes, 1994; Nwana & Ndumu, 1996, 1997; Bailey & Bakos, 1997; Soltysiak & Crabtree, 1998) to fulfill the needs of consumers patronizing online e-commerce stores. This includes intelligent filtering services (Chanan, 2000) and product brokering services to understand users' needs before alerting users of suitable products according to their needs and preferences.

We present a generic approach to capture individual user responding towards product attributes including non-quantifiable ones. The proposed solution does not generalize or stereotype user preference, but captures the user's unique taste and recommends a list of products to the user. Under the proposed generic approach, the system is able to handle the inclusion of any unaccounted attribute that is not predefined in the system, without re-programming the system. The system is able to cater for any unaccounted attribute through a general description field found in most product databases. This is extremely useful as hundreds of new attributes of products emerge each day, making any complex analysis impossible. In addition, the system is self-adjusting in nature and can adapt to changes in a user's preference.

BACKGROUND

Although there is a tremendous increase in e-commerce activities, technology in enhancing consumers' shopping experience remains primitive. Unlike real-life department stores, there are no sales assistants to aid consumers in selecting the most appropriate product for users. Consumers are further confused by the large options and varieties of goods available. Thus there is a need to provide, in addition to the provided filtering and search services (Bierwirth, 2000), an effective piece of software in the form of a product brokering agent to understand their needs and assist them in selecting suitable products.

Definitions

A user's choice in selecting a preferred product is often influenced by the product attributes that range from price to brand name. This research shall classify attributes as accounted, unaccounted, and detected. The same attributes may also be classified as quantifiable or non-quantifiable attributes.

Accounted attributes are predefined attributes that the system is specially catered to handle. A system may be designed to capture the user's choice in terms of price and brand name, making them accounted attributes. *Unaccounted attributes* have the opposite definition, and such attributes are not predefined in the ontology of the system. The system does not understand whether an unaccounted attribute represents a model or a brand name. Such attributes merely appear in the product description field of the database. The system will attempt to detect the unaccounted attributes that affect the user's preference and consider them as *detected attributes.* Thus detected attributes are unaccounted attributes that are detected to be vital in affecting the user's preference.

Quantifiable attributes contain specific numeric values (e.g., hard disk size), and thus their values are well defined. Non-quantifiable attributes on the other hand do not have any logical numeric values, and their valuation may differ from user to user (e.g., brand name).

The proposed system shall define price and quality of a product in the ontology and consider it to be quantifiable, accounted attributes. All other attributes defined in the system and considered as unaccounted attributes will be detected by the system.

Related Work

A lot of research and work has been done to aid transactions in electronic commerce. One of the research aims found is to understand a user's needs before recommending products through the use of product brokering services. Due to the difference in complexity, different approaches are proposed to handle quantifiable and non-quantifiable attributes.

One of the main approaches to handle quantifiable attributes is to compile these attributes and assign weights representing their relative importance to the user (Guan, Ngoo, & Zhu, 2002; Zhu & Guan, 2001; Sheth & Maes, 1993). The weights are adjusted to reflect the user's preference.

Much research is aimed at creating an interface to understand user preference in terms of non-quantifiable attributes. This represents a more complex problem as attributes are highly subjective with no discrete quantity to measure their

Figure 1. System flow diagram

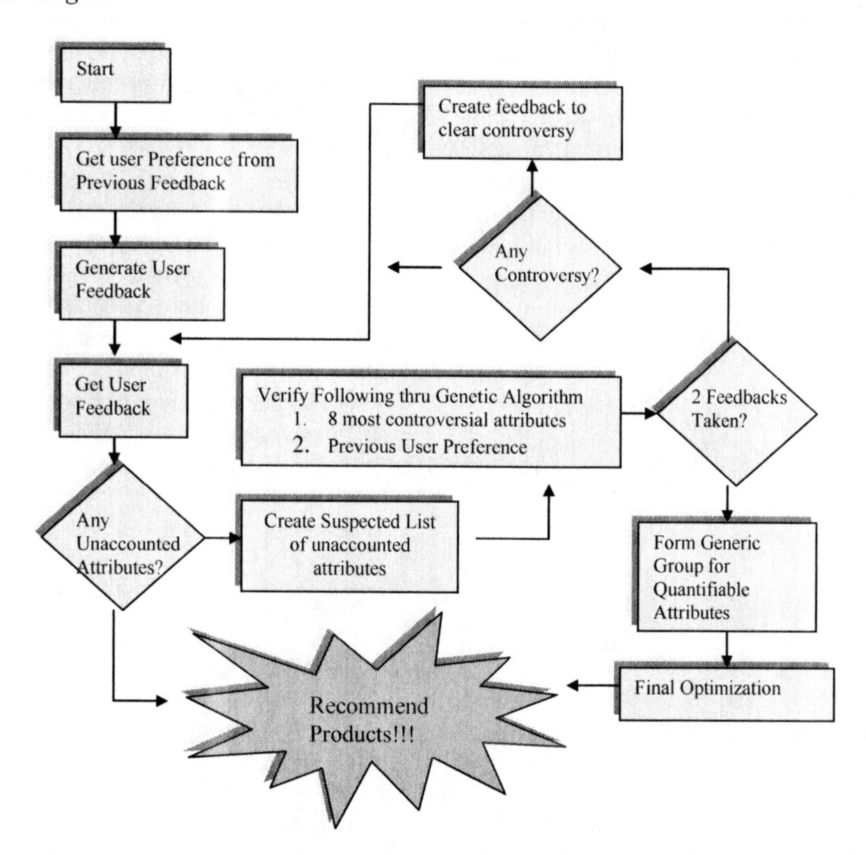

values. Different users will give different values to a particular attribute. MARI (multi-attribute resource intermediary) proposed a "word-of-mouth approach" to solve this problem. The project split up users into general groups and estimated their preference to a specific set of attributes through the group the user belongs to.

Another approach in the handling of non-quantifiable attributes involves specifically requesting the user for the preferred attributes. Shearin and Liberman (2001) provided a learning tool for the user to explore his or her preference before requesting him or her to suggest desirable attributes.

Some of the main problems in related work lie in the handling of non-quantifiable attributes, as the approaches are too general. Most work so far only attempts to understand user preference through generalization and stereotyping instead of understanding specific user needs. Another main problem is that most works are only able to handle a specific set of attributes. The attributes that they are able to handle are hard-coded into the design of the system, and the consequence is that they are not able to handle attributes that are unaccounted and beyond the pre-defined list. However, the list of product attributes is often large, possibly infinite. The approach used in related research may not be able to cover all the attributes, as they need to classify them into the ontology.

DESCRIPTION OF INTELLIGENT USER PREFERENCE DETECTION

The proposed approach attempts to capture user preference on the basis of two quantifiable accounted attributes, price and quality. It incrementally learns and detects any unaccounted attribute that affects the user's preference. If any unaccounted attribute is suspected, the system attempts to come up with a list of highly suspicious attributes and verify their importance through a genetic algorithm (Haupt & Haupt, 1998). Thus vital attributes that are unaccounted for previously will be considered. The unaccounted attributes are derived from the general description field of a product. The approach is therefore generic in nature, as the system is not restricted by the attributes it is designed to cater to.

Overall Procedure

The overall procedure is as shown in Figure 1. As the system is able to incrementally detect the attributes that affect user preference, it first retrieves any information captured regarding the user from some previous feedback and generates feedback in the form of a list of products for the user to rank, and attempts to investigate the presence of any unaccounted attribute affecting the user's preference. The system shall compile a list of possible attributes that are unaccounted for by analyzing the user feedback and rank them according to their suspicion levels. The most suspicious attributes and any information captured from previous feedback are then verified through a genetic algorithm. If two cycles of feedback are completed, the system attempts to detect any quantifiable attributes that are able to form a generic group of attributes. The system finally optimizes all information accumulated by a genetic algorithm and recommends a list of products for the user according to the preference captured.

Tangible Score

In our application, we shall consider two quantifiable attributes, price and quality, as the basis in deriving the tangible score. The effects of these two attributes are always accounted for. The equation to derive this score is as shown in equations 1-3.

Equation 1 measures the price competitiveness of the product. PrefWeight is the weight or importance the user places on price competitiveness as compared to quality with values ranging from 0 to 1.0. A value of 1.0 shows that 100% of the user's preference is based on price competitiveness. A product with a price close to the most expensive product will have a low score in terms of price competitiveness and vice versa.

Equation 2 measures the score given to quality. The quality attribute measures the quality of the product and takes a value ranging from 0.0 to 1.0. The value of "1.0 – PrefWeight" measures the importance of quality to the user.

The final score given to tangible attributes are computed by adding equations 1 and 2 as shown in equation 3 as follows:

$$\text{TangibleScore} = \text{Score}_{\text{PriceCompetitive}} + \text{Score}_{\text{Quality}} \quad (3)$$

$$\text{Modification Score} = \left(\sum_{i=1}^{\text{NoOfAttributes}} K_i - 1\right) * \text{TangibleScore} \quad (4)$$

Modification Score for Detected Attributes

The modification score is the score assigned to all detected attributes by the system. These detected attributes are previously unaccounted attributes, but had been detected by the system to be a vital attribute in the user's preference. These include all other attributes besides price and quality. As these attributes may not have a quantifiable value, the score is taken as a factor of the TangibleScore derived earlier. The modification score is as shown in equation 4 whereby the modification factor K is introduced.

The values of each modification factor K range between 0.0 and 2.0. A value of K shall be assigned for each newly detected attribute (e.g., each brand name will have a distinct value of K). The modification factor K takes a default value of 1.0 that gives a modification score of 0. Such a situation arises when the detected attribute does not affect the user's choice. When K<1.0, we will have a negative or penalty score for the particular attribute. This will take place when the user dislikes products from a certain brand name or other detected attributes. When K >1.0 we will have a bonus score to the attributes, and it takes place when the user has special positive preference towards certain attributes. By using a summation sign as shown in equation 4, we are considering the combined effects of all the attributes that are previously unaccounted by the system.

With all new attributes captured, the final score for the product is as shown in equation (5) as the summation of tangible and modification scores.

$$\text{Final Score} = \text{TangibleScore} + \text{Modification Score} \quad (5)$$

A Ranking System for User Feedback

As shown in equations 1 and 2 earlier, there is a need to capture the user's preference in terms of the PrefWeight in equation 1 and the various modification factor K in equation 4. The system shall request the user to rank a list of products as shown in Figure 2. The user is able to rank the products according to his or her preference with the up and down button and submit when done. The system shall make use of this ranked list to assess a best value for PrefWeight in equations 1 and 2. In a case whereby no unaccounted attributes affect the user's feedback, the agents will be evolved along the PrefWeight gradient to optimize a value for the PrefWeight.

Fitness of Agents

The fitness of each agent shall depend on the similarity between the agent's ranking of the product and the ranking made by the user. It reflects the fitness of agents in capturing the user's preference.

Figure 2. Requesting the user to rank a list of products

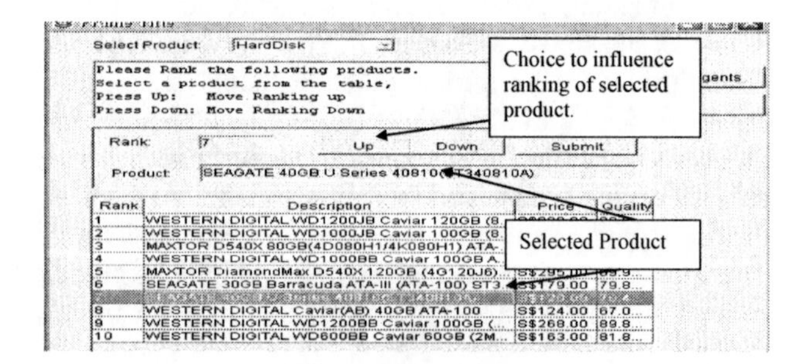

Unaccounted Attribute Detection

To demonstrate the system's ability to detect unaccounted attributed, the ontology shall contain only price and quality, while all other attributes are unaccounted and remain to be detected, if they are vital to the user. These unaccounted attributes include non-quantifiable attributes that are subjective in nature (e.g., brand name). The unaccounted attributes can be retrieved by analyzing the description field of a product database, thus allowing new attributes to be included without the need of change in ontology or system design.

The system firstly goes through a detection stage where it comes up with a list of attributes that affect the user's preference. These attributes are considered as unaccounted attributes as the system has not accounted for them during this stage. A *confidence score* is assigned to each attribute according to the possibility of it being the governing attribute influencing the user's preference.

The system shall request the user to rank a list of products and analyze the feedback according to the process shown in Figure 3. The agents shall attempt to explain the rankings by optimizing the PriceWeight value and various K values. The fittest agent shall give each product a score.

The system shall loop through the 10 products that are ranked by the user and compare the score given to products. If the user ranks a product higher than another, this product should have a higher score than a lower ranked product. However if the agent awards a higher score to a product ranked lower than another (e.g., product ranked 2nd has higher score than 3rd), the product is deemed to contain an unaccounted attribute causing an illogical ranking. This process shall be able to identify all products containing positive unaccounted attributes that the user has preference for.

The next step is to identify the unaccounted attributes inside these products that give rise to such illogical rankings. The products with illogical rankings are tokenized. Each word in the product description field is considered as a possible unaccounted attribute affecting the user's preference. Each of the tokens is considered as a possible attribute affecting the user's taste. The system shall next analyze the situation and modify the confidence score according to the cases as shown.

1. The token appears in other products and shows no illogical ranking: deduction of points.
2. The token appears in other products and shows illogical ranking: addition of points.

The design above only provides an estimate on the Confidence Points according the two cases described and may not be 100% reliable.

Confirmation of Attributes

Attributes captured in previous feedback may be relevant in the current feedback as the user may choose to provide more than one set of feedback. The system thus makes a hypothesis that the user's preference is influenced by attributes affecting him in previous feedback if available and eight other new attributes with the highest confidence score. The effect of these attributes on the user's preference is verified next.

Each agent in the system shall make estimation on the user's preference by randomly assigning a modification factor (or "K" value) for each of the eight attributes with high confidence score. Attributes identified to be positive are given K values greater than 1.0, while negative attributes have K values less than 1.0. The K values and PrefWeight are optimized by a genetic algorithm to improve the fitness level of the agents.

The remaining attributes undergo another filtering process whereby redundant attributes that do not affect the agent's fitness are filtered off.

Optimization Using Genetic Algorithm

The status of detected attributes perceived by the agents and the most suspicious attributes should be verified here. The PrefWeight and various K values shall be optimized here to

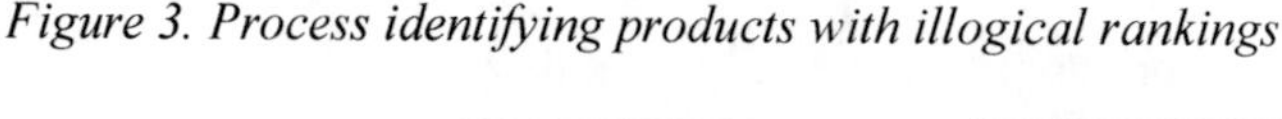

Figure 3. Process identifying products with illogical rankings

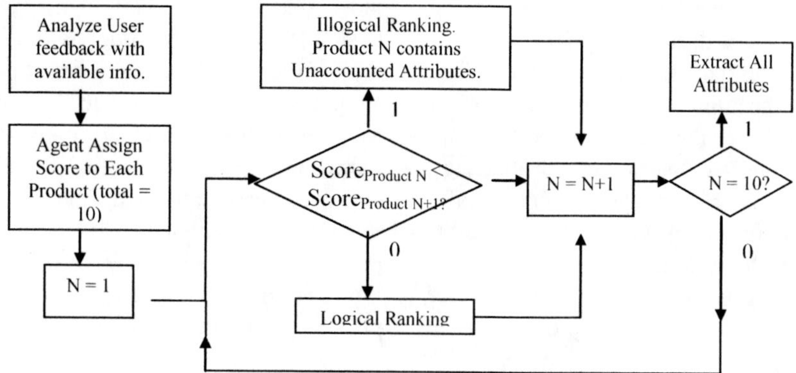

Figure 4. Chromosome encoding

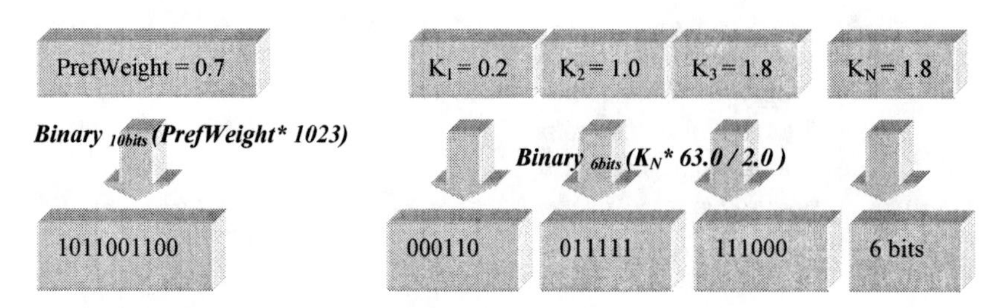

Figure 5. One set of feedback consisting of 2 feedback cycles

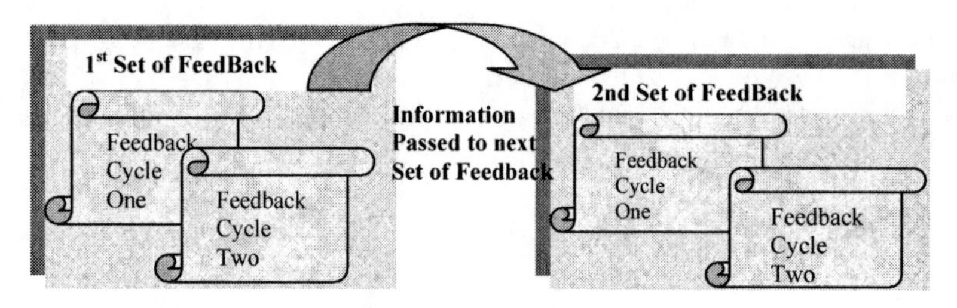

produce maximum agent fitness. As this represents a multi-dimensional problem (Osyczka, 2001) with each new K value introducing a new dimension, we shall optimize using a genetic algorithm that converts the attributes into binary strings. The agents shall evolve under the genetic algorithm to optimize the fitness of each agent. The PrefWeight and various K values are optimized in this algorithm.

The attributes of each agent are converted into a binary string as shown in Figure 4, and each binary bit represents a chromosome. In the design, 10 bits of data are used to represent the PrefWeight, while five bits of data are used to represent the various K values.

An Incremental Detection System and Overall Feedback Design

The system takes an incremental detection approach in understanding user preference, and the results show success in analyzing the complex user preference situation. The system acknowledges that not all vital attributes may be captured within one set of feedback and thus considers the results of previous sets. The attributes that affect a user's preference in one feedback become the prime candidates in the next set of feedback. In this way, the attributes that are detected are preserved and verified, while new unaccounted attributes are being detected, allowing the software agents incrementally to learn about the attributes that affect the user's preference. However some of the information captured by the system may be incorrect or no longer valid as the number of feedback cycles increase. This creates a problem in the incremental detection system, as the information may not be relevant. To solve this problem, the system checks the validity of past attributes influencing the user's preference and deletes attributes that are no longer relevant in the current feedback. Every set of feedback contains two feedback cycles as shown in Figure 6.

Both feedback cycles will attempt to detect the presence of any unaccounted attributes. In addition, the first cycle shall delete any attributes that are passed from previous feedback and no longer relevant. These attributes should have a K value of 1.0 after we apply the genetic algorithm discussed earlier. Any controversial attributes detected by the first cycle shall be clarified using the second feedback cycle.

IMPACT OF INTELLIGENT USER PREFERENCE DETECTION

A prototype was created to simulate the product broker. An independent program is written and run in the background to simulate a user. This program is used to provide feedback to the system and ranks the list of products on behalf of a simulated user who is affected by price and quality, as well as a list of unaccounted attributes. The system is also affected by some generic groups of quantifiable attributes. It was observed that the performance of the system is closely related to the complexity of the problem. More complex problems will give a lower overall performance as shown in the various cases. However, this is greatly alleviated by providing multiple sets of feedback. The system incrementally detected attributes affecting the user's preference, and in the cases shown, the gap in performance was negligible.

The system also demonstrated its ability to adapt to changes in consumer preference. This is extremely important when multiple sets of feedback are involved, as the user's preference may vary between feedback cycles. It also demonstrated the system's ability to correct its own mistakes and search for a better solution.

CONCLUSION

In this article, we demonstrated a solution in the handling of previously unaccounted attributes without the need for change in the ontology or database design. The results showed that the system designed is indeed capable of understanding the user's needs and preferences even when previously unknown or unaccounted attributes were present. The system is also able to handle the presence of multiple unaccounted attributes and classify quantifiable attributes into a generic group of unaccounted attributes.

In addition, the system demonstrated the power of incremental detection of unaccounted attributes by passing the detected attributes from within one feedback to the other.

FUTURE TRENDS

The current system generated user feedback to clarify its doubts on suspicious attributes. However, more than half of the feedback was generated in random to increase the chances of capturing new attributes. This random feedback was generated with products of different brand names having equal chances of being selected to add to the variety of the products used for feedback. This could be improved by generating feedback to test certain popular attributes to increase the detection capabilities.

REFERENCES

Bailey, J. P., & Bakos, Y. (1997). An exploratory study of the emerging role of electronic intermediates. *International Journal of Electronic Commerce, 1.1*(3), 7-20.

Bierwirth, C. (2000). *Adaptive search and the management of logistic systems—base models for learning agents.* Kluwer Academic.

Chanan, G., & Yadav, S. B. (2001). A conceptual model of an intelligent catalog search system. *Journal of Organizational and Electronic Commerce, 11*(1), 31-46.

Guan, S. U., Ngoo, C. S., & Zhu, F. (2002). HandyBroker—An intelligent product-brokering agent for m-commerce applications with user preference tracking. *Electronic Commerce and Research Applications, 1*(3-4), 314-330.

Haupt, R. L., & Haupt, S. E. (1998). *Practical genetic algorithm.* New York: John Wiley & Sons.

Maes, P. (1994). Agents that reduce work and information overload. *Communications of the ACM, 37*(7), 30-40.

MIT. (n.d.). Retrieved from http://www.media.mit.edu/get-wari/MARI/

Nwana, H. S., & Ndumu, D. T. (1996). An introduction to agent technology. *BT Technology Journal, 14*(4), 55-67.

Nwana, H. S., & Ndumu, D. T. (1997). Research and development challenges for agent-based systems. *IEEE Proceedings on Software Engineering, 144*(1), 2-10.

Osyczka, A. (2001). *Evolutionary algorithms for single and multicriteria design optimization.* Physica-Verlag.

Shearin, S., & Lieberman, H. (2001). Intelligent profiling by example. In *Proceedings of the International Conference on Intelligent User Interfaces* (Vol. 1, pp. 145-151), Santa Fe, NM.

Sheth, B., & Maes, P. (1993). Evolving agents for personalized information filtering. *Proceedings of the 9th Conference on Artificial Intelligence for Applications* (pp. 345-352). IEEE Press.

Soltysiak, S., & Crabtree, B. (1998). Automatic learning of user profiles—Towards the personalization of agent services. *BT Technology Journal, 16*(3), 110-117.

Zhu, F. M., & Guan, S. U. (2001). Evolving software agents in e-commerce with GP operators and knowledge exchange. *Proceedings of the 2001 IEEE Systems, Man and Cybernetics Conference.*

KEY TERMS

E-Commerce: Consists primarily of the distributing, buying, selling, marketing, and servicing of products or services over electronic systems such as the Internet and other computer networks.

Genetic Algorithm: Search technique used in computer science to find approximate solutions to optimization and search problems. Genetic algorithms are a particular class of evolutionary algorithms that use techniques inspired by evolutionary biology such as inheritance, mutation, natural selection, and recombination (or crossover).

Ontology: Studies being or existence and their basic categories and relationships, to determine what entities and what types of entities exist.

Product Brokering: A broker is a party that mediates between a buyer and a seller.

Software Agent: An abstraction, a logical model that describes software that acts for a user or other program in a relationship of agency.

Interactive Multimedia File Sharing Using Bluetooth

Danilo Freire de Souza Santos
Federal University of Campina Grande, Brazil

José Luís do Nascimento
Federal University of Campina Grande, Brazil

Hyggo Almeida
Federal University of Campina Grande, Brazil

Angelo Perkusich
Federal University of Campina Grande, Brazil

INTRODUCTION

In the past few years, industry has introduced cellular phones with increasing processing capabilities and powerful wireless communication technologies. These wireless technologies provide the user with mechanisms to easily access services, enabling file sharing among devices with the same technology interfaces (Mallick, 2003). In the context of electronic commerce, which demands new techniques and technologies to attract consumers, these wireless technologies aim to simplify the shopping process and provide up-to-date information about available products.

In order to exemplify the application of mobile and wireless technologies to satisfy these new commerce functionalities and needs, we present in this article the interactive multimedia system (IMS). IMS is a system for sharing multimedia files between servers running on PCs, and client applications running on mobile devices. The system was conceived initially to be deployed in CDs/DVDs and rental stores to make available product information in a simple and interactive way.

In a general way, the system allows a user to obtain information about available products through a mobile device. Then, a user can listen or watch parts (stretches) of available videos or songs. For that, the user needs to enter the store, choose a product in the store shelf, and type its identity code in the mobile device, choosing which music (or video) to listen to (or watch).

The IMS system has a client/server architecture, where the server was developed in C++ for the Windows operating system and the client application was developed in C++ for the Symbian operating system, which is a mobile device operating system mainly used in smart phones. Client/server communication is performed based on Bluetooth wireless technology. Bluetooth is suitable for this kind of application because it has a satisfactory transmission rate with enough range, and it is also supported by more than 500 million mobile devices (Bluetooth Official Website, 2006).

The rest of this article is organized as follows. In next section we present a background of the main technologies used in this project. We then present the architecture of the proposed system and describe how the system works, before discussing future trends of mobile multimedia systems and offering final remarks.

BACKGROUND

This section provides an overview of the main technologies used in the IMS development. More specifically, we outline the Bluetooth wireless technology and the programming language used with the Symbian Operational System.

Bluetooth

To provide communication between devices, the IMS client/server architecture uses the Bluetooth wireless technology. Bluetooth is a short-range wireless technology present in a large number of smart phones of the Symbian OS Series 60 platform. It is suitable for fast file exchange, including text files, photo files, and short video files. Bluetooth technology covers a distance of about 10 meters for class 2 devices (most common devices), and each server (or master) can be connected to up to seven slaves in its coverage area (Mallick, 2003). Another important feature of Bluetooth is its lower power consumption, around 2.5 mW at most, which reinforces its use in embedded devices.

With Bluetooth, it is possible to use two kinds of connections: ACL (asynchronous connectionless) and SCO (synchronous connection oriented) (Andersson, 2001). ACL links are defined for data transmission, supporting sym-

Figure 1. Symbian OS architecture

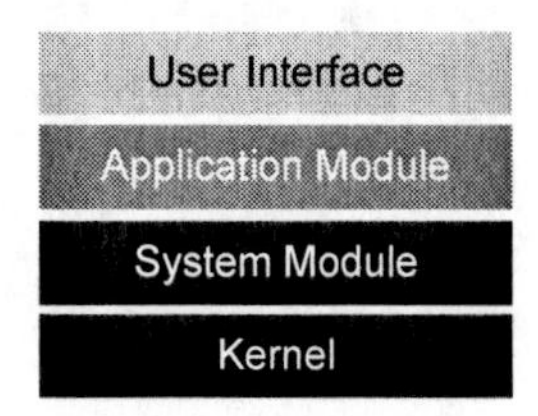

Figure 2. IMS general view

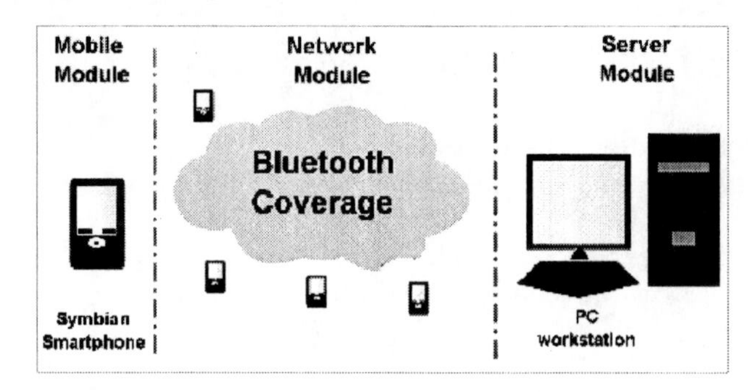

metrical and asymmetrical packet-switched connections. In this mode, the maximum data rate could be 723 kbps in one direction and 57.6 kbps in the other direction, and these rates are controlled by the master of a cell. SCO links support only a symmetrical, circuit-switched, point-to-point connection for primarily voice traffic. The data rate for SCO links is limited to 64 kbps, and the number of devices connected at the same time with the master is restricted to three devices.

In the Bluetooth protocol stack, some profiles that implement some kind of particular communication partner are defined. The general profiles in Bluetooth stack are: GAP (generic access profile), SDAP (service discovery application profile), SPP (serial port profile), and GOEXP (generic object exchange profile) (Forum Nokia, 2003). GAP defines generic procedures related to discovery of Bluetooth devices and links management aspects of connecting Bluetooth devices. SDAP defines features and procedures to allow an application in a Bluetooth device to discover services of another Bluetooth device. With SPP used in ACL links, it is possible to emulate serial cable connections using RFCOMM (RS232 Serial Cable Emulation Profile) between two peer devices. RFCOMM emulates RS-232 (Serial Cable Interface Specification) signals and can thus be used in applications that are formerly implemented with a serial cable. The GOEXP profile defines protocols and procedures that should be used by applications requiring object exchange capabilities.

Symbian

Symbian is an operating system specifically designed for mobile devices with limited resources, such as memory and processor performance. The programming language C++ for Symbian provides a specific API (application program interface), with new features for the programmer that allow access to services such as telephony and messaging (Stichbury, 2004). Also, the Symbian C++ API enables programmers to efficient deal with multitasking and memory functions. These functions reduce memory-intensive operations. Symbian OS is event driven rather than multi-thread. Although multi-thread operations are possible, they potentially create kilobytes of overhead per thread. Services in Symbian OS are provided by servers through client/server architectures. For developing applications, Symbian offers an application framework, which constitutes a set of core classes that are the basis and structure of all applications.

The Symbian OS architecture can be described by a layered approach, as illustrated in Figure 1.

The layers can be defined as follows:

- **User Interface (UI):** Can be specifically defined per vendor or per family of mobile devices, such as Series 60 platform devices;
- **Application Module:** Allows access to applications' built-in functionality concerned with data processing and not how it is presented for the user;
- **System Module:** Contains the set of OS APIs; and
- **Kernel:** The core of the operational system and cannot be directly accessed by user programs.

The mobile application was implemented using a Series 60 platform (Series 60 Website, 2006; Edwards & Barker, 2004). Series 60 is a complete smart phone-based UI design reference. It completes the Symbian OS architecture with a configurable graphical user interface library and a suite of applications, besides other general-purpose engines.

SYSTEM ARCHITECTURE

This section presents the IMS architecture. As introduced earlier, the IMS system has a client/server architecture, where the server was developed in C++ for the Microsoft Windows OS and the client application was developed in C++ for Symbian OS. Client/server communication is performed based on Bluetooth wireless technology, which can provide connection to up to seven users at the same time. Figure 2 illustrates a general system view.

According to this general view, the system can be divided into three specific modules: mobile, network, and server. The mobile module is composed of the software running on mobile devices, such as smart phones. It offers a friendly and intuitive user interface, which is responsible

Figure 3. IMS screenshots

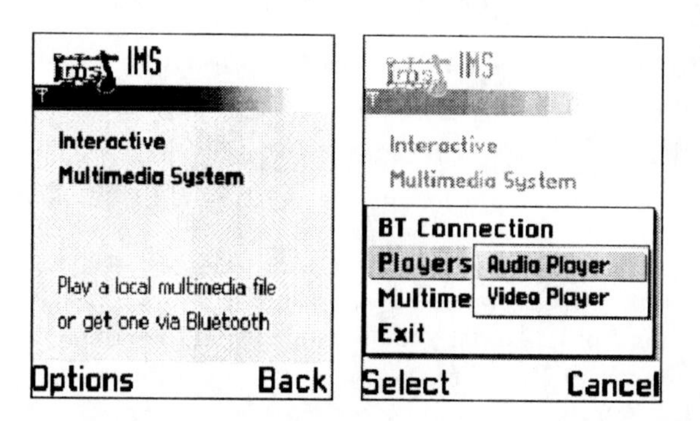

for showing relevant information about available products. This information is obtained by typing the product code into the device. Then, the product description is returned with an option to run the multimedia file related to the product. Also, the mobile module controls a multimedia player installed together with the IMS software. Screenshots of the IMS mobile application are presented in Figure 3.

The network module controls the mobile network, in this case the Bluetooth wireless technology. To handle the connection, the network module offers an application layer protocol to manage the exchange of messages between the other two modules. This protocol is text based. Also, this module is responsible to control the transfer of files using the Bluetooth Serial Port Profile (SPP). The SPP profile is supported by Symbian (and Series 60 Platform) Bluetooth-enabled devices (Jipping, 2003). The system can support more than seven users by sharing the available Bluetooth communication ports at server side. This way, after using one communication port with one user, the network module switches the port to another user if necessary.

Finally, the server module is responsible for database and system management. Files and relevant information (description of video or song, time, author, etc.) are stored into a database. When required, the server sends relevant information about a specific product. After confirmation from the mobile software, the server sends the multimedia file.

SYSTEM EXECUTION

In this section we describe how IMS works. For this execution scenario, consider that the client has entered the IMS Bluetooth coverage area. Then, he/she has chosen a product to retrieve information (such as a DVD or CD) available at the store shelf. After that, as illustrated in Figure 4, the following steps are performed:

1. The user types the product identification code into the mobile device software. Then, the software running on the mobile device searches for the IMS server using Bluetooth technology. After finding the IMS server, the IMS network module establishes the connection between the mobile device and the server. Using a specific application protocol, the IMS software asks for information about that product code, such as audio/video stretches, release notes, and so forth.
2. If the information is available, the server module retrieves it from the database and returns it to the mobile device through the Bluetooth link. Then, the IMS software receives that information and displays it to the client, who will have available the description and the option to run a multimedia file.
3. When the client chooses to run the file, it is downloaded through the Bluetooth link and executed using a multimedia player managed by the IMS software.

Figure 4. IMS operation

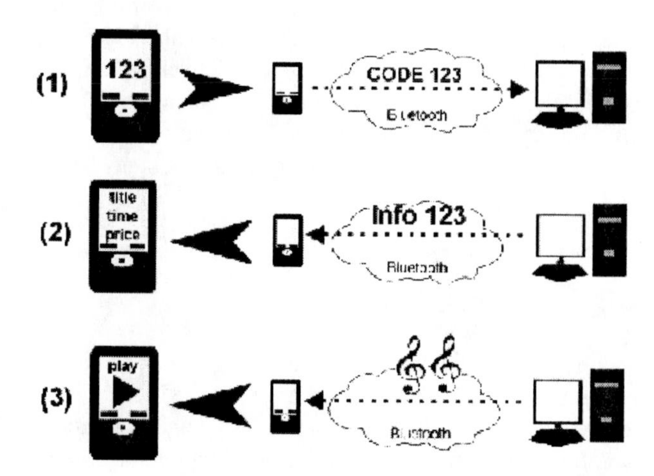

FUTURE TRENDS

Today, the mobile multimedia area tends to investigate and develop mechanisms for carrying multimedia streams over wireless links, as described by Chen, Kapoor, Lee, Sanadidi, and Gerla (2004). Mobile multimedia systems are focused on providing multimedia streams directly to the mobile device in an efficient way. Future interactive multimedia systems will offer real-time multimedia streams, with more general services such as IPTV (television over IP) (Santos, Souto, Almeida, and Perkusich, 2006), radio, and so forth. Also, with the advent of smart phones with different wireless interfaces such as Wi-Fi and Bluetooth, those systems will support heterogeneous technologies offering different kinds of services for different kinds of devices.

Another prominent future trend is related to pervasive computing (Weiser, 1991; Saha & Mukherjee, 2003; Satyanarayanan, 2001). Within this context, users with mobile and wireless technology access can define their personal preferences to adapt systems and environments according to it. IMS, in the future, could be used in a pervasive infrastructure enabling file sharing according to personal preferences of users. IMS can be primarily used to delivery file in the pervasive environment, and in the future can be associated with variable bit rate multimedia streaming according to battery life of a mobile device or any kind of strategy based on the user's profile.

CONCLUSION

Mobile multimedia systems are becoming reality, and more and more accessible to ordinary users. These users, as potential consumers, are attracting industry and commerce attention, motivating research and development of several solutions in this area. Within this context, one of the relevant efforts is to promote multimedia file sharing.

To illustrate the large application of multimedia file sharing to support consumer activities, we presented in this article an interactive multimedia system. IMS provides attractive and interactive ways for users to obtain product information, which can be used in several commerce domains.

Interactive multimedia systems, such as IMS, are very useful and have great relevance for multimedia commerce. They could substitute traditional multimedia players installed in stores, offering a new way of interacting with clients using mobile devices.

REFERENCES

Andersson, C. (2001). *GPRS and 3G wireless applications.* New York: John Wiley & Sons.

Bluetooth Official Website. (2006). *Bluetooth technology benefits.* Retrieved April 28, 2006, from http://www.bluetooth.com/Bluetooth/Learn/Benefits/

Chen, L., Kapoor, R., Lee, K., Sanadidi, M.Y., & Gerla, M. (2004). Audio streaming over Bluetooth: An adaptive ARQ timeout approach. *Proceedings of the 24th International Conference on Distributed Computing Systems Workshops (ICDCSW'04).*

Edwards, L., & Barker, R. (2004). *Developing Series 60 applications—a guide for Symbian OS C++ developers.* Nokia Mobile Developer Series. Boston: Addison-Wesley.

Forum Nokia. (2003, April 4). *Bluetooth technology overview.* Retrieved March 1, 2005, from http://www.forum.nokia.com

Jipping, M. (2003). *Symbian OS communications programming.* New York: John Wiley & Sons.

Mallick, M. (2003). *Mobile and wireless design essentials.* New York: John Wiley & Sons.

Saha, D., & Mukherjee, A. (2003, March). Pervasive computing: A paradigm for the 21st century. *IEEE Computer,* 25-31.

Santos, D.F.S., Souto, S.F., Almeida, H., & Perkusich, A. (2006, April). An IPTV architecture using free software (in Portuguese). *Proceedings of the Brazilian Computer Society's Free Software Workshop in the International Free Software Forum,* Porto Alegre, Brazil.

Satyanarayanan, M. (2001). Pervasive computing: Vision and challenges. *IEEE Personal Communications, 8*(4).

Series 60 Web Site. (2006). *S60—about Series 60.* Retrieved April 28, 2006, from http://www.s60.com/about

Stichbury, J. (2004). *Symbian OS explained—effective C++ programming for smart phones.* New York: John Wiley & Sons.

Weiser, M. (1991). The computer for the 21st century. *Scientific American, 265*(3), 94-104.

KEY TERMS

ACL: Asynchronous connectionless.

API: Application program interface.

Bluetooth Link: Connection between two Bluetooth peers.

GAP: Generic access profile.

GOEXP: Generic object exchange profile.

IMS: Interactive multimedia system.

OS: Operational system.

SCO: Synchronous connection oriented.

SDAP: Service discovery application profile.

Smart Phone: Cell phone with special computer-enabled features.

SPP: Serial port profile.

UI: User interface.

Interactive Product Catalog for M-Commerce

Sheng-Uei Guan
Brunel University, UK

Yuan Sherng Tay
National University of Singapore, Singapore

INTRODUCTION

We propose a product catalog where browsing is directed by an integrated recommender system. The recommender system is to take incremental feedback in return for browsing assistance. Product appearance in the catalog will be dynamically determined at runtime based on user preference detected by the recommender system. The design of our hybrid m-commerce catalog-recommender system investigated the typical constraints of m-commerce applications to conceptualize a suitable catalog interface. The scope was restricted to the case of having a personal digital assistant (PDA) as the mobile device. Thereafter, a preference detection technique was developed to serve as the recommender layer of the system.

BACKGROUND

M-commerce possesses two distinctive characteristics that distinguish it from traditional e-commerce: the mobile setting and the small form factor of mobile devices. Of these, the size of a mobile device will remain largely unchanged due to the tradeoff between size and portability. Small screen size and limited input capabilities pose a great challenge for developers to conceptualize user interfaces that have good usability while working within the size constraints of the device.

In response to the limited screen size of mobile devices, there has been unspoken consensus that certain tools must be made available to aid users in coping with the relatively large volume of information. Recommender systems have been proposed to narrow down choices before presenting them to the user (Feldman, 2000).

Catalog Browsing

In one study, a new user behavior, termed *opportunistic exploration,* has been identified, where users have multiple, ill-defined overlapping interests (Bryan & Gershman, 1999). Throughout the course of browsing, exposure to items affect interests, and interest may evolve due to exposure or whim. In Tateson and Bonsma (2003), the emphasis was that the paradigm of online shopping is fundamentally different from that of information retrieval.

Despite the importance of having a well-designed online catalog that supports the shopping behavior of users, the challenge of including such browsing capabilities in m-commerce is great, given that the small screen size of mobile devices severely limits the number of products that may be presented on-screen.

The predominant strategy of organizing products into narrow categories has many problems (Lee, Lee, & Wang, 2004). The alternative solution of interactive catalogs (Tateson & Bonsma, 2003) allows for fluid navigation of the product space, whereby users are given the freedom to redirect the browsing process as and when their interests change.

Recommender System

Recommender systems perform the role of sales agents by first understanding a user's preferences through querying and profiling, and subsequently presenting information or products of relevance to the user (Schafer, Konstan, & Riedl, 2001). Recommender systems have long been regarded as a highly desirable feature of e-commerce.

Currently there are numerous ongoing studies to improve recommender technology in the context of e-commerce (Konsten, 2004; Montaner, Lopez, & Lluis, 2003). However, the approaches of such studies are seldom directly applicable to the domain of m-commerce. With respect to the m-commerce constraints, a "best effort" recommender system that makes do with whatever information is available will serve as an interesting alternative to the "best quality" emphasis of current recommendation technology.

DESCRIPTION OF INTERACTIVE CATALOG

The interface of a catalog is divided into three components: visual presentation, browsing process, and feedback mechanism.

Figure 1. Screenshot

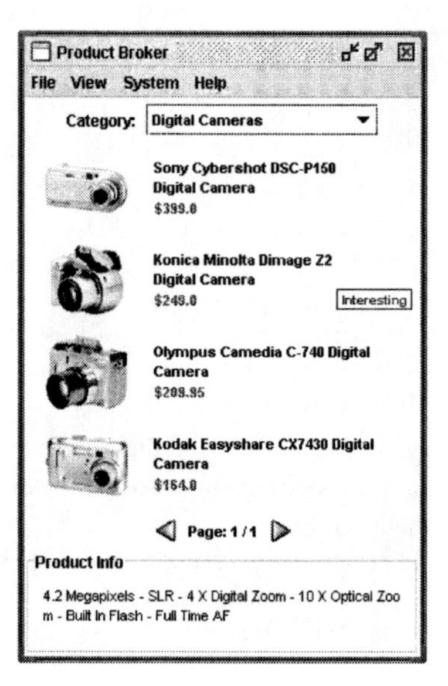

Presentation

Given the constraint of a PDA screen, the main concern of our design is to maximize emphasis on product presentation while simplifying the control elements. Human cognition is more adapted to the processing of visual images as compared to textual information (Lee et al., 2004). Visual elements are thus useful mechanisms to improve the usability of a catalog. To save space while facilitating easy examination of products, we incorporate a product information panel. Figure 1 shows a screenshot of the implemented user interface.

Browsing Process

Browsing naturally induces a sense of flow, which may be imagined as a navigation process through the product space. The main challenge in the design of such a navigation system is to define the relation of products with respect to one another. Differing standpoints of people dictate that each individual sees the product relations from a different perspective. One method of custom defining product relations doing so is through interactive critiquing of products (Burke, 2002). Interactive critiquing involves allowing a user to express the goals that are not satisfied by current items. Another method to understand the preference of a user is through clustering. In our case, clustering may be used to group items that receive similar feedback from a user in an attempt to identify the underlying pattern that matches the preference of the user.

While the sharp focus on a single point in the product space, a feature of interactive critiquing, makes it unsuitable for expansive browsing, in our catalog, one desirable feature is to have an adaptable focus that allows the user to glance at the entire product range as well as zoom in on a few products of interest. We define two parameters in our browsing: breadth and preference. Breadth is a measure of diversity in the product presentation, whereas preference is the inferred interest of the user.

Breadth needs to be changed according to the state of browsing. As the user increasingly grasps some understanding of the available choices, breadth should be narrowed down to focus on recommended products based on the user's preference, allowing the user to discover products of increasing interest and at the same time facilitate a comparison of close alternatives to aid in the purchase decision. At any time, should a shift be detected in the user interest, breadth has to be relaxed accordingly to allow the user the possibility to explore again products of differing nature.

To implement such a mechanism, we divided each page of the catalog into two portions, the first containing products recommended based on the detected preference of the user and the second containing randomly sampled products. Breadth is defined as the size of the latter portion.

Feedback Mechanism

In our case, we note that the most intuitive and compact feedback method is for a user to comment directly on the products on display, as proposed by Burke, Hammond, and Young (1997), in this using a case-based critiquing approach and adopting a bipolar rating system for simplification.

Using the bipolar rating system, we obtain a set of selected products and its complement. The selected set is derived through explicit feedback by the user. This establishes it as a strong indicator of user interest. The converse however is not necessarily true for the complementary set of non-selected products.

The usefulness of non-selected products is the relativistic nature of product selection. A user initially selects what appears to be the best available option. With greater exposure to relevant products, it is natural for a user to become more discerning in making a choice. It is thus inaccurate to conclude that non-selected products are disliked by the user. In view of the ambiguity in interpreting the set of non-selected products, the approach adopted in this article is to analyze only the selected set.

Prototype

A prototype of the catalog was developed for testing purposes. Though fully implemented in Java, the interface was designed to be easily presentable in HTML format. In an actual implementation, the catalog software is intended to reside on a Web server and be remotely accessed via PDA.

Figure 2. Product ontology

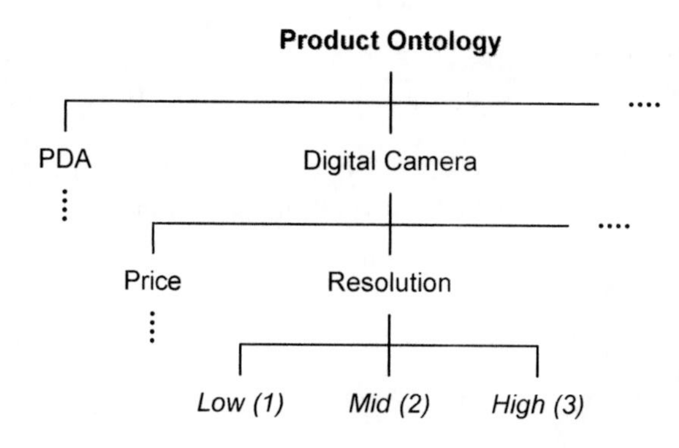

Preference Detection

Clustering is the conceptual grouping of similar products. For our case, we seek to identify a few dominant areas of interest associated with a user so to find relevant products for recommendation. To do so, we perform clustering on the set of positive examples volunteered by the user.

Product Ontology

The products are represented through the specification of an encoding scheme that maps products from the same category into a conceptual product space. The encoding scheme is responsible for the enumeration of product attributes, and in so doing, determines the relationship between products.

We adopted a static encoding scheme in the form of product ontology (Smith, 2003). In our context, product ontology is simply a descriptive tree that defines the key attributes of each product category as well as their relevant enumeration schemes. Figure 2 shows an example of a product ontology.

Product Definition

Let p denote a product and P the product space such that $p \in P$.

A product is characterized by a set of attributes as well as their associated value. We define an attribute as a particular aspect of a product's characteristics (e.g., weight, color), while an attribute instance is a value taken by a product attribute (e.g., 100g, red).

Let α denote an attribute instance and A the domain that the attribute belongs to such that $\alpha \in A$.

A product space P is defined as a vector space of η dimensions where η is the total number of unique attributes possessed by products in P.

$$P : A_1 \times A_2 \times \cdots \times A_\eta$$

Products are mapped into the product space through a predefined product ontology. Products may then be represented by ordered η-tuples with the ith value representing the attribute instance for the ith attribute of the product. We shall refer to this η-tuple as the product characteristic.

$$p : \{\alpha_1, \alpha_2, \ldots, \alpha_\eta\},\ \alpha_i \in A_i$$

A product is assumed to be entirely characterized by the set of ordered attribute instances it is associated with.

Cluster Definition

To facilitate the clustering of products, we adopt the concept of a schema proposed by John Holland (1975) in his schema theorem. In our context, a schema is a template that partially specifies a set of product characteristics. This is possible with the introduction of wildcards that match with any value. A schema effectively defines a subset of the product space for all products that match with the schema.

Let χ denote a schema and X the schematic domain such that $\chi \in X$.

$$X : G_1 \times G_2 \times \cdots \times G_\eta \text{ where } G_j = A_j \cup *$$

$$\chi : \{ \gamma_1, \gamma_2, \ldots, \gamma_\eta \} \text{ where } \gamma_j \in G_j$$

To determine if a product p matches with a schema χ, we define the following functions:

$$\delta(\alpha, \gamma) = \begin{cases} 1 & \alpha = \gamma \text{ or } \gamma = * \\ 0 & else \end{cases} \tag{1}$$

$$\delta_{match}(p, \chi) = \prod_{j=1}^{\eta} \delta(\alpha_j, \gamma_j) \tag{2}$$

A schema serves as a useful means to define a cluster, providing both a signature to determine membership to the cluster as well as a definition of product similarity. Products within a cluster are similar in the sense that they match with the schema representative of the cluster.

In this article, we shall adopt the schema as the sole definition of a cluster $\chi \equiv C$. We term such an approach schematic clustering.

$$p \in C \Leftrightarrow \delta_{match}(p, \chi) = 1$$
$$p \notin C \Leftrightarrow \delta_{match}(p, \chi) = 0$$

Scoring

With the definition of cluster in place, the best cluster that generalizes a sequence of user selection S has to be found.

For this purpose, we need to be able to evaluate the relative quality of each possible cluster as a generalization of *S*.

Span

Let *S* be mapped into an $n \times \eta$ matrix $\{\alpha_{ij}\}$, such that α_{ij} denotes the jth attribute instance of the ith product.

Adapting the match function (2) for use on a matrix:

$$\delta_{match}(p_i, \chi) = \prod_{j=1}^{\eta} \delta(\alpha_{ij}, \gamma_j) \quad (3)$$

We define span as the number of matches a schema has on a set of products:

$$\sigma(S, \chi) = \sum_{i=1}^{n} \delta_{match}(p_i, \chi) \quad (4)$$

Given two clusters with different span, we derive greater confidence in the cluster with a larger span as an area of interest with greater significance. For example if a user selected six products, of which five belong to cluster A while only one belongs to cluster B, we naturally conclude that cluster A serves as a better representation of the user's area of interest. Span thus serves as an important measure of quality.

Order

Given a schema, we define order as the number of non-wildcard values present in the schema.

$$\overline{\delta}_{wildcard}(\gamma) = \begin{cases} 1 & \gamma \neq * \\ 0 & \gamma = * \end{cases} \quad (5)$$

$$d(\chi) = \sum_{j=1}^{\eta} \overline{\delta}_{wildcard}(\gamma_j) \quad (6)$$

Considering the definition of span, it is clear that the number of wildcards present in a schema is proportionate to the chances of the schema having a large span. However, having too many wildcards may not be a desirable because it dilutes the interpretation of the area of interest.

For example, the null schema [*,*,…,*] is undoubtedly the schema with the largest span in any situation, for it encompasses the entire product space. However, the null schema does not give any inference as to where the actual area of interest may lie. Assuming that a product fits the cluster [1,*,*,*,*] as well as the cluster [1,2,3,*,*], we see that the latter is a more precise interpretation of the area of interest because it has a more exclusive membership. Order thus serves as an equally important measure of quality as compared to span.

Span: Order Tradeoff

Having established that span and order are two competing objectives, it is not possible to maximize both measures simultaneously.

To distinguish better the quality of a schema from another, we introduce another measure called coverage:

$$\textbf{Coverage: } \kappa(S, \chi) = \sigma(S, \chi) \cdot d(\chi) \quad (7)$$

$$\textbf{Score}_2\textbf{: } \Gamma_2(S, \chi) = \kappa(S, \chi) \quad (8)$$

Coverage eliminates schemas with extreme span or order to give preference to those with a balance of the two. However, in certain cases it is still not possible to discern schemas with equally good balance. To do so, we have to decide whether to give greater priority to span or order. Since span represents a measure of the level of confidence in an area of interest, we adopt a prudent approach by giving it a higher priority.

$$\textbf{Score}_3\textbf{: } \Gamma_3(S, \chi) = \kappa(S, \chi) + \mu \cdot \sigma(S, \chi) \quad (9)$$

where $0 < \mu < 1$.

Noise Correction

In the context of data processing, it is usually inevitable that the data be distorted by a certain level of noise due to uncontrollable factors. In our case, noise may be introduced either due to ignorance on the part of the user or the lack of appropriate choices for the user to express freely a preference. To overcome this limitation, we introduce a noise threshold K to relax the condition for a match between a schema and a product.

$$\gamma(p_i, \chi) = \sum_{j=1}^{\eta} (1 - \delta(\alpha_{ij}, \gamma_j)) \quad (10)$$

$$\delta'_{match}(p_i, \chi) = \begin{cases} 1 & \gamma(p_i, \chi) \leq K \\ 0 & \gamma(p_i, \chi) > K \end{cases} \quad (11)$$

where $0 \leq K < \eta$.

$$\textbf{Span': } \sigma'(S, \chi) = \sum_{i=1}^{n} \delta'_{match}(p_i, \chi) \quad (12)$$

With such an allowance given for noise, the scoring system will be able to pick up the optimum schema that matches the user preference. This is because the noise threshold allows schemas to be credited for partial matches with the selected products.

Figure 3. Genetic encoding

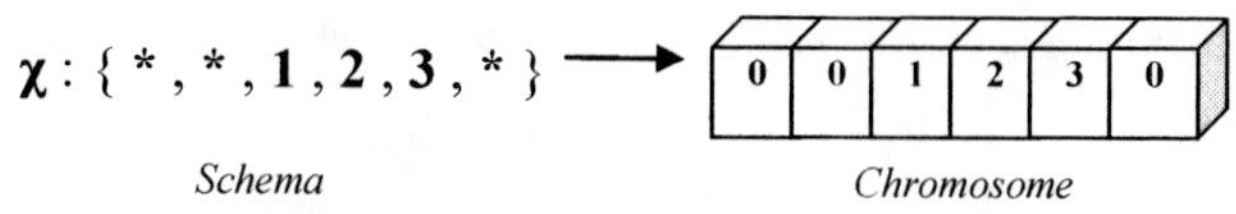

$\chi : \{ *, *, 1, 2, 3, * \} \longrightarrow$ | 0 | 0 | 1 | 2 | 3 | 0 |

Schema *Chromosome*

Owing to the noise threshold, ambiguity appears in the assessment of schemas. A schema that takes advantage of the threshold term in an unwarranted context stands to gain a higher coverage. One main reason is the simple definition of coverage as a product of span and order, which gives unnecessary credit to schema values that do not match the actual attribute instance value.

$$\delta_{pt}(\alpha,\gamma) = \begin{cases} 1 & \alpha = \gamma \ \ and \ \ \gamma \neq * \\ 0 & else \end{cases} \quad (13)$$

$$\textbf{Coverage:} \kappa'(S,\chi) = \sum_{i=1}^{n}\sum_{j=1}^{\eta} \delta'_{match}(\alpha_{ij},\gamma_j)\delta_{pt}(\alpha_{ij},\gamma_j) \quad (14)$$

With the redefinition of coverage, there is an improvement in the score to give less emphasis to matches that makes use of the noise threshold. Despite having a more equitable score, the redefined coverage is still incapable of differentiating between the sensible use of the noise threshold to accommodate noise or the abuse of it to increase coverage. To correct this error, the approach adopted is the inclusion of a penalty term to penalize the usage of the noise threshold.

$$\textbf{Penalty:} \ \pi(S,\chi) = -\sum_{i=1}^{n} \delta'_{match}(p_i,\chi)\gamma(p_i,\chi) \quad (15)$$

$$\textbf{Score:} \ \Gamma(S,\chi) = \sigma'(S,\chi) + \mu \cdot \kappa'(S,\chi) + \lambda \cdot \pi(S,\chi) \quad (16)$$

where $0 < \mu < 1, \lambda > 1$.

Emphasis

Finally, we recognize that a user's preference may evolve in the course of browsing. Products that were selected more recently are thus likely to be more in line with the current preference of the user. To take this factor into account, we allow a progressive emphasis to be set on more recent selection.

We define *E(i)* as the emphasis factor on a product p_i as a function of the product index in the sequence of user selection *S*. The function may follow either a linear or a geometric progression depending on the desired degree of emphasis. The emphasis factor is then applied to all application of the match function (12).

Figure 4. GA pseudocode

```
INITIALIZE population with random candidate solutions
repeat until TERMINATION CONDITION
    1.  EVALUATE chromosomes
    2.  SELECT parents
    3.  RECOMBINE pairs of parents
    4.  MUTATE offspring
    5.  EVALUATE offspring
    6.  SELECT survivors to next generation
```

The optimal emphasis varies in different contexts. Though a high degree of emphasis improves the responsiveness of the system, the tradeoff is poorer overall generalization. It is thus advisable to use moderate values of *E(i)* in most circumstances. Empirical trial tests must be carried out to investigate the effect of a chosen emphasis.

Global Optimization

Having defined a scoring function to evaluate the relative superiority of each schema, we seek to design an algorithm to search for the best schema given a sequence of user selection. EA was found to be a more appropriate choice in our context. In particular, we chose a genetic algorithm (GA), which is a form of EA for the optimization of our scoring function.

Genetic Algorithm

By assigning a value of zero to the wildcard, the η-tuple of positive integer values of a schema is encoded directly into a chromosome as an array of integers. Figure 3 illustrates the encoding process.

Having defined the chromosomes, we apply the typical genetic algorithm as summarized in Figure 4. Evaluation is done using the scoring function defined in the previous section.

Performance

To determine the performance of the algorithm, we define accuracy and efficiency as the performance measures. Accuracy is the frequency that results produce when the genetic algorithm matches the actual global optimum. We calculate accuracy as the average percentage of such matches.

On the other hand, efficiency is the amount of computational effort required to execute the algorithm. We thus calculate efficiency as the average number of generations.

CONCLUSION

The approach in this study focused on realizing the possibility for a more complete m-commerce environment. This outlook is shared by other researchers who attempt to tackle the same problem with different strategies (Guan, Ngoo, & Zhu, 2000). To the best of our knowledge, a customized catalog for m-commerce has not been conceived. This study shares the same intent to make shopping a more pleasant experience for users.

Our approach differs in the absence of a passive viewing mode, as the context of m-commerce makes it unfeasible for users to concentrate on the screen for an extended period of time. Interaction control was greatly simplified in our catalog. Through the usage of recommender technology, we streamlined the browsing process by using a reduced form of feedback.

In summary, this article highlighted the need for specialized applications in the domain of m-commerce. In particular, the need for expansive browsing as a complement to existing search and filter functions has been emphasized. As a possible solution, a novel method of product catalog navigation with the aid of a recommender system has been proposed. This approach emphasizes a minimal-attention user interface that allows a user to browse through a catalog quickly with as little cognitive effort as possible. The associated recommender system that has been conceived adopts a best effort strategy that accommodates any level of user participation. It has been shown to be capable of detecting non-linear preferences in a set of incremental feedback, as well as tolerate noisy input produced by a user. One drawback of this design is the danger of using predefined product ontology in the enumeration of attribute instances. This leads to stereotypical preference interpretation whose relevance depends largely on how the product ontology is defined.

FUTURE TRENDS

For future improvement, it may be worth investigating the possibility of having the recommender generate the ontology from the collective feedback of an ensemble of users. Another enhancement to the existing system would be to incorporate fuzzy logic into the enumeration process. Doing so eliminates the problem around segment boundaries where similar attribute values may be arbitrarily classified into different clusters.

REFERENCES

BizRate. (2004). Retrieved from www.bizrate.com

Bryan, D., & Gershman, A. (1999). Opportunistic exploration of large consumer product spaces. *Proceedings of the 1st ACM Conference on Electronic Commerce* (pp. 41-47).

Burke, R. D. (2002). Interactive critiquing for catalog navigation in e-commerce. *Artificial Intelligence Review, 18,* 245-267.

Burke, R. D., Hammond, K. J., & Young, B. C. (1997). The FindME approach to assisted browsing. *IEEE Expert: Intelligent Systems and Their Applications, 12*(4), 32-40.

Feldman, S. (2000). Mobile commerce for the masses. *IEEE Internet Computing, 4,* 75-76.

Guan, S. U., Ngoo, C. S., & Zhu, F. M. (2000). Handy broker: An intelligent product-brokering agent for m-commerce applications with user preference tracking. *Electronic Commerce Research and Applications, 1,* 314-330.

Holland, J. H. (1975). *Adaptation in natural and artificial systems.* Ann Arbor: The University of Michigan Press.

Lee, J. Y., Lee, H. S., & Wang, P. (2004). An interactive visual interface for online product catalogs. *Electronic Commerce Research, 4,* 335-358.

Montaner, M., Lopez, B., & Lluis, J. (2003). A taxonomy of recommender agents on the Internet. *Artificial Intelligence Review, 19,* 285-330.

Schafer, J. B., Konstan, J., & Riedl, J. (2001). E-commerce recommendation applications. *Data Mining and Knowledge Discovery, 5,* 115-153.

Tateson, R., & Bonsma, E. (2003). ShoppingGarden—Improving the customer experience with online catalogs. *BT Technology Journal, 21*(4), 84-91.

KEY TERMS

E-Commerce: Consists primarily of the distributing, buying, selling, marketing, and servicing of products or services over electronic systems such as the Internet and other computer networks.

Feedback Mechanism: Process whereby some proportion or, in general, function of the output signal of a system is passed (fed back) to the input.

Genetic Algorithm: Search technique used in computer science to find approximate solutions to optimization and search problems. Genetic algorithms are a particular class of evolutionary algorithms that use techniques inspired by evolutionary biology such as inheritance, mutation, natural selection, and recombination (or crossover).

Global Optimization: A branch of applied mathematics and numerical analysis that deals with the optimization of a function or a set of functions to some criteria.

M-Commerce: Electronic commerce made through mobile devices.

Product Catalog: Organized, detailed, descriptive list of products arranged systematically.

Product Ontology: Studies' being or existence, and their basic categories and relationships, to determine what entities and what types of entities exist.

An Interactive Wireless Morse Code Learning System

Cheng-Huei Yang
National Kaohsiung Marine University, Taiwan

Li-Yeh Chuang
I-Shou University, Taiwan

Cheng-Hong Yang
National Kaohsiung University of Applied Sciences, Taiwan

Jun-Yang Chang
National Kaohsiung University of Applied Sciences, Taiwan

INTRODUCTION

Morse code has been shown to be a valuable tool in assistive technology, augmentative and alternative communication, and rehabilitation for some people with various conditions, such as spinal cord injuries, non-vocal quadriplegics, and visual or hearing impairments. In this article, a mobile phone human-interface system using Morse code input device is designed and implemented for the person with disabilities to send/receive SMS (simple message service) messages or make/respond to a phone call. The proposed system is divided into three parts: input module, control module, and display module. The data format of the signal transmission between the proposed system and the communication devices is the PDU (protocol description unit) mode. Experimental results revealed that three participants with disabilities were able to operate the mobile phone through this human interface after four weeks' practice.

BACKGROUND

A current trend in high technology production is to develop adaptive tools for persons with disabilities to assist them with self-learning and personal development, and lead more independent lives. Among the various technological adaptive tools available, many are based on the adaptation of computer hardware and software. The areas of application for computers and these tools include training, teaching, learning, rehabilitation, communication, and adaptive design (Enders, 1990; McCormick, 1994; Bower et al., 1998; King, 1999).

Many adapted and alternative input methods now have been developed to allow users with physical disabilities to use a computer. These include modified direct selections (via mouth stick, head stick, splinted hand, etc.), scanning methods (row-column, linear, circular) and other ways of controlling a sequentially stepping selection cursor in an organized information matrix via a single switch (Anson, 1997). However, they were not designed for mobile phone devices. Computer input systems, which use Morse code via special software programs, hardware devices, and switches, are invaluable assets in assistive technology (AT), augmentative-alternative communication (AAC), rehabilitation, and education (Caves, 2000; Leonard et al., 1995; Shannon et al., 1981; Thomas, 1981; French et al., 1986; Russel & Rego, 1998; Wyler & Ray, 1994). To date, more than 30 manufactures/developers of Morse code input hardware or software for use in AAC and AT have been identified (Anson, 1997; http://www.uwec.edu/Academic/Outreach/Mores2000/morse2000.html; Yang, 2000; Yang, 2001; Yang et al., 2002; Yang et al., 2003a; Yang et al., 2003b). In this article, we adopt Morse code to be the communication method and present a human interface for persons with physical disabilities.

The technology employed in assistive devices has often lagged behind mainstream products. This is partly because the shelf life of an assistive device is considerably longer then mainstream products such as mobile phones. In this study, we designed and implemented an easily operated mobile phone human interface device by using Morse code as a communication adaptive device for users with physical disabilities. Experimental results showed that three participants with disabilities were able to operate the mobile phone through this human interface after four weeks' practice.

SYSTEM DESIGN

Morse code is a simple, fast, and low-cost communication method composed of a series of dots, dashes, and intervals

Figure 1. System schematics of the mobile phone human-interface

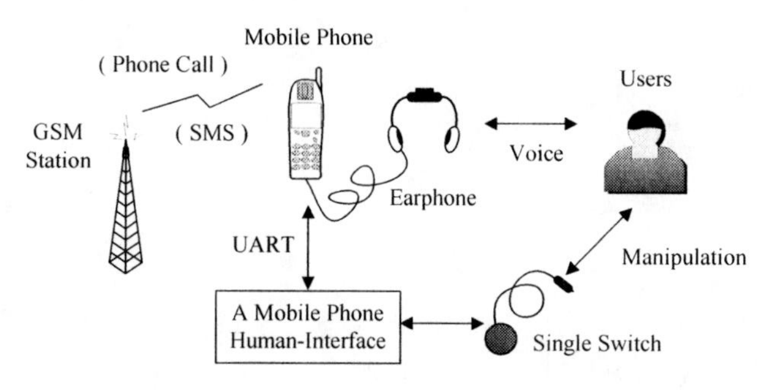

in which each character entered can be translated into a pre-defined sequence of dots and dashes (the elements of Morse code). A dot is represented as a period ".", while a dash is represented as a hyphen, or minus sign, "-". Each element, dot or dash, is transmitted by sending a signal for a standard length of time. According to the definition of Morse code, the tone ratio for dot to dash must be 1:3. That means that if the duration of a dot is taken to be one unit, then that of a dash must be three units. In addition, the silent ratio for dot-dash space to character-space also has to be 1:3. In other words, the space between the elements of one character is one unit while the space between characters is three units (Yang et al., 2002).

In this article, the mobile phone human interface system using Morse code input device is schematically shown in Figure 1. When a user presses the Morse code input device, the signal is transmitted to the key scan circuit, which translates the incoming analog data into digital data. The digital data are then sent into the microprocessor, an 8051 single chip, for further processing. In this study, an ATMEL series 89C51 single chip has been adopted to handle the communication between the press-button processing and the communication devices. Even though the I/O memory capacity of the chip is small compared to a typical PC, it is sufficient to control the device. The 89C51 chip's internal serial communication function is used for data transmission and reception (Mackenzie, 1998). To achieve the data communication at both ends, the two pins, TxD and RxD, are connected to the TxD and RxD pins of a RS-232 connector. Then the two pins are connected to the RxD and TxD of an UART (Universal Asynchronous Receiver Transmitter) controller on the mobile phone device. Then, persons with physical disabilities can use this proposed communication aid system to connect their mobile communication equipment, such as mobile phones or GSM (global system for mobile communications) modems, and receive or send their messages (SMS, simple message service). If they wear an earphone, they might be able to dial or answer the phone. SMS is a protocol (GSM 03.40 and GSM 03.38), which was established by the ETSI (the European Telecommunications Standards Institute) organization. The transmission model is divided into two models: text and PDU (protocol description unit). In this system, we use the PDU model to transmit and receive SMS information through the AT command of the application program (Pettersson, 2000). Structurally the mobile phone human-interface system is divided into three modules: the input module, the control module, and the display module. The interface framework is graphically shown in Figure 2. A detailed explanation is given below.

INPUT MODULE

A user's input will be digitized first, and then the converted results will be sent to the micro controller. From the signal processing circuit can monitor all input from the input device, the Morse code. The results will be entered into the input data stream. When the user presses the input key, the micro-operating system detects new input data in the data stream, and then sends the corresponding characters to the display module. Some commands and/or keys, such as *OK, Cancel, Answer, Response, Send, Receive, Menu, Exit,* and so forth, have been customized and perform several new functions in order to accommodate the Morse code system. These key modifications facilitate the human interface use for a person with disabilities.

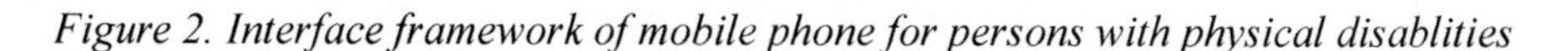
Figure 2. Interface framework of mobile phone for persons with physical disablities

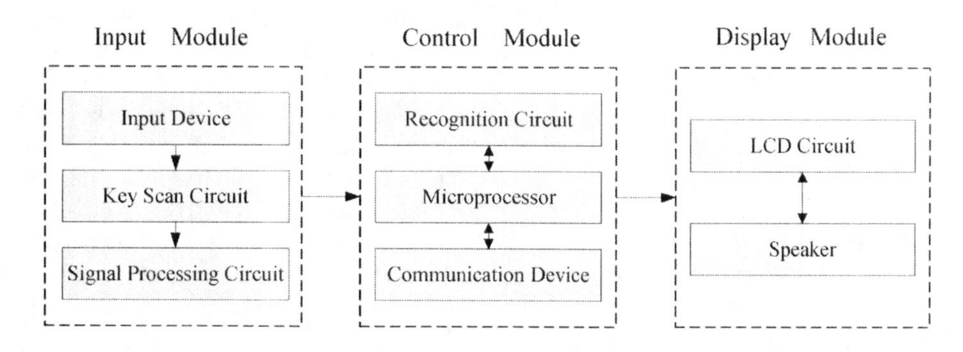

Figure 3. Block diagram of the Morse code recognition system

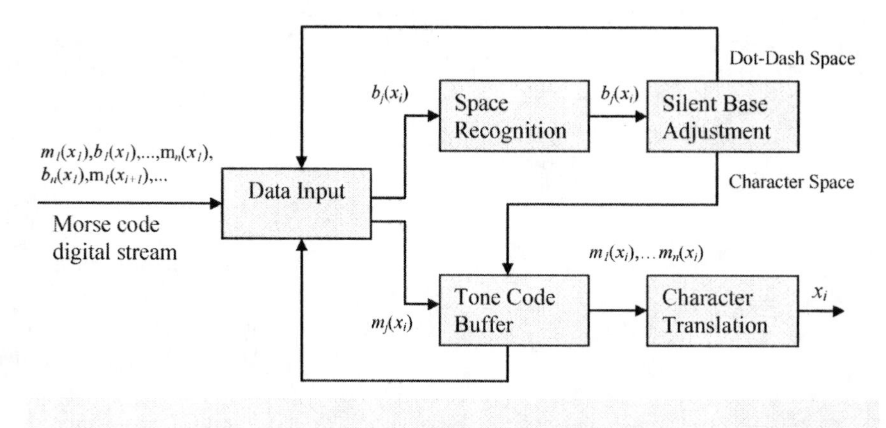

CONTROL MODULE

The proposed recognition method is divided into three modules (see Figure 3): space recognition, adjustment processing, and character translation. Initially, the input data stream is sent individually to separate tone code buffer and space recognition processes, which are based on key-press (Morse code element) or key-release (space element). In the space recognition module, the space element value is recognized as a dot-dash space or a character space. The dot-dash space and character space represent the spaces existing between individual characters and within isolated elements of a character respectively. If a character space is identified, then the value(s) in the code buffer is (are) sent to character translation. To account for varying release speeds, the space element value has to be adjusted. The silent element value is sent into the silent base adjustment process. Afterwards, the character is identified in the character translation process.

A Morse code character, x_i, is represented as follows:

$$m_1(x_i),\ b_1(x_i),\ \ldots,\ m_j(x_i),\ b_j(x_i),\ \ldots,\ m_n(x_i),\ b_n(x_i)$$

where

$b_j(x_i)$: jth silent duration in the character x_i.

n: the total number of Morse code elements in the character x_i.

$m_j(x_i)$: the jth Morse code element of the input character x_i.

DISPLAY MODULE

Since users with disabilities have, in order to increase the convenience of user operations, more requirements for system interfaces than a normal person, the developed system shows selected items and system condition information on an electronic circuit platform, which is based on LCD (liquid crystal display). The characteristics of the proposed system can be summarized as follows: (1) easy operation for users with physical disabilities with Morse code input system, (2) multiple operations due to the selection of different modes, (3) highly tolerant capability from adaptive algorithm recognition, and (4) system extension for customized functions.

RESULTS AND DISCUSSIONS

This system provides two easily operated modes, the phone panel and LCD panel control mode, which allow a user with disabilities easy manipulation. The following shows how the proposed system sends/receives simple message service (SMS) message or make /respond to a phone call.

SMS Receiving Operation

First, when users receive a message notification and want to look at the content, this system will provide "phone panel" and "LCD panel" control modes to choose from. In the phone panel mode, users can directly key-in Morse code "..." (as character 'S'). The interface system will go through the message recognition process, then exchange the message into AT command "AT+CKPD='S', 1", to execute the "confirm" action of the mobile phone. The purpose of this process is the same as users keying-in "yes" on the mobile phone keyboard, then keying-in Morse code ". - - . ." (as key '↓'). The system will recognize the message, then automatically send the "AT+CKPD='↓', 1" instruction. The message cursor of the mobile phone is moved to the next line, or key-in Morse code ". - . . -" (as key '↑') for moving it to the previous data line. Finally, if users want to exit and return to the previous screen, they only need to key-in Morse code ". . - ." (as character 'F'), and start the c key function on the mobile phone keyboard. If LCD panel mode is selected, one can directly follow the selected items on the LCD crystal, to execute the reception and message reading process.

SMS Transmitting Operation

Message transmission services are provided in two modes: phone panel and LCD panel. In the phone panel mode, continually type two times the Morse code ". - . - ." (as key '→'). The system will be converted into AT Command and transferred into mobile phone to show the selection screen of the message functions. Then continuing to key-in three times the Morse code ". . ." (as character 'S'), one can get into the editing screen of message content, and wait for users to input the message text data and receiver's phone number. The phone book function can be used to directly save the receiver's phone number. After the input, press the "yes" key to confirm that the message sending process has been completed. In addition, if the LCD panel mode is selected, one can follow the LCD selection prompt input the service selection of all the action integrated in the LCD panel. Then go through the interface and translate to a series of AT command orders, and batch transfer these into the mobile phone to achieve the control purpose.

The selection command "Answer a phone," displays on the menu of the LCD screen, and can be constructed using Morse code. The participants could press and release the switch, and input the number code ". - - - -" (as character '1') or hot key ". -" (as character 'A'). The mobile phone is then answered automatically. Problems with this training, according to participants, are that the end result is limited typing speed and users must remember all the Morse code set of commands.

Three test participants were chosen to investigate the efficiency of the proposed system after practicing on this system for four weeks. Participant 1 (P1) was a 14-year-old male adolescent who has been diagnosed with cerebral palsy. Participant 2 (P2) was a 14-year-old female adolescent with cerebral palsy, athetoid type, who experiences involuntary movements of all her limbs. Participant 3 (P3) was a 40-year-old male adult, with a spinal cord injury and incomplete quadriparalysis due to an accident. These three test participants with physical impairments were able to make/respond to phone calls or send/receive SMS messages after practice with the proposed system.

FUTURE TRENDS

In the future, Morse code input device could be adapted to several environmental control devices, which would facilitate the use of everyday appliances for people with physical disabilities considerably.

CONCLUSION

To help some persons with disabilities such as amyotrophic lateral sclerosis, multiple sclerosis, muscular dystrophy, and other conditions that worsen with time and cause the user's abilities to write, type, and speak to be progressively lost, it requires an assistive tool for purposes of augmentative and alternative communication in their daily lives. This article presents a human interface for mobile phone devices using Morse code as an adapted access communication tool. This system provides phone panel and LCD panel control modes to help users with a disability with operation. Experimental results revealed that three physically impaired users were able to make/respond to phone calls or send/receive SMS messages after only four weeks' practice with the proposed system.

ACKNOWLEDGMENTS

This research was supported by the National Science Council, R.O.C., under grant NSC 91-2213-E-151-016.

REFERENCES

Anson, D. (1997). *Alternative computer access: A guide to selection.* Philadelphia, PA: F. A. Davis.

Bower, R. et al. (Eds.) (1998). *The Trace resource book: Assistive technology for communication, control, and computer access.* Madison, WI: Trace Research & Development Center, Universities of Wisconsin-Madison, Waisman Center.

Caves, K. (2000). *Morse code on a computer—really?* Keynote presentation at the First Morse 2000 World Conference, Minneapolis, MN.

Enders, A., & Hall, M. (Ed.) (1990). *Assistive technology sourcebook.* Arlington, VA: RESNA Press,.

French, J. J., Silverstein, F., & Siebens, A. A. (1986). An inexpensive computer based Morse code system. In *Proceedings of the RESNA 9th Annual Conference, Minneapolis* (pp. 259-261). Retrieved from http://www.uwec.edu/Academic/Outreach/Mores2000/morse2000.html.

King, T. W. (1999). *Modern Morse code in rehabilitation and education.* MA: Allyn and Bacon.

Lars Pettersson. (n.d.). *Dreamfabric.* Retrieved from http://www.dreamfabric.com/sms

Leonard, S., Romanowski, J., & Carroll, C. (1995). Morse code as a writing method for school students. *Morsels, University of Wisconsin-Eau Claire*, *1*(2), 1.

Mackenzie, I. S. (1998). *The 89C51 Microcontroller* (3rd ed.). Prentice Hall.

McCormick, J. A. (1994). Computers and the Americans with disabilities act: A manager's guide. Blue Ridge Summit, PA: Wincrest/McGraw Hill.

Russel, M., & Rego, R. (1998). A Morse code communication device for the deaf-blind individual. In *Proceedings of the ICAART, Montreal* (pp. 52-53).

Shannon, D. A., Staewen, W. S., Miller, J. T., & Cohen, B. S. (1981). Morse code controlled computer aid for the nonvocal quadriplegic. *Medical Instrumentation*, *15*(5), 341-343.

Thomas, A. (1981). Communication devices for the non-vocal disabled. *Computer*, *14*, 25-30.

Wyler, A. R., & Ray, M. W. (1994). Aphasia for Morse code. *Brain and Language*, *27*(2), 195-198.

Yang, C.-H. (2000), Adaptive Morse code communication system for severely disabled individuals. *Medical Engineering & Physics*, *22*(1), 59-66.

Yang, C.-H. (2001). Morse code recognition using learning vector quantization for persons with physical disabilities. *IEICE Transactions on Fundamentals of Electronics, Communication and Computer Sciences*, *E84-A*(1), 356-362.

Yang, C.-H., Chuang, L.-Y. Yang, C.-H., & Luo, C.-H. (2002). An Internet access device for physically impaired users of Chanjei Morse code. *Journal of Chinese Institute of Engineers*, *25*(3), 363-369.

Yang, C.-H. (2003a). An interactive Morse code emulation management system. *Computer & Mathematics with Applications*, *46*, 479-492.

Yang, C.-H., Chuang, L.-Y., Yang, C.-H., & Luo, C.-H. (2003b, December). Morse code application for wireless environmental control system for severely disabled individuals. *IEEE Transactions on Neural System and Rehabilitation Engineering*, *11*(4), 463-469.

KEY TERMS

Morse Code: Morse code is a transmission method, implemented by using just a single switch. The tone ratio (dot to dash) in Morse code has to be 1:3 per definition. This means that the duration of a dash is required to be three times that of a dot. In addition, the silent ratio (dot-space to character-space) also has to be 1:3.

Adaptive Signal Processing: Adaptive signal processing is the processing, amplification and interpretation of signals that change over time through a process that adapts to a change in the input signal.

Augmentative and Alternative Communication (AAC): Support for and/or replacement of natural speaking, writing, typing, and telecommunications capabilities that do not fully meet communicator's needs. AAC, a subset of AT (see below), is a field of academic study and clinical practice, combining the expertise of many professions. AAC may include unaided and aided approaches.

Assistive Technology (AT): A generic term for a device that helps a person accomplishes a task. It includes assistive, adaptive and rehabilitative devices, and grants a greater degree of independence people with disabilities by letting them perform tasks they would otherwise be unable of performing.

Simple Message Service (SMS): A service available on digital mobile phones, which permits the sending of simple messages between mobile phones.

Global System for Mobile Communications (GSM): GSM is the most popular standard for global mobile phone communication. Both its signal and speech channels are digital and it is therefore considered a 2nd generation mobile phone system.

Interworking Architectures of 3G and WLAN

Ilias Politis
University of Patras, Greece

Tasos Dagiuklas
Technical Institute of Messolonghi, Greece

Michail Tsagkaropoulos
University of Patras, Greece

Stavros Kotsopoulos
University of Patras, Greece

INTRODUCTION

The complex and demanding communications needs of modern humans led recently to the deployment of the 3G/UMTS mobile data networks and the wireless LANs. The already established GSM/GPRS radio access technology can easily handle the voice and low-rate data traffic such as short messages (SMS); however, it is inadequate for the more challenging real-time multimedia exchanges that require higher data rates and ubiquitous connectivity. The UTRAN radio access technology provides wide area coverage and multimedia services up to 2Mbps, while the recently deployed WLANs offer radio access at hotspots such as offices, shopping areas, homes, and other Internet/intranet-connected networks, with very high data rates up to 54Mbps (IEEE 802.11g). Hence, there is a strong need to integrate WLANs and 3G access technologies, and to develop a heterogeneous network based on an all-IP infrastructure that will be capable to offer ubiquitous and seamless multimedia services at very broadband rates.

The major benefits that drive towards an all-IP based core network are the following (Wisely et al., 2002):

- **Cost Saving on Ownership and Management:** Network operators need to own and manage one single network, instead of multiple.
- **Cost Saving on Transport:** For example, the cost to provide IP transport is lower.
- **Future Proof:** It can be claimed that the future of backbone network, both for voice and data, is IP based. An IP-based network allows smooth interworking with an IP backbone and efficient usage of network resources.
- Smooth integration of heterogeneous wireless access technologies.
- The IP multimedia domain can support different access technologies and greatly assist towards fix/mobile convergence.
- **Capacity Increase:** The capacity enhancement of an IP-based transport network is quicker and cheaper. The same is also true to service capacity, thanks to the distributed nature of the service architecture.
- **Rich Services:** The benefits of VoIP are available for improved and new services, for example, voice/multimedia calls can be integrated with other services, providing a powerful and flexible platform for service creation.
- Enable peer-to-peer networking and service model.

This hybrid network architecture would allow the user to benefit from the high throughput IP-connectivity in 'hotspots' and to attain service roaming across heterogeneous radio access technologies such as IEEE 802.11, HiperLan/2, UTRAN, and GERAN. The IP-based infrastructure emerges as a key part of next-generation mobile systems since it allows the efficient and cost-effective interworking between the overlay networks for seamless provisioning of current and future applications and services (De Vriendt et al., 2002). Furthermore, IP performs as an adhesive, which provides global connectivity, mobility among networks, and a common platform for service provisioning across different types of access networks (Dagiuklas et al., 2002). The development of an all-IP interworking architecture, also referred to as fourth-generation (4G) mobile data network, requires specification and analysis of many technical challenges and functions, including seamless mobility and vertical handovers between WLAN and 3G radio technologies, security, authentication and subscriber administration, consolidated accounting and billing, QoS, and service provisioning (Tafazolli, 2005).

This article discusses the motivation, interworking requirements, and different architectures regarding 3G/WLAN interworking towards an all-IP hybrid networking environment. Five common interworking techniques and architectures that effectively can support most of the issues addressed previously are presented and discussed. These are namely: open coupling, loose coupling, tight coupling, very tight coupling (3GPP, 2004), and the recently developed interworking technology named unlicensed mobile access (UMA), which arises as a very competitive solution for the interworking environment (3GPP-UMAC, 2005). The focus of the article is on a comparison and qualitative analysis of the above architectures.

3G AND WLAN INTERWORKING

Motivation

The main motivation for mobile operators to get involved in the WLAN business (Dagiuklas & Velentzas, 2003) is the following:

- Public WLANs provide the opportunity for mobile operators to increase their revenues significantly from mobile data traffic.
- WLANs can be considered as an environment for testing new applications at the initial stage.
- High-demand data traffic from hotspot areas can be diverted from 3G to WLAN, relieving potential network congestion.
- Location-based services in hotspot areas could be based on WLAN technology rather than using more-complex GPS-like systems.

On the other hand, a shift from WLAN to 3G could take place due to the following reasons:

- **Poor Coverage:** Users may be able to use WLAN services at the airport of departure, but not at the airport of arrival or at the hotel.
- **Lack of Brand Recognition:** The service operators are often new start-ups, which causes end-users to hesitate to use the service.
- **Lack of Roaming Agreements:** End users are forced to locate different service providers at the places they roam to.

The service provider value proposition for utilizing integrated WLANs with cellular networks includes the following benefits for carriers as well as their subscribers:

- Extension of current service offering by:
  - integrating cellular data and WLAN solutions,
  - positioning for voice phone service in hotspots, and
  - engaging enterprises with in-building solutions.
- Improve bottom line with new revenue and lower churn:
  - The carrier provides improved in-building coverage by using intranet bandwidth instead of in-building cell sites to provide coverage.
  - Cross system/service integration features become a competitive advantage for the carriers offering seamless mobility services.
  - The cellular provider derives service revenue for authentication services, mobility services, and calls that do not use cellular bearer channels.
  - The cellular handset becomes an indispensable element.
  - The handset can operate with more functionality, for example, even as gateway.
  - The subscriber increases his dependency on the handset.
- Payload traffic trade-off:
  - Some calls will hand over from cellular channels to WLAN connections when subscribers enter these coverage areas.
  - Other calls will hand over to cellular bearer channels when people leave WLAN coverage areas.
  - A more integrated approach to data traffic will probably increase the use of data transferred over cellular networks.

It becomes evident that as subscribers become more dependent on their much more useful handsets, they will call and be called more and everywhere.

3G and WLAN Architectures

The interworking between 3G and WLAN is a trivial issue that is under study by international standardization fora, namely, ETSI, 3GPP, and the UMTS Forum. The undergoing investigation has provided specific requirements that interworking solutions need to meet. The demands include the establishment of some kind of partnership between the 3G operator and the wireless Internet service provider (WISP), a common billing and accounting policy between roaming partners, and a shared subscriber database for authentication, authorization, and accounting (AAA) and security provisioning (Nakhjiri & Nakhjiri, 2005).

The work in this article refers to four already established interworking scenarios (Salkintzis, 2004) regarding 3G and WLAN, which are presented and compared to the recently developed UMA architecture.

In *open coupling* interworking architecture, there is no requirement for specific WLAN access, while each of the networks—3G and WLAN—follows separate authentication procedures. This architecture does not support seamless services, while the user performs a vertical handover from 3G towards WLAN and vice versa.

Figure 1. Open coupling scenario

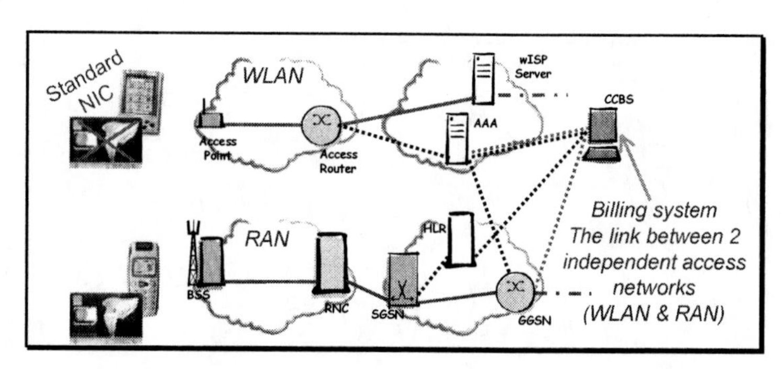

Figure 2. Loose coupling scenario

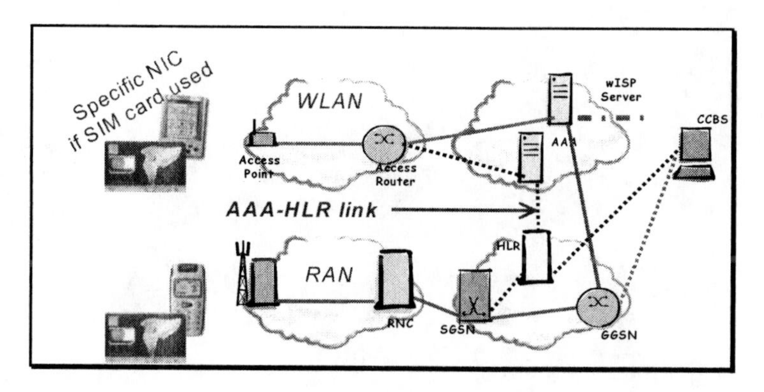

Figure 3. Tight coupling scenario

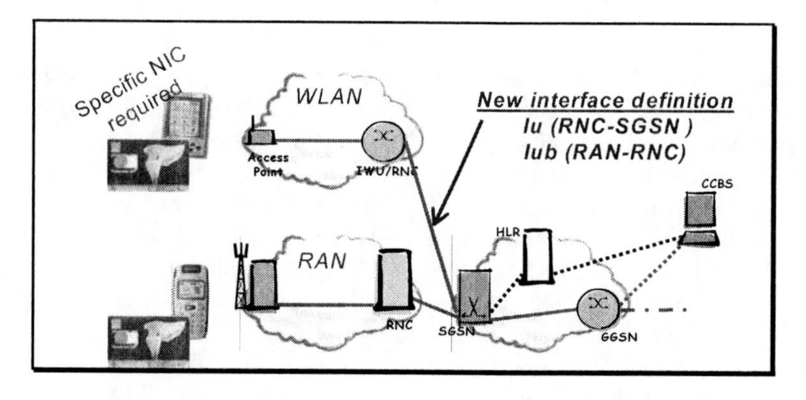

In a *loose coupling* scenario, 3G and WLAN share a common customer database and authentication process. There is no requirement for specific WLAN access. In addition, no load balancing is provided for applications with specific QoS requirements, and the architecture does not support seamless services. Similar to previous architecture, loose coupling does not support seamless services.

On the other hand, a *tight coupling* scenario supports seamless service provisioning for vertical handovers and load balancing for QoS demanding applications. However, there is need for definition of the interface interconnection between the WLAN and SGSN node.

Similar to the previous scenario, *very tight coupling* offers the same advantages, although in this case WLAN is considered as part of the UTRAN and a new interface interconnecting the WLAN and the UTRAN/RNC needs specification.

Figure 4. Very tight coupling scenario

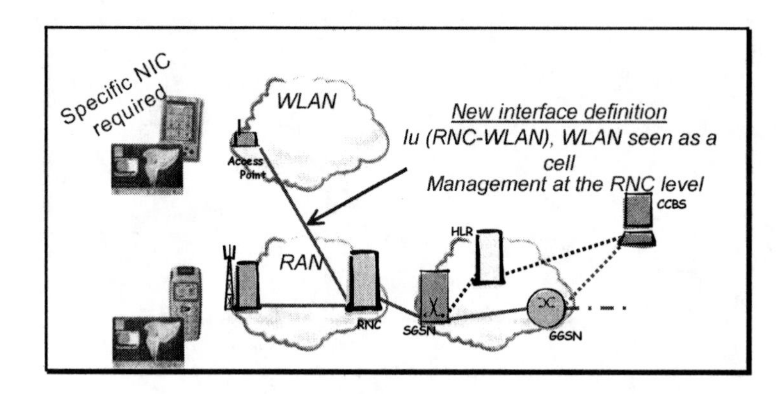

Figure 5. UMA architecture

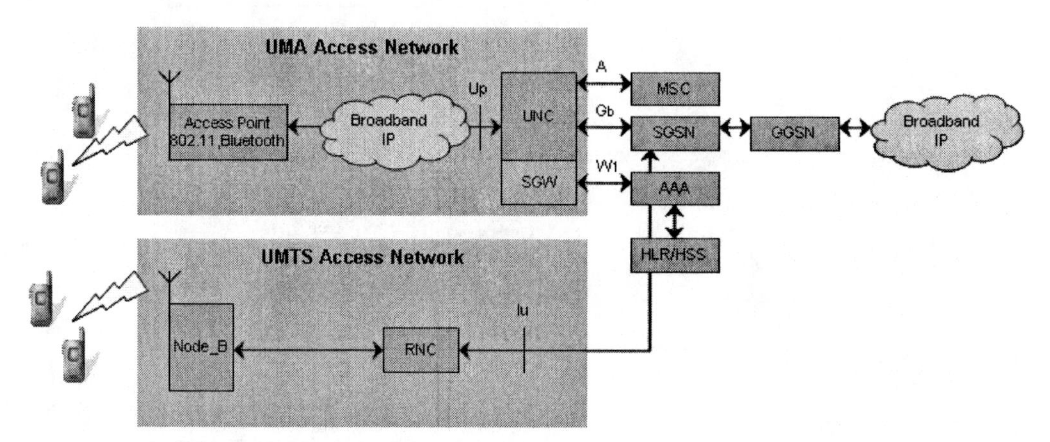

The above interworking scenarios are constrained by the fact that they cover only three wireless technologies. The B3G vision considers the employment of new emerging wireless technologies (e.g., WiMAX, 4G, etc.). The limitation of the approach mentioned above can be alleviated through the approach envisioned by the Unlicensed Mobile Access Consortium (UMAC), where all wireless technologies can be smoothly integrated towards an all-IP-based heterogeneous network. Recently, UMAC produced a multi-access architecture known as unlicensed mobile access (UMA), which rises as an exiting prospect. In the next section, the UMA architecture is presented.

UNLICENSED MOBILE ACCESS

The UMA technology provides access to circuit and packet-switched services over the unlicensed spectrum, through technologies which include Bluetooth and IEEE 802.11 WLANs. The standardization procedure has been initiated by the Unlicensed Mobile Access Consortium (UMAC) and is currently continued by the 3rd Generation Partnership Project (3GPP) under the work item "Generic Access to A./Gb interface." UMA deployment offers to the subscribers a ubiquitous connectivity and consistent experience for mobile and data services as they transition between cellular and public or private unlicensed wireless networks. Compared to other cellular-WLAN interworking solutions, UMA is superior in both the technology aspects and the cost effectiveness for customers and providers (Velentzas & Dagiuklas, 2005).

In more detail, UMA technology does not affect the operation of current cellular radio access networks (UTRAN/GERAN), like cell planning. Hence the investments in existing and future mobile core network infrastructures are preserved. UMA utilizes standard all-IP infrastructure, which provides access to standard packet-switched services and applications. Additionally, the UMA network (UMAN) is independent of the technology used for unlicensed spectrum access, thus it is open to new wireless technologies with no extra requirements. It enables seamless handover between the heterogeneous access technologies, while it ensures service continuity across the network coverage area. Compared to a loose-coupling WLAN-3G interworking scenario, UMA offers greater control over the authentication, authorization, and accounting procedures and tighter end-to-end security. Moreover, UMA technology supports load balancing for efficient allocation of bandwidth and data rates according to the customer requirements, and provides higher throughput and network capacity to the operator that it is translated to more connected customers, thus producing higher revenues.

The UMA technology introduces a new network element, the UMA Controller (UNC), and associates protocols that provide secure transport of GPRS signaling and user plane traffic over IP. UNC acts as a GERAN base station controller (BSC) and includes a security gateway (SGW) that: (1) terminates secure remote access tunnels between the UNC and the mobile station (MS); and (2) provides authentication, encryption, and data integrity for signaling and media traffic. The UNC is connected to a unique mobile switching center (MSC) through the standard A-interface and a serving GPRS support node (SGSN) through the Gb-interface in order to relay GSM and GPRS services respectively to the core network, while through the Wm-interface allows authentication signaling with the corresponding AAA server. The Up-interface between the UNC and MS operates over the IP transport network, and relays circuit and packet-switched services and signaling among the mobile core network and the MS. The MS, which is capable of switching between cellular radio access networks (RANs) and unlicensed, also has an IP interface to the access point that extends the IP access from the UNC to the MS.

COMPARISON OF DIFFERENT ARCHITECTURES AND QUALITATIVE ANALYSIS

The following table showcases the comparison between the proposed interworking architectures. The comparison is based on already mentioned parameters that affect not only the technical efficiency of each of the proposed solutions, but also the cost efficiency, as this is perceived from the operator's and from the user's point of view.

Table 1. Qualitative comparison of interworking scenarios

Qualitative Parameters	Open Coupling	Loose Coupling	Tight Coupling	Very Tight Coupling	UMA
Service Continuity	The running application will not continue across 3G and WLAN once vertical handover takes place	It is not supported and time sensitive services will be interrupted during handover	Service continuity is supported, although QoS may be degraded during handover	Similar to Tight Coupling	Service continuity is fully supported
Simplicity	The user for the same service may have to subscribe to at least two service providers	The user for the same service may have to subscribe to at least two service providers	One service, one mailbox	One service, one mailbox	One service, one mailbox
Seamlessness	Seamless vertical handovers are not supported	Seamless services are not supported by this architecture	Seamless handovers and mobility are supported	Seamless handovers and mobility are supported	Seamless mobility for circuit and packet switched services
Load Balancing	The architecture has no capability to divert the services according to their QoS demands	It is no supported, the system cannot select the network that is suitable for the QoS requests of the service	It is not supported, network selection is based on network coverage at current user's location	It is not supported, network selection is based on network coverage at current user's location	Supports load balancing decisions based on the required QoS of the application and the application's requested bandwidth
Security and Authentication	Separate authentication procedure and security provisioning	Common authentication procedure, the 3G/HLR database is shared between the networks	Common authentication procedures, SGSN is the point of decision, 3G like security scheme	Common authentication procedures and security provisioning	The architecture supports SIM/EAP-AKA authentication and IPSec protocol for the unlicensed mobile part and common GPRS/3G authentication procedure from UNC towards SGSN and MSC

Table 1. Qualitative comparison of interworking scenarios, continued

Qualitative Parameters					
Openness	Additional radio access schemes may be added to the architecture	The architecture may support other heterogeneous wireless technologies	The architecture is proprietary and depends on the WLAN technology used	The architecture is proprietary and depends on the WLAN technology used	The architecture supports interworking between 3G and heterogeneous wireless technologies
Access Control	No specific WLAN access is required, the access control is WLAN based	Independent of access technology used, the access control is WLAN based	Dependent on access technology used due to the new IWU/RNC-SGSN interface, access control is 3G based	Dependent on access technology, management at RNC, 3G based access control	Access is WLAN based over unlicensed radio spectrum
MT Complexity	Standard MT are used with some kind of WLAN interface	Standard MT are used with some kind of WLAN interface	MT with specific NIC	MT with specific NIC	MT are required to support mechanism (software module) for switching between UMA and 3G radio interface
Standardization Complexity	No further standardization effort is required	ISP/AAA-3G/HLR link standardization, translation from MAP to RADIUS/ DIAMETER	Significant standardization effort, new interface definition, 'Iu' (RNC-SGSN), 'Iub' (RAN-RNC)	'Iu' interface between RNC and WLAN	Further standardization is required, however leverages many already defined 3G and IP based protocols
Cost Efficiency	WLAN and 3G networks run separately and there is no further financial burden for the operator	A very cost efficient solution since it is based on the implementation of well established technologies	The operator is required to install new infrastructure at hotspots to interconnect WLAN in the SGSN	The operator is required to install new infrastructure at hotspots to interconnect WLAN in the RNC	New infrastructure needs to be installed, UNC, however utilisation of unlicensed spectrum and openness to new technologies compensates for the cost.

In comparison with the loose and tight coupling scenarios, the UMA approach for WLAN 3G interworking has the advantage in both technology and cost efficiency for operators and customers. Specifically, the installation of a UNC controller, although it requires further standardization and protocol definitions, arises as a better solution. It allows easy openness to new wireless technologies with no extra requirements and is able to provide QoS guarantees for multimedia real-time applications and time-sensitive services in general, seamless mobility, and uninterrupted services across the network coverage area. Hence, the operator is able to cover the demands of customers and the customers have access to all services anywhere and anytime. Authentication, authorization, accounting, and security are common between 3G and WLAN and based on already established protocols and standards. Compared to the other solutions, UMA offers greater control on AAA procedures and tighter end-to-end security. Finally, UNC supports load balancing, which provides—in addition to the best possible solution in terms of connectivity, bandwidth, and data rates, according to the service that the customer requires—higher throughput and network capacity to the operator so that it is translated to more connected customers, thus producing higher revenues.

FUTURE TRENDS

There is no industry consensus on what next generation networks will look like, but as far as the next generation networks are concerned (Kingston, Morita, & Towle, 2005), ideas and concepts include:

- transition to an "All-IP" network infrastructure;
- support of heterogeneous access technologies (e.g., UTRAN, WLANs, WiMAX, xDSL, etc.);
- VoIP substitution of the pure voice circuit switching;
- seamless handovers across both homogeneous and heterogeneous wireless technologies;
- mobility, nomadicity, and QoS support on or above IP layer;
- need to provide triple-play services creating a service bundle of unifying video, voice, and Internet;
- home networks are opening new doors to the telecommunication sector and network providers;
- unified control architecture to manage application and services; and
- convergence among network and services.

Two important factors have been considered to satisfy all these requirements. The first one regards the interworking of existing and emerging access network under the umbrella of a unified IP-based core network and unified control architecture supporting multimedia services. A proposed solution towards this direction is the unlicensed mobile access (UMA), allowing heterogeneous wireless technologies to interconnect to a core network through a network controller. The second requirement regards IP multimedia subsystem (IMS) evolution in order to cope with requirements imposed by NGN architecture (Passas & Salkintzis, 2005). The initial release of 3GPP IMS was developed only for mobile networks. The increasing demand of interworking between different access devices and technologies led to subsequent releases that defined IMS as a core independent element and a key enabler for applying fixed mobile convergence (FMC). FMC comprises two attributes: using one number, voice/mail and seamless handover of multimedia sessions. In the B3G/4G vision, IMS is required to become the common architecture for both fixed and mobile services. Towards this end the ETSI Telecoms and Internet converged services and protocols for advanced networks (TISPAN) is also producing new functionality extensions for the IMS (ETSI TISPAN, n.d.).

CONCLUSION

The conclusion of the qualitative analysis relates the most suitable solution for an interworking architecture of 3G and WLAN radio access technologies based on an all-IP core network with the UMA network. The most important characteristics of UMA, as they have been discovered during the analysis, are among others: the seamless support for vertical handovers and the QoS guarantees for multimedia and time-sensitive applications due to the load balance capability; and the network continuity, scalability, and cost efficient openness. Moreover, the UMA network solution for integrated 3G and WLAN technologies enables network operators to leverage cost and performance benefits of VoIP, broadband, and Wi-Fi, while it supports all mobile services voice, packet, and IMS/SIP, and utilizes standard interfaces into the all-IP core network.

REFERENCES

Dagiuklas, T. et al. (2002). Seamless multimedia services over all-IP network infrastructures: The EVOLUTE approach. *Proceedings of the IST Summit 2002* (pp. 75-78).

Dagiuklas, T., & Velentzas, S. (2003, July). *3G and WLAN interworking scenarios: Qualitative analysis and business models.* IFIP HET-NET03, Bradford, UK.

De Vriendt, J. et al. (2002). Mobile network evolution: A revolution on the move. *IEEE Communications Magazine, 4,* 104-111.

ETSI TISPAN. (n.d.). *NGN functional architecture: Resource and admission control subsystems, release 1.*

Kingston, K., Morita, N., & Towle, T. (2005). NGN architecture: Generic principles, functional architecture and implementation. *IEEE Communications Magazine,* (October), 49-56.

Nakhjiri, M., & Nakhjiri, M. (2005). *AAA and network security for mobile access* (pp. 1-23). New York: John Wiley & Sons.

Passas, N., & Salkintzis, A. (2005). WLAN/3G integration for next generation heterogeneous mobile data networks. *Wireless Communication and Mobile Computing Journal,* (September).

Salkintzis, A. (2004). Interworking techniques and architectures for WLAN/3G integration towards 4G mobile data networks. *IEEE Wireless Communications,* (June), 50-61.

Tafazolli, R. (2005). *Technologies for the wireless future.* New York: John Wiley & Sons.

3GPP. (2004a, September). *Technical specification group services and system aspects: 3GPP system to wireless local area network (WLAN) interworking: System description, 3 TS 23.234.* Retrieved from http://www.3gpp.org

3GPP. (2004b, September). IP multimedia subsystem version 6. 3G TS 22.228.

3GPP-UMAC. (2005, June). *UMA architecture (stage 2).* Retrieved from http://www.3gpp.org

Unified Mobile Access Consortium. (n.d.). Retrieved from http://www.uma.org

Velentzas, S., & Dagiuklas, T. (2005, July). *Tutorial: 4G/wireless LAN interworking.* IFIP HET-NET 2005, Ilkley, UK.

Wisely, D. et al. (2002). *IP for 3G: Networking technologies for mobile communications.* New York: John Wiley & Sons.

KEY TERMS

Authentication Authorization Accounting (AAA): Provides the framework for the construction of a network architecture that protects the network operator and its customers from attacks and inappropriate resource management and loss of revenue.

B3G/4G: Beyond 3G and 4G mobile communications that provide seamless handover between heterogeneous networks and service continuity.

IP Multimedia Subsystem (IMS): Provides a framework for the deployment of both basic calling and enhanced multimedia services over IP core.

NGN: An ITU standard for Next Generation Networks where cellular mobile 3G systems, WLANs, and fixed networks are integrated over IP protocol.

3G: Third generation of cellular mobile communications (GPRS/UMTS).

iPod as a Visitor's Personal Guide

Keyurkumar J. Patel
Box Hill Institute, Australia

Umesh Patel
Box Hill Institute, Australia

INTRODUCTION

Over the past few years, use of mobile devices for various purposes has increased. Apple released its first iPod on October 23, 2001, a breakthrough MP3 player. Today, Apple's fifth-generation iPod is available which can be considered as a portable media player that focuses on the playback of digital video, as well as storing and displaying pictures and video (see apple.com). Since then the iPod has been successfully and effectively used for various purposes including as a media player, bootable drive, external data storage device, PDA replacement, and for podcasting.

Academia and tourism are two areas where the use of mobile devices are encouraged to gain benefits from the technology. For academic use, the iPod's recording and storage capabilities have been explored by some educational institutes across the United States. According to the Duke University iPod First-Year Experience Final Evaluation Report, the iPod supports individual learning preferences and needs, and easy-to-use tools for recording interviews, field notes, and small-group discussions. The tourism industry is also identified as a potential area to use mobile technologies. Recently, Dublin Tourism, Ireland discovered the use of the iPod as a portable tourist guide; Ireland's neighbor Scotland followed (see Physorg, 2006).

Sales of interactive portable MP3 players have increased explosively in the last few years. Information Media Group predicts that sales will continue to increase at the rate of 45% for next six years (Macworld UK, 2005). The iPod is currently the world's best-selling digital audio player and increased its popularity in Australia sevenfold in 2004 (see apple.com). Greg Joswiak, the worldwide vice president of iPod marketing, said: "As of August 2005, market share in Australia is 68% of [the] digital player market."

With the increasing use of digital media together with the handheld devices, this iPod application will eliminate the need for human guides and will provide an entertaining experience to visitors. It will be very useful for landmark tourist destinations such as aquariums and museums, and there will be a huge demand with the increasing flow of tourists in Australia, which according to Tourism Australia (2005) was an increase of 5.4% from 2004 to 2005, with tourists numbering 5.5 million in the latter year.

BACKGROUND

Tourism is an important activity for human life as a source of pleasure and during the holidays. We visit various places every now and then, including tourist destinations such as a museum, commercial destinations such as a stock market, educational institutes such as a university, or public places such as a shopping mall.

Every new visitor suffers from preconceptions and anxiety from their lack of knowledge about the visiting site. This acts as a barrier and must be overcome before an effective visit can take place. As for visitors' preconceptions, the authors of this article encourage visitors to address their anxiety and introduce excitement before they start the tour. The tourism industry so far has promoted the various communication mediums such as maps on the board, written information about specific locations, and now display video screens. Tourism has been a popular area for mobile information systems (Cheverst, Davies, Mitchell, Friday, & Efstratiou, 2000) and other PDA-based systems.

Audio and video has been neglected or underused as a leaning medium in recent years (Scottish Council for Educational Technology, 1994.). The general view is that video is a better tool for leaning than audio. Animation and interactive media like simulations can attract attention, but they proved to be expensive. Hearing is an astoundingly efficient skill according to Clark and Walsh (2004).

Portable media players such as PDAs and iPods can provide information anytime and anywhere. These devices come with their own hard drives and eliminate transportation of storage devices, which is a requirement for video communication. The iPod, with built-in speakers and microphone, makes it easier to record and playback information stored into its hard drive. Clark and Walsh (2004) stated that besides its popularity and ease of use, listening to an iPod and similar devices at public places is socially accepted.

At Box Hill Institute of TAFE, we realized the use of an iPod as a part of our "Innovation Walk" project. The "Innovation Walk" is developed with the aim of showcasing the institute's prized innovations. Career teachers, overseas visitors, students, industry and government dignitaries, and member of the community can undertake the walk independently or as a guided tour.

Figure 1. Designed main menu of a prototype device

A prototype device is being developed using an Apple iPod (see Figure 1).

PERSONAL GUIDE DESIGN

The visitor's personal guide itself will be in the form of an iPod, which can be programmed to give details of a defined list of locations, as well as playing an audible narration of each featured location. This will allow visitors to navigate the visitor's site on their own with the use of the iPod. The following technologies were considered initially to program the iPod as per the requirements:

- creating an application in J2ME on the Java platform porting it to the iPod;
- installing a variant of Linux (more on this later), and modifying its operation to create the system from this platform; and
- creating a text-based guide on the iPod.

The text-based option is the easiest way, with some limitations and given preference on the basis of the estimated time and skills available. To get multiple text pages to run is a fairly simple concept. It requires a specifically named file located under the "Notes" folder that acts as an index page, from which the menu would be created and all other notes will be created. As discussed earlier, iPods as storage devices can easily be connected to a computer via the USB port, and the drive that is mounted for the iPod can be navigated easily from "My Computer."

Creating a Content Page

Open up the "Notes" folder and create a new file called "Main.linx." This file name is required for two reasons. The first reason is that by naming the file "Main," the iPod will automatically display the page as an index, rather than providing a list of available files to open. Secondly, the extension ".linx" of the filename defines the method used to display on the iPod screen a link to another text file.

The iPod has two methods of displaying a link. The default is to have a link created within a text file appear as a hyperlink, similar to that of an html Web page with the word or sentence underlined. The second method is to display the link as an actual menu item on the iPod. This method would be ideal for the contents of our visitor's guide.

Once the "Main.linx" file has been created and located correctly, the next step will be the contents of the main page. This will create the major links to each of the locations that will contain information. This is achieved by opening up the "Main.linx" file in the Notepad and entering the following:

<title>

Alternative Operating Systems for Apple's iPod

Currently there are two main alternatives to Apples' iPod Operating System: iPodLinux (an open source venture into porting Linux onto the iPod) and Rockbox (an open source replacement firmware for mp3 players).

iPodlinux (www.ipodlinux.org) and Rockbox Operating System (www.rockbox.org) are able to replace Apple's Operating System and still maintain the same functionality. The alternative operating systems are capable of playing mp3s and other audio formats, videos, and reading notes. The main difference between Apple's Operating System is that with iPodlinux and Rockbox you can:

- play video games,
- run applications,
- simply develop your own applications without requiring commercial development tools, and
- programmers can develop their own applications or modify/customize existing iPodlinux GPL (General Public Licenxe) applications.

Certain Linux applications are recompiled to run on the iPod without modification. Both alternatives to Apple's iPod Operating System have a following of enthusiastic programmers and developers who have figured out the workings of the five generations of the iPod. Developers and programmers of the iPodlinux have contributed a lot to an open source operating system by setting up Internet relay chat rooms, news groups, forums, wikis, and Web sites. Sourceforge hosts the source code and development comma separated value (CSV) tree, which is maintained by the iPodlinux core developers. Documentation of the iPod hardware components such as the microcontroller, display, memory, battery, and so on is now accessible to everyone. Rockbox Operating System developers thank the hard work of the iPodlinux project because if it was not for iPodlinux documentation and developers, the Rockbox Operating System may have never been ported to the iPod.

Why Choose an Alternative iPod Operating System

The iPod as a visitor's personal guide project initially was looking at the bleeding edge mobile Java J2ME application technology to fulfill its requirements. After research it was discovered that there are other ways to implement a tour guide on an iPod. The research found iPod Notes, iPodlinux, and Rockbox.

- The iPod Apple Operating System is proprietary and therefore a close source.
- iPodLinux and Rockbox are open source operating systems written under the GNU General Public License.
- iPod Apple OS only supports a crippled html language in "Notes" which allows the development of interactive Notes that can contain pictures, video, and text.

How iPod Can be Programmed for iPodLinux

Programming for iPodLinux is done in C, and as a prerequisite the standard functions and libraries must be used. Here is an example of the "Hello World" code using the print function from the stdio.h library.

Using notepad or a C application, do the following:

- Start off by including the precompiler derivative includes statement: #include <stdio.h>.
- Next create the main function from which we will put in the code to print Hello World (see Figure 2).
- Now save the code you have entered into the notepad using the filename of hworld.c. The step is to compile hworld.c using the arm complier tools, arm-elf-gcc hworld.c -o hworld -elf2flt.
- Executing hworld on the iPod running iPodlinux will display "Hello World" to the iPod screen. Once you are happy with your application, it can be packaged as a module and inserted into the Podzilla menu structure.

Figure 2. Sample main function created using programming language "C"

```
int main (int argc, char **argv)
{
printf ("Hello World!\n" );
return 0;
}
```

ADVANTAGES

This new application and use of iPod will:

- eliminate the need for a human guide;
- provide a self-guided tour with entertainment to visitors;
- lead to interactive customer service;
- provide flexibility to tourists to tour the area per their own need, time, and interest, which is an important perspective; and
- achieve great tourist turnaround, as there is no need to wait for some predefined number of tourists.

FUTURE TRENDS

This concept can further be considered to provide visitor information in other languages than English, with possible navigation for the use of different languages in a multicultural environment. Also our next application will discover the possibilities of porting iPodLinux platform on the fifth-generation iPod which is not done so far. Further, the possibility of using an iPod—similar to a PDA—in a commercial environment will be investigated. For now, this cost-effective solution can be implemented at various landmark tourist destinations such as mines, aquariums, and museums, and in the near future it will replace existing expensive technologies.

CONCLUSION

We have demonstrated that the iPod can be used as an innovative and cost-effective tool. To realize the use of the iPod as a visitor's personal guide, iPod's simple user interface designed around a central scroll wheel can be explored for the navigation and recording/ playback facility. It provides the latest information to visitors. Furthermore, iTunes can add an entertaining experience with preferable music while using it as a personal guide.

ACKNOWLEDGMENTS

We would like to thank our final-year students Andrew Obersnel, Ben Coster, Randima Sampath Ratnayake, and Pathum Wickasitha Thamawitage for their contributions toward this project. We would also like to thank Rob McAllister, general manager of teaching, innovation, and degrees, Box Hill Institute; Stephen Besford, center manager, Center for Computer Technology, Box Hill Institute; and John Couch, administrative officer, Center for Computer Technology, Box Hill Institute.

REFERENCES

Cheverst, K., Davies, N., Mitchell, K., Friday, A., & Efstratiou, C. (2000). Developing a context-aware electronic tourist guide: Some issues and experiences. *Proceedings of CHI 2000* (pp. 17-24). The Hague: ACM Press.

Clark, D., & Walsh, S. (2004). *iPod learning.* White Paper, Epic Group, Brighton, UK.

Duke University. (2005). *Duke iPod first year experience.* Retrieved February 17, 2006, from http://www.duke.edu/ipod/

Macworld UK. (2005). Music player sales 'set to double'. *Macworld UK,* (July 22). Retrieved April 26, 2006, from http://www.macworld.co.uk/news/index.cfm?NewsID=92 18&pagePos=5

Physorg. (2006). *Portable tourist guide now on service.* Retrieved February 19, 2006, from http://www.physorg.com/news10338.html

Scottish Council for Educational Technology. (1994). Audio. In *Technologies and learning* (pp. 24-25). Glasgow: SCET.

Tourism Australia. (2005). *Visitors arrival data.* Retrieved February 19, 2006, from http://www.tourism.australia.com/Research.asp?sub=0318&al=2100

KEY TERMS

Java2 Macro Edition (J2ME): A Java platform especially for programming mobile devices such as PDAs.

Linux: An open source operating system for computers.

MP3: MPEG-1 Audio Layer 3, of sound or music recordings stored in the MP3 format on computers.

Personal Digital Assistant (PDA): A mobile device that can be used both as a mobile phone and a personal organizer primarily.

Universal Serial Bus (USB) Port: Port used to connect devices to computers such as PCs, laptops, and Apple Macintosh computers.

Keyword-Based Language for Mobile Phones Query Services

Ziyad Tariq Abdul-Mehdi
Multimedia University, Malaysia

Hussein M. Azia Basi
Multimedia University, Malaysia

INTRODUCTION

A mobile system is a communications network in which at least one of the constituent entities—that is, user, switches, or a combination of both—changes location relative to another. With the advancements in wireless technology, mobile communication is growing rapidly. There are certain aspects exuded by mobile phones, which make them a high potential device for mobile business transactions. Firstly, there is a growing statistic on the number of users who own at least one mobile phone. In 2003 alone, the numbers of mobile phone users were as high as 1.3 billion, and this number is growing steadily each year. Secondly, more and more mobile phones are equipped with much better features and resources at a considerably lower price, which make them affordable to a larger number of users. And thirdly, and most importantly, the small size of mobile phones makes them easily transportable and can truly be the device for anywhere and anytime access (Myers & Beigl, 2003).

Database querying, which is the interest of this article, is a kind of business transaction that can benefit from mobile phones. In general, database querying concerns the retrieval of information from stored data (kept in a database) based on the query (request) posed by the users. This aspect of the database transactions had been the focus of many database researchers for a long time. The mobile phone aspect of the transaction had only recently gained interest from the database communities, and these interests were mainly targeted to the "fatter" mobile devices. The work on mobile database querying can be grouped into those focused on the technology and techniques to handle the limitations of the mobile transmissions due to the instability of the mobile cellular networks, which were concentrated on developing applications that involved access to databases for the mobile environment, and those that handled both of the above issues. For example, caching (Cao, 2002) and batching (Tan & Ooi, 1998) are some of the popular techniques that were and still are investigated in detail to handle the problems of the mobile transmissions. On the other hand, Hung and Zhang's (2003) telemedicine application, Koyama, Takayama, Barolli, Cheng, and Kamibayashi's (2002) education application, and Kobayashi and Paungma's (2002) Boonsrimuang transportation application are some examples of the work on mobile database application. These works were observed to be application-specific and supported a very limited and predefined number and type of possible queries. Each of the possible queries, in turn, requires several interactions with the server before a full query can be composed.

This article will highlight the framework opted by the authors in developing a database query system for usage on mobile phones. As the development work is still in progress, the authors will share some of the approaches taken in developing a prototype for a relationally complete database query language. This work concentrates on developing an application-independent, relationally complete database query language. The remainder of this article is organized as follows. The next section presents some of the existing work related to the study. We then introduce and describe the framework undertaken in order to develop a database query system for mobile phones, and discuss the prototype of the database query language used by the system. We end with our conclusion.

RELATED WORK

Query languages are specialized languages for asking questions, or queries, which involve data in a database (Ramakrishnan & Gehrke, 2000). Query languages for relational databases originated in the 1970s with the introduction of relational algebra and relational calculus by E.F. Codd. Both relations are equivalent in their level of expressiveness or query completeness. These two formal relations had interchangeably been used as the benchmarks for measuring the completeness of the later query languages. Codd originally proposed eight operations to be included in the relational algebra, but out of the eight, five were considered fundamentals as they allowed most of the data retrieval operations. These operations are known as selection, projection, cartesian product, union, and set difference. If a query language supports the five operations, then it is considered as being relationally complete (Connolly, Begg, & Strachan, 1997). Throughout the years, several other measures of query completeness were proposed such as datalog, stratified, computable, and others (Chandra, 1988). However, in the authors' opinion, these later measures might be too extensive to be considered for mobile phones and their users' application.

Although both relational algebra and relational calculus are complete, they are hard to understand and use. This resulted in a search for other easy-to-use languages that are at least compatible to relational algebra and calculus. Some of these query languages are transform-oriented non-procedural-based languages, which use relations to transform input data into required outputs. Structured Query Language (SQL) is an example of such a language. Besides non-procedural languages, visual query languages have also gained much acceptance in the database community. Some of the work on visual query languages found in the literature, such as Czejdo, Rusinkiewicz, Embley, and Reddy (1989), Catarci (1991), and Polyviou, Samaras, and Evripidou (2005), used the entity relationship diagram and other data modeling as the basis for query formulation, and some used icons to present pre-defined queries (Massari, Weissman, & Chrysanthis, 1996). Query languages are textual languages that caught the interest of some database query language researchers. Some of these languages were represented in the form of natural language (Kang, Bae, & Lee, 2002; Hongchi, Shang, & Ren, 2001), and some were represented in the form of keywords (Calado, da Silva, Laender, Ribeiro-Neto, & Vieira, 2004; Agrawal, Chaudhuri, & Das, 2001). This type of languages is less restrictive compared to the other types of languages. However, they need extra work in approximating the meaning of the terms or keywords used in a query. Thesaurus and ontology are few approaches used to approximate meanings of terms or keywords (Kimoto & Iwadera, 1991; Weibenberg, Voisard, & Gartmann, 2006) in this type of query language.

Even though each type of query language mentioned above has its own advantages, very few of them, except for Polyviou et al. (2005) and Massari et al. (1996), were tested on small devices. SQL, for example, would be too tedious to be entered using mobile phones and too complicated for ordinary users. Visual query languages, on the other hand, would require considerable screen space as well as resource consuming in order to be rendered. Natural language and keyword language would also be difficult to be textually keyed in using mobile phones. There were, however, attempts to use spoken method for query languages (Chang et al., 2002; Bai, Chen, Chien, & Lee, 1998). But this approach leads to another problem in matching the intonation of users. The textual form of query languages (keyword method in particular) might be the most suitable language to be used on mobile phones since they are the least resource consuming and easily extensible. However, there must be a method to ease the input part of the query formulation process. To date, the authors have not been able to find any publication of the investigations of the same method as applied to mobile phones. Therefore, we believe that the keyword-based language is worth some investigation.

Framework Model

Polyviou, Samaras, and Evripiou (Kang et al., 2002) laid down several challenges that must be dealt with in order to develop a modern search interface. The challenges specified were: the search interface must be usable, powerful, flexible, and scalable. These challenges are adopted in our approach while developing the database query system for mobile phones. The concept of usable is implemented in our design by providing a language that supports menu-based guidance for the users to form valid queries. The concept of powerful is implemented by making sure that the language is relationally complete. The concept of flexible, on the other hand, is implemented in the language by allowing the language to work with any type of relational databases and any type of applications. Finally, the scalability aspect is handled by allowing the language to accept a database of any size, but at the same time filtering the data to be presented to the users according to some form of user grouping and access patterns. The keyword-based language is developed to answer the above challenges. The reasons for choosing such a language, among others, are due to the ability of such a language to present complex relationships with a minimal number of keywords, and it takes lesser space for presentation. For example, it is possible to access information from two indirectly linked relations, no matter how far apart the relations might be, by simply providing the name of the two relations as query keywords. This ability makes the language scalable and easily extensible. However, keyword-based language does have constraints. Firstly, it is in textual form and therefore is tedious and prone to typing error. Secondly, it requires users to have exact knowledge of the keywords to be entered in order to form valid queries. Therefore, the authors have modified the keyword-based approach by providing users with selectable keywords in a menu form. This approach has another advantage: it allows users to point and click the keywords needed without having to type them manually, which is a way to handle the input mechanism problem of mobile phones (most phones only have keypads as an input mechanism). This approach requires lengthy display space, which is lacking for mobile phones. Therefore, the authors intend to handle this problem by providing only selected keywords to users based on their personal profiles and preferred queries. Figure 1 shows the general framework of the query language, and Figure 2 shows the position of the query language with respect to the rest of the whole database query system. As shown in both figures, the query language basically resides in two locations: in the mobile phones as the query interface, and in the application server as the query engine that transformed the keyword

Figure 1. General framework for query language

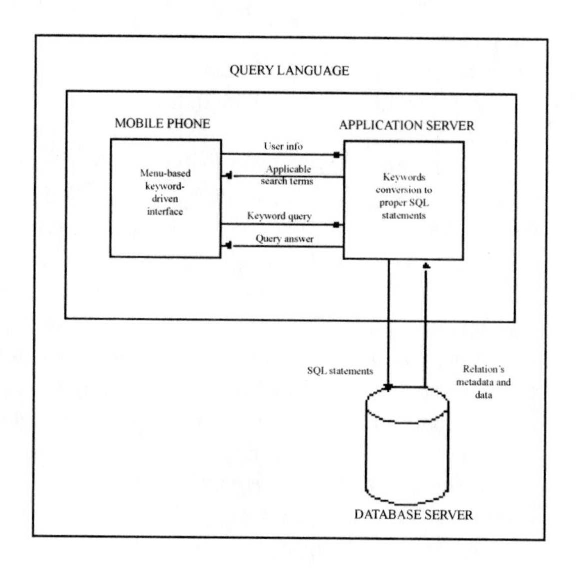

Figure 3. Activities during pre-language operational stage

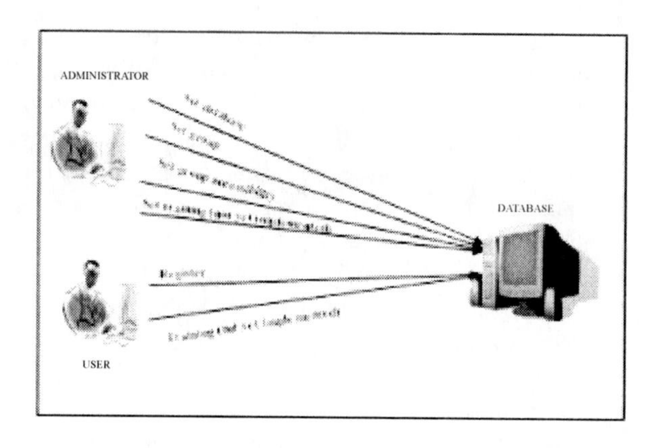

queries into their relevant Structured Query Language (SQL) queries. The keywords that are presented as options during query formulation are selected metadata (i.e., relations' names and attributes' names) of the relations in the database. The selected metadata is based on the user information provided during login (currently personal information; later we will consider results from group training as well) and the accessibility information set by the database administrator prior to the query language becoming operational.

Figure 3 shows the activities that are done by the database administrator and users during the pre-language operational stage. The operation of the database query system is depicted in the system architecture, as shown in Figure 4. The pre-language operational stage will be conducted using a normal computing terminal over the Internet, while the query operations will be handled through query formulation using

Figure 2. The position for database query system

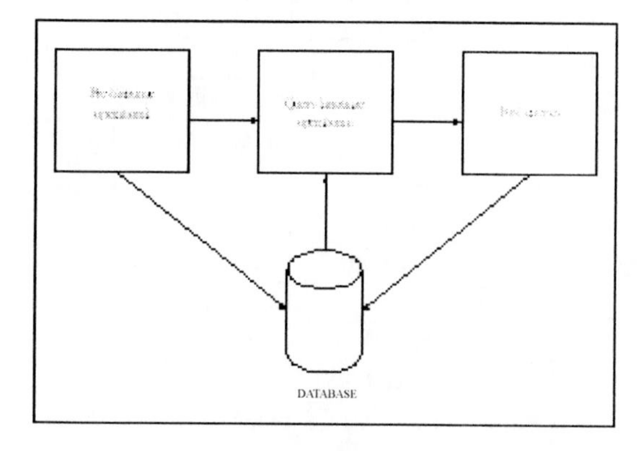

Figure 4. The operational stage of network architecture

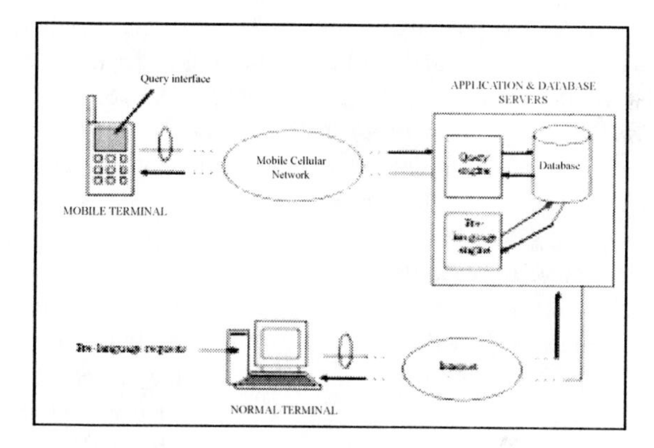

the mobile phones and query processing by the application and database servers.

EXPERIMENTAL STUDY

The prototype of the query language is developed to identify the query interface midlet on the mobile phones and J2SE for the query engine servlets on the application server. The database consists of data on students, subjects, and staff of the university. Students enroll in many subjects that are conducted in many sessions at several venues. The subjects are taught by many lecturers who are of different ranks. The students stay in hostels managed by wardens, and their outings must be approved by a staff member. The schema in each relation in the database is shown in Figure 5 and their relationships in Table 1 respectively.

The database is used to prove that the developed query language is relationally complete. As mentioned earlier, there are five fundamental operations that must be satisfied in order for a language to be considered relationally complete.

Table 1. The relationships between attributes

ATTRIBUTE	REFER TO
Hostelwarden	Staff ID
Staff position	StaffPost code
Staff ranking	StaffRank code
Student hostel Code	Hostel code
Subject lecturer	Staff ID
Session subjectCode	Subject code
Session venue	Venue code
Outing studentID	Student ID
Outing approver	Staff ID
Enrolment studentID	Student ID
Enrolment subjectCode	Subject code

In order to explain the operations, let us assume that there are two relations *R* and *S* which contain m and n numbers of attributes respectively. The five operations are: *Selection, Projection, Cartesian Product, Union,* and *Set Difference.* Here we describe how the five operations are handled by the developed query language:

- **Selection:** Mathematically denoted as $\sigma_{predicate}(R)$, works on a single relation *R* and defines a relation that contains only those tuples of *R* that satisfy the specified condition (predicate). For example, a *Selection* query of $\sigma_{studYear=3}(Student)$ will produce as output, tuples from the *Student* relation which have a value 3 in their *studYear* attribute.
  The developed query language handles the *Selection* operation by allowing a user to select the proper relation name from the list of keywords. This action will allow the user to later choose the name of the attribute that he/she wants to check the value of, to choose the operation he/she wants to perform, and to provide the value he/she is looking for. Figure 6 show the screen shots of a sample *Selection* operation. The query language also allows multiple conditions to be implemented.
- **Projection:** Mathematically denoted as $\pi_{a1,a2,..,am}(R)$, works on a single relation *R* and defines a relation that contains a vertical subset of *R*, extracting the values of specified attributes and eliminating duplicates. For example, a *Projection* query of $\pi_{subjCode,subjName}(Subject)$ will produce as output all tuples in the *Subject* relation. For each tuple, the only values associated with the attributes of *subjCode* and *subjName* will be returned. The developed query language handles the *Projection* operation by allowing a user to select the proper relation name which later gives as a list all of the attributes of the relation. A user can then select as many attributes as he/she likes to view as output. Figure 7 shows the screen shots of a sample *Projection* operation.
- **Cartesian Product:** Mathematically denoted as *R X S*, defines a relation that is the concatenation of every tuple of relation *R* with every tuple of relation *S*. For example, if a *Staff* relation contains 100 tuples with 10 attributes each and a *Subject* relation contains 100 tuples with five attributes each, a *Cartesian product* query of *Staff X Subject* will produce as output 1,000 times 100 tuples, since each tuple of the *Staff* relation will be concatenated with each one of the tuples from the *Subject* relation. Furthermore, each tuple of the output relation will have ten plus five attributes. The result of a *Cartesian product* operation is less meaningful. Therefore, a more restrictive form of the operation, called *Join*, is more preferable. A *Join* operation, mathematically denoted as $R \rhd\lhd_{predicate} S$, includes only the combinations of both relations that satisfy certain conditions. For example, a query of $Staff \rhd\lhd_{staffID=lectID} Subject$ will produce tuples which combine a tuple from the *Staff* with its associated *Subject* tuple. The number of the output tuples will be equal to the number of the tuples of *Staff*. There are several variations to the *Join* operation such as *left-outer join,*

Figure 5. The database schema

Hostel (code, name, warden)
Staff (ID, name, gender, DOB, mobile#, position, ranking, area)
StaffPost (code, name)
StaffRank (code, name)
Student (ID, name, gender, DOB, year, hostelCode, room)
Subject (code, name, creditHour, lecturer)
Session (subjectCode, day, timeStart, timeEnd, venue)
Venue (code, name, capacity)
Enrolment (studentID, subjectCode)
Outing (studentID, dateOut, tiemout, destination, dateIn, timeIn, approver)

Figure 6. Selection operation

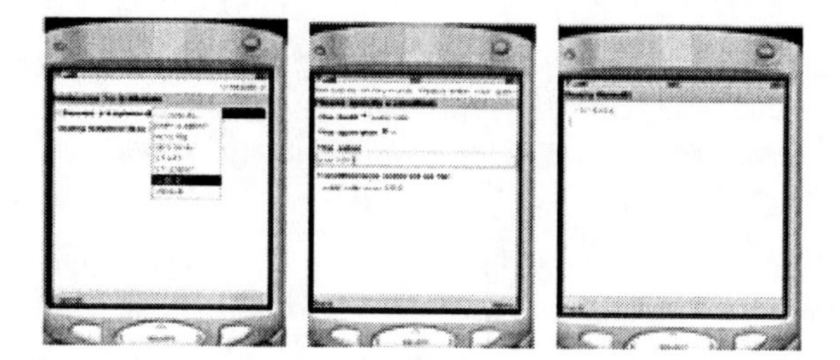

Figure 7.

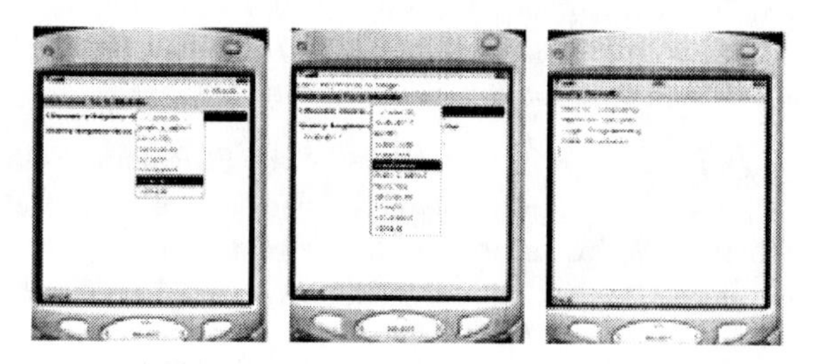

Figure 8.

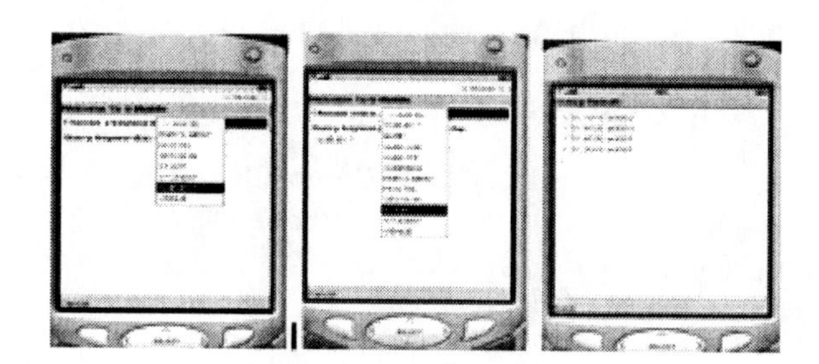

right-outer join, equijoin, and so on. The developed query language handles the *Join* operation by allowing a user to select as many relations as he/she wants to join and the relations can be indirectly related as well. Figure 8 shows screen shots of a sample *Join* operation.

- **Union:** Mathematically denoted as $R \cup S$, concatenates all tuples of R and all tuples of S into one relation with duplicate tuples being eliminated. However, R and S must be union-compatible (i.e., both relations contain the same number of attributes, and each corresponding attribute is of the same domain) in order for the operation to be valid. For example, the query of *Lecturing-Staff* $\cup$ *Administrative-Employee* is valid if they both have the same schema type. This operation is not yet implemented by the query language. But the concept would be possibly implemented by allowing a user to select two relations and the union operator. The query engine will then check for the compatibility of their schema types.
- **Set Difference:** Mathematically denoted as $R - S$, defines a relation consisting of the tuples that are in relation R, but not in S. R and S again must be union-compatible. For example, the query of *Staff–Lecturing-Staff* will produce all administrative staff, assuming lecturers and administrators are the only two types of staff in the university. This operation is not yet implemented by the query language. Similarly, the concept would be possibly implemented by allowing a user to select two relations and a difference operator. The query engine will then check for the compatibility of their schema types. Besides the five operations, the query language is capable of combining multiple operations in one query, and it can also be easily extended to include other operations as needed.

CONCLUSION

The use of a keyword-based query language with menu-based guidance for formulating database queries using mobile phones is feasible due to its usability, powerfulness, flexibility, and scalability. With the physical constraints of the mobile phones, this type of query language uses minimal space for presentation and a lesser number of interactions to form complex queries. Furthermore, the keyword-based language is robust since it enables users to enter all possible queries by combining relevant keywords. Therefore, it is able to accept unplanned queries; it can be extended, and it is adaptable to other database applications. With further research, especially in the method for recommending the keywords relevant to a user, the keyword-based language could be the answer to access a full-scale database from mobile phones.

REFERENCES

Agrawal, S., Chaudhuri, S., & Das, G. (2002, February 26-March 1). DBXplorer: A system for keyword-based search over relational databases. *Proceedings of the IEEE 18th International Conference on Data Engineering (ICDE'02)* (pp. 5-16).

Bai, B. R., Chen, C. L., Chien, L. F., & Lee, L. S. (1998). Intelligent retrieval of dynamic networked information from mobile terminals using spoken natural language queries. *IEEE Transactions on Consumer Electronics, 44*(1), 62-72.

Boonsrimuang, P., Kobayashi, H., & Paungma, T. (2002, July 3-5). Mobile Internet navigation system. *Proceedings of the 5th IEEE International Conference on High Speed Networks and Multimedia Communications* (pp. 325-328).

Calado, P., da Silva, A.S., Laender, A. H. F., Ribeiro-Neto, B. A., & Vieira, R. C. (2004). A Bayesian Network approach to searching Web databases through keyword-based queries. *Information Processing and Management, 40*(5), 773-790.

Cao, G. (2002). On improving the performance of cache invalidation in mobile environments. *Mobile Networks and Applications, 7*(4), 291-303.

Catarci, T. (1991). On the expressive power of graphical query languages. *Proceedings of the 2nd IFIP W.G. 2.6 Working Conference on Visual Databases* (pp. 404-414).

Chandra, A. (1988). Theory of database queries. *Proceedings of the 7th ACM Symposium on Principles of Database Systems* (pp. 1-9).

Chang, E., Seide, F., Meng, H. M., Chen, Z., Shi, Y., & Li, Y. C. (2002). A system for spoken query information retrieval on mobile devices. *IEEE Transactions on Speech and Audio Processing, 10*(8), 531-541.

Connolly, T., Begg, C., & Strachan, A. (1997). *Database systems—A practical approach to design, implementation and management.* Boston: Addison-Wesley.

Czejdo, B., Rusinkiewicz, M., Embley, D., & Reddy, V. (1989, October 4-6). A visual query language for an ER data model. *Proceedings of the IEEE Workshop on Visual Languages* (pp.165-170).

Hongchi, S., Shang, Y., & Ren, F. (2001, October 7-10). Using natural language to access databases on the Web. *Proceedings of the IEEE International Conference on Systems, Man, and Cybernetics* (Vol. 1, pp. 429-434).

Hung, K., & Zhang, Y.-T. (2003). Implementation of a WAP-based telemedicine system for patient monitoring. *IEEE Transactions on Information Technology in Biomedicine, 7*(2), 101-107.

Kang, I.-S., Bae, J.-H. J., & Lee, J. H. (2002, November 6-8). Database semantics representation for natural language access. *Proceedings of the 1st International Symposium on Cyber Worlds* (pp. 127-133).

Kimoto, H., & Iwadera, T. (1991, July 8-14). A dynamic thesaurus and its application to associated information retrieval. *Proceedings of IJCNN-91, the Seattle International Joint Conference on Neural Networks* (Vol. 1, pp. 19-29).

Koyama, A., Takayama, N., Barolli, L., Cheng, Z., & Kamibayashi, N. (2002, November 6-8). An agent based campus information providing system for cellular phone. *Proceedings of the 1st International Symposium on Cyber Worlds* (pp. 339-345).

Massari, A., Weissman, S., & Chrysanthis, P. K. (1996). Supporting mobile database access through query by icons. *Distributed and Parallel Databases (Special Issue on Databases and Mobile Computing), 4*(3), 47-68.

Myers, B. A., & Beigl, M. (2003). Handheld computing. *Computer, 36*(9), 27-29.

Polyviou, S., Samaras, G., & Evripidou, P. (2005). A relationally complete visual query language for heterogeneous data sources and pervasive querying. *Proceedings of IEEE 2005.*

Ramakrishnan, R., & Gehrke, J. (2000). *Database management systems.* New York: McGraw-Hill.

Tan, K. L., & Ooi, B. C. (1998). Batch scheduling for demand-driven servers in wireless environments. *Journal of Information Sciences, 109*(1-4), 281-298.

Weibenberg, N., Voisard, A., & Gartmann, R. (2006). Using ontologies in personalized mobile applications. *Proceedings of the12th Annual International Workshop on GIS.* ACM Press.

Knowledge Representation in Semantic Mobile Applications

Pankaj Kamthan
Concordia University, Canada

INTRODUCTION

Mobile applications today face the challenges of increasing information, diversity of users and user contexts, and ever-increasing variations in mobile computing platforms. They need to continue being a successful business model for service providers and useful to their user community in the light of these challenges.

An appropriate representation of information is crucial for the agility, sustainability, and maintainability of the information architecture of mobile applications. This article discusses the potential of the Semantic Web (Hendler, Lassila, & Berners-Lee, 2001) framework to that regard.

The organization of the article is as follows. We first outline the background necessary for the discussion that follows and state our position. This is followed by the introduction of a knowledge representation framework for integrating Semantic Web and mobile applications, and we deal with both social prospects and technical concerns. Next, challenges and directions for future research are outlined. Finally, concluding remarks are given.

BACKGROUND

In recent years, there has been a proliferation of affordable information devices such as a cellular phone, a personal digital assistant (PDA), or a pager that provide access to mobile applications. In a similar timeframe, the Semantic Web has recently emerged as an extension of the current Web that adds technological infrastructure for better knowledge representation, interpretation, and reasoning.

The goal of the mobile Web is to be able to mimic the desktop Web as closely as possible, and an appropriate representation of information is central to its realization. This requires a transition from the traditional approach of merely presentation to *representation* of information. The Semantic Webprovides one avenue towards that.

Indeed, the integration of Semantic Webtechnologies in mobile applications is suggested in Alesso and Smith (2002) and Lassila (2005). There are also proof-of-concept semantic mobile applications such as MyCampus (Gandon & Sadeh, 2004) and mSpace Mobile (Wilson, Russell, Smith, Owens, & Schraefel, 2005) serving a specific community. However, these initiatives are limited by one or more of the following factors: the discussion of knowledge representation is one-sided and focuses on specific technology(ies) or is not systematic, or the treatment is restricted to specific use cases. One of the purposes of this article is to address this gap.

Table 1. Knowledge representation tiers in a semantic mobile application

Semiotic Level	Semantic Mobile Web Concern and Technology Tier	Decision Support
Social	Trust	Feasibility
Pragmatic	Inferences	
Semantic	Metadata, Ontology, Rules	
Syntactic	Markup	
Empirical	Characters, Addressing, Transport	
Physical	Not Directly Applicable	

UNDERSTANDING KNOWLEDGE REPRESENTATION IN SEMANTIC MOBILE APPLICATIONS

In this section, our discussion of semantic mobile applications is based on the knowledge representation framework given in Table 1.

The first column addresses semiotic levels. Semiotics (Stamper, 1992) is concerned with the use of symbols to convey knowledge. From a semiotics perspective, a representation can be viewed on six interrelated levels: physical, empirical, syntactic, semantic, pragmatic, and social, each depending on the previous one in that order. The physical level is concerned with the representation of signs in hardware and is not directly relevant here.

The second column corresponds to the Semantic Web"tower" that consists of a stack of technologies that vary across the technical to social spectrum as we move from bottom to top, respectively. The definition of each layer in this technology stack depends upon the layers beneath it.

Finally, in the third column, we acknowledge that there are time, effort, and budgetary constraints on producing a representation and include feasibility as an all-encompassing factor on the layers to make the framework practical. For example, an organization may choose not to adopt a technically superior technology as it cannot afford training or processing tools available that meet the organization's quality expectations. For that, analytical hierarchy process (AHP) and quality function deployment (QFD) are commonly used techniques. Further discussion of this aspect is beyond the scope of the article.

The architecture of a semantic mobile application extends that of a traditional mobile application on the server-side by: (a) expressing information in a manner that focuses on *description* rather than presentation or processing of information, and (b) associating with it a knowledge management system (KMS) consisting of one or more domain-specific ontologies and a reasoner.

We now turn our attention to each of the levels in our framework for knowledge representation in semantic mobile applications.

Empirical Level of a Semantic Mobile Application

This layer is responsible for the communication properties of signs. Among the given choices, the Unicode Standard provides a suitable basis for the signs themselves and is character-by-character equivalent to the ISO/IEC 10646 Standard Universal Character Set (UCS). Unicode is based on a large set of characters that are needed for supporting internationalization and special symbols. This is necessary for the aim of universality of mobile applications.

The characters must be uniquely identifiable and locatable, and thus addressable. The uniform resource identifier (URI), or its successor international resource identifier (IRI), serves that purpose.

Finally, we need a transport protocol such as the hypertext transfer protocol (HTTP) or the simple object access protocol (SOAP) to transmit data across networks. We note that these are limited to the transport between the mobile service provider that acts as the intermediary between the mobile client and the server. They are also layered on top of and/or used in conjunction with other protocols, such as those belonging to the Institute of Electrical and Electronics Engineers (IEEE) 802 hierarchy.

Syntactic Level of a Semantic Mobile Application

This layer is responsible for the formal or structural relations between signs. The eXtensible Markup Language (XML) lends a suitable syntactical basis for expressing information in a mobile application.

The XML is supported by a number of ancillary technologies that strengthen its capabilities. Among those, there are domain-specific XML-based markup languages that can be used for expressing information in a mobile application (Kamthan, 2001).

The eXtensible HyperText Markup Language (XHTML) is a recast of the HyperText Markup Language (HTML) in XML. XHTML Basic is the successor of compact HTML (cHTML) that is an initiative of the NTT DoCoMo, and of the Wireless Markup Language (WML) that is part of the wireless application protocol (WAP) architecture and an initiative of the Open Mobile Alliance (OMA). It uses XML for its syntax and HTML for its semantics. XHTML Basic has native support for elementary constructs for structuring information like paragraphs, lists, and so on. It could also be used as a placeholder for information fragments based on other languages, a role that makes it rather powerful in spite of being a small language.

The Scalable Vector Graphics (SVG) is a language for two-dimensional vector graphics that works across platforms, across output resolutions, across color spaces, and across a range of available bandwidths; SVG Tiny and SVG Basic are profiles of SVG targeted towards cellular phones and PDAs, respectively.

The Synchronized Multimedia Integration Language (SMIL) is a language that allows description of temporal behavior of a multimedia presentation, associates hyperlinks with media objects, and describes the layout of the presentation on a screen. It includes reusable components that can allow integration of timing and synchronization into XHTML and into SVG. SMIL Basic is a profile that meets the needs of resource-constrained devices such as mobile phones and portable disc players.

Namespaces in XML is a mechanism for uniquely identifying XML elements and attributes of a markup language, thus making it possible to create heterogeneous (compound) documents (Figure 1) that unambiguously mix

Figure 1. The architecture of a heterogeneous XML document for a mobile device

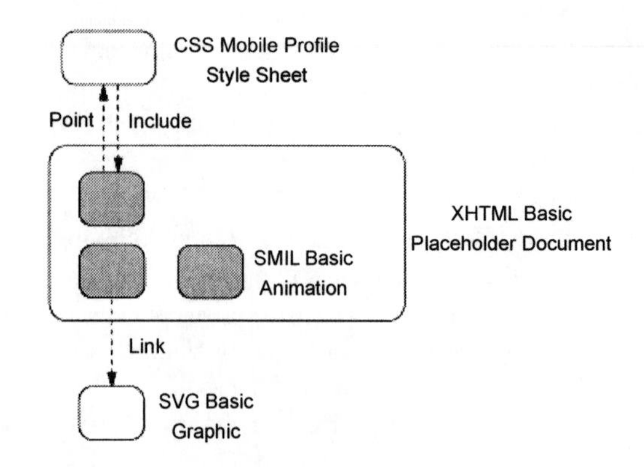

elements and attributes from multiple different XML document fragments.

Appropriate presentation on the user agent of information in a given modality is crucial. However, XML in itself (and by reference, the markup languages based on it) does not provide any special presentation semantics (such as fonts, horizontal and vertical layout, pagination, and so on) to the documents that make use of it. This is because the separation of the structure of a document from its presentation is a design principle that supports maintainability of a mobile application. The cascading style sheets (CSS) provides the presentation semantics on the client, and CSS mobile profile is a subset of CSS tailored to the needs and constraints of mobile devices.

With the myriad of proliferating platforms, information created for one platform needs to be adapted for other platforms. The eXtensible Stylesheet Language Transformations (XSLT) is a style sheet language for transforming XML documents into other, including non-XML, documents. As an example, information represented in XML could be transformed on-demand using an XSLT style sheet into XHTML Basic or an SVG Tiny document, as appropriate, for presentation to users accessing a mobile portal via a mobile device.

Representing information in XML provides various advantages towards archival, retrieval, and processing. It is possible to down-transform and render a document on multiple devices via an XSLT transformation, without making substantial modifications to the original source document. However, XML is not suitable for completely representing the knowledge inherent in information resources. For example, XML by itself does not provide any specific mechanism for differentiating between homonyms or synonyms, does not have the capabilities to model complex relationships precisely, is not able to extract implicit knowledge (such as hidden dependencies), and can only provide limited reasoning and inference capabilities, if at all.

The combination of the layers until now forms the basis of the mobile Web. The next two layers extend that and are largely responsible for what could be termed as the semantic mobile Web.

Semantic Level of a Semantic Mobile Application

This layer is responsible for the relationship of signs to what they stand for. The resource description framework (RDF) is a language for metadata that provides a "bridge" between the syntactic and semantic layers. It, along with RDF Schema, provides elementary support for *classification* of information into classes, properties of classes, and means to model more complex relationships among classes than possible with XML only. In spite of their usefulness, RDF/RDF Schema suffer from limited representational capabilities and non-standard semantics. This motivates the need for additional expressivity of knowledge.

The declarative knowledge of a domain is often modeled using ontology, an explicit formal specification of a conceptualization that consists of a set of concepts in a domain and relations among them (Gruber, 1993). By explicitly defining the relationships and constraints among the concepts in the universe of discourse, the *semantics* of a concept is constrained by restricting the number of possible interpretations of the concept.

In recent years, a number of initiatives for ontology specification languages for the semantic Web, with varying degrees of formality and target user communities, have been proposed, and the Web Ontology Language (OWL) has emerged as the successor. Specifically, we advocate that OWL DL, one of the sub-languages of OWL, is the most suitable among the currently available choices for representation of domain knowledge in mobile applications due to its compatibility with the architecture of the Web in general; and the Semantic Webin particular benefits from using XML/RDF/RDF Schema as its serialization syntax, its agreement with the Web standards for accessibility and internationalization, well-understood declarative semantics from its origins in description logics (DL) (Baader, McGuinness, Nardi, & Schneider, 2003), and provides the necessary balance between computational expressiveness and decidability.

Pragmatic Level of a Semantic Mobile Application

This layer is responsible for the relation of signs to interpreters. There are several advantages of an ontological representation. When information is expressed in a form that is oriented towards presentation, the traditional search engines usually return results based simply on a string match. This can be ameliorated in an ontological representation where the search is based on a *concept* match. An ontology also allows the logical means to distinguish between homonyms and synonyms, which could be exploited by a reasoner conforming to the language in which it is represented. For example, Java in the context of coffee is different from that in the context of an island, which in turn is different from the context of a programming language; therefore a search for one should not return results for other. Further, ontologies can be applied towards precise access of desirable information from mobile applications (Tsounis, Anagnostopoulos, & Hadjiefthymiades, 2004). Even though resources can be related to one another via a linking mechanism, such as the XML Linking Language (XLink), these links are merely structural constructs based on author discretion that do not carry any special semantics.

Explicit declaration of all knowledge is at times not cost effective as it increases the size of the knowledge base, which

Example 1. Ontological Inferences

```
<Region rdf:ID=“MontTremblant”>
 <subRegionOf rdf:resource=“#Laurentides”/>
</Region>
<Region rdf:ID=“Laurentides”>
 <subRegionOf rdf:resource=“#Qu&eacute;bec”/>
</Region>
<owl:TransitiveProperty rdf:ID=“subRegionOf”>
 <rdfs:domain rdf:resource=“#Region”/>
 <rdfs:range rdf:resource=“#Region”/>
</owl:TransitiveProperty>
```

becomes rather challenging as the amount of information grows. However, an ontology with a suitable semantical basis can make implicit knowledge (such as hidden dependencies) *explicit.* A unique aspect of ontological representation based for instance on OWL DL is that it allows logical constraints that can be reasoned with and enables us to *derive* logical consequences—that is, facts not literally present in the ontology but *entailed* by the semantics.

We have a semantic mobile portal for tourist information. Let Mont Tremblant, Laurentides, and Québec be defined as regions, and the subRegionOf property between regions be declared as transitive in OWL (see Example 1.)

Then, an OWL reasoner should be able to derive that if Mont Tremblant is a sub-region of Laurentides, and Laurentides is a sub-region of Québec, then Mont Tremblant is also a sub-region of Québec. This would give a more complete set of search results to a semantic mobile application user.

In spite of its potential, ontological representation of information presents certain domain-specific and human-centric challenges (Kamthan & Pai, 2006). Query formulations to a reasoner for extracting information from an ontology can be rather lengthy input on a mobile device. It is currently also difficult both to provide a sound logical basis to aesthetical, spatial/temporal, or uncertainty in knowledge, and represent that adequately in ontology.

Social Level of a Semantic Mobile Application

This layer is responsible for the manifestation of social interaction with respect to the representation. Specifically, ontological representations are a result of consensus, which in turn is built upon trust.

The client-side environment in a mobile context is constrained in many ways: devices often have restricted processing capability and limited user interface input/output facilities. The Composite Capabilities/Preference Profiles (CC/PP) Specification, layered on top of XML and RDF, allows the expression of user (computing environment and personal) preferences, thereby informing the server side of the delivery context.

Example 2. Device Profile

```
<rdf:RDF xmlns:rdf=“http://www.w3.org/1999/02/22-rdf-syntax-ns#”
xmlns:ccpp=“http://www.w3.org/2002/11/08-ccpp-schema#”
xmlns:prf=“http://a.com/schema#”>
...
<ccpp:component>
<rdf:Description rdf:about=“http://a.com/HardwareDevice”>
<rdf:type rdf:resource=“http://a.com/schema#HardwarePlatform”/>
<ccpp:defaults rdf:resource=“http://a.com/HardwareDefault”/>
<prf:vendor>MyMobileCompany</prf:vendor>
<prf:cpu>ABC</prf:cpu>
<prf:displayHeight>200</prf:displayHeight>
<prf:displayWidth>320</prf:displayWidth>
<prf:memoryMb>16</prf:memoryMb>
</rdf:Description>
</ccpp:component>
...
</rdf:RDF>
```

In Example 2, CC/PP markup for a device whose processor is of type ABC and the preferred default values of its display and memory as determined by its vendor are given. The namespace in XML is used to disambiguate elements/attributes that are native to CC/PP or RDF from those that are specific to the vendor vocabulary.

CC/PP can be used as a basis for introducing context-awareness in mobile applications (Sadeh, Chan, Van, Kwon, & Takizawa, 2003; Khushraj & Lassila, 2004).

One of the major challenges to the personalization based on profile mechanism is the user concern for privacy. The Platform for Privacy Preferences Project (P3P) allows the expression of privacy preferences of a user, which can be used by agents to decide if they have the permission to process certain content, and if so, how they should go about it. This ensures that users are informed about privacy policies of the mobile service providers before they release personal information. Thus, P3P provides a balance to the flexibility offered by the user profiles in CC/PP.

The Security Assertion Markup Language (SAML), XML Signature, and XML Encryption provide assurance of the sanctity of the message to processing agents. We note that an increasing number of languages to account for may place an unacceptable load, if it is to be processed exclusively, on the client side. We also acknowledge that these technologies alone will not solve the issue of trust, but when applied properly, could contribute towards it.

FUTURE TRENDS

The transition of the traditional mobile applications to semantic mobile applications is an important issue. The previ-

ous section has shown the amount of expertise and level of skills required for that. Although up-transformations are in general difficult, we anticipate that the move will be easier for the mobile applications that are well-structured in their current expression of information and in their conformance to the languages deployed.

The production of mobile applications, and by extension semantic mobile applications, is becoming increasingly complex and resource (time, effort) intensive. Therefore, a systematic and disciplined approach for their development, deployment, and maintenance, similar to that of Web engineering, is needed. Related to that, the issue of quality of represented and delivered information will continue to be important. The studies of specific attributes such as usability (Bertini, Catarci, Kimani, & Dix, 2005) and "best practices" for mobile applications from the World Wide Web Consortium (W3C) Mobile Web Initiative are efforts that could eventually be useful in an "engineering" approach for producing future semantic mobile applications.

The process of aggregation and inclusion of information in a mobile application is primarily manual, which can be both tedious and error prone. This process could be, at least partially, automated via the use of Web services where mobile applications could be made to automatically update themselves with (candidate) information. Therefore, manifestations of mobile applications through Semantic Webservices (Wagner & Paolucci, 2005; Wahlster, 2005) are part of a natural evolution.

CONCLUSION

For mobile applications to continue to provide a high quality-of-service (QoS) to their user community, their information architecture must be evolvable. The incorporation of Semantic Webtechnologies can be much more helpful in that regard. The adoption of these technologies does not have to be an "all or nothing" proposition: the evolution of a mobile application to a semantic mobile application could be gradual, transcending from one layer to another in the aforementioned framework. In the long term, the benefits of transition outweigh the costs.

Ontologies form one of the most important layers in semantic mobile applications, and ontological representations have certain distinct advantages over other means of representing knowledge. However, engineering an ontology is a resource-intensive process, and an ontology is only as useful as the inferences (conclusions) that can be drawn from it.

To be successful, semantic mobile applications must align themselves to the Semantic Webvision of inclusiveness for all. For that, the semiotic quality of representations, particularly that of ontologies, must be systematically assured and evaluated.

REFERENCES

Alesso, H. P., & Smith, C. F. (2002). *The intelligent wireless Web.* Boston: Addison-Wesley.

Baader, F., McGuinness, D., Nardi, D., & Schneider, P. P. (2003). *The description logic handbook: Theory, implementation and applications.* Cambridge University Press.

Bertini, E., Catarci, T., Kimani, S., & Dix, A. (2005). A review of standard usability principles in the context of mobile computing. *Studies in Communication Sciences, 1*(5), 111-126.

Gandon, F. L., & Sadeh, N. M. (2004, June 1-3). Context-awareness, privacy and mobile access: A Web semantic and multiagent approach. *Proceedings of the 1st French-Speaking Conference on Mobility and Ubiquity Computing* (pp. 123-130), Nice, France.

Gruber, T.R. (1993). Toward principles for the design of ontologies used for knowledge sharing. *Formal ontology in conceptual analysis and knowledge representation.* Kluwer Academic.

Hendler, J., Lassila, O., & Berners-Lee, T. (2001). The semantic Web. *Scientific American, 284*(5), 34-43.

Kamthan, P. (2001, March 20-22). Markup languages and mobile commerce: Towards business data omnipresence. *Proceedings of the WEB@TEK 2001 Conference,* Québec City, Canada.

Kamthan, P., & Pai, H.-I. (2006, May 21-24). Human-centric challenges in ontology engineering for the Semantic Web: A perspective from patterns ontology. *Proceedings of the 17th Annual Information Resources Management Association International Conference (IRMA 2006)*, Washington, DC.

Khushraj, D., & Lassila, O. (2004, November 7). CALI: Context Awareness via Logical Inference. *Proceedings of the Workshop on Semantic Web Technology for Mobile and Ubiquitous Applications,* Hiroshima, Japan.

Lassila, O. (2005, August 25-27). Using the Semantic Web in ubiquitous and mobile computing. *Proceedings of the 1st International IFIP/WG 12.5 Working Conference on Industrial Applications of the Semantic Web (IASW 2005)*, Jyväskylä, Finland.

Sadeh, N. M., Chan, T.-C., Van, L., Kwon, O., & Takizawa, K. (2003, June 9-12). A Semantic Web environment for context-aware m-commerce. *Proceedings of the 4th ACM Conference on Electronic Commerce* (pp. 268-269), San Diego, CA.

Stamper, R. (1992, October 5-8). Signs, organizations, norms and information systems. *Proceedings of the 3rd Australian*

Conference on Information Systems (pp. 21-55), Wollongong, Australia.

Tsounis, A., Anagnostopoulos, C., & Hadjiefthymiades, S. (2004, September 13). The role of Semantic Web and ontologies in pervasive computing environments. *Proceedings of the Workshop on Mobile and Ubiquitous Information Access (MUIA 2004)*, Glasgow, Scotland.

Wagner, M., & Paolucci, M. (2005, June 9-10). Enabling personal mobile applications through Semantic Web services. *Proceedings of the W3C Workshop on Frameworks for Semantics in Web Services*, Innsbruck, Austria.

Wahlster, W. (2005, June 3). Mobile interfaces to intelligent information services: Two converging megatrends. *Proceedings of the MINDS Symposium,* Berlin, Germany.

Wilson, M., Russell, A., Smith, D. A., Owens, A., & Schraefel, M. C. (2005, November 7). mSpace mobile: A mobile application for the Semantic Web. *Proceedings of the 2nd International Workshop on Interaction Design and the Semantic Web,* Galway, Ireland.

KEY TERMS

Delivery Context: A set of attributes that characterizes the capabilities of the access mechanism, the preferences of the user, and other aspects of the context into which a resource is to be delivered.

Knowledge Representation: The study of how knowledge about the world can be represented and the kinds of reasoning can be carried out with that knowledge.

Ontology: An explicit formal specification of a conceptualization that consists of a set of terms in a domain and relations among them.

Personalization: A strategy that enables delivery that is customized to the user and user's environment.

Semantic Web: An extension of the current Web that adds technological infrastructure for better knowledge representation, interpretation, and reasoning.

Semiotics: The field of study of signs and their representations.

User Profile: An information container describing user needs, goals, and preferences.

Location-Based Multimedia Content Delivery System for Monitoring Purposes

Athanasios-Dimitrios Sotiriou
National Technical University of Athens, Greece

Panagiotis Kalliaras
National Technical University of Athens, Greece

INTRODUCTION

Advances in mobile communications enable the development and support of real-time multimedia services and applications. These can be mainly characterized by the personalization of the service content and its dependency to the actual location within the operational environment. Implementation of such services does not only call for increased communication efficiency and processing power, but also requires the deployment of more intelligent decision methodologies.

While legacy systems are based on stationary cameras and operational centers, advanced monitoring systems should be capable of operating in highly mobile, ad-hoc configurations, where overall situation and users roles can rapidly change both in time and space, exploiting the advances in both the wireless network infrastructure and the user terminals' capabilities. However, as the information load is increased, an important aspect is its filtering. Thus, the development of an efficient rapid decision system, which will be flexible enough to control the information flow according to the rapidly changing environmental conditions and criteria, is required. Furthermore, such a system should interface and utilize the underlying network infrastructures for providing the desired quality of service (QoS) in an efficient manner.

In this framework, this article presents a *location-based multimedia content delivery system* (LMCDS) for monitoring purposes, which incorporates media processing with a decision support system and positioning techniques for providing the appropriate content to the most suitable users, in respect to user profile and location, for monitoring purposes. This system is based on agent technology (Hagen & Magendanz, 1998) and aims to promote the social welfare, by increasing the overall situation awareness and efficiency in emergency cases and in areas of high importance. Such a system can be exploited in many operational (public or commercial) environments and offers increased security at a low cost.

SERVICES

The LMCDS provides a platform for rapid and easy set up of a monitoring system in any environment, without any network configurations or time-consuming structural planning. The cameras can be installed in an ad hoc way, and video can be transmitted to and from heterogeneous devices using an Intelligent decision support system (IDSS) according to the user's profile data, location information, and network capabilities.

Users can dynamically install ad-hoc cameras to areas where the fixed camera network does not provide adequate information. The real-time transmission of still images or video in an emergency situation or accident to the available operational centers can instantly provide the necessary elements for the immediate evaluation of the situation and the deployment of the appropriate emergency forces. This allows the structure of the monitoring system to dynamically change according to on-the-spot needs.

The IDSS is responsible for overviewing the system's activity and providing multimedia content to the appropriate users. Its functionality lies in the following actions:

- identifying the appropriate user or group of users that need access to the multimedia content (either through user profile criteria or topological criteria);
- providing the appropriate multimedia content in relevance to the situation and the location; and
- adapting the content to the user's needs due to the heterogeneity of the users devices—that is, low bit rate video to users with portable devices.

The LMCDS can evaluate users' needs and crisis events in respect to the topological taxonomy of all users and provide multimedia content along with geographical data. The location information is obtained through GPS or from GPRS through the use of corresponding techniques (Markoulidakis, Desiniotis, & Kypris, 2004). It also provides intelligent methodologies for processing the video and image content according to network congestion status and terminal devices. It can handle the necessary monitoring management mechanisms, which enable the selection of the non-congested network elements for transferring the appropriate services (i.e., video streaming, images, etc.) to the concerned users. It also delivers the service content in the most appropriate

Figure 1. System functionality and services

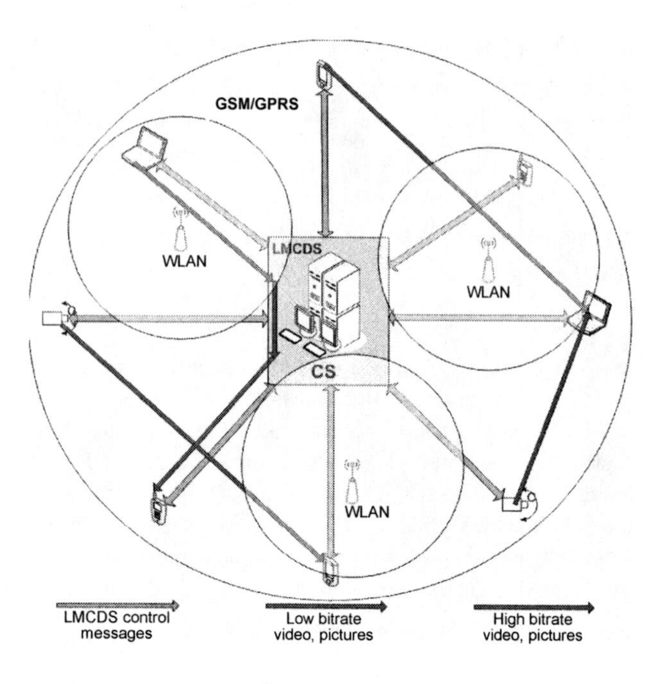

format, thus allowing the cooperation of users equipped with different types of terminal devices.

Moreover, the LMCDS provides notification services between users of the system for instant communication in case of emergency through text messaging or live video feed.

All of the above services outline the requirements for an advanced monitoring system. The LMCDS functionality meets these requirements, since it performs the following features:

- location-based management of the multimedia content in order to serve the appropriate users;
- differentiated multimedia content that can be transmitted to a wide range of devices and over different networks;
- lightweight codecs and decoders that can be supported by devices of different processing and network capabilities;
- IP-based services in order to be transparent to the underlying network technology and utilize already available hardware and operating systems platforms;
- intelligent delivery of the multimedia content through the LMCDS in order to avoid increased traffic payload as well as information overload; and
- diverse localization capabilities through both GPS and GPRS, and generation of appropriate topological data (i.e., maps) in order to aid users.

However, the system architecture enables the incorporation of additional location techniques (such as WLAN positioning mechanisms)) through the appropriate, but simple, development of the necessary interfaces with external location mechanisms.

In order to describe the above services in a more practical way, a short list of available options and capabilities of the system is given below. The target group of the LMCDS consists of small to medium businesses that seek a low-cost solution for monitoring systems or bigger corporations that need an ad hoc extension to their current system for emergency cases and in order to enhance their system's mobility. Even though the users of the system consist mainly of security staff and trained personnel that are in charge of security, the system's ease of use, low user complexity, and device diversity allow access even to common untrained users.

The system offers a range of capabilities, most of which are summarized in Figure 1, such as:

- User registration and authentication.
- User profile (i.e., device, network interface).
- Location awareness:
  - User is located through positioning techniques.
  - User is presented with appropriate topographical information and metadata.
  - User is aware of all other users' locations.
  - User can be informed and directed from a Center of Security (CS) to specified locations.
- Multimedia content:
  - Video, images, and text are transmitted to user in real time or off-line based on situation or topological criteria.
  - User can provide feedback from his device through camera (laptop, PDA, smart phone) or via text input.
  - Content is distributed among users from the CS as needed.
- Ad hoc installation of cameras that transmit video to CS and can take advantage of wireless technology (no fixed network needed).
- Autonomous nature of users due to agent technology used.

LMCDS ARCHITECTURE

The LMCDS is designed to distribute system functionality and to allow diverse components to work independently while a mass amount of information is exchanged. This design ensures that new users and services can be added in an ad hoc manner, ensuring future enhancements and allowing it to support existing monitoring systems.

Multi-agent systems (MASs) (Nwana & Ndumu, 1999) provide an ideal mechanism for implementing such a heterogeneous and sophisticated distributed system in contrast to traditional software technologies' limitations in communication and autonomy.

Figure 2. Architecture overview

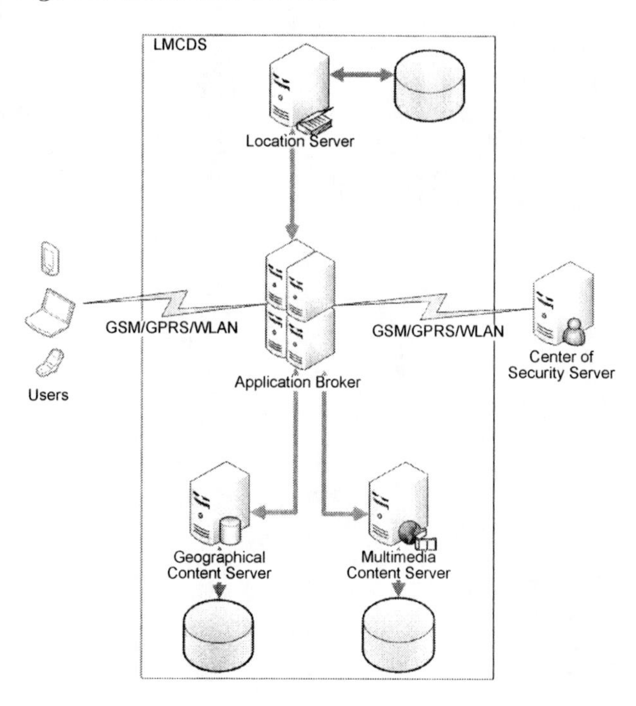

The system is developed based on MAS and allows diversified agents to communicate with each other in an autonomous manner, resulting in an open, decentralized platform. Tasks are being delegated among different components based on specific rules, which relate to the role undertaken by each agent, and information is being composed to messages and exchanged using FIPA ACL (FIPA, 2002a). An important aspect for the communication model is the definition of the content language (FIPA, 2002b). Since LMCDS targets to a variety of devices, including lightweight terminals, the LEAP language (Berger, Rusitschka, Toropov, Watzke, & Schlichte, 2002) of the JADE technology has been exploited.

In addition to the security mechanisms supported by the underlying network components, the JADE platform offers a security model (Poggi, Rimassa, & Tomaiuolo, 2001) that enables the delegation of tasks to respective agent components by defining certificates, and ensures the authentication and encryption of TCP connections through the secure socket layer (SSL) protocol (http://www.openssl.org/).

The general architecture of the LMCDS is shown in Figure 2. The platform is composed of different agents offering services to the system which are linked by an application broker agent, acting as the coordinator of the system. These agents are the location server, the center of security server, the application broker, the geographical content server, and the multimedia content server. The latter two are discussed in later sections in more detail, while a brief description of the functionality of the others is given as follows.

The location server agent is responsible for the tracking of all users and the forwarding of location-based information to other components. The information is gathered dynamically and kept up-to-date according to specific intervals. The intelligence lies in the finding of the closest users to the demanded area, not only in terms of geographical coordinates, but also in terms of the topology of the environment. More information on location determination is given in a following section.

The center of security server agent monitors all users and directs information and multimedia content to the appropriate users. It is responsible for notifying users in emergency situations, and also performs monitoring functions for the system and its underlying network.

The User agent components manage information, including the transmission and reception of image or video, the display of location information, critical data, or other kinds of displays. They are in charge of several user tasks, such as updating users' preferences and location, decoding the response messages from the application broker, and performing service level agreement (SLA) (ADAMANT, 2003) monitoring functionalities. The user agent can reside in a range of devices, since it is Java based.

Finally, the application broker agent acts as a mediator that provides the interface between all of the components. It is responsible for prioritizing user requests according to users' profiles, performing authentication functions, and acting as a facilitator for SLA negotiations. It also coordinates the communication process between users in order for the multimedia content to be delivered in the appropriate format according to the user's processing and network capabilities, and also the network's payload.

A closer look into the components of each agent, along with the interaction between them, is shown in Figure 3. Each agent is composed of three main components: the graphical display, which is responsible for user input and information display; the communication unit, in charge of agent communication through ACL messages; and the decision processing unit, which processes all received information. In addition, the user and the CS include a multimedia content component for the capture, playback, and transmission of multimedia data through RTP or FTP channels.

MULTIMEDIA PROCESSING AND ENCODING

Video/Image Formats

One of the novelties is the ability of the system to perform real-time format conversions of the image or video data and transmit to several heterogeneous recipients, ranging from large PC servers to small personal devices like PDAs or mobile phones.

Figure 3. Agent components

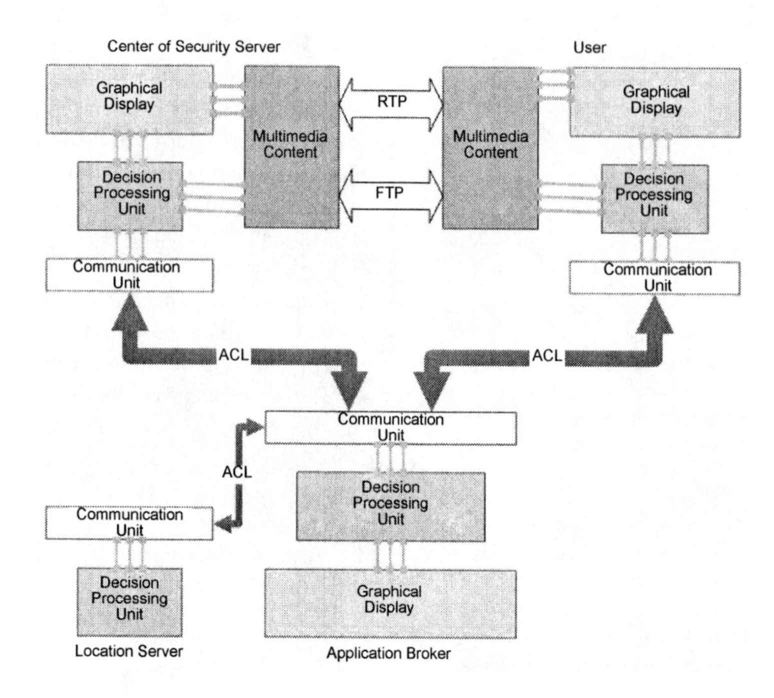

The LMCDS enables the adoption and support of different video formats, according to the partial requirements of the user terminals and the available networks status. The most commonly used is the M-JPEG (http://www.jpeg .org/index.html) format. It was preferred over other common video formats suitable for wireless networks like MPEG (http://www. m4if.org/) and H.263 (http://www.itu.int/rec/ recommendation.asp), which provide higher compression ratio, because using them can require intense processing power both for the encoder and the decoder. Also, frames in MPEG or H.263 streams are inter-related, so a single packet loss during transmission may degrade video quality noticeably. On the contrary, M-JPEG is independent of such cascaded-like phenomena, and it is preferable for photo-video application temporal compression requirements for smoothness of motion.

It is a lossy codec, but the compression level can be set at any desired quality, so the image degradation can be minimal. Also, at small data rates (5-20Kbps) and small frame rates, M-JPEG produces better results than MPEG or H.263. This is important, as the photos or video can be used as clues in legal procedures afterwards, where image quality is more crucial than smooth motion. Another offering feature is the easy extraction of JPEG (http://www.jpeg .org/index.html) images from video frames.

The video resolution can be set in any industry-standard (i.e., subQCIF) or any other resolution of width and height dividable of 8, so the track is suitable for the device it is intended for. Video is streamed directly from the camera-equipped terminals in a peer-to-peer manner. Transmission rates for the video depend on the resolution and the frame rate used. Some sample rates are given in Table 1.

Apart from M-JPEG, another set of video formats have been adopted, such as H263 and MPEG-4. The development of these formats enables the testing and evaluation of the LMCDS, based on the network congestion and the current efficiency of the supported video formats and the crisis situations in progress. This means that for a specific application scenario, the encoding with M-JPEG format can lead to better quality on the user terminal side, while the MPEG 4 format can be effective in cases that the network infrastructure is highly loaded, so the variance in bit rate can keep the quality in high values.

Image compression is JPEG with resolution of any width and height dividable of 8. For the real-time transmission of video stream, the real-time transfer protocol (RTP, http:// www.ietf.org/rfc/rfc3550.txt) is used, while for stored images and video tracks, the file transfer protocol (FTP) is used.

Video Processing

It is important to point out that the output video formats can be produced and transmitted simultaneously with the use of the algorithm shown in Figure 4.

Note that the image/video generator can be called several times for the same captured video stream as long as it is fed with video frames from the frame grabber. So, a single user can generate multiple live video streams with variations,

Table 1. Output video formats for the application

Resolution	Frame Rate	Suitable Network	Trans. Rate (kbps)	Target Device
160 x 120	1	GPRS WLAN	2 – 3	Smartphone PDA, PC
160 x 120	5	GPRS WLAN	10 – 15	Smartphone, PDA,PC
232 x 176	2	GPRS WLAN	10 – 15	PDA,PC
320 x 240	5	WLAN	30 – 40	PC
320 x 240	15	WLAN	90 – 120	PC
640 x 480	5	WLAN	45 – 55	PC

not only in frame rate and size, but also in JPEG compression quality, color depth, and even superimpose layers with handmade drawings or text. The algorithm was implemented in Java with the use of the JMF API (http://java.sun. com/products/java-media/jmf/).

For a captured stream at a frame rate of n frames per second, the frame grabber component extracts from the raw video byte stream n/A samples per second, where A is a constant representing the processing power of the capturing device. Depending on how many different qualities of video streams need to be generated, m Image/Video Generator processes are activated, and each process i handles K_i fps. The following relationship needs to be applied:

$$K_i = B_i * n / A * m \text{ , where}$$

$$B_i = (K_i * A * m)/n \text{ and}$$

$$\sum B_i = m$$

There is also a latency of $m*C_i$ seconds per video stream, where C_i depends on the transcoding time of the image to each final format.

So, this tendency of the LMCDS to keep the frame rate low is inevitable due to this sampling process. However, using low frame rates is quite common in surveillance systems. It also allows long hours of recording, where video size is optimal when minimum both for storage and transmission, and is less demanding in processing power for use with video players running on small devices.

Figure 4. LMCDS video processing overview

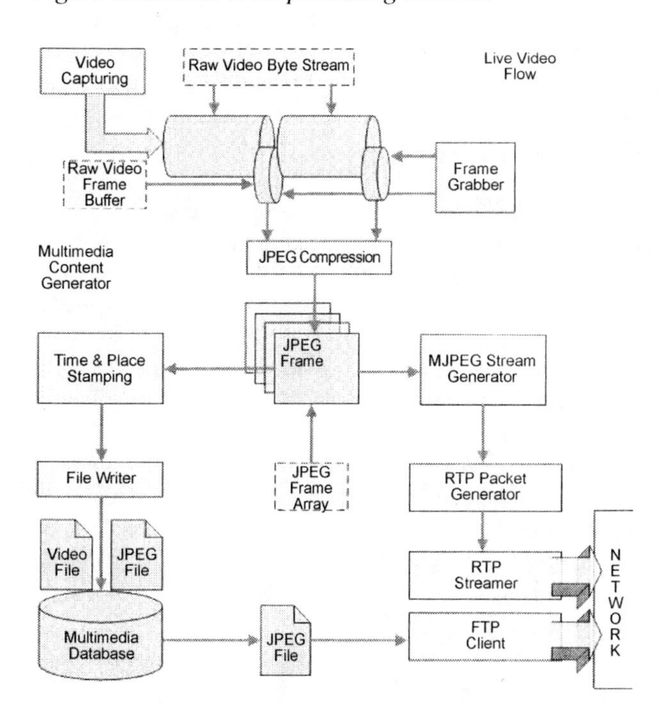

GEOGRAPHICAL CONTENT DELIVERY

Positioning Methods

The LMCDS uses the following techniques for locating the users' positions inside the served environment.

- *GPS (Global Positioning System)* is a global satellite system based on a group of non-geostatic satellites in middle altitude orbit (12,000 miles). The GPS-enabled devices have the ability to locate their position with a high degree of accuracy by receiving signals from at least six satellites.
- *A GSM/GPRS subscriber can be located upon request, depending on the received power from adjacent cells. More information about these mechanisms can be found in* Markoulidakis et al. (2004).
- The *Ekahau* position engine (http://www.ekahau.com/pdf/EPE_2.1_datasheet.PDF) is designed to locate 802.11b (at present) wireless LAN users positions in the indoor environment. In the context of radio resource management, the software could apply the access point's radio coverage map in the indoor environment. It uses a calibration method. Initially it measures a set of sample points' radio strength. Based on these sample points, the engine can estimate a client WLAN station's approximate location, given an arbitrary combination of signal strengths.

Geographical Database

The geographical database of the LMCDS is storing information about the positions of the users that are registered in the system. The information is obtained regularly by scheduled queries. When the users perform service request messages to the system, their position coordinates are automatically retrieved (through any of the above methods), so their location in the geographical DB is also updated. Special importance has been given to the ability of the system to serve queries about the relative position of its users and estimation of distances. So, the scheduled queries can be of the following types:

- Find my 3 nearest users and send them a video.
- Estimate the time that I need to get to Building A.
- *Find the users closest to point B and send pictures to those that operate in GPRS network and high-quality video to those in UMTS or WLAN.*
- *When user C enters a specified area, send him a message.*

The user is displayed visual information in the form of maps to its device. The map consists of raster data—that is, the plan of the area and also several layers of metadata, showing points of interest, paths, as well as the position of relative users.

FUTURE TRENDS

Future steps involve the exploitation of video streaming measurements for providing guaranteed QoS of the video content to the end user, as well as the better utilization of the available radio resources. Furthermore, the incorporation of new trends in video streaming in conjunction with a markup language for multimedia content, such as MPEG-7 or MPEG-21, can offer a higher level of personalized location-based services to the end user and are in consideration for future development.

CONCLUSION

This article presented a location-based multimedia content system enabling real-time transfer of multimedia content to end users for location-based services. Based on the general architecture of multi-agent systems, it focused on fundamental features that enable the personalization of the service content and the intelligent selection of the appropriate users for delivering the selected content.

REFERENCES

Berger, M., Rusitschka, S., Toropov, D., Watzke, M., & Schlichte, M. (2002). Porting distributed agent-middleware to small mobile devices. *Proceedings of the Workshop on Ubiquitous Agents on Embedded, Wearable, and Mobile Devices,* Bologna, Italy.

FIPA (Foundation for Intelligent Physical Agents). (2002a). *FIPA ACL message structure specification.* SC00061G.

FIPA. (Foundation for Intelligent Physical Agents). (2002b). *FIPA SL content language specification.* SC00008I.

Hagen, L., & Magendanz, T. (1998). Impact of mobile agent technology on mobile communication system evolution. *IEEE Personal Communications, 5*(4).

IST ADAMANT Project. (2003). *SLA management specification.* IST-2001-39117, Deliverable D6.

Markoulidakis, J. G., Desiniotis, C., & Kypris, K. (2004). Statistical approach for improving the accuracy of the CGI++ mobile location technique. *Proceedings of the Mobile Location Workshop , Mobile Venue '04.*

Nwana, H., & Ndumu, D. (1998). A perspective on software agents research. *The Knowledge Engineering Review, 14*(2).

Poggi, A., Rimassa, G., & Tomaiuolo, M. (2001). Multi-user and security support for multi-agent systems. *Proceedings of WOA 2001 Workshop,* Modena, Italy.

KEY TERMS

Agent: A program that performs some information gathering or processing task in the background. Typically, an agent is given a very small and well-defined task.

Application Broker: A central component that helps build asynchronous, loosely coupled applications in which independent components work together to accomplish a task. Its main purpose is to forward service requests to the appropriate components.

IDSS: Intelligent decision support system.

LMCDS: Location-based multimedia content delivery system.

MAS: Multi-agent system.

Media Processing: Digital manipulation of a multimedia stream in order to change its core characteristics, such as quality, size, format, and so forth.

Positioning Method: One of several methods and techniques for locating the exact or relative geographical position of an entity, such as a person or a device.

Location-Based Multimedia Services for Tourists

Panagiotis Kalliaras
National Technical University of Athens, Greece

Athanasios-Dimitrios Sotiriou
National Technical University of Athens, Greece

P. Papageorgiou
National Technical University of Athens, Greece

S. Zoi
National Technical University of Athens, Greece

INTRODUCTION

The evolution of mobile technologies and their convergence with the Internet enable the development of interactive services targeting users with heterogeneous devices and network infrastructures (Wang et al., 2004). Specifically, as far as cultural heritage and tourism are concerned, several systems offering location-based multimedia services through mobile computing and multimodal interaction have already appeared in the European research community (LOVEUS, n.d.; Karigiannis, Vlahakis, & Daehne, n.d.).

Although such services introduce new business opportunities for both the mobile market and the tourism sector, they are not still widely deployed, as several research issues have not been resolved yet, and also available technologies and tools are not mature enough to meet end user requirements. Furthermore, user heterogeneity stemming both from different device and network technologies is another open issue, as different versions of the multimedia content are often required.

This article presents the AVATON system. AVATON aims at providing citizens with ubiquitous user-friendly services, offering personalized, location-aware (GSM Association, 2003), tourism-oriented multimedia information related to the area of the Aegean Volcanic Arc. Towards this end, a uniform architecture is adopted in order to dynamically release the geographic and multimedia content to the end users through enhanced application and network interfaces, targeting different device technologies (mobile phones, PDAs, PCs, and TV sets). Advanced positioning techniques are applied for those mobile user terminals that support them.

SERVICES

AVATON is an ambient information system that offers an interactive tour to the user (visitor) in the area of the Aegean Volcanic Arch (see http://www.aegean.gr/petrified_forest/). The system can serve both as a remote and as an onsite assistant for the visitor, by providing multimedia-rich content through various devices and channels:

- over the Internet, via Web browsers with the use of new technologies such as rich-clients and multi-tier architecture in order to dynamically provide the content;
- with portable devices (palmtops, PDAs) and 2.5G or 3G mobile phones, which are capable of processing and presenting real-time information relevant to the user's physical position or areas of interest; and
- via television channels—AVATON allows users to directly correlate geographic with informative space and conceivably pass from one space to the other, in the context of Worldboard (Spohrer, 1999).

With the use of portable devices equipped with positioning capabilities, the system provides:

- dynamic search for geographical content, information related to users' location, or objects of interest that are in their proximity;
- tours in areas of interest with the aid of interactive maps and 3-D representations of the embossed geography;
- search for hypermedia information relative to various geographic objects of the map;
- user registration and management of personal notes during the tour that can be recalled and reused during later times; and

Figure 1. The AVATON architecture

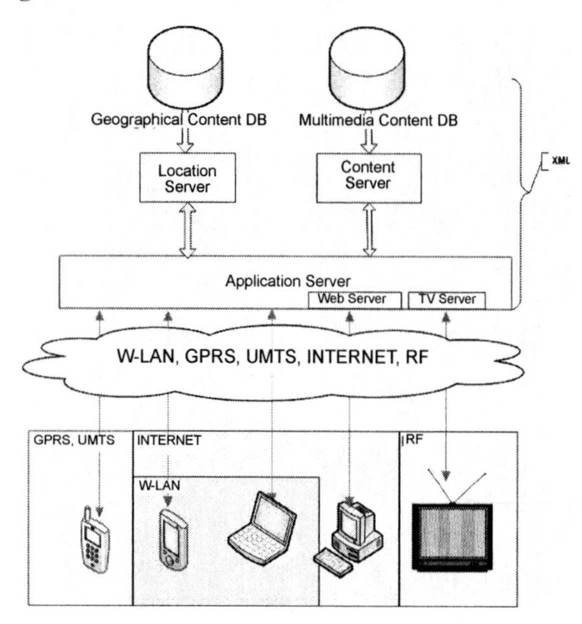

Figure 2. The XML-based technologies in the client side

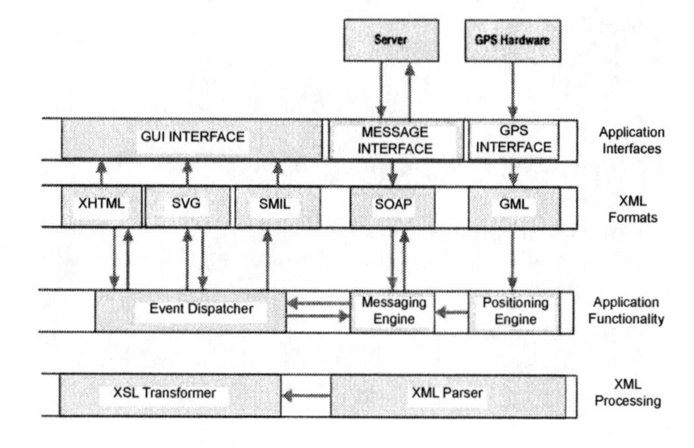

- interrelation of personal information with natural areas or objects for personal use or even as a collective memory relative to visited areas or objects.

THE AVATON ARCHITECTURE

Overview

The AVATON system is based on a client-server architecture composed of three main server components: the application server, the content server, and the location server. The application server combines information and content from the content and location servers, and replies to client requests through different network technologies. The content is retrieved from two kinds of databases, the geographical and multimedia content DBs. The above architecture is shown in Figure 1.

In more detail:

- **Multimedia Content Database:** This database contains the multimedia content such as images, video, audio, and animation.
- **Geographical Content Database:** A repository of geographical content such as aerial photos, high-resolution maps, and relevant metadata.
- **Content Server:** The content server supplies the application server with multimedia content. It retrieves needed data from the multimedia content database according to user criteria and device capabilities, and responds to the application server.
- **Location Server:** Serves requests for geographical content from the application server by querying the geographical content database. The content retrieved is transformed into the appropriate format according to user device display capabilities and network bandwidth available.
- **Application Server:** The application server receives requests from different devices through GPRS, UMTS (third-generation mobile phone), W-LAN, Internet (PDA, laptop, PC), and RF (television). The server identifies each device and transmits data in an appropriate format. More precisely, the application server incorporates a Web server and a TV server in order to communicate with PCs and televisions respectively.

Client

This section focuses on the mobile-phone and PDA applications. The scope of the AVATON system includes Java-enabled phones with color displays and PDAs with WLAN or GPRS/UMTS connectivity. While all the available data for the application can be downloaded and streamed over the network, data caching is exploited for better performance and more modest network usage.

When the users complete their registration in the system, they have in their disposal an interactive map that initially portrays the entire region as well as areas or individual points of interest. For acquiring user position, the system is using GPS. The client also supports multi-lingual implementation, as far as operational content is concerned, for example menus, messages, and help. These files are maintained as XML documents. XML is extensively used in order to ease the load of parsing different data syntaxes. A single process, the XML Parser is used for decoding all kinds of data and an XSL Transformer for transcoding them in new formats. The different XML formats are XHTML, SVG, SMIL, SOAP, and GML, as shown in Figure 2.

Geographical Info Presentation

In order to render the geographical data, the client receives raster images for the drawing of the background map, combined with metadata concerning areas of interest and links to additional textual or multimedia information. The raster data are aerial high-resolution photographs of the region on two or three scales. Because of the high resolution of the original images, the client is receiving small portions, in the form of tiles from the *raster data processing engine* in the server side, which are used to regenerate the photorealistic *image layer* in a resolution that is suitable for the device used. The attributes of the geographical data are generated in vector ShapeFile (ESRI ShapeFile) format, which is quite satisfactory for the server side but not for lightweight client devices. So, a *SHP TO SVG converter* at the server side is regenerating the metadata in SVG format that can be viewed properly from a handheld device. As soon as the metadata is downloaded to the client device, a final filtering (XSLT transformation) is done and the additional layer is opposed to the image layer in the *SVG viewer.* On the *SVG data layer,* the user can interact with points of interest and receive additional information in the form of text or multimedia objects. The above are shown in Figure 3.

Multimedia Info Presentation

The presentation of multimedia information mainly depends on user position. The system is designed to provide audio and video clips, 3-D representations, and also textual information concerning each place of interest. Not all devices, though, receive the same content, since they differ in display, processor, or network speed. For that purpose, for each registered user, the system decides what kind of content is more suitable for them to receive and the multimedia content server generates the appropriate script. Depending on the available memory of the client's device, media objects stay resident in the cache memory so that frequently requested content is accessed without delays that occur due to network latency. In Figure 4 the components that are involved in the multimedia presentation are shown. The *TourScript Data* contains the script which describes the multimedia presentation. It is transcoded inside the *SMIL generator* to a SMIL message that follows the XML syntax, so that it can be incorporated seamlessly to the messages that are exchanged in the AVATON system. At the client side, the SMIL message is received by the *SMIL processor* which coordinates the process of fetching the *multimedia objects* from the *client cache memory* to the suitable renderer, so that the *multimedia presentation* can be completed.

Location Server

The location server is the component that handles the geographical content of the AVATON system. It provides a storage system for all geographical data and allows querying of its contents through location criteria, such as global position and areas of interest. Content management is based on a PostgreSQL (http://www.postgres.org) relational database. A JDBCInterface uses the JDBC APIs in order to provide support for data operations. A GISExtension is also present, based on PostGIS (http://www.postgis.org), in order to enable the PostgreSQL server to allow spatial queries. This feature is utilized through a GISJDBCInterface, a PostGIS layer on top of PostgreSQL.

Figure 3. Client-side map rendering

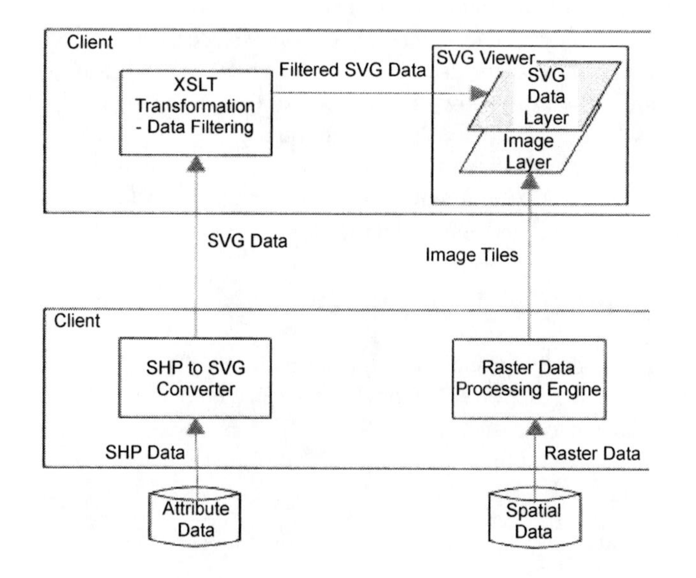

Figure 4. Client-side map rendering

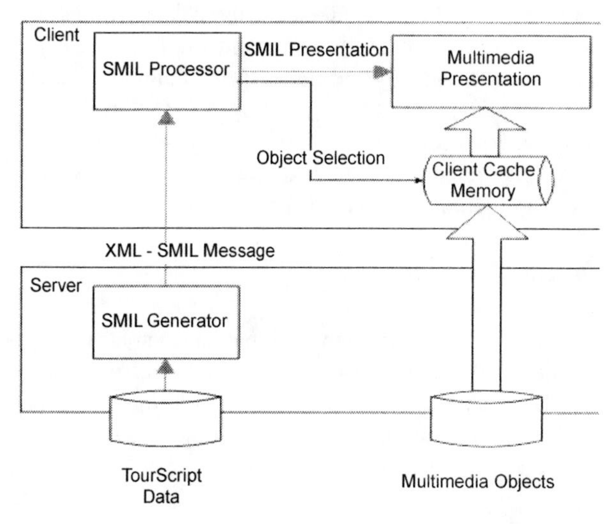

Cartographic Data

Concerning the photorealistic information, the user can choose from several distinct zoom levels. The mobiles phones and PDAs in the market that support GPRS or WLAN have displays of different resolutions that, in most cases, are multiples of 16 pixels. Hence, the location server can generate tiles with a multiple of 16p x 16p, which can be presented in the user's mobile device. The server always holds multiple resolutions for every level of cartographic (photographic) information. The levels of cartographic information define the degree of focus.

Apart from the photographic layer, additional layers of vector information also exist, and their size is approximately 5% or 10% of the corresponding photographic. Therefore, in practice, every device will initially request from the server the cartographical information with the maximum resolution this device can support. Hence, the server decides the available resolution that best corresponds to the requested resolution from the device. The size of every tile is approximately 1K. The devices with greater resolution per tile receive more files, with greater magnitude, for every degree of focus due to the higher resolution.

Multimedia Content Server

The multimedia content server component comprises the major unit that controls the mixing and presentation of different multimedia objects. Its purpose is to upload all the objects necessary and present them in a well-defined controlled order that in general depends on the user position, interactions, and tracking information available. The multilingual audiovisual information scheduled for presentation is coordinated so that several objects may be presented simultaneously. The multimedia content server component is also responsible for choosing "relevant" objects for the user to select among in the case the user requires more information on a topic. The multimedia content server interfaces with the multimedia content database, a relational database storing the multimedia content. The database is organized thematically and allows the creation of hierarchical structures. It also contains a complete list of multimedia material, covering all content of the physical site, such as 3D reconstructed plants, audio narration, virtual 3D models, avatar animations, and 2D images.

Media Objects

As mentioned already, the multimedia content server is responsible for mixing the basic units of multimedia information. These elements are hierarchically ordered. At the finest level of granularity, there are atomic objects called MediaObjects with specializations such as AudioMediaObject,

Figure 5. Hierarchy formulation of media objects at the multimedia content server

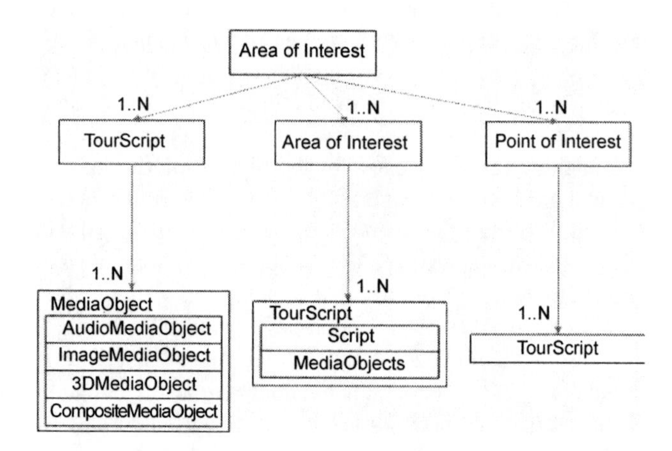

ImageMediaObject, 3DMediaObject, and CompositeMediaObject. These objects contain the actual data to be rendered along with additional profile metadata characterizing them. At a higher level of complexity, a TourScript represents an ordered sequence of MediaObjects, all of which are to be presented if the script is chosen.

According to user requirements, the user will be able to navigate through the site in a geographically based tree. This is made possible through the use of points of interest (PoI) and areas of interest (AoI). A PoI can only contain TourScripts and can be viewed as the end node of the site tree. In contrast, an AoI may contain either another PoI, an AoI, or TourScripts. This allows the system to map the actual site into a hierarchy model containing PoI at the top and MediaObject components at the leaf level.

The multimedia content server is also responsible for managing this site-tree for the entire site. Moreover it is responsible for traversing it. The use of the site tree is quite interesting: when a media object, for instance *audio* object, is presented to the user, it belongs to a node in the site hierarchy. Figure 5 shows the structure of the site in a tree view as described previously. The multimedia content server is responsible for coordinating the rendering components in order to provide a synchronized presentation to the user, according to user preferences, position, and commands.

Deployment and Usage

Based on the proposed architecture, the AVATON services are being deployed to physical sites within the Aegean Volcano Arc (such as Santorini and Lesvos islands) and evaluated by real end users under different scenarios. The main air interfaces that will be used by the system (along

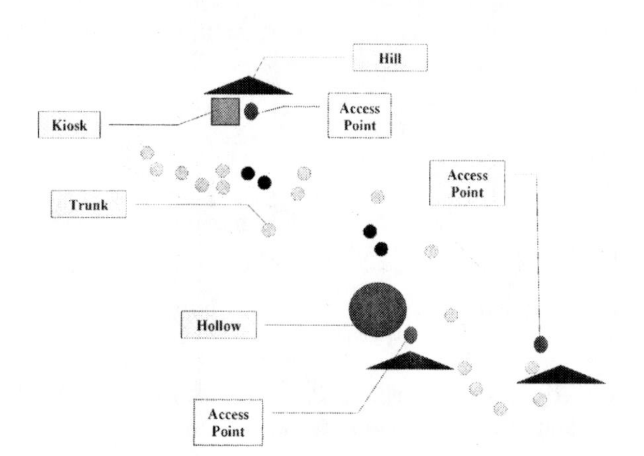

Figure 6. Implementation plan for the Lesvos island site

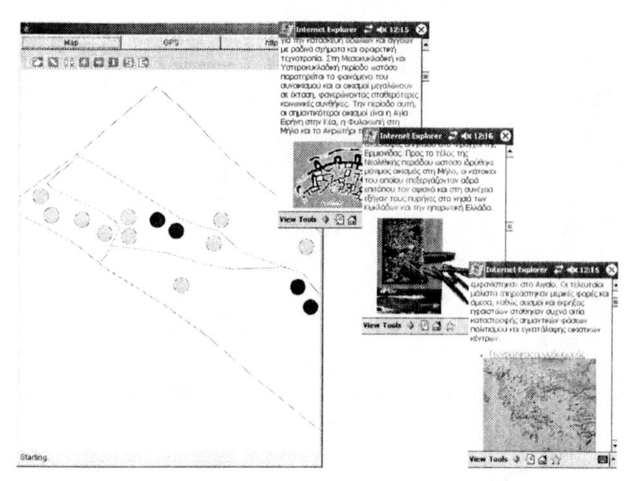

Figure 7. SVG map and interface

with standard wired access through common LANs or the Internet) are:

- **WLAN:** A standard 802.11b wireless access network provides connectivity for users equipped with portable devices (such as PDA with built-in WLAN cards or laptops).
- **GPRS/UMTS:** For mobile users and smart phones, access will be provided through the GSM network, using GPRS (Bettsteter, Vögel, & Eberspächer, 1999). This restricts the system from providing video or 3-D animations to such users, and the services offered are focused on text, images (including map information), and short audio. As GPRS is already packet oriented, our implementation can be easily transferred to UMTS, if available.

In Figure 6, the actual implementation plan is given for the location of the Sigri Natural History Museum on Lesvos island. The area consists of an open geological site, the Petrified Forest, where the ash from a volcanic eruption some 15 to 20 million years ago covered the stand of sequoia trees, causing their petrification. Wireless access is provided by the use of a 3 Netgear 54Mbps access point equipped with additional Netgear antennas in order to overcome the physical limitations of the area (hills, trunks, and hollows).

Visitors to the site are equipped with PDAs or smart phones (provided at the entrance kiosk) and stroll around the area. A typical scenario consists of the following: The users enter the archaeological site and activate their devices. They then perform a login and provide personal details to the server, such as username, language selection, and device settings. The client then requests from the server and loads the map of the area in SVG (http://www.w3.org/TR/SVG) format, as seen in Figure 7. The circles on the map present distinct points of interest (yellow indicating trees and blue indicating leafs). The application monitors the users' location and updates the SVG map in real time, informing the users of their position. The SVG map is interactive, and when the users enter the vicinity of a point of interest, the application automatically fetches and displays (via their browser or media player) the corresponding multimedia content (in the form of HTML pages, audio, or video) at the requested language. The users are also able to navigate manually through the available content and receive additional information on topics of their interest. During the tour, the users' path is being tracked and displayed (the red line on Figure 7) in order to guide them through the site. They are also able to keep notes or mark favorite content (such as images), which can be later sent to them when they complete the tour.

FURTHER WORK

The system is currently being deployed and tested in two archeological sites. Users are expected to provide useful feedback on system capabilities and assist in further enhancements of its functionality. Also, as 3G infrastructure is being expanded, incorporation of the UMTS network in the system's access mechanisms will provide further capabilities for smart phone devices and also use of the system in areas where wireless access cannot be provided.

Towards commercial exploitation, billing and accounting functionalities will be incorporated into the proposed architecture. Finally, possible extensions of the system are considered in order to include other cultural or archeological areas.

REFERENCES

Bettsteter, C., Vögel, H.-J., & Eberspächer, J. (1999). GSM Phase 2+ General Packet Radio Service GPRS: Architecture, protocols, and air interface. *IEEE Communication Surveys,* (3rd Quarter).

ESRI ShapeFile Technical Description. (1998, July). *An ESRI white paper.*

GSM Association. (2003, January). Location based services. *SE.23, 3.10.*

Karigiannis, J. N., Vlahakis, V., & Daehne, P. (n.d.). ARCHEOGUIDE: Challenges and solutions of a personalized augmented reality guide for archeological sites. *Computer Graphics in Art, History and Archeology, Special Issue of the IEEE Computer Graphics and Application Magazine.*

LOVEUS. (n.d.). Retrieved from http://loveus.intranet.gr/documentation.htm

Spohrer, J. C. (1999). Information in places. *IBM System Journal, 38*(4).

Wang, Y., Cuthbert, L., Mullany, F. J., Stathopoulos, P., Tountopoulos, V., Sotiriou, D. A., Mitrou, N., & Senis, M. (2004). Exploring agent-based wireless business models and decision support applications in an airport environment. *Journal of Telecommunications and Information Technology,* (3).

KEY TERMS

Cartographic Data: Spatial data and associated attributes used by a geographic information system (GIS).

Content Provider: A service that provides multimedia content.

Location-Based Services: A way to send custom advertising and other information to cell phone subscribers based on their curent location.

Multimedia: Media that usesmultiple forms of information content and information processing (e.g., text, audio, graphics, animation, video, interactivity) to inform or entertain the (user) audience.

Network: A network of telecommunications links arranged so that data may be passed from one part of the network to another over multiple links.

Tourism: The act of travel for predominantly recreational or leisure purposes, and the provision of services in support of this act.

Location-Based Services

Péter Hegedüs
Budapest University of Technology and Economics, Hungary

Mihály Orosz
Budapest University of Technology and Economics, Hungary

Gábor Hosszú
Budapest University of Technology and Economics, Hungary

Ferenc Kovács
Budapest University of Technology and Economics, Hungary

INTRODUCTION

The basically two different technologies, the location-based services in the mobile communication and the well-elaborated multicast technology, are joined in the multicast over LBS solutions. As the article demonstrates, this emerging and new management area has many possibilities that have not been completely utilized.

Currently an important area of mobile communications is the *ad-hoc computer networking,* where mobile devices need base stations; however, they form an overlay without any Internet-related infrastructure, which is a virtual computer network among them. In their case the selective, location-related communication model has not been elaborated on completely (Ibach, Tamm, & Horbank, 2005). One of the various communication ways among the software entities on various mobile computers is the one-to-many data dissemination that is called *multicast.* Multicast communication over mobile ad-hoc networks has increasing importance. This article describes the fundamental concepts and solutions, especially focusing on the area of *location-based services* (LBSs) and the possible multicasting over the LBS systems. This kind of communication is in fact a special case of the multicast communication model, called *geocast,* where the sender disseminates the data to that subset of the multicast group members in a specific geographical area. The article shows that the geocast utilizes the advantages of the LBS, since it is based on the location-aware information being available in the location-based solutions (Mohapatra, Gui, & Li, 2004).

There are several unsolved problems in LBS, in management and low surfaces. Most of them are in quick progress, but some need new developments. The *product managers* have to take responsibility for the software and hardware research and development part of the LBS product. This is a very important part of the design process, because if the development engineer leaves the product useful out of consideration, the whole project could possibly be led astray. Another import question is that of LBS-related *international* and *national laws,* which could throw an obstacle into LBS's spread. These obstacles will need to be solved before LBS global introduction.

The article presents this emerging new area and the many possible management solutions that have not been completely utilized.

BACKGROUND

Location-based services are based on the various distances of mobile communications from different base stations. With advances in automatic position sensing and wireless connectivity, the application range of mobile LBS is rapidly developing, particularly in the area of geographic, tourist, and local travel information systems (Ibach et al., 2005). Such systems can offer maps and other area-related information. The LBS solutions offer the capability to deliver location-aware content to subscribers on the basis of the positioning capability of the wireless infrastructure. The LBS solutions can push location-dependent data to mobile users according to their interests, or the user can pull the required information by sending a request to a server that provides location-dependent information.

LBS may have many useful applications in homeland security (HLS). A few of the more significant of these applications are security and intelligence operations, notification systems for emergency responders, search and rescue, public notification systems, and emergency preparedness (Niedzwiadek, 2002). Mobile security and intelligence operatives can employ LBS to aid in monitoring people and resources in space and time, and they can stay connected with emergency operations centers to receive the necessary updates regarding the common operating picture for a situation. Emergency operations centers can similarly coordinate search and rescue operations. Call-down systems can be employed to notify the public in affected disaster areas. In this application the multicast communication is preferable, since it uses in an efficient way the communication channels, which can be partly damaged after a disaster. Location-based public information services can give time-sensitive details concerning nearest available shelters, safe evacuation routes, nearest health services, and other public safety information (Niedzwiadek, 2002).

Location-based services utilize their ability of location-awareness to simplify user interactions. With advances in wireless connectivity, the application range of mobile LBSs is rapidly developing, particularly in the field of tourist information systems—telematic, geographic, and logistic information systems. However, current LBS solutions are incompatible with each other since manufacturer-specific protocols and interfaces are applied to aggregate the various system components for positioning, networking, or payment services. In many cases, these components form a rigid system. If such system has to be adapted to another technology, for example, moving from *global positioning system* (GPS)-based positioning to in-house *IEEE 802.11a*-based *wireless local-area network* (WLAN) or *Bluetooth*-based positioning, it has to be completely redesigned (Haartsen, 1998). In such a way the ability of interoperation of different resources under changeable interconnection conditions becomes crucial for the end-to-end availability of the services in mobile environments (Ibach, & Horbank, 2004).

There are a lot of location determination methods and technologies, such as the satellite-based GPS, which is widely applied (Hofmann-Wellenhof, Lichtenegger, & Collins, 1997). The three basic location determination methods are *proximity, triangulation* (lateration), and *scene analysis* or *pattern recognition* (Hightower & Borriello, 2001). Signal strength is frequently applied to determine proximity. As a proximity measurement, if a signal is received at several known locations, it is possible to intersect the coverage areas of that signal to calculate a location area. If one knows the angle of bearing (relative to a sphere) and distance from a known point to the target device, then the target location can be accurately calculated. Similarly, if somebody knows the range from three known positions to a target, then the location of the target object can be determined. A GPS receiver uses range measurements to multiple satellites to calculate its position. The location determination methods can be *server based* or *client based* according to the locus of computation (Hightower & Borriello, 2001).

Chen et al. (2004) introduce an enabling infrastructure, which is a middleware in order to support location-based services. This solution is based on a *location operating reference model* (LORE) that solves many problems of constructing location-based services, including location modeling, positioning, tracking, location-dependent query processing, and smart location-based message notification. Another interesting solution is the mobile yellow page service.

The LBS is facing technical and social challenges, such as location tracking, privacy issues, positioning in different environments using various locating methods, and the investment of location-aware applications.

An interesting development is the *Compose* project, which aims to overcome the drawbacks of the current solutions by pursuing a service-integrated approach that encompasses *pre-trip* and *on-trip services,* considering that on-trip services could be split into *in-car* and *last-mile services* (Bocci, 2005). The pre-trip service means the 3D navigation of the users in a city environment, and the on-trip service means the in-car and the last-mile services together. The in-car service is the location-based service and the satellite broadcasting/multicasting. In this case, the user has wireless-link access by PDA to broadcast or multicast. The last-mile service helps the mobile user with a PDA to receive guidance during the final part of the journey.

In order to create applications that utilize multicast over LBS solutions, the middleware platform of LocatioNet Systems provides all required service elements such as end-user devices, service applications, and position determination technologies, which are perfectly integrated (LocatioNet, 2006). The middleware platform of LocatioNet has an open API and a *software development kit* (SDK). Based on these, the application developers are able to easily implement novel services, focusing on comfortable user interface and free from complex details of the LBS (LocatioNet, 2006).

The article focuses on the multicast solutions over the current LBS solutions. This kind of communication is in fact a special case of the multicast communication model, called geocast, where the sender disseminates the data to a subset of the multicast group members that are in a specific geographical area.

MULTICASTING

The models of the multicast communication differ in the realization of the multiplication function in the intermediate nodes. In the case of the datalink level, the intermediate nodes are switches; in the network level, they are routers; and in the application level, the fork points are applications on hosts.

The datalink-level-based multicast is not flexible enough for new applications therefore it has no practical importance. The *network-level multicast* (NLM)—named IP-multicast—is well elaborated, and sophisticated routing protocols are developed for it. However, it has not been deployed widely yet since routing on the whole Internet has not been solved perfectly. The application-level solution gives less efficiency compared to the IP-multicast, however its deployment depends on the application itself and it has no influence on the operation of the routers. That is why the *application-level multicast* (ALM) currently has an increasing importance.

There are a lot of various protocols and implementations of the ALM for wired networks. However, the communication over wireless networks further enhances the importance of the ALM. The reason is that in the case of mobile devices, the importance of ad-hoc networks is increasing. Ad-hoc is a network that does not need any infrastructure. Such networks are *Bluetooth* (Haartsen, 1998) and *mobile ad hoc network* (MANET), which comprise a set of wireless devices that can move around freely and communicate in relaying packets on behalf of one another (Mohapatra et al., 2004).

In computer networking, there is a weaker definition of this ad-hoc network. It says that ad-hoc is a computer network that does not need routing infrastructure. It means that the mobile devices that use base stations can create an *ad-hoc computer network.* In such situations, the usage of *application-level networking* (ALN) technology is more practical than IP-multicast. In order to support this group communication, various multicast routing protocols are developed for the mobile environment. The multicast routing protocols for ad-hoc networks differ in terms of state maintenance, route topology, and other attributes.

The speed and reliability of sharing the information among the communication software entities on individual hosts depends on the network model and the topology. Theory of *peer-to-peer* (P2P) networks has gone through a great development in past years. Such networks consist of peer nodes. Usually, registered and reliable nodes connect to a grid, while P2P networks can tolerate the unreliability of nodes and the quick change of their numbers (Uppuluri, Jabisetti, Joshi, & Lee, 2005). Generally the ALN solutions use the P2P communication model, and multicast services overlay the P2P target created by the communicating entities.

The simplest ad-hoc multicast routing methods are *flooding* and *tree-based routing.* Flooding is very simple, which offers the lowest control overhead at the expense of generating high data traffic. This situation is similar to the traditional IP-multicast routing. However, in a wireless ad-hoc environment, the tree-based routing fundamentally differs from the situation in wired IP-multicast, where tree-based multicast routing algorithms are obviously the most efficient ones, such as in the *multicast open shortest path first* (MOSPF) routing protocol (Moy, 1994). Though tree-based routing generates optimally small data traffic on the overlay in the wireless ad-hoc network, the tree maintenance and updates need a lot of control traffic. That is why the two simplest methods are not scalable for large groups.

A more sophisticated ad-hoc multicast routing protocol is the *core-assisted mesh protocol* (CAMP), which belongs to the mesh-based multicast routing protocols (Garcia-Luna-Aceves & Madruga, 1999). It uses a shared mesh to support multicast routing in a dynamic ad-hoc environment. This method uses cores to limit the control traffic needed to create multicast meshes. Unlike the core-based multicast routing protocol as the traditional *protocol independent multicast-sparse mode* (PIM-SM) multicast routing protocol (Deering et al., 1996), CAMP does not require that all traffic flow through the core nodes. CAMP uses a receiver-initiated method for routers to join a multicast group. If a node wishes to join to the group, it uses a standard procedure to announce its membership. When none of its neighbors are mesh members, the node either sends a join request toward a core or attempts to reach a group member using an expanding-ring search process. Any mesh member can respond to the join request with a join *acknowledgement* (ACK) that propagates back to the request originator.

In contrast to the mesh-based routing protocols, which exploit variable topology, the so-called gossip-based multicast routing protocols exploit randomness in communication and mobility. Such multicast routing protocols apply gossip as a form of randomly controlled flooding to solve the problems of network news dissemination. This method involves member nodes talking periodically to a random subset of other members. After each round of talk, the gossipers can recover their missed multicast packets from each other (Mohapatra et al., 2004). In contrast to the deterministic approaches, this probabilistic method will better survive a highly dynamic ad hoc network because it operates independently of network topology, and its random nature fits the typical characteristics of the network.

THE LOCATION-AWARE MULTICAST

An interesting type of ad-hoc multicasting is the *geocasting.* The host that wishes to deliver packets at every node in a certain geographical area can use such a method. In such case, the position of each node with regard to the specified geocast region implicitly defines group membership. Every node is required to know its own geographical location. For this purpose they all can use the *global positioning system* (GPS). The geocasting routing method does not require any explicit join and leave actions. The members of the group tend to be clustered both geographically and topologically. The geocasting routing exploits the knowledge of location.

The geocasting can be combined with flooding. Such methods are called *forwarding zone* methods, which constrain the flooding region. The forwarding zone is a geographic area that extends from the source node to cover the geocast zone. The source node defines a forwarding zone in the header of the geocast data packet. Upon receiving a geocast packet, other machines will forward it only if their location is inside the forwarding zone. The *location-based multicast* (LBM) is an example for such *geocasting-limited flooding* (Ko & Vaidya, 2002).

LBM GEOCASTING AND IP MULTICAST

Using LBM in a network where routers are in fixed locations and their directly connected hosts are in a short distance, the location of the hosts can be approximated with the location of their router. These requirements are met by most of the GSM/UMTS, WIFI/WIMAX, and Ethernet networks, therefore a novel IP layer routing mechanism can be introduced.

This new method is a simple *geocasting-limited flooding*, extending the normal multicast RIB with the geological location of the neighbor routers. Every router should know its own location, and a routing protocol should be used to spread location information between routers. The new IP protocol is similar to the UDP protocol, but it extends it with a source location and a radius parameter. The source location parameter is automatically assigned by the first router. When a router receives a packet with empty source location, it assigns its own location to it. The radius parameter is assigned by the application itself, or it can be an administratively defined parameter in routers.

Routers forward received packets to all their neighbors except the neighbor the packet arrived from, and the neighbors outside the circle defined by the source location and radius parameters. If a packet arrives from more than one neighbor, only the first packet is handled; the duplicates are dropped.

This method requires changes in routing operating systems, but offers an easy way to start geocasting services on existing IP infrastructure without using additional positioning devices (e.g., GPS receiver) on every sender and receiver. The real advantages of the method are that geocasting services can be offered for all existing mobile phones without any additional device or infrastructure.

PRODUCT MANAGEMENT

The product managers have to take responsibility for the software and hardware part of the LBS product. This is a very important part of the process because, if the development engineer leaves a useful product out of consideration, the whole project could possibly be led astray. There are several product-management systems that help to coordinate the whole process.

FUTURE TRENDS

The multicast communication over mobile ad-hoc networks has increasing importance. This article has described the fundamental concepts and solutions. It especially focused on the area of *location-based services* (LBS) and the possible multicasting over the LBS systems. It was shown that a special kind of multicast, called *geocast* communication model, utilizes the advantages of the LBS since it is based on the location-aware information being available in the location-based solutions.

There are two known issues of this IP-level geocasting. The first problem is the scalability, the flooding type of message transfer compared to multicast tree-based protocols is less robust, but this method is more efficient in a smaller environment than using tree allocation overhead of multicast protocols. The second issue is a source must be connected directly to the router, being physically in the center position, to become source of a session. The proposed geocasting-limited flooding protocol should be extended to handle those situations where the source of a session and the target geological location are in different places.

CONCLUSION

The basically two different technologies, the location-based services in the mobile communication and the well-elaborated multicast technology, are jointed in the multicast over LBS solutions. As it was described, this emerging new area has a lot of possibilities that have not been completely utilized.

As a conclusion it can be stated that despite the earlier predicted slower development rate of the LBS solutions, nowadays the technical possibilities and the consumers' demands have already met. Furthermore, based on the latest development of the multicast over P2P technology, the one-to-many communication can be extended to the LBS systems. Also, in the case of the emerging homeland security applications, the multicast over LBS is not only a possibility, but it became a serious requirement as well. The geospatial property of the LBS provides technical conditions to apply a specialized type of the multicast technology, called geocasting, which gives an efficient and users' group targeted solution for the one-to-many communication.

REFERENCES

Bocci, L. (2005). *Compose project web site.* Retrieved from http://www.newapplication.it/compose

Chen, Y., Chen, Y. Y., Rao, F. Y., Yu, X. L., Li, Y., & Liu, D. (2004). LORE: An infrastructure to support location-aware services. *IBM Journal of Research & Development, 48*(5/6), 601-615.

Deering, S.E., Estrin, D., Farinacci, D., Jacobson, V., Liu, C-G., & Wei, L. (1996). The PIM architecture for wide-area multicast routing. *IEEE/ACM Transactions on Networking, 4*(2), 153-162.

Garcia-Luna-Aceves, J. J., & Madruga, E. L. (1999, August). The core-assisted mesh protocol. *IEEE Journal of Selected Areas in Communications,* 1380-1394.

Haartsen, J. (1998). The universal radio interface for ad hoc, wireless connectivity. *Ericsson Review, 3.* Retrieved 2004 from http://www.ericsson.com/review

Hightower, J., & Borriello, G. (2001). Location systems for ubiquitous computing. *IEEE Computer,* (August), 57-65.

Hofmann-Wellenhof, B., Lichtenegger, H., & Collins, J. (1997). *Global positioning system: Theory and practice* (4th ed.). Vienna/New York: Springer-Verlag.

Hosszú, G. (2005). Current multicast technology. In M. Khosrow-Pour (Ed.), *Encyclopedia of information science and information technology* (pp. 660-667). Hershey, PA: Idea Group Reference.

Ibach, P., Tamm, G., & Horbank, M. (2005). Dynamic value webs in mobile environments using adaptive location-based services. *Proceedings of the 38th Hawaii International Conference on System Sciences.*

Ibach, P., & Horbank, M. (2004, May 13-14). Highly-available location-based services in mobile environments. *Proceedings of the International Service Availability Symposium,* Munich, Germany.

Ko, Y-B., & Vaidya, N. H. (2002). Flooding-based geocasting protocols for mobile ad hoc networks. *Proceedings of the Mobile Networks and Applications,* 7(6), 471-480.

LocatioNet. (2006). *LocatioNet and Ericsson enter into global distribution agreement.* Retrieved from http://www.locationet.com

Mohapatra, P., Gui, C., & Li, J. (2004). Group communications in mobile ad hoc networks. *Computer, 37*(2), 52-59.

Moy, J. (1994, March). *Multicast extensions to OSPF.* Network Working Group RFC 1584.

Niedzwiadek, H. (2002). *Location-based services for homeland security.* Retrieved March 2006 from http://www.jlocationservices.com/LBSArticles

Uppuluri, P., Jabisetti, N., Joshi, U., & Lee, Y. (2005, July 11-15). P2P grid: Service oriented framework for distributed resource management. *Proceedings of the IEEE International Conference on Web Services,* Orlando, FL.

KEY TERMS

Ad-Hoc Computer Network: Mobile devices that require base stations can create the ad-hoc computer network if they do not need routing infrastructure.

Ad-Hoc Network: A network that does not need any infrastructure. One example is Bluetooth.

Application-Level Multicast (ALM): A novel multicast technology that does not require any additional protocol in the network routers, since it uses the traditional unicast IP transmission.

Application-Level Network (ALN): The applications, which are running in the hosts, can create a virtual network from their logical connections. This is also called overlay network. The operations of such software entities are not able to understand without knowing their logical relations. In most cases these ALN software entities use the *P2P model,* not the *client/server* for the communication.

Client/Server Model: A communicating method, where one hardware or software entity (server) has more functionalities than the other entity (the client), whereas the client is responsible to initiate and close the communication session towards the server. Usually the server provides services that the client can request from the server. Its alternative is the *P2P model.*

Geocast: One-to-many communication method among communicating entities, where an entity in the root of the multicast distribution tree sends data to that certain subset of the entities in the multicast dissemination tree, which are in a specific geographical area.

Multicast: One-to-many and many-to-many communication method among communicating entities on various networked hosts.

Multicast Routing Protocol: In order to forward the multicast packets, the routers have to create multicast routing tables using multicast routing protocols.

Peer-to-Peer (P2P): A communication method where each node has the same authority and communication capability. The nodes create a virtual network, overlaid on the Internet. The members organize themselves into a topology for data transmission.

Product Management: A function within a corporation dealing with the continuous management and welfare of the products at all stages of the production procedure in order to ensure that the products profitably meet the needs of customers.

M-Advertising

Michael Decker
University of Karlsruhe, Germany

INTRODUCTION

According to our comprehension, mobile advertising (also called "wireless advertising" or "mobile marketing") is the presentation of advertising information on mobile handheld devices with a wireless data link like cellular phones, personal digital assistants and smartphones; however notebooks/laptops and tablet PCs are not considered as mobile devices in this sense, because they are used like stationary devices at different locations. For example SMS-messages with product offers would be a simple form of m-advertising. In this article we discuss the special features of m-advertising, but also the problems involved. Afterwards we name basic methods of m-advertising and compare their general strengths and weaknesses using a set of criteria.

M-ADVERTISING COMPARED TO TRADITIONAL FORMS OF ADVERTISING AND INTERNET-BASED ADVERTISING

Conventional media for advertising are newspapers, advertising pillars, TV and radio commercials. Relative new media for advertising are the Internet and mobile devices. Both have some features in common:

- **Individually Addressable:** The user can be addressed individually, so a high degree of personalization is possible: it is possible to tailor the content of each advert according to the profile of the consumer (mass customization).
- **Interactive:** if end users receive an advert, they can immediately request further information, participate in a sweepstake or forward an advert to friends. The last one is especially interesting in terms of "viral marketing."
- **Multimedia-Capable:** Multimedia elements (e.g., pictures, movies, jingles, tunes, sounds) are important to realize entertaining adverts and to generate brand awareness.
- **Countable:** Each impression of an ad can be counted; for most conventional methods of advertising like TV/radio/cinema commercials or adverts in print media this can not be done and thus the advertisers are billed according to a rough estimate of the number of generated contacts.

But there are additional features of m-advertising [see also Barnes (2002)]:

- **Context:** In the sense of mobile computing, context is "[...] any information that can be used to characterize the situation of an entity" (Dey, 2001). This information helps to support a user during an interaction with an application. For mobile terminals with their limited user interface context-awareness is especially important. The most prominent example of context is the location of a user, because it changes often and there are a lot of useful scenarios of how to exploit that information. The location information for these "location-based services" can be retrieved based on the position of the currently used base station, the runtime difference when using more than two base stations (TDOA: Time Differential of Arrival) or using a GPS-receiver (Zeimpekis, Giaglis, & Lekakos, 2003). Other examples of context information also used for m-advertising are "weather" and "time" (Salo & Tähtinen, 2005).
- **Reachability:** People carry their mobile terminal along with them most of the day and rarely lend it away or share it with other people, because it is a personal device. Therefore marketers can reach people almost anywhere and anytime.
- **Convenience:** Mobile terminals are much simpler to handle than personal computers because they are preconfigured by the mobile network operator and have no boot-up time, so they are a medium for electronic advertising to reach people who don't want to use a computer.
- **Penetration Rates:** Mobile terminals—especially cellular phones—have very high penetration rates, which exceed those of fixed line telephones and personal computers. Mobile terminals are more popular than PCs because they are more affordable and simpler to handle. The current global number of cellular phones is beyond one billion, there are even countries with penetration rates over 100 % (Netsize, 2005).

At first glance, m-advertising seems to be a direct continuation of Internet-based advertising: instead of a fixed computer with a wired data link a mobile terminal with wireless data link is used. But most forms of advertising in the Internet are based on the idea of showing additional advertising information on the user interface (banners on Web pages, sponsored links in search engines). Due to the limited display size of mobile terminals these forms of advertising can't be used for m-advertising, so new methods have to be developed.

Mobile terminals are much more personal devices than personal computers, so a higher degree of personalization can be obtained than with Internet advertising, which leads to better response rates than with other forms of direct marketing (Kavassalis et al., 2003).

CHALLENGES

As shown in the last section, m-advertising has some unique advantages when compared to other forms of advertising. But one shouldn't conceal the challenges associated with it:

- **Unsolicited Messages:** Unsolicited mass-mailing with commercial intention ("Spam") as well as malware (viruses, trojan horses, spyware, etc.) are a great worriment in the fixed-line Internet; the portion of spam messages in e-mail communication exceeds the 50% mark by far. Unsolicited messages on mobile terminals are a much bigger problem, because mobile terminals have limited resources to handle them and are personal devices.
- **Limited Usability:** Due to the limited dimensions of mobile terminals they have only a small display and no real keyboard. This has to be considered when designing adverts for mobile terminals. One cannot ask the user for extensive data input, for example, about his/her fields of interests or socio-demographic particulars.
- **Limited Resources:** Mobile terminals have very limited resources, for example, memory, CPU-power and available bandwidth. These have to be considered when designing adverts for mobile terminals, for example, transmission of adverts with a lot of data volume is not adequate,
- **Privacy Concerns:** As already mentioned, mobile terminals are personal devices with personal data stored on them; it is also possible to track the location of the users.
- **Cost of Mobile Data Communication:** Mobile data communication is still very expensive in some regions, so no consumer wants to cover the costs caused by the transmission of adverts.
- **Technical Heterogeneity:** The underlying network infrastructure and the capabilities of mobile terminals are much more heterogeneous than for ordinary fixed-line computers: one advert might look great on one type of terminal, but isn't displayable on another one. It might cause significant costs when the creator of an m-advertising campaign has to consider many different types of mobile terminals.

Due to the problems with unsolicited e-mails and telephone calls "permission-based marketing" is a generally accepted principle when designing systems and campaigns for m-advertising (Barwise & Strong, 2002): A consumer will only receive advertising-messages on his mobile device if he explicitly gave his permission. The adverts sent to a user will be chosen according to the profile of interests of the user. But there is one problem with this principle: a consumer has to know about an m-advertising campaign or system to give his explicit agreement, so one has to "advertise for m-advertising." Because of this m-advertising is very often integrated into bigger campaigns along with traditional media, see Kavassalis et al. (2003) for examples.

DIFFERENT APPROACHES FOR M-ADVERTISING

We distinguish different approaches for m-advertising by the underlying mode of wireless communication used:

- When using broadcast communication, messages are sent to all ready-to-receive terminals in the area covered by the radio waves according to their natural propagation. If the area covered is rather restricted we talk about a local broadcast, which allows realizing a certain degree of location-aware adoption; the opposite case is global broadcast. Examples: cell-broadcast to all terminals in a certain network-cell (local broadcast) or digital television standards like DVB-H or DMB (global broadcast).
- Mobile ad-hoc networks (MANETs) are wireless networks without a dedicated infrastructure or a central administration (Murthy & Manoj, 2004). Two terminals of such a network exchange messages when the distance between them is short enough, a message can also be routed via several terminals to the recipient (multi-hop).
- Unicast communication provides a dedicated point-to-point connection between a base station and a mobile terminal in an infrastructure-based network like GSM, WLAN or UMTS.

To compare different advertising approaches based on these modes we apply the following set of criteria:

- Which degree of personalization and location-aware adoption can be achieved?

- Will there be costs for the data transmission which have to be borne by the end-user?
- How reliable is the transmission of the advert? Can it be guaranteed that the advert is received indeed within a certain time span? The last point might be crucial for campaigns with temporary and "last minute" offers.
- Is the mode capable of direct user interaction?
- Unsolicited messages aren't bearable on mobile devices, so a system for m-advertising should be designed to make the dispatching of unsolicited messages impossible.
- Is it possible to detect the exact number of contacts generated?

BROADCASTING

Using broadcast communication, adverts are addressed undirected to an anonymous crowd of people, so there is no possibility of personalization. If the broadcast is a local one at least a certain degree of location-aware adoption can be achieved, for example adverts about shopping facilities and sights located in one cell of a GSM-network. Receiving broadcast messages is free of charge for the end-users, but if the user hasn't turned on his device or isn't in the area covered by the broadcast he may miss a message. If a user wants to respond to an advert or forward it to other people he has to resort to unicast communication. Unsolicited messages can only occur if the user is in ready-to-receive state.

Using broadcasting-based advertising on mobile terminals we lose a lot of the specific opportunities of m-advertising, for example, the high personalization and the interactive nature, but there are no costs for data transmission for the end-user and his anonymity is protected. Without further measures it is not possible to find out how many users received an advertisement.

MANETs

The idea of MANETs can be applied for the distribution of adverts (Ratsimor, Finin, Joshi, & Yesha, 2003; Straub & Heinemann, 2004): following the idea of "word-of-mouth" recommendations mobile terminals exchange adverts when they approach each other, whereas the initial transmission of adverts is performed by fixed "information sprinklers." If these are positioned at an appropriate place we can obtain a certain degree of location awareness at least for those users whose devices receive the message from the sprinklers directly. A client application installed on the mobile terminal will only display adverts that match the profile of the user, so a certain degree of personalization is possible. There are also incentive schemes: when an advert leads to a purchase a user whose device participated in the recommendation chain may receive a bonus.

The multi-hop case requires a special client application on the device. At present only the single hop-case can be found in practice: "sprinklers" installed at appropriate places (e.g., entrance halls of shopping or conference centers or even at billboards) submit adverts to mobile devices using infrared [e.g., "marketEye™" by Accinity (2006)] or bluetooth [e.g., "BlipZones ™" by BLIP Systems (2006)] communication, but the adverts are not transmitted to other devices automatically.

M-advertising based on MANETs doesn't cause costs for the end-user. The reliability of this form of advertising is not very high, because one cannot give guarantees how long it takes until a consumer receives—if at all—an ad. For interaction the user has to resort to Unicast communication. In the multi-hop case the local client applications can decide which adverts to display and which not to, so the user won't be harassed by unsolicited messages; in the single-hop case the user can disable his infrared or bluetooth interface if he doesn't want to receive ads. The advertisers can only count contacts that led to certain defined actions (e.g., if a digital voucher is redeemed). Receiving ads from unknown mobile terminals all the time may also be very energy consuming and there is the danger of receiving malware which exploits flaws of the mobile device.

UNICAST

Unicast-communication can be further divided into "push" and "pull:" in push-mode the consumer receives a message without a direct request, for example, SMS; in pull-mode the consumer has to perform an explicit request for each message, for example, request of WAP-pages. Since Unicast communication provides a dedicated point-to-point connection, the advertiser can deliver a different ad for each user, so there is a high degree of personalization possible and the advertiser can calculate the exact number of contacts generated.

The most obvious form of pull-advertising for mobile terminals is Web-pages in special formats like WML or cHTML. When viewing such pages the user has to pay for the data volume, so there is the idea that the advertiser covers the costs if the user's profile meets certain criteria (Figge, Schrott, Muntermann, & Rannenberg, 2002). Pull-advertising isn't vulnerable for unsolicited messages, but the reliability is also restricted, because the user might miss offers he is interested in or obtain offers too late if he doesn't request the right page at the right time.

SMS as form of mobile push-messaging is also the most successful data service for mobile phones, but the messages can only consist of text and are bound to a maximum size of 140 bytes or 160 letters for a 7-bit-encoding—and Barwise and Strong (2002) even recommend to use much shorter messages when using SMS for advertising. But it is the data service which is supported by almost all cellular phones, so it is an interesting opportunity for m-advertising.

The multimedia messaging service (MMS) is a further development of SMS, which is also capable of displaying multimedia content, but the creation of an appealing message on a mobile device is challenging; also in many countries sending MMS is very expensive. "V-Card" (Mohr, Nösekabel, & Keber, 2003) is a platform for the creation of MMS and offers templates and multimedia content, so the user can design easily a personalized message. To keep the service free of charge for the end-user a sponsor bears the costs and thus has his advertising message included in the MMS.

The "MoMa" system (Bulander, Decker, Schiefer, & Kölmel, 2005) is a system for context aware push advertising for mobile terminals with a special focus on privacy aspects. The system acts as mediator between end-users and advertisers: end-users put "orders" into the system using a special client application to express that they are interested in advertising concerning a certain product or service (e.g., restaurants or shoe shops in their current surrounding); the possible products and services are listed in a hierarchical catalogue. Depending on their type the orders can be refined through the specification of attributes. On the other side of the system the advertisers put "offers" into the system. A matching component tries to find fitting pairs of orders and offers, in case of a hit a notification via SMS/MMS, e-mail or text-to-speech call will be dispatched. When appropriate the matching process also considers context-parameters like the location of the user or the weather.

Push advertising in general provides a high degree of reliability but is also vulnerable with regard to unsolicited messages.

CONCLUSION

We highlighted the special features of mobile and wireless terminals as medium for advertising, but saw also the considerable challenges. M-advertising might be one of the first successful m-business applications in the business-to-consumer sector, because it doesn't require m-payment. The current market share of m-advertising in relation to the total advertising market seems to be less than one percent in most countries, but is expected to reach a few percent-points (like nowadays Internet-advertising) in the medium term.

REFERENCES

Accinity. (n.d.). Retrieved March 28, 2006, from http://www.accinity.com

Barwise, P., & Strong, C. (2002). Permission-based mobile advertising. *Journal of Interactive Marketing, 16*(1), 14-24.

Barnes, J. (2002). Wireless digital advertising: Nature and implications. *International Journal of Advertising, 21*(3), 399-420.

BLIP Systems. (n.d.). Retrieved March 28, 2006, from http://www.blipsystems.com

Bulander, R., Decker, M., Schiefer, G., & Kölmel, B. (2005). Advertising via mobile terminals. In *Proceedings of the 2nd International Conference on E-Business and Telecommunication Networks (ICETE '05)*. Reading, UK.

Dey, A.K. (2001). Understanding and using context. *Personal and Ubiquitous Computing Journal, 5*(1), 4-7.

Figge, S., Schrott, G., Muntermann, J., & Rannenberg, K. (2002). Earning M-Oney—A situation based approach for mobile business models. In *Proceedings of the 11th European Conference on Information Systems (ECIS)*. Naples, Italy.

Kavassalis, P., Spyropoulou, N., Drossos, D., Mitrokostas, E., Gikas, G., & Hatzistamatiou, A. (2003). Mobile permission marketing: Framing the market inquiry. *International Journal of Electronic Commerce, 8*(1), 55-79.

Mohr, R., Nösekabel, H., & Keber, T. (2003). V-Card: Sublimated message and lifestyle services for the mobile mass market. In *Proceedings of the 5th International Conference on Information and Web-Based Applications and Services*. Jakarta, Indonesia.

Murthy, C., & Manoj, B. (2004). *Ad hoc wireless networks—Architectures and protocols*. Upper Saddle River, NJ: Prentice Hall.

Netsize. (2005). *The Netsize guide 2005*. Paris, France.

Ratsimor, O., Finin, T., Joshi, A., & Yesha, Y. (2003). eNcentive: A framework for intelligent marketing in mobile peer-to-peer environments. In *Proceedings of the 5th ACM International Conference on Electronic Commerce* (ICEC 2003). Pittsburgh, Pennsylvania.

Salo, J., & Tähtinen, J. (2005). Retailer use of permission-based mobile advertising. In I. Clarke, III & T.B. Flaherty (Eds.), *Advances in Electronic Marketing* (pp. 139-155). Hershey, PA: Idea Group Publishing.

Straub, T., & Heinemann, A. (2004). An anonymous bonus point system for mobile commerce based on word-of-mouth-recommendation. In *Proceedings of the ACM Symposium on Applied Computing (SAC '04)*. Nicosia, Cyprus.

Zeimpekis, V., Giaglis, G., & Lekakos, G. (2003). A taxonomy of indoor and outdoor positioning techniques for mobile location services. *ACM SIGecom Exchanges, 3*(4), 19-27.

KEY TERMS

Broadcast: sending data using wireless communication to an anonymous audience according to the natural propagation of radio waves.

Context: information available at runtime of a computer system to support the user when interacting with the system.

Location-Based Services: most prominent case of context-aware services, which adapt themselves according to the current location of the user.

Mobile Advertising (M-Advertising): adverts displayed on mobile and wireless terminals like cellular phones, smartphones or PDAs.

Mobile Ad-Hoc Network (MANET): Wireless network without infrastructure and central administration, the nodes (devices) pass messages to other nodes in reach.

Pull: A user receives a message only as direct response.

Push: A user receives a message without directly requesting it.

Unicast: Communication using an infrastructure-based network which provides dedicated point-to-point-connection. Examples are GSM/GPRS, WLAN or UMTS.

Viral Marketing: Dissemination of adverts by consumers themselves ("tell a friend!"); is expected to generate exponential rates of contacts and a higher trustability than adverts received from firms directly.

Man–Machine Interface with Applications in Mobile Robotic Systems

Milan Kvasnica
Tomas Bata University, Zlin, Czech Republic

INTRODUCTION

This article focuses on the current state-of-the-art assistive technologies in man-machine interface and its applications in robotics. This work presents the assistive technologies developed specifically for disabled people. The presented devices are as follows:

- The head joystick works on a set of instructions derived from intended head movements. Five laser diodes are attached to the head at specific points whose light rays' spots are scanned by a set of CCD cameras mounted at strategic locations (on the ceiling, on the wall, or on a wheelchair).
- Automatic parking equipment has two laser diodes attached at the back of the wheelchair, and their light rays' spots are scanned by the CCD cameras.
- A range-inclination tracer for positioning and control of a wheelchair works on two laser diodes attached onto the front of the wheelchair. A CCD camera mounted on the front of the wheelchair detects the light rays' spots on the wall.
- The body motion control system is based on a set of instructions derived from intended body motion detected by a six component force-torque transducer, which is inserted between the saddle and the chassis of the wheelchair.
- An optoelectronic handy navigator for blind people consists of four laser diodes, the 1-D CCD array (alternatively PSD array), a microprocessor, and a tuned pitch and timbre sound source. The functionality of this system is based on the shape analysis of the structured lighting. The structured lighting provides a cutting plane intersection of an object, and a simple expert system can be devised to help blind people in classification and articulation of 3-D objects. There are two parameters involved: the distance and the inclination of the object's articulation. The time-profiles of the distance and inclination are used to adjust the frequency and amplitude of the sound generator. The sound representation of a 3-D object's articulation enables the skill-based training of a user in recognizing the distance and ambient articulation.

The head joystick, the automatic parking equipment, the range-inclination tracer, and the body motion control system for the wheelchair control are suitable for people who have lost the ability to use their own lower limbs to walk or their upper limbs for quadriplegics. The optoelectronic handy navigator is suitable for blind people. The mentioned sensory systems help them to perform daily living tasks, namely to manage independent mobility of electrical wheelchair or to control a robot manipulator to handle utensils and other objects. The customization of described universal portable modules and their combinations enable convenient implementation in rooms and along corridors, for the comfort of the wheelchair user. Smart configuration of the optoelectronic handy navigator for blind people enables the built-in customization into a handy phone, handheld device, or a white stick for blind people.

BACKGROUND

Significant progress in human-computer interfaces for elderly and disabled people has been reported in recent years. Some examples for such devices are the eye-mouse tracking system, hand gesture systems, face gesture systems, head controller, head joystick, and human-robot shoulder interface, all presented at recent international conferences. The aim of this article is to publish further progress in the field of assistive technologies like the head joystick, automatic parking equipment, range-inclination tracer, body motion control system, and optoelectronic handy navigator for blind people.

HEAD JOYSTICK AND AUTOMATIC PARKING EQUIPMENT

Following parts of the modular sensory system enables the processing of multi-DOF information for the control and the positioning of a wheelchair by means of two types of modules for alternative use, as shown in Figure 1.

The module Apc (the module of four laser diodes) is designed for tracking the head motion of the wheelchair user. The ceiling-mounted CCD cameras detect the Apc laser rays. The fifth, auxiliary laser diode with redundant light

Figure 1. The modules Apc and App positioned relative to the plain of the ceiling and the module Bp of the CCD camera mounted in perpendicular view or in perspective view against the light spots on the ceiling

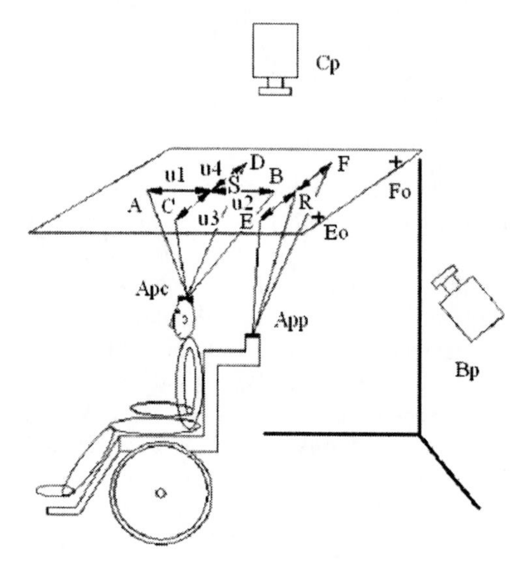

spot is used for the verification of accurate functionality. The module App (the module of two laser diodes) is designed for automatic parking of the wheelchair into a predefined position in the room. The third auxiliary laser diode with redundant light spot is used for the verification of accurate functionality. The modules Apc and App have the presetting control 1 of the angle 2s contained by mutual opposite light rays 2, as shown in Figures 1 and 2. The auxiliary fifth or third laser diode is centered in the axis. The camera with 2-D CCD array can be arranged in two ways:

- The perpendicular view downwards against the translucent screen, which is mounted parallel to the ceiling. The module Cp for direct sampling is shown in Figure 3. The light spots reflected by light rays of Apc, or App respectively, are sampled by the camera with a 2-D CCD array. The module Cp for direct sampling of the light spots (no. 3) consists of the camera with 2-D CCD array (no. 6) with focusing optics (no. 5) and the flange (no. 1) mounted perpendicular to the translucent screen. The light spots from the laser light rays are projected onto the translucent screen (no. 4). The translucent screen spans the entire ceiling of the room. In larger rooms, four Cp modules are attached onto the ceiling in front of the laser rays.
- The perspective view of the ceiling and light spots of the laser ray images spots from the modules Apc and App are shown in Figure 1. The module Bp, depicted in Figure 4, is shown in Figure 1 in perspective view on the wall. The camera with 2-D array makes the sampling of the light spot position from the laser light rays on the ceiling plane. The X-Y coordinate system on the ceiling is used to monitor the parking position of the wheelchair.

The module Cp is mounted on the ceiling against the modules Apc and App, respectively. The Apc module is attached to the head of the wheelchair user, and the rays are intersecting the ceiling plain of the Bp or Cp modules respectively, in light spots A, B, C, and D. The intersection of abscises AB and CD is the point S centered by auxiliary laser diode. The lengths of abscises AS, SB, CS, and SD are u_1, u_2, u_3, and u_4. The App module is attached to the wheelchair, and the light rays intersect the ceiling plain in light spots E, F in equal distance u from the middle point R centered by auxiliary laser diode. The light spots position and configuration is analyzed and processed for the navigation of a wheelchair.

Purposeful head motion of the module Apc represented by the light spots configuration is sampled by means of the modules Bp, or Cp from the ceiling. The following commands are used for three degrees-of-freedom control of the wheelchair with an adjustable operating height of the wheelchair perpendicular to the 2-D coordinate frame on the ground:

- **Start:** The head movement with the module Apc outwards the dead zone position of the light spots.
- **Stop:** The head movement with the module Apc inwards the dead zone position of the light spots.
- The dead zone is defined by the ratio u_1/u_2 and u_3/u_4, for example (only for perpendicular view of the CCD Camera) by the interval <0,8; 1,2> and for the angle Ω by the interval <-15°;+15°>. Inside these intervals, following commands are not valid because of physiological trembling of the head. Outwards these intervals are valid following commands.
- **Forward, Backward for the First DOF:** The dividing ratio of diagonals $u_1/u_2 > 1{,}2$ respectively $u_1/u_2 < 0{,}8$.
- **Up, Down for the Second DOF:** The dividing ratio of diagonals $u_3/u_4 > 1{,}2$ respectively $u_3/u_4 < 0{,}8$.
- **Turn to the Left – Turn to the Right for the Third DOF:** Last increment of the angle Ω is positive $\Omega > +15^0$, respectively, and negative $\Omega < -15^0$ oriented.
- The magnitude of the dividing ratios u_1/u_2, u_3/u_4, and the angle Ω is assigned to the velocity of the wheelchair motion for each DOF.
- The motion control system of the wheelchair enables parallel control of all three degrees-of-freedom.
- The light spot A from the module Apc is recognized by means of the enhanced intensity, color, or shape in contrast with light spots B, C, and D. This is needed for the orientation of the wheelchair against the basic light spot position.

Figure 2. The multi-laser configuration modules App, and Apc

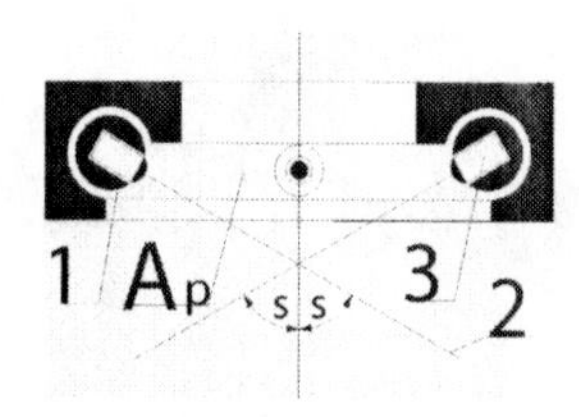

Figure 3. The Cp module for direct sampling

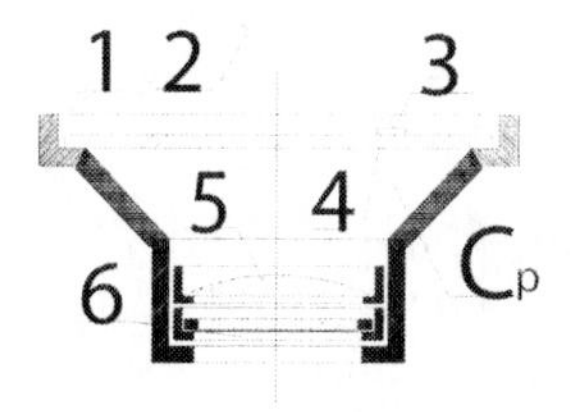

Figure 4. The module Bp for the ground plane or for perspective ceiling sampling

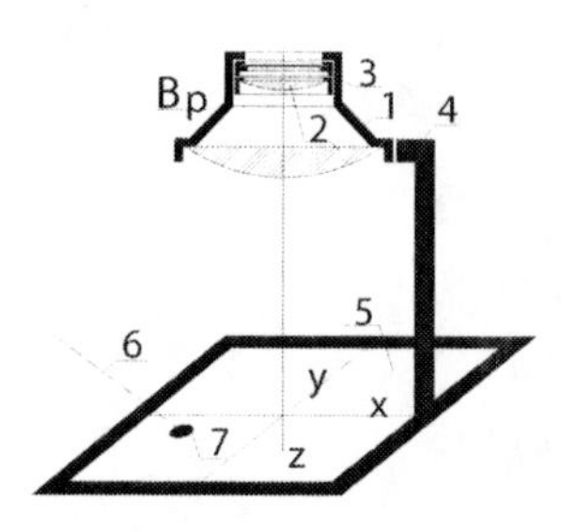

Figure 5. Navigation by automatic parking system

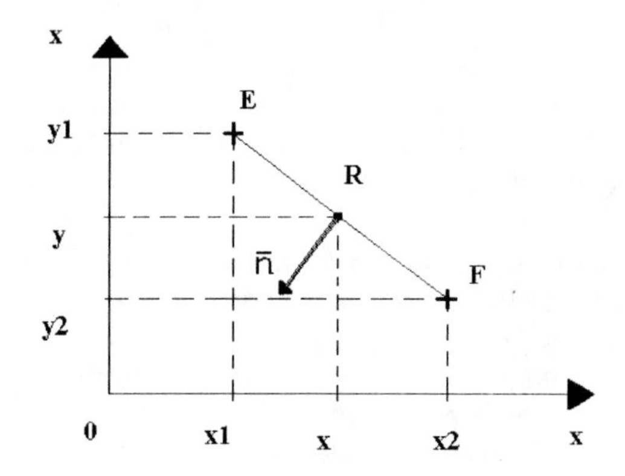

The automatic parking equipment of a wheelchair is devised on the module App, which is attached at the back of the wheelchair:

- The module App is used for automatic return into the parking position using controls in forward-backward, up-down, and turns right-left.
- The module App and the module Apc operate independently and sequentially into common modules Bp or Cp.
- The position $E(x_1,y_1)$ and $F(x_2,y_2)$ of the light spots from the module App on the ceiling are sampled by the camera. The information about the light spots position $E(x_1,y_1)$ and $F(x_2,y_2)$ is sufficient for the navigation into the points $E_0(x_{01},y_{01})$ and $F_0(x_{02},y_{02})$ of the basic position highlighted from the reflexive material on the ceiling.
- The light spot E from the module App is recognized by means of the intensity, color, or shape in contrast with light spot F. This is needed for the orientation of the wheelchair against the basic light spot parking position.
- Another technique of recognizing the orientation is to sample every light spot in separate picture synchronized with sequentially switched light rays.

The coordinates of the position x,y of the wheelchair at point R of the ceiling rectangular coordinate frame is given by the relationship (1), and the normal vector n in the middle point R determines the direction of the wheelchair trajectory.

$$x = \frac{1}{2}(x_1 + x_2);\ \ y = \frac{1}{2}(y_1 + y_2); \tag{1}$$

- The direction **n** of the wheelchair motion in the ceiling rectangular coordinate frame is given by the relationship (3) according to Figure 5.
- All coordinates x_i, y_i, z are with respect to the geometrical center of the light spots.
- The z coordinate is used for calibration of the magnitude x_i, y_i according to the position of the light spot image center of the 2-D CCD coordinate frame.

The distance z of the wheelchair from the ceiling in the rectangular coordinate frame (recommended the value s = arctg(1/2)) according to Figures 2 and 5 is given by the relationship (2).

$$z = \sqrt{(y_2 - y_1)^2 + (x_2 - x_1)^2}\ ; \tag{2}$$

$$n = -\left[\frac{y_2 - y_1}{x_2 - x_1}\right]^{-1}. \tag{3}$$

The coordinates of the wheelchair position x,y and the normal vector direction n of the wheelchair motion are used for the planning of the wheelchair trajectory as well as for collision avoidance.

THE RANGE-INCLINATION TRACER

The range-inclination tracer consists of the camera with 1-D CCD array or linear PSD element and two laser diodes radiating two intersecting light rays against the wall, as shown in Figures 6 and 7. Two light spot positions of the light rays are sampled by a simple sampling algorithm, in order to evaluate the distance D and the inclination β of the wall against the wheelchair. The sampling of every light spot coordinate in separate picture synchronized with sequentially switched light rays enables recognition of the varying orientation of the light spots before and after the crossing point of the laser light rays.

Control algorithms are derived in the rectangular coordinate system x,y with an origin in zero point of mutual intersection of light rays (no. 4), so that the y coordinate fuses with optical axis (no. 5) of the camera (no. 3) with 1-D CCD array, and its positive orientation leads into the camera (see Figure 7). The coordinates of the light spot on the wall (no. 6) emitted by the light source (no. 1) are designated by $A(r_1,s_1)$, and the coordinates of the light spot emitted by the light source (no.,2) are designated by $B(r_2,s_2)$. Every light ray creates the angle $\alpha/2$ with optical axis of the camera. The magnitude of the coordinates r_1,r_2 is derived according to the calibration of the light spot's position in the image coordinate frame of the 1-D CCD array. Coordinates s_1,s_2 are computed from the relationship (4).

Figure 6. Range-inclination tracer attached to the wheelchair

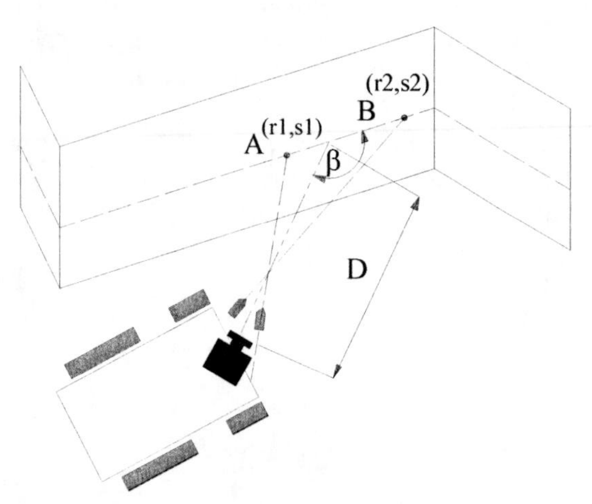

Figure 7. Geometrical approach of the activity of the range-inclination tracer

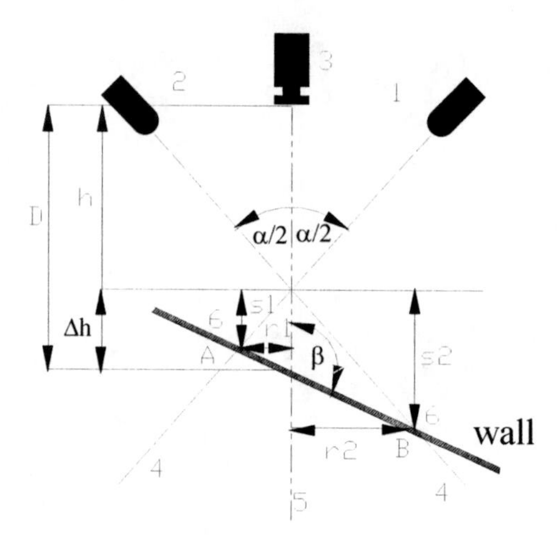

$$s_1 = \frac{r_1}{tg\frac{\alpha}{2}}\ ; \quad s_2 = \frac{r_2}{tg\frac{\alpha}{2}} \tag{4}$$

where h is the distance between the camera and the point of the intersection of light rays. The Δh is the difference between the point of the light rays' intersection and the intersection of the optical axis of the camera with the wall, given by the relationship (5).

The distance D between the camera and the wall is given by the relationship (6), and the inclination β of the wall against the wheelchair is given by the relationship (7).

$$D = h \pm \Delta h \tag{5}$$

$$\Delta h = -\frac{s_1 - s_2}{r_2 - r_1} r_1 + s_1 \tag{6}$$

$$\beta = arcctg\frac{r_2 - r_1}{s_1 - s_2} \tag{7}$$

The described sensory system is alternatively used for independent control of the wheelchair trajectory in the vicinity of the walls, namely in long corridors, and enables the implementation of a low-cost collision avoidance system.

HUMAN-ROBOT INTERFACE USING THE BODY MOTION

Many people with limited mobility prefer to enjoy life. They particularly enjoy a wheelchair ride at a popular tourist spot with typical leisure activities, and sometimes go on high adventures and wilderness expeditions. These expeditions are usually long trips, and active compliant assistance by way of feedback control based on the damping device with a shock absorber is required. This active compliant assistance is able to overcome some barriers and to predict hazardous scenarios in unexpected situations like the quick halt on the stone or the raid on a sharp slope. A basic part of compliant assistance is the six-component force-torque transducer inserted between the saddle and the chassis as shown in Figures 8 and 9. The force-torque transducer used for dynamic weighing of the user and the load on the wheelchair improves the dynamic stability at maneuvering and the comfort of the ride. In addition this force-torque transducer is possible to use for the sampling of the information about the user's body motion.

The control system based on the user's body motion is based on a set of instructions derived from the body inclination and sampled by means of the six-component force-torque transducer inserted between the saddle and the chassis of the wheelchair. The six-component force-torque transducer is also used for the active compliant assistance.

The explanation of the activity is introduced on a simple modification of the six-component force-torque transducer. An example of the six-component force-torque transducer with the acting force -Fz is depicted in Figure 8. Laser diodes (no. 1) emit intersecting light rays (no. 2) creating the edges of a pyramid, intersecting the 2-D CCD array (no. 4) in light spots (no. 3). The beginning of the 3D rectangular pyramid coordinate frame x,y,z is chosen in the crossing point of the light rays (the apex of a pyramid shape). Unique light spots configuration on the 2-D CCD array changes under the force-torque acting between the flanges (no. 5 and no. 6) connected by means of elastic deformable medium (no. 7). The beginning of a floating 2D coordinates frame x_{CCD}, y_{CCD} is chosen in the geometrical center of the 2-D CCD array. An even number of four light rays simplifies and enhances the accuracy of the algorithm for the evaluation of three axial shiftings and three angular displacements. In addition it removes the ambiguity of the imagination at the rotation of the 2-D CCD array around the straight line passing through

Figure 8. Six-component force-torque transducer

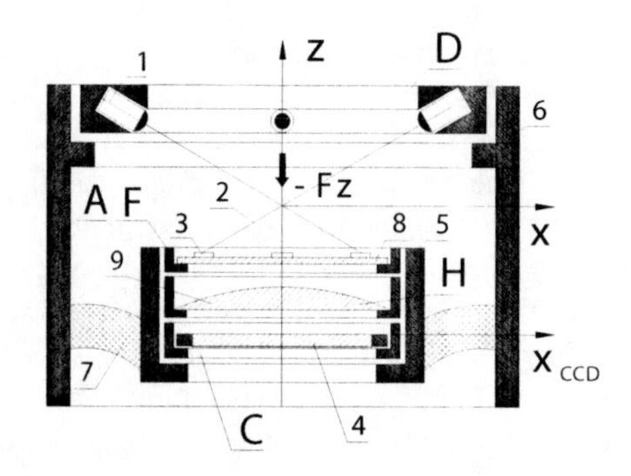

two light spots, where for two inclines of opposite orientation belongs one position of the third light spot.

The module of five laser diodes is attached on the outer flange of the force-torque sensor, and the rays intersect the translucent screen in light spots A, B, C, and D. The intersection of the abscises AB and CD is the point S. The length of the abscises AS, SB, CS, and SD is u_1, u_2, u_3, and u_4. The light spot configuration is analyzed and processed for the navigation and positioning of the wheelchair.

Intentional body inclinations against the saddle are represented by the light spots (no. 3) configuration in the 2-D CCD array (no. 4) of the force-torque transducer, as shown in Figure 8, which is analyzed like the force-torque acting in six components. The following commands are used for two-degrees-of-freedom control of the wheelchair as shown in Figure 9:

- **Start:** The body inclination in chosen direction outwards the light spots dead zone position.
- **Stop:** The body inclination inwards the light spots dead zone position.
- The dead zone is defined for the dividing ratio u_1/u_2 for example by the interval <0.8; 1.2>, u_3/u_4 is symmetric <0.8; 1.2>.
- **Forwards, Backwards for the First DOF:** The dividing ratio of diagonals $u_1/u_2 > 1.2$ respectively $u_1/u_2 < 0.8$.
- **Turn to the Left—Turn to the Right:** Similar for the u_3/u_4. Final direction of the wheelchair motion is the vector sum of components u_1/u_2 and u_3/u_4.
- The magnitude of the dividing ratio u_1/u_2, u_3/u_4 is assigned to the velocity of the wheelchair motion for two degrees of freedom (DOF).
- The motion control system of the wheelchair enables parallel control of two degrees-of-freedom.
- The light spot A is oriented in front direction. This is needed for the orientation of the wheelchair against the basic light spot position.

Some applications of the signal filtering for the elimination of the body inclination are needed for users with muscular trembling at neurological diseases and at the motion of the wheelchair on rough surface.

Figure 9 depicts the CCD camera for the sampling of the head joystick's light spots configuration on the floor. The difference between the sampling from the CCD camera attached to the wall or to the ceiling against the one attached to the wheelchair is in the level of navigation. The CCD camera attached to the wall or to the ceiling enables the positioning of the wheelchair with respect to the shape of the room. The CCD camera attached to the wheelchair enables only the relative positioning of the wheelchair.

Figure 9. The force-torque transducer inserted between the saddle and the chassis of the wheelchair and the CCD camera for the sampling of the head joystick's light spots configuration on the floor

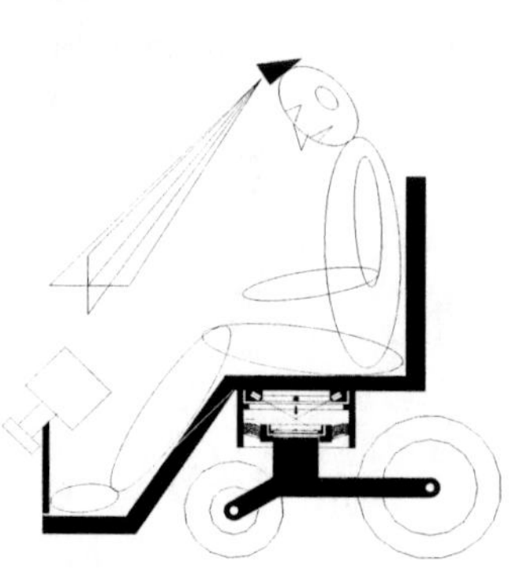

OPTOELECTRONIC HANDY NAVIGATOR FOR BLIND PEOPLE

Safe and effective mobility of blind people depends largely upon reliable orientation about the articulation of the ambient. The current situation of people who suffer severe visual impairments is that they mostly require active assistance from relatives or occasionally assistance of passing by strangers in order to travel or to use transportation or to manage street traffic in crowded areas. Nevertheless, blind people, just like sighted people, do not like to be dependent on others. There are restricted sources of sound information at the traffic light control or acceptable spoken information at landmarks and significant places in information stations. Independent travel and transportation for blind people involves orienting oneself and finding a safe path through known and unknown articulated environments. Most efforts have been to solve the mobility part of the problem to help the blind traveler detect irregularities on the floor such as boundaries, objects located near or alongside his or her path in order to avoid collisions, and steer a straight and safe course through the immediate environment. Ultrasonic and laser devices for the navigation using active methods for the identification of the environment have been reported.

One way for the classification of 3-D object articulation destined for blind people is based on the principle of the twofold range-inclination tracer. The range-inclination tracer enables the shape analysis by means of the structured light-cutting plane intersection with an object. This navigation system consists of four laser diodes with focusing optics used for the light spots imagination 1-D CCD array (alternatively PSD array), microprocessor, and tuned pitch and timbre of sound source. Two parameters on the intersection are computed and evaluated: the distance and the inclination of the object.

The time-profiles of the distance and inclination, and their combinations are used to tune the sound generator's frequency and intensity. The sound representation of a 3-D object's articulation enables the skill-based training of the user in recognizing the distance and ambient articulation. Smart configuration enables the customizing of navigation device into:

- a handy phone that uses only one line from the built-in 2-D CCD array, or
- a white stick for blind people using 1-D CCD array (alternatively PSD array) or the handheld device.

The twofold range-inclination tracer consists of the camera with 1-D CCD array or linear PSD element and four laser diodes radiating four intersecting light rays against the wall, as shown in Figure 10. Four light spots are sampled by a simple sampling algorithm, which then evaluates the distance D and the inclination β of the plane against the range-inclination tracer. The sampling of left and right light spot coordinates in subsequent pictures synchronized with sequentially switched light rays enables the recognition of the changing difference and orientation of the light spots and their orientation before and after the crossing point. Control algorithms are derived in the rectangular coordinate system x,y with the origin in the point of mutual intersection of four light rays, as described in Figure 7 using the relationships (4), (5), (6), and (7). The described range-inclination tracer is alternatively used for navigation:

- in single mode, using two crossing light rays with light spots $A(r_1,s_1)$ and $B(r_2,s_2)$, with the evaluation of the distances and inclination of the ambient; or
- in double mode, using four crossing light rays with light spots $A(r_1,s_1)$, $B(r_2,s_2)$, $E(r_3,s_3)$, and $F(r_4,s_4)$, with the evaluation of the distances and inclination of the ambient including the evaluation of the invariant symptoms of the ambient articulation, like simple plain, parallelism, or perpendicularity of two plains.

The navigation system for blind people is based on the cooperation of four laser diodes with focusing optics used to determine light spot in the 1-D CCD array (alternatively PSD array), microprocessor, and tuned pitch and timbre of sound source. The first couple of laser rays with light spots A,B form an angle α, and the second couple of laser rays with light spots E,F form an angle σ; both are symmetric with the axis of the CCD array. Generally the navigation system consists of two independently working range inclination tracers. The configuration of the navigation system enables, by means of purposeful swept hand motion, simple evalu-

Figure 10. Navigation system based on the twofold range-inclination tracer fastened on the handy device

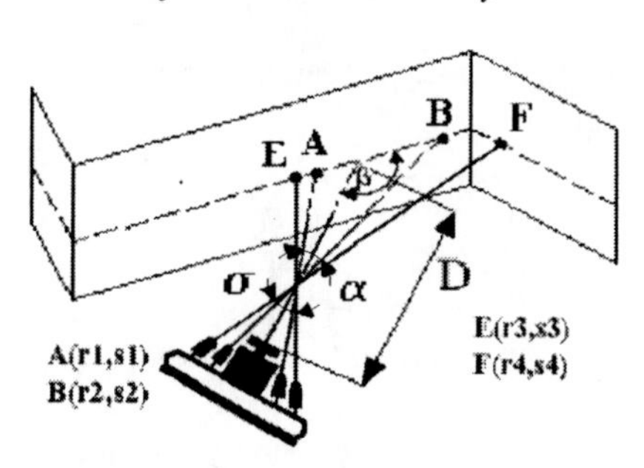

ation of invariant symptoms of the object for the indication of the following:

- simple plane, when the points A,B,E,F belong to one straight line as shown in Figure 11;
- parallel planes, when the points A,E and B,F create two parallel straight lines as shown in Figure 12; or
- two plains forming an angle with pointing up of the right angle, when the points A,E and B,F create two mutually intersecting lines, as shown in Figure 13.

The navigation device can be used in two ways:

1. The single mode using only two light spots A,B or E,F on the way of the range-inclination tracer. This way is effective to assign a safe and acceptable range of the distance and inclination using two-tones. For example, when both light spots are on the pavement, a unique tone is played. The loss of safe inclination is signaled by the dissonance, for example at the positions of the light spots on the pavement-wall. The loss of the front light spot is signaled by the interrupted tone of enhanced intensity.
2. The navigation system using two pairs of the light spots A,B and E,F enables, by means of purposeful swept hand motion, a simple indication of the object articulation. This method is effective to assign various chord combinations (melodies) for different shapes such as simple plane, parallel planes, or two plains forming an angle. Simultaneously the range inclination resonates in the background.

Figure 11. Indication of a simple plain

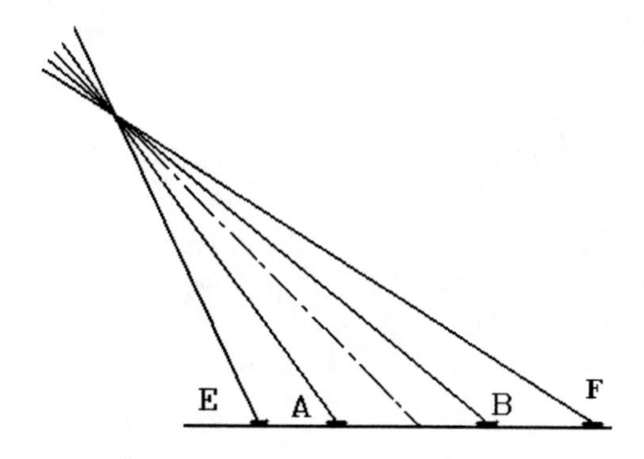

Figure 12. Indication of parallel plains

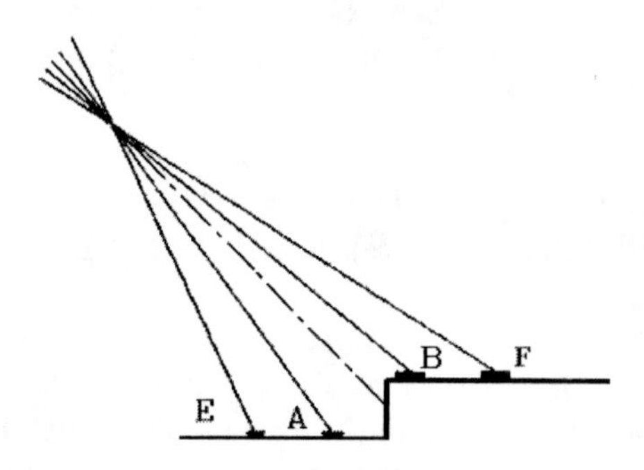

Figure 13. Indication of two plains forming an angle with pointing up of the right angle

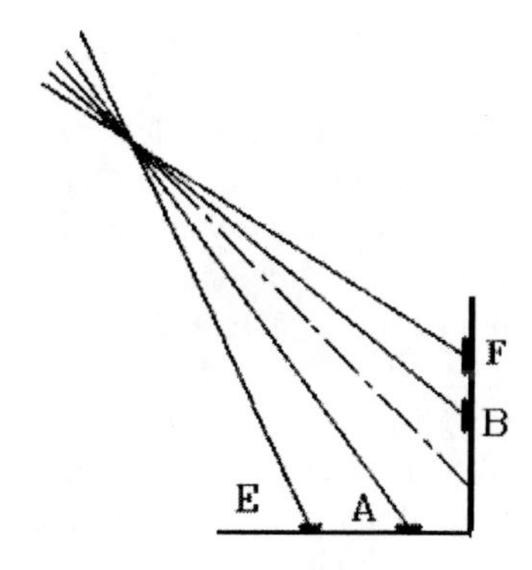

Smart configuration of the handy navigator needs simple built-in customizing into a handy phone, handheld device, or white stick for blind people. Video signal from the 1-D or the 2-D CCD array is processed by a simple method at low cost, and in the form of an embedded system, with a single cheap microprocessor in order to decrease the size of opto-electronic devices. The video signal output is preprocessed in the comparator, which is set up on the white level. It causes the signal from every pixel of the light spot to be indicated like an impulse. Every impulse is assigned to the video dot information sequence in the range between 1 and the maximal number of pixels in the CCD array in order to determine its position in the picture coordinate frame. The run of the dot video information is switched for every picture separately by vertical (picture) synchronization impulse. Every couple of light spots A,B and E,F is evaluated in separate pictures. It enables recognition of the varying orientation of the light spot before and after the crossing point of the light rays. An alternative application of the 1-D PSD array is based similarly on the sampling of only one light spot's position.

EXPERIMENTAL HARDWARE FOR COMPOSING OF SIMPLE TASKS

The automatic parking equipment, the range-inclination tracer, the optoelectronic handy navigator, and the wheel-chair control by means of the head motion and by means of body motion were implemented by means of following experimental hardware:

- **The Data Translation High-Accuracy, Programmable, Monochrome Frame Grabber Board DT3155 for the PCI Bus:** This is suitable for both image analysis and machine vision applications.
- **The Microprocessor-Controlled Programmable Timer PIKRON ZO-CPU2:** This used for the timing of the light exposure and asynchronous switching of the laser diodes configuration.
- **The Configuration of Miniature Laser Diodes F-LASER 5mW:** This includes focusing optics radiating structured light rays.
- **Digital B/W Video Camera SONY KC-381CG:** This includes a digital signal processor and high-resolution 795Hx596V 1/3" CCD sensor with high sensitivity (0.02lux at F0.75) and interline transfer, digital light level control system for the back-light compensation, with auto-exposure or manual exposure system, aperture correction, and internal or gen-lock/line-lock external transfer.
- **Zoom Lens Computar MLH-10X:** This includes 10x macro zoom, maximal magnification 0.084~0.84x, maximal aperture 1:5.6, maximal image format 6.4×4.8mm (average 8mm), and focus 0.18~0.45m.

The image processing of the light spots from the CCD camera including the control algorithms for the first step of skill-based education was developed on MATLAB. The application of the frame grabber is used only for experiments.

The second step of the education is oriented on the processing of the video signal from the 1-D or the 2-D CCD array for the user's application by a single cheap microprocessor.

The activity of the PSD element is based on the sampling of only one light spot's position. This means that for every light beam, it assigns a separate PSD element. In this way the subsequent sampling of the six-DOF information by means of the PSD element is guaranteed, but causes dynamic distortion dependent on the sampling frequency. Correct activity of the six-DOF sensor based on the CCD element depends on the continuity of sampling. This indicates the verification of every light spot position in regard to responding basic position A,B,C,D,S. The continuous motion of the first light spot can be recognized in various ways, for example by means of:

- different shape of the light spot,
- different color of the light spot,
- different intensity of the light spot,
- dividing of the picture frame into four quadrants for every light spot, and
- minimal distance of preliminary position of the light spot.

Having the identity of the first light spot, the other light spots can be recognized clockwise or counter-clockwise. The third step of the educational applications is oriented toward embedded systems in order to decrease the size of developed optoelectronic devices. The structure of the embedded system is oriented to the use of microprocessors in order to enable flexibility in the proof of various algorithmic modules for the enhancement of the accuracy, like the approximation of the light spot center or for the elimination of the nonlinearity, and on the tasks of the dynamic control using C language with subroutines in assembler.

CONCLUSION

The modular sensory system design presented here enables easy adaptation of various applications of human-machine interfaces for assistive technologies and mobile robotic systems. In general, this modular sensory system concept is appropriate for a low-cost design and in addition enables the understanding of basic problems concerning the interaction between human and mobile robotic systems.

ACKNOWLEDGMENTS

The support from the grant Vyzkumne zamery MSM 7088352102 "Modelovani a rizeni zpracovatelskych procesu prirodnich a syntetickych polymeru" is gratefully acknowledged.

REFERENCES

Akira, I. (2002). An approach to the eye contact through the outstaring game Nirammekko. *Proceedings of the 11th IEEE International Workshop on Robot and Human Interactive Communication (RO-MAN 2002)*, Berlin, Germany.

Fukuda, T., Nakashima, M., Arai, F., & Hasegawa, Y. (2002). Generalized facial expression of character face based on deformation model for human-robot communication. *Proceedings of the 11th IEEE International Workshop on Robot and Human Interactive Communication (RO-MAN 2002)*, Berlin.

Humusoft. (n.d.). Retrieved from http://www.humusoft.cz

Kawarazaki, N., Hoya, I., Nishihara, K., & Yoshidome, T. (2003). Welfare robot system using hand gesture instructions. *Proceedings of the 8th International Conference on Rehabilitation Robotics (ICORR 2003)*, Daejeon, Korea.

Kim, D.-H., Kim, J.-H., & Chung, M. J. (2001). An eye-gaze tracking system for people with motor disabilities. *Proceedings of the 7th International Conference on Rehabilitation Robotics (ICORR 2001)*, Evry Cedex, France.

Kim, J.-H., Lee, B. R., Kim, D.-H., & Chung, M. J. (2003). Eye-mouse system for people with motor disabilities. *Proceedings of the 8th International Conference on Rehabilitation Robotics (ICORR 2003)*, Daejeon, Korea.

Kvasnica, M. (1999, July). Modular force-torque transducers for rehabilitation robotics. *Proceedings of the IEEE International Conference on Rehabilitation Robotics*, Stanford, CA.

Kvasnica, M. (2001, April). A six-DOF modular sensory system with haptic interface for rehabilitation robotics. *Proceedings of the 7th International Conference on Rehabilitation Robotics (ICORR 2001)*, Paris-Evry, France.

Kvasnica, M. (2001). Algorithm for computing of information about six-DOF motion in 3-D space sampled by 2-D CCD array. *Proceedings of the 7th World Multi-Conference (SCI'2001-ISAS)* (Vol. XV, Industrial Systems, Part II), Orlando, FL.

Kvasnica, M. (2002). Six DOF measurements in robotics, engineering constructions and space control. *Proceedings of the 8th World Multi-Conference on Systemics, Cybernetics and Informatics (SCI'2002-ISAS 2002)* (Ext. Vol. XX), Orlando, FL.

Kvasnica, M. (2003, September). Head joystick and interactive positioning for the wheelchair. *Proceedings of the 1st International Conference on Smart Homes and Health Telematics (ICOST 2003)*, Paris.

Kvasnica, M. (2003). Six-DOF sensory system for interactive positioning and motion control in rehabilitation robotics. *Proceedings of the 8th International Conference on Rehabilitation Robotics (ICORR 2003)*, Daejeon, Korea.

Kvasnica, M. (2003). Six-DOF sensory system for interactive positioning and motion control in rehabilitation robotics. *International Journal of Human-Friendly Welfare Robotic Systems, 4*(3).

Kvasnica, M. (2004, September). Six-DOF force-torque transducer for wheelchair control by means of body motion. *Proceedings of the 2nd International Conference on Smart Homes and Health Telematics (ICOST 2004)*, Singapore.

Kvasnica, M. (2005). Assistive technologies for man-machine interface and applications in education and robotics. *International Journal of Human-Friendly Welfare Robotic Systems, 6*(3). Daejeon, Korea: KAIST Press.

Kvasnica, M., & Vasek, V. (2004, May). Mechatronics on the human-robot interface for assistive technologies and for the six-DOF measurements systems. *Proceedings of the 7th International Symposium on Topical Questions of Teaching Mechatronics,* Rackova Dolina, Slovakia.

Kvasnica, M., & Van der Loos, M. (2000). Six-DOF modular sensory system with haptic interaction for robotics and human-machine interaction. *Proceedings of the World Automation Congress (WAC 2000)*, Maui, HI.

Mathworks. (n.d.) Retrieved form http://www.mathworks.com/

Min, J. W., Lee, K., Lim, S.-C., & Kwon, D.-S. (2003). Human-robot interfaces for wheelchair control with body motion. *Proceedings of the 8th International Conference on Rehabilitation Robotics (ICORR 2003)*, Daejeon, Korea.

Moon, I., Lee, M., Ryu, J., Kim, K., & Mun, M. (2003). Intelligent robotic wheelchair with human-friendly interfaces for disabled and the elderly. *Proceedings of the 8th International Conference on Rehabilitation Robotics (ICORR 2003)*, Daejeon, Korea.

Neovision. (n.d.). Retrieved from http://www.neovision.cz

KEY TERMS

$\alpha/2$: The angle of the optical axis of the camera with the laser light ray.

B: The inclination of the wall against the wheelchair.

D: The distance between the camera and the wall.

Δh: The difference between the point of the light rays' intersection and the intersection of the optical axis of the camera with the wall.

h: The distance between the camera and the point of the intersection of light rays.

N: The direction of the wheelchair motion.

(r_1,s_1), (r_2,s_2): The coordinates of the light spots A, B.

2s: Presetting control of the angle contained by mutual opposite light rays of the modules Apc and App.

x,y: The coordinates of the position of the wheelchair.

(x_1,y_1), (x_2,y_2): Coordinates of the points E, F.

z: The distance of the wheelchair from the ceiling in the rectangular coordinate frame.

M-Commerce Technology Perceptions on Technology Adoptions

Reychav Iris
Bar-Ilan University, Israel

Ehud Menipaz
Ben-Gurion University, Israel

INTRODUCTION

This article presents a tool for assessing the probability of adopting a new technology or product before it is marketed. Specifically, the research offers managers in firms dealing with mobile electronic commerce a way of measuring perceptions of technology usage as an index for assessing the tendency to adopt a given technology. The article is based on an ongoing study dealing with m-commerce in Israel and internationally. It is centered on creating a research tool for predicting the usage of m-commerce in Israel, based on the PCI model. The suggested model is based on a questionnaire presented to the potential consumer, containing questions linking the consumer's perception of the various aspects of the technological innovation offered, together with his tendency to buy and therefore adopt it. The tool was found to possess high reliability and validity levels. The average score in the questionnaire is used to predict the probability of adoption of the mobile electronic commerce technology. Implications related to m-commerce technology in Israel and worldwide are discussed.

BACKGROUND

The main purpose of this study is to assess the tendency to adopt mobile electronic commerce technologies, prior to actually launching a new product or service based on cellular technology. The focus in the present study is on the general population and not on organizations.

Contrary to products or services sold to end users, the mobile electronic commerce field offers an innovative system of business ties with the client, by utilizing mediating tools such as the cellular device. The focus in the mobile electronic commerce field is on influencing consumers' preferences. This study presents a tool that examines the perceived utilization of technology advances in the field.

The focus in this study on the characteristics of using cellular phones for mobile electronic commerce is based on findings from a wide range of studies dealing with the characteristics of the perception of innovativeness itself.

Rogers (1983) studied thousands of cases of diffusion and managed to define five characteristics of innovativeness affecting its diffusion: relative advantage, compatibility, complexity, visibility, and trialability. While Rogers' characteristics were based on the perception of innovativeness itself, Ajzen and Fishbein (1980, p. 8) claimed that the attitude toward the object is different in essence from the attitude towards a certain behavior related to the object. Innovation penetrates because of accumulating decisions by individuals to adopt it. Therefore, not perceiving the efforts of innovation itself, but the perception of using innovation is the key to its diffusion. In the diffusion studies, the subject of perceptions was treated in relation to innovation itself. Nevertheless, the characteristics of the perceptions of innovation can be redesigned in terms of perceiving the use of the innovation (Moore, 1987). Rewriting the characteristics of *perceptions of innovativeness* into characteristics of the *perceptions of using the innovation* was the basis for the PCI (perceived characteristics of innovating) model, developed by Moore and Benbasat (1991) and used as a tool for studying the adoption of information technologies. The PCI model expands the conceptual framework designed by Rogers, by adding additional characteristics that may influence the decision to adopt a new technology. The tool was presented as reliable and valid.

MOTIVATION AND PURPOSE OF THE STUDY

The motivation for creating a tool for measuring the perceptions regarding cellular phone usage for mobile electronic commerce of the potential adopters of the technology originated from three main factors.

First were findings from previous research, which focused on adoption patterns of Internet and cellular technologies in various countries, in an attempt to present a methodology for analyzing diffusion (Reychav & Menipaz, 2002). Second, while carrying out the above-mentioned study, the researchers realized there was a lack in theoretical background for studying the initial adoption process of innovative technologies such as m-commerce, as well as the understanding of how to successfully assimilate innovative technology. Third, an opportunity presented itself to examine the research model in Israel, a country in which the usage of cellular technology is widespread.

METHOD

The study took place in Israel, where the knowledge constraint towards cellular technology does not exist and apprehension on the part of the general population from adopting unknown technologies due to this constraint is nearly unknown. Outsland (1974, p. 28) suggested that perceptions of innovations by potential adopters of innovative technology might be an effective prediction measure for adoption of the innovation, more than personal factors. Based on this assumption, the current research focused on the university student population, which represents a segment in society that is essentially aware of computer and Internet technologies, and therefore its apprehension from adopting unknown technologies is relatively low. In addition, in Israel the penetration percentage of cellular technology has already reached its full potential, and the interest of the current research is to examine the tendencies to adopt usage of cellular phones for mobile electronic commerce. In order to do so, a research questionnaire was constructed, including 35 items dealing with perceptions regarding the use of mobile electronic commerce technology. Each item in the questionnaire was assessed on three time scales—perceptions of usage in the past, at present, and an estimate of usage perceptions in future.

The questionnaire was distributed amongst students from various departments at Ben Gurion University in the Negev. The distribution included most university departments. A total of 1,300 questionnaires were distributed. They were completed in the presence of the researcher and handed in directly.

CHARACTERISTICS OF THE MODEL

The model is based on the characteristics of the perceptions of innovativeness, which have been identified in previous studies, with a change in wording from "perception of innovativeness" to "perception of the use of innovation," as suggested in the PCI model (Moore & Benbasat 1991). The characteristics are as follows:

- **Relative Advantage:** The extent to which the use of an innovation is perceived as better than the use of its predecessor (based on work by Roger, 1983).
- **Compatibility:** The extent to which the use of an innovation is perceived as being persistent with other existing values, needs, and experiences of the potential adopters (Roger, 1983).
- **Ease of Use:** The extent to which individuals believe that the use of a specific system does not require investment of physical and emotional efforts (Davis, 1986).
- **Results Demonstrability:** The extent to which the results of using an innovation are tangible and presentable (Rogers, 1983, p. 232). Research has shown that merely being exposed to a product can in itself create a positive attitude toward it among individuals (Zajonc & Markus, 1982).
- **Image:** The extent to which the use of an innovation is perceived as improving the individual's status in society.
- **Visibility:** The extent to which the results of the use of an innovation are visible to others. The characteristic "Observability," which was mentioned by Rogers (1983), is presented in the PCI model via two variables (Results Demonstrability and Visibility).
- **Trialability:** The extent to which the use of an innovation can be experienced prior to its adoption.

RESULTS

Out of 1,300 distributed questionnaires, 1,005 were completed correctly (11.49% missing data). The study results point to 55.6% potential users of cellular phones for m-commerce in two years' time, compared to 34.8% two years ago and 38.9% users today. This is indicative of the fact that the questionnaire reflects the target population studied in the current research.

The percentage of explained variance obtained in the model runs is 67.732% in the past, 65.896% at present, and 61.470% in the future.

It is safe to say that the model for this study has been validated and verified. Therefore, we can conclude that in order to assess the probability of actual usage of cellular phone for m-commerce, the model for perceptions of use of cellular phone for m-commerce presented in this study may be used.

Predicting the Probability of Using M-Commerce

After verifying the model, a test was carried out to identify which variables assist in predicting the probability for *using cellular phones for mobile electronic commerce.*

The testing method used was Forward Stepwise Logistical Regression (Hosmer & Lemeshow, 2000), which first brings into the equation variables having the highest level of significance, and then re-calculates the level of significance

Table 1. Logistical regression equation constants in test run on perceptions categories

Variable	*B (past)*	*B (current)*	*B (future)*
Relative Advantage	0.117	0	0
Image	0.096	0	0
Compatibility	0	0.182	0.385
Visibility	0.305	0.273	0.185
Cellular Phone Use	0	0	1.021
Method of Connecting with Internet Provider	0	0	-0.533
Equation Constant	-1.770	-1.501	-2.310

of each of the variables in the equation. For the past timeframe, the results were given after running the first step of the regression, while for the present and future timeframes, the results were given after three steps of the regression.

The results show that for the period of two years ago, the *average perceptions* of potential adopters of the technology is a significant variable ($p<0.01$) and can therefore serve as a tool for predicting the probability of using cellular phones for mobile electronic commerce at that time, and for present and future times as well. For the period of the next two years (future), the variables which were found to be significant are: *estimate of the average perceptions in two years' time* ($p<0.01$), *estimate of cellular phone use,* and *method of connecting to an Internet provider in two years' time* ($p=0.037$).

An assessment of the predictions for using cellular phones for mobile electronic commerce was also done by using each of the categories in the model.

The findings show that for the timeframe of two years ago, the significant perception categories are: *relative advantage* ($p=0.039$), *image* ($p=0.033$), and *visibility* ($p=0.001$). In the present timeframe, the significant categories are: *compatibility* ($p=0.001$) and *visibility* ($p=0.000$), with the *image* category having only a slight significance ($p=0.055$). In the future timeframe, the significant categories are: *compatibility* ($p=0.001$), *visibility* ($p=0.002$), and the additional variables *use of cellular phone* ($p=0.016$) and *method of connecting to an Internet provider* ($p=0.010$).

For each of the significant variables, the logistical regression odds parameters present the probability for prediction of the dependent variable, *use of cellular phone for mobile electronic commerce.* The results show that the categories *relative advantage* and *Image* serve as a prediction tool only for the past timeframe. The *visibility* category serves as a prediction tool for all timeframes, although for each point in the average of the *visibility* category, the probability along the timeframes is lower than 1.357 in the past, 1.314 at present, and 1.203 in future. The *compatibility* category serves as a prediction tool only at present and in the future. Also in the future, additional variables are added which may influence the probability for prediction: *use of cellular phone* (2.775) and *method of connecting to an Internet provider* (0.587). However, the probability presented by the variable *method of connecting to an Internet provider* is lower than 1, therefore reducing the probability for using mobile phones for mobile electronic commerce if the *method of connecting to an Internet provider* is by way of a telephone line.

Given regression coefficients for each one of the test runs performed across the three timeframes, the regression equation can be created so as to predict the probability for *using cellular phone for mobile electronic commerce* for each questionnaire respondent. The results of the regression coefficients are presented in Table 1.

In this case, we can also insert the regression coefficients obtained in each test run in the logistical regression equation, and receive the probability for *using cellular phones for mobile electronic commerce* in each timeframe and for every score received in each of the categories.

For a presentation of the probability to *use cellular phones* according to the average score obtained by every potential adopter of the technology who answers the questionnaire, see Figure 1.

Figure 1 presents a trend in the change in attitudes towards *using cellular phones for mobile electronic commerce* from non-usage to usage. As the *average score of perception of potential adopter* grows from past to present to future, so grows the probability to adopt the technology, from 63%, 79%, and 90% respectively for the maximal score given by the respondent. The variable *method of connecting to an Internet provider,* which was also found to influence the probability of predicting use in the future, heightens the probability of adopting the technology from 90% (when using a telephone line) to 93.485% (when using ADSL).

Internet, Cellular, and Credit Card Technology Usage Habits

In addition to the research hypothesis, which dealt with the perceptions model presented in the study, additional research hypotheses were tested, relating to the usage habits of Internet and cellular technologies such as the *use of cellular phones, method of connecting to an Internet provider,* and *use of credit card.*

Cellular Technology

Cellular phone usage rate was 87.1% two years ago. This rate has risen to 95.4% today, and is estimated to change by 0.4%, to reach 95.8% in two years' time.

Figure 1. The probability of using cellular phones for mobile electronic commerce, dependent on the average score of perceptions of using cellular phones for mobile electronic commerce

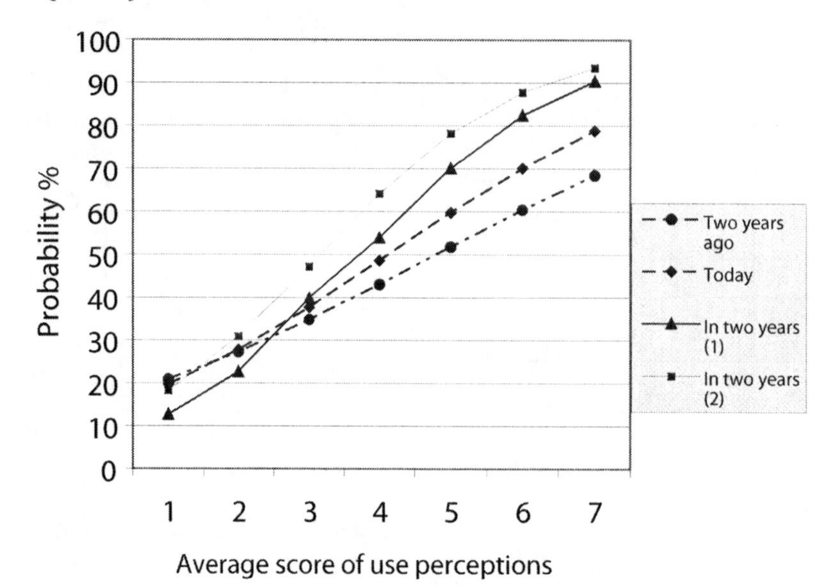

The results presented here point to the existence of a trend from past to present time only (Sig. (McNemar) p<0.01).

Internet Technology

Regarding the *method of connecting to an Internet provider*, results show that the use of a telephone line as a connecting method has been in decline, from 85.7% two years ago to 53.3% today. This trend is expected to continue with a further decline to 12.8% two years from now. In comparison, the use of ADSL as a connection method has grown by 694.8% over the past two years (from 5.8% two years ago to 46.1% today). It is estimated that this trend will continue in the near future and will reach 84.1% in two years' time (a growth rate of 82.4% compared to the present rate).

The rate of Internet use two years ago was 79.4%. Today it has grown to 97.9%. This rate is expected to grow to 98.4% two years from now.

The results presented here point to the existence of a trend in Internet use from past to present (Sig. (McNemar) p<0.01) and from present to future (Sig. (McNemar) p=0.039). A significant correlation was found, using the chi-square test, between *Internet use* and *cellular phone use* for all timeframes (Sig. (2 tailed) p<0.01). The percentage of cellular phone users found in the study out of the overall Internet users shows a 6% growth in the past two years and an additional 0.8% expected growth in the next two years.

In comparison, the percentage of Internet users out of the overall cellular phone users showed a growth of 16.6% in the past two years, but is expected to decline by 13.1% in the next two years, from a rate of 98.8% users to 85.7% in two years. A further interesting finding is the non-significant correlation between *Internet use* and *use of cellular phones for mobile electronic commerce* for all timeframes.

Credit Card Use

Concerning *credit card use,* the results of the study show that the usage rate of credit cards has risen from 86.1% two years ago to 93.8% today, with an expected rate of 97.3% in two years.

The results presented here point to the existence of a trend in *credit card use* for all timeframes (Sig. (McNemar) p<0.01). The results also point to a significant correlation, by using the chi-square test, between *use of credit cards for Internet use* for all timeframes (Sig. (2 tailed) p<0.01) and the *use of credit cards for cellular phone use* in the past and present times (p<0.01), and also in the future (p=0.015). Similarly to the non-significant correlations obtained between the use of basic technologies for electronic commerce, such as Internet and cellular technologies, the study results also show that the *use of cellular phones for mobile electronic commerce* is not necessarily an indication of the *use of credit cards.*

CONCLUSION

Following are the main conclusions that can be deducted from the study so far:

a. The research tool presented in this study was found to have a high reliability level (Nunnally, 1967), and it may serve as a tool for predicting the probability of *using cellular phones for mobile electronic commerce* in the general population, at an initial adoption environment. The present study attempts to analyze respondents' responses when the decision to adopt a technology is voluntary for the general population. The prediction is deducted from the average score obtained by the respondent in the questionnaire, or from the average score obtained by the respondent in the *visibility* and *compatibility* categories, which were found to be significant in this study.

b. Predicting the *use of cellular phones for mobile electronic commerce* two years from now was found to be influenced not only by perceptions, but by a number of additional factors such as *the use of cellular phone* and *method of connecting to an Internet provider.* In Israel, the *use of cellular phones* has reached its full potential. This finding is also supported by the results of the present study (a 95.4% usage rate today).

c. A significant trend was found in the past, present, and future for transferring from *using a phone line,* to *using ADSL* as the *method of connecting to an Internet provider.* This result points to the fact that consumers who use relatively advanced technologies will also tend to do so regarding new advanced technologies. The logical foundation for this argument is based on the findings obtained from the model of perceptions of using cellular phones for mobile electronic commerce, which related the two categories *compatibility* and *relative advantage* to the same factor in the factor analysis. Hence, ensuring the compatibility of an innovative technology to the needs of potential adopters may also constitute a relative advantage, thus increasing the level of use. The *compatibility* category was also found to have a high level of significance in other diffusion studies (Hurt & Hubbard, 1987). Testing the perceptions of use categories has presented a significant difference between past and present, as well as between present and future times.

d. The findings specified above (c) may also assist in explaining the trend of the innovation of mobile electronic commerce two years ago, which was seen as superior to other available alternatives and was thus identified as having a relative advantage.

e. Additional interesting results found in the study point to a non-significant correlation between *credit card use*, together with the *use of Internet and cellular phone technologies* and *using cellular phones for mobile electronic commerce.* The *use of credit cards* for mobile electronic commerce is perceived as unsafe and requires a high level of functional interaction.

f. The study points to a significant correlation between the *use of Internet* and *the use of cellular technologies.* These two technologies are perceived as complementing each other.

g. The decline in the number of Internet users out of the overall cellular user population, together with the finding that points to a non-significance in the relationship between the *use of Internet* and the *use of cellular phones for mobile electronic commerce,* presents a change towards adopting a more innovative technology, such as the mobile electronic commerce technology. A further verification of this claim can be found in the finding showing a significant relationship between the *method of connecting to an Internet provider* and the *use of cellular phones for mobile electronic commerce.*

h. Similarly to the diffusion theory presented by the Bass model (1969), in which the adopter's population is divided into two groups—innovators and imitators, also in the present research, the population, which tends to use an innovative technology, does not limit itself to using a specific technology, but will continue to do so regarding other advanced technologies such as mobile electronic commerce.

RESEARCH IMPLICATIONS

The research and practical implications of this study include a tool for assessing the probability to adopt an innovative technology or product before it is marketed. Specifically, this study offers managers in companies dealing with mobile electronic commerce to base their assessment of the technologies' adoption chances on the perceptions of use of the technology by potential adopters.

The tool presented in the present study enables predicting the probability of adopting a new technology and modeling the factors influencing its diffusion. From the practical aspect, the study may help decision makers in cellular companies, both as regards infrastructure and service providers, to understand the characteristics of perceptions of using the m-commerce technology in Israel and worldwide, and as a result, to be able to focus their time and efforts on defining appropriate marketing strategies. Performing an analysis from the point of view of the end user, as in absorbing a new information system, is an important factor for assessing the future level of use of a new technology.

This study is one of the first attempts to deal with the diffusion of innovations in the mediating technologies environment. The results of the study open a number of additional research opportunities, such as assessing the usage perceptions of potential adopters of a specific product or service in the m-commerce environment (for instance, designated content services). This future study may serve as an indication to marketers regarding the perceptions of using a product or service while it is in its initial development phase, so that the required adjustments can be made before it is actually launched. An additional interesting research direction would be to examine the tool suggested in the present study on international markets, so as to broaden its scope of generalizations and significance.

REFERENCES

Ajzen, I., & Fishbein, M. (1980). *Understanding attitudes and predicting behavior.* Englewood Cliffs, NJ: Prentice-Hall.

Bass, F. M. (1969). A new-product growth model for consumer durables. *Management Science, 15*(January), 215-227.

Davis, F. D. (1986). *A technology acceptance model for empirically testing new end user information systems: Theory and result.* Unpublished doctoral dissertation, Massachusetts Institute of Technology, USA.

Hosmer, D. W, & Lemeshow, J. R. S. (2000). *Applied survival analysis regression modeling of time to event data.* New York: Wiley-Interscience.

Hurt, H. T., & Hubbard, R. (1987, May). The systematic measurements of the perceived characteristics of information technologies: Microcomputers as innovations. *Proceedings of the ICA Annual Conference,* Montreal Quebec.

Moore, G. C. (1987). End user computing and office automation: A diffusion of innovation perspective. *INFOR, 25*(3), 214-235.

Moore, G. C., & Benbasat, I. (1991). Development of an instrument to measure the perceptions of adopting an information technology innovation. *Information Systems Research, 2*(3), 192-222.

Nunnally, J. C. (1967). *Psychometric theory.* New York: McGraw Hill.

Olshvsky, R. W. (1980). Time and the rate of adoption of innovation. *Journal of Consumer Research, 6*(March), 425-428.

Ostlund, L. E. (1974). Perceived innovation attributes as predictors of innovativeness. *Journal of Consumer Research, 1*(2), 23-29.

Reychav, I., & Menipaz, E. (2002). M-business diffusion and use: Global perspective. *Proceedings of the 12th Industrial Engineering Conference,* Israel.

Rogers, E. M. (1983). *Diffusion of innovations.* New York: The Free Press.

Zajonc, R. B., & Markus, H. (1982). Affective and cognitive factors in preferences. *Journal of Consumer Research, 9*(September), 123-131.

KEY TERMS

ADSL: Method of connecting to an Internet provider.

Compatibility: The extent in which the use of an innovation is perceived as persistent with other existing values, needs, and experiences of the potential adopters (Roger, 1983).

Ease of Use: The extent in which individuals believe that the use of a specific system does not require investment of physical and emotional efforts (Davis, 1986).

Image: The extent in which the use of an innovation is perceived as improving the individual's status in society.

Perceived Characteristics of Innovating (PCI) Model: Developed by Moore and Benbasat (1991), and is used as a tool for studying the adoption of information technologies.

Relative Advantage: The extent in which the use of an innovation is perceived as better than the use of its predecessor (based on work by Roger, 1983).

Results Demonstrability: The extent in which the results of using an innovation are tangible and presentable (Rogers, 1983, p. 232). Research has shown that merely being exposed to a product can in itself create a positive attitude toward it among individuals (Zajonc & Markus, 1982).

Trialability: The extent in which the use of an innovation can be experienced prior to its adoption.

Visibility: The extent in which the results of the use of an innovation are visible to others. The characteristic "observability," which was mentioned by Rogers (1983), is presented in the PCI model via two variables (results demonstrability and visibility).

M-Learning with Mobile Phones

Simon So
Hong Kong Institute of Education, Hong Kong

INTRODUCTION

The Internet is a major driver of e-learning advancement and there was an estimate of over 1000 million Internet users in 2004. The ownership of mobile devices is even more astonishing. ITU (2006) reported that 77% of the population in developed countries are mobile subscribers. The emergence of mobile, wireless and satellite technologies is impacting our daily life and our learning. New Internet technologies are being used to support small-screen mobile and wireless devices. In a field marked by such rapid evolution, we cannot assume that the Web as we know it today will remain the primary conduit for Internet-based learning (Bowles, 2004, p.12). Mobile and wireless technologies will play a pivotal role in learning. This new field is commonly known as mobile learning (m-learning).

In this article, the context of m-learning in relation to e-learning and d-learning is presented. Because of the great importance in Web-based technologies to bridge over mobile and wireless technologies, the infrastructure to support m-learning through browser-based technologies is described. This concept represents my own view on the future direction of m-learning. An m-learning experiment, which implemented the concept, is then presented.

Figure 1. M-learning space as part of e-learning and d-learning spaces

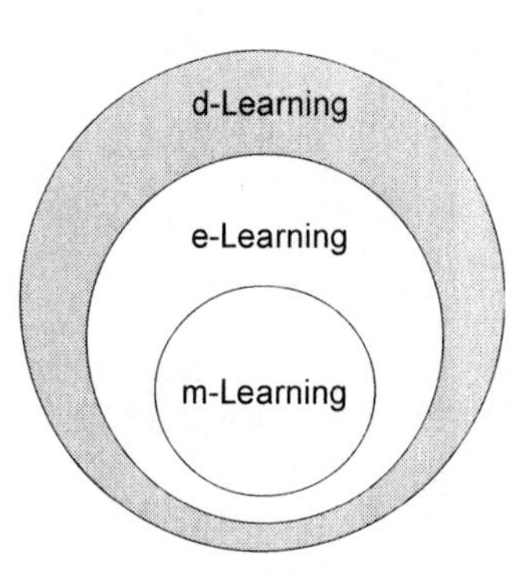

BACKGROUND

Many researchers and educators view that m-learning is the descendant of e-learning and originates from d-learning (Wikipedia M-Learning, 2006; Georgiev, Georgieva, & Smrikarov, 2004). The m-learning space is subsumed in the e-learning space and, in turn, in the d-learning space, as shown in Figure 1. This may be true chronologically. D-learning has more than hundred years of evolution starting from the printed media of correspondence (signified by carefully designed and produced materials by specialists to support the absence of instructors and independent study [Charles Wedemeyer] and the industrialization of teaching [Otto Peters]), to mass and broadcast media (marked by the opening of British Open University in 1961 [Daniel, 2001]), and to the telecommunication technologies supporting asynchronous and synchronous learning through teleconferencing, computer mediated communication and online interactive environments for students to create and re-create knowledge individually or collaboratively. In d-learning, the teacher and students are separated quasi-permanently by time, location, or both (Keegan, 2002; ASTD Glossary, 2006). With the advent of computer and communication technologies, e-learning covers a wide set of applications and processes, such as Web-based learning, computer-based learning, virtual classrooms, and digital collaboration (ASTD Glossary, 2006). The delivery of content is through a media-rich and hyperlinked environment utilizing internetworking services. M-learning can be considered as learning taking place where the learner is not at a fixed, predetermined location, or where the dominant technologies are handheld devices such as mobile phones, PDAs and palmtops, or tablet PCs. It can be spontaneous, personal, informal, contextual, portable, ubiquitous and pervasive (Kukulska-Hulme & Traxler, 2005, p. 2).

In my view, new concepts in teaching and learning can be generated from m-learning. For example, mobile phones can be used as voting devices for outdoor learning activities or in classrooms without computer supports, as interactive devices in museums, positioning or data logging devices at field trips or in many pedagogical situations. The justification of m-learning being descendent of e-learning and d-learning is rather thin, and Figure 2 is better represented. Furthermore, not everything can be delivered through m-learning. The small form factor, one-finger operation in some cases—slow computational and communication speed, short battery life and limited multimedia capabilities in contrast with computers do not really suit applications requiring heavy reading, high over-the-air communication and a lot of typing or texting.

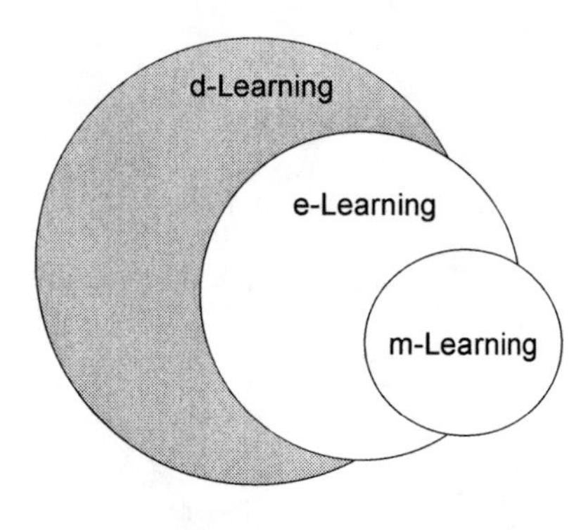

Figure 2. Overlapping and differential spaces of m-learning, e-learning and d-learning

Table 1. Different teaching and learning contexts

	Salient characteristics
d-learning	• Separation of teachers and learners • Learning normally occurs in a different place from teaching • Formal educational influence and organization
e-learning	• Multimedia-rich • Hypermedia • Independent • Collaborative
m-learning	• Mobile • Portable • Ubiquitous • Pervasive

In summary, m-learning is restricted and expedited by its nature. Different teaching and learning applications require different approaches, whether it is in d-learning, e-learning or m-learning. We must keep in mind their salient characteristics in different teaching and learning contexts, as shown in Table 1.

M-LEARNING INFRASTRUCTURE

In order to support m-learning, mobile devices such as PDAs, mobile phones and tablet PCs, together with servers such as Web servers, streaming servers and database servers on top of applications such as specific adaptation of LMS must be employed (Horton & Horton, 2003; Chen & Kinshuk, 2005). Despite the rapid development in mobile technologies, Figure 3 provides a typical browser-based architecture to support m-learning. It represents a full-scale implementation of any learning system formally. Processing and logic are controlled from the server-side and the mobile devices act as interfaces (Hodges, Bories, & Mandel, 2004, p. 2).

It is also possible that the learning applications are run locally on mobile devices with or without accessing network resources. Applications can be built using Java, such as mobile information device profile (MIDP), C++ on Symbian or native OSs, and Adobe Flash for mobile devices. Feature-rich applications can be implemented to take advantage or avoid limitations of the hardware.

Many researchers believe that, in order to support m-learning, a mobile learning management system (mLMS) is necessary. The logical derivation of mLMS is through the extension of conventional LMS (Trifonova & Ronchetti, 2003; Trifonova, Knapp, Ronchetti, & Gamper, 2004). Direct presentation of materials from computers to mobile devices is likely not legible, aesthetically pleasant, or technically not feasible. Adaptation according to the hardware and device profiles is required. This view is also supported by Goh & Kinshuk (2004). CSS, XSLT and XSL transformation in XML technologies are used to support WML, XHTML and HTML through server pages (Shotsberger & Vetter, 2002). Open standards, including e-learning standards such as SCORM (Fallon & Brown 2003), are the keys for the success of any mLMS.

M-LEARNING WITH MOBILE PHONES

To illustrate the concept discussed above, an m-learning experiment using phone simulators with one of my classes

Figure 3. Browser-based support for m-learning

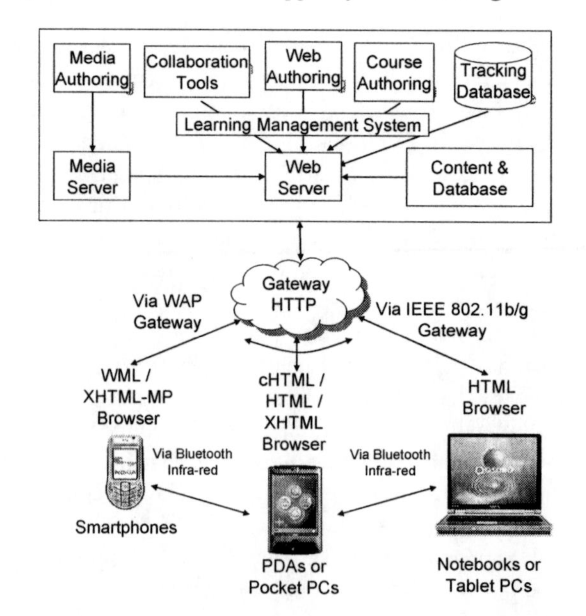

Figure 4. An m-learning experiment using phone simulators

Figure 5. The system architecture for the m-learning experiment

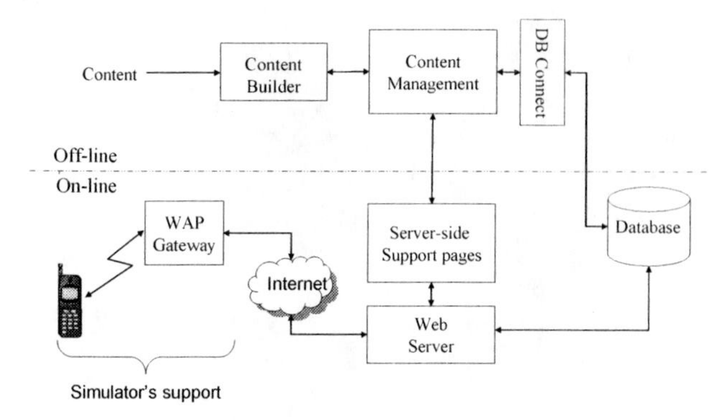

Figure 6. Voting activity

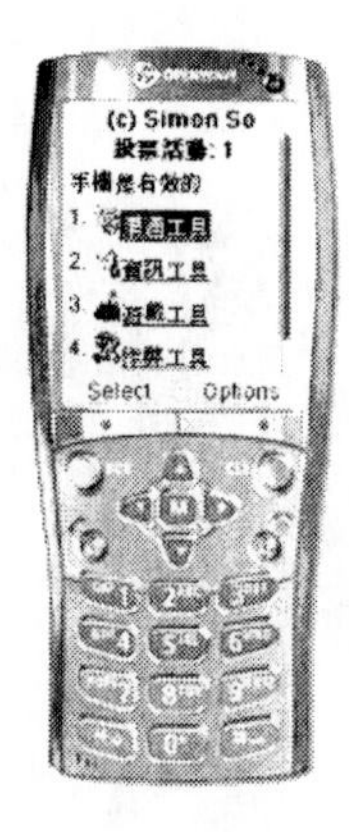

Figure 7. The corresponding voting result

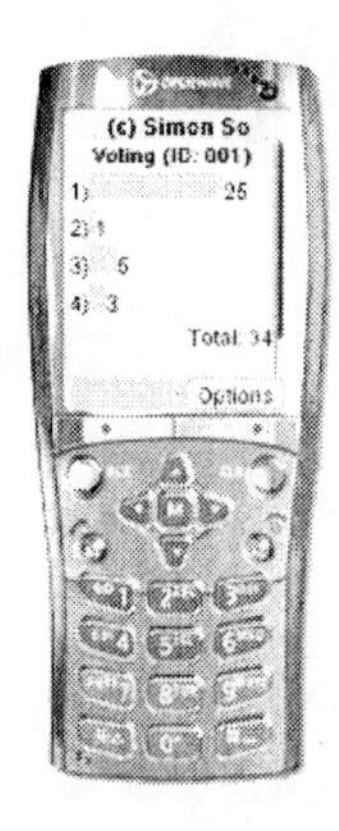

in a computer lab was conducted, as shown in Figure 4. The purpose of the experiment is to find out how my students react to the concept of m-learning. Three activities were developed to address different applications of mobile phones for teaching and learning. Simulators developed to execute in real mobile phones are used for this study (Openwave, 2006). There are three reasons for this. Firstly, the chosen software has been implemented in a number of real phone models. It behaves like a real phone. Secondly, some students may not have mobile phones with advanced features to support WAP 2.0 (Wapforum, 2006) and XTHML-MP, or connect to the mobile service providers with the features turned on. Some students may still have text-based mobile phones! Thirdly, as long as students operate the simulator (e.g., one-finger operation) as the experiment intended, I have a much better controlled environment to answer my research questions.

To support this experiment, a WAP gateway connected to a Web server is needed. Figure 5 outlines a practical and partial implementation of the architecture described in the previous section. Apache, PHP and MySQL are chosen as the Web server, server-side programming and database support respectively.

Among the three applications developed for this experiment, the first application is a voting system. Students can cast their votes on their simulators and teachers can interactively check the voting results as illustrated in Figure 6 and Figure 7. Students can use the quick access keys ("1" to "X") on the keypad to cast their votes. This acts as if the voter has a simple voting machine at hand. Teachers can retrieve the voting results from the database onto their handsets as well.

The second application is an interactive game called "15/16" which is a popular game on Hong Kong's television. Instead of two players per game, it was modified that the whole class can participate in each game. Students make their selections and the teacher (or any student) suggests the explanation. Students can change their mind depending on whether they believe the teacher/students or not. Figure 8 illustrates two questions. Teachers can show or refresh the selections at anytime. Figure 9 shows the students' selections for Question #1 in Figure 8.

The third application is a system to administrate tests. Students attempt the questions stored in the database. The overall score can be sent to the students at the end of the test, as shown in Figure 10 and Figure 11. The scores are kept in the database as well.

Figure 8. Two questions for the game

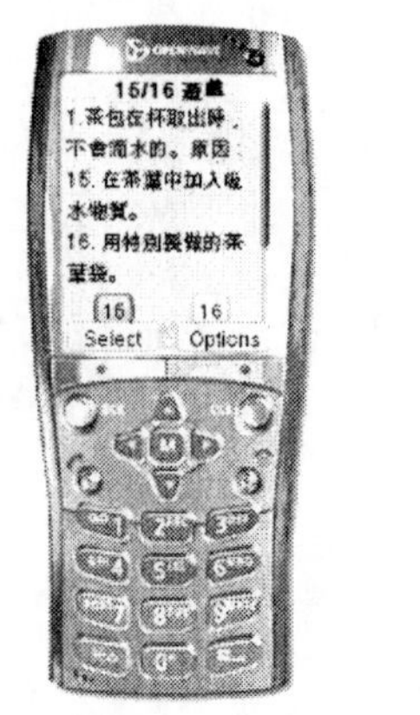

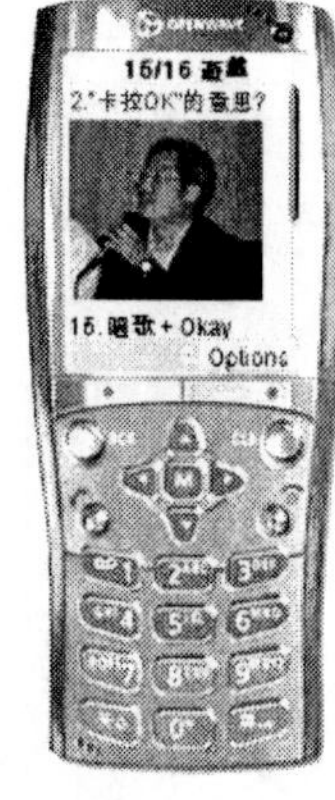

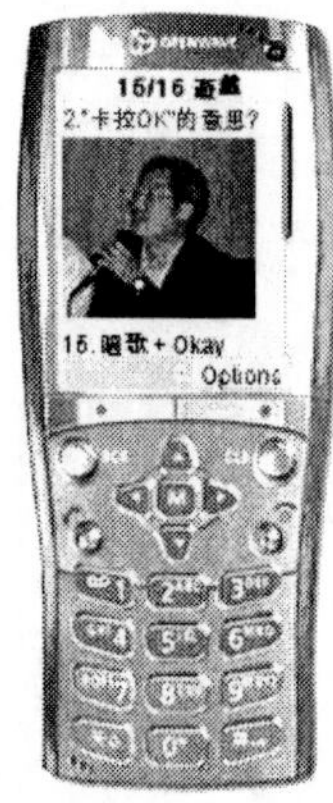

Figure 9. Students' selections

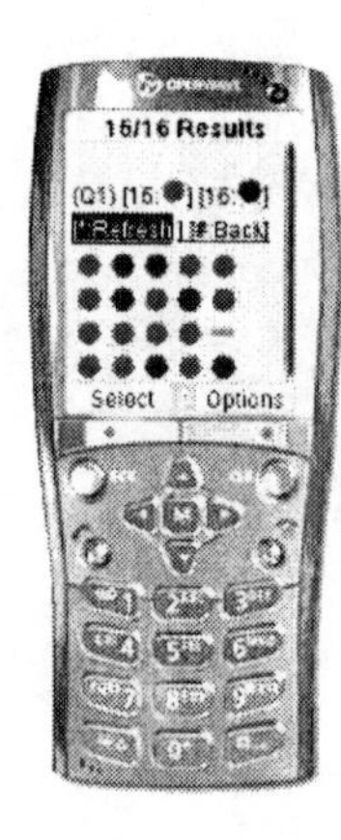

Figure 10. A test on mobile phones

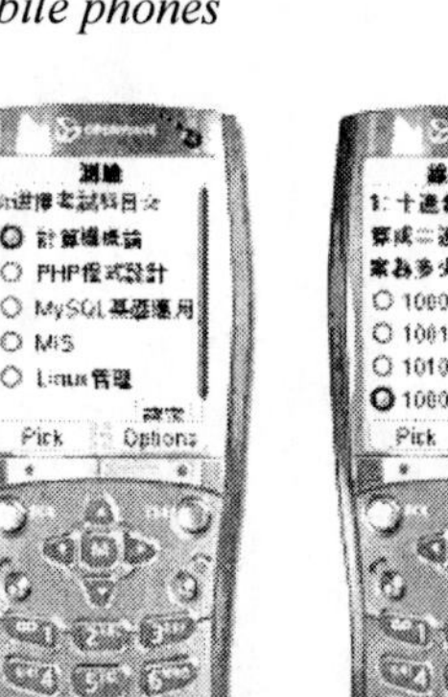

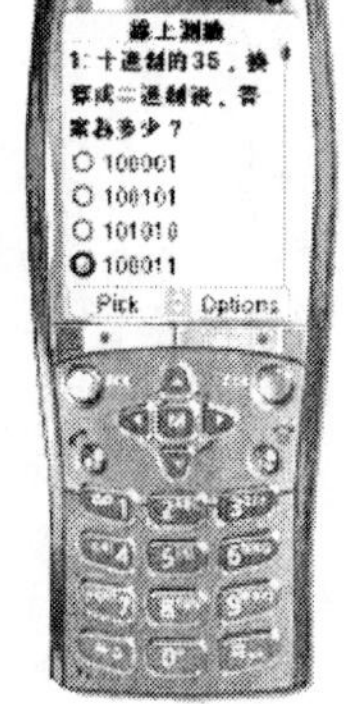

(login) (subject selection) (questions)

Figure 11. The score

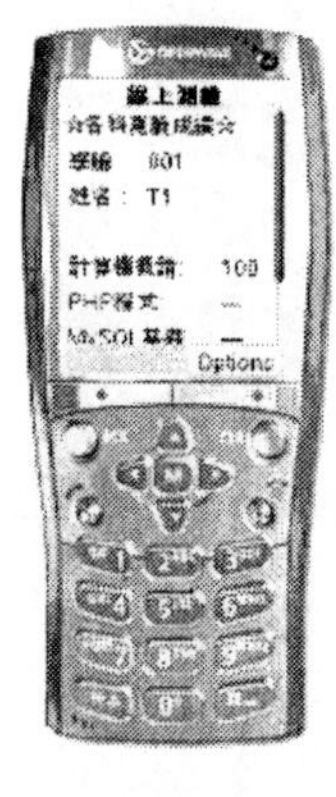

(student's score)

Figure 12. Nokia's handset implementation corresponding to Figures 8 and 9

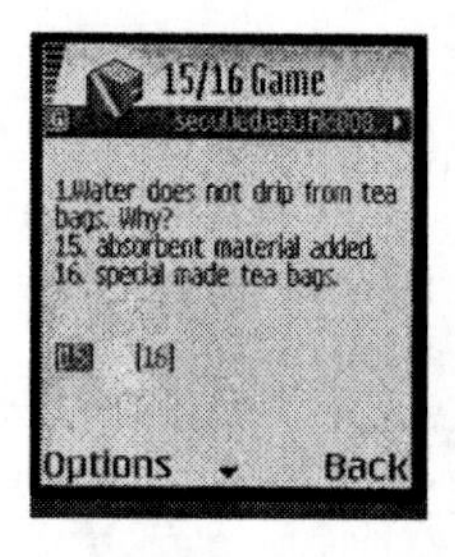

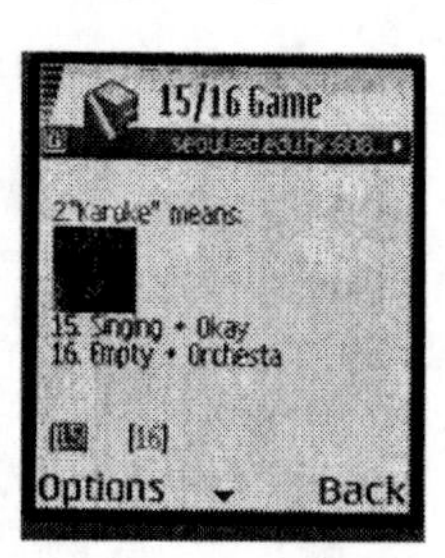

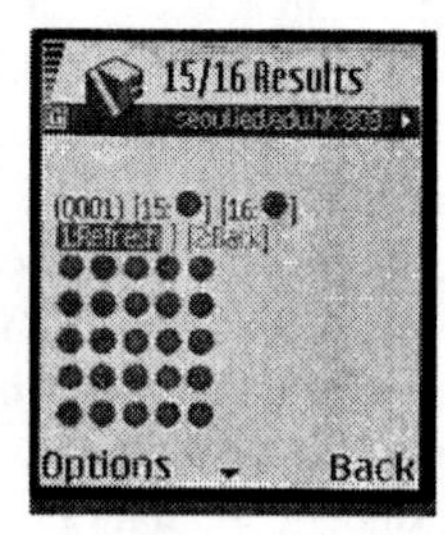

The applications described above are currently being rewritten in English and implemented for Nokia handsets. Figure 12 provides some of the snapshots.

CONCLUSION

M-learning has attracted a lot of research interest recently. It is a fashionable term in education. We can expect a lot of research work in this area will emerge for years to come. It is an exciting field. It also poses a lot of challenges to educators, instructional designs, software engineers and network specialists.

The main concept of m-learning has been highlighted in the article. The browser-based approach to m-learning is presented. It is illustrated by the experiment conducted with my students. This article serves as an example for those researchers to pursue further studies in this direction.

REFERENCES

ASTD Glossary. (2006). *ASTD's source for e-learning*. Retrieved on June 30, 2006, from http://www.learningcircuits.org/glossary.html

Bowles, M. S. (2004). Relearning to e-learn: Strategies for electronic learning and knowledge. Melbourne: Melbourne University Press.

Chen, J., & Kinshuk. (2005). Mobile technologies in educational services. *AACE Journal of Educational Multimedia and Hypermedia, 14*(1), 89-107

Daniel, J. (2001). The UK Open University: Managing success and leading change in a mega-university. In C. Latchem & D. Hanna (Eds.), *Leadership for 21st Century: Global Perspectives from Educational Innovators*. London: Kogan Page.

Fallon C., & Brown S. (2003). *E-learning standards: A guide to purchasing, developing, and deploying standards-conformant e-learning*. FL.: St. Lucie Press

Georgiev, T., Georgieva, E., & Smrikarov, A. (2004). M-learning: A new stage of e-learning. In *Proceedings of International Conference on Computer Systems and Technologies, CompSysTech'2004*. Retrieved on June 30, 2006, from http://ecet.ecs.ru.acad.bg/cst04/Docs/sIB/428.pdf

Goh, T., & Kinshuk. (2004). Getting ready for mobile learning. In *Proceedings of the 2004 World Conference on Educational Multimedia, Hypermedia and Telecommunications (ED-MEDIA2004)* Lugano, Switzerland (pp.56-63).

Hodges, A., Bories, J., & Mandel, R. (2004). Designing applications for 3G mobile devices. In R. Longoria (Ed.), *Designing Software for the Mobile Context: A Practitioner's Guide*. London: Springer.

Horton, W., & Horton, K. (2003). *E-learning tools and technologies: A consumer's guide for trainers, teachers, educators, and instructional designers*. New York: Wiley.

ITU. (2006). Executive summary. In *World Telecommunication/ICT Development Report 2006: Measuring ICT for Social and Economic Development*. Retrieved on June 30, 2006, from http://www.itu.int/ITU-D/ict/publications/wtdr_06/material/WTDR2006_Sum_e.pdf

Keegan, D. (2002). *The future of learning: From eLearning to mLearning*. Retrieved on June 30, 2006, from http://learning.ericsson.net/mlearning2/project_one/book.html

Kukulska-Hulme, A., & Traxler, J. (2005). *Mobile learning: A handbook for educators and trainers*. London: Routledge.

Openwave. (2006). *V7 simulator*. Retrieved on April 15, 2006, from http://www.openwave.com

Shotsberger, P., & Vetter, R. (2002). The handheld Web: How mobile wireless technologies will change Web-based instruction and training. In Allison Rossett (Ed.), *The ASTD E-Learning Handbook: Best Practices, Strategies, and Case Studies for Emerging Field*. New York: McGraw-Hill

Trifonova, A., & Ronchetti, M. (2003). A general architecture for m-learning. In *Proceedings of the Second International Conference on Multimedia and ICTs in Education*. Badajoz, Spain.

Trifonova, A., Knapp, J., Ronchetti, M., & Gamper, J. (2004). Mobile ELDIT: Transition from an e-Learning to an m-Learning. In *Proceedings of the 2004 World Conference on Educational Multimedia, Hypermedia and Telecommunications (ED-MEDIA2004)*, Lugano, Switzerland (pp.188-193).

Wapforum. (2006). *WAP 2.0 Standard*. Retrieved on June 30, 2006, from http://www.wapforum.org

Wikipedia M-learning. (2006). *M-learning*. Retrieved on June 30, 2006, from http://en.wikipedia.org/wiki/M-learning

KEY TERMS

Distance Learning (D-Learning): The teacher and students are separated quasi-permanently by time, location, or both. The content can be delivered synchronously or asynchronously.

Electronic Learning (E-Learning): Processes of learning through Web-based learning, online learning, computer-based learning and/or virtual classrooms. The delivery of content is through media-rich and hyperlinked environment utilizing internetworking services.

Learning Management System (LMS): LMS allows the tracking of learner's needs and achievement over periods of time.

Mobile Learning (M-Learning): Learning takes place where the learner is not at a fixed, predetermined location, or where the dominant technologies are handheld devices such as mobile phones, PDAs and palmtops, or tablet PCs.

Wireless Application Protocol 2.0 (WAP 2.0): This is a new version of WAP, which facilitates a cut-down version of XHTML and makes it work in mobile devices.

XHTML Mobile Profile (XHTML-MP): XHTML-MP is a markup language for mobile phones and the like.

Mobile Ad-Hoc Networks

Moh Lim Sim
Multimedia University, Malaysia

Choong Ming Chin
British Telecommunications (Asian Research Center), Malaysia

Chor Min Tan
British Telecommunications (Asian Research Center), Malaysia

INTRODUCTION

Within the coming years, it is inevitable that mobile computing will flourish, evolving toward integrated and converged next generation wireless technology (Webb, 2001), and an important role to play in this technological evolution is mobile ad hoc networks (Liu & Chlamtac, 2004). In short, mobile ad hoc network (MANET) is a self-configuring network that consists of a number of mobile communication nodes that are interconnected by wireless links. The communication nodes are free to move in random manners, which include stationary as a special case. Due to the dynamic movement of the nodes, the ad hoc network topology is normally formed in a decentralized manner and on an ad hoc basis. MANET is in fact a peer-to-peer wireless network that transmits from a client node to another without the use any preexisting network infrastructure-based centralised base stations to coordinate the communications. This type of network is particularly favourable when the information is to be transmitted to other nodes in the locality or the individual communication node has limited radio ranges. The self-configuration, stand alone and quick deployment nature make MANETs suitable for emergency situations like disasters, wars, sporting events, and so forth. Other examples of MANETs with other functionalities include wireless sensor networks and vehicular ad hoc networks.

A wireless sensor network (WSN) is a network formed by a collection of small computers, which are employed in the processing of sensor data (Hać, 2003). These small computers have limited capabilities in terms of the processing and communication power. They usually consist of sensors, a communication device, and a power supply. WSNs find many and varied applications in various fields ranging from industrial monitoring of dangerous environments to agriculture monitoring.

A vehicular ad hoc network (VANET) is a network formed by a collection of vehicles with communications capabilities and also having the potential to support various intelligent transport services. This class of traffic telematics applications ranges from emergency warnings, for example, in the case of accidents, via floating car data gathering and distribution, to more advanced applications, like platooning and co-operative driving (Festag et al., 2004).

In the future, communication devices, communication-capable devices or sensors and home electronic appliances will have the capability to form various MANETs, and interoperate with the global communication networks. These MANETs play an important role in supporting various visions toward the creation of a world of ubiquitous computing where computation is integrated into the environment, rather than having computers that are distinct objects. One of the goals of ubiquitous computing is to enable devices to sense changes in their respective surroundings and to automatically adapt and act on these changes based on user needs and preferences. With ubiquitous computing, people can move around and interact with computers, devices and home appliances more naturally than they currently do.

BACKGROUND

The earliest MANETs were called packet radio networks, and were sponsored by Defense Advanced Projects Agency (DARPA) in the early 1970s (Mobile ad-hoc network, 2006). It is interesting to note that these early packet radio systems predated the Internet, and indeed were part of the motivation of the original Internet Protocol suite. Later DARPA experiments included the Survivable Radio Network (SURAN) project, which took place in the 1980s (Mobile ad-hoc network, 2006). The third wave of academic activity started in the mid-1990s with the advent of inexpensive wireless sensor devices, and Wi-Fi or IEEE 802.11 family of radio cards for personal computers, notebooks and smartphones.

The existing cellular-based broadband access for mobile communications is foreseen to be inefficient due to a number of reasons. Firstly, as the bandwidth required is getting higher approaching hundreds of MHz or tens of GHz range, higher carrier frequency (at least ten times the bandwidth as a rule of thumb) is expected. For a same transmit power level, the wireless channel suffers from greater attenuation as a results of using higher carrier frequency (Etoh, 2005). This calls for the research into the use of multihop communication for the provisioning of broadband access where each hop can support high bandwidth transmission over a short range. Hence MANETs, which is multihop in nature, promise to be one of the most innovative and challenging areas of wireless networking in the future. As mobile technologies are growing at an ever-faster rate, therefore higher reliability and capacity, better coverage and services are required.

The future MANETs will likely evolve along the following directions:

- Different MANETs such as wireless sensor networks, VANETs and infrastructure ad hoc networks are interconnected to form a bigger MANET for better exchange of information.
- The emergence of various radio technologies, such as Bluetooth, UWB, ZigBee, Wi-Fi, WiMAX, and so forth, which are optimized for different functions and with the affordable price of radio cards due to economic of scale, made it practical to install more than one radio card, either of the same type or different, or in a single device. When the communication nodes communicate with each other using more than one radio interface type or channel frequency, it is called multi-radio communication or multi-channel communication.

CHALLENGES OF MANETS

There are a number of technical challenges that need to be addressed in order to ensure good connectivity and quality of service (QoS) for the end-users or client nodes in future MANETs. In the following paragraphs we discuss a few of them.

Dynamic Routing Protocol

Basically, routing protocols with different characteristics may be required for different types of MANETs or under different operating environments. Alternatively, a dynamic routing protocol that can adapt itself to different operating environment is required. For example, conventional routing protocols that have been proven to work fine in MANETs with communication nodes in random movement patterns may not be optimum to support inter-vehicular communications (e.g., VANET) within close proximity that may be moving in cluster form in a specific direction but with micro randomness. Meanwhile, in the case of WSN, conventional MANET protocols such as AODV and DSR may not scale well as the network size increases due to the reservation of large bandwidth for control messages. In addition, the energy limitation of the communication nodes has not been considered (Hać, 2003). The presence of multiple or heterogeneous network interfaces posed a need for an efficient routing mechanism such as multi-radio routing when multiple types of radio technologies are used, or multi-channel routing when different channels of a common radio technology is used. Meanwhile, an appropriate rewarding scheme that helps to accelerate the sharing of resources, which include bandwidth and processing time, among communication nodes in a client ad-hoc network is required.

Network Topology Control

In contrast to wired networks, which typically have fixed network topologies, each communication node in a MANET can change the network topology by adjusting its transmit power or selecting specific nodes to forward its messages, thus controlling its neighbor list. In conjunction with the use of optimum routing algorithms, the challenges of topology control in MANETs are to maintain network connectivity, and optimized network lifetime, throughput, and delay with high scalability, minimum overhead, and high fault tolerance.

In order to achieve high scalability and reduce overhead, formation of sub-groups of nodes among the MANET nodes that perform the routing has been proposed in many algorithms (Perkins et al., 2001; Stojmenovic et al., 2002). In this method, a virtual backbone is formed by using the connected dominating set. The relatively smaller sub-network size helps to reduce the amount of routing information.

A good fault tolerance network may require a fully integrated mesh solution among all communication nodes or through the use of higher transmit power. However, the interference generated may correspondingly degrade the overall network performance in terms of throughput and delay. Thus, there is a trade-off between fault tolerance and capacity performance. In a heterogeneous network environment, the problem becomes worst, as the network topology is governed by the capabilities of diverse types of radio interfaces.

Radio Resource Management

The decentralized nature of MANETs makes it difficult for coordinating the sharing and utilization of radio resources. For each communication node, a large amount of physical parameters are involved, which include the number of radio channels, the type and capability of radio interfaces, the channel conditions (channel quality) that determine the performance of the radio transmission, current communication state (busy or not), and so forth. The collection of communication nodes within a MANET may have different amount of resources for use. Hence, a scheme for the discovery, optimum utilization and scheduling of available resources is required, where further information can be found in Chin et al., 2006.

Power Control and Antenna Beamforming

In a MANET environment, communication nodes tend to interfere with each other due to the decentralized nature of the network formation. This calls for effective methods for interference mitigation. The application of power control in cellular communications has been proven to be able to improve system capacity by reducing unnecessary interference and prolong battery life through reducing the transmit power (Sim et al., 1998; Sim et al., 1999). It can be divided into open loop and closed-loop power control and is widely used in cellular communication systems where base stations will coordinate the power control operations. Adaptation of existing power control algorithms for MANETs is required due to the fact that no centralized node is available for the coordination of power control operations. Without such coordination, all the mobile nodes may greedily transmit at maximum power level for its own sake and hence will cause undue interference to other existing nodes. For a detailed discussion see Olafsson et al. (2005).

Antenna beamforming technique offers a significantly improved solution to reduce interference levels and improve the signal-to-noise ratio through the use of narrower beam in the direction toward the receiving node (Alexiou & Haardt, 2004). With this technology, each mobile node's signal is transmitted and received by the dedicated pair of transmitting and receiving nodes only. The main challenges in using beamforming antennas for MANETs are the cost and physical size. Currently there are efforts on achieving inexpensive beamforming methods, but further works are still required to improve the performance (Liberti & Rappaport, 1999).

Mobility Management

It is anticipated that radio traffics in future wireless networks will be mostly generated by multimedia applications and, hence, next generation networks are expected to provide adequate supports for mobile entertainment with extended geographic coverage. However, multimedia applications often require a dynamic amount of bandwidth and in order to guarantee QoS for such bandwidth-greedy applications when used over a wireless link, current schemes for supporting such services in conventional networks have to be reviewed and new resource management solutions have to be proposed. One of the most critical aspects of guaranteeing QoS support in providing seamless access under dynamic radio conditions is handoff (or handover). A handoff process is either mobile station-triggered or network-triggered, and it involves four successive phases: (1) measurement, (2) handoff initiation, (3) channel assignment, and (4) network connection reconfiguration. Given the prevalent trend toward higher order heterogeneous networks, such a MANET requires advanced schemes to coordinate handoff and reservation of radio resources to ensure the continuity of services to mobile client nodes (Chin et al., 2006).

Security Issues

The MANETs also pose unique challenges in security implementation. This is mostly due to the following properties of some MANETs: resource-constraint mobile nodes, uncontrollable environment and large dynamic network topology. For example, wireless sensor nodes are characterized as severely resource-constraint devices in terms of available power, memory, bandwidth and computational capability. These mobile node-specific factors have set several constraints on the design of security architecture (Sajal et al., 2004). As only a fraction of the total memory may be used by the cryptographic algorithms, the security architecture demands relatively lightweight cryptographic algorithms with a reasonable execution time. The extra overhead required in providing the security service should not substantially degrade the overall efficiency of the MANET.

CONCLUSION

The widespread use of computers and the advancement in embedded system design, microwave techniques and VLSI techniques has stimulated the emergence of various MANETs. The integration of various types of MANET is foreseen in the near future due to the need for pervasive computing. However, the success of such integrated hybrid MANET is very much dependent on the issues discussed herein. Proper solutions are therefore necessary to ensure a successful deployment of MANETs in future wireless environments.

So far the decentralized nature of MANETs has been assumed in various studies. Owing to the complexity of various challenges encountered, there may be a need to look into locally centralized and coordinated MANETs in the future. For example, the gateway node to Internet in a WSN may coordinate the work of topology formation and sensor data aggregation. In addition, the open environments where MANETs will operate in many occasions are uncontrollable and not trustworthy. Hostile circumstances could be envisioned in some situations. Generally, the MANET consists of numerous mobile client nodes organized in a flat or hierarchical structure. Considering the node mobility, authentication and key exchanges must not generate too much overhead messages, since the topology is subject to frequent changes (Schmidt et al., 2005). Additionally, all necessary cryptographic functions and keying material must reside and be executable in the mobile client nodes. Finally, the security architecture needs to be scalable to accommodate a large number of mobile client nodes.

REFERENCES

Alexiou, A., & Haardt, M. (2004). Smart antenna technologies for future wireless systems: Trends and challenges. *IEEE Communications Magazine, 42*(9), 90-97.

Chin, C. M., Tan, C. M., & Sim, M. L. (2007). Emerging solutions for optimal utilization of future wireless resources. In D. Taniar (Ed.), *Encyclopedia of mobile computing and commerce.* Hershey, PA: Idea Group Reference.

Etoh, M. (2005). *Next generation mobile systems 3G and beyond.* West Sussex, UK: John Wiley & Sons.

Festag, A., Fubler, H., Hartenstein, H., Sarma, A., & Schmitz, R. (2004). FleetNet: Bringing car-to-car communication into the real world. In *Proceedings of 11th World Congress on ITS.* Nagoya, Japan.

Hać, A. (2003). *Wireless sensor network designs.* West Sussex, UK: John Wiley & Sons.

Liberti, J. C., & Rappaport, T. S. (1999). *Smart antennas for wireless communications.* Prentice Hall.

Liu, J. J.-N., & Chlamtac, I. (2004). Mobile ad-hoc networking with a view of 4G wireless: Imperatives and challenges. In S. Basagni, M. Conti, S. Giodano, & I. Stojmenovic (Eds.), *Mobile ad hoc networking* (pp. 46). IEEE Press Wiley-Interscience.

Mobile ad-hoc network. (2006). *Wikipedia.* Retrieved from http://en.wikipedia.org/wiki/Mobile_ad-hoc_ network

Olafsson, S., Freysson, G., Chin, E., & Sim, M.L. (2005, October 6-9). The relevance of adaptive power control for connectivity in de-centralized wireless systems. Paper presented at the the 2005 Networking and Electronic Commerce Research Conference (NAEC 2005), Lake Garda, Italy.

Perkins, C. E., Royer, E. M., Das, S. R., & Marina, M. K. (2001). Performance comparison of two on-demand routing protocols for ad hoc networks. *IEEE Personal Communications, 8*(1), 16-28.

Sajal, K. D., Afrand, A., & Kalyan, B. (2004). Security in wireless mobile and sensor networks. In *Wireless communications systems and networks* (pp. 531-557). Plenum Press.

Schmidt, S., Krahn, H., Fischer, S., & Wätjen, D. (2005). *A security architecture for mobile wireless sensor networks (LNCS).* Springer Verlag.

Sim, M. L., Gunawan, E., Soh, C. B., & Soong, B. H. (1998). Characteristics of closed loop power control algorithms for a cellular DS/CDMA system. *IEEE Proceedings Communications, 145*(5), 355-362.

Sim, M. L., Gunawan, E., Soong, B. H., & Soh, C. B. (1999). Performance study of close-loop power control algorithms for a cellular CDMA system. *IEEE Transactions on Vehicular Technology, 48*(3), 911-921.

Stojmenovic, I., Seddigh, M., & Zunic, J. (2002). Dominating sets and neighbor elimination-based broadcasting algorithms in wireless networks. IEEE *Transactions on Parallel and Distributed Systems, 13*(1), 14-25.

Webb, W. (2001). The future of wireless communications. Norwood, US: Artech House.

KEY TERMS

AODV: Ad-hoc on demand distance vector.

DARPA: Defense Advanced Projects Agency.

DSR: Dynamic source routing.

MANET: Mobile ad hoc network.

Multihop Communications: A communication mode in which traffic is forwarded to the destination through a number of intermediate communication nodes or routers.

Packet Radio Network: A wireless network that employs packet switching.

Ultra Wide Band (UWB): It is a wireless access technology that uses low power and provides higher speed than Wi-Fi or Bluetooth. It is developed to provide wireless video transmission for home theater systems, cable TV, auto safety and navigation, medical imaging and security surveillance.

VANET: Vehicular ad hoc network.

Very Large Scale Integration (VLSI): It is a microelectronic technology where large-scale systems of transistor-based circuits are integrated into circuits on a single chip.

Wireless Fidelity (Wi-Fi): It is also referred as IEEE802.11. It is a set of standards that set forth the specifications for transmitting data over a wireless network.

WiMAX: World interoperability for microwave access. It is frequently referred to as the IEEE 802.16 wireless broadband standard. It was initially designed to extend local Wi-Fi networks across greater distances, such as a campus, as well as to provide last mile connectivity.

WSN: Wireless sensor network.

ZigBee: A wireless networking technology conforming to the IEEE 802.15.4 standards used for home, building and industrial control and monitoring. It supports the deployment of wireless sensor network. With a slow maximum speed of 250 kbps at 2.4 GHz, ZigBee is slower than Wi-Fi and Bluetooth, but is designed for low power so that batteries can last for months and years. The ZigBee transmission range is short and roughly about 50 meters, but that can vary greatly depending on channel conditions, temperature, humidity and air quality.

Mobile Agent Protection for M-Commerce

Sheng-Uei Guan
Brunel University, UK

INTRODUCTION

The introduction of the mobile Internet is probably one of the most significant revolutions of the 20th century. With a simple click, one can connect to almost every corner of the world thousands of kilometers away. This presents a great opportunity for m-commerce. Despite its many advantages over traditional commerce, m-commerce has not taken off successfully. One of the major hindrances is security. The focus of this article is secure transport of mobile agents. A mobile agent is useful for handheld devices like a palmtop or PDA. Such m-commerce devices usually have limited computing power. It would be useful if the users of such devices could send an intelligent, mobile agent to remote machines to carry out complex tasks like product brokering, bargain hunting, or information collection.

When it comes to online transactions, security becomes the primary concern. The Internet was developed without too much security in mind. Information flows from hubs to hubs before it reaches its destination. By simply tapping into wires or hubs, one can easily monitor all traffic transmitted. For example, when Alice uses her VISA credit card to purchase an album from Virtual CD Mall, the information about her card may be stolen if it is not carefully protected. This information may be used maliciously to make other online transactions, thus causing damage to both the card holder and the credit card company.

Besides concerns on security, current m-commerce lacks the intelligence to locate the correct piece of information. The Internet is like the world's most complete library collections unsorted by any means. To make things worse, there is no competent librarian that can help readers locate the book wanted. Existing popular search engines are attempts to provide librarian assistance. However, as the collection of information is huge, none of the librarians are competent enough at the moment.

An intelligent agent is one solution to providing intelligence in m-commerce. But having an agent that is intelligent is insufficient. There are certain tasks that are unrealistic for agents to perform locally, especially those that require a large amount of information. Therefore, it is important to equip intelligent agents with roaming capability.

Unfortunately, with the introduction of roaming capability, more security issues arise. As the agent needs to move among external hosts to perform its tasks, the agent itself becomes a target of attack. The data collected by agents may be modified, the credit carried by agents may be stolen, and the mission statement on the agent may be changed. As a result, transport security is an immediate concern to agent roaming. The SAFE (*s*ecure roaming *a*gent *f*or *e*-commerce) transport protocol is designed to provide a secure roaming mechanism for intelligent agents. Here, both general and roaming-related security concerns are addressed carefully. Furthermore, several protocols are designed to address different requirements. An m-commerce application can choose the protocol that is most suitable based on its need.

BACKGROUND

There has been a lot of research done on the area of intelligent agents. Some literature (Guilfoyle, 1994; Johansen, Marzullo, & Lauvset, 1999) only propose certain features of intelligent agents, some attempt to define a complete agent architecture. Unfortunately, there is no standardization in the various proposals, resulting in vastly different agent systems. Efforts are made to standardize some aspects of agent systems so that different systems can inter-operate with each other. Knowledge representation and exchange is one of the aspects of agent systems for which KQML (Knowledge Query and Manipulation Language; Finin, 1993) is one of the most widely accepted standards. Developed as part of the *Knowledge Sharing Effort,* KQML is designed as a high-level language for runtime exchange of information between heterogeneous systems. Unfortunately, KQML is designed with little security considerations because no security mechanism is built to address common security concerns, not to mention specific security concerns introduced by mobile agents. Agent systems using KQML will have to implement security mechanisms on top of KQML to protect themselves.

While KQML acts as a sufficient standard for agent representation, it does not touch upon the security aspects of agents. In an attempt to equip KQML with built-in security mechanisms, Secret Agent is proposed by Thirunavukkarasu, Finin and Mayfield (1995).

Another prominent transportable agent system is Agent TCL developed at Dartmouth College (Gray, 1997; Kotz et al., 1997). Agent TCL addresses most areas of agent transport by providing a complete suite of solutions. It is probably one of the most complete agent systems under research. Its security mechanism aims at protecting resources and the agent itself. In terms of agent protection, the author acknowledges that "it is clear that it is impossible to protect an agent from the machine on which the agent is executing ... it is equally clear that it is impossible to protect an agent from a resource that willfully provides false information" (Gray, 1997). As a result, the author "seeks to implement a verification mechanism so that each machine can check whether an agent was modified unexpectedly after it left the home machine" (Gray, 1997). The other areas of security, like non-repudiation, verification, and identification, are not carefully addressed.

Compared with the various agent systems discussed above, SAFE is designed to address the special needs of m-commerce. The other mobile agent systems are either too general or too specific to a particular application. By designing SAFE with m-commerce application concerns in mind, the architecture will be suitable for m-commerce applications. The most important concern is security, as discussed in previous sections. Due to the nature of m-commerce, security becomes a prerequisite for any successful m-commerce application. Other concerns are mobility, efficiency, and interoperability. In addition, the design allows certain flexibility to cater to different application needs.

MAIN FOCUS OF THE ARTICLE

As a prerequisite, each SAFE entity must carry a digital certificate issued by SAFE Certificate Authority, or SCA. The certificate itself is used to establish the identity of a SAFE entity. Because the private key to the certificate has signing capability, this allows the certificate owner to authenticate itself to the SAFE community. An assumption is made that the agent private key can be protected by function hiding (Thomas, 1998). Other techniques were also discussed in the literature (Bem, 2000; Westhoff, 2000), but will not be elaborated in this article.

From the host's viewpoint, an agent is a piece of foreign code that executes locally. In order to prevent a malicious agent from abusing the host resources, the host should monitor the agent's usage of resources (e.g., computing resources, network resources). The agent receptionist will act as the middleman to facilitate and monitor agent communication with the external party.

General Message Format

In SAFE, agent transport is achieved via a series of message exchanges. The format of a general message is as follows:

SAFE Message = Message Content + Timestamp + Sequence Number + MD(Message Content + Timestamp + Sequence Number) + Signature(MD)

The main body of a SAFE message comprises message content, a timestamp, and a sequence number. The message content is defined by individual messages. Here MD stands for the Message Digest function. The first MD is the function applied to Message Content, Timestamp, and Sequence Number to generate a message digest. The second MD in the equation is the application of digital signature to the message digest generated. A timestamp contains the issue and expiry time of the message.

To prevent replay attack, message exchanges between entities during agent transport are labeled according to each transport session. A running sequence number is included in the message body whenever a new message is exchanged.

In order to protect the integrity of the main message body, a message digest is appended to the main message. The formula of the message digest is as follows:

Message Digest = MD5(SHA(message_body) + message_body)

Here SHA (Secure Hash Algorithm) stands for a set of related cryptographic hash functions. The most commonly used function, SHA-1, is employed in a large variety of security applications. The message digest alone is not sufficient to protect the integrity of a SAFE message. A malicious hacker can modify the message body, and recalculate the value of message digest using the same formula and produce a seemingly valid message digest. To ensure the authenticity of the message, a digital signature on the message digest is generated for each SAFE message. In addition to ensuring message integrity, the signature serves as a proof for non-repudiation as well.

If the message content is sensitive, it can be encrypted using a symmetric key algorithm (e.g., Triple DES). The secret key used for encryption will have to be decided at a higher level.

To cater for different application concerns, three transport protocols are proposed: supervised agent transport, unsupervised agent transport, and bootstrap agent transport. These three protocols will be discussed in the following sections in detail.

Supervised Agent Transport

Supervised agent transport is designed for applications that require close supervision of agents. Under this protocol, an agent must request a roaming permit from its owner or butler before roaming. The owner has the option to deny the roaming request and prevent its agent from roaming to undesirable hosts. Without the agent owner playing an active role in the transport protocol, it is difficult to have tight control over agent roaming.

The procedure for supervised agent transport is shown in Figure 1.

Figure 1. Supervised agent transport

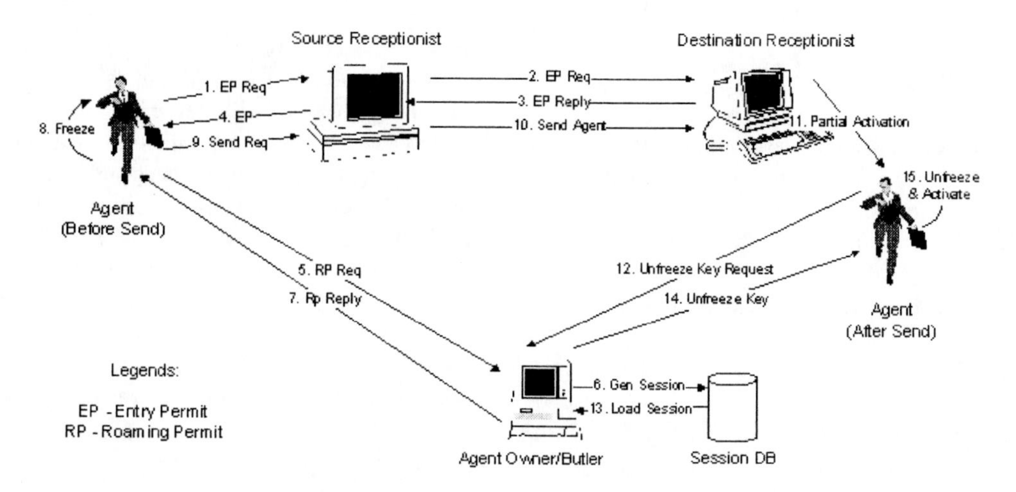

Agent Receptionist

Agent receptionists are processes running at every host to facilitate agent transport. If an agent wishes to roam to a host, it should communicate with the agent receptionist at the destination host to complete the transport protocol. Every host will keep a pool of agent receptionists to service incoming agents. Whenever an agent roaming request arrives, an idle agent receptionist from the pool will be activated to entertain the request. In this way, a number of agents can be serviced concurrently.

Request through Source Receptionist for Entry Permit

To initiate supervised agent transport, an agent needs to request for an entry permit from the destination receptionist. Communication between a visiting agent and foreign parties (other agents outside the host, agent owner, etc.) is done using an agent receptionist as a proxy.

Request for Roaming Permit

Once the source receptionist receives the entry permit from the destination receptionist, it simply forwards it to the requesting agent. The next step is for the agent to receive a roaming permit from its owner/butler. The agent sends the entry permit and address of its owner/butler to the source receptionist. Without processing, the source receptionist forwards the entry permit to the address as specified in the agent request.

The agent owner/butler can decide whether the roaming permit should be issued based on its own criteria. If the agent owner/butler decides to issue the roaming permit, it will have to generate a session number, a random challenge, and a freeze/unfreeze key pair. The roaming permit should contain the session number, random challenge, freeze key, timestamp, entry permit, and a signature on all of the above from the agent owner/butler.

In order to verify that the agent has indeed reached the intended destination, a random challenge is generated into the roaming permit. A digital signature on this random challenge is required for the destination to prove its authenticity.

Agent Freeze

With the roaming permit and entry permit, the agent is now able to request for roaming from the source receptionist. In order to protect the agent during its roaming, sensitive function and codes inside the agent body will be frozen. This is achieved using the freeze key in the roaming permit. Even if the agent is intercepted during its transmission, the agent's capability is restricted such that it cannot be run due to the freezing of agent functions. Not much harm can be done to the agent owner/butler. To ensure a smooth roaming operation, the agent's life support systems cannot be frozen. Functions that are critical to the agent's roaming capability must remain functional when the agent is roaming.

Agent Transport

Once frozen, the agent is ready for transmission over the Internet. To activate roaming, the agent sends a request containing the roaming permit to the source receptionist. The source receptionist can verify the validity and authenticity of the roaming permit.

If the agent's roaming permit is valid, the source receptionist will transmit the frozen agent to the destination receptionist as specified in the entry permit. Once the transmission is completed, the source receptionist will terminate the execution of the original agent and make itself available to other incoming agents.

Agent Pre-Activation

When the frozen agent reaches the destination receptionist, it will inspect the agent's roaming permit and the entry permit (contained in the roaming permit) carefully. By doing so, the destination receptionist can establish the following:

1. The agent has been granted permission to enter the destination.
2. The entry permit carried by the agent has not expired.
3. The agent has obtained sufficient authorization from its owner/butler for roaming.
4. The roaming permit carried by the agent has not expired.

If the destination receptionist is satisfied with the agent's credentials, it will activate the agent partially and allow it to continue agent transport process.

Request for Unfreeze Key and Agent Activation

Although the agent has been activated, it is still unable to perform any operation since all sensitive codes/data are frozen. To unfreeze the agent, it has to request for the unfreeze key from its owner/butler. To prove the authenticity of the destination, the destination receptionist is required to sign the random challenge in the roaming permit. The request for unfreeze key contains the session number, the certificate of destination, and the signature on the random challenge.

The direct agent transport process is completed.

Unsupervised Agent Transport

Supervised agent protocol is not a perfect solution to agent transport. Although it provides tight supervision to an agent owner/butler, it has its limitations. Since the agent owner/butler is actively involved in the transport, the protocol inevitably incurs additional overhead and network traffic. This results in lower efficiency of the protocol. This is especially significant when the agent owner/butler is located behind a network with lower bandwidth, or the agent owner/butler is supervising a large number of agents. In order to provide flexibility between security and efficiency, unsupervised agent transport is proposed. The steps involved in unsupervised agent transport are shown in Figure 2.

Request for Entry Permit

In supervised agent transport, session ID and key pair are generated by the agent butler. However, for unsupervised agent transport, these are generated by the destination receptionists because agent butler is no longer online to the agents.

Pre-Roaming Notification

Unlike supervised agent transport, the agent does not need to seek for explicit approval to roam from its owner/butler. Instead, a pre-roaming notification is sent to the agent owner/butler first. It serves to inform the agent owner/butler that the agent has started its roaming. The agent does not need to wait for the owner/butler's reply before roaming.

Agent Freeze

Agent freeze is very close to the same step under supervised agent transport, only that the encryption key is generated by destination instead of agent butler.

Figure 2. Unsupervised agent transport

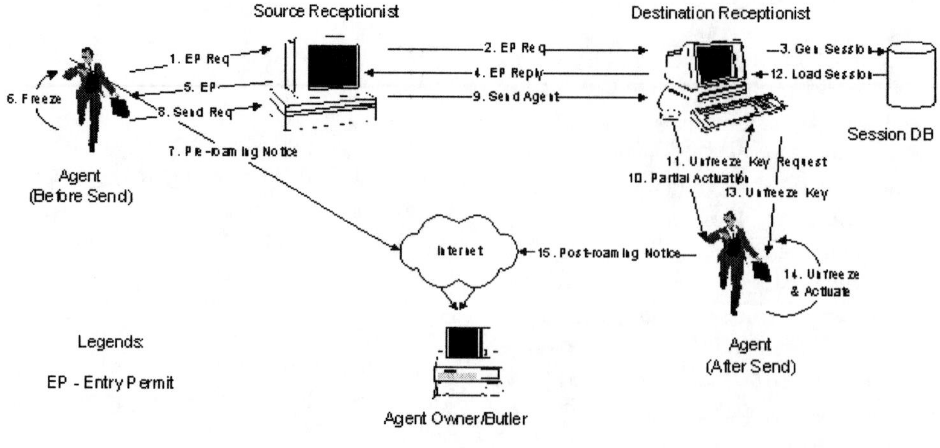

Figure 3. Bootstrap agent transport

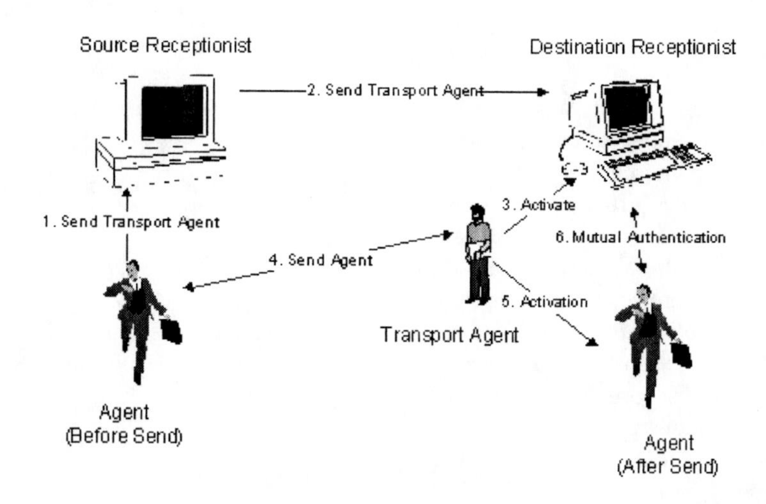

Agent Transport

This step is the same as that in supervised agent transport protocol.

Request for Unfreeze Key

The identification and verification processes are the same as compared to supervised agent transport, the exception being that the unfreeze key comes from destination receptionist.

Agent Activation

This step is the same as that in supervised agent transport.

Post-Roaming Notification

Upon full activation, the agent must send a post-roaming notification to its owner/butler. This will inform the agent owner/butler that the agent roaming has been completed successfully. Again, this notification will take place through an indirect channel so that the agent does not need to wait for any reply before continuing with its normal execution.

Bootstrap Agent Transport

Both supervised and unsupervised agent transport make use of a fixed protocol for agent transport. The procedures for agent transport in these two protocols have been clearly defined without much room for variations. It is realized that there exist applications that require a special transport mechanism for their agents. In order to allow this flexibility, SAFER provides a third transport protocol, bootstrap agent transport. Under bootstrap agent transport, agent transport is completed in two phases. Bootstrap agent transport is illustrated in Figure 3.

In the first phase, the transport agent is sent to the destination receptionist using either supervised or unsupervised agent transport with some modifications. The original supervised and unsupervised agent transport requires agent authentication and destination authentication to make sure that the right agent reaches the right destination. Under bootstrap agent transport, the transmission of transport agent does not require both agent authentication and destination authentication.

Once the transport agent reaches the destination, it starts execution in a restricted environment. It is not given the full privilege as a normal agent because it has yet to authenticate itself to the destination. This is to prevent the transport agent from hacking attempts to local host. Under the restricted environment, the transport agent is not allowed to interact with local host services. It is only allowed to communicate with its parent until the parent reaches destination. SAFER allows individual transport agents be customized to use any secure protocol for parent agent transmission.

When the parent agent reaches the destination, it can continue the handshake with the destination receptionist and perform mutual authentication directly. The authentication scheme is similar to that in supervised/unsupervised agent transport.

FUTURE TRENDS

As an evolving effort to deliver a more complete architecture for agents, SAFER (secure agent fabrication, evolution, and roaming) architecture is being proposed to extend the SAFE architecture. In SAFER, agents not only have roaming capability, but can make electronic payments and can evolve to perform better.

CONCLUSION

SAFE is designed as a secure agent transport protocol for m-commerce. The foundation of SAFE is the agent transport protocol, which provides intelligent agents with roaming capability without compromising security. General security concerns as well as security concerns raised by agent transport have been carefully addressed. The design of the protocol also takes into consideration differing concerns for different applications. Instead of standardizing on one transport protocol, three different transport protocols are designed, catering to various needs. Based on the level of control desired, one can choose between supervised agent transport and unsupervised agent transport. For applications that require a high level of security during agent roaming, bootstrap agent transport is provided so that individual applications can customize their transport protocols. The prototype of SAFE agent transport protocol has been developed and tested.

REFERENCES

Bem, E. Z. (2000). Protecting mobile agents in a hostile environment. *Proceedings of the ICSC Symposia on Intelligent Systems and Applications (ISA 2000).*

Corley, S. (1995). The application of intelligent and mobile agents to network and service management. *Proceedings of the 5th International Conference on Intelligence in Services and Networks,* Antwerp, Belgium.

Finin, T. (1994). *KQML—A language protocol for knowledge and information exchange.* Technical Report CS-94-02, University of Maryland, USA.

Finin, T., & Weber, J. (1993). *Draft specification of the KQML agent communication language.* Retrieved from http://www.cs.umbc.edu/kqml/kqmlspec/spec.html

Gray, R. (1997). *Agent TCL: A flexible and secure mobile-agent system.* PhD Thesis, Department of Computer Science, Dartmouth College, USA.

Guan, S. U., & Yang, Y. (1999). SAFE: Secure-roaming agent for e-commerce. *Proceedings of CIE'99,* Melbourne, Australia (pp. 33-37).

Guilfoyle, C. (1994). *Intelligent agents: The new revolution in software.* London: OVUM.

Johansen, D., Marzullo, K., & Lauvset, K.J. (1999). An approach towards an agent computing environment. *Proceedings of the ICDCS'99 Workshop on Middleware.*

Kotz, D., Gray, R., Nog, S., Rus, D., Chawla, S., & Cybenko, C. (1997). Agent TCL: Targeting the needs of mobile computers. *IEEE Internet Computing, 1*(4), 58-67.

Odubiyi, J. B., Kocur, D. J., Weinstein, S. M., Wakim, N., Srivastava, S., Gokey, C., & Graham, J. (1997). SAIRE—A Scalable Agent-Based Information Retrieval Engine. *Proceedings of the Autonomous Agents 97 Conference* (pp. 292-299), Marina Del Rey, CA.

Rus, D., Gray, R., & Kotz, D. (1996). Autonomous and adaptive agents that gather information. *Proceedings of the AAAI '96 International Workshop on Intelligent Adaptive Agents.*

Rus, D., Gray, R., & Kotz, D. (1997). Transportable information agents. In M. Huhns & M. Singh (Eds.), *Readings in agents.* San Francisco: Morgan Kaufmann.

Sander, T., & Tschundin, C.F. (1998). Protecting mobile agents against malicious hosts. *Mobile Agents and Security, LNCS 1419,* 44-60.

Schneider, F. B. (1997). Towards fault-tolerant and secure agentry. *Proceedings of the 11th International Workshop on Distributed Algorithms,* Saarbrücken, Germany.

Schneier, B. (1996). *Applied cryptography: Protocols, algorithms, and source code in C* (2nd ed.). New York: John Wiley & Sons.

Schoonderwoerd, R., Holland, O., & Bruten, J. (1997). Ant-like agents for load balancing in telecommunications networks. *Proceedings of the 1997 1st International Conference on Autonomous Agents* (pp. 209-216). Marina Del Rey, CA.

Thirunavukkarasu, C., Finin, T., & Mayfield, J. (1995). Secret agents—a security architecture for the KQML agent communication language. *Proceedings of the CIKM'95 Intelligent Information Agents Workshop,* Baltimore, MD.

Westhoff, D. (2000). On securing a mobile agent's binary code. *Proceedings of the ICSC Symposia on Intelligent Systems and Applications (ISA 2000).*

White, D. E. (1998). *A comparison of mobile agent migration mechanisms.* Senior Honors Thesis, Dartmouth College, USA.

KEY TERMS

Agent: A piece of software that acts to accomplish tasks on behalf of its user.

Digital Certificate: Certificate that uses a digital signature to bind together a public key with an identity—information such as the name of a person or an organization, his or her address, and so forth. The certificate can be used to verify that a public key belongs to an individual

Electronic Commerce (E-Commerce): Consists primarily of the distributing, buying, selling, marketing, and servicing of products or services over electronic systems such as the Internet and other computer networks.

Encryption: The art of protecting information by transforming (*encrypting*) it into an unreadable format, called cipher text. Only those who possess a secret *key* can decipher (or *decrypt*) the message into plain text.

Flexibility: The ease with which a system or component can be modified for use in applications or environments other than those for which it was specifically designed.

Mobile Commerce (M-Commerce): Electronic commerce made through mobile devices.

Protocol: A convention or standard that controls or enables the connection, communication, and data transfer between two computing endpoints. Protocols may be implemented by hardware, software, or a combination of the two. At the lowest level, a protocol defines a hardware connection

Security: The effort to create a secure computing platform, designed so that agents (users or programs) can only perform actions that have been allowed.

Mobile Agent-Based Discovery System

Rajeev R. Raje
Indiana University Purdue University Indianapolis, USA

Jayasree Gandhamaneni
Indiana University Purdue University Indianapolis, USA

Andrew M. Olson
Indiana University Purdue University Indianapolis, USA

Barrett R. Bryant
University of Alabama at Birmingham, USA

INTRODUCTION

For reasons of economy and scalability, many of the current distributed computing systems (DCSs) are realized as an integration of prefabricated and deployed components offering specific services. A critical task that the assembler of such a system needs to address is to locate and select appropriate components scattered over a network. This requires solving many research challenges. These include: (a) deployment of components and their specifications, (b) efficient searching for and gathering of appropriate specifications, (c) representation of queries, and (d) semantics of matching between queries and specifications. UniFrame (Raje, Auguston, Bryant, Olson, & Burt, 2001) is a framework that allows the seamless discovery and integration of such distributed software components. It addresses three key research issues: (1) architecture-based interoperability, (2) distributed discovery of resources, and (3) quality validation. This article presents a mobile-agent-based discovery service, which is one of the alternatives developed under research issue (2).

BACKGROUND

There have been many attempts at creating discovery services. This section reviews only a few prominent ones for the sake of brevity.

Jini (Waldo, 1999) is based on the underlying Java Remote Method Invocation infrastructure (Sun Microsystems, 1994), and thus provides a simplified interoperability. Services register themselves in Lookup Registries, which clients search to download their required services. The matching used in Jini is based on attribute comparisons.

The model used in the Ninja secure service discovery service (SSDS) (Czerwinski, Zhao, Hodes, Joseph, & Katz, 1999) to locate an appropriate service for a request is based on the concept of advertisement. SSDS tracks services in a network and allows authenticated users to locate them through expressive queries. It uses XML to describe the services and to allow complex queries. It supports the possibility of describing various attributes, such as the quality of service (QoS) parameters and associated costs, which are used in the matching process.

CORBA® (Common Object Request Broker Architecture) includes the Trader service (OMG, 2000), which uses a standardized Interface Definition Language to describe service interfaces. These interfaces provide the basis on which lookup and client invocations take place. The trader provides a simple attribute matching.

The aim of Agora (Seacord, Hissam, & Wallnau, 1998) is to provide an automatically generated, indexed, worldwide database of software products classified by their types. Agora combines introspection with Web search engines to reduce the costs of seeking components in the software marketplace. The query terms used for finding components are compared against the index collected by the search engines. The result is inspected by the user so the search can be broadened or refined based on the number and quality of matches.

Universal description, discovery, and integration (UDDI) (OASIS Consortium, 2000) defines a set of services supporting the description and discovery of businesses, organizations, and other Web service providers, as well as the Web services they provide. It utilizes Web Services Description Language (WSDL) for describing the capabilities of the services. UDDI provides a simple textual matching process by comparing each search term with various fields in a service's description.

Web services peer-to-peer discovery service (Banaei-Kashani, Chen, & Shahabi, 2004) is a decentralized discovery service with a matching capability that extends up to the semantic level. It is used to locate Web services that are geographically dispersed across a network. It uses keywords and semantically annotated WSDL to describe Web service interfaces. Each entity, called a "Servent," in this environment serves as both client and server. When a Servent receives a query for a Web service that is not available locally, it shifts its capacity from server to client and queries the network for that specific request. For discovery purposes, a Servent formulates a query encapsulated in a simple object access protocol (SOAP) message (W3C, 2004) and propagates it over the network based on a probabilistic flooding dissemination mechanism.

SLP (Guttman, 1999) provides hosts with access to information about the existence, location, and configuration of networked services. In this framework, user agents model client applications, service agents advertise services, and directory agents cache service information. A user agent can issue service requests to specify the requirements of the client application. It can transmit a request to service agents or a directory SLP. The SLP supports matching only at the syntactical level.

The monitoring and discovery service (MDS) (Globus Alliance, n.d.; Kandagatla, 2003), a part of the Globus Toolkit, is used for discovering computational resources deployed in a Grid environment. The resources are described using a standard schema made up of keywords and can be discovered using specific characteristics. The MDS is made up of two components: the Grid information resource service (GRIS) and the Grid index information service (GIIS). The GRIS runs on resources deployed on the Grid and is an information provider framework for specific information sources. A GIIS is a user-accessible directory server at a higher level that accepts information from child GIIS and GRIS instances and aggregates it for use at a higher level.

THE MOBILE UNIFRAME RESOURCE DISCOVERY SERVICE

A majority of the previously mentioned approaches for discovering service-providing components use relatively simple schemes for describing and matching services against a request. Also, none of these alternatives uses mobile agents in the discovery process. This section provides details about the Mobile UniFrame Resource Discovery Service (MURDS), which has a hierarchical architecture and uses mobile agents to discover services deployed over a network. MURDS is an enhancement of the UniFrame Resource Discovery Service (URDS) (Siram, 2002).

URDS is a hierarchical discovery service that supports the proactive discovery of component specifications, resolves technological heterogeneity, and allows multi-level matching. URDS is one of the entities in the UniFrame approach for developing DCS from heterogeneous, distributed software components. The core concept behind UniFrame is the unified meta-component model (UMM). UMM, as described in Raje (2000), consists of: (a) component, (b) services, and (c) infrastructure.

A component in UniFrame is developed by following a specification in a standardized knowledgebase (KB) (Raje et al., 2001) and implemented in any distributed-component technology. In addition to the implementation of a component, its developer must create, following the specification format described in the KB, a comprehensive specification for it. This is the UMM specification for that component. This, as indicated in Olson, Raje, Bryant, Burt, and Auguston (2005), consists of multiple levels—syntax, semantics, synchronization, and quality of service. Such a complex specification supports multi-level matching while seeking components for a specific query.

Each component in UniFrame offers a service whose UMM description provides its specification. In addition to the functionality of the service, UniFrame emphasizes the service's QoS aspect. Each component indicates its QoS using parameters described in the UniFrame QoS catalog (Brahnmath, 2002).

The infrastructure part of the UMM defines the necessary computational fabric on which components can be deployed and their specifications can be advertised. This allows a proactive discovery of the components for specific queries. URDS provides this infrastructure in UniFrame.

Incorporating the use of mobile agents in the discovery process into URDS creates MURDS. Its architecture, shown in Figure 1, comprises the following entities: (a) Internet Component Broker (ICB), (b) Headhunters (HHs), (c) Meta-Repositories (MR), (d) Active Registries (AR), (e) Components (C_1..C_n), and (f) mobile agents (MA). Figure 1 also depicts their interactions. Gandhamaneni (2004) discusses these entities in detail. A brief description follows below.

Internet Component Broker (ICB)

The ICB is a collection of the following entities: *query manager (QM), domain security manager (DSM), link manager (LM),* and *adapter manager (AM).* The ICB is a component broker that is pervasive in an interconnected environment. It is expected that there will be a number of ICBs deployed at well-known locations hosted by organizations supporting the UniFrame paradigm for developing distributed systems. The functions of the entities that make up the ICB are as follows:

- **Domain Security Manager (DSM):** The DSM serves as an authorization controller that handles the generation and distribution of the secret keys needed for communication between various constituents of MURDS. It enforces group memberships and performs access control checks on HHs on behalf of the ARs. For performing access control checks, the DSM has a repository of valid users (i.e., HHs, MAs acting on behalf of HHs, and the ARs) and the policies that

Figure 1. MURDS architecture

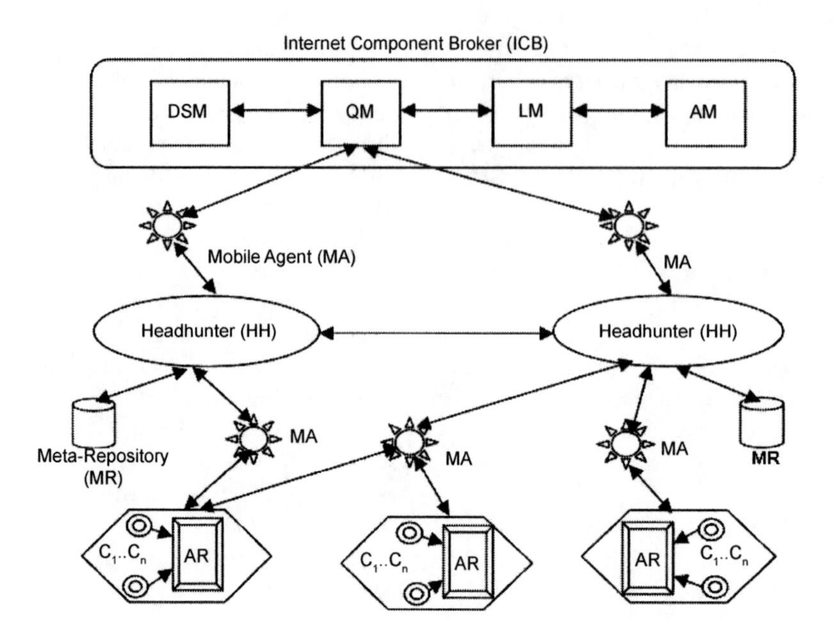

regulate the associations among them (i.e., the ARs and the HHs).

- **Query Manager (QM):** The QM is responsible for propagating the component selection queries it receives from a user to 'appropriate' HHs. The QM accomplishes this by sending a mobile agent on its behalf to select a list of service provider components that match the search criteria in the query. The current MURDS prototype bases *appropriateness* on the application domain specified in the search requirements. However, more complex schemes, say that use past performance, can be employed to decide to which HHs to send the queries.
- **Link Manager (LM):** The LM establishes links between different ICBs to form a federation of ICBs. Such a federated approach provides a much larger search space for discovering components. An ICB administrator configures the LM with the location information of other ICBs with which links are to be established. Then the QM and the LM can propagate the queries to the other linked ICBs as necessary.
- **Adapter Manager (AM):** The AM acts as a lookup service for clients needing adapter components. These adapter components assist in resolving technological heterogeneity that may occur between two communicating components.

Headhunters (HHs)

The HHs are responsible for proactively detecting, with the help of mobile agents, the presence of components offering services and registering their functionalities in their respective meta-repositories. After receiving a query from a QM, an HH searches its meta-repository and returns a list of components that match the query. The component selection performed by the HH can be based on the concepts of multi-level matching, which aims to match the query requirements with different levels present in the UMM specification. Thus, the matching that the HHs perform (and hence, the MURDS) is much more comprehensive than simple attribute-based matching—a scenario utilized by various discovery services described earlier.

Active Registry (AR)

The AR is an enhancement of the native registry/lookup mechanism that is present in a distributed computing model. For example, in the case of Java-RMI, the AR is a modification of the built-in naming service so that it can listen to broadcasts from HHs and permit mobile agents to access the information about its registered components. ARs have introspection capabilities so that they can provide the specifications of the components registered with them.

Meta-Repository (MR)

The MR is a database belonging to an HH for storing the UMM specifications of the various components the HH finds. Each HH continually attempts to discover new components available on the network to populate its MR.

Mobile Agents (MA)

Mobile agents act as proxies for headhunters in discovering components and for query managers in propagating queries. The MAs that a headhunter sends carry and present that headhunter's credentials to the active registries and seek components from them to be sent back to the meta-repository of that headhunter. The MAs that a query manager sends carry the incoming query to a set of headhunters (or ICBs via the link manager). The addition of the MAs distinguishes the MURDS from the URDS.

Components (C_1...C_n)

The components offering services, which are deployed on the network, may be implemented in different distributed component models. Each of these (and hence, its service) registers itself with its corresponding AR by providing its type name and associated UMM specification.

MURDS' Method of Operation

Developers of a DCS are users, or clients, of the MURDS system. Their goal is to obtain components that match certain functional and non-functional requirements for use in their development process. MURDS receives an incoming query from a user via its QM. Once the QM receives the query, it determines a subset of HHs to which to propagate the query. The MURDS prototype described later selects this subset randomly. After the QM identifies the subset, it sends MAs to the HHs in it. On receiving the query, each HH checks its MR for matching components. It returns any present via the MA to the QM. Also, periodically an HH sends MAs to a set of ARs in order to discover components newly registered with these ARs. An AR, after acknowledging an MA from an HH, then decides which specific type of access, if any, to grant the MA based on the HH's credentials the MA carries. The type of access provided depends upon the policy decisions enforced by the owner of that AR. These policy decisions can be dynamic, and hence the components revealed by an AR to a MA can vary over time. After gathering the desired component specifications from an AR, an MA may choose to send them back to the HH immediately, or continue to other ARs and send a consolidated collection at the end of its journey.

DESIGN, IMPLEMENTATION, AND EXPERIMENTATION

The design of a prototype of MURDS follows a multi-tier architecture, typical of distributed applications, as Figure 2 exhibits. The architecture consists of three tiers, namely the client, middle, and database. The prototype's *client tier* supports application clients. The *middle tier* supports client services (i.e., DSM, HH, QM, and AR) and Grasshopper™ version 2.2.4-enabled mobile agents (Magedanz, Bäumer, & Choy, n.d.). The Java 2 Platform, Enterprise Edition (J2EE)™ version 1.4.2 (Sun Microsystems, 2001) software environment implements various components of the MURDS prototype. The core architectural components (DSM, QM, HH, and AR) are Java-RMI-based services. The *database*

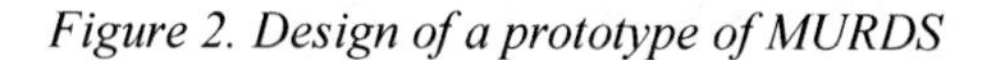

Figure 2. Design of a prototype of MURDS

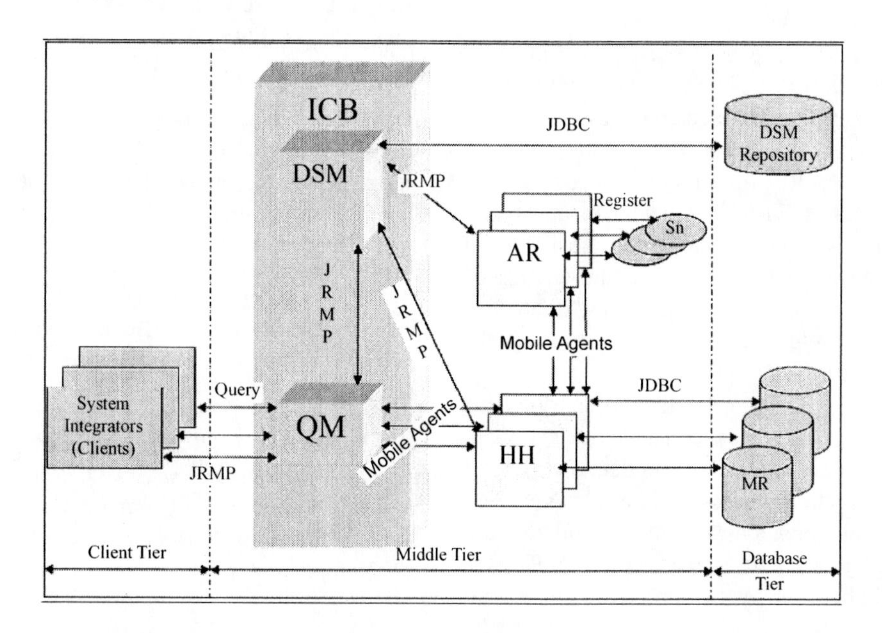

Figure 3. Effect of the number of HHs on CQRRT

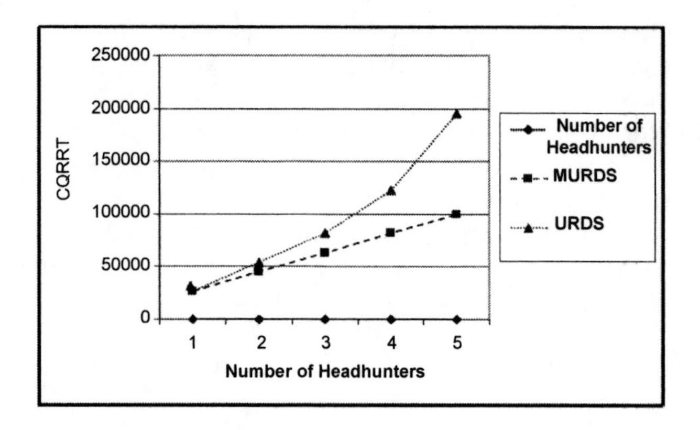

tier supports access to the repositories by means of standard APIs. The repositories (DSM-Repository and Meta-Repositories) are Oracle ™ v. 9.2 databases.

Various experiments were conducted to evaluate the prototype of MURDS. Due to space constraints, this article describes only one. Gandhamaneni (2004) presents the details about the others. An important metric these studies used to judge the performance of the prototype was client query result retrieval time (CQRRT). CQRRT is defined as the time taken from the instant a client issues a query to MURDS until the instant the results return back to the client. The experiments investigated the impact of various parameters, such as the number of components, of HHs, and of ARs on the CQRRT.

The result of one such experiment carried out to investigate the effect of increasing the number of HHs on the CQRRT appears in Figure 3. It also compares the results from the MURDS prototype (mobile agents) with the results from the URDS prototype (stationary agents). The configuration used for this experiment consisted of one QM, one client, and seven ARs. As Figure 3 reveals, increasing the number of headhunters increases the CQRRT for both the mobile-agent-based and the stationary-agent-based component selection processes. The reason is that the agents in both processes contact one headhunter at a time to retrieve component information from that HH. Each headhunter consumes time to retrieve components from its local meta-repository and to return them to the contacting agents. As the number of headhunters specified in the search list increases, the amount of time that it takes to return results to the query manager increases, which increases the CQRRT.

Figure 3 indicates, in addition, that the stationary-agent-based component selection process consumes more time to retrieve results from a set of headhunters than the mobile-based-component selection process. This is attributable to the synchronous communication mode of the former vs. the asynchronous mode of the latter, because these modes are the only differences between the two processes.

FUTURE TRENDS

Many extensions to the architecture of MURDS are possible. These could involve comprehensive study of its scalability and classifying the HHs and ARs into categories for deciding the mobile agents' itineraries, or using multi-level matching criteria in selecting appropriate services. The ARs could provide differentiated service and cost frameworks to the agents. An incorporation of these features into MURDS would provide a more comprehensive discovery service that uses mobile agents to search for components deployed over a network. In this way, it would anticipate demands that discovery services will face in the near future.

CONCLUSION

During the assembly of a distributed computing system, the importance of discovering the most appropriate components from those available over a network cannot be overstated. Such discovery, due to its inherent complexity, presents many interesting challenges. This article has briefly described one possible approach, which uses the concepts of UniFrame and mobile agents to identify appropriate components that are scattered over a network. The prototype's results have demonstrated the feasibility of this approach, and further investigations are currently underway as a part of ongoing UniFrame research.

REFERENCES

Banaei-Kashani, F., Chen, C., & Shahabi, C. (2004). *WSPDS: Web services peer-to-peer discovery service.* Retrieved March 22, 2006, from http://infolab.usc.edu/DocsDemos/isws2004_WSPDS.pdf

Brahnmath, G. (2002). *The UniFrame quality of service framework.* Unpublished MS thesis, Department of Computer and Information Science, Indiana University Purdue University, USA. Retrieved March 22, 2006, from http://www.cs.iupui.edu/uniFrame/

Czerwinski, S., Zhao, B., Hodes, T., Joseph, A., & Katz, R. (1999). An architecture for a secure service discovery service. *Proceedings of ACM Mobicom'99* (pp. 24-35). Retrieved March 22, 2006, from http://bnrg.cs.berkeley.edu/~czerwin/publications/sds-mobicom.pdf

Gandhamaneni, J. (2004). *UniFrame mobile, agent-based resource discovery system (MURDS).* Unpublished MS project, Department of Computer and Information Science, Indiana University Purdue University, USA. Retrieved March 22, 2006, from http://www.cs.iupui.edu/uniFrame/

Globus Alliance. (n.d.). Towards open grid services architecture. *Proceedings of the Open Grid Forum,* Chicago, IL. Retrieved March 22, 2006, from http://www.globus.org/ogsa/

Guttman, E. (1999). Service location protocol: Automatic discovery of IP network services. *IEEE Internet Computing, 3*(4), 71-80.

Kandagatla, C. (2003). *Survey and taxonomy of Grid resource management systems.* Retrieved March 22, 2006, from http://www.cs.utexas.edu/users/browne/cs395f2003/projects/KandagatlaReport.pdf

Magedanz, T., Bäumer, M., & Choy, S. (n.d.). *Grasshopper—A universal agent platform based on OMG MASIF and FIPA standards.* Retrieved March 22, 2006, from http://www.cordis.lu/infowin/acts/analysys/products/thematic/agents/ch4/ch4.htm

OASIS Consortium. (2000). *UDDI technical white paper.* Retrieved March 22, 2006, from http://www.uddi.org/pubs/Iru_UDDI_Technical_White_Paper.pdf

OMG (Object Management Group). (2000). *Trading object service specification.* Retrieved March 22, 2006, from http://www.omg.org/docs/formal/00-06-27.pdf

Olson, A., Raje, R., Bryant, B., Burt, C., & Auguston, M. (2005). UniFrame: A unified framework for developing service-oriented, component-based, distributed software systems. In Z. Stojanovic & A. Dahanayake (Eds.), *Service oriented software system engineering: Challenges and practices* (pp. 68-87). Hershey, PA: Idea Group Publishing.

Raje, R. (2000). UMM: Unified Meta-object Model for open distributed systems. *Proceedings of the 4th IEEE International Conference on Algorithms and Architecture for Parallel Processing* (pp. 454-465). Los Alamitos, CA: IEEE Press. Retrieved March 22, 2006, from http://www.cs.iupui.edu/uniFrame

Raje, R., Auguston, M., Bryant, B., Olson, A., & Burt, C. (2001). A unified approach for the integration of distributed heterogeneous software components. *Proceedings of the Workshop on Engineering Automation for Software Intensive System Integration* (pp. 109-119). Monterey, CA: U.S. Naval Postgraduate School. Retrieved March 22, 2006, from http://www.cs.iupui.edu/uniFrame

Seacord, R., Hissam, S., & Wallnau, K. (1998). *Agora: A search engine for software components.* Technical Report, CMU/SEI-98-TR-011, ESC-TR-98-011, Carnegie Mellon University, USA.

Siram, N. (2002). *An architecture for discovery of heterogeneous software components.* Unpublished MS thesis, Department of Computer and Information Science, Indiana University Purdue University Indianapolis, USA. Retrieved March 22, 2006, from http://www.cs.iupui.edu/uniFrame/

Sun Microsystems. (1994). *Remote method invocation.* Retrieved March 22, 2006, from http://java.sun.com/products/jdk/rmi

Sun Microsystems. (2001). *Designing enterprise applications with the J2EE™ platform.* Retrieved March 22, 2006, from http://java.sun.com/blueprints/guidelines/designing_enterprise_applications/

Waldo, J. (1999). The Jini architecture for network-centric computing. *Communications of ACM, 42*(7), 76-82.

W3C. (2004). *SOAP versions & reports.* Retrieved March 22, 2006, from http://www.w3.org/TR/soap

KEY TERMS

Active Registry: An enhanced version, capable of listening to broadcasts, of the basic registration mechanism present in a component model.

Client Query Result Retrieval Time (CQRRT): The time taken from the instant a client issues a query to MURDS until the instant the results return back to the client.

Distributed Computing System (DCS): A system made up of networked processors, each with its own memory, that communicate with each other by sending messages.

Headhunter: A critical piece of MURDS whose task is to accept queries and send mobile agents to discover the requested components, which are deployed over the network.

Internet Component Broker (ICB): A collection of services that provides a secure infrastructure for accepting incoming queries from distributed system developers, propagating the queries to the headhunters, and collecting results from the headhunters.

Mobile Agent: A software agent that can migrate from one point of connection to another on a network.

Mobile Agent-based Resource Discovery System (MURDS): An enhancement of URDS. It uses mobile agents for locating deployed services in a network and for propagating the discovery queries.

UniFrame: A unifying framework that supports a seamless integration of distributed and heterogeneous components.

UniFrame Resource Discovery System (URDS): Provides an infrastructure for proactively discovering components deployed over a network.

Mobile Business Applications

Cheon-Pyo Lee
Carson-Newman College, USA

INTRODUCTION

As an increasing number of organizations and individuals are dependent on mobile technologies to perform their tasks, various mobile applications have been rapidly introduced and used in a number of areas such as communications, financial management, information retrieval, and entertainment. Mobile applications were initially very basic and simple, but the introduction of higher bandwidth capability and the rapid diffusion of Internet-compatible phones, along with the innovations in the mobile technologies, allow for richer and more efficient applications.

Over the years, mobile applications have primarily been developed in consumer-oriented areas where products such as e-mail, games, and music have led the market (Gebauer & Shaw, 2004). According to the ARC group, mobile entertainment service will generate $27 billion globally by 2008 with 2.5 billion users (Smith, 2004). Even though mobile business (m-business) applications have been slow to catch on mobile applications for consumers and are still waiting for larger-scale usage, m-business application areas have received enormous attention and have rapidly grown. As entertainment has been a significant driver of consumer-oriented mobile applications, applications such as delivery, construction, maintenance, and sales of mobile business have been drivers of m-business applications (Funk, 2003).

By fall of 2003, Microsoft mobile solutions partners had registered more than 11,000 applications including e-mail, calendars and contacts, sales force automation, customer relationship management, and filed force automation (Smith, 2004). However, in spite of their huge potential and benefits, the adoption of m-business applications appears much slower than anticipated due to numerous technical and managerial problems.

BACKGROUND

M-business applications can be classified into two distinct categories in terms of target groups: vertical and horizontal target group (Paavilainen, 2002). Vertical targets are typically narrow user segments, such as filed service engineers or sales representatives. On the other hand, horizontal targets are a massive number of users. For example, mobile e-mail, mobile bulletin board, and mobile calendar are applications for a horizontal target group, while mobile recruitment tools, mobile sales reporting, and mobile remote control represent vertical applications (see Table 1). Generally, the goal of horizontal applications is to improve communication and streamlined processes in horizontal procedures, such as travel management and time entry. In contrast, the goal of vertical applications is to improve and solve business processes in more detailed and specific areas such as the needs of sales departments. Various vertical and horizontal applications are currently used in a number of industries. Table 2 provides examples of m-business applications in various industries.

THE IMPACTS OF MOBILE BUSINESS APPLICATIONS ON BUSINESSES

The advantages of using m-business applications are mobility, flexibility, and dissemination of m-business applications (Nah, Siau, & Sheng, 2005). Mobility allows users to conduct business anytime and anywhere, and flexibility allows users to capture data at the source or point of origin. In addition, m-business applications offer an efficient means of disseminating real-time information to a larger user population, which consequently enhances and improves customer service. According to Gebauer and Shaw (2004), users valued two

Table 1. Examples of vertical and horizontal mobile business applications (Paavilainen, 2002)

Vertical Mobile Applications	Horizontal Mobile Applications
• Mobile e-mail	• Mobile recruitment tools
• Mobile bulletin board	• Mobile tools for filed engineers
• Mobile time entry	• Mobile sales reporting
• Mobile calendar	• Mobile supply chain tools
• Mobile travel management	• Mobile fleet control
• Mobile pay slips	• Mobile remote control
	• Mobile job dispatch

Table 2. Examples of various mobile business applications (Sources: Chen & Nath, 2004; Collett, 2003; Dekleva, 2004)

Hotel	• Embassy Suite: Maintenance and housekeeping crews are equipped with mobile text messaging devices, so the front desk can inform the crew of the location and nature of the repair without physically locating them. • Las Vegas Four Seasons: Customer food orders are wirelessly transmitted from the poolside to the kitchen. • Carlson hotels: Managers use Pocket PCs to access all of the information they need to manage the properties in real-time.
Hospital & Healthcare	• Johns Hopkins Hospital: Pharmacists use a wireless system for accessing critical information on clinical interventions, medication errors, adverse drug reactions, and prescription cost comparisons. • St. Vincent's Hospital: Physicians can retrieve a patient's medical history from the hospital clinical database to their PDA. • ePocrates: Healthcare professionals receive drug, herbal, and infections disease information via handheld devices.
Insurance	• Producer Lloyds Insurance: Field agents can assess the company's Policy Administration & Services System (PASS) and Online Policy Updated System (OPUS).
Government	• Public safety agencies can access federal and state database and file reports.
Manufacture	• General Motors: Workers can receive work instructions wirelessly • Celanese Chemicals Ltd.: Maintenance workers are able to arrange for repair parts and equipment to be brought to the site using wireless Pocket PCs. • Roebuck: Technicians can communicate and order parts directly from their job location instead of first walking back to their truck.
Delivery Service	• UPS & FedEx : Drivers can access GPS and other important information in real-time

things most in m-business applications use: notification, especially in connection with high mobility, and support for simple activities like tracking. The study suggested that the combination of mobility and the frequency with which each task occurred is a primary indicator of the usage of m-business applications.

M-business applications have shown significant impacts and created enormous business values. For example, m-business applications have improved operational efficiency as well as flexibility and the ability to handle situations to current operations (Chen & Nath, 2004; Gebauer & Shaw, 2004). In addition, m-business applications allow users to have access to critical information from anywhere at anytime, resulting in greater abilities to seize business opportunities.

It is very difficult to measure the direct impact of mobile business applications in *productivity* statistics, but according to an OMNI (2005) consulting report, financial services agents executed approximately 11.4% more trade options on an annualized basis with mobile business applications and achieved an average nominal improvement of 3.1% in overall portfolio performance. Also, health care and pharmaceutical filed sales representatives conducted an additional 8.3 physical briefings per week due to mobile business applications. Finally, insurance-filed claims adjusters handled an additional 7.4 claims per worker per week and improved payout ratios by an annual yield of 6.4% per adjuster using mobile business applications. Table 3 provides a list of values created by mobile business applications.

FACILITATORS AND INHIBITORS OF MOBILE BUSINESS APPLICATIONS GROWTH

Several factors are expected to contribute to the continued growth of m-business applications. Across the globe, mobile devices such as Internet-enabled mobile phones and personal digital assistants (PDAs) are gaining rapid popularity among businesses and consumers. This rapid penetration of mobile devices can provide strong support for mobile business applications. Employees' demand to access critical business processes and services from anywhere at any time is also a significant driving factor for m-business applications (Chen

Table 3. Values of mobile business applications (Sources: Chen & Nath, 2004)

	Value
Efficiency	Reduce business process cycle time
	Capture information electronically
	Enhance connectivity and communication
	Track and surveillance
Effectiveness	Reduce information float
	Access critical information anytime-anywhere
	Increased collaboration
	Alert and m-marketing campaigns
Innovation	Enhance service quality
	React to problems and opportunities anytime-anywhere
	Increase information transparency to improve supply chain
	Localize

& Nath, 2004). The traditional methods of wired communication, which have a limited reach and range, are no longer suitable for the fast-paced business environment. Finally, corporate and individual customers, who are demanding more channels for interaction and services, also contribute to the growth of m-business applications.

However, in spite of their huge potential and benefits, the adoption of m-business applications appears much slower than anticipated. Various factors have been offered as explanations for this slow growth, including the immaturity of the wireless technology, the existence of a chaotic array of competing technologies and standards, and the lack of killer applications (Chen & Nath, 2004). According to Gebauer and Shaw (2004), poor technology characteristics have inhibited application usage to a great extent. In addition, according to Nah et al. (2005), security, cost, and employee acceptance are also significant barriers of the growth of m-business applications. Companies have been concerned about the loss or theft of mobile devices, which are easily misplaced or stolen, and their likelihood to contain sensitive or confidential data that can be accessed by unauthorized persons. Huge cost is also a concern to companies. To implement mobile applications, the company must invest in mobile devices, pay service fees for wireless access, and train employees. According to Lucas (2002), some U.S. firms are spending between $5 million and $50 million for mobile business applications. Finally, employee acceptance is also a big barrier. Not every employee is willing to embrace new technology, and some employees accustomed to standard operation procedures resist adoption of m-business applications.

FUTURE TRENDS

In the future, more customized and personalized business applications will be introduced. These applications are called context-aware or situation-dependent m-business applications (Figge, 2004; Heer, Peddemors, & Lankhorst, 2003). Currently, the majority of context-aware computing has been restricted to location-aware computing for mobile applications. However, more contextual information including spatial (e.g., speed and acceleration), temporal (e.g., time of the day), environmental (e.g., temperature), and social situation (e.g., office nearby) information will be added to increase the value of mobile business applications. In context mobile business applications, the most necessary information for the user to perform tasks will be provided in advance without the user's involvement. Therefore, in most cases, the user simply presses a single button rather than making several text inputs.

However, for m-business applications to grow, current limitations in technical and managerial issues should be resolved. Current technical limitations are mainly related to mobile devices such as small multi-function keypads, less computation power, and limited memory and disk capacity (Siau, Lim, & Shen, 2001). Other technical issues such as the lack of network standards and security problems also must be resolved (Chen & Nath, 2004). In addition, a clear understanding of the value of m-business applications is also very important to grow m-business applications. The m-business development and adoption decision should always be based on clearly identified needs and business requirements (Paavilainen, 2002).

CONCLUSION

M-business applications have shown significant impacts on business processes. M-business applications not only increase productivity, but also develop new business processes that yield increased customer and job satisfaction as well as competitive advantage. In the future, richer and more efficient m-business applications will be introduced to attract more businesses. However, current technical and managerial limitations should be resolved to support continued growth of m-business applications. Especially, it is very important to understand the fundamental value derived from m-business applications before developing and adopting them.

REFERENCES

Chen, L.-D., & Nath, R. (2004). A framework for mobile business applications. *International Journal of Mobile Communications, 2,* 368-381.

Collett, S. (2003). Wireless gets down to business. *Computerworld, 37*(18), 31.

Dekleva, S. (2004). M-business: Economy driver or a mess? *Communications of the Association for Information Systems, 13,* 111-135.

Figge, S. (2004). Situation-dependent services: A challenges for mobile network operators. *Journal of Business Research, 57*(12), 1416-1422.

Funk, J. (2003). *Mobile disruption: Key technologies and applications that are driving the mobile Internet.* New York: John Wiley & Sons.

Gebauer, J., & Shaw, M.J. (2004). Success factors and impacts of mobile business applications: Results from a mobile e-procurement study. *International Journal of Electronic Commerce, 8*(3), 19-41.

Heer, J.D., Peddemors, A.J.H., & Lankhorst, M.M. (2003). *Context-aware mobile business applications.* Retrieved October 29, 2005, from https://doc.telin.nl/dscgi/ds.py/Get/File-25810/coconet.pdf

Lucas, M. (2002). Wireless financial apps grow slowly. *Computerworld, 36,* 14.

Nah, F.F.-H., Siau, K., & Sheng, H. (2005). The value of mobile applications: A utility company study. *Communications of the ACM, 48,* 85-90.

Omni. (2005). *Study finds 13.4 percent increase in worker productivity.* Retrieved October 10, 2005, from http://newsroom.cisco.com/dlls/2005/prod_020905.html

Paavilainen, J. (2002). *Mobile business strategies.* London: Wireless Press.

Siau, K., Lim, E.P., & Shen, Z. (2001). Mobile commerce: Promises, challenges, and research agenda. *Journal of Database Management, 12*(3), 4-13.

Smith, B. (2004). Business apps: Going for the tried and true. *Wireless Week, 10,* 22.

KEY TERMS

Horizontal Mobile Business Application: Mobile business application developed for a massive number of users to improve communication and streamline processes.

Location-Aware Computing: The capability of computing to recognize and react to location context. Global Positioning System (GPS) is the most widely known location-aware computing system.

Mobile Business Application: Mobile application used to perform business tasks such as sales force automation, customer relationship management, and filed force automation.

Situation-Dependent Mobile Application: Mobile application using various contextual information such as spatial, temporal, environmental, and social.

Vertical Mobile Business Application: Mobile business application developed for a specific target group such as filed service engineers and sales representatives.

Mobile Cellular Traffic with the Effect of Outage Channels

Hussein M. Aziz Basi
Multimedia University, Malaysia

M. B. Ramamurthy
Multimedia University, Malaysia

INTRODUCATION

The designer of the cellular network must evaluate the possible configurations of the system components and their characteristics in order to develop a system with greater efficiency. This article studies the grade of service (GOS) degradation in the presence of outage for a mobile cellular network where the number of channels in outage can be used as an indicator of the traffic load for two models, namely fixed outage rate and traffic dependent outage rate. The performance parameters considered for this article are: the probability of delay, waiting time for priority and non-priority calls, mean waiting time, and priority gain; each is estimated for both models. The system is evaluated and compared under different conditions.

BACKGROUND

The mobile user behavior has a higher traffic impact (in both space and time) than in the fixed network line. The call initiation sites are scattered and dynamically changing over a geographical area, while the bandwidth associated with a connection may have to be provided to different sites throughout the call; the radio signal will change from one cell to another following the user call movement, and in such environments, the efficient allocation of wireless channels for communication sessions is of vital importance as the bandwidth allotted for cellular communication is limited. The number of wireless communication service users as well as the frequency of the available services increased with an unexpected rate. The analysis of traffic deployed with wireless communication networks is important for determining the operation for a mobile user's status. Mobile cellular traffic varies greatly from one period to another and not in any uniform manner, but according to the cellular user's needs. Teletraffic theory is used to specify the methods to ensure that the actual GOS is fulfilling the requirements. The calls in the cellular network are made by individual customers according to their habits, needs, and so forth, and the overall pattern of calls will vary throughout the day. The cellular network equipment should be sufficient in quantity to cope satisfactorily for the period of maximum demand in the busy hour, depending on availability of free channel. In order to determine the optimal channel loading, it is necessary to relate the GOS to traffic characteristics. Traffic modeling is necessary for cellular network provisioning, for predicting utilization of cellular network resources, and for cellular network planning and developments (Vujicic, Cackov, Vujicic, & Trajkovic, 2005) to specify emergency actions when systems are overloaded or technical faults occur. GOS could be defined as the number of unsuccessful calls relative to the total number of attempted calls (Nathan, Ran, & Freedman, 2002; Zhao, Shen, & Mar, 2002; Hong, Malhamd, & Gerald, 1991).

Voice traffic network has been modeled by the Erlang C formula (Yacoub, 1993), which is used in cases where all users have access to all channels in the mobile network and where there are a large number of users using the available channels (Nathan et al., 2002). The number of required channels is used as a fraction of user traffic intensity and desired GOS. The GOS in a cellular system is affected not only by the system's traffic but also by co-channel interference. The cellular system presence of co-channel interference can cause the carrier-to-interference ratio (C/I) to drop below a specified threshold level (Annamalai, Tellambura, & Bhargava, 2001; Aguirre, Munoz, Molina, & Basu, 1998; Yang & Alouini, 2002, 2006; Zhang, 1996); such an event is known as outage. In some cases, outage can cause the loss of the communication system.

Aguirre et al. (1998) estimated the effect of an outage channel for many models where there are no available channels and the call is blocked or dropped. In this case they did not consider the aspect of buffering the dropped calls (outage calls) and the mean waiting time with priority calls. Another researcher evaluated the performance of mobile systems with priority concept, where no channel is available when the call is queued through to when the available channel has been assigned, and the priority calls are placed in a queue before all non-priority calls but never interrupt a call in progress (Barcelo & Paradells, 2000); however they did not consider the concept of outage.

Figure 1. Outage parameter

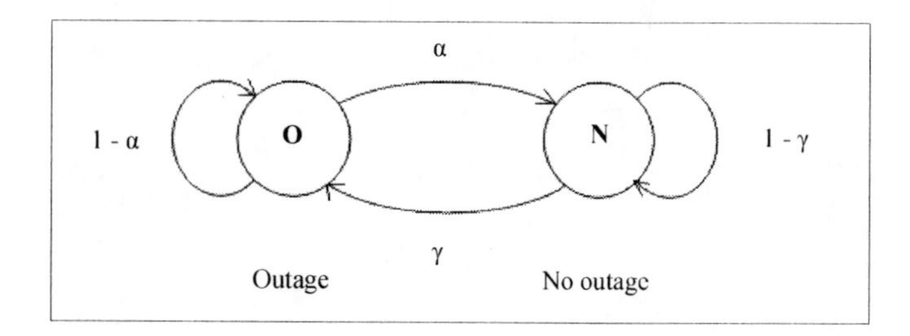

The major contributions of this article are to analyze the performance of a mobile communication system including GOS degradation due to the outage, when the calls as well as the outage channels in the cellular system are queued in the same buffer according to their priority, and for two different models to evaluate the performance of outage channel on the cellular network.

THE OUTAGE PARAMETERS

When the C/I is dropped below a certain quality threshold (ϑ) in a given channel, it becomes unusable and it affects the GOS in the cell. While two subscribers are communicating in the cellular network, the user could experience an absence of the desired signal and some noise or crosstalk. Even if link outages are very short, they collectively degrade the system performance, although they may not be individually recognized. Generally only outages listing longer than tens of milliseconds are recognized and can cause the dropout of the communication (Caini, Immovilli, & Merani, 2002). When the new calls come to the cellular system (by arrival rateλ) and there is a free channel in the cell, the call will engage one of the free channels and the channel becomes busy. The channel can go into outage with outage arrival rate γ, and the outage channel may recover by the outage recovery rate α. Thus, the numbers of available channels for service become a random variable due to the stochastic nature of the outage.

A channel from a normal working condition may become unavailable (or move into the outage state) due to drop in C/I. Thus, the two-state simple model shown in Figure 1 can represent its behaviors. The state $\boldsymbol{O}$ represents a channel in outage, while state $\boldsymbol{N}$ represents a state in it normal condition. The parameters γ and α represent failure and recovery rates. These rates can be represented in terms of steady-state probabilities $\boldsymbol{O}$ and $\boldsymbol{N}$ by the following analysis.

The probabilities of being in states $\boldsymbol{O}$ and $\boldsymbol{N}$ are:

$$\boldsymbol{O} = \gamma \boldsymbol{N} + (1-\alpha)\boldsymbol{O} \tag{1}$$

$$\boldsymbol{N} = (1-\gamma)\boldsymbol{N} + \alpha \boldsymbol{O} \tag{2}$$

In addition, they satisfy $\boldsymbol{O}+\boldsymbol{N} = 1$. After solving this system of equations, the following is obtained:

$$\boldsymbol{O} = \frac{\gamma}{\gamma + \alpha} \tag{3}$$

$$\boldsymbol{N} = \frac{\alpha}{\gamma + \alpha} \tag{4}$$

By sorting out $\gamma + \alpha$ in both previous equations and equalizing them, it is found that

$$\alpha = \boldsymbol{N}\gamma / \boldsymbol{O} \tag{5}$$

With the above equation, the outage arrival rate γ or the outage recovery rate α, assuming one of them, a value of the outage probability can be obtained. The relations for the outage γ and α are used to find the probability of delay for a different cellular system under different conditions. The design is extended for the normal cellular system by considering the outage channels where the outage channel calls as well as the normal incoming calls are queued in the same buffer as shown in Figure 2.

Queues occur wherever an unbalance occurs between requests for a limited resource and the ability of a service facility to provide that resource. The size of the buffer depends on the amount of the resource available and the demand for it by subscriber. The most common service discipline in real life is called first in first out (FIFO) or first come first served (FCFS); non-preemptive priority calls are used in this article where the priority calls have been affected by outage and thus the priority calls will be re-queued in the head of the buffer as it is assigned as high priority. Queuing of new calls and waiting for requests to be served can generally improve channel utilization at the expense of time spent in the queue.

PRESENT MODELS

Outage can cause the loss of the communication system and affects the GOS. Study of the GOS degradation due to

Figure 2. Queuing the outage and the normal calls

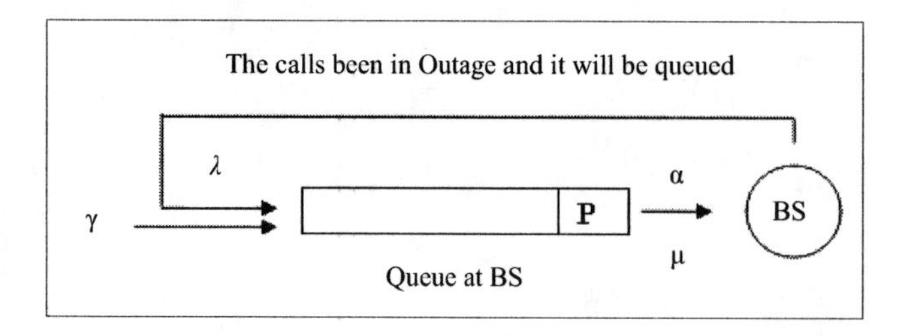

BS: Base Station, P: Priority calls, λ: Arrival rate, μ: Departure rate, γ : Outage arrival rate, α : Outage recovery rate

outage will be useful in evaluating the system performance. For a mobile communication system with channels in outage, a modification for the Erlang C formula is proposed by the authors for two cases, namely fixed outage rate and traffic dependent outage rate model (Basi & Murthy, 2004, 2005). In fixed outage rate model, the outage arrival rate (γ) is independent of the state, while the outage recovery rate (α) increases with the number of channels in outage. In a traffic-dependent outage rate model, both outage arrival rate (γ) and outage recovery rate (α) are state dependent. The most important queuing system, called Erlang C, has been widely used to evaluate the queuing system behavior as shown below (Yacoub, 1993). This equation is known as the Erlang C formula:

$$C(N,A)=\frac{A^N}{N!}\frac{1}{1-A/N}p_0 \tag{6}$$

where

$$P_0=\left[\sum_{k=0}^{N-1}\frac{A^k}{k!}+\frac{A^N}{N!}\frac{1}{1-A/N}\right]^{-1} \tag{7}$$

The modified formula is carried out in this article to evaluate the probability of delay for both models. Then using this probability of delay, the mean waiting time is calculated. Considering a priority option for some calls, the priority gain is estimated in view of the modification to the Erlang C formula. Thus analysis is carried out for two situations, namely priority and non-priority cases. The modification formula for the Erlang C model is proposed by Basi and Murthy (2004, 2005). The effective traffic intensity is taken to be $A_e = A + A_o$, replacing A in the Erlang C formula. In a system when k channels are in outage, the available channels will be N-k. Thus using the effective rates and number of channels as (N-k) instead of N in Erlang's C formula, the modified expression for probability of delay is:

$$P_D=C(N-k,A_e)=\frac{A_e^{N-k}}{(N-k)}\frac{p_0}{1-\frac{A_e}{(N-k)}} \tag{8a}$$

where

$$p_o=\left[\sum_{i=0}^{N-k-1}\frac{Ae^i}{i!}+\frac{Ae^{N-k}}{(N-k)}\frac{1}{1-\frac{Ae}{N-k}}\right]^{-1}$$

where N > k and $A_o < 1$ (8b)

These expressions are used in the present work. The probability of delay is estimated with varying call duration, number of calls/hour, and number of outage channels. The probability of delay is calculated according to the proposed formula, and it is used to evaluate the waiting times when a channel is in outage. The total duration (d) of call is known as call duration plus the recovery rate time; using this concept to calculate mean waiting time of priority and non-priority calls, the relations of Barceló and Paradells (2000) are used with modification, where N is replaced by N-k and d is taken as $d_r + \alpha$; d_r as duration time and α as outage recovery rate used in the modified relations will be:

$$WT_1=\frac{(P_D*d)}{(N-k)(1-p\rho)} \tag{9}$$

$$WT_2=\frac{(P_D*d)}{(N-k)(1-p\rho)(1-\rho)}=\frac{WT_1}{(1-\rho)} \tag{10}$$

WT_1 and WT_2 are the waiting time for calls with priority and non-priority (regular calls) respectively, ρ is overall system load, and p is the priority proportion. The load due to priority calls will be ρ x p. The evaluation conditions of channel system are heavy traffic (high ρ) and low priority propagation (low p) to maintain the effectiveness of the

Figure 3. P_D vs. outage channels (k) where $\alpha = 0.2$ and $\gamma = 0.00664$

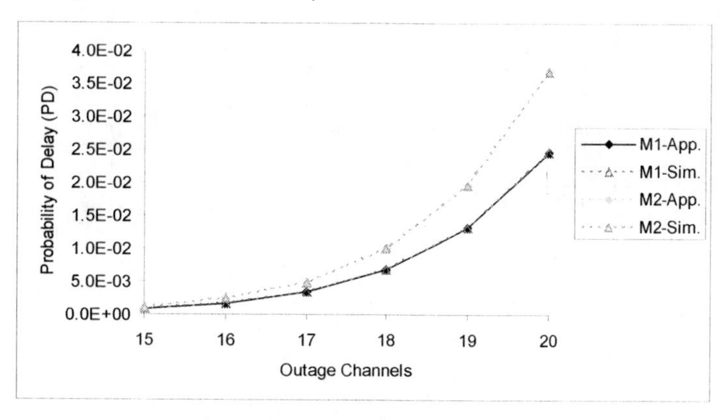

priority system as statues (Barceló & Paradells, 2000). It can easily be checked that the following relation holds for the average mean waiting time for all calls:

$$W_m = p\,WT_1 + (1-p)\,WT_2 \qquad (11)$$

The priority gain is convenient ration. It is the quotient between the mean waiting time for all calls (as if there was no priority) and mean waiting time for priority calls:

$$P_G = W_m \,/\, WT_1 \qquad (12)$$

The parameters are calculated using the original Erlang C formula, and unmodified relations of Barceló and Paradells (2002) are given as results for normal system. The simulation system has been designed to evaluate the above models, where the coming calls as well as the calls of outage channels in the cellular system are queued with consideration of the following assumption (Hussein & Murthy, 2006):

1. All the channels are fully available for servicing calls until all channels are occupied.
2. The offered traffic is uniformly distributed in the cell.
3. The number of subscribers is assumed infinite.
4. The call is initiation as a Poisson process with a mean call arrival of λ calls/hour.
5. The call holding time is exponentially distributed with a mean of 120 s.
6. The threshold level for new calls>19dB and for dropped calls <17.3 dB.
7. The interference channel (outage channels) should be limited.
8. The outage recover rate is 0.00664.
9. The buffer is assumed infinite.
10. Sorting the calls according to their priority without interrupting calls in progress.

The analytical as well as the simulation results are obtained for these parameters and plotted as graphs for the two proposed models.

RESULTS AND DISCUSSION

The results are presented in tables and plotted as graphs for the two models. The testing is carried out with 40 channels and a call rate of 360 calls per hour, while the outage arrival rate is 0.00664. From the results of both models reported, it is found that probability of delay (P_D) is much lower in the first model (fixed outage model) than the second model (traffic-dependent outage rate model). This means that the call has to wait for shorter time in the buffer in the case of a fixed outage rate model, as shown in Figure 3 and Table 1.

The waiting time for priority calls (WT_1) and non-priority calls (WT_2) for the first model is less than those of second model under different conditions for different priority call percentages as shown in Figures 4 and 5. The mean waiting time (W_m) for model one is less than in model two as shown in Figure 6; this means that the time for the call to wait in the first model is less than that in the second model. Figure 7 shows that the priority gain (P_G) is higher in case of the second model for heavy load and low priority call percentage.

The comparison of the two proposed models with the Erlang C model and the normal system (Barceló & Paradells, 2002) for the waiting time and the priority gain are present with the value of k = 1, as given in Tables 2, 3, 4, and 5. The P_D for the proposed models gives a higher delay than Erlang C because of the effect of the channel in outage on the proposed models as shown in Table 2. With the WT_1, WT_2, and W_m, it is found that the waiting time for the proposed models is higher than the normal system as given in Tables 3 and 4. For P_G, it is found that proposed models are given a higher delay for heavy load and a lower delay for priority

Table 1. Comparison of P_D between model one and two, $\alpha = 0.2$ & $\gamma = 0.00664$

k	Model-1(A)	Model-1(S)	Model-2(A)	Model-2(S)
0	1.58116E-10	1.58115E-10	1.58116E-10	1.58115E-10
1	5.34530E-10	5.34527E-10	5.36156E-10	5.36152E-10
2	1.76274E-09	1.76273E-09	1.78350E-09	1.78349E-09
3	5.66726E-09	5.66722E-09	5.81274E-09	5.81271E-09
4	1.77525E-08	1.77524E-08	1.85380E-08	1.85379E-08
5	5.41465E-08	5.41462E-08	5.77766E-08	5.77763E-08
6	1.60702E-07	1.60701E-07	1.75742E-07	1.75741E-07
7	4.63780E-07	4.63777E-07	5.21010E-07	5.21007E-07
8	1.30057E-06	1.30056E-06	1.50338E-06	1.50337E-06
9	3.54127E-06	3.54125E-06	4.21635E-06	4.21633E-06
10	9.35532E-06	9.35526E-06	1.14773E-05	1.14772E-05
11	2.39600E-05	2.39598E-05	3.02800E-05	3.02799E-05
12	5.94415E-05	5.94412E-05	7.73161E-05	7.73157E-05
13	1.42728E-04	1.42727E-04	1.90794E-04	1.90793E-04
14	3.31424E-04	3.31422E-04	4.54396E-04	4.54394E-04
15	7.43628E-04	7.43624E-04	1.04302E-03	1.04302E-03
16	1.61094E-03	1.61093E-03	2.30458E-03	2.30457E-03
17	3.36691E-03	3.36689E-03	4.89585E-03	4.89583E-03
18	6.78465E-03	6.78461E-03	9.99031E-03	9.99027E-03
19	1.31747E-02	1.31746E-02	1.95669E-02	1.95668E-02
20	2.46444E-02	2.46443E-02	3.67676E-02	3.67675E-02

Figure 4. WT_1 vs. the percentage of priority calls (p) where $k = 20$, $\alpha = 0.2$, and $\gamma = 0.00664$

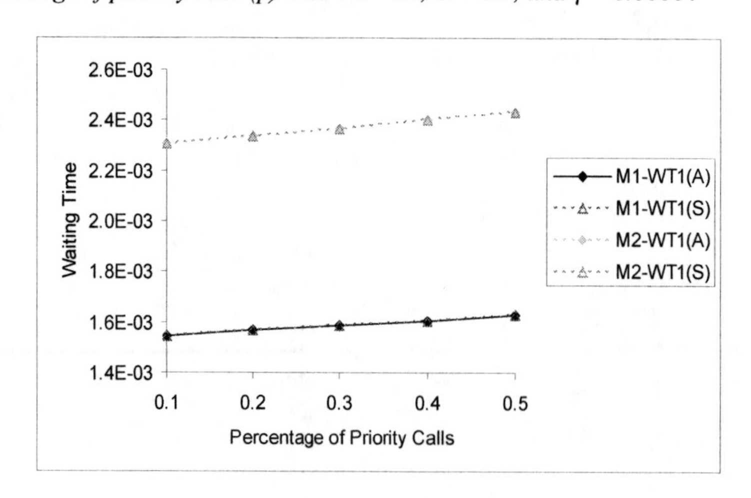

Figure 5. WT_2 vs. the priority call percentage where $k = 20$, $\alpha = 0.2$, and $\gamma = 0.00664$

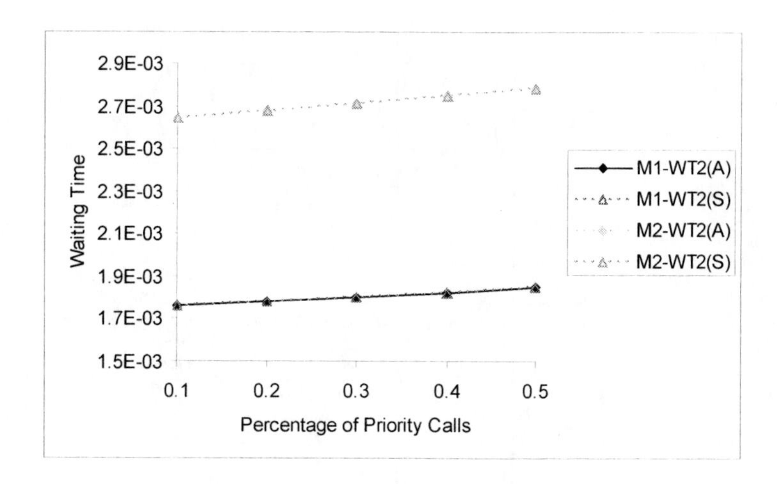

Figure 6. W_m vs. k with $\alpha = 0.2$ and $\gamma = 0.00664$

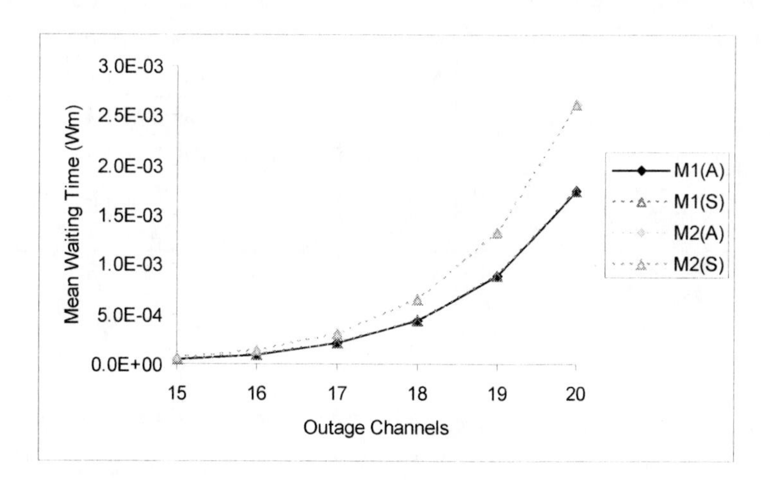

Figure 7. P_G vs. priority percentage with $k = 20$, $\alpha = 0.2$, and $\gamma = 0.00664$

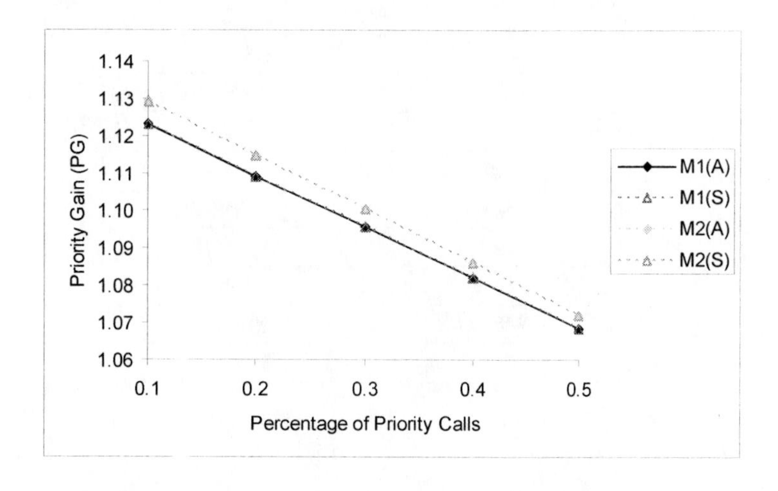

Table 2. Comparison of P_D between model one, two, and the Erlang C model; $\alpha = 0.2$, $\gamma = 0.00664$, and $k=1$

Channels	Erlang C	Model 1	Model 2
35	5.345E-08	1.587E-07	1.591E-07
36	1.756E-08	5.359E-08	5.373E-08
37	5.620E-09	1.761E-08	1.766E-08
38	1.752E-09	5.635E-09	5.651E-09
39	5.329E-10	1.758E-09	1.763E-09
40	1.581E-10	5.345E-10	5.362E-10
41	4.580E-11	1.586E-10	1.591E-10
42	1.296E-11	4.595E-11	4.610E-11
43	3.583E-12	1.300E-11	1.305E-11
44	9.686E-13	3.595E-12	3.608E-12
45	2.562E-13	9.721E-13	9.756E-13

Table 3. Comparison of waiting times between model one, two, and normal system; $\alpha = 0.2$, $\gamma = 0.00664$, and $k = 1$

p	WT_1	M1- WT_1	M2- WT_1	WT_2	M1- WT_2	M2- WT_2
0.1	4.801E-12	1.667E-11	1.673E-11	5.456E-12	1.895E-11	1.901E-11
0.2	4.860E-12	1.688E-11	1.693E-11	5.523E-12	1.918E-11	1.924E-11
0.3	4.921E-12	1.709E-11	1.714E-11	5.592E-12	1.942E-11	1.948E-11
0.4	4.983E-12	1.731E-11	1.736E-11	5.662E-12	1.967E-11	1.973E-11
0.5	5.046E-12	1.753E-11	1.758E-11	5.734E-12	1.992E-11	1.998E-11

Table 4. Comparison of mean waiting time between model one, two, and normal system; $\alpha = 0.2$, $\gamma = 0.00664$, $k = 1$, and priority percentage is 0.3

Channels	W_m	Model 1- W_m	Model 2- W_m
35	2.082E-09	6.376E-09	6.392E-09
36	6.652E-10	2.091E-09	2.097E-09
37	2.071E-10	6.681E-10	6.699E-10
38	6.289E-11	2.080E-10	2.086E-10
39	1.863E-11	6.318E-11	6.336E-11
40	5.390E-12	1.872E-11	1.878E-11
41	1.523E-12	5.416E-12	5.434E-12
42	4.207E-13	1.531E-12	1.536E-12
43	1.136E-13	4.228E-13	4.243E-13
44	3.002E-14	1.142E-13	1.146E-13
45	7.762E-15	3.018E-14	3.029E-14

Table 5. Comparison of priority gain between model one, two, and normal system; α = 0.2, γ = 0.00664, and k = 1

p	P_G	Model 1- P_G	Model 2 - P_G
0.1	1.12273	1.12274	1.12276
0.2	1.10909	1.10910	1.10912
0.3	1.09545	1.09547	1.09548
0.4	1.08182	1.08183	1.08184
0.5	1.06818	1.06819	1.06820

call percentage as in Table 5. The results of the mathematical model are indicated by (A) and those of simulation by (S). M1 refers to model one (fixed outage model) and M2 to model two (traffic-dependent outage rate model).

CONCLUSION

The mobile cellular system has appeared more recently as a consequence of the high demand for mobile services; the Erlang C model is used in cases where all users have access to all channels in the mobile network and where there are a large number of users using the available channels. Co-channel interference can cause the carrier-to-interference ratio (C/I) to drop below a specified threshold level, and such an event is known as outage. The calls of outage channels are queued along with normal arriving calls into the same buffer. The call duration, number of calls per hour, and number of channels in outage affect the probability of delay as observed. The study helps in understanding the performance of a mobile link and may help in deciding the number of channels for given traffic. With increasing probability of delay to meet given traffic demands, one has to select a memory of suitable size so that all waiting calls can be queued up. According to the scheme proposed by authors, no call is lost, but they may be delayed. The above results can be successfully used in designs of cellular systems to decide the number of channels needed for satisfactory, reliable operation of the cellular system.

REFERENCES

Annamalai, A., Tellambura, C., & Bhargava, V. K. (2001). Simple and accurate methods for outage analysis in cellular mobile radio system—a unified approach. *IEEE Transactions in Communications, 49*(2), 303-308.

Aguirre, A., Munoz, D., Molina, C., & Basu, K. (1998). Outage—GOS relationship in cellular systems. *IEEE Communications Letters, 2*(1), 5-7.

Barcelo, F., & Paradells, J. (2000, September). Performance evaluation of Public Access Mobile Radio (PAMR) systems with priority calls. *IEEE Proceedings of the 11th PIMRC* (pp. 979-983), London.

Basi, H. M. A., & Murthy, M.B.R. (2004). A simple scheme for improved performance of fixed outage rate cellular system. *American Journal of Applied Sciences, 1*(3), 190-192.

Basi, H. M. A., & Murthy, M. B. R. (2005). Improved performance of traffic dependent outage rate cellular system. *Journal of Computer Sciences, 1*(1), 72-75.

Basi, H. M. A., & Murthy, M. B. R. (2006). The simulation study on the effect of outage channels on mobile cellular network. *International Journal of Computer Science & Network Security, 6*(4), 146-150.

Caini, C., Immovilli, G., & Merani, M. L. (2002). Outage probability for cellular mobile radio systems: Simplified analytical evaluation and simulation results. *Electronics Letters, 28*(7), 669-671.

Hong, H.H., Malhamd, R., & Chen, G. (1991, May 19-22). Traffic engineering of trunked land mobile radio dispatch system. *Proceedings of the 41st IEEE Vehicular Technology Conference: Gateway to the Future Technology in Motion* (pp. 251-256).

Nathan, B., Ran, G., & Freedman, A. (2002). Unified approach of GOS optimization for fixed wireless access. *Vehicular Technology, IEEE Transactions, 51*(1), 200-208.

Vujicic, B., Cackov, N., Vujicic, S., & Trajkovic, L. (2005). Modeling and characterization of traffic in public safety wireless networks. *Proceedings of SPECTS 2005* (pp. 214-223), Philadelphia, PA.

Yang, L., & Alouini, M.-S. (2006). Performance comparison of different selection combining algorithms in presence of co-channel interference. *Vehicular Technology, IEEE Transactions, 55*(2), 559-571.

Yang, L., & Alouini, M.-S. (2002). Outage probability of dual-branch diversity system in presence of co-channel interference. *IEEE Transactions on Wireless Communication, 2*(2), 310-319.

Yacoub, M. D. (1993). *Foundations of mobile radio engineering.* CRC Press.

Zhang, Q. T. (1996). Outage probability in cellular mobile radio due to Nakagami signal and interferers with arbitrary parameters. *Vehicular Technology, IEEE Transactions, 45*(2), 364-372.

Zhao, D., Shen, X., & Mar, J. W. K. (2002). Performance analysis for cellular system supporting heterogeneous services. *Proceedings of ICC 2002* (vol. 5, pp, 3351-3355).

M

Mobile Commerce

JiaJia Wang
University of Bradford, UK

Pouwan Lei
University of Bradford, UK

INTRODUCTION

The rapid development and deployment in wireless networks and mobile telecommunication systems are leading to a phenomenal growth of innovative and intelligent mobile applications generally referred to as mobile commerce (m-commerce). Mobile devices like the mobile phone become a necessity for everyone. M-commerce makes networks more productive by seamlessly bringing together voice, data communication, and multimedia services. There is an increasing demand in mobile applications or m-commerce. The objective of this short article is to discuss the reasons for the growth of m-commerce. First, variety of wireless and mobile telecommunication technologies will be reviewed. Second, the evolution of m-commerce application architecture will be studied. Third, we will examine the landscape of m-commerce. Finally, we conclude the article.

BACKGROUND

The recent phenomenal convergence of the Internet and mobile telecommunication has accelerated the demand for "Internet in the pocket" on light, low-cost terminals, as well as for radio technologies that boost data throughput and reduce the cost per bit. This trend to higher data rates over wireless networks will culminate in the introduction of 3G IMT-2000 (International Mobile Telecommunications-2000) systems. This revolution continues to 3.5G, which is HSDPA (High-Speed Downlink Packet Access) spreading in Europe and Japan currently, and further will get to 3.75G-HSUPA for solving uplink problems. In addition to these wide area cellular networks, a variety of wireless transmission technologies are being deployed, including DAB (Digital Audio Broadcast), DVB (Digital Video Broadcast), and DMB (Digital Multimedia Broadband) for wide area broadcasting; LMDS (Local Multipoint Distribution System) and MMDS (microwave multipoint distribution system) for fixed wireless access; and IEEE 802.11b, a, g, h, and the new standard i for WLAN (Wireless Local Area Networking), as well as WiMAX (Worldwide Interoperability for Microwave Access) extending from the enterprise world into the public and residential domains.

M-commerce, which refers to access to the Internet via a handheld device such as a cell phone or a PDA, is becoming a leading driver for the successful rollout of the current cellular systems, and will influence the relations between existing and emerging players (Paavilainen, 2001). It is expected to be one of the most important applications for nearly all social classes, as the UMTS Forum predicted the significant potential of the mobile Internet for m-commerce in 3G with the expectation about 50% of mobile subscribers (UMTS Forum, 2003), with a further 1.5 billion mobile users worldwide. The target m-commerce applications imaginable today are ranging from telemetry and credit card applications to electronic postcards, Web browsing, audio or video on demand, and even videoconferences. This will result in an estimated m-commerce global revenue of US$88 billion, and the ticket purchased and phone-based retail POS sales will result US$39 billion and US$299 million respectively in 2009 (Juniper Research, 2004).

This rapid development of m-commerce technologies has opened up hitherto unseen business opportunities. It has increased an organization's ability to reach its customers regardless of location and distance, and has also been successful to a certain extent in creating a consumer demand for more advanced mobile devices with interactive features. While the distinctive e-commerce is characterized by e-marketplaces, an explosion in m-commerce innovative applications has presented the business world with a fresh set of strategy based on personalized and location-based services (Buvat, 2005).

THE EVOLUTION OF M-COMMERCE ARCHITECTURE

M-commerce is enabled by a combination of technologies such as networking, embedded systems, databases, and security. Mobile hardware, software, and wireless networks enable m-commerce systems to transmit data more quickly, locate a user's position more accurately, and conduct business with better security and reliability. In this section, three areas of technologies that are fundamental for m-commerce will be examined which are wireless networks, wireless protocol(s), and mobile devices.

Wireless Networks

Wireless networks provide the backbone of m-commerce activities. The evolution of wireless networks continued with the implementation of 2G (Second-Generation) systems such as TDMA (Time Division Multiple Access), CDMA (Code Division Multiple Access), and GSM (Global System Of Mobile Communication), which were also used primarily for voice applications, with the exception of the SMS (Short Message Service) capability offered by the GSM network. An upgrade of the 2G networks is referred to as 2.5G wireless networks such as high-speed circuit-switched data, GPRS (General Packet Radio Service), and EDGE (Enhanced Data Rates For Global Evolution). Being either circuit-switched or packet-switched, these networks are primarily intended to allow for increases in data transmission rates and, in the case of packet-switched networks, an "always-on" connection.

3G networks are commonly referred as IMT-2000 on a global scale. Along with voice functionality, 3G networks support higher-speed transmission for high-quality audio and video enabled through high-bandwidth data transfers, as well as provide a global "always on" roaming capability. Better modulation methods and smart antenna technology are two of the main research areas that enable fourth-generation wireless systems to outperform third-generation wireless network (PriceWaterhouseCoopers, 2001).

Wireless Protocol(s)

Wireless networks are evolving, similar to the communication protocols; WAP and iMode are the two main wireless protocols that are implemented in m-commerce. The following "information exchange technology" for these two protocols is described:

- Hyper-Text Markup Language (HTML) is not a suitable format for information exchange in the wireless domain, while the compact version of HTML, known as cHTML, has been used in the NTT DoCoMo's iMode services.
- eXtensible Markup Language (XML) is a meta-language, designed to communicate the meaning of the data through a self-describing mechanism. It tags data and puts content into context, therefore enabling content providers to encode semantics into their documents. For XML-compliant information systems, data can be exchanged directly, even between organizations with different operation systems and data models, as long as the organizations agree on the meaning of the data they exchange.
- Wireless Markup Language (WML), which has been derived from XML, has been developed especially for WAP (Wireless Application Protocol). It allows information to be represented as cards suitable for display on mobile devices. So WML is basically to WAP what HTML is to the Internet.

Of course, iMode is a serious competitor of WAP 2.0 (NTTCoCoMo, 2005). It has been suggested that WAP may push ahead of iMode in popularity because WAP has a large community of developers, whereas the tightly NTT-controlled iMode may be stifled by lack of development blood (Frank, 2001). As iMode evolves towards support of XHTML and TCP (Transmission Control Protocol), with the current WAP evolution, these two technologies will probably converge. It has been rumored that the iMode supporters are evolving their platforms to support WAP users by enabling WAP phones to access iMode content. This is being done in Japan, and it is one way for iMode manufacturers and service providers to sell more equipment and services. By enabling a WAP user to get iMode content, an iMode service provider could use the product as a way of convincing the WAP user to buy his or her primary service from the iMode carrier. More than likely, a gateway function will be used to act as a mediation and conversion access point.

CHTML will likely become the common markup language for both iMode and WAP. XHTML is a combination of HTML and XML, and the combined format will define the data and the presentation of the data. This convergence for the technologies will create more opportunities to content providers and the Internet industry between the wireless Internet and the wired Internet, which in turn can offer more applications to m-commerce users and further expand the subscriber base in order to grow the revenue stream.

SMS enables sending and receiving text messages to and from mobile phones. Up to 160 alphanumeric characters can be exchanged in each SMS message. Widely used in Europe, SMS messages are mainly voicemail notification and simple person-to-person messaging. It also provides mobile information services, such as news, stock quotes, sports, weather, SMS chat, and downloading of ringing tones.

In mobile communication, knowledge of the physical location of a user at any particular moment is central to offering relevant service. Location identification technologies are important to certain types of mobile commerce applications, particularly those whose content is varied depending on location. GPS (Global Positioning System), a useful location technology, uses a system of satellites orbiting the earth.

Mobile Terminal(s)

The development of mobile terminals is partly dependent on the evolution of the networks. Bandwidth is an advanced feature, while it is not the only feature that narrows down potential applications. Network-based location services are also dependent on the equipment installed by the mobile operator. Location technologies are especially important with the evolution of car navigation systems, which use network and satellite-dependent positioning. Mobile terminals inside the car are able to use both technologies in order to provide driving directions and information on special points of interest.

Another factor affecting the evolution of mobile handsets is consumer adoption. It remains to be seen how the advanced features are welcomed by the end users. For example, consumers in Europe are more concerned about ease-of-use of their handsets, while Asian consumers are more concerned about the appearance and the size of their mobile phones, and more and more members of the younger generation already regard their mobile phones as a fashion statement.

The evolution of mobile terminals will be characterized by customer segmentation. Handsets focusing on a narrow target group, such as teens, construction workers, or business professionals, have specific requirements in terms of functions and applications. Business professionals require efficient time management and team working capabilities. Teens may choose a handset with a built-in game console. Construction workers may rely on a water-resistant phone covered with rubber. Therefore, customer segmentation will be a crucial part of the future, and device manufactures have to develop several models in order to stay in the business (Paavilanimen, 2001).

As can be seen, consumer electronics and mobile communication come closer to each other by integrating new technologies with handsets. Mobile handset owners can already use their device as a calculator, MP3 player, radio, remote controller, game console, and digital camera. Naturally, devices with the biggest potential are those integrating a mobile phone with another mobile device such as a digital camera or game console. In this way, the portability of the two devices is used to create totally new service concepts.

Now, with the emergence of data services, it is likely that the size of a mobile terminal is going to be increased, as the use of Internet applications requires a bigger screen and more flexible character input methods. A 3G consumer in the UK said, "It was like going back 10 years to when mobiles where the size of a brick" (BBC News, 2004a). For this reason, some mobile carriers like AT&T Wireless now provide users with shortcuts that allow the consumer to access Web content and services through voice-activated dialing (BBC News, 2004b). Most researchers are researching and developing to compensate this defect. The latest news shows that V920, the "world first" video eyewear, can be used as a portable high-resolution display or as the ultimate viewer in the rapidly growing mobile video markets with portable DVD players, "in-car" video systems, video-enabled cell phones, game consoles, and the new personal digital media/video players. This revolutionary device overcomes the limitations of traditional direct view displays and creates big-screen images from micro displays, providing users with an unparalleled solution for mobile entertainment and information applications (3G Newsletter, 2005).

THE LANDSCAPE OF M-COMMERCE APPLICATIONS

The evolution of m-commerce applications will be driven by the user's preference for new high-speed services and their demands, while on the move, to replicate their experience of broadband at home and work. To analyze the impact of the future m-commerce applications is challenging. All of the potential services are unforeseen, difficult to say which are really going to be the "killer applications." For this reason the selected approach has been to broaden the granularity and use a classification that can help in assessing the penetration, usage, bandwidth, and other requirements, and thus revenue potential of the forthcoming m-commerce applications.

M-commerce operation modes can be generalized in four categories: (1) content distribution mode, (2) financial transaction mode, (3) interaction mode, and (4) communication mode—all described in detail as follows. Also, some prototypes are shown in Figure 1.

Content distribution services are concerned with real-time information notification (e.g., bank overdraft) and using positioning systems for intelligent distribution of personalized information by location (e.g., selective advertising of locally available services and entertainment). Real-time information such as news, traffic reports, stock prices, and weather forecasts can be distributed to mobile phones via the Internet. The information is personalized to users' interests. Users also can retrieve local information such as restaurant and shopping information, as well as traffic reports. Content distribution services with a greater degree of personalization and localization can be effectively provided through a mobile portal (Tsalgatidou & Veijalainen, 2000). Localization means to supply information relevant to the current location of the user. A user's profile—such as past behavior, situation, and location—should be taken into account for personalization and localized service provision. Notification can be sent to the mobile device too.

Figure 1. Mobile commerce operation modes

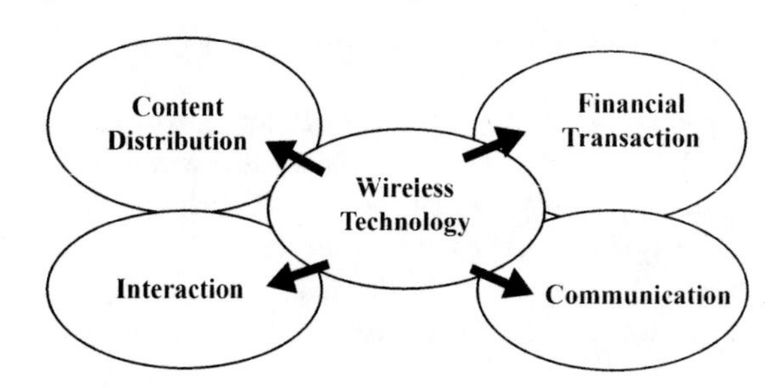

In the financial transaction mode, companies use the wireless Internet to run business transactions. M-commerce consumers can browse through the catalog and order products online. Although there are still some hidden obstacles such as transaction security, speed, and ease of use, it seems that most companies are likely to benefit directly from transactions on the wireless Internet, especially for small and medium-sized enterprises (Das, Wang, & Lei, 2006). The micro-payment m-commerce system, which is capable if executing transactions from external online merchants, includes vending machines, tickets, gasoline, and tax fares. In other words, the mobile phone is used as an ATM card or debit card. Time-sensitive and simple procedure transactions are the key success factors to this operation mode.

The mobile phone has also become a new personal entertainment medium. A wide range of interactive entertainment services are available which consist of playing online games, downloading ring tones, watching football video clips, watching live TV broadcasts, downloading music, and so on. According to *Screen Digest* estimates, Korea and Japan accounted for 80% of worldwide games download revenues of Euro 380 million (Screen Digest, 2005). Unsurprisingly, adult mobile services and mobile gambling services are among the fast-growing services. According to Juniper Research, the total revenue from adult mobile services and mobile gambling services could be worth US$1billion and US$15billion respectively by 2008 (Kowk, 2004). Law regulators have to stay ahead of the fast-growing development.

Community tools also generate a large amount of revenue. It evolves from voice and SMS messaging service in the early stage, to the current messenger chatting tools and the distribution of broadband multimedia messaging. Messaging and chatting allow a mobile user to keep contact with the others while he or she is on the move. M-commerce is one of the most important means to communicate in the society.

With different operation modes, each of these can be further classified by the bandwidth utilization (Cherry, 2004):

- **Higher Interactive Multimedia:** Data rate lower than 144kb/s.
- **Narrowband (NB):** Designed as the applications with data rates in the range [144, 384] kb/s.
- **Wideband (WB):** With data rates in the range [384, 2048] kb/s.
- **Broadband (BB):** With data rates higher than 2Mb/s.

As can be seen from this classification, the broadband class is available only when WLAN and other wireless technology access is possible. The NB and WB classes can be distinctive by circuit-switched and packet-switched, therefore the clear evolution path can be drawn as follows: beginning from circuit-switched NB services like basic voice service gradually to the packet-switched WB services then towards to the purely packet-switched BB services.

Figure 2. M-commerce mode prototypes (designed by authors using NMIT 4.0)

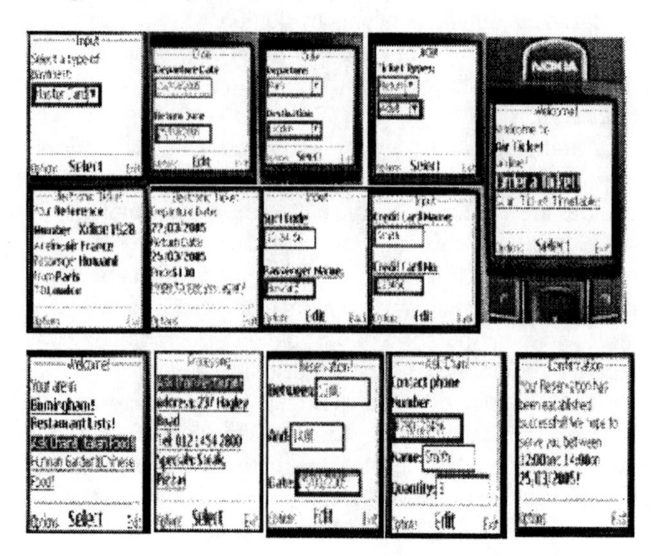

RESEARCH FINDINGS

In this research, the Nokia WAP emulator version 4.0 is used to develop the m-commerce application scenarios. Two m-commerce execution scenarios are designed for prototype. The first prototype is concerned with an LBS (location-based service), which provides a list of restaurants that are located near the current location and match a set of user preferences. The second prototype presents a whole financial transaction procedure by purchasing a mobile air ticket. It is more complicated than the first scenario, involving money transaction and payment procedure. Figure 2 shows the entire procedure. The previous steps are similar for both scenarios, which input a set of user preferences, such as departure time, destination, and ticket type. Following this information, the user comes to a secure domain, which can be a financial institution or bank. This step is a significant part in m-commerce applications. The money transaction will be performed in this secure channel by selecting the payment type.

This kind of mobile ticket service creates an extra purchase possibility for public transportation tickets via the mobile phone. Even though the scenarios described above are complicated, from mobile users' point of view, it is transparent and the benefit for them is purchasing goods and request services at anytime, anywhere without constraint of opening hour and physical distribution points, and most importantly it is a cashless payment (Wang, Song, Lei, & Sheriff, 2005).

From the procedures we presented in the two scenarios, it is obvious that there are two critical procedures urgently needing to be solved: user input usability between the client and server, and credit card payment security as performed in financial institutions. As the mobile phone user scrolls the information categories available to be requested and selects the category by pressing a key on the phone pad, a wireless device is dramatically easier to use such that the usability seems to be the critical limitation, one that the user is anxious to solve. And the security relative to a money transaction is still the main concern of business to adapt m-commerce for its intranet and extranet applications.

CONCLUSION

As mobile and wireless technologies are evolving rapidly and sophisticated mobile devices becomes affordable, m-commerce will become a part of our daily lives. The mobile Internet is ideal for particular applications and has useful characteristics that offer a range of services and contents. The widespread adoption of m-commerce is fast approaching.

REFERENCES

BBC News. (2004a). Retrieved March 8, 2004, from http://bbc.co.uk

BBC News. (2004b). Retrieved December 6, 2004, from http://bbc.co.uk

Buvat, J. (2005). Two disruptive technologies. *Land Mobile, 12*(4), 20-21.

Cherry, S. M. (2004). WiMax and Wi-Fi: Separate and unequal. *IEEE Spectrum,* (March).

Das, R., Wang, J. J., & Lei, P. (2006). A social-cultural analysis of the present and the future of the m-commerce industry. In B. Unhelkar (Ed.), *Handbook of research in mobile business: Technical, methodological and social perspective.* Hershey, PA: Idea Group Reference.

Frank, P.C. (2001). *Wireless Web, a manager's guide* (1st ed., pp. 115-132). Boston: Addison-Wesley.

Garber, L. (2002). Will 3G really be the next big wireless technology? *IEEE Computer, 35*(1), 26-32.

Juniper Research. (2006). Retrieved April 20, 2006, from http://www.epaynews.com/statistics/mcommstats.html#49

Kwok, B. (2004). Watershed year for mobile phones. *Companies and Finance in South China Morning Post,* (January 3).

Lamont, D. (2001). *Conquering the wireless world: The age of m-commerce.* New York: Capstone/John Wiley & Sons.

NTT DoCoMo. (2004). *iMode, an overview: Mobile communication and mobile computing.* Retrieved from http://www.rn.inf.tu-dresden.de/scripts_lsrn/Lehre/mobile/print_en/18_en.pdf

Paavilainen, J. (2001). *Mobile business strategies: Understanding the technologies and opportunities* (pp. 32-79). London: Wireless Press.

PriceWaterhouseCoopers. (2001). *2001 global forest & paper industry survey.*

Screen Digest. (2005, February 9). *Mobile gaming gets its skates on.* Retrieved from http://www.theregister.com/2005/02/09/mobile_gaming_analysis

3G Newsletter. (2005). Retrieved January 4, 2005, from http://www.3g.co.uk/PR/Jan2005/8904.htm

Tsalgatidou, A., & Veijalainen, J. (2000, September). Mobile electronic commerce: Emerging issues. *Proceedings of the 1st International Conference on E-commerce and Web Technologies (EC-WEB 2000)* (pp. 477-486), London. Berlin: Springer-Verlag (LNCS 1875).

UMTS Forum. (2003). *Mobile evolution shaping the future.* A UMTS forum white paper.

Wang, J. J., Song, Z., Lei, P., & Sheriff, R. E. (2005, October 3-5). Design and evaluation of m-commerce applications. *Proceedings of the 2005 Asia-Pacific Conference on Communications,* Perth, Australia.

KEY TERMS

Application: An application program (sometimes shortened to application) is any program designed to perform a specific function directly for the user or, in some cases, for another application program. For example, software for project management, issue tracking, file sharing, and so forth.

Bandwidth: A measure of frequency range, measured in hertz, of a function of a frequency variable. Bandwidth is a central concept in many fields, including information theory, radio communications, signal processing, and spectroscopy. Bandwidth also refers to data rates when communicating over certain media or devices. Bandwidth is a key concept in many applications.

Micro-Payment: Means for transferring money in situations where collecting money with the usual payment systems is impractical, or very expensive, in terms of the amount of money being collected.

Mobile Commerce (M-Commerce): The buying and selling of goods and services through wireless handheld devices such as cellular telephones and personal digital assistants (PDAs). Known as next-generation e-commerce, m-commerce enables users to access the Internet without needing to find a place to plug in.

Mobile Internet: Internet access over wireless devices.

Mobile Commerce Adoption Barriers

Pruthikrai Mahatanankoon
Illinois State University, USA

Juan Garcia
Illinois State University, USA

INTRODUCTION

Mobile commerce (m-commerce) emerged as one of the technologies that could change the way consumers engage in electronic business. Consumers have envisioned it as the mobile "electronic commerce," which allows them to purchase goods and services using their wireless mobile devices anywhere, anytime. This mobility, supported by a mobile telecommunications infrastructure, is the major characteristic that differentiates mobile computing from other forms of information technology applications.

Although the widespread use of mobile commerce has been intermingled with advanced telecommunications infrastructure, perceived benefits, and consumer demands, the industry is continuously searching for new and innovative mobile applications. Many consumers are still reluctant to make use of various mobile commerce applications. Technological hype and unreal consumer expectations have generated high hopes for innovative mobile applications that cannot be conceptualized during their initial stages. In many cases, unfilled gaps exist between the potential applications and the actual services provided by leading mobile carriers.

The purpose of this article is to identify and explain different socio-psychological drivers and barriers affecting consumers' motivations to use mobile commerce applications. These determinants are based on our literature reviews and exploratory consumer-based research. We later suggest a research framework to which researchers and practitioners can refer.

BACKGROUND: CONSUMER-BASED DRIVERS OF MOBILE COMMERCE

Mobile computing has two major characteristics that differentiate it from other forms of computing: mobility and broad reach (Turban, Rainer, & Potter, 2006). These two characteristics have created several value-added attributes that drive the demands for mobile-based computing, such as convenience, instant connectivity, and personalization. Wen and Mahatanankoon (2004) capture these demands through their 'aspects of mobility' concept, suggesting that the main driving forces of mobile applications are based on consumers' perception that: (1) their mobile devices are 'always on'; (2) they have the ability to customize their usage according to their lifestyle and social-psychological needs; (3) their location-based services (LBSs) can recognize where they are and then personalize the available services accordingly; and (4) their mobile devices have built-in authentication procedures that support secure mobile transactions. These aspects of mobility have tremendous impact on how consumers perceive various mobile applications.

The success of mobile commerce relies on the synergy of technology innovation, evolution of new value chains, and active customer demand (Zhang, Yuan, & Archer, 2003). These interrelated factors shift the telecommunications industry from being the provider of products or services to being the facilitator of customers' socio-psychological needs. Some practitioners suggest a consumer-centric approach to design effective mobile portals (Chen, Zhang, & Zhou, 2005). A good mobile application not only needs to be ergonomically easy to use, but it also has to provide consumers with sufficient, relevant, and personalized information. The industry should exploit these demand drivers and strengthen them by creating unique sets of innovative mobile applications that interact seamlessly between consumers and their surroundings. To ensure critical mass of mobile commerce adoption, we suggest further development of these existing applications and services to support consumer socio-psychological needs.

Integrated Mobile Devices

These devices are evolving from being a simple telephone with some extra features to an integration of the functionality of a personal digital assistant (PDA) with cellular telephones. The result of such integration creates a device that is able not only to connect to wireless networks, but also to manage organizer features. Speech recognition has become increasingly popular to support mobile commerce activities and will change the nature of user interface design (Fan, Saliba, Kendall, & Newmarch, 2005). In the near future, consumer perspective will change as these technologies are integrated into miniature wearable devices.

Ultramodern Mobile Applications

New and innovative services should exceed today's conventional usage of mobile devices. Enticing future applications will not only blur the boundary between work and play, but will also permit various ubiquitous services to take place simultaneously based on consumer demands (Varshney & Vetter, 2002). Ubiquitous services should be built based on the social network of mobile users.

Geographic-Oriented Applications

Location-based services will be the fastest growing enabler of mobile commerce applications. To consumers, the idea of conducting commercial transactions based on their current location is very appealing (e.g., consumers receiving a coupon for their favorite drink while walking past the coffee shop). In the foreseeable future, various industry consortiums will seek new ways to improve consumers' satisfaction by mapping their usage behaviors to specific locations, times, and events while providing them with options to customize their experience.

Advance Security and Privacy Applications

Two major factors exist concerning security issues: network security and storage security (i.e., securing the information stored in the mobile device). A mobile network needs to protect its users by continuously authenticating its subscribers (Patiyoot & Shepherd, 1999). Biometrics-ready phones can identify and enable authorized users to access the devices' full capabilities while preventing malicious individuals from accessing important personal information. Consumers can also locate lost mobile devices via location-based services by using the embedded global positioning system (GPS) capability.

However, most consumers are not totally convinced that mobile commerce would be a satisfactory experience. Consumers think twice before engaging in mobile commerce since most mobile commerce functionalities and services are not similar to those of electronic commerce. Mobile commerce is not intended to replace electronic commerce, but rather supplement it. Various factors, such as user interface, network speed, and users' self-efficacy, hinder many potential mobile applications.

BARRIERS TO MOBILE COMMERCE ADOPTION

Based on our preliminary findings, we are able to identify six main consumer-based barriers to mobile commerce adoption. These socio-psychological barriers are tightly integrated and include unawareness, device inefficiency, personalization/customization, nice-to-have/must-have, roaming, and electronic commerce perception. A successful solution to these interrelated factors will most likely result in mobile commerce reaching its critical mass. Figure 1 suggests a research framework on mobile commerce applications. These integrated barriers directly impact the mobile commerce adoption, as well as moderate the strength of external industry, technological, and consumer-based drivers. The external industry and technological drivers directly influence the consumer-based drivers and vice versa.

Unawareness Barrier

Awareness of mobile commerce existence implies that the individual has heard of it and has some idea of the kind of services it provides. Consumers are not always aware of their wireless devices' mobile commerce capabilities or their carrier's pricing scheme. Sometimes mobile carriers fail to communicate the mobile commerce capabilities to consumers. Only a few active users explore their mobile devices beyond voice communications and information-seeking activities. With various third-party electronic commerce vendors joining the bandwagon, it is often up to the users to discover how to connect to the Internet, download applications, or figure out how to use such applications. Therefore, in many mobile usage settings, consumer self-efficacy generally plays a significant role in exploring ground-breaking functionalities.

Figure 1. Consumer-based mobile commerce adoption framework

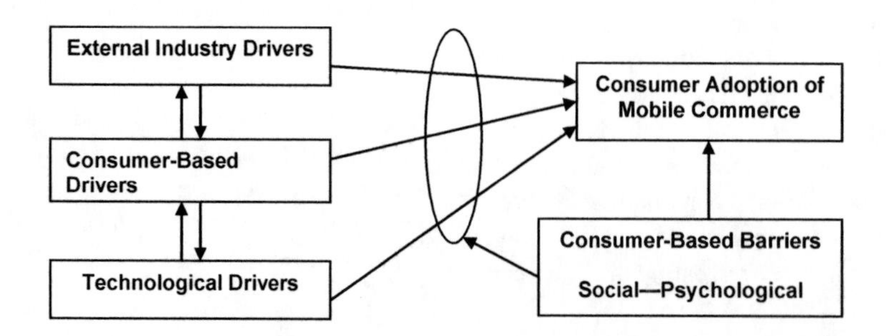

Device Inefficiency Barrier

The inefficiency of small mobile devices continues to be a problem. Every extra navigational input reduces the possibility of a transaction by 50% (Clarke, 2001). In addition to the limitations of screen size, power, and processing capability, device manufacturers need to be aware that consumers have a variety of multi-tasking activities (Lee & Benbasat, 2003). Personalization can compensate for the drawbacks of a small user interface (Ho & Kwok, 2003), but it may negatively affect other aspects, such as privacy and security. In many aspects, mobile user interfaces need to be designed to support users' limited but ever-shifting tasks.

Personalization/Customization Barrier

Customization services are context-specific applications that target each individual. These operations range from customized ring tones to location-based services. Since most customizable systems typically store users' essential information, issues related to privacy will be a major concern for consumers. These concerns for individual privacy negatively impact the adoption of mobile commerce. Consumers fear that they can be profiled, and their purchase history and navigation behaviors analyzed and abused (Pitkow et al., 2002). The lack of trust also leads to consumer avoidance of personalization/customization mobile applications.

Nice-to-Have vs. Must-Have Attitudinal Barrier

The industry must move beyond nice-to-have services and devise new 'must-have' services that positively affect people's lives (Jarvenpaa, Lang, Takeda, & Tuunainen, 2003). A nice-to-have feature may tempt consumers into buying a mobile device, but it cannot sustain a steady stream of revenue for the industry. Mobile application developers are searching for their killer application without trying to understand the socio-psychological aspects of hedonic mobile usage activities. Despite more than 40 inventive mobile applications, only five consumer-based applications are considered must-have applications (Mahatanankoon, Wen, & Lim, 2005); these are location-based, banking, entertainment, Internet, and emergency applications. Mobile designers must take into consideration usage environments that are relatively unstable and dynamic, potentially changing in a matter of seconds (Tarasewich, 2003).

Geographical Roaming Barrier

Consumers should be able to use the same mobile devices and services from anywhere in the world. Interoperability relates to the ability to use the same mobile device anywhere in the world. However, interoperability that accrues significant charges hinders the rapid adoption of mobile applications. Due to the competitive nature of the telecommunications industry, third-party providers and mobile carriers generally design their mobile applications based on device characteristics and specific network standards, which do not support communication and information sharing across mobile device platforms. Open Mobile Alliance (OMA) and the World Wide Web Consortium (W3C) have set their goal to create a global and interoperable mobile commerce market. It is hopeful that these consortiums will more closely connect worldwide consumers.

Perceptual Barrier

Many characteristics of traditional and electronic commerce can impact the way consumers perceive mobile commerce. Prior exposure to electronic commerce applications can have a significant impact on consumers' tendencies to modify their behaviors to fit the nature of small handheld devices (Orlikowski & Gash, 1994). Trust and trustworthiness of mobile commerce are still questionable. The idea of not dealing with somebody face to face at a physical location, or not being able to touch the merchandise, may sound unattractive to consumers. Many users simply do not like the idea of entering personal information and credit card numbers into their mobile transactions, fearing unsecured wireless networks or becoming potential victims of identity theft when the devices are stolen. Siau and Shen (2003) recommend that building customer trust in mobile commerce is a continuous process. Nevertheless, unlike electronic commerce's virtual communities, mobile commerce still lacks the sense of *virtualness* among consumers (e.g., customers cannot interact with other customers and gain feedback about a merchant from other customers).

These foremost barriers suggest that the mobile commerce buying experience is totally different than the traditional or electronic commerce buying experience. Electronic commerce customers may decide to buy products from a trusted vendor just by looking at its reliability and reviews, but for mobile commerce consumers, this functionality still remains a challenge. The industry is obligated to assist customers to overcome such barriers before it can reap any potential revenue from mobile commerce.

FUTURE TRENDS OF MOBILITY RESEARCH

Enormous potential exists for mobile commerce applications in the future, although the industry is still searching for distinct and profitable business models. Its current hype has surpassed its usefulness, but the joint efforts of industry players are pushing mobile commerce to become a widely used consumer-based technology. Practitioners and researchers can examine these drivers/barriers and their impact on consumers' socio-psychological behaviors. As mobile commerce evolves and sets its goal on changing the way consumers interact with the world, it is necessary to explore and take into account the obstacles that prevent the technology from reaching its true potential.

CONCLUSION

This article discusses the most salient characteristics of mobile commerce based on its mobility and success attributes. These interrelated factors help identify different consumer-based drivers of mobile commerce adoption, such as ultramodern, geographical-oriented, and advance security/privacy applications. The article then discusses the socio-psychological barriers of mobile commerce adoption. Various factors such as device inefficacy, interoperability, users' perceptions, and self-efficacy hinder many potential mobile applications. Given these limitations, mobile commerce applications should exploit the demand drivers and strengthen them by creating their own set of unique and innovative mobile applications.

REFERENCES

Clarke, I., III. (2001). Emerging value propositions from mobile commerce. *Journal of Business Strategies, 18*(2), 133-148.

Fan, Y., Saliba, A., Kendall, E. A., & Newmarch, J. (2005) Speech interface: An enhancer to the acceptance of mobile commerce applications. *Proceedings of the International Conference on Mobile Business.* Los Alamitos, CA: IEEE Computer Society Press.

Jarvenpaa, S. L., Lang, K. L., Takeda, Y., & Tuunainen, V. K. (2003). Mobile commerce at a crossroads. *Communications of the ACM, 46*(12), 41-44.

Lee, Y. E., & Benbasat, I. (2003). Interface design for mobile commerce. *Communications of the ACM, 46*(12), 49-52.

Mahatanankoon, P., Wen, H. J., & Lim, B. (2005). Consumer-based m-commerce: Exploring consumer perception of mobile applications. *Computer Standards and Interfaces, 27*(4), 347-357.

Orlikowski, W. J., & Gash, D. (1994). Technological frames: Making sense of information technology in organizations. *ACM Transactions on Information Systems, 12*(2), 174-207.

Patiyoot, D., & Shepherd, S. J. (1999). Cryptographic security techniques for wireless networks. *ACM SIGOPS Operating Systems Review, 33*(2), 36-50.

Pitkow, J., Schutze, H., Cass, T., Cooley, R., Turnbull, D., Edmonds, A., et al. (2002). Personalized search. *Communications of the ACM, 45*(9), 50-55.

Siau, K., & Shen, Z., (2003). Building consumer trust in mobile commerce. *Communications of the ACM, 46*(4), 91-94.

Tarasewich, P. (2003). Designing mobile commerce applications. *Communications of the ACM, 46*(12), 57-60.

Turban, E., Rainer, K., & Potter, R. (2006). *Introduction to information technology* (3rd ed.). New York: John Wiley & Sons.

Varshney, U., & Vetter, R. (2002). Mobile commerce: Framework, applications and networking support. *Mobile Networks and Applications, 7*(3), 185-198.

Wen, H., & Mahatanankoon, P. (2004). Mobile commerce operation modes and applications. *International Journal of Electronic Business, 2*(3), 301-315.

Zhang, J., Yuan, Y., & Archer, N. (2003). Driving forces for mobile commerce success. In M.J. Shaw (Ed.), *E-business management: Integration of Web technologies with business models* (pp. 51-76). Boston: Kluwer Academic.

KEY TERMS

Global Positioning System (GPS): A satellite-based tracking system that enables the determination of a GPS device's location.

Location-Based Service (LBS): One of several mobile services and applications offered to consumers via the utilization of GPS technology via the mapping of existing spatial information.

Open Mobile Alliance (OMA): An alliance of leading mobile operators, device and network suppliers, information technology companies, and content providers to create a universal mobile interoperability standard.

Personal Digital Assistant (PDA): A small handheld organizer, sometimes equipped with operating systems and wireless Internet capability.

Mobile Commerce (M-Commerce): The process of conducting electronic commerce activities through small mobile devices, such as mobile phones, pocket PCs, or PDAs.

Smart Phone: Internet-enabled cell phone that can support mobile applications and generally having advanced microprocessors to support various mobile applications.

Socio-Psychology (Social Psychology): A study of psychology related to the behaviors of groups and the influence of social factors on an individual.

Wireless Mobile Computing: The combination of mobile devices used in a wireless environment.

A Mobile Computing and Commerce Framework

Stephanie Teufel
University of Fribourg, Switzerland

Patrick S. Merten
University of Fribourg, Switzerland

Martin Steinert
University of Fribourg, Switzerland

INTRODUCTION

This encyclopedia on mobile computing and commerce spans the entire nexus from mobile technology over commerce to applications and end devices. Due to the complexity of the topic, this chapter provides a structured approach to understand the interrelationship in-between the mobile computing and commerce environment. A framework will be introduced; the approach is based on the Fribourg ICT Management Framework, elaborated at our institute with input from academics and practitioners, which has been tried and tested in papers, books, and lectures on ICT management methods. For published examples, please consult Teufel (2001, 2004), Steinert and Teufel (2002, 2004), or Teufel, Götte, and Steinert (2004).

THE MOBILE CONVERGENCE CHALLENGE

The information revolution has drastically reshaped global society and is pushing the world ever more towards the information-based economy. In this, information has become a commodity good for companies and customers. From an economical perspective, the demand for information at the right time and place, for the right person, and with minimal costs has risen. The transformation towards this information-driven society and economy is based on the developments of modern *information and communication technology (ICT)*. Different industries are able to generate enormous synergy effects from the use of ICT and the *information systems (IS)* building on these technologies, especially the Internet. It is a possible instrument to chance the structure and processes of entire markets.

As shown in Figure 1, information and communication technology can be differentiated in its infrastructure, the technologies themselves, and the information systems running on these technologies. In general, the infrastructure consists of all hardware- and software-related aspects as well as human resources. Consequently, the technologies themselves enable the collection, storage, administration, and communication of all data. These data can be used to synthesize information in respective systems, supporting the decision process and enabling computer-supported cooperative work.

Figure 1. Information and communication technology, infrastructure, and systems

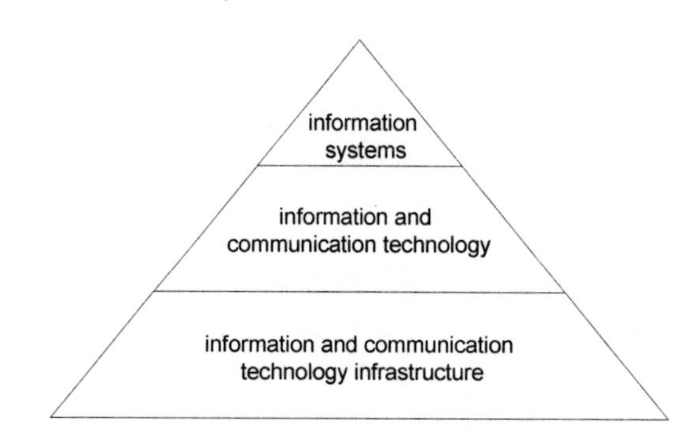

The term information and communication technology (ICT) appeared in recent years. Due to the harmonization of *information technology (IT)* and the digitalization of the *telecommunications (CT)* infrastructure and the liberalization of the latter business sector, the ICT market established itself (see Figure 3). Consequently, the development and convergence of ICT became increasingly complex. Figure 2 illustrates the associated technology convergence.

Nowadays, a new aspect has entered the arena: mobility. Mobility is perhaps the most important trend on the ICT market. The fundamental characteristic of mobile technologies is the use of the radio frequency band for (data) communication, which is often referred to as "wireless." The "wireless trend" has influenced not only the telecommunications and IT sector, but also most traditional markets, in the same way wired ICT did before. In addition, a convergence of wired and wireless, respectively fixed and mobile ICT can be observed.

Figure 2. Technology convergence (Teufel, 2004, p. 17)

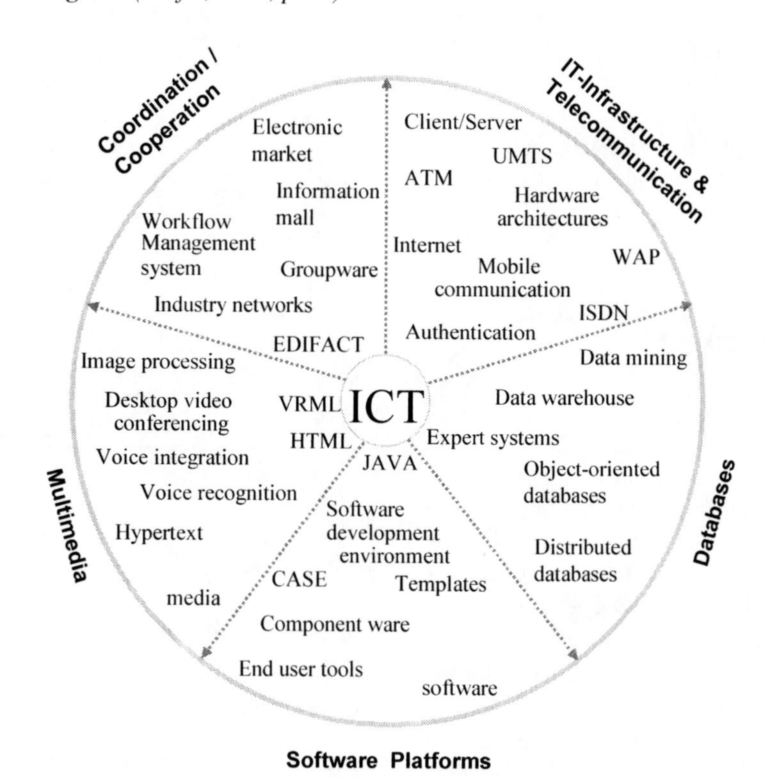

Figure 3. Mobile and fixed-line ICT convergence

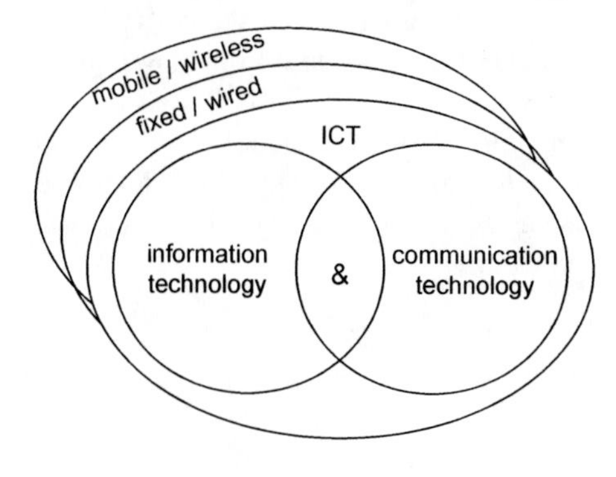

Figure 4. ICT and multimedia entertainment convergence (Teufel, 2004, p. 14)

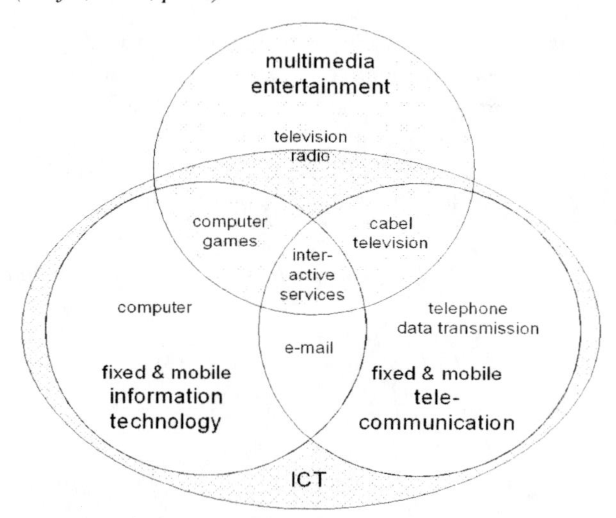

As shown in Figure 3, the convergence of information technology and communication technology to ICT can be seen as the first phase of convergence. This was caused by the digitalization and liberalization in the telecommunications sector. The next phase of convergence was the success of mobile ICT, initializing a competition between wireless and fixed ICT. Meanwhile, information and communication as well as mobile and wired technologies have not only coexisted; they have merged, generating enormous synergy effects for both business and customer. In addition, another not just technological convergence can be observed. The entertainment and multimedia branch has entered the ICT market and vice versa, as illustrated in Figure 4.

The trend shown in Figure 4 becomes obvious when looking at the boom in interactive games or home cinema computerized equipment—again accelerated by the digitalization in a sector, this time the television (DVB) and radio

(DAB). Again, the Asian market is leading edge. In South Korea, they are already running a fully functional system, based on the digital mobile broadcasting standard (DMB), bringing video broadcasting directly to the mobile end-device via satellite (tu4u, 2006). Finally, the three dimensions, fixed and mobile ICT convergence plus entertainment/multimedia, form the core of this encyclopedia's topic: the challenges of mobile computing and commerce.

THE MOBILE COMPUTING AND COMMERCE FRAMEWORK

Mobile computing and commerce comprises all business processes between administration, business, and customer via public or private wireless communication networks and with value creation. To understand the actual trends, recognizing the possibilities and threats and coping with the challenges of mobile computing and commerce are complex tasks. It becomes obvious that mobile computing and commerce consists of multiple dimensions, which are, in addition, interrelated. In order to structure the discussion, a framework for mobile computing and commerce is introduced. Using the classical scientific engineering approach, the framework allows a detailed analysis of single aspects and a reintegration of the diverse solutions in the synthesis. Furthermore, it covers the main issues, controversies, and problems from a market and business perception. Figure 5 features this ,mobile computing and commerce framework.

The four different dimensions of the framework as demonstrated in Figure 5 in addition show a common underlying scope. The strategic scope covers issues of long-term influence (more than five years of impact), as the tactical scope deals with all aspects in a timeframe of one to five years. Finally, all short-term topics are subject of the operational scope and handled within a year's period. The individual

Figure 5. Mobile computing and commerce framework

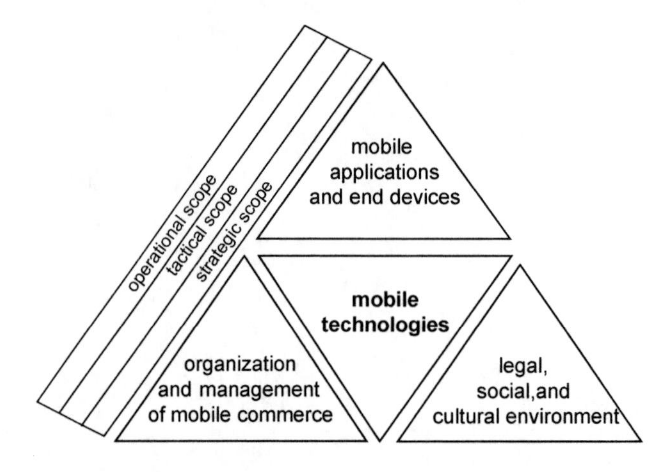

four main parts of the framework are examined in the following sections.

Mobile Technologies

The origin and foundation of every case of mobile computing and commerce are mobile technologies. They are the centerpiece of the framework and comprise the different technological aspects. They are building the foundation for discussing all other aspects of the framework. Mobile technologies have evolved rapidly in the last decade, not only gaining market penetration, but in terms of bandwidth and relative speed. Figure 6 presents today's available wireless access technologies—also introducing a physical and economic border.

As such, this dimension includes aspects which are dealt with in the categories mobile information systems, mobile service technologies, and enabling technologies of this encyclopedia.

Figure 6. Wireless access technologies (adapted from Schiller, 2003, p. 450)

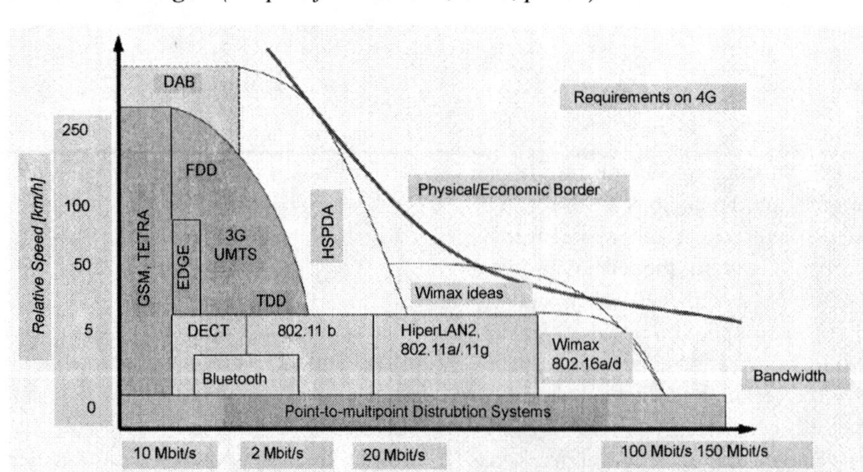

Legal, Social, and Cultural Environment

In a mobile environment, corporate social responsibility (CSR) is a fairly new field of increasing attention. It deals with the consequences of globalization, economic and ecological disaster, as well as financial affairs and others. Referring to the Global Compact Program 2000 from the United Nations and the Green Paper on CSR from the European Union, principles and guidelines are available today. These have led to programs that enable companies to continuously analyze and handle the versatile influences and effects on society and vice versa (Teufel et al., 2004).

Furthermore, the existence, use, and diffusion of mobile technologies are also strongly influenced by environmental aspects, especially from legal, social, and culture sub-environments. Examples are data protection issues, surveillance discussions, and radiation concerns, respectively. Other issues may also include important aspects such as standardization and regulation.

Furthermore, mobile technologies also change the way of living—introducing new concepts like mobile working. Especially the new work-life-(un)balance is subject to heated debates. Mobile technologies, applications, and end devices not only represent new opportunities in a business environment, but also create an interconnected and virtual world. In this, the digital divide more and more becomes a critical threat. Therefore topics of the categories like "mobile enterprise implications for society, business, and security" are to be considered.

Organization and Management of Mobile Commerce

To cope with the business challenges of mobile computing and commerce, all company internal aspects of the organization and the management of mobile commerce form a particular topic space. First of all, the classical roles of the CIO and the CTO have to be re-evaluated, taking mobile ICT into account. Furthermore, mobility also affects a whole set of management issues, which have already been previously influenced by fixed ICT. For example, the information management has to consider the aspect of mobile working when planning information system architectures. This in turn results in an adoption of current business processes and workflow implementations. Especially the procurement and distribution processes go through a fundamental change. In addition, mobile ICT offers new possibilities in the customer relationship management.

This dimension for example includes aspects described in "Mobile Commerce and E-Business."

Mobile Applications and End Devices

The focus point of every examination of mobile computing and commerce is the actual applications and end devices it is running onto. Again the Asian market can be consulted, to give an example of cutting-edge end device research. NTT DoCoMo is working on a future mobile phone device that uses human fingers as receiver. For this, a wristwatch-like bone conduction terminal is used in contact with the human arm (NTT DoCoMo, 2006). Above all, the Asian market is leading the way towards an all IP-based mobile network environment. Thus this last but probably most important framework dimension features topics such as: mobile to "consumer applications", "mobile applications for the extended enterprise", and "enabling applications."

CONCLUSION

Throughout the previous sections, it has been shown that coping with the challenges of mobile computing and commerce is a complex problem. Therefore, and to structure this encyclopedia, the framework has been introduced. It aims to provide managers, engineers, and practitioners with a profound approach to handle fixed and mobile information and communication technology. In such a mobile computing and commerce environment, the different market players themselves can be subsequently differentiated as shown in Figure 7, following the EITO on their special on "Entering the UMTS era—mobile applications for pocket devices and services."

Figure 7 illustrates the mobile data value net. In this interconnected environment, the nine different market players experience an enforced competition, due to the fact that every action influences the entire business network. As a result, a duality of competitive and cooperative business strategies established itself (Steinert & Bult, 2004, p. 31) to generate network effects, introducing the phenomenon of co-opetition (Brandenburger & Nalebuff, 1996). On a more general level, this again shows the complexity of a mobile computing and commerce environment.

Figure 7. Mobile data value net (EITO, 2002, p. 205)

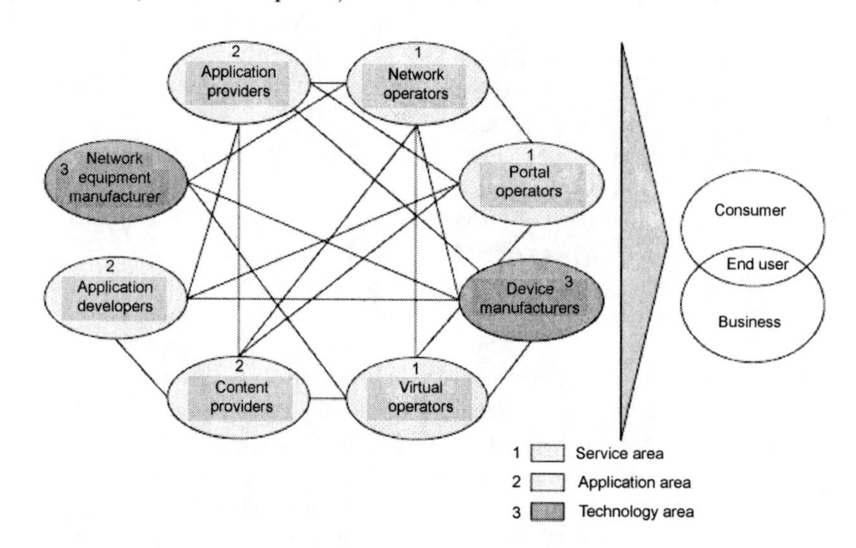

REFERENCES

Brandenburger, A., & Nalebuff, B. (1996). *Co-opetition* (1st ed.). New York: Doubleday.

EITO (European Information and Technology Observatory). (2002). *Eito report 2002.*

NTT DoCoMo. (2006). *R&D.* Retrieved January 12, 2006, from http://www.nttdocomo.com/corebiz/rd/index.html

Schiller, J. (2003). *Mobile communication* (2nd ed.). London: Addison-Wesley.

Steinert, M., & Bult, A. (2004). Strategische unternehmensführung von hightech-unternehmen—insights von swisscom-fixnet. In S. Teufel, S. Götte, & M. Steinert (Eds.), *Managementmethoden für ICT-unternehmen* (p. 12).

Steinert, M., & Teufel, S. (2002). The Asian lesson for mobile provider—An all-out strategic paradigm shift. *Proceedings of ITU Telecom Asia 2002* (pp. 25-44), Hong Kong.

Steinert, M., & Teufel, S. (2004, September 17-19). Beyond e-business—why e-commerce and Web organizations should monitor the mobile dimension. *Proceedings of the 2nd International Conference on Knowledge Economy and Development of Science and Technology (KEST2004)* (pp. 446-454), Beijing, China.

Teufel, S. (2001, August 6-12). ICT-management framework. *Proceedings of the International Conference on Advances in Infrastructure for Electronic Business, Science, and Education on the Internet (SSGRR 2001)* (pp. 9-24), L'Aquila, Italy.

Teufel, S. (2004). Managementmethoden für ICT-unternehmen—dargestellt mittels dem Fribourg ICT management framework. In S. Teufel, S. Götte, & M. Steinert (Eds.), *Managementmethoden für ICT-unternehmen.* Zurich: Verlag Industrielle Organisation/Orell Füssli.

Teufel, S., Götte, S., & Steinert, M. (Eds.). (2004). *Managementmethoden für ICT-unternehmen: Aktuelles wissen von forschenden des iimt der Université Fribourg und spezialisten aus der praxis.* Zürich: Verlag. Industrielle Organisation.

tu4u. (2006). *TU media corporation.* Retrieved January 12, 2006, from http://www.tu4u.com/

KEY TERMS

Co-Opetition: Following Brandenburger and Nalebuff (1996), co-opetition is the economic situation between a company and a competing company that provides complementary products and services. Following game theory, a differentiated approach strategic than the generic competitive strategies are necessary (see also *ValueNet*).

Fribourg ICT Management Framework: The framework has been elaborated at the International Institute of Management in Technology (IIMT) of the University of Fribourg (Switzerland) with input of academics and practitioners. It provides an integrated approach to cope with the business challenges of the information-based economy.

Information and Communication Technology (ICT): The result of developments in the fields information technology (IT) and communication technology (CT), and their convergence caused by the digitalization and liberalization in the telecommunication sector.

Legal, Social, and Cultural Environment: This framework dimension covers all aspects and implications of Mobile ICT for Society and Business.

Mobile Application and End Device: Mobile applications running on mobile end devices are the topic of this framework dimension.

Mobile Computing and Commerce Framework: The framework is based upon the Fribourg ICT Management Framework and presents an integrated view on the different fields to be considered, while examining the issues and controversies of mobile computing and commerce.

Mobile ICT Convergence: As ICT can be seen as the first phase of convergence, mobile ICT convergences introduce wireless technologies next to wired ICT.

Mobile Technology: Wireless mobile access technology and the centerpiece of the framework.

Network Effect: Following Katz and Shapiro (1985), each new network participant directly increases the benefit of all other actors in a network, for example, by offering a new communication possibility (primary or direct network effect); an increased size of a network also indirectly increased the value of the entire network indirectly, for example by pushing an industry standard (secondary or indirect network effect).

Organization and Management of Mobile Commerce: All company internal aspects of the organization and the management issues, which are influenced by mobile ICT, also including aspects such as mobile business.

ValueNet or ValueWeb: Instead of a linear value chain, the company, its suppliers, and customers, and also its complementors and competitors, form a ValueNet or ValueWeb. Co-opetition, reciprocal actions, and network effects must be taken into account in the economics of such a value net.

Mobile E-Commerce as a Strategic Imperative for the New Economy

Mahesh S. Raisinghani
TWU School of Management, USA

INTRODUCTION

A new form of technology is changing the way commerce is being done globally. This article provides an overall description of mobile commerce and examines ways in which the Internet will be changing. It explains the requirements for operating mobile commerce and the numerous ways of providing this wireless Internet business. While the Internet is already a valuable form of business that has already changed the way the world is doing business, it is about to change again. Telecommunications, the Internet, and mobile computing are merging their technologies to form a new business called *mobile commerce* or the *wireless Internet.* This is being driven by consumer demand for wireless devices and the desire to be connected to information and data available through the Internet. There are many new opportunities that have only begun to be explored, and for many this will become a large revenue source for those who capitalize upon this new form of technology. However, like other capital ventures, these new opportunities have their drawbacks, which may limit growth of the mobile commerce market if not dealt with. Mobile e-commerce technology is changing our world of business just as the Internet alone has changed business today.

BACKGROUND

Mobile commerce is the delivery of electronic commerce capabilities directly into the consumer's hand via wireless technology and putting a retail outlet in the customers' hand anywhere. This form of e-commerce allows businesses to reach consumers directly regardless of their location. The term mobile commerce or m-commerce is a variation of the e-commerce or electronic commerce term used for business being done over the Internet. Known as next-generation technology, m-commerce enables users to access the Internet without the need to find a place to plug in. There are signs that m-commerce is growing in popularity. Gartner Research (2004) forecasts that in six years time, 60% of people aged 15 to 50 in the European Union and the United States will wear an always-on wireless communications device for at least six hours a day, and more than 75% will do so by the year 2010.

Mobile commerce is the integration of technologies using wireless devices for conducting business over the Internet. M-commerce can be done by computer solutions, such as laptops and palm pads, with wireless devices attached to connect to the Internet or by using newly adapted cellular phones to receive digital transmissions of Internet material to these phones. These are all linked by software and service providers which provide the platform to conduct these operations.

A new business model is emerging: the integration of wireless networks with data communications, combined with electronic commerce, to create wireless e-commerce. Wireless e-commerce will generate significant revenues within the next several years from such services as wireless banking, wireless stock trading, and a variety of wireless-based shopping ventures. Wireless communications and e-commerce already are multi-billion-dollar global businesses. The integration of mobile communications with e-commerce has already started. For years companies in the vertical markets, such as field repair, have been utilizing mobile communications networks to enable their technicians to order parts and check inventories. The opportunities for wireless e-commerce in the horizontal markets, such as traveling executives and the consumer markets, is generating much appeal (Reiter, 1999).

M-commerce is a Quantum leap of technology applications and will not be limited simply to banking and brokerages. Other market uses will emerge. Payment options are one example that are being tested now in which products in a store may be scanned as one walks out and automatically deducted from a *smart card* which stores cash on it from your local bank. Airline and rail connections will be enhanced with ticket reservation and payment facilities. Mobile phone users will also have access to new online auction houses to submit bids and check developments by use of the cellular phone (Brokat, 2000).

MOBILE COMMERCE: STATE OF THE INDUSTRY

According to the Strategis Group (2005), by the year 2010, there will be one billion wireless subscribers worldwide on 3G (third-generation) networks. ARC Group estimates that

Table 1. Payback period for wireless LAN

	Retail	Manufacturing	Healthcare	Office Automation	Education
Benefits per company (millions $)	5.6	2.2	.94	2.5	.5
Costs per company (millions $)	4.2	1.3	.90	1.3	.3
Payback (# of months)	9.7	7.2	11.4	6.3	7.1

Table 2. Wireless Internet users

Region		2001	2004	2010
USA	Internet Users (#M)	149	186	247
	Wireless Internet Users (#M)	5.5	23	91
	Wireless Internet User Share (%)	3.7	12.5	35.1
Worldwide	Internet Users (#M)	552	941	1,781
	Wireless Internet Users (#M)	79	200	779
	Wireless Internet User Share (%)	14.4	21.2	43.8

by 2007 approximately 546 million users will spend close to $40 billion on mobile commerce (Schone, 2004). The reasons for this phenomenal growth are attributed to business factors such as substantial increase in remote workers and the telecommuters' need for improved customer service; the economic justification of mobile computing solutions through productivity gains and competitive advantages gained by early implementers; availability of inexpensive hardware with pre-packaged vertical industry application solutions, and less expensive and faster wireless networks; convergence of the Internet, wireless, and e-commerce technologies; and emergence of location-specific and mobile commerce applications, especially by a socially upscale and mobile population. As illustrated in Table 1, the data from the Wireless LAN Association (2005) shows that across all industries, with all economic benefits such as increased productivity, organizational efficiency, and extra revenue/profit gain considered, the wireless LAN paid for itself within 12 months time.

By 2008, a tenth of the world's mobile phone users will use their handsets as video players and cameras, and to download news, sports, and entertainment news (Gartner Research, 2004). Table 2 lists the wireless Internet users from 2001 and 2004, and forecasts the statistics for 2010 (eTForecasts, 2006).

Figure 1 summarizes the expected growth in m-commerce revenues over the period of 2001-2006.

International Data Corporation's (IDC's) forecast shows that total mobile data revenues are expected to be increasing by more than 31% per annum, whereby the CAGR for revenues generated by m-commerce is estimated to be more than 265% per annum. As seen in Figure 1, despite rapid growth, income from m-commerce will remain a very small portion of total data revenues (highest value 5.2% in 2006). This implies that in the short run m-commerce will not turn in profits to justify the investments in the new technologies that were initially believed to boost it. This applies to Europe and the United States. Asia, on the other hand, has developed more quickly with respect to m-commerce. Figure 2 illustrates the mobile operator data services revenue from 2001 to 2006 in Hong Kong and China.

The combined figures for Hong Kong and China show that the total mobile data services revenues are expected to increase for the period 2001-2006 by more than 70% per annum. The growth in m-commerce revenues in Hong Kong and China is currently 26.3% and is expected to outstrip 45% by 2006.

This outstanding acceptance of m-commerce in Asia is only partially due to the currently used mobile technologies that require fewer investments for the upgrade to 3G. The major drivers behind this trend are the habits of the Asians who are keener on innovative technologies, and the variety of content providers that attracts an increasing number of mobile services users.

Interestingly in Asia, the investments were restricted to the lower priced 3G licenses. The service providers, therefore, are able to offer their services at a much lower cost. Especially in China and Hong Kong, where mobile technology was introduced at a later stage and the 2G CDMA standard was adopted initially, the transition to 3G required little investments for the upgrade of the networks. This allows them to offer the services at a lower cost compared to the potential

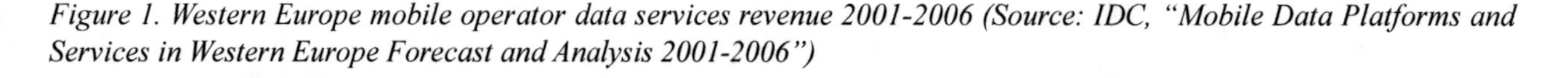

Figure 1. Western Europe mobile operator data services revenue 2001-2006 (Source: IDC, "Mobile Data Platforms and Services in Western Europe Forecast and Analysis 2001-2006")

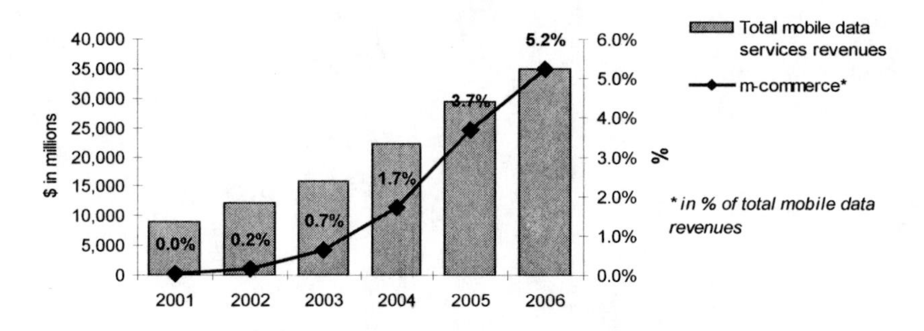

Figure 2. Hong Kong and China mobile operator data services revenue 2001-2006 (Source: IDC, "Asia/Pacific M-Commerce Forecast and Analysis: Opportunities Await")

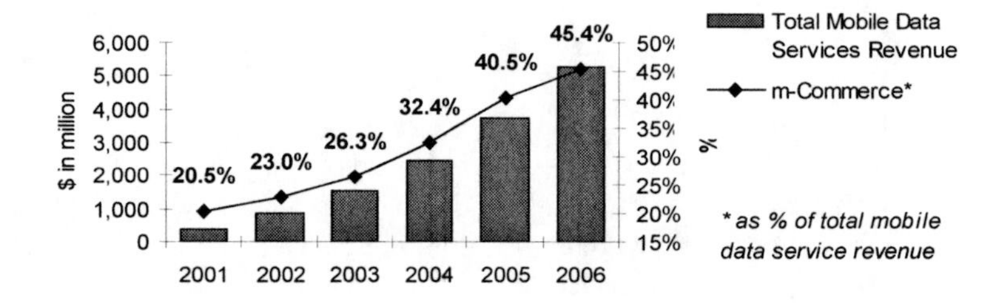

costs to the European consumers. The lower costs make it easier for consumers to try this new technology, become accustomed to, and ultimately adopt it.

The mobile market penetration in the U.S. is around 41%, far less than countries like Finland with 75%, Hong Kong with 89%, or the United Kingdom with 74% (Magura, 2003). Besides, only 6% of users in the U.S. use their mobile phones to access the Internet, and this is a much lower percentage compared with other countries like Japan with 72%, Germany with 16%, or the United Kingdom with 10% (Beal et al., 2001, p. 6).

CONDITIONS SPARKING DEVELOPMENT OF MOBILE COMMERCE

A large reason for the high level of acceptance for m-commerce is the large number of mobile phone users and Internet users around the world. The Internet has promoted electronic services, and m-commerce is another means of using the Internet. Customers now want to take advantage of Internet services from mobile end devices so that they can conduct business from any location in the world.

The highly lucrative industry of Internet commerce and mobile communication is a driving force in bringing many companies to develop this technology. The second catalyst is that many mobile phone users, especially in Europe and then in South East Asia, will be using smart cards, and this technology is another way to use cellular phones for business. Much of today's wireless e-commerce technology is a result of technology being developed by many of the mobile phone makers. The Europeans have led this charge since they have some of the highest numbers of cellular phone users. This is a result of the global economic and political environment during the 1980s, which promoted greater unification and collaboration, which helped the new telecommunications industries in Europe to flourish. As the need for better communication facilities grew, due to increased trade and investment flows, the solutions provided by the new technological developments become more viable.

EQUIPMENT USED IN MOBILE COMMERCE

This is an overview of the equipment necessary to conduct mobile commerce. It consists of digital cellular phones, smart cards, laptops, or palm pads, and the software to operate and communicate with this hardware.

Digital Cellular Phone

Dual-slot mobile phone technology was developed in Europe. Two key things are required for it to work: phones with the capability and chip-based cards. Dual-slot mobile phones offer a suite of value-added services, including mobile banking. It can be reprogrammed in the field and has substantial free memory for further applications. This allows subscribers to turn their mobile phones into tools to support their business and leisure lifestyle (Brokat, 2000).

To understand digital cellular technology, we must understand the background of the cellular phones as it relates to speed of data transmissions. Analog technology is considered to be the first generation of cellular technologies. The second generation of digital cellular technology is high-speed circuit-switched and packet-switched data technology. High-speed circuit-switched data technology uses a single voice channel and delivers data at a rate of 9.6 Kbps. Packet-switched means the computer that is connected to the cell phone sends and receives bursts, or packets, across the radio channel. The channel is occupied only for the duration of the data transmission instead of continuous transmissions, making it more efficient than circuit switched. The third generation (3G) of digital cellular technology refers to a much higher data transmission speed in the range of 14.4 Mbps. It will enable wireless multimedia applications such as videoconferencing. 3G is the collective term used for several engineering proposals to make wireless networks more data capable than first-generation analog and second-generation digital cellular networks. Some of the challenges of network speed and volume capability are addressed by these networks which must be able to transmit wireless data at 144 kilobits per second at mobile user speeds, 384 kbps at pedestrian user speeds, and 2 megabits per second in fixed locations (Schone, 2004).

Smart Cards

Smart cards are a cross between an ID card and electronic wallet. They can be used to store and exchange money from banks as well as support the payment functions of digital cellular phones. Already used in parts of Europe, this card provides many attributes that will enable technology to better serve consumers. Some present mobile communication in Europe relies on dual smart card technology. It consists of one smart card, internal to the cell phone, and one external card, which can hold personal information and be used as a cash card or electronic wallet, as well as phone card (Rundgren, 1999b).

Smart cards are a plastic ID card containing an integrated circuit chip that is capable of reading, writing, storing, and processing information. The size and shape of the plastic, the positioning of the chip, and its resilience to attack are defined by international standards. They cost between $2 and $20 depending on their capabilities. Multiple applications include contactless smart cards that can be read by radio signal from a card reader.

Smart cards have won the battle with magnetic-strip cards because of their security, reliability, capability, and lifetime cost. A contactless smart card ticketing solution is much cheaper in the long run. Capital investment can be 90% lower, revenues can be increased by 5% to 10% through lower fraud, and maintenance can be 30 times lower with a contactless smart card system. Smart cards are already used for public transport and parking services in cities in Europe and Asia. The most important advantages of smart cards are the capability and security that they offer.

One advantage of smart card IDs is they are extremely hard to forge. A PIN-code is added as an extra security

Figure 3. Dual slot mobile phone

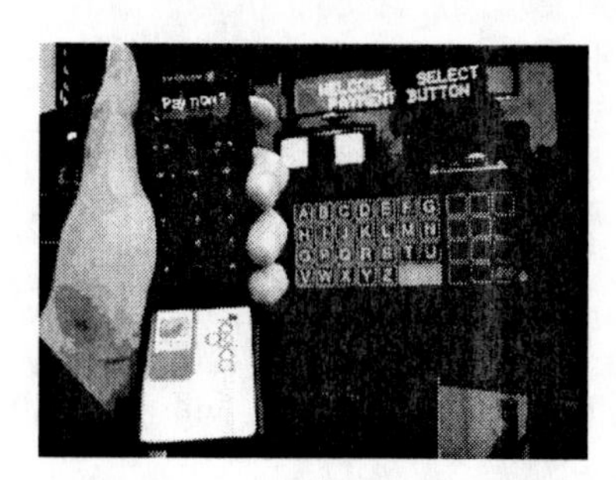

Figure 4. Smart card

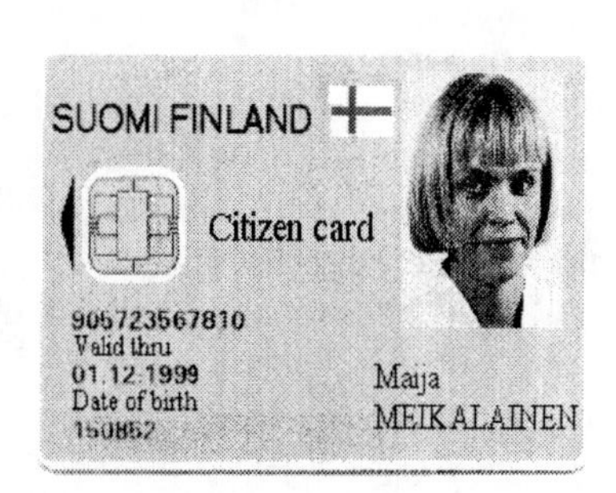

measure to avoid abuse if the card gets stolen or lost. An ordinary ID card can only be used for identification, while a smart-card-based ID card can also be used to digitally sign documents and transactions in a non-repudiateable way (Rundgren, 1999a).

Palm Pad Computer and Laptop Computers

Palm pad and laptop computers have become another means of doing business over the Internet. Primarily developed to be portable and used for computing on the go, they are now being used for communication and access of information from databases at other locations. They originally could connect to the Internet via a mobile phone and conduct business. While laptops are fairly expensive, the palm pads are less expensive in price and are becoming more common in mobile commerce.

Software

There are four basic components that make up a wireless Web service: browser phones, WML, link server, and services. Browser phones are handheld devices with special software that replaces conventional Web browsers. The WML (Wireless Markup Language) is a programming language consisting of a set of statements that defines what the browser phone displays in its window and how it interacts with the user. Instead of Web pages, the wireless world uses decks consisting of cards (Vujosesevic & Laberge, 2000).

In 1995, European telecommunication companies wanted a common platform and decided on Java, which has today become the development language of choice for advanced cellular mobile phone services under the global system for mobile communication (GSM) digital communication platform. The advent of Java for the smart card computing environment, standardized as JavaCard API (application programming interface), now offers the prospect of an open mobile platform: one that can store multiple applications, as well as delete, replace, and upgrade them over the air, at the point-of-sale, or via the Internet. This technology gives operators new freedom to forge links with content providers, as well as develop their own unique applications and services (Brokat, 2000).

Wireless application protocol (WAP) is a leading global standard for delivering information over wireless devices. WAP bridges the gap between mobile devices and the Internet, delivering a wide range of mobile services to subscribers independent of their network, bearer, and terminal. The WAP-framework will be useful in digital cell phones in two ways: as a low-level communication protocol, and as an application environment supporting a "mini-browser." WAP is similar to the combination of HTML and HTTP, but includes optimization for low-bandwidth, low-memory, and low-display capability environments necessary to deliver information to mobile devices (Schone, 2004).

ISSUES AND CHALLENGES

Wireless Constraints

Developing content for wireless devices requires rethinking the Web experience. Wireless content developers need to begin from the ground up developing content for these new devices. These devices tend to have very little real estate available for viewing content—often as small as 14 × 7 characters. Wireless devices also tend to be monochromatic, so images do not render well. Keyboards are difficult to use. Wireless devices tend to have limited CPU, memory, and battery life. Developers and designers need to find new, intuitive navigational techniques to overcome these constraints. Today, the most common navigational technique on wireless is the drill-down capabilities (Gutzman, 2000).

Another constraint of wireless capabilities is the amount of bandwidth available for use of data transmission. This new technology would put a greater burden on current bandwidths available for wireless transmissions. Alternate bandwidths must be opened for transmission.

Wireless User Behavior

Wireless users will not be expected to "surf the Web" in the traditional sense. This is due to the viewing and input constraints of using a wireless device and the relative inconvenience of performing any but the most straightforward, time-critical tasks. More likely, wireless users are expected to use their devices to execute small, specific tasks that they can take care of quickly, such as finding the time of local events, purchasing tickets, looking up news, or checking e-mail. Content developers need to develop with these motives in mind. Rather than just translating a content-rich site into WML, developers need to think in terms of surgical access to content and drilling down capabilities to detailed information in the site (Gutzman, 2000).

Larger screens have been developed for viewing, but use of magnification or projection techniques would be easier for users to view Internet content. Keypads designed for smaller appliances should be developed with small typing ability in mind.

Infrastructure for Wireless Internet

Currently, the infrastructure to handle smart cards is not generally established (except in Europe). Most industry analysts believe that smart cards will eventually become mainstream for paying in shops and on the Internet together with a PC. In many countries, smart ID cards will also become fairly widespread. One of the problems is that the cost for shops, banks, companies, homes, and PC owners to convert to smart cards makes the process fairly slow. There is an obvious risk that consumers, banks, and companies, after the initial WAP euphoria is gone, start to question the rationale behind having multiple payment systems and could begin to put pressure on the mobile-phone makers to force them to adapt their systems to the rest of the world. This is a very awkward solution because it sets unnecessary physical constraints for mobile phones and is also likely to need "software fixes" for each new card variant. Even when used over GSM, operators will simply be supplying a gateway to the Internet, which will be regarded as a standard part of a subscription (Rundgren, 1999c).

Security

Security of data transmissions and commerce being conducted by wireless devices is a great concern for businesses and individuals today. The wired Internet is vulnerable to hacking attacks. Individuals have been wary of using Internet commerce for fear of having their credit card being used improperly. A prerequisite for the success of m-commerce applications is the legal recognition and non-disputability of any transactions affected. The mobile digital signature may be an answer to this problem (Brokat, 2000).

New smart cards, available for wireless communications applications, will enable secure transactions via the Internet. The wireless identity module (WIM) will guarantee a new level of security by giving mobile Internet users the ability to safeguard their transactions through encryption and digital signatures. Compliant with the wireless application protocol (WAP), the WIM device will allow mobile network operators and service and content providers to begin implementing mobile commerce services such as secure information access, online banking, and the purchase of goods and services. The WAP-powered identity module supports "logical channels," enabling users to pass from one application to another without losing transactions that have already been carried out. The card offers two forms of protection: client-to-server authentication using ultra-long keys, and the ability to generate the digital signature required to secure the application. Unlike an encryption-enabled browser, the secret keys handling the encryption remain in the user's smart card, by definition a tamper-resistant device, and allow it to be removed and transferred to other devices (Electronic, 1999).

One advantage of smart card IDs is they are extremely hard to forge. To crack a private key stored in such a smart card or guess its value based on corresponding public key is very difficult. A PIN-code is added as an extra security measure to avoid abuse if the card gets stolen or lost. An ordinary ID card can only be used for identification, while a smart-card-based ID card can also be used to digitally sign documents and transactions in a non-repudiateable way (Rundgren, 1999a).

Privacy

Privacy is another issue not resolved by the growth of mobile commerce. The new connectivity of consumers to the Internet is a great convenience for consumers, but it also comes at a price. The price is the value of privacy that individuals lose as they become hooked up to the Internet. One part of privacy is that the development of smart cards for use with cell phones is convenient for consumers wanting to buy or sell. However, much personal data is enclosed on the card, and it could be used for the wrong purposes. Many cell phones can be equipped with a global positioning chip, which can identify the location of the user. This new technology would be good for emergencies, but could also be used against the individual for monitoring purposes or other activity. These are issues that still need to be addressed and have been downplayed by current technology developers. Privacy is one of several issues that complicate the long-term timetable for developing location-based m-commerce. Another issue is the level of direct access marketers will have to customers since the Internet will be located with the individual customer and can be contacted by voice, e-mail, or Internet (Vujovesic & Laberge, 2000).

FUTURE TRENDS

The value of transactions conducted over mobile phones in Europe is set to reach 23 billion euros ($23.7 billion) by 2003, according to a new study from Durlacher Research. Europe's mobile phone service operators are poised to increasingly derive revenue from Internet content and services, and will become leading Internet portals in the future. Europe has adopted a clear lead in usage and application development, fueled by its high penetration of mobile phones and successful adoption of GSM as the single digital phone standard. The U.S. has not been able to reach a single standard nor to settle on a generic type of terminal, thus slowing the establishment of a critical mass of handsets in the market needed for introduction of new services (Uimonen, 1999).

According to a recent study by IDC, the number of people using wireless devices to connect to the Internet will increase by some 728% by 2003. That is an increase from 7.4 million users in 1999 to 61.5 million users in 2003 (Blackwell, 2000).

Forecasters predict that in 2003 over half of all Web access will be from a mobile device, by which time consumers will be comfortable with m-commerce. In 2003 around a quarter of all mobile Internet users are likely to use their mobile phone to access travel services such as booking flights,

Figure 5. Worldwide mobile commerce use (Cahners In-Stat Group)

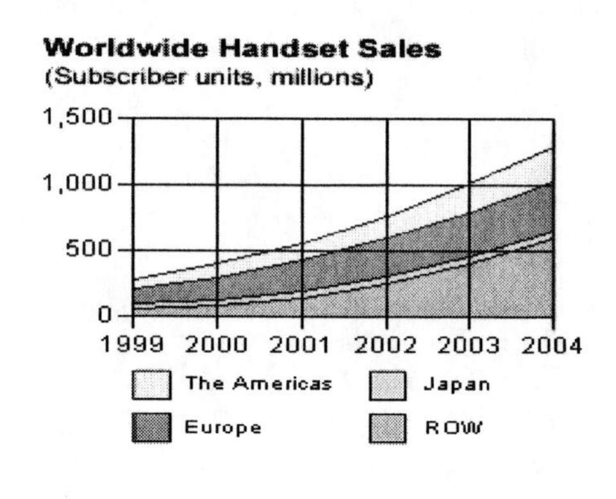

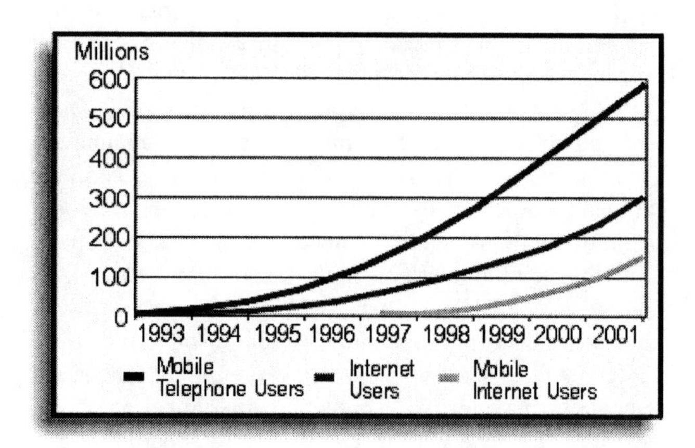

finding local hotel accommodations, sourcing last-minute holidays, or purchasing rail tickets. These are all services that are particularly suited to both business and experienced leisure travelers. This mix of mobile transactions is likely to result in mobile travel commerce revenues overtaking online travel (Cross, 2000).

Smart cards have always offered the potential to radically change and automate the way mass consumer business operates. The huge business potential presented by Java-based smart cards to service providers has not been lost to GSM-based mobile communications operators. Today, the cost of mobile communications is almost the same as fixed-line communications. Mobile phone penetration in Asia is expected to reach 35-40%. With the merging of electronic tools, such as the palmtop with the mobile phone, and more and more using the mobile phone for accessing data, the mobile phone complements the Internet to enable Web surfers to access information without needing a PC. Eventually, e-commerce will be a hot application for mobile phone users. The same should be expected for GSM-based mobile e-commerce. The capability of mobile phones is expected to increase and this will accelerate more developments (Brokat, 2000).

IMPLICATIONS FOR MANAGEMENT

Consumer-oriented m-commerce is becoming a reality today. Many businesses and consumers are taking up the wireless Web through many services such as AT&T, Sprint PCS, Verizon Wireless, Motorola, Nokia, Ericsson, and other wireless service providers. It will still take a few years for consumer-oriented m-commerce applications to become as generally available as the wired Web is today. In the near term, the most promising opportunities for mobile wireless transactions are those built for industrial use. These involve the development of software for vertical applications that allow delivery agents, salespeople, and mobile workers to perform logistical and other data-driven duties. Doctors are using wireless-enabled palmtops to access and update patient records or write prescriptions. The most visible wireless developments are consumer oriented. These include delivery of time-critical information to mobile banking and travel-ticket purchase. The current crop of consumer-oriented services will become the foundation for more advanced services that will deliver time- and location-critical data to consumers (Vujovesic & Laberge, 2000).

Mobile commerce is a reality now and will not be going away in the near term. Problems with these systems are being addressed and new applications are being developed rapidly. Most of the market is behind this new technology, and it will likely change business by making it easier and more accessible to individuals. Mobile commerce is a tool of telecommunication and Internet industries. Those who are involved in it now may become the giants that Microsoft and Intel have been in the computing industry. Those who follow may be able to capitalize on leading m-commerce mistakes and perfect this technology. One should measure the risk involved and understand it must still be a carefully planned strategy to implement this new technology into corporate future goals.

CONCLUSION

The development of mobile commerce is the evolution of several different technologies to make the Internet more accessible and commerce easier for the consumer. While the Internet is already a valuable form of business, which has already changed the way the world is doing business, the format in which we will view it is changing. There are many new opportunities that have only begun to be explored. This will become a large opportunity for those who capitalize upon this technology. The growth trends are impressive, and the public interest and large companies are behind this technology.

Time will tell whether it is the treasure that most have touted it to be. If it is the "cash cow" most are looking for, then for those who trail or lag behind, the leaders may not be in business in a few years. If it turns out to be a bust, then those who invested so heavily in this technology will find their efforts wasted. The risk is very high in this new development, but the benefit is that consumers seem to be driving the demand for mobile commerce.

The application of this technology is the true seller. Its success is contingent upon a majority of Internet browsers using mobile digital phones. To be fully accepted, all these technologies must overcome their current drawbacks. Technology is being developed to overcome the security drawbacks, but enhanced viewing devices and input devices for controlling the data must be developed. Also, the infrastructure to control smart card payments may be a few years off for the U.S., but it will need to be accepted at shops and businesses throughout the U.S. to make it useful. Mobile e-commerce will change our world of business to a similar degree that the Internet alone has changed business today.

REFERENCES

Blackwell, G. (2000). *Wireless to outstrip wired Net access.* Retrieved from http:/www.isp-planet.com/research/more_wireless.html

Beal, A., Beck, J. C., Keating, S. T., Lynch, P. D., Tu, L., Wade, M., et al. (2001). *The future of wireless: Different than you think, bolder than you imagine.* Retrieved June 4, 2004, from http://www.accenture.com/xd/xd.asp?it=enWeb&xd=_isc/iscresearchreportabstract_134.xml

Cross, T. (2000). *Mobile travel commerce—A bigger deal than online travel?* Retrieved from http:/www.gmcforum.com/PressRelease/PressRelease_110500.htm

Durlacher Research. (1998). *Mobile electronic commerce.* Retrieved from http://network365.com/mobilecommerce.html

eTForecasts. (2006). *Internet user forecast by country: Wireless Internet users.* Retrieved on March 26, 2006, from http://www.etforecasts.com/products/ES_intusersv2.htm#1.0

Electronic Buyer's News. (1999). *Schlumberger says new smart card will ensure secure mobile-Internet transactions.* Retrieved from http:/www.ebns.com/ecomponents/commnews/story/OEG19991116S0008

Gartner Research. (2004, October 29). *Predictions 2005: Mobile and wireless technologies.* Retrieved July 3, 2005, from http://www.analysphere.com/13Aug01/wireless.htm

Gutzman, A. (2000). *The who, what and why of WAP.* Retrieved from http:/www.allnetdevices.com/wireless/opinions/2000/06/20/the_who.html

Hansen, C. (2000). *GSM-based mobile e-commerce will be hot.* Retrieved from http:/www.globalsources.com/MAGAZINE/TS/9909/SLB.HTM

Intel. (1999). *The future GSM data knowledge.* Retrieved from http:/www.gsmdata.com/Future.html

Magura, B. (2003). What hooks m-commerce customers? *MIT Sloan Management Review,* (Spring), 9.

Mobile Business. (2000). *Brokat global e-commerce services.* Retrieved from http:/www.brokat.com/int/mobile/index.html

Muller, J., & Schnoring, T. (1995). *Mobile telecommunications: Emerging European markets* (p. 247). Artech House Publishers.

Reiter, A. (1999a). *Dynamics of wireless e-commerce, conditions sparking the development of international wireless e-commerce.* Retrieved from http:/www.wirelessinternet.com/dynamics.htm

Reiter, A. (1999b). *Wireless e-commerce: A new business model.* Retrieved from http:/www.wirelessinternet.com/wireless2.htm

Rundgren, A. (1999a). *ID-cards: Yesterday, today and in the future.* Retrieved from http:/www.mobilephones-tng.com/papers/idcards.html

Rundgren, A. (1999b). *The cyber ID card.* Retrieved from http:/www.mobilephones-tng.com/v100/cyberphonecards.html

Rundgren, A. (1999c). *The new Swiss Army Knife? (Smart cards vs. smart terminals).* Retrieved from http./www.mobilephones-tng.com/papers/thenewswissarmyknife.htm

Rundgren, A. (1999d). *WAP—Wireless Application Protocol.* Retrieved from http:/www.mobilephones-tng.com/v100/wap.htm

Schone, S. (2004). *Computer Technology Review, 24*(10), 1, 38.

Strategis Group. (2005). *Mobile computing outlook.* Retrieved July 5, 2005, from http://www.mobileinfo.com/Market/market_outlook.htm

Uimonen, T. (1999). *European mobile commerce to hit $24 billion.* Retrieved from http:/www.durlacher.com

Varshney, U., & Vetter, R. (2000). Emerging mobile & wireless networks. *Communications of the ACM, 43*(6).

M

Vujosevic, S., & Laberge, R. (2000). *Info on the go: Wireless Internet database connectivity with ASP, XML, and SQL server.* Retrieved from http:/www.msdn.microsoft.com/msdnmag/issues/0600/wireless/wireless.asp

Walker, M. (2000). *M-commerce tricks emerge from tech magician's bag.* Retrieved from http:/www.bizjournals.com/houston/stories/2000/06/26/focus6.html

Wireless LAN Association. (2005). *Wireless LAN ROI.* Retrieved June 25, 2005, from http://wlana.org/learn/roi.htm

KEY TERMS

Mobile Commerce (M-Commerce): The delivery of electronic commerce capabilities directly into the consumer's hand via wireless technology and putting a retail outlet in the customers' hand anywhere.

Smart Card: A cross between an ID card and an electronic wallet. It can be used to store and exchange money from banks, as well as support the payment functions of digital cellular phones.

Wireless Application Protocol (WAP): A leading global standard for delivering information over wireless devices. It bridges the gap between mobile devices and the Internet, delivering a wide range of mobile services to subscribers independent of their network, bearer, and terminal.

Wireless Identity Module (WIM): Guarantees a new level of security by giving mobile Internet users the ability to safeguard their transactions through encryption and digital signatures.

Wireless Web Service: The four basic components that make up a wireless Web service are browser phones, WML, link server, and services.

Mobile Enterprise Readiness and Transformation

Rahul C. Basole
Georgia Institute of Technology, USA

William B. Rouse
Georgia Institute of Technology, USA

INTRODUCTION

Recent studies claim that mobile information and communication technologies (ICT) offer a plethora of new value propositions and promise to have a significant transformational impact on business processes, organizations, and supply chains (Kornak et al., 2004). However, despite its potential contributions, enterprise adoption of mobile ICT has not been as widespread as initially anticipated. Previous research has argued that successful adoption and implementation of any emerging ICT, such as mobile ICT, often requires fundamental changes across an enterprise and its current business practices, organizational culture, and workflows (Taylor & McAdam, 2004; Rouse, 2006). Hence, in order to minimize organizational risks and maximize the potential benefits of mobile ICT, enterprises must be cognizant of the value of enterprise mobility to their organization and accurately evaluate their level of "readiness" for mobile ICT adoption (Hartman & Sifonis, 2000; Ward & Peppard, 2002). This paper reviews the transformational value and impact of enterprise mobility and explores the critical dimensions for determining an enterprise's readiness for mobile ICT. Both theoretical and managerial implications are discussed.

BACKGROUND

Over the past few years mobile ICT have advanced at a tremendous pace making an always-on connection, anywhere and anytime, a growing reality. The rapid proliferation of mobile devices has led to an increasingly mobile society in which users now expect to have instant communication means, data access, and commerce capabilities. A similar trend has also seeped into the enterprise domain. The use of mobile ICT in enterprises has evolved from being simplistic point solutions and small projects focused on productivity improvements and costs savings to strategic and large-scale enterprise-wide implementations that enable organizations to create new core competencies, gain and sustain competitive advantages, and define new markets (Davidson, 1999; Kornak et al., 2004).

The Mobile Enterprise

So what is a mobile enterprise? Simply deploying laptops so employees can take work home does not constitute a mobile enterprise. Pundits have argued that a slight increase in mobility that a laptop affords amounts to little more than a very small geographic extension of the existing static enterprise. Similarly, a mobile enterprise is not merely a collection of people with handheld devices, smart phones, tablet PCs, and pagers. Many enterprises already have such a workforce, however, it often does not change how those people work with each other and the rest of the organization. Therefore, bolting a group of mobile workers onto an organizational chart does not create a new organization and often does very little to enhance the existing one. However, the more mobile workers an organization has, the greater will be the need to transform at least part of that company into a mobile enterprise. More specifically, it will require a rethinking of how business is organized, how people interact and collaborate, how corporate resources are accessed, and how adaptable an enterprise is (Barnes, 2003; Rouse, 2005). Building on this notion, we propose that mobile enterprises exhibit higher levels of access, interaction, and adaptability than their static counterparts do. In visual terms, static enterprises tend to exist in spheres closer to the origin (see Figure 1). The further the sphere is from the origin, the higher the

Figure 1. The dimensions of the mobile enterprise

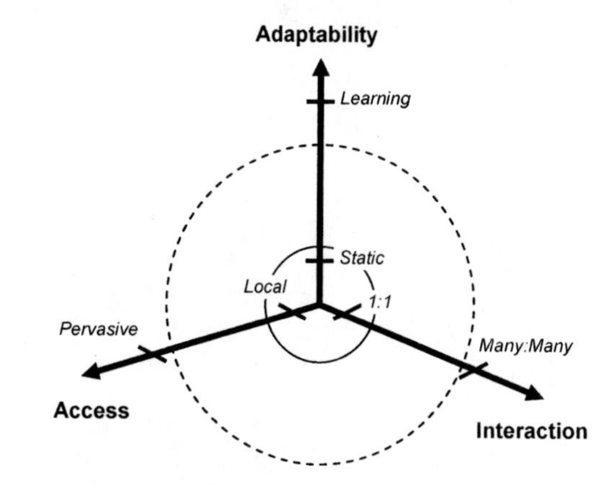

level of enterprise mobility. Thus, independent of location, the mobile enterprise is built on a foundation of processes and technologies allowing full access to organizational resources, which results in improved adaptability, access, and interaction among employees, customers, partners, and suppliers (Basole, 2005).

Benefits of Enterprise Mobility

With this understanding of the mobile enterprise, it therefore becomes more transparent what benefits mobile ICT can offer. The ability to access the corporate network and resources anywhere and anytime is one of the primary benefits and key drivers to adopting mobile enterprise solutions. Field workers are no longer tied to desktop computers to check mission- and task-critical data. The use of mobile ICT enables workers to receive timely answers, which in turn can lead to timely decisions. Enterprise mobility solutions also offer the potential of achieving significant cost savings. Expensive computing equipment can be replaced with smaller, more portable, and less expensive handheld devices. Field workers can use these devices to be immediately connected to all the sources they need. Furthermore, replacing paper-based processes with mobilized applications reduces the potential for errors in transferring information to a call report or clinical chart, leading to a higher level of data accuracy and integrity, which in turn can be harvested for overall business intelligence use. Better access to corporate resources—both data and people—naturally leads to a higher level of productivity, as mobile workers are able to view data that allows them to respond and execute faster to changing market conditions.

ENTERPRISE TRANSFORMATION THROUGH MOBILE ICT

Mobilizing enterprise applications and providing business professionals access to information anywhere and anytime is clearly an important first step in gaining business value (Barnes, 2003; Kornak et al., 2004); however these gains are only the beginning. We argue that enterprises can realize a much broader range of benefits over time by pursuing a multi-stage mobile transformation process. Research has shown that ICT have the ability to change and fundamentally transform enterprises in a number of ways (Basole & DeMillo, 2006; Rouse, 2006). This transformational impact can be primarily experienced and realized at the strategic, operational and organizational culture level (Taylor & McAdam, 2004). Indeed, the impact of mobile ICT is far beyond mere business process improvements and enhancements (Davidson, 1999; Kornak et al., 2004; Basole, 2005). Extending this previous work, four distinct stages of mobile enterprise transformations are proposed (see Figure 2).

Mobilization (Stage 1)

The first stage of the transformation process begins with the mobilization of existing data and applications. Mobilization refers to the process of making current business data, pro-

Figure 2. Stages of enterprise transformation through mobile ICT

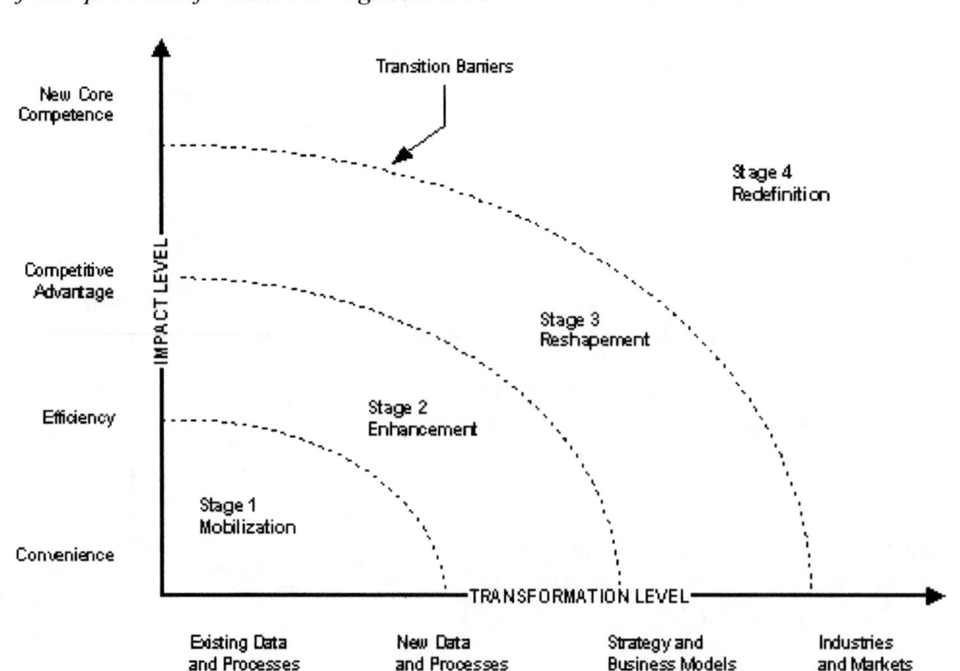

cesses, and applications available for use on mobile/wireless devices. The first stage aims to provide end-users with a new level of convenience by enabling access to resources anywhere and anytime. Examples include access to corporate e-mail, the Intranet, and other data and human resources. Generally, Stage 1 solutions will lead to higher levels of convenience and generate significant performance gains in productivity, speed, efficiency, quality, and customer service (Kornak et al., 2004).

Enhancement (Stage 2)

The second stage shifts its focus from mobilizing existing data and applications to enhancing existing and creating new business processes that leverage the unique functionalities and capabilities of mobile ICT (Barnes, 2003). Characteristics of these business processes generally include two elements, namely (1) mobility (do it anywhere) and (2) immediacy (do it now), all with the user's context in mind. While solutions in the enhancement stage may affect working practices and modify business processes, they seldom change the business in a fundamental manner. This level of transformation occurs in Stage 3 of the mobile transformation process.

Reshapement (Stage 3)

As enterprises transition to Stage 3, mobile ICT begin to reshape business models and strategies. The creation of innovative new mobile processes and services provide enterprises with a source of competitive advantage. In this stage, mobile ICT often enable a business capability and become a critical element in the overall business model. For example, wireless sensors could enable a pharmaceutical company to shift from selling only medication to a business model in which the company provides both medication and sensors, and enters into a contract with a medical practitioner to perform continuous monitoring and keep a patient's blood pressure within an agreed range.

Redefinition (Stage 4)

In the fourth and final stage of the transformation process, mobile ICT create entirely new core enterprise competencies. Business models and strategies are based and revolve around enterprise mobility and in turn lead to a redefinition of entire markets and industries. Concrete examples for this stage of the mobile transformation process have not emerged yet; however, as enterprises continue to embrace mobility and mobile ICT mature, mobile redefinition is expected to become an increasingly common business phenomenon.

The four stages of mobile enterprise transformation are not purely sequential. Activities performed during Stage 1 continue during Stages 2-4. Some companies may elect transitioning directly from Stage 1 to Stage 3. New ventures may begin their business models based on Stage 2 philosophies. Stage 4 examples are still scarce, but are poised to emerge as mobile ICT continue to mature and new business models take shape. Yet, all four stages are inextricably linked in significant ways. Diligent pursuit of Stage 1 initiatives will lead to many Stage 2 and 3 opportunities. Similarly, Stage 4 opportunities will emerge as enterprises realize the full transformational potential of mobile ICT solutions.

Adoption and Transition Barriers

Enterprises that undergo significant organizational changes generally encounter a number of transition barriers. Empirical evidence suggests that these barriers can be broadly categorized as economic/strategic, technological, organizational, and environmental-related issues (Taylor & McAdam, 2004).

Despite tremendous advances, mobile ICT are still in their infancy stage. Evolving standards, lack of technology maturity, and issues of compatibility with existing systems and infrastructure are causing organizations to delay mobile ICT implementation (Basole, 2005). Another prevalent barrier is related to the ongoing debate of business value and cost. Investments in emerging technologies such as mobile ICT often require significant financial commitments by the enterprise. With shrinking IT budgets, it becomes critical to understand what value enterprise mobility can deliver now, and in the future. Mobile ICT implementations must thus be aligned with the overall business strategy and support enterprises' current and future business objectives. Similarly, the availability of other organizational resources – such as human and technical support—must be in place in order to successfully adopt mobile ICT and transition across the stages. From an organizational perspective, enterprise culture, size and structure also play a critical role in the adoption and transition process. As with most new ICT implementations, end-users often show resistance to new processes and change. Mobile ICT will have a radical, and potentially transformational, impact on the way work is done; hence, particular attention to end-user needs, education, motivation and incentives must be provided to ensure a successful adoption, implementation, and transition. Lastly, unfavorable market conditions, strong regulatory influences, lack of customer and supplier pressure, and inadequate vendor support may also inhibit organizational adoption of mobile ICT.

In summary, in order to avoid a "fragmented" mobile ICT adoption and transformation, enterprises should determine the fit between the value of mobile ICT and the overall business strategy, and ensure that a common vision, leadership support, and strategic path for implementing enterprise mobility solutions is in place (Ward & Peppard, 2002).

ENTERPRISE READINESS FOR MOBILE ICT

The previous discussion highlights the complexity of mobile ICT adoption and implementation, as well as the transition across the four transformation stages. It also underlines the fact that successful assimilation will require significant organizational changes, its current business practices, culture, processes, and workflows. While previous studies identified the potential benefits, challenges, and drivers for enterprise adoption of mobile ICT, exact enterprise benefits and ROI of mobile ICT implementations may not be known for the near future. As ICT budgets have decreased and failure rates for new ICT implementations have continued to rise over the past years, a smaller tolerance to ICT failures has emerged. The hype of potential benefits often drives enterprises to jump onto the "fad" bandwagon and rush into ineffective implementations of new ICT. In contrast, however, a range of studies have shown that many potentially successful IT projects fail due to a lack of assessment of potential barriers and organizational risks associated to the implementation of new ICT.

In order to minimize the associated risks and maximize the potential benefits of enterprise mobility solutions, organizations must thus not only understand the value and economics of enterprise mobility solutions, but also carefully evaluate and measure their level of "enterprise readiness" for mobile ICT (Hartman & Sifonis, 2000; Basole, 2005). Readiness assessment enables decision makers to become more knowledgeable about the characteristics of mobile ICT, form attitudes about it, and make a decision regarding the fit between the technology and the organization (Hartman & Sifonis 2000). It also enables decision makers to determine whether enterprises can truly benefit from mobile ICT and take appropriate measures to steer the organization towards a successful adoption and mobile transformation transition.

Defining Enterprise Readiness for Mobile ICT

The concept of enterprise readiness for ICT has received very limited attention in both the academic and business press literature. Preparedness, agility, and maturity are often some terms commonly associated with enterprise readiness. However, anecdotal evidence has shown that higher levels of enterprise readiness generally lead to lower levels of innovation risk and more successful implementation outcomes. A similar argument can be transposed to the context of mobile ICT: higher levels of mobile ICT readiness leads to lower organizational risks and implementations that are more successful.

So what constitutes enterprise readiness for mobile ICT? Extending previous theories of organizational readiness and technology adoption (Hartman & Sifonis, 2000; Taylor & McAdam, 2004), we postulate the following definition:

Enterprise readiness for mobile ICT is an assessment of an organization's (1) preparedness, (2) potential, and (3) willingness to adopt and implement mobile ICT.

More specifically, preparedness refers to an organization's ability to adopt, diffuse, and assimilate mobile ICT; potential refers to an organization's processes, employee, and strategy that could benefit from mobile ICT; and willingness reflects the attitudinal orientation of leadership and employee towards adopting mobile ICT.

Dimensions of Enterprise Readiness

We further argue that enterprise readiness for mobile ICT is comprised of eight dimensions: (1) technology, (2) data and information, (3) process, (4) resource, (5) knowledge, (6) leadership, (7) employee, and (8) values and goals. A complete enterprise readiness assessment will thus involve an evaluation across the three layers—preparedness, potential, and willingness—and along all eight readiness dimensions (see *Figure 3*). Preparedness is assessed for all eight dimensions; potential is evaluated along the process, employee, and value and goals dimensions; and, willingness is assessed along the employee and leadership dimensions.

Theoretical and practical support for each of the eight dimensions and associated assessment indicators is provided as follows:

- **Technology Readiness:** Technology readiness refers to the ability of the underlying technology infrastructure (network services, hardware, software, and security) to support the adoption and implementation of mobile ICT. A robust, comprehensive, and open-standards oriented technological infrastructure, flexible and scalable to accommodate any change and emerging requirements, leads to a higher level of technology readiness.
- **Data and Information Readiness:** Data and information readiness refers to the ability to federate data from multiple sources, provide a single view of enterprise data, and make it available to any system at the time when it is needed. Higher levels of data and information readiness are achieved through a consistent, reliable, and secure data and information infrastructure that provides both synchronization and data recovery capabilities for highly disconnected and variable environments.
- **Process Readiness:** Process readiness refers to the ability of organizational processes (e.g., human processes, information processes, organizational change processes, etc.) to facilitate the adoption and imple-

Figure 3. Dimensions and layers of enterprise readiness for mobile ICT

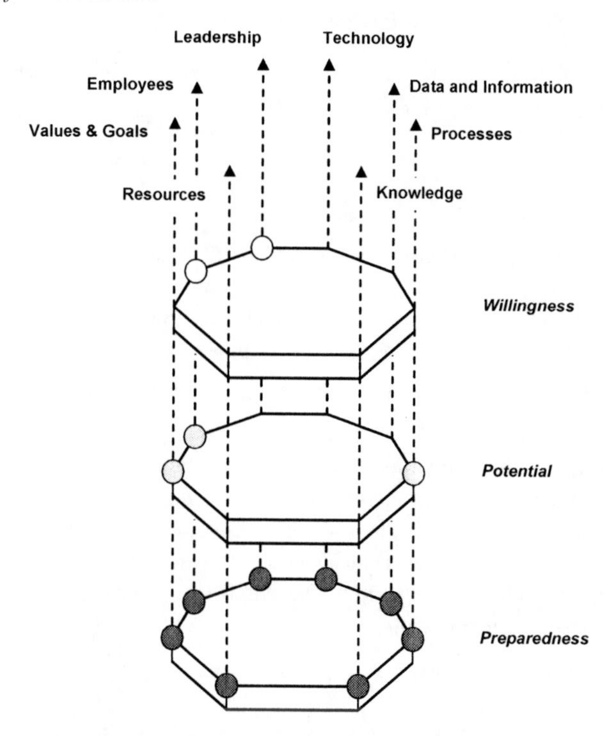

mentation of mobile ICT. Well-defined, documented, managed, repeatable and optimized processes indicate a high level of readiness along this dimension.

- **Resource Readiness:** Resource readiness represents an organization's ability to support mobile ICT adoption and implementation. These resources may include (1) financial, (2) human, and (3) technical assets. The availability of resources for current and future plans is an important aspect in successful assimilations of mobile ICT.
- **Knowledge Readiness:** Knowledge readiness reflects both the general and specific knowledge required by decision makers for mobile ICT adoption and implementation. General knowledge includes awareness and understanding of the state of emerging ICT, ICT-related decision-making processes, and previous experiences with ICT adoptions and implementations. Specific knowledge encompasses an awareness and understanding of the opportunities, challenges, barriers, and opportunities that come with the adoption and implementation of mobile ICT. This will includes an understanding of mobile ICT characteristics, its potential impact on strategy, processes, and people, and the changing enterprise mobility market.
- **Leadership Readiness:** Previous studies have shown that one of the most critical factors in technology adoption decisions is the support and vision of top management. Leadership readiness, hence, reflects an appropriate level of skills, innovativeness, knowledge, and risk orientation of top management. It also indicates the level of support and strategic vision that management offers in association to the adoption and implementation of mobile ICT. Leadership needs to ensure that mobile strategies fit with the way they are doing business rather than changing their ways of doing business to fit the strategy.
- **Employee Readiness:** Employee readiness reflects the end-users attitude towards change, their level of skills, and perceived benefits by the end-users. A high level of employee readiness can lead to a faster adoption and diffusion of mobile ICT.
- **Values and Goals Readiness:** Values and goals readiness reflects the fit between existing structural and nonstructural enterprise characteristics and mobile ICT characteristics. Structural characteristics may include organizational size, centralization, formalization, autonomy, specialization, functional differentiation, strategic objectives and goals. Nonstructural characteristics may include culture, bureaucracy, task environment, and political climate.

It should be noted that all of these dimensions have an influence on each other and must therefore be considered as a whole. A lack in one dimension may influence the overall enterprise readiness for mobile ICT. Similarly, a lack of readiness in one of the three layers will also result in a lower degree of enterprise readiness. As such, a comprehensive assessment of all dimensions on all layers should be conducted.

FUTURE TRENDS AND CONCLUSION

Enterprise mobility is not merely a fad; it has become a reality in a wide-range of organizations and industries. Mobile ICT clearly offers a plethora of lucrative value propositions that will impact and fundamentally transform business processes, organizations, and supply chains. As mobile ICT continues to evolve and mature, enterprises must prepare themselves for a more "mobile" future. In order to minimize organizational risks and maximize the potential benefits of mobile ICT adoption and implementation, is therefore of utmost importance in order to assess the level of enterprise readiness for mobile ICT. We further argue that an understanding of the transformational stages can provide decision makers with a strategic map of the potential impact of mobile ICT. An assessment of an enterprise's preparedness, potential, and willingness in conjunction with an understanding of the transformative influence will enable decision makers to make more objective judgments on why and when to adopt mobile ICT, aid in the formulation of appropriate mobility strategies, and enable successful transformations.

REFERENCES

Barnes, S. J. (2003). Enterprise mobility: Concept and examples. *International Journal of Mobile Communications, 1*(4), 341-359.

Basole, R.C. (2005). Mobilizing the enterprise: A conceptual model of transformational value and enterprise readiness. In *Proceedings of the 26th American Society of Engineering Management*. Virginia Beach, VA.

Basole, R. C., & DeMillo, R. A. (2006). Enterprise IT and transformation. In W. B. Rouse (Ed.), *Enterprise Transformation: Understanding and Enabling Fundamental Change* (pp. 223-237). New York: Wiley.

Davidson, W. (1999). Beyond re-engineering: The three phases of business transformation. *IBM Systems Journal, 38*(2/3), 485.

Hartman, A., & Sifonis, J. (2000). *Net ready: Strategies for success in the e-conomy*, McGraw-Hill.

Kornak, A., et al. (2004). *Enterprise guide to gaining business value from mobile technologies*. New York: Wiley.

Rouse, W. B. (2005). A theory of enterprise transformation. *Systems Engineering, 8*(4), 279-295.

Rouse, W. B. (2006). *Enterprise transformation: Understanding and enabling fundamental change*. New York: Wiley.

Taylor, J., & McAdam, R. (2004). Innovation adoption and implementation in organizations: A review and critique. *Journal of General Management, 30*(1), 17-38.

Ward, J., & Peppard, J. (2002). *Strategic planning for information systems*. New York: Wiley.

KEY TERMS

Enterprise Readiness for Mobile ICT: Enterprise readiness for mobile ICT is an assessment of an organization's (1) preparedness, (2) potential, and (3) willingness to adopt and implement mobile ICT, and is an essential element for successful enterprise transformation through mobile ICT.

Enterprise Transformation: Enterprise transformation concerns fundamental change that substantially alters an organization's relationships with one or more key constituencies, and can involve new value propositions in terms of products and services, how these offerings are delivered and supported, and/or how the enterprise is organized to provide these offerings.

Mobile ICT: Mobile ICT refers to all mobile information and communication technologies, including network infrastructure (e.g., WiFi), devices (e.g., smart phones, laptops, PDAs), and mobile applications.

Mobile Entertainment

Chin Chin Wong
British Telecommunications (Asian Research Center), Malaysia

Pang Leang Hiew
British Telecommunications (Asian Research Center), Malaysia

INTRODUCTION

Mobile commerce is forecasted to be a significant growth market in leading countries. This high growth estimate of mobile phones is leading investors to take special interest in device manufacturing, provision for future innovations, and system management areas. Mobile commerce services can be adopted through different wireless and mobile networks, with the aid of several mobile devices (Andreou et al., 2002). Mobile commerce opens a new evolutionary era in global business (Maharramov, 1999). In mobile business there will be no need for international custom regulations that vary from country to country, therefore it is business without borders (Maharramov, 1999).

Mobile entertainment is a newly emerging subset of mobile commerce. A primary difficulty when researching mobile entertainment is that of definition (Moore & Rutter, 2004). It is recognized that, as mobile entertainment is a social and commercial process as well as a technical one, a diversity of other definitions for mobile entertainment is held by numerous industry producers, manufacturers, and end users, as well as researchers of dissimilar background (Moore & Rutter, 2004). It is noteworthy to rethink and redefine mobile entertainment, as it is more complex than other subsets of mobile commerce.

The problem of producing common understandings of mobile entertainment has been previously highlighted by the Mobile Entertainment Forum (MEF) when stating that two different industries make up the mobile entertainment industry: entertainment and telecommunications (Wiener, 2003). Mobile entertainment is created as the convergence of both industries. Each of these worlds speaks a different language and holds different assumptions about the nature of its work. Recent research demonstrates that many consumers are unclear about the mobile entertainment and related wireless technology options available to them. For example, a Packard Bell-sponsored survey of nearly 1,000 British home personal computer users found that 70% of the respondents did not know what Wi-Fi was (MORI, 2003).

Mobile entertainment represents one of the few mobile services that has mass market potential that will drive the adoption of the next generation of mobile devices (Ollila, Kronzell, Bakos, & Weisner, 2003). Proper classification of mobile entertainment services enable players in the value web to adopt suitable business models to bring services to market and how they should cooperate, share revenue, and jointly create competitive advantages.

This article presents a framework to examine mobile entertainment from multiple points of views concerning the service, network, and device-related sectors. This allows future research to be conducted with the clarity of distinguishing mobile entertainment services of different domains. The article also tries to collate and rationalize possibilities and restrictions of existing and emerging mobile entertainment technologies with respect to this framework. The study explores a number of scenarios to reflect the understanding on the value web. This study serves as a foundation for further studies in the area of mobile entertainment.

BACKGROUND

Travish and Smorodinsky (2002) as well as Kalyanaraman (2002) define mobile entertainment as services that offer gaming experiences on-par with those to be had in other mediums such as Xbox and PlayStation 2. On the contrary, it is of the authors' opinion that mobile entertainment services are more than merely games. Besides, the definition does not cover what constitutes mobile games. For example, if one considers games deployed on laptop and Game Boy as mobile games, a similar development approach could not be taken to launch mobile games on mobile phones because, generally, mobile games development on mobile devices should take into consideration key characteristics such as short session time, fresh content, continuous and reliable availability, cultural compliance, and so forth (Kalyanaraman, 2002). Furthermore, a game that is installed on a laptop cannot be installed on a mobile phone due to dissimilar platforms.

In other literature, Ollila et al. (2003) assume mobile entertainment includes any leisure activity undertaken via a personal technology, which is or has the potential to be networked, and facilitates transfer of data over geographic distance either on the move or at a variety of discrete locations. While workable, the definition does not cover whether mobile entertainment services must interact with service providers. It does not cover whether such service would

Table 1. Terminology of mobile entertainment (Wiener, 2003)

Terminology	Definition
Platform Vendor	Develops, implements, manufactures, supplies, and supports standard or customized platforms to the platform operator.
Service Provider	Brings content to the end user, undertakes the commercial and regulatory obligations that accompany the provision of service; does not involve the operator of the service.
Mobile Network Operator	Provides the infrastructure for mobile communications: the service, billing, and customer care.
Publisher	Refers to any company or individual that allows for the "publishing" of a piece of content; typically assumes the financial risk for the creation of the content; maintains control of all aspects of the entertainment service, including rights management and payment, user-service interaction, multi-user interaction, and user-per-service preferences.
Retailer	Delivers services to end users. In the mobile industry the retailers are either specialized for mobile services or mass retailers. Entertainment retailers are usually mass retailers.
Developer	Performs application development.
Subscriber	Refers to the end user or consumer of mobile entertainment services.

incur a cost upon usage. If mobile entertainment was said to be a subset of mobile commerce, it must therefore involve transaction of an economic value. The social aspects of mobile entertainment are hidden within the phrase "any leisure activity" (Moore & Rutter, 2004).

From a business perspective, various literatures attempt to classify the mobile entertainment value web by referring to its players within the industry. For example, Wiener (2003) asserts that to help all participants in this industry collaborate, clarification of how each industry defines the nature of its work is necessary. The goal is to offer a set of common definitions of typical industry players and various mobile entertainment roles for the interfaces between the businesses (Wiener, 2003). In another paper, Camponovo (2002) classifies the players in the value web based on technology, services, network, regulation, and user. A summary of the findings is concluded in Table 1.

A search on Google on the term *mobile entertainment* reveals that even everything portable, including DVD player, television, radio, external player, MP3 player, amplifiers, speakers, as well as woofers and so forth, are considered devices of mobile entertainment. This proves that confusion with regards to the definition of mobile entertainment is common among stakeholders of the value web.

Mobile entertainment comprises a range of activities including but not limited to downloading ring tone, logo, music, and movies; playing games; instant messaging; gambling; accessing location-based entertainment services; and Internet browsing. Hitherto, the list is constantly expanding.

REDEFINING MOBILE ENTERTAINMENT

In this section, the authors briefly explain the three different segments and come up with a model that is believed to be useful in the development of end user models and consumer scenarios. Subsequently, players in the mobile entertainment value web may improve their understanding of the consumers and their usage scenarios. This will make them perform better evaluations of the likelihood of adoption, and will improve their foundation for designing, evaluating, and timing mobile entertainment end user services (Pedersen, Methlie, & Thorbjørnsen, 2002).

In essence, taxonomy is a system of classifications. To put the framework into use, a few examples will be discussed in this section. The purpose of this section is to present a classification of these segments to identify relevant categories of mobile entertainment services for this study.

Scenario 1: Downloading Music onto Mobile Devices

A mobile user connects to the Internet via his 3G-enabled mobile phone, searches for a particular song, and downloads it onto his mobile phone. This falls under segment 1 where this activity utilizes wireless telecommunication networks, incurs a cost upon file download, interacts with the service provider, and is a form of leisure activity. If he transfers the music file to his friend via Bluetooth or infrared, this falls under segment 2 where such activity still utilizes the wireless network, yet does not incur a cost upon file transfer or involve any interaction with service providers. However, if he records his own singing (provided if the mobile device supports voice recording functionality), such activity is still considered as mobile entertainment, but it does not utilize the wireless network or incur a cost upon usage. Therefore, this activity falls under segment 3.

Scenario 2: Downloading Pictures onto Mobile Devices

A mobile user connects to the Internet via his 3G-enabled mobile phone, searches for a particular picture on the content provider's site, and downloads it onto his mobile phone as wallpaper. This activity falls under segment 1. If the mobile user transfers the image to his friend via Bluetooth or infrared, this activity falls under segment 2. On the other hand, if the mobile user snaps a picture with his camera phone, this would fall under segment 3. He did not download the picture, nor did he transfer the picture from another device.

Scenario 3: Playing Games on Mobile Devices

A mobile user connects to the Internet via his WAP-enabled mobile phone, searches for a particular Java game, and downloads it onto his mobile phone. This activity falls under segment 1. Assuming that the downloaded game can be played as either a single player or multiplayer game, if the mobile user competes against his friend via Bluetooth, this activity falls under segment 2. In the third scenario, a mobile user plays preinstalled games on his mobile phone. He did not download the game nor did he transfer the game from another device. Therefore, the third scenario falls under segment 3.

Mobile entertainment does not exclude portfolio technologies such as Apple's iPod or Palm's Zire, which are not wireless but rely on being networked to other devices between periods of mobility (MGAIN, 2004). However, in the authors' opinion, such service falls under segment 3. Besides, such activity does not incur a cost and does not interact with service providers.

Hence, in these scenarios, the players in the value web vary in all three segments. The definitions of mobile entertainment for all three segments differ as well. Hence, the model in Figure 1 aids the industries to determine an appropriate business model to adopt in order to target the right audience. By classifying mobile entertainment service in its appropriate segment, it is then possible to determine the stakeholders involved, the network- and device-related requirements, and the business model required to develop and market the service (Wong & Hiew, 2005).

Figure 1. Mobile entertainment model: Position in the value web

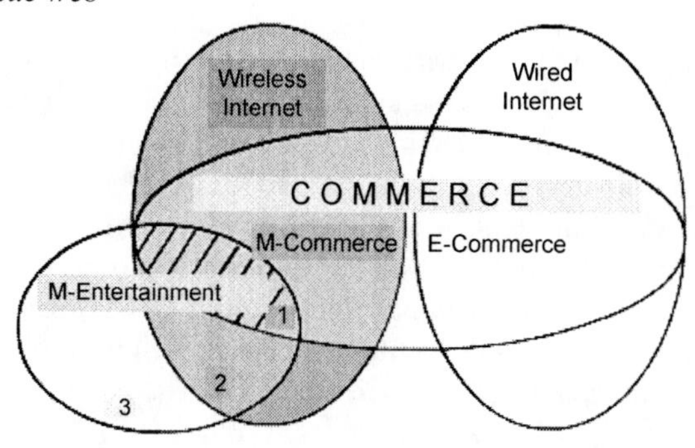

FUTURE TRENDS

Mobile entertainment services have come a long way since the introduction of games like 'Snake' on mobile handsets (Drewitt & Bell, 2002). Consumers now have access to a host of services including online games, betting, and messaging, and the advent of 3G networks is poised to bring further developments (Drewitt & Bell, 2002).

Key players in the industry are taking the "wait-and-see" approach, as it is difficult to determine what the "killer applications" are in mobile entertainment arena. Some have already tested the waters with promotional music videos, sponsored ring tones, and song downloads (Williamson, 2005). But mobile video advertising could potentially be even bigger (Williamson, 2005).

However, there is a long list of challenges that will make mobile entertainment a difficult market to crack:

- **Bandwidth Issues:** Cellular systems like 2G and 3G have since been viewed as technologies capable of offering full mobility support due to their advanced roaming, handover, and network management capabilities, with relatively much lower data rate however. Yet, the so-called 3G mobile technology still lags in most of the United States (Williamson, 2005). Sprint and Verizon are the only two carriers to offer 3G phones (Williamson, 2005). Speed is the key driver for most mobile music and video applications.
- **Competing Technologies**: Mobile video today is an alphabet soup of technologies, and wireless operators have yet to agree on a standard (Williamson, 2005). That means content must be customized for each wireless network and consumers will, for the near term, select carriers based on the media content they offer (Williamson, 2005).

Finally, the mobile entertainment value web needs to carry on the innovation, development, and deployment of new technology that can be exploited for the improvement and expansion of value for mobile entertainment. This could be the means to finally make the consumers interested, as long as this is done from the consumers' perspective and not technology for its own sake. The future looks promising in this regard, with interesting new devices having additional functionality and improved user interfaces; and eventually, appealing 3G networks, security, and digital rights management (DRM) solutions will support a trustworthy use for both consumers and content providers.

CONCLUSION

The mobile revolution is changing the way people live and work. Mobile phones are already pervasive in all major developed economies and in an increasing number of developing ones as well. Prediction, based on both anecdotal and empirical information, on the future popularity and volume of mobile commerce has been widely presented in academic literature, as well as business and technological press.

In short, the authors define mobile entertainment as any leisure activity undertaken via a mobile device, interacting with service providers, incurring a cost upon usage, and utilizing wireless communication networks. Mobile entertainment services and applications can be adopted through different wireless and mobile networks, with the support of various mobile devices. An important factor in designing mobile entertainment services and applications is the necessity for apt identification of consumers' requirements, as well as mobile device and technology constraints. Services and applications are designed and developed based on these requirements and limitations.

Foresight is a series of methods and tools for creating future-orientated scenarios at national, regional, and sectoral levels. However, consumers are rarely consulted in traditional foresight exercises.

While consumer foresight is no more a game of forecasting the future than other forms of foresight research, it is clear that knowledge of consumer expectations can aid the mobile entertainment industry in focusing on those issues that are of concern to their market base. Directing technical and market developments towards fulfilling consumers' expectations of the future will support diffusion of new mobile technologies and services.

REFERENCES

Andreou, A. S., Chrysostomou, C., Leonidou, C., Mavromoustakos, S., Pitsillides, A., Samaras, G., et al. (2002). *Mobile commerce applications and services: A design and development approach, M-Business 2002.* Athens, July 8-9.

Booz, Allen, & Hamilton. (2003). Future mobile entertainment scenarios. *Proceedings of the Mobile Entertainment Forum,* (pp. 4-16).

Camponovo, G. (2002). *Mobile commerce business models, presented at International workshop on Business Models.* Lawsanne, Switzerland.

Drewitt, A., & Bell, P. (2002). *Play away: The future of mobile entertainment: BWCS.*

Kalyanaraman, R. (2002). *Mobile entertainment services—A perspective.* White Paper Series, Wipro Technologies.

Maharramov, S. (1999). M-commerce: Evolution in business. *Proceedings of the E-Business Forum,* pp. 1-6.

MGAIN. (2004). *Mobile entertainment industry and culture.* UK: MGAIN.

Moore, K., & Rutter, J. (2004). Understanding consumers' understanding of mobile entertainment. *Proceedings of Mobile Entertainment: User-Centred Perspectives,* Manchester, UK. pp. 113-148.

MORI. (2003). *Knowledge of WiFi hotspots.* MORI, London.

Ollila, M., Kronzell, M., Bakos, M., & Weisner, F. (2003). *Mobile entertainment industry and culture: Barriers and drivers.* UK: MGAIN.

Pedersen, P. E., Methlie, L. B., & Thorbjørnsen, H. (2002). Understanding mobile commerce end-user adoption: A triangulation perspective and suggestions for an exploratory service evaluation framework. *Proceedings of the 35th Hawaii International Conference on System Sciences,* Hawaii.

UK Trade and Investment. (2002). *Communication market in Malaysia, UK.*

Wiener, S. N. (2003). Terminology of mobile entertainment: An introduction. *Proceedings of the Mobile Entertainment Forum.*

Williamson, D. A. (2005). Mobile video: Present and future. *iMedia Connection,* (December 16). Available at http://www.imediaconnection.com/content/7587.asp

Wong, C. C., & Hiew, P. L. (2005). Mobile entertainment: Review and redefine. *Proceedings of the IEEE 4th International Conference on Mobile Business,* Sydney, Australia, pp. 187-192.

KEY TERMS

Bluetooth: A standard developed by a group of electronics manufacturers that allows any sort of electronic equipment to make its own connections, without wires or any direct action from a user. The name Bluetooth was derived from the tenth-century king of Denmark, known as King Harold Bluetooth, who engaged in diplomacy that led warning parties to negotiate with one another. The inventors of the Bluetooth technology thought this a fitting name for their technology, which allowed different devices to talk to each other.

Business Model: The mechanism by which a business intends to generate revenue and profits. It is a summary of how a company plans to select, serve, and keep its customers; define and differentiate its product offerings; position itself in the market; as well as capture profit. Also called a business design.

Consumer Scenario: Describes how consumers use a particular service or product in their everyday lives.

Digital Rights Management (DRM): The umbrella term referring to any of several technologies used to enforce predefined policies controlling access to software, music, movies, or other digital data. In more technical terms, DRM deals with the description, layering, analysis, valuation, trading, and monitoring of the rights held over a digital work.

Foresight: The providence by virtue of planning prudently for the future. In other words, it is the discipline of developing a forward view in time, the link-theme between spiritual dimension and mental dimension.

Infrared: Infrared data transmission is employed in short-range communication among computer peripherals and personal digital assistants. These devices usually conform to standards published by IrDA, the Infrared Data Association.

Mobile Commerce: A mobile commerce transaction is defined as any type of transaction of economic value that is conducted through a mobile device that uses a wireless telecommunications network for communication with the electronic commerce infrastructure.

Taxonomy: Initially taxonomy was only the science of classifying living organisms, but later the word was applied in a wider sense, and may also refer to either a classification of things or the principles underlying the classification.

3G: 3G is a short term for third-generation wireless, and refers to developments in personal and business wireless technology, especially mobile communications. 3G networks were conceived from the Universal Mobile Telecommunications Service (UMTS) concept for high-speed networks for enabling a variety of data-intensive applications.

Mobile File-Sharing over P2P Networks

Lu Yan
Åbo Akademi, Finland

INTRODUCTION

Peer-to-peer (P2P) computing is a networking and distributed computing paradigm which allows the sharing of computing resources and services by direct, symmetric interaction between computers. With the advance in mobile wireless communication technology and the increasing number of mobile users, peer-to-peer computing, in both academic research and industrial development, has recently begun to extend its scope to address problems relevant to mobile devices and wireless networks.

The mobile ad hoc network (MANET) and P2P systems share key characteristics including self-organization and decentralization, and both need to solve the same fundamental problem: connectivity. Although it seems natural and attractive to deploy P2P systems over MANET due to this common nature, the special characteristics of mobile environments and the diversity in wireless networks bring new challenges for research in P2P computing.

Currently, most P2P systems work on wired Internet, which depends on application layer connections among peers, forming an application layer overlay network. In MANET, overlay is also formed dynamically via connections among peers, but without requiring any wired infrastructure. So the major differences between P2P and MANET that concern us in this article are:

a. P2P is generally referred to the application layer, but MANET is generally referred to the network layer, which is a lower layer concerning network access issues. Thus, the immediate result of this layer partition reflects the difference of the packet transmission methods between P2P and MANET: the P2P overlay is a unicast network with virtual broadcast consisting of numerous single unicast packets, while the MANET overlay always performs physical broadcasting.
b. Peers in P2P overlay are usually referred to static node though no priori knowledge of arriving and departing is assumed, but peers in MANET are usually referred to mobile node since connections are usually constrained by physical factors like limited battery energy, bandwidth, computing power, and so forth.

BACKGROUND

Since both P2P and MANET are becoming popular only in recent years, the research on P2P systems over MANET is still in its early stage. The first documented system is Proem (Kortuem et al., 2001), which is a P2P platform for developing mobile P2P applications, but it seems to be a rough one, and only IEEE 802.11b in ad hoc mode is supported. 7DS (Papadopouli & Schulzrinne, 2001) is another primitive attempt to enable P2P resource sharing and information dissemination in mobile environments, but it is rather a P2P architecture proposal than a practical application. In a recent paper, Lindemann and Waldhorst (2002) proposed *passive distributed indexing* for such kinds of systems to improve the search efficiency of P2P systems over MANET, and in ORION (Klemm, Lindemann & Waldhorst, 2003), a broadcast-over-broadcast routing protocol was proposed. The above works were focused on either P2P architecture or routing schema design, but how efficient the approach is and what the performance experienced by users is—these are still in need of further investigation.

Previous work on performance study of P2P over MANET was mostly based on simulative approach, and no concrete analytical mode was introduced. Performance issues of these kinds of systems were first discussed in Goel, Singh, and Xu (2002), but it simply shows the experiment results and no further analysis was presented. There is a survey of such kinds of systems in Ding and Bhargava (2004), but no further conclusions were derived. A sophisticated experiment and discussion on P2P communication in MANET can be found in Hsieh and Sivakumar (2004). However, all above works fall into a practical experience report category, and no performance models are proposed.

There have been many routing protocols in P2P networks and MANET respectively. For instance, one can find a very substantial P2P routing scheme survey from HP Labs in Milojicic et al. (2002), and U.S. Navy Research publishes ongoing MANET routing schemes (MANET, n.d.); but all of the above schemes fall into two basic categories: broadcast-like and DHT-like. More specifically, most early P2P search algorithms, such as in Gnutella (www.gnutella.com), Freenet (freenet.sourceforge.net), and Kazaa (www.kazaa.com), are

Figure 1. Broadcast over broadcast

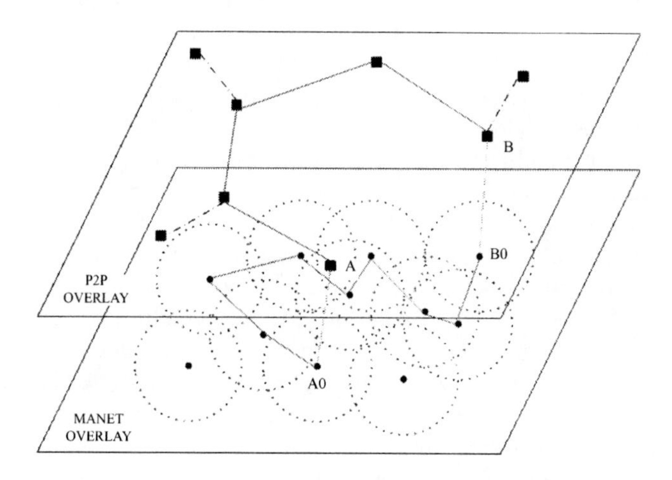

broadcast-like, and some recent P2P searching, like in eMule (www.emule-project.net) and BitTorrent (http://bittorrent.com/), employs more or less some feathers of DHT. On the MANET side, most on-demand routing protocols such as DSR (n.d.) and AODV (n.d.) are basically broadcast-like. Therefore, we here introduce different approaches to integrate these protocols in different ways according to categories.

BROADCAST OVER BROADCAST

The most straightforward approach is to employ a broadcast-like P2P routing protocol at the application layer over a broadcast-like MANET routing protocol at the network layer. Intuitively, in the above settings, every routing message broadcasting to the virtual neighbors at the application layer will result in a full broadcast to the corresponding physical neighbors at the network layer.

The scheme is illustrated in Figure 1 with a searching example: peer A in the P2P overlay is trying to search for a particular piece of information, which is actually available in peer B. Due to broadcast mechanism, the search request is transmitted to its neighbors, and recursively to all the members in the network, until a match is found or it times-out. Here we use the blue lines to represent the routing path at this application layer. Then we map this searching process into the MANET overlay, where node A0 is the corresponding mobile node to the peer A in the P2P overlay, and B0 is related to B in the same way. Since the MANET overlay also employs a broadcast-like routing protocol, the request from node A0 is flooded (broadcast) to directly connected neighbors, which in turn flood their neighbors and so on, until the request is answered or a maximum number of flooding steps occur. The route establishing lines in that network layer are highlighted in red, where we can find that there are few overlapping routes between these two layers, though they all employ broadcast-like protocols.

We have studied a typical broadcast-like P2P protocol, Gnutella (Clip2, 2001), in previous work (Yan & Sere, 2003). This is a pure P2P protocol, in which no advertisement of shared resources (e.g., directory or index server) occurs. Instead, each request from a peer is broadcast to directly connected peers, which themselves broadcast this request to their directly connected peers and so on, until the request is answered or a maximum number of broadcast steps occur. It is easy to see that this protocol requires a lot of network bandwidth, and it does not prove to be very scalable. The complexity of this routing algorithm is O(n) (Ripeanu, Foster, & Iamnitch, 2002; Chawathe, Ratnasamy, Breslau, & Shenker, 2003).

Generally, most on-demand MANET protocols, like DSR (Johnson & Maltz, 1996) and AODV (Perkins & Royer, 2000), are broadcast-like in nature (Kojima, Harada, & Fujise, 2001). Previously, one typical broadcast-like MANET protocol, AODV, was studied (Yan & Ni, 2004). In that protocol, each node maintains a routing table only for active destinations: when a node needs a route to a destinations, a path discovery procedure is started based on a RREQ (route request) packet; the packet will not collect a complete path (with all IDs of involved nodes) but only a hop count; when the packet reaches a node that has the destination in its routing table, or the destination itself, a RREP (route reply) packet is sent back to the source (through the path that has been set up by the RREQ packet), which will insert the destination in its routing table and will associate the neighbor from which the RREP was received as the preferred neighbor to that destination. Simply speaking, when a source node wants to send a packet to a destination, if it does not know a valid route, it initiates a route discovery process by flooding the RREQ packet through the network. AODV is a pure on-demand protocol, as only nodes along a path maintain routing information and exchange routing tables. The complexity of that routing algorithm is O(n) (Royer & Toh, 1999).

This approach is probably the easiest one to implement, but the drawback is also obvious: the routing path of the requesting message is not the shortest path between source and destination (e.g., the red line in Figure 1), because the virtual neighbors in the P2P overlay are not necessarily also the physical neighbors in the MANET overlay, and actually these nodes might be physically far away from each other. Therefore, the resulting routing algorithm complexity of this broadcast-over-broadcast scheme is unfortunately $O(n^2)$, though each layer's routing algorithm complexity is O(n) respectively.

It is not practical to deploy such a scheme for its serious scalability problem due to the double broadcast; and taking the energy consumption portion into consideration, which is somehow critical to mobile devices, the double broadcast will

Figure 2. DHT over broadcast

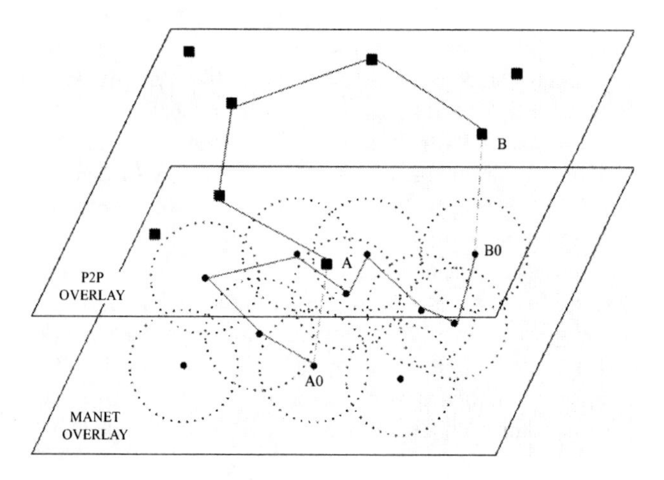

also cost a lot of energy consumption and make it infeasible in cellular wireless data networks.

DHT OVER BROADCAST

The scalability problem of broadcast-like protocols has long been observed, and many revisions and improvement schemas are proposed (Lv, Ratnasamy, & Shenker, 2002; Yang & Garcia-Molina, 2002; Chawathe et al., 2003). To overcome the scaling problems in broadcast-like protocols where data placement and overlay network construction are essentially random, a number of proposals are focused on structured overlay designs. The distributed hash table (DHT) (Stoica, Morris, Karger, Kaashoek, & Balakrishnan, 2001) and its varieties (Ratnasamy, Francis, Handley, Karp, & Schenker 2001; Rowstron & Druschel, 2001; Zhao et al., 2004) advocated by Microsoft Research seem to be promising routing algorithms for overlay networks. Therefore it is interesting to see the second approach: to employ a DHT-like P2P routing protocol at the application layer over a broadcast-like MANET routing protocol at the network layer.

The scheme is illustrated in Figure 2 with the same searching example. Compared to the previous approach, the difference lies in the P2P overlay: in a DHT-like protocol, files are associated to keys (e.g., produced by hashing the file name); each node in the system handles a portion of the hash space and is responsible for storing a certain range of keys. After a lookup for a certain key, the system returns the identity (e.g., the IP address) of the node storing the object with that key. The DHT functionality allows nodes to put and get files based on their key, and each node handles a portion of the hash space and is responsible for a certain key range. Therefore, routing is location-deterministic distributed lookup (e.g., the blue line in Figure 2).

DHT was first proposed in Plaxton, Rajaraman, and Richa (1997) without intention to address P2P routing problems. DHT soon proved to be a useful substrate for large distributed systems, and a number of projects are proposed to build Internet-scale facilities layered above DHTs; among them are Chord, CAN, Pastry, and Tapestry. All take a key as input and route a message to the node responsible for that key. Nodes have identifiers, taken from the same space as the keys. Each node maintains a routing table consisting of a small subset of nodes in the system. When a node receives a query for a key for which it is not responsible, the node routes the query to the hashed neighbor node towards resolving the query. In such a design, for a system with n nodes, each node has O(log n) neighbors, and the complexity of the DHT-like routing algorithm is O(log n) (Ratnasamy, Shenker, & Stoica, 2002).

Additional work is required to implement this approach, partly because DHT requires a periodical maintenance (i.e., it is just like an Internet-scale hash table or a large distributed database); since each node maintains a routing table (i.e., hashed keys) to its neighbors according to DHT algorithm, following a node join or leave, there is always a nearest key reassignment between nodes.

This DHT-over-broadcast approach is obviously better than the previous one, but it still does not solve the shortest path problem as in the broadcast-over-broadcast scheme. Though the P2P overlay algorithm complexity is optimized to O(log n), the mapped message routing in the MANET overlay is still in the broadcast fashion with complexity O(n); the resulting algorithm complexity of this approach is as high as O(n log n).

This approach still requires a lot of network bandwidth and hence does not prove to be very scalable, but it is efficient in limited communities such as a company network.

CROSS-LAYER BROADCAST

A further step of the broadcast-over-broadcast approach would be a cross-layer broadcast. Due to similarity of broadcast-like P2P and MANET protocols, the second broadcast could be skipped if the peers in the P2P overlay would be mapped directly into the MANET overlay, and the result of this approach would be the merge of application layer and network layer (i.e., the virtual neighbors in P2P overlay overlaps the physical neighbors in MANET overlay).

The scheme is illustrated in Figure 3, where the advantage of this cross-layer approach is obvious: the routing path of the requesting message is the shortest path between source and destination (e.g., the blue and red lines in Figure 3), because the virtual neighbors in the P2P overlay are de facto physical neighbors in the MANET overlay due to the merge of two layers. Thanks to the nature of broadcast, the algorithm complexity of this approach is O(n), making it

M

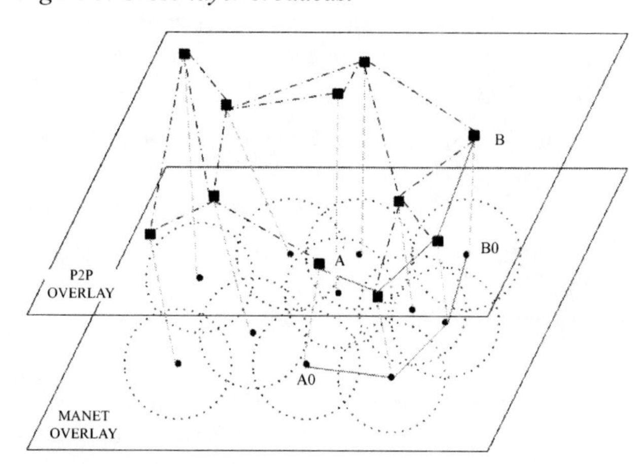

Figure 3. Cross-layer broadcast

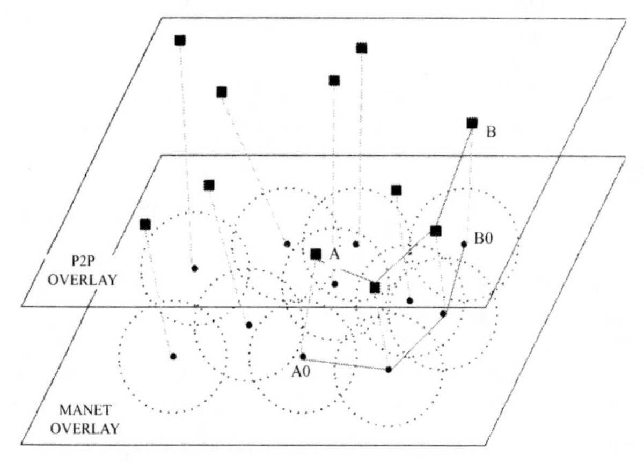

Figure 4. Cross-layer DHT

suitable for deployment in relatively large-scale networks, but still not feasible for Internet-scale networks.

CROSS-LAYER DHT

It is also possible to design a cross-layer DHT in Figure 4 with the similar inspiration, and the algorithm complexity would be optimized to O(log n) with the merit of DHT, which is advocated to be efficient even in Internet-scale networks. The difficulty in that approach is implementation: there is no off-the-shelf DHT-like MANET protocol as far as we know, though recently, some research projects like Ekta (Pucha, Das, & Hu, 2004) towards a DHT substrate in MANET are proposed.

CONCLUSION

In this article, we studied the peer-to-peer systems over mobile ad hoc networks with a comparison of different settings for the peer-to-peer overlay and underlying mobile ad hoc network. We show that the cross-layer approach performs better than separating the overlay from the access networks in Table 1. Our results would potentially provide useful guidelines for mobile operators, value-added service providers, and application developers to design and dimension mobile peer-to-peer systems.

REFERENCES

AODV. (n.d.). *AODV IETF draft v1.3.* Retrieved from http://www.ietf.org/internet-drafts/draft-ietf-manet-aodv-13.txt

Chawathe, Y., Ratnasamy, S., Breslau, L., & Shenker, S. (2003). Making Gnutella-like P2P systems scalable. *Proceedings of ACM SIGCOMM.*

Clip2. (2001). *The Gnutella protocol specification v0.4* (document revision 1.2). Retrieved from http://www9.limewire.com/developer /gnutella protocol 0.4.pdf

DSR. (n.d.). *DSR IETF draft v1.0.* Retrieved from http://www.ietf.org/internet-drafts/draft-ietf-manet-dsr-10.txt

Ding, G., & Bhargava, B. (2004). Peer-to-peer file-sharing over mobile ad hoc networks. *Proceedings of the 2nd IEEE Conference on Pervasive Computing and Communications Workshops.*

Goel, S. K., Singh, M., & Xu, D. (2002). Efficient peer-to-peer data dissemination in mobile ad-hoc networks. *Proceedings of the International Conference on Parallel Processing.*

Table 1. How efficient does a user try to find a specific piece of data?

	Efficiency	Scalability	Implementation
Broadcast over Broadcast	O(n^2)	n/a	Easy
DHT over Broadcast	O(n log n)	Bad	Medium
Cross-Layer Broadcast	O(n)	Medium	Difficult
Cross-Layer DHT	O(log n)	Good	n/a

Hsieh, H.Y., & Sivakumar, R. (2004). On using peer-to-peer communication in cellular wireless data networks. *IEEE Transactions on Mobile Computing, 3*(1).

Johnson, D. B., & Maltz, D. A. (1996). *Dynamic source routing in ad-hoc wireless networks. Mobile computing.* Kluwer.

Klemm, A., Lindemann, C., & Waldhorst, O. (2003). A special-purpose peer-to-peer file sharing system for mobile ad hoc networks. *Proceedings of the IEEE Vehicular Technology Conference.*

Kojima, F., Harada, H., & Fujise, M. (2001). A study on effective packet routing scheme for mobile communication network. *Proceedings of the 4th Symposium on Wireless Personal Multimedia Communications.*

Kortuem, G., Schneider, J., Preuitt, D., Thompson, T. G. C., Fickas, S., & Segall, Z. (2001). When peer-to-peer comes face-to-face: Collaborative peer-to-peer computing in mobile ad hoc networks. *Proceedings of the 1st International Conference on Peer-to-Peer Computing.*

Lindemann, C., & Waldhorst, O. (2002). A distributed search service for peer-to-peer file sharing in mobile applications. *Proceedings of the 2nd IEEE Conference on Peer-to-Peer Computing.*

Lv, Q., Ratnasamy, S., & Shenker, S. (2002). Can heterogeneity make Gnutella scalable? *Proceedings of the 1st International Workshop on Peer-to-Peer Systems.*

MANET. (n.d.). *MANET implementation survey.* Retrieved from http://protean.itd.nrl.navy.mil/manet/survey/survey.html

Milojicic, D. S., Kalogeraki, V., Lukose, R., Nagaraja, K., Pruyne, J., Richard, B., Rollins, S., & Xu, Z. (2002). *Peer-to-peer computing.* Technical Report HPL-2002-57, HP Labs.

Papadopouli, M., & Schulzrinne, H. (2001). A performance analysis of 7DS, a peer-to-peer data dissemination and prefetching tool for mobile users. *IEEE Sarnoff Symposium Digest.*

Perkins, C. E., & Royer, E. M. (2000). *The ad hoc on-demand distance vector protocol. Ad hoc networking.* Boston: Addison-Wesley.

Plaxton, C., Rajaraman, R., & Richa, A. (1997). Accessing nearby copies of replicated objects in a distributed environment. *Proceedings of ACM SPAA.*

Pucha, H., Das, S. M., & Hu, Y. C. (2004). Ekta: An efficient DHT substrate for distributed applications in mobile ad hoc networks. *Proceedings of the 6th IEEE Workshop on Mobile Computing Systems and Applications.*

Ratnasamy, S., Francis, P., Handley, M., Karp, R., & Schenker, S. (2001). A scalable content-addressable network. *Proceedings of the Conference on Applications, Technologies, Architectures, and Protocols for Computer Communications.*

Ratnasamy, S., Shenker, S., & Stoica, I. (2002). Routing algorithms for DHTs: Some open questions. *Proceedings of the 1st International Workshop on Peer-to-Peer Systems.*

Ripeanu, M., Foster, I., & Iamnitch, A. (2002). Mapping the Gnutella network: Properties of large-scale peer-to-peer systems and implications for system design. *IEEE Internet Computing, 6*(1).

Rowstron, A., & Druschel, P. (2001). Pastry: Scalable, distributed object location and routing for large-scale peer-to-peer systems. *Proceedings of the IFIP/ACM International Conference on Distributed Systems Platforms.*

Royer, E. M., & Toh, C. K. (1999). *A review of current routing protocols for ad-hoc mobile wireless networks.* IEEE Personal Communications.

Stoica, I., Morris, R., Karger, D., Kaashoek, F., & Balakrishnan, H. (2001). Chord: A scalable peer-to-peer lookup service for Internet applications. *Proceedings of ACM SIGCOMM.*

Yan, L., & Ni, J. (2004). Building a formal framework for mobile ad hoc computing. *Proceedings of the International Conference on Computational Science.*

Yan, L., & Sere, K. (2003). Stepwise development of peer-to-peer systems. *Proceedings of the 6th International Workshop in Formal Methods.*

Yang, B., & Garcia-Molina, H. (2002). Improving search in peer-to-peer networks. *Proceedings of the International Conference on Distributed Systems.*

Zhao, B. Y., Huang, L., Stribling, J., Rhea, S. C., Joseph, A. D., & Kubiatowicz, J. (2004). Tapestry: A Resilient global-scale overlay for service deployment. *IEEE Journal on Selected Areas in Communications.*

KEY TERMS

AODV: Ad hoc On-demand Distance Vector routing.

DHT: Distributed hash table.

DSR: Dynamic source routing.

MANET: Mobile ad hoc network.

P2P: Peer-to-peer.

Mobile Gaming

Krassie Petrova
Auckland University of Technology, New Zealand

INTRODUCTION

A number of multifunctional handheld devices with Internet and multimedia capabilities are currently available on the market. The mobile network technologies implemented make it possible for a range of value-added mobile services known as "mobile entertainment" to be offered to paying subscribers (Carlsson, Hyvonen, Repo, & Walden, 2005). Examples include watching streamed news, downloading music and images, or playing a game on one's mobile phone (alone, or in an interaction with other players).

Mobile game development depends on the choice of a middleware platform, as the application needs to be portable across a wide spectrum of handheld devices and technologies (Yuan, 2004; Hagleitner & Mueck, 2002). A business model for offering mobile gaming as a service has been successfully trialled in Japan where playing games is one of the main components of the popular Japanese entertainment platform iMode (Natsuno, 2003, pp.88-90).

Research in the area of mobile gaming adoption has focused on the investigation of the value generation process and on identifying the critical factors for mobile gaming acceptance. A number of critical success factors have been identified (e.g., Shchglick, Barnes, Scornavacca, & Tate, 2004; Moore & Rutter, 2004; Yoon, Ha, & Choi, 2005), adapting and extending existing mobile business frameworks and models (e.g., Siau, Lim, & Shen, 2001; Varshney & Vetter, 2002; Lee, Hu, & Yeh, 2003; Barnes, 2003). This short article investigates the relationship between mobile gaming customers and the mobile gaming value chain, and discusses the implications from mobile gaming supply and demand perspectives.

BACKGROUND

Playing mobile games ("mobile gaming") is classified as a "mobile entertainment" application (Barnes & Huff, 2003; Van de Kar, Maitland, de Montalvo, & Bouwman, 2003). Mobile entertainment includes personal leisure activities undertaken via a network technology. Entertainment services might feature data transfer including voice and video over significant geographic distance while the user of the service is either on the move or has the potential to move without interrupting the activity (Ollila, Kronzell, Bakos, & Weisner, 2003).

Mobile gaming is also an example of a mobile commerce (m-commerce) application, provided through a specially designed m-commerce service. Typically, an m-commerce service involves payment: a monetary transaction which the customer conducts using the mobile payment mechanism provided with the service (Paavilainen, 2002). In the case of mobile gaming, the player's subscriber account with the mobile network operating the service is used to collect the revenue. Subsequently the network operator makes payments to other service providers who might be involved: content developers and publishers, portal aggregators, and retailers (Wiener, 2003).

Some mobile games are simply downloaded and played off-line, paying once or with every update. Such games might be suitable for "low-end" handheld devices and might use text messaging. Other mobile games need to be played on smartphones, as interactivity among multiple players needs to be supported. These real-time mobile games require a persistent network connection to a dedicated game server. Advancements in mobile game development include the inclusion of location-based features into the game (Moore & Rutter, 2004; Maitland, van de Kar, de Montalvo, & Bouwman, 2005). In all cases, the mobile game player does not need to be stationary – he or she is "released" from the need to use a stationary networked device (Finn, 2005) .The assumed mobility of the mobile game player is one of the defining features of mobile gaming.

A leader in mass mobile entertainment, the Japanese company NTT DoCoMo developed a comprehensive mobile service and platform: iMode. Entertainment applications and specifically mobile gaming are seen as the catalyst for the increased use of the range of other iMode services (Baldi & Thaung, 2002; Natsuno, 2003, p.92; Barnes & Huff, 2003; Funk, 2003, p. 28). In the global market, mobile gaming is seen as a viable business opportunity (Kleijnen, de Ruyter, & Wetzels, 2003, p. 205; Anckar & D'Incau, 2002; Paavilainen, 2002). According to some predictions, by 2010 the revenue from downloading mobile games might reach US$8.4 billion (Graft, 2006).

THE MOBILE GAMING VALUE CHAIN

A number of value chain models (e.g., Buellingen & Woerter, 2004; Siau et al., 2001, Barnes, 2003) and mobile frameworks (e.g., Varshney & Vetter, 2002; Stanoevska-Slabeva, 2003)

Figure 1. Customer relationships in mobile gaming (Derived from Petrova, 2005)

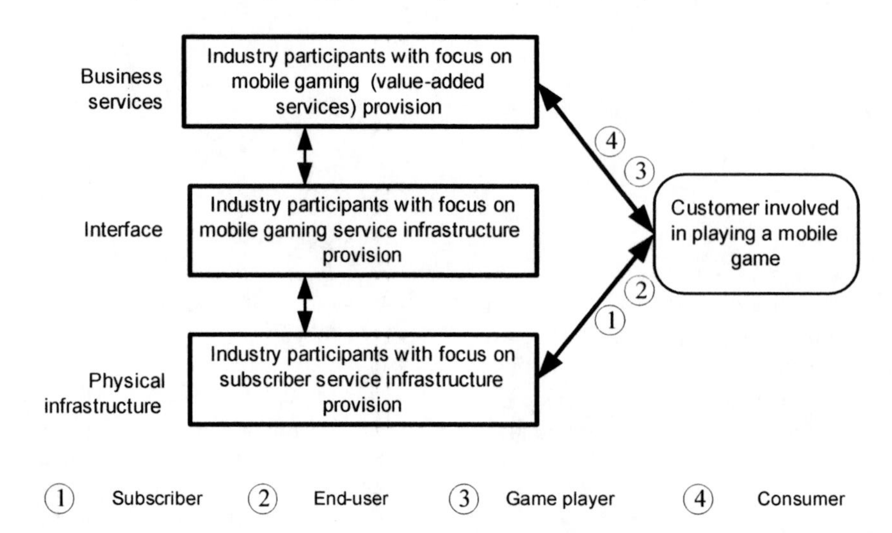

for m-commerce have been suggested in the literature and used to map industry players, roles and functions. A multiple value-chain model representing the relationships between a customer and the mobile gaming industry is shown in Figure 1. It applies the m-commerce reference model proposed by Petrova (2005) to the mobile entertainment value Web (Ollila et al., 2003) to identify the relationships between the customer (a mobile game player) and the mobile industry participants.

The network developers and providers, along with device developers, create the physical foundation needed for mobile gaming, and together with the mobile network providers, form a physical infrastructure layer. The interface layer includes developers of middleware platforms, which serve as game developing and servicing environments and enable the use of the networks and technologies for game service provision. The top layer represents the category of value-added services where mobile games are offered to customers directly or as a part of a mobile entertainment package. The main industry participants are mobile game developers, publishers and aggregators.

A company involved in mobile gaming can be categorised under more than one category: *Vodafone,* for example, provides a subscriber network and subscriber service, as well as mobile game downloading through its portal Vodafone Live! (Harmer, 2003). The revenue model of the company depends on its position in the value chain (Ollila et al., 2003).

MOBILE VALUE CHAIN RELATIONSHIPS

Mobile network operators are part of the physical infrastructure by building and maintaining the network. In most cases they also provide subscriber services and access to the network (e.g., *Vodafone, Orange).* The 2G-2.5G technologies currently implemented include CDMA, GSM, and GPRS, and are capable of maintaining real-time, persistent network connection (Mobile Games, 2001). Customers interact with mobile network operators as subscribers to the network service (Relationship 1 in Figure 1).

Mobile device supply manufacturers (e.g., *Nokia*, *Siemens*, *Motorola*) provide customers with the devices needed to connect to a network and to access the services provided. In the case of mobile gaming, devices might need extended functionality such as a very fast processor (Leavitt, 2003). Customers interact with devices as information technology (IT) end-users (Relationship 2 in Figure 1).

A feature of the mobile gaming industry sector is the coupling between device and network technology: a handheld device might be designed to be used only with a particular technology and for access to networks based on that technology. The IT end-user experience will depend on the characteristics of the device and the underlying wireless network technology.

In the interface category, some widely used middleware products for game development are WAP and iMode (Baldi & Thaung, 2002; Lee et al., 2003). New software platforms (J2ME and BREW) and operating systems (Symbian) allow for the development of portable mobile games. The industry participants in this category provide the service infrastructure needed by game developers and other related content providers, but typically do not interact directly with customers (although *Nokia* have developed a mobile phone microbrowser). Rather they serve as a link between the physical infrastructure and the business services that bring mobile gaming to the customer.

Table 1. Critical success factors, customer roles and industry players (extended from Petrova & Qu, 2006)

Critical Success Factor		Customer Role	Industry Participants
Facilitating conditions	Providing payment and billing mechanisms	Consumer	Aggregators/publishers/ mobile service infrastructure providers
Trialability	Providing opportunities for a free trial of a service		Aggregators/publishers
Compatibility	Meeting the gaming needs of specific customer segments	Game player	Game developers/ publishers
Observability	Emphasizing the social importance of playing a mobile game		Aggregators/ publishers
Image	Emphasising the status of the game player		Aggregators/ publishers
Normative beliefs	Building a critical mass to create social pressure		Aggregators/ publishers
Complexity	Ensuring easy to use gaming applications	IT End-User	Device manufacturers/ game developers
Trust	Alleviating security and privacy concerns		Mobile network operators/application service infrastructure providers
Relative advantage	Providing a ubiquitous and accessible mobile gaming service		Mobile network operators/application service infrastructure providers
Self-efficacy	Technical services matching different customer segment needs	Subscriber	Device manufacturers/ game developers

In the business services category, mobile game developers (for example *In-Fusio*) and mobile game aggregators and publishers (for example *MFORMA*, *Digital Bridges*) provide games to customers. Customers act as game players (Relationship 3 in Figure 1) and as consumers who pay for the valued added service (Relationship 4 in Figure 1).

CUSTOMER ROLES AND CRITICAL SUCCESS FACTORS

Understanding the priorities of a customer would lead to a better understanding of the criticality of the success factors of mobile gaming adoption and to more informed decision-making when offering a mobile gaming service to a targeted audience. From the discussion above, the question that arises is: What is the significance of customer roles in mobile gaming adoption?

Well known models and theories have been used to investigate customer adoption of different mobile applications (Barnes & Huff, 2003; Kleijnen et al., 2003; 2004; Pagani & Schipani, 2003; Carlsson et al., 2005; Pedersen, 2005; Yoon, Ha, & Choi, 2005). Key contributing factors based on customer attitudes and perspectives (Moore & Rutter, 2004; Pedersen, Methlie, & Thorbjornsen, 2002) and potential adoption drivers (Baldi & Thaung, 2002) have been identified. Petrova and Qu (2006) extracted a set of critical success factors for mobile gaming adoption based on findings from the literature, and proposed a framework for studying the adoption process (Table 1, first two columns). Extending this work further, in Table 1 (the last two columns) each factor is matched to a specific customer role based on the relationships discussed in the previous section.

Most of the critical success factors (seven) relate to two customer roles: "game player" and "IT end-user." They represent a relationship with the physical infrastructure layer and the business services layer of the mobile gaming industry, respectively, and constitute a "first priority" group of factors. The remaining three factors (matching the roles of "consumer" and of "subscriber") form the "second priority" group—also related to both infrastructure and business

Table 2. Future trends in mobile game development

Game Development Trends	Example
Games for 3G mobile networks	"Virtual Girlfriend" (*Artificial Life Inc.*, Hong Kong)
Multi-player, networked, interactive games	"MLSN Sports Picks" (*Digital Chocolate Inc.*, Finland)
Location–based games	"The Shroud" (*Your World Games*, U.S.)
Games for casual game players	Puzzle games (*Future Platforms*, UK). Backgammon (*Nokia* and *Octopi*, U.S.)
Highly personalised mobile gaming services.	"Vomitron" (*Ninemsn*, Australia). "PrizePlay" (*Sennari*, U.S.A). iMode in Japan, Europe and Australia.
3D games.	"V-Rally 3D" (*Fishlabs*, Germany)
Games which can be used for learning, or for awareness development	"FreedomHIV/AIDS" (*ZMQ*, India)

services provision. The significance of these results and their implications are discussed next.

Customer Priorities

Four factors are related to the role of the customer as a "game player." The implications are that mobile gaming will be successfully adopted by customers who are likely to enjoy mobile gaming as a social activity. Therefore mobile game content needs to be tailored to match the needs of particular social groups; developers will need to provide content of greater diversity—for example, multi-player games, building a virtual world (Raghu, Ramesh, & Whinston, 2002) and location-based gaming (Finn, 2005). The importance of the social context is confirmed by the success of mobile gaming in Japan (Barnes & Huff, 2003) and in other Asian countries (Kymalainen, 2004, p. 131). Game developers, aggregators and publishers should continue to focus on market segments that display a strong, pre-existing disposition towards mobile gaming as innovation. According to Baldi and Thaung (2002) and Kleijnen et al. (2004), Internet usage is such a predictor. Non-core gamer markets might also be penetrated with a range of casual games.

Three success factors relate to the role of the customer as an IT end-user. The implication for industry participants involved in the provision of physical infrastructure is that the usability of handheld devices will continue to be important as it brings "compelling" value to the customer (Venkatesh, Ramesh, & Massey, 2003). Network operators might need to consider "opening up their networks" to facilitate the development and provisioning of interactive games, operable across provider networks (Smorodinsky, 2002), while game developers will need to develop more portable applications (Yuan, 2004). All industry participants will need to address users' concerns for privacy and security, possibly by adapting solutions developed for e-commerce (Yiliant-tila, 2004, p. 71).

In summary, the adoption factors related to the consumer roles of "game player" and "IT end-user" are the ones that will ultimately determine the success of mobile gaming application and services.

Customer Support

Two factors relate to the customer role of "consumer" and one to the role of "subscriber." The implication is that that mobile game customers will expect a transaction and trading environment similar to other commercial environments, where customer and technical support are readily available (Kleijnen, de Ruyter, & Wetzels, 2003, p. 213). Industry participants from both the infrastructure layer and the business services layer will need to explore different cooperation strategies to be able to meet these requirements and so gain sustainable competitive advantage (Feldmann, 2002).

In summary, while adoption factors related to the customer roles of "consumer" and "subscriber" are not crucial, they will have a strong influence on customer choice and loyalty.

FUTURE TRENDS

The game development effort is focused on providing games targeting different customer segments, games exploring new technologies and games enhancing personalization. A summary of the future trends in mobile game development is provided in Table 2 along with some examples from the industry to illustrate them.

An important global industry trend is the move to define mobile game interoperability standards, so that game development can focus on producing games deployable across multiple game servers, wireless networks, and handheld devices. An example is the work of the Open Mobile Alliance (OMA) Games Services working group.

Academic work will continue to focus on multi-platform development and middleware architecture, on factors influencing adoption and on the development of innovative business models, studying customers from diverse cultural backgrounds and their needs and priorities.

CONCLUSION

The article reviews the literature on mobile games provision and adoption and outlines the different roles played by customers involved in mobile gaming. A mobile gaming framework is introduced, including critical success factors, customer roles and industry participants. Two priority groups of success factors are identified. The factors related to customers acting as "game players" and "IT end-users" are crucial for selecting and penetrating a market segment that is ready to adopt mobile gaming as a social activity; the factors related to customers acting as "consumers" and "subscribers" play an important but secondary role and might confer competitive advantage in the selected market. The future trends in mobile gaming will be towards the development of multi-player real-time games, and providing interoperability across platforms and networks, which will facilitate new market segment penetration.

REFERENCES

Anckar, B., & D'Incau, D. (2002). Value-added services in mobile commerce: An analytical framework and empirical findings from a national consumer survey. In *Proceedings of the 35th Annual Hawaii International Conference on System Sciences* (pp. 1087-1096).

Baldi, S., & Thaung, H. P-P. (2002). The entertaining way to m-commerce: Japan's approach to the mobile Internet—A model for Europe? *Electronic Markets, 12*(1), 6-13.

Barnes, S. J. (2003). The mobile commerce value chain in consumer markets. In *m-Business: The strategic implications of wireless technologies* (1st ed.) (pp. 13-37). Burlington MA: Butterworth-Heinemann.

Barnes, S. J., & Huff, S. L. (2003). Rising sun: iMode and the wireless Internet. *Communications of the ACM, 46*(11), 79-84.

Buellingen, F., & Woerter, M. (2004). Development perspectives, firm strategies and applications in mobile commerce. *Journal of Business Research, 57*(12), 1402-1408.

Carlsson, C., Hyvonen, K., Repo, P., & Walden, P. (2005). Asynchronous adoption patterns of mobile services. In *Proceedings of the 38th Annual Hawaii International Conference on Systems Sciences* (p. 189a).

Feldmann, V. (2002). Competitive strategy for media companies in the mobile Internet. *Schmalenbach Business Review, 54*, 351-371. Retrieved January 5, 2006, from http://www.vhb.de/sbr/pdfarchive/einzelne_pdf/sbr_2002_oct-351-371.pdf

Finn, M. (2005). Gaming goes mobile: Issues and Implications. *Australian Journal of Emerging Technologies and Society, 39*(1), 32-42

Funk, J. L. (2003). *Mobile disruption: The technologies and applications driving the mobile Internet*. New Jersey: John Wiley & Sons.

Graft, K. (2006, January 22). Analysis: A history of cellphone gaming. *BusinessWeek Online*. Retrieved January 25, 2006, from http://www.businessweek.com/innovate/content/jan2006/id20060122_077129.htm

Hagleitner, M., & Mueck, T. A. (2002). WAP-G: A case study in mobile entertainment. In *Proceedings of the 35th Hawaii International Conference on System Sciences* (Vol. 3) (p. 88).

Harmer, J. A. (2003). Mobile multimedia services. *BT Technology Journal, 21*(3), 169-180.

Kleijnen, M. D., de Ruyter, K., & Wetzels, M. G. M. (2003). Factors influencing the adoption of mobile gaming services. In B. E. Mennecke & T. J. Strader (Eds.), *Mobile Commerce: Technology, Theory, and Applications* (pp. 202-217). Hershey, PA: Idea Group Publishing.

Kleijnen, M. D., de Ruyter, K., & Wetzels, M. G. M. (2004). Customer adoption of wireless services: Discovering the rules, while playing the game. *Journal of Interactive Marketing, 18*(2), 51-60.

Kymalainen, P. (Ed.). (2004). Mobile entertainment industry and culture. *European Commission User-Friendly Information Society.* Retrieved January 10, 2006, from http://www.mgain.org/mgain-wp8-D823_book-delivered.pdf

Leavitt, N. (2003). Will wireless gaming be a winner? *Computer, 36*(1), 24-27.

Lee, C.-W., Hu, W.-C, & Yeh, J.-H. (2003). A system model for mobile commerce. In *23rd International Conference on Distributed Computing Systems Workshops* (pp. 634-639).

Maitland, C. F., van de Kar, E. A. M., de Montalvo, U. W., & Bouwman, H. (2005). Mobile information and entertainment services: Business models and service networks. *International Journal of Management and Decision Making, 6*(1), 47-64.

Mobile Games (2001, June). *Mobile games. Mobile Entertainment,* pp. 4-50.

Moore, K., & Rutter, J. (2004). Understanding consumers' understanding of mobile entertainment. In K. Moore & J. Rutter (Eds.), *Mobile Entertainment: User-Centred Perspectives* (pp. 49-65). Manchester, UK: University of Manchester.

Natsuno, T. (2003). *I-mode strategy*. Chichester, UK: Wiley & Sons.

Ollila, M., Kronzell, M., Bakos, N., & Weisner, F. (2003). *Mobile entertainment business*. European Commission User-Friendly Information Society. Retrieved December 23, 2005, from http://www.mgain.org/mgain-wp5-D542-delivered3.pdf

Paavilainen, J. (2002). Consumer mobile commerce. In *Mobile Business Strategies: Understanding the Technologies and Opportunities* (pp. 69-121). London: IT Press.

Pagani, M., & Schipani, D. (2003). Motivations and barriers to the adoption of 3G mobile multimedia services: An end user perspective in the Italian market. In M. Khosrow-Pour (Ed.), *Proceedings of the 2003 Information Resources Management Association International Conference* (pp. 957-960). Hershey, PA: IRM Press.

Pedersen, P. E., Methlie, L. B., & Thorbjornsen, H. (2002). Understanding mobile commerce end-user adoption: A triangulation perspective and suggestions for an exploratory service evaluation framework. In *Proceedings of the 35th Annual Hawaii International Conference on System Sciences* (pp. 1079-1086).

Pedersen, P. E. (2005). Adoption of mobile Internet services: An exploratory study of mobile commerce early adopters. *Journal of Organizational Computing and Electronic Commerce, 15*(3), 203-221.

Petrova, K. (2005). A study of the adoption of mobile commerce applications and of emerging viable business models. In M. Khosrow-Pour (Ed.), *Managing Modern Organizations with Information Technology. Proceedings of the 2005 Information Resources Management Association International Conference* (pp. 1133-1135). Hershey, PA: IRM Press.

Petrova, K., & Qu. H. (2006). Mobile gaming: A reference model and critical success factors. In M. Khosrow-Pour (Ed.), *Emerging Trends and Challenges in Information Technology Mangement: Proceedings of the 2006 Information Resources Mangement Association International Conference* (pp. 228-231). Hershey, PA: IRM Press.

Raghu, T. S., Ramesh, R., & Whinston, A. B. (2002). Next steps for mobile entertainment portals. *Computer, 35*(5), 63-70

Shchiglik, C., Barnes, S., Scornavacca, E., & Tate, M. (2004). Mobile entertainment service in New Zealand: An examination of consumer perceptions towards games delivered via the wireless application protocol. *International Journal of Services and Standards, 1*(2), 155-171.

Siau, K., Lim, E., & Shen, Z. (2001). Mobile commerce: Promises, challenges and research agenda. *Journal of Database Management, 12*(3), 4-13.

Smorodinsky, R. (2002) Harnessing the potential of mobile entertainment. In *Digital Infrastructure Technology* (pp. 55-56). Retrieved January 5, 2006, from http://www.m-e-f.org/pdf/Harnessing%20the%20Potential.pdf

Stanoevska-Slabeva, K. (2003). Towards a reference model for m-commerce applications. In *Proceeding of the 2003 European Conference on Information Systems*. Retrieved March 1, 2004, from http://inforge.unil.ch/yp/Terminodes/papers/03ECISSG.pdf

Van de Kar, E., Maitland, C. F., de Montalvo, U. W., & Bouwman, H. (2003). Design guidelines for mobile information and entertainment services: Based on the Radio 538 ringtunes I-mode service case study. In *Proceedings of the 5th International Conference on Electronic Commerce* (pp. 413-417).

Varshney, U., & Vetter, R. (2002). Mobile commerce: Framework, applications and networking support. *Mobile Networks and Applications, 3*(7), 185-187.

Venkatesh, V., Ramesh, V., & Massey, A.P. (2003). Mobile commerce opportunities and challenges: Understanding usability in mobile commerce. *Communications of the ACM, 46*(12), 30-32.

Wiener, S. N. (2003, August). *Terminology of mobile entertainment*. Mobile Entertainment Forum. Retrieved November 25, 2005, from http://www.m-e-f.org/pdf/GlossaryRelease_new%20logo.pdf

Yilianttila, M. (2004). *Emerging and future mobile entertainment technologies*. European Commission User-Friendly Information Society. Retrieved January 10, 2006, from http://www.mgain.org/mGain-wp4-d421-delivered-revised.pdf

Yoon, Y. S., Ha, I. S., & Choi, M.-K. (2005). Nature of potential mobile gamers' behaviour under future wireless mobile environment. In *Proceedings of the 7th International Conference on Advanced Communication Technology* (pp. 551-558).

Yuan, M. (2004, September 28). *Challenges and opportunities in mobile games*. IBM. Retrieved December 23, 2005, from http://www-128.ibm.com/developerworks/wireless/library/wi-austingameconf.html

KEY TERMS

Aggregator: A content provider who maintains relationships with several mobile network operators and delivers mobile game content to their users in a transparent way (i.e., using the same mobile number). Example: *MFORMA* (U.S.).

Binary Runtime Environment for Wireless (BREW): An application development platform for client applications designed for mobile networks using CDMA, based on the "C++" programming language.

Code Division Multiple Access (CDMA): A wireless technology used in 2G mobile networks, which is also developed for 3G services (e.g., W-CDMA, CDMA2002)

Developer: Typically a team of designers, artists, and engineers designing and creating game content (including graphics and sound). Example: *Future Platforms* (UK).

General Packet Radio Services (GPRS): An advanced technology for mobile data transmission based on GSM. Supports mobile multimedia applications and Web interaction, and enables pay-per-data models.

Global System for Mobile Communication (GSM): A standard for digital mobile telephony especially popular in Europe, but used worldwide. Provides across-border compatibility and maintains services at the 2G level.

Java 2 Platform Micro Edition (J2ME): An application development platform used primarily for client applications designed for mobile networks using GSM, based on the "Java" programming language.

Publisher: A company involved in controlling and marketing a mobile game, often funding its development. Example: *Your World Games* (U.S.).

Second Generation (2G) Mobile Telephony: Includes a range of technologies for wireless communication such as GSM and CDMA. Provides digitised voice communication, short messaging service (SMS) and some special features.

Symbian: An operating system for mobile devices with data processing capabilities (an open industry standard).

Third Generation (3G): "third generation." A range of advanced wireless communication technologies capable of supporting multimedia, video streaming, and video-conferencing.

Wireless Application Protocol (WAP): A set of open standards to enable Web and e-mail access and data display on handheld devices.

Mobile Healthcare Communication Infrastructure Networks

Phillip Olla
Madonna University, USA

INTRODUCTION

M-health is defined as "mobile computing, medical sensor, and communications technologies for healthcare" (Istepanian & Zhang, 2004). The use of the m-health terminology relates to applications and systems such as telemedicine and biomedical sensing systems (Budinger, 2003). The rapid advances in information and communication technology (ICT) (Godoe, 2000), nanotechnology, bio-monitoring (Budinger, 2003), mobile networks (Olla, 2005a), pervasive computing (Akyildiz & Rudin, 2001), wearable systems, and drug delivery approaches (Amy & Richards, 2004) are transforming the healthcare sector. The insurgence of innovative technology into healthcare practice is not only blurring the boundaries of the various technologies and fields, but is also causing a paradigm shift that is blurring the boundaries between public health, acute care, and preventative health (Hatcher & Heetebry, 2004). These developments have not only had a significant impact on current e-health and telemedical systems (Istepanian & Zhang, 2004), but they are also leading to the creation of a new generation of m-health systems with convergence of devices, technologies, and networks at the forefront of the innovation.

The phenomenon to provide care remotely using ICT can be placed into a number of areas such as m-health, telemedicine, and e-health. Over the evolution of telemedicine, new terminologies have been created, as new health applications and delivery options became available and the application areas extended to most healthcare domains. This resulted in confusion, and identification of what falls under telemedicine and what falls under telehealth or e-health became more complicated as the field advanced. New concepts such as pervasive health and m-health are also adding to this confusion. The first section of this article provides the background of telemedicine and the advancements of mobile networks, which are collectively the foundation of m-health. The evolution and growth of telemedicine is highly correlated with ICT advancements and software development. Telemedicine advancements can be categorized into three eras (Bashshur, Reardon, & Shannon, 2000; Tulu & Chatterjee, 2005) discussed in the next section.

There are numerous wireless infrastructures available for healthcare providers to choose from. Mobile networks that provide connectivity within buildings use different protocols from the standard digital mobile technologies such as global mobile systems (GSMs), which provide wide area connectivity. The second section of this article provides a summary of these mobile technologies that are having a profound impact on the healthcare sector. This section is then followed by the conclusion.

ERAS OF TELEMEDICINE

The *first era* of telemedicine solely focused on the medical care as the only function of telemedicine. This era can be named the telecommunications era of the 1970s. The applications in this era were dependent on broadcast and television technologies in which telemedicine applications were not integrated with any other clinical data. The *second era* of telemedicine was a result of digitalization in telecommunications, and it grew during 1990s. The transmission of data was supported by various communication mediums ranging from telephone lines to integrated service digital network (ISDN) lines. During this period there was a high costs attached to the communication mediums that provided higher bandwidth. The bandwidth issue became a significant bottleneck for telemedicine in this era. Resolving the bandwidth constraints has been a critical research challenge for the past decade, with new approaches and opportunities created by the Internet revolution; now more complex and ubiquitous networks are supporting the telemedicine. The *third era* of telemedicine was supported by the networking technology that was cheaper and accessible to an increasing user population. The improved speed and quality offered by Internet2 is providing new opportunities in telemedicine. In this new era of telemedicine, the focus shifted from an technology assessment to a deeper appreciation of the functional relationships between telemedicine technology and the outcomes of cost, quality, and access.

This article proposes a *fourth era,* which is characterized by the use of Internet protocol (IP) technologies, ubiquitous networks, and mobile/wireless networking capabilities, and can be observed by the proliferation of m-health applications that perform both clinical and non-clinical functions. Since the proliferation of mobile networks, telemedicine has attracted a lot more interest from both academic researchers and industry (Tachakra, Wang, Istepanian, & Song, 2003). This has resulted in many mobile/wireless telemedicine applications being developed and implemented. Critical healthcare information regularly travels with patients and clinicians, and therefore the need for information to become securely and accurately available over mobile telecommunication networks is key to reliable patient care and reliable medical systems.

The telecommunication industry has progressed significantly over the last decade. There has been significant innovation in digital mobile technologies. The mobile telecommunication industry has advanced through three generations of systems and is currently on the verge of designing the fourth generation of systems (Olla, 2005b). The recent developments in digital mobile technologies are reflected in the fast-growing commercial domain of mobile telemedical services. Specific examples include mobile ECG transmissions, video images and tele-radiology, wireless ambulance services to predict emergency and stroke morbidity, and other integrated mobile telemedical monitoring systems (Istepanian & Zhang, 2004; New Scientist, 2005; Istepanian & Lacal, 2003; Warren, 2003). There is no doubt that mobile networks can introduce additional security concerns to the healthcare sector.

As security is a major concern, it is important to implement a mobile trust model that will ensure that a mobile transaction safely navigates multiple technologies and devices without compromising the data or the healthcare systems. M-health transactions can be made secure by adopting practices that extend beyond the security of the wireless network used and implementing a trusted model for secure end-to-end mobile transactions. The mobile trust model proposed by Wickramasinghe and Misra (2005) utilizes both technology and adequate operational practices to achieve a secure end-to-end mobile transaction. The first level highlights the application of technologies to secure elements of a mobile transaction. The next level of the model shows the operational policies and procedures needed to complement technologies used. No additional activity is proposed for the mobile network infrastructure since this element is not within the control of the provider or the hospital.

The next section will discuss the mobile network technologies and infrastructure which are key components of any m-health system; the network infrastructure acts as a channel for data transmission and is subject to the same vulnerabilities, such as sniffing, as in the case of fixed network transaction. The mobile networks discussed in the next section are creating the growth and increased adoption of m-health applications in the healthcare sector.

MOBILE HEALTHCARE COMMUNICATION INFRASTRUCTURE

The implementation of an m-health application in the healthcare environment leads to the creation of a mobile healthcare delivery system (MHDS). An MHDS can be defined as the carrying out of healthcare-related activities using mobile devices such as a wireless tablet computer, personal digital assistant (PDA), or a wireless-enabled computer. An activity occurs when authorized healthcare personnel access the clinical or administrative systems of a healthcare institution using mobile devices (Wickramasinghe & Misra, 2005). The transaction is said to be complete when medical personnel decide to access medical records (patient or administrative) via a mobile network to either browse or update the record.

Over the past decade there has been an increase in the use of new mobile technologies in healthcare such as Bluetooth and wireless local area networks (WLANs) that use different protocols from the standard digital mobile technologies such as 2G, 2.5, and 3G technologies. A summary of these technologies are presented below, and an overview of the speeds and range is presented in Table 1.

These mobile networks are being deployed to allow physicians and nurses easy access to patient records while on rounds, to add observations to the central databases, and to check on medications, among a growing number of other functions. The ease of access that wireless networks offer is matched by the security and privacy challenges presented by the networks. This serious issue requires further investigation and research to identify the real threats for the various types of networks in the healthcare domain.

Second-Generation (2G/2.5G) Systems

The second-generation cellular systems were the first to apply digital transmission technologies such as time division multiple access (TDMA) for voice and data communication. The data transfer rate was on the order of tens of kbit/s. Other examples of technologies in 2G systems include frequency division multiple access (FDMA) and code division multiple access (CDMA).

The 2G networks deliver high-quality and secure mobile voice and basic data services such as fax and text messaging, along with full roaming capabilities around the world. 2G technology is in use by more than 10% of the world's population, and it is estimated that 1.3 billion customers across more than 200 countries and territories around the world use this technology (GSM, 2005). The later advanced technological applications are called 2.5G technologies and include networks such as general packet radio service (GPRS) and EDGE. GPRS-enabled networks provide functionality such as: 'always-on', higher capacity, Internet-based content and packet-based data services enabling services such as color Internet browsing, e-mail on the move, visual communications, multimedia messages, and location-based services. Another complimentary 2.5G service is enhanced data rates for GSM evolution (EDGE), which offers similar capabilities to the GPRS network.

Table 1. Mobile networks

Networks	Speed	Range and Coverage	Main Issues for M-Health
2nd-Generation GSM	9.6 kilobits per second (KBPS)	World wide coverage, dependent on network operators' roaming agreements	Bandwidth limitation, interference
High-Speed Circuit-Switched Data (HSCSD)	Between 28.8 KBPS and 57.6 KBPS	Not global, only supported by service providers network.	Not widely available, scarcity of devices
General Packet Radio Service (GPRS)	171.2 KBPS	Not global, only supported by service providers network	Not widely available
EDGE	384 KBPS	Not global, only supported by service providers network	Not widely available, scarcity of devices
UMTS	144 KBPS—2 MBPS depending on mobility	When fully implemented, should offer interoperability between networks, global coverage	Device battery life, operational costs
Wireless Local Area	54 MBPS	30-50 m indoors and 100-500 m outdoors; must be in the vicinity of hot spot	Privacy, security
Personal Area Networks—Bluetooth	400 KBPS symmetrically, 150-700 KBPS asymmetrically	10-100m	Privacy, security, low bandwidth
Personal Area Networks—ZigBee	20 kb/s-250 KBPS depending on band	30m	Security, privacy, low bandwidth
WiMAX	Up to 70MBPS	Approx. 40m from base station	Currently no devices and networks cards
RFID	100 KBPS	1 m; non-line-of-sight and contactless transfer of data between a tag and reader	Security, privacy
Satellite Networks	400 to 512 KBPS new satellites have potential of 155 MBPS	Global coverage	Data costs, shortage of devices with roaming capabilities; bandwidth limitations

Third-Generation (3G) Systems

The most promising period is the advent of 3G networks, which are also referred to as the Universal Mobile Telecommunications System (UMTS). A significant feature of 3G technology is its ability to unify existing cellular standards, such as code-division multiple-access (CDMA), global system for mobile communications (GSM, 2005), and time-division multiple-access (TDMA), under one umbrella. Over 85% of the world's network operators have chosen 3G as the underlying technology platform to deliver their third-generation services (GSM, 2004). Efforts are underway to integrate the many diverse mobile environments in addition to blurring the distinction between the fixed and mobile networks. The continual roll out of advanced wireless communication and mobile network technologies will be the major driving force for future developments in m-health systems (Istepanian & Zhang, 2004). Currently the GSM version of 3G alone saw the addition of more than 13.5 million users, representing an annual growth rate of more than 500% in 2004. As of December 2004, 60 operators in 30 countries were offering 3GSM services. The global 3GSM customer base is approaching 20 million and has already been commercially launched in Africa, the Americas, Asia Pacific, Europe, and the Middle East (GSM, 2005), thus making this technology ideal for developing affordable global m-health systems.

Fourth Generation (4G)

The benefits of fourth-generation network technology include (Istepanian, Laxminarayan, & Pattichis, 2005; Olla, 2005a; Qiu, Zhu, & Zhang, 2002): voice-data integration, support for mobile and fixed networking, and enhanced services through the use of simple networks with intelligent terminal devices. 4G also incorporates a flexible method of payment for network connectivity that will support a large number of network operators in a highly competitive environment. Over the last decade, the Internet has been dominated by non-real-time, person-to machine communications (UMTS, 2002). The current developments in progress will incorporate real-time person-to-person communications, including high-quality voice and video telecommunications, along with extensive use of machine-to-machine interactions to simplify and enhance the user experience.

Currently the Internet is used solely to interconnect computer networks. IP compatibility is being added to many types of devices such as set-top boxes to automotive and home electronics. The large-scale deployment of IP-based networks will reduce the acquisition costs of the associated devices. The future vision is to integrate mobile voice communications and Internet technologies, bringing the control and multiplicity of Internet application services to mobile users (Olla, 2005b). 4G advances will provide both mobile patients and citizens the choices that will fit their lifestyle and make it easier for them to interactively get the medical attention and advice they need, when and where it is required, and how they want it, regardless of any geographical barriers or mobility constraints.

Worldwide Interoperability for Microwave Access (WiMAX)

WiMAX is considered to be the next generation of wireless fidelity (WiFi), wireless networking technology that will connect you to the Internet at faster speeds and from much longer ranges than current wireless technology allows (http://wimaxxed.com/). WiMax has been undergoing testing and is expected to launch commercially by 2007. The research firm Allied Business Research predicts that by 2009, sales of WiMax accessories will top US$1 billion (Taylor & Kendall, 2005), and Strategy Analytics predicts a market of more than 20 million WiMAX subscriber terminals and base stations per year in 2009 (ABI, 2005).

The technology holds a lot of potential for m-health applications, with the capabilities of providing data rates of up to 70 mbps over distances of up to 50 km. The benefits to both developing and developed nations are immense. There has been a gradual increase in popularity of this technology. Intel recently announced plans to mass produce and release processors aimed to power WiMax-enabled devices (WiMax, 2005). Other technology organizations investing to further the advancement of this technology include Qwest, British Telecom, Siemens, and Texas Instruments. They aim to get the prices of the devices powered by WiMax to affordable levels so that the public can adopt them in large numbers, making it the next global wireless standard. There are already Internet service providers in metropolitan areas offering pre-WiMAX service to enterprises in a number of cities including New York, Boston, and Los Angeles (WiMax, 2005).

Wireless Local Area Networks

Wireless local area networks (WLANs) use radio or infrared waves and spread spectrum technology to enable communication between devices in a limited area. WLAN allows users to access a data network at high speeds of up to 54 Mb/s as long as users are located within a relatively short range (typically 30-50 meters indoors and 100-500 meters outdoors) of a WLAN base station (or antenna). Devices may roam freely within the coverage areas created by wireless "access points," the receivers and transmitters connected to the enterprise network. WLANs are a good solution for healthcare today, plus they are significantly less expensive to operate than wireless WAN solutions such as 3G (Daou-Systems, 2001).

Personal Area Networks

A wireless personal area network (WPAN) (IBM Research, 2006; Istepanian & Zhang, 2004) is the interconnection of information technology devices within the range of an individual person, typically within a range of 10 meters. For example, a person traveling with a laptop, a PDA, and a portable printer could wirelessly interconnect all the devices, using some form of wireless technology. WPANs are defined by IEEE Standard 802.15 (IEEE Working Group, 2006). The most relevant enabling technologies for m-health systems are Bluetooth (http://www.bluetooth.org/) and ZigBee (http://www.zigbee.org/). ZigBee is a set of high-level communication protocols designed to use small, low-power digital radios based on the IEEE 802.15.4 standard for wireless personal area networks. ZigBee is aimed at applications with low data rates and low power consumption. ZigBee's current focus is to define a general-purpose, inexpensive, self-organizing network that can be shared by industrial controls, medical devices, smoke and intruder alarms, building automation, and home automation. The network is designed to use very small amounts of power, so that individual devices might run for a year or two with a single alkaline battery, which is ideal for use in small medical devices and sensors. The Bluetooth specification was first developed by Ericsson and later formalized by the Bluetooth Special Interest Group established by Sony Ericsson, IBM, Intel, Toshiba, and Nokia, and later joined by many other companies. A Bluetooth WPAN is also called a *piconet* and is composed of up to eight active devices in a master-slave relationship. A piconet typically has a range of 10 meters, although ranges of up to 100 meters can be reached under ideal circumstances. Implementations with Bluetooth versions 1.1 and 1.2 reach speeds of 723.1 kbit/s. Version 2.0 implementations feature Bluetooth Enhanced Data Rate (EDR) and thus reach 2.1 Mbit/s (http://www.bluetooth.org/; http://en.wikipedia.org/wiki).

Radio Frequency Identification (RFID)

RFID systems consist of two key elements: a tag and a reader/writer unit capable of transferring data to and from the tag. An antennae linked to each element allows power to be transferred between the reader/writer and remotely sited tag through inductive coupling. Since this is a bi-directional process, modulation of the tag antenna will be reflected back to the reader's/writer's antenna, allowing data to be transferred in both directions. Some of the advantages of RFID that makes this technology appealing to the healthcare sector are:

- no line-of-sight required between tag and reader;
- non-contact transfer of data between a tag and reader;
- tags are passive, which means no power source is required for the tag component;
- data transfer range of up to 1 meter is possible; and
- rapid data transfer rates of up to 100 Kbits/sec.

The use of RFID in the healthcare environment is set to rise and is currently being used for drug tracking. RFID technology is expected to decrease counterfeit medicines and make obtaining drugs all the more difficult for addicts (Weil, 2005). There are also applications that allow tagging of patients, beds, and expensive hospital equipment.

Satellite Technologies

Satellite broadband uses a satellite to connect customers to the Internet. Two-way satellite broadband uses a satellite link to both send and receive data. Typical download speeds are 400 to 512 kbps, while upload speeds on two-way services are typically 64 to 128 kbps. Various organizations (Inmarsat Swift64, 2000) have been investigating the development of an ultra-high-data-rate Internet test satellite for use for making a high-speed Internet society a reality (JAXA, 2005). Satellite-based telemedicine service will allow a real-time transmission of electronic medical records and medical information anywhere on earth. This will make it possible for doctors to diagnose emergency patients even from remote areas, and also will increase the chances of saving lives by receiving early information as ambulance data rates of 155mbps are expected. One considerable drawback associated with using this technology is cost (Olla, 2004).

CONCLUSION

The first section of this article provides the background of telemedicine and the advancements of mobile networks, which is fuelling the increase of m-health applications in the healthcare domain. The evolution and growth of telemedicine is highly correlated with ICT advancements and software development. This article has also summarized the various mobile network technologies that are being used in the healthcare sector. The mobile technologies described above have a significant impact on the ability to deploy mobile healthcare applications and systems.

REFERENCES

ABI. (2005). WIFI-WIMAX. Retrieved from http://www.abiresearch.com/category/Wi-Fi_WiMAX

Akyildiz, I., & Rudin, H. (2001). Pervasive computing. *Computer Networks—The International Journal of Computer and Telecommunications Networking, 35*(4), 371-371.

Amy, C., & Richards G. (2004). A BioMEMS review: MEMS technology for physiologically integrated devices. *Proceedings of IEEE* (Vol. 92, no. 1, pp. 6-21).

Bashshur, R. L., Reardon, T. G., & Shannon, G. W. (2000). Telemedicine: A new health care delivery system. *Annual Review of Public Health, 21,* 613-637.

Budinger, T. F. (2003). Biomonitoring with wireless communications. *Annual Review of Biomedical Engineering, 5,* 412.

Daou-Systems. (2001). *Going mobile: From eHealth to mHealth.* Retrieved from http://www.daou.com/emerging/pdf/mHealth_White_Paper_April_2001.PDF

Godoe, H. (2000). Innovation regimes, R&D and radical innovations in telecommunications. *Research Policy, 29*(9), 1033-1046.

GSM. (2005). *Homepage.* Retrieved from http://www.gsmworld.com/index.shtml

GSM. (2004). *Information.* Retrieved from http://www.gsmworld.com/index.shtml

Hatcher, M., & Heetebry, I. (2004). Information technology in the future of health care. *Journal of Medical Systems Issue, 28*(6), 673-688.

IBM Research. (2006). Retrieved from http://www.research.ibm.com/topics/popups/smart/mobile/html/phow.html

IEEE Working Group. (2006). *IEEE 802.15 Working Group for WPAN.* Retrieved from http://www.ieee802.org/15/

Imrasat Swift64. (2000). *Inmarsat announces availability of the 64kbit/s mobile office in the sky.* Retrieved from http://www.inmarsat.com/swift64/press_1.htm

Istepanian, R. S. H., & Lacal, J. (2003). M-health systems: Future directions. *Proceedings of the 25th Annual IEEE International Conference on Engineering Medicine and Biology,* Cancun, Mexico.

Istepanian, R. S. H., Laxminarayan, S., & Pattichis, E. (2005). *M-health: Emerging mobile health systems.* New York.

Istepanian, R. S. H., & Zhang, Y. T. (2004). Guest editorial introduction to the special section on m-health: Beyond seamless mobility and global wireless health-care connectivity. *IEEE Transactions on Information Technology in Biomedicine, 8*(4).

JAXA. (2005). *Aerospace Exploration Agency*. Retrieved February, 2006, at www.jaxa.jp/missions/projects/sat/index_e.html

New Scientist. (2005). Future trends, convergence IS. *New Scientist,* (March 12), 53.

Olla, P. (2004). A convergent mobile infrastructure: Competition or Co-operation. *Journal of Computing and Information Technology: Special Issue on Information Systems: Healthcare and Mobile Computing, 12*(4), 309-322.

Olla, P. (2005a). Evolution of GSM network technology. In M. Pagani (Ed.), *Encyclopedia of multimedia technology and networking*. Hershey, PA: Idea Group Reference.

Olla, P. (2005b). Incorporating commercial space technology into mobile services: Developing innovative business models. In M. Pagani (Ed.), *Mobile and wireless systems beyond 3G: Managing new business opportunities*. Hershey, PA: IRM Press.

Qiu, R. C., & Zhang, Y. Q. (2002, May 26-29). Third-generation and beyond (3.5G) wireless networks and its applications. *Proceedings of the IEEE International Symposium on Circuits and Systems (ISCS)*, Scottsdale, AZ.

Tachakra, S., Wang, X. H., Istepanian, R. S. H., & Song, Y. H. (2003). Mobile e-health: The unwired evolution of TelemedicineMobile. *Telemedicine Journal and E-Health, 9*(3), 247.

Taylor, C., & Kendall, P. (2005, June 1). *Strategy analytics: WiMAX 3G killer or fixed broadband wireless standard?* Retrieved from www.strategyanalytics.net/default.aspx?mod=ReportAbstractViewer&a0=2393

Tulu, B., & Chatterjee, S. (2005). A taxonomy of telemedicine efforts with respect to applications, infrastructure, delivery tools, type of setting and purpose. *Proceedings of the 38th Hawaii International Conference on System Sciences*, Hawaii.

UMTS. (2002). *Support of third generation services using UMTS in a converging network environment*. Retrieved from http://www.umts-forum.org/servlet/dycon/ztumts/umts/Live/en/umts/Resources_Reports_index: UMTS

Warren, S. (2003). Beyond telemedicine: Infrastructures for intelligent home care technology. *Proceedings of the Pre-ICADI Workshop Technology for Aging, Disability, and Independence,* London.

Weil, N. (2005). Companies announce RFID drug-tracking project: Unisys and SupplyScape plan to track Oxycontin through the supply chain. *Computerworld*.

Wickramasinghe, N., & Misra, S. K. (2005). A wireless trust model for healthcare. *International Journal of Electronic Healthcare, 1*(1), 62.

WiMax. (2005). Retrieved from http://wimaxxed.com

KEY TERMS

Bluetooth: Worldwide industrial specification for wireless personal area networks (PANs). Bluetooth provides a way to connect and exchange information between devices like personal digital assistants (PDAs), mobile phones, laptops, PCs, printers, and digital cameras via a secure, low-cost, globally available short range radio frequency.

Global System for Mobile Communications (GSM): A digital worldwide mobile phone and data standard. GSM service is used by over 1.5 billion people across more than 210 countries and territories. The ubiquity of the GSM standard makes international roaming very common between mobile phone operators, enabling subscribers to use their phones in many parts of the world.

M-Health: Mobile computing, medical sensor, and communications technologies for healthcare.

Personal Area Network (PAN): A computer network with a reach of a meters used for communication among computer devices such as telephones and personal digital assistants or medical sensors close to the human body.

Radio Frequency Identification (RFID): An automatic identification system that transmits, stores, and remotely retrieves data using mobile devices called RFID tags or transponders. The purpose of an RFID system is to enable data to be transmitted by a mobile device, called a tag, which is read by an RFID reader and processed according to the needs of a particular application.

Mobile Hunters

Jörg Lonthoff
Technische Universität Darmstadt, Germany

INTRODUCTION

Growing Internet mobility due to various transmission methods such as broadband data transmission get mobile service providers interested in providing services that offer more than voice telephony. Modern cellular phones support general packet radio service (GPRS) have a color display and are usually Java-compliant. This meets the device's requirements for context-based services. As global system for mobile communications (GSM)-based cellular phones are widely used and, at least in Europe, the GSM-network is available almost everywhere, the context variable "location" seems useful for extending the relevant value-added services (Rao & Minakakis, 2003, p. 61). For example, there are services for finding friends in the vicinity (Buddy Alert by Mobiloco, www.mobiloco.de) and mobile navigation systems for cellular phones (NaviGate by T-Mobile, www.t-mobile.de/navigate). But there has not yet been a real breakthrough for location-based services (LBSs) (Lonthoff & Ortner, 2006).

Supported by T-Systems International, we developed the adventure game "Mobile Hunters." The game demonstrates what is possible with LBS and uses the currently available infrastructure mobile network providers offer for creating a virtual playing field. This playing field will be adapted to the real world. The object of the game is a hunt. Players can either be a hunter who must find a fugitive or a fugitive who has to make sure he is not getting caught. Of course, this hunt will become eventful as there are a variety of obstacles. Playing a so-called mobile location-based game (MLBG) could increase the acceptance of further LBSs.

This article will first summarize the current state of research in this field and then present *mobile location-based gaming*. Then the reader will get to know the game "Mobile Hunters." After that, we will discuss the lessons learned from the game, and possible further developments will be considered. At the end of this, we draw a short conclusion.

BACKGROUND

MLBGs are a special category of location-based games, as follows:

A MLBG is a location-based game running on a mobile device. By using a communication channel the game exchanges information with a game server or other players. (Lonthoff & Ortner, 2006)

Figure 1. Taxonomy for location-based services

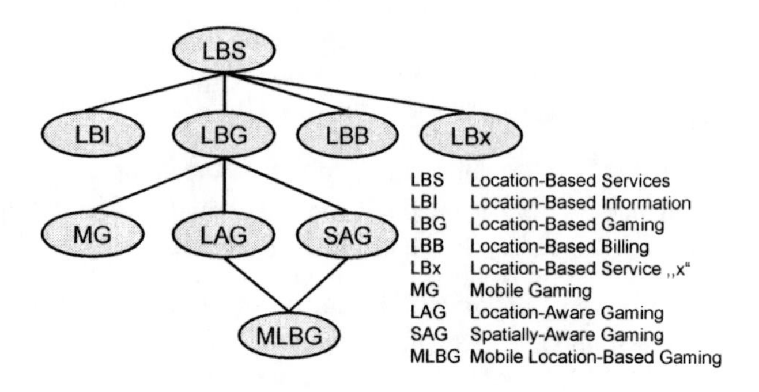

Applying this definition, the fields *location-aware games* and *spatially aware games* become relevant. Figure 1 shows all terms relevant in MLBG.

Location-Based Services

The added value of mobile services opens up opportunities for service providers to address a new dimension of the user: the user's spatiotemporal position. Such services are called location-based services. LBSs are based on a variety of localization methods for determining a user's position. In the field of so-called *context-aware computing* (Schilit, Adams, & Want, 1994, p. 85), LBSs provide location information as context references (Dey, 2001). There are many possibilities of using the location reference in an application system (Unni & Harmon, 2003, p. 417; Schiller & Voisard, 2004). All of these services are based on mobile positioning. Mobile positioning comprises all technologies for determining the location of mobile devices. A position can be determined in two different ways: using network-based technology (the network provides the position) or using terminal-based technology (the device provides the position).

Network-Based Positioning Technologies

GSM-networks offer basically six different methods of network-based localization (Röttger-Gerigk, 2002). Cell of origin (COO) is the simplest mobile positioning technique. It identifies the cell (cell ID) in which a cellular phone is logged on. The cell ID is connected with the radiation range of a mobile base station. The cellular coverage area has a certain range around the position of the mobile base station. One mobile base station can have several radiation areas (cell IDs), whereby these cells always refer to the same geographic position of the mobile base station's location (Hansmann, Merk, Nicklous, & Stober, 2001, p. 243). The positioning accuracy that can be achieved depends on the size of the cellular coverage area; it may range between 25 m and 35 km in diameter. In addition, there are more complex techniques such as angle of arrival (AOA), time of arrival (TOA), time difference of arrival (TDOA), signal attenuation (SA), and the radiocamera system.

Terminal-Based Positioning Technologies

Cell of origin can also be considered a terminal-based technique, as the desired cell ID can be read out directly from the device (terminal). To do this, however, a reference database is needed that contains the geographic coordinates stored for each cell ID. The following further terminal-based techniques are available: enhanced observed time difference (E-OTD), as well as the satellite-based systems such as the global positioning system (GPS), or assisted-GPS (A-GPS), which works without modifications to the cellular phone network infrastructure, except that the mobile device must possess a GPS receiver.

Gaming

Mobile games for cellular phones are currently experiencing a growing demand. There is a trend towards more complicated 3D games. This trend is supported by the current hardware development that is the availability of high-performance cellular phones or smart phones, respectively. Games that allow direct communication with remote participants are of great interest (multi-player games). Multi-player games on offer can be played using a wireless application protocol (WAP) portal or locally by two people (infrared) or by several players (Bluetooth).

Games for personal digital assistants (PDAs) are also very interesting. Such games are usually intended for one player. But games for several players become possible, if infrared, Bluetooth, or wireless local area network (WLAN) are used.

In location-based games the movements of a player (in the sense of a geographical change of location) influence the game. Nicklas, Pfisterer, and Mitschang (2001, pp. 61-62) suggest a classification of location-based games into mobile games, location-aware games, and spatially aware games.

Mobile games require as a location reference only one more player who is in the vicinity. The location information itself is not considered in the game. A typical example of this kind of game is Snake, a game of dexterity for two delivered with the older Nokia cellular phone models that can be played using infrared or Bluetooth. Location-aware games include information about the location of a player in the game. A typical example would be a treasure quest whereby a player must reach a particular location. Spatially aware games adapt a real-world environment to the game. This creates a connection between the real world and the virtual world. The MLBG "Mobile Hunters" presented in the following belongs to this category of games.

CHARACTERISTICS AND CHALLENGES OF MLBG

Important for MLBG are the type of device used, the communication and network infrastructure it is based on, the way positions are determined, and the kind of game.

Devices such as cellular phones, smart phones, and PDAs can be used, possibly laptop also. In addition to this rough classification, the device properties can serve for further distinction: the operating system, client programming (Java virtual machine, Web-client/WAP-client), the types of available user interfaces, as well as battery life and processor power.

The relevant communication media are wide area networks such as WLAN, GSM, and universal mobile telecommunication system (UMTS). These technologies vary in range and bandwidth. The accuracy of a determined position depends on the technique used and on the network structure.

When looking at the type of game, two dimensions are of interest: the number of players and the type of game. There are single-player games and multi-player games. You can also play multi-player games alone, if players are simulated. Massive-multi-player games are a special type of game in which the end of the game is not defined. Players can actively participate in the game for some time and improve their ranking in the community associated with the game. Relevant genres of game would be role-playing games, scouting games, real-time strategy games, and first-person-shooter games.

Users can find a variety of game collections on the Internet that include MLBGs. For example:

- www.smartmobs.com/archive/2004/12/28/location-based_.html
- www.we-make-money-not-art.com/archives/001653.php
- www.in-duce.net/archives/locationbased_mobile_phone_games.php

Newer publications offer studies (Jegers & Wiberg, 2006) and overviews (Magerkurth, Cheok, Nilsen, & Mandryk, 2005; Rashid, Mullins, Coulton, & Edwards, 2006) of pervasive games, including MLBGs.

What is fascinating about MLBG is: "This ability for you to actually use your real-world movements to play the game means that you are no more playing a game... You are in the Game!" (Mikoishi, 2004).

If you want to turn a "classic" game into an MLBG, there are four central problem areas, identified by Nicklas et al. (2001, p. 62): adaptation of the playing field, adaptation of the pawn in a game, representation of cards or objects, and adaptation of the moves in the game.

Another challenge results from the characteristics of mobile networks. In MLBG it may happen that some of the players interrupt the connection for short periods of time. These interruptions must be covered.

MOBILE HUNTERS

The Game

The idea of the adventure game "Mobile Hunters" was inspired by the well-known board game "Scotland Yard" (Ravensburger, 1983). The creation of a players' community reflects partly the notion of massive multi-player games. To be able to start a game session, at least two players must participate. There are two possible roles in the game: one or more hunters and one fugitive. The hunters want to catch the fugitive before the specified playing time is over. The fugitive must try to escape the hunters or, to prove that he is innocent, collect a number of items as proof of his innocence.

After logging onto the "Mobile Hunters" server with user name and password as authentication, a player can initiate a game. Once a game has been initiated, several more players can enter into the game. If a new player enters into the game, the game server checks whether the potential player's location is within an appropriate distance to the center of the playing field. This ensures that the spatial distance between the players does not become too large. The person who initiated the game decides when the number of players is sufficient and then starts the game. The maximum number of players can be specified.

When the game is started, the game server randomly assigns the roles and informs the players about whether they are a hunter or a fugitive. When the game has been started, the game server distributes all the items for hunters and fugitives randomly on the playing field. After this initialization phase the game begins synchronously on all participating clients and the countdown for the playing time starts. In previously specified intervals (default 1.5 minutes), the current position of the fugitive appears on a map in the hunters' display. In the playing field there are a number of locked (virtual) boxes that players can open if their position is the same as the position of the box (geo-coordinate with specified radius = playing field). Some of the boxes are visible only for hunters. These boxes contain offensive weapons. The fugitive can only see the boxes that contain items for him. These are items for defense or proof of innocence. Table 1 gives an overview of the various items.

A player can attack, if one player is located in the same position and the attacker has a weapon. If a hunter attacks another hunter, the person attacked will become incapable of action for some time (default 30 seconds)-his client's menu will be hidden. In this attack the attacker loses his weapon. If a hunter attacks a fugitive, the fugitive can defend himself using a matching item for defense. If this happens, the attacker will become incapable of action for some time (default 30 seconds). If the fugitive cannot defend himself, the hunter wins.

The fugitive wins if he is able to find a certain number (default 3) of items that prove his innocence or if no attack

Table 1. Overview of the items available for the players

Recipient	Type	Instance	Purpose
Hunter	Attack	K.O.	In the boxes the hunter will find offensive weapons he can use to attack the fugitive or for attacking another hunter to make him incapable of action (e.g., handcuffs).
Fugitive	Proof		In the boxes the fugitive will find items that can prove his innocence (e.g., a theater ticket).
Fugitive	Defense		In the boxes the fugitive will find items he can use to defend himself against attacks by a hunter (e.g., paper clip for defense against handcuffs)

Figure 2. Overall architecture

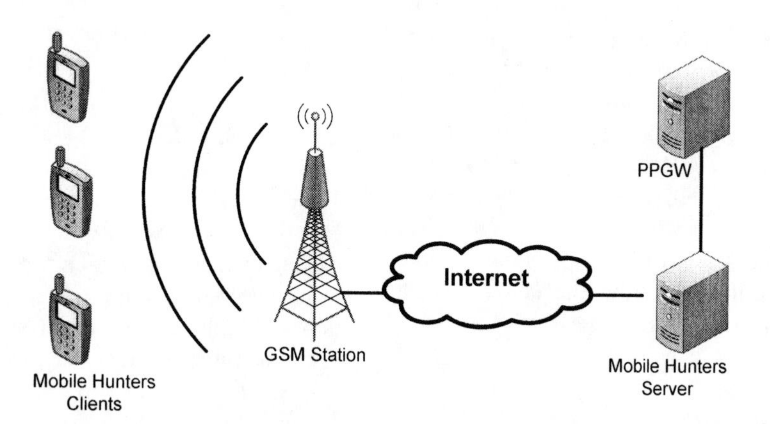

on him was successful during the duration (default 30 minutes) of the game.

Architecture

In the development of "Mobile Hunters," we only used technologies that have already been accepted in the market and are, or will most certainly be, widely spread, to facilitate acceptance of the game. We chose the GSM network, which is currently the most widely used (Nicklas et al., 2001, p. 61), to ensure a wide range of application. We chose a Java-compliant cellular phone (Nokia 6680) as the playing device.

It was essential that the positioning technique would only use the available cellular phone network infrastructure and that it would not be necessary to make modifications to it. This is why COO seemed best suited.

The implementation is based on a client/server architecture. Communication between the components is realized using the German T-Mobile GSM network. For data transfer we chose GPRS; communication takes place on an application level via secure hypertext transfer protocol (HTTPS). The "Mobile Hunters" server is a Java-based game server, which provides the user management and the game management. For determining the current position of each player, T-Systems' research platform "Permission and Privacy Gateway" (PPGW) is used. PPGW also provides the "Mobile Hunters" server with maps of the areas in question. Figure 2 shows the overall architecture schematically.

Client

The "Mobile Hunters" client (see Figure 3) is the game's user interface. It displays the current map segments; here, interaction with the other players takes place. Players also use the "Mobile Hunters" client to register and log on.

Current prerequisites are a cellular phone with Symbian operating system version 7.0 and a Java 2 Micro Edition (J2ME) framework. A class implemented in Symbian C++ is needed for reading out the cell ID the cellular phone uses. The cell ID is transmitted to the "Mobile Hunters" server. The server requests the current status in specified intervals (default every 10 seconds). The necessary unique identification of each player is the user's name he or she entered at the beginning of the game.

Server

The "Mobile Hunters" server's task is to centrally control the game. On the server, identification and authentication of the users takes place, as well as the game logic and the communication between PPGW and the clients. Every initialization as well as the specification of the playing field happens dynamically on the game server. The server was implemented in Java as an Apache Tomcat application on a Microsoft Windows 2003 Server platform and connected

Figure 3. "Mobile Hunters" client on Nokia 6680 at Luisenplatz, Darmstadt, Germany

to a Microsoft SQL Server via Java database connectivity (JDBC)/open database connectivity (ODBC). This ensures a clear separation of application level and data level. The server establishes the connection to the PPGW for requesting the geographic data of each cell ID used in the game at any time. The PPGW has a reference database where the relevant geo-reference-data for each cell ID are stored. The geographic data that correspond to the cell IDs are returned by the PPGW via an eXtensible Markup Language (XML)-based interface. The server keeps a high-score list to challenge the gaming community. Using a configuration file, the parameters that can be controlled are passed to the server.

LESSONS LEARNED

The adaptation of the game's concept to the virtual world turned out to be difficult because of the inaccuracy of positioning. Cell switches occur even if a player does not move. This may happen, for example, if several participants are logged on to a cell and, due to the limits of bandwidth or for reasons of optimization, a neighboring cell is allocated to the cellular phone.

Cellular coverage areas have different sizes: in an urban area approximately 25 m, in rural areas up to 35 km. Therefore "Mobile Hunters" is suited for playing in high-density urban areas. The cellular coverage areas of single cells overlap with those of other cells. Large cells cover smaller cells almost completely. It is still under discussion whether to exclude cells from the game if they are too large and from which size on they should be banned. To increase the accuracy of positioning, GPS or A-GPS offers interesting opportunities for mobile network providers and for the MLBG.

When a "classic" game is adapted to an MLBG, it is important to make sure the point of the game is kept and that the duration of the game is adequate. The hype phases we observe are much shorter with games than with any other service offers. Once a game is considered boring or error prone, acceptance drops. For the game "Mobile Hunters" to be funny and exciting, a game duration of 30 minutes is recommended. The interval in which a fugitive's current position is displayed should be 1.5 minutes, the time a player is incapable of action should be 30 seconds, and the interval in which the cell ID is read ought to be less than one second. A powerful processor such as the Nokia 6680 (220 MHz) is necessary for playing this game. It is very important to develop the client's code in a way best suited for cellular phones (Lonthoff, Ortner, & Wolf, 2006). Session handling is essential for playing an MLBG. The user interface must be designed quite simply, so that no explanations will be necessary.

Our mobile-specific modeling resulted in a very economic consumption of resources and of handling communication data. In a 30-minute game, only an average of 300 kB data is transferred. Half of that data volume is used for loading the graphics (maps and icons) at the beginning of the game.

CONCLUSION

The conclusion we draw is that we have partly succeeded in adapting the real world to a game's virtual world. It is possible to read out the cell ID from a variety of mobile devices using different techniques. However, no standard application programming interface (API) is currently available to offer this functionality. This means that anyone who wants to create a game needs to develop a suitable API for every single cellular phone, if the game is to be widely used. This problem may be solved by deploying a middleware such as BREW from Qualcomm Inc. (www.qualcomm.com/brew; Tarumi, Matsubara, & Yano, 2004, p. 546).

Addressing the human play instinct, MLBGs increase the acceptance of LBSs. Playing, the use of LBSs will become effortless and people will get interested in such systems. Location-based games are becoming a mass market. If this mass market can be served, prices for location requests will go down.

The experience gained in the field of gaming also applies to situations in private and business life. Advertising and interactive marketing (Han, Cho, & Choi, 2005, p. 103) are potential application domains. Possibly, brand-new application systems useful in everyday life can be developed and implemented. In this context "application system" is understood in a comprehensive way for all tasks in user and computer-based information processing (Ortner, 2005, p. 34).

REFERENCES

Dey, A. (2001). Understanding and using context. *Personal and Ubiquitous Computing, 5*(1), 4-7.

Han, S., Cho, M., & Choi, M. (2005). Ubitem: A framework for interactive marketing in location-based gaming environment. *IEEE Proceedings of the International Conference on Mobile Business (ICMB'05)* (pp. 103-108), Sydney, Australia.

Hansmann, U., Merk, L., Nicklous, M., & Stober, T. (2001). *Pervasive computing handbook.* Berlin: Springer-Verlag.

Jegers, K., & Wiberg, M. (2006). Pervasive gaming in the everyday world. *Pervasive Computing, 5*(1), 78-85.

Lonthoff, J., & Ortner, E. (2006). Mobile location-based gaming (MLBG) as enabler for location-based services (LBS). *Proceedings of the E-Society,* Dublin, Ireland (pp. 485-492).

Lonthoff, J., Ortner, E., & Wolf, M. (2006). Implementierungsbericht Mobile Hunters. In J. Roth, J. Schiller, & A. Voisard (Hrsg.): *3. GI/ITG KuVS Fachgespräch Ortsbezogene Anwendungen und Dienste* (pp. 26-31), Technical Report, FU Berlin, Germany.

Magerkurth, C., Cheok, A.D., Nilsen, T., & Mandryk, R. (Eds.). (2005). *Proceedings of PerGames 2005,* Munich, Germany.

Mikoishi. (2004). *gunslingers.* Retrieved from http://www.gunslingers.mikoishi.com

Nicklas, D., Pfisterer, C., & Mitschang, B. (2001). Towards location-based games. *Proceedings of the International Conference on Applications and Development of Computer Games in the 21st Century (ADCOG 21)* (pp. 61-67), Hong Kong Special Administrative Region, China.

Ortner, E. (2005). *Sprachbasierte Informatk—Wie man mit Wörtern die Cyber-Welt bewegt.* Leipzig, Germany: Eagle-Verlag.

Ravensburger. (1983). *Scotland Yard.* Ravensburg, Germany.

Rao, B., & Minakakis, L. (2003). Evolution of mobile location-based services. *Communications of the ACM, 46*(12), 61-65.

Rashid, O., Mullins, I., Coulton, P., & Edwards, R. (2006). Extending cyberspace: Location based games using cellular phones. *ACM Computers in Entertainment (CIE), 4*(1), 1-18.

Röttger-Gerigk, S. (2002). Lokalisierungsmethoden. In *Handbuch mobile-commerce* (pp. 419-426). Berlin: Springer-Verlag.

Schilit, B., Adams, N., & Want, R. (1994). Context-aware computing applications. *Proceedings of the Workshop on Mobile Computing Systems and Applications* (pp. 85-90), Santa Cruz, CA.

Schiller, J., & Voisard, A. (2004). *Location-based services.* San Francisco: Morgan Kaufmann.

Tarumi, H., Matsubara, K., & Yano, M. (2004). Implementations and evaluations of location-based virtual city system for mobile phones. *IEEE Proceedings of the Global Telecommunications Conference Workshops* (pp. 544-547), Dallas, TX.

Unni, R., & Harmon, R. (2003). Location-based services: Models for strategy development in m-commerce. *IEEE Proceedings of the Portland International Conference on Management of Engineering and Technology* (PICMET'03) (pp. 416-424), Portland, OR.

KEY TERMS

General Packet Radio Service (GPRS): An IP-based data communication service for cellular phones using GSM networks.

Global System for Mobile Communications (GSM): The most popular standard in the world for second-generation (2G) cellular phone systems.

Java 2 Micro Edition (J2ME): A special Java environment for small but smart devices, using a minimum of resources.

Location-Based Service (LBS): A context-based service using the context variable "location." In the area of m-commerce and m-business, LBSs enable new value-added services by using the customers' location.

Mobile Location-Based Game (MLBG): A MLBG is a location-based game running on a mobile device. By using a communication channel, the game exchanges information with a game server or other players.

Mobile Hunter: A pretty nice MLBG developed by the Department of Commercial Information Technology I at Technische Universität Darmstadt, Germany, in cooperation with T-Systems International.

Mobile Positioning: Summarizes technologies for providing the location of a mobile device.

Permission and Privacy Gateway (PPGW): A T-Systems research and development platform that provides a location interface. The PPGW was developed as a location brokerage platform that aggregates location sources (cell ID, WLAN, A-GPS, RFID) and offers this information to service providers using a unified interface.

Index

E

G

H

I

M

N

O

S

W

X

Z